GENERAL CHEMISTRY

. . . truth, which only doth judge itself, teacheth that the inquiry of truth, which is the love-making or wooing of it, the knowledge of truth, which is the presence of it, and the belief of truth, which is the enjoying of it, is the sovereign good of human nature.

Francis Bacon
Essays, 1597

GENERAL CHEMISTRY

Second Edition

THEODORE L. BROWN
University of Illinois

CHARLES E. MERRILL PUBLISHING COMPANY
COLUMBUS, OHIO
A Bell & Howell Company

Library of Congress Catalog Card Number: 68-11553

2 3 4 5 6 7 8 9 10 11 12 13 14 15 -76 75 74 73 72 71 70 69 68

PRINTED IN THE UNITED STATES OF AMERICA

The Merrill Physical and Inorganic Chemistry Series

About the Cover

The cover shows crystals of Cortisone* as seen under a polarizing microscope. Cortisone is a corticosteroid hormone, a compound secreted by the cortex or outer layer of the adrenal glands, situated at the upper end of each kidney. Cortisone and related compounds are widely used in treatment of rheumatoid arthritis. These compounds are members of a large family of organic compounds called steroids, which include Vitamin D and the sex hormones. The empirical formula of cortisone is $C_{21}H_{28}O_5$. The structural formula is

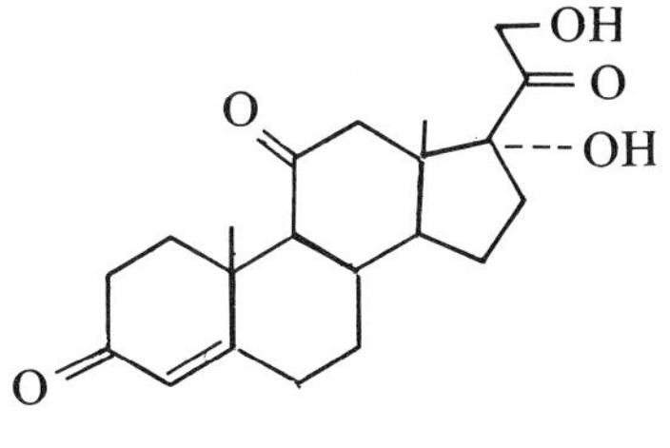

OH
O
O
---OH
O

*We are indebted to Merck and Co., Inc. for making available the photograph of crystals of Cortone, the trade name under which cortisone is distributed by Merck and Co., Inc.

Author's Preface

I have decided to write a new preface to this second edition rather than merely add a few remarks to the old preface. This decision results in part from my conviction that this is really a much different book from the first edition. The constructive criticisms of many users of the first edition were extremely helpful to me in rewriting the book, and I am grateful for all those assistances. Aside from organizational changes to be mentioned later, the second edition differs from the first in some important areas of content, and in style. The first edition proved to be too concise for the tastes of many teachers and students. In this edition, the text has been rewritten in a more ruminative style, with considerably expanded discussions of many topics, but with no change in the general level of the material. On the whole, the book should be easier for both students and teachers to use.

It has been assumed, in writing the book, that the students will have a reasonable facility with ordinary algebraic methods, but not that they will necessarily have had previous work in physics or chemistry. Some of the mathematics necessary for proper use of the text is reviewed in Appendix C, which includes a discussion of linear relationships for students who have no acquaintance with analytic geometry. It is not expected that the students will have any knowledge of calculus.

Almost every chapter has been substantially rewritten. The chapters on ionic and covalent bonding have been very considerably expanded. Covalent bonding is treated in terms of molecular orbital theory, although the elements of the valence bond approach have not been neglected. Thermodynamics is now treated in a single chapter, and material dealing with the second law has been completely rewritten. In the light of further experience in teaching thermodynamics at the freshman level, I have come to conclude that most students can more easily acquire a feeling for the concepts of entropy and free energy when entropy is treated in terms of statistics and probabilities. Since it is very difficult, and probably unwise, to span in one book a wide range of material in terms of level of difficulty, I have elected not to treat thermodynamics at a level which requires calculus. The difficulties which students experience in their first encounter with thermodynamics are usually conceptual rather than mathematical. An introduction to thermodynamics in which the emphasis is on concepts, as in this text, is important even for students who are prepared to deal with the subject in terms of calculus.

The general chemistry course should provide an adequate introduction to chemistry for students who may receive no further formal training in chemistry. At the same time, however, it must provide a good foundation for students who will go on to more advanced work in chemistry, other sciences, and engineering. It is quite possible, in a properly designed course, to meet the needs of both groups. To be a successful venture in education, the general chemistry course must arouse in students an enthusiasm for chemistry as an intellectual activity worthy of their interest and best efforts. With this in mind, I have attempted to make a choice of subject matter which exposes a realistic view of chemistry to the student. He learns something about the kinds of problems with which chemists have to deal, and also something of the approaches taken to solve these problems. After all, chemistry is, in the best semantic sense, that which chemists *do*. The training of a chemist consists in equipping him to perform, as well as possible, in the way that chemists are required to perform in daily work — whether at the laboratory bench, in the library, or at a desk. In presenting chemistry to students who will act in the future as informed laymen, one can do no better than to lead them along the path that the chemist himself must travel. To re-emphasize the point, they must learn something about chemistry as an intellectual activity, rather than solely as a collection of factual material which may soon become outdated.

In organizing the material of the book, I have made every effort to allow for flexibility in both the order and choice of subject matter. A few special remarks on certain aspects of the text are in order:

1. The order of chapters, which is not the order which needs to be followed

by the teacher, is such as to introduce gases very early in the book. Such fundamental considerations as the mole concept, Avogadro's number, combining weight, etc., can therefore be introduced early and made the basis for laboratory work. Aside from this practical reason for treating gases early, the phenomenological treatment of gases provides a proper logical development leading to a discussion of the modern theory of atomic structure. Portions of Chapter 10 could also be dealt with early in the course if this should seem advisable in terms of the laboratory program.

2. I have placed the important chapter on thermodynamics as early in the book as feasible, to allow for maximum application of thermodynamic concepts in the course. So much of what is taught in general chemistry is based on thermodynamic principles, even though not often identified as such, that it is really quite important to introduce thermodynamics as early as possible.
3. I have introduced a good deal of material dealing with solids into this edition. An adequate appreciation of the solid state is of growing importance in a variety of scientific and engineering disciplines. The notions of non-stoichiometry and defect structure as seen from a chemical viewpoint are, I believe, useful additions to a general chemistry course.
4. The material dealing with the "descriptive" chemistry of the elements has been greatly expanded. In discussing this material, I have tried to make as much use as possible of the principles elaborated upon in the early part of the text. The sections on The Atmosphere, Reactions of Transition Metal Complexes, and Metal Carbonyls are among the new topics, and should be of special interest.
5. The number of exercises has been considerably increased over the first edition. While many of the new problems are of a more or less routine character, I have tried to keep to interesting problems. Some of them are, admittedly, rather difficult. The number for which answers are given in the appendix is increased over the first edition. Some teachers like to have at their disposal a larger number of exercises for assignment than is available in this, or any other single text. Fortunately, a number of quite good problems manuals are now available as supplementary material.

I should like to acknowledge the helpful criticisms of the first edition of the text by Professor Durward Shriver. Professors A. C. Breyer and H. Eick were very helpful in commenting upon the manuscript for this edition. I wish to acknowledge again many helpful criticisms, notices of error, *etc.*, from students and teachers who used the first edition. Although there were very few actual errors in the first edition, it would have been better if there had been none. I hope there are none in this edition, but I am prepared to receive with gratitude notice of errors which come to the attention of the reader.

Finally, I wish to express my appreciation to the staff of the Charles E. Merrill Publishing Company for their interest, and for their efforts in producing an attractive text.

Theodore L. Brown

Champaign, Illinois
January, 1968

Table of Contents

Chapter 3

Chapter 4

Chapter 5

Chapter 12

Acids and Bases 293

Chapter 13

Chemical Kinetics 307

Chapter 14

Chemical Equilibrium I. Principles 331

Chapter 15

Chemical Equilibrium II. Applications 351

Chapter 19

Chemistry of the Non-Metals 441

Chapter 20

Metallic Elements; Metal Ion Complexes 485

Chapter 21

Chemistry of the Metallic Elements 525

Chapter 1

Matter and Energy

1-1 Introduction

In the past few decades, the demarcations of the various branches of physical and natural science have become so vague that it is unwise to attempt a precise definition of chemistry. Indeed, one of the purposes of this book is to show how the most general concepts and theories of science find application in a wide variety of situations and problems. As a more or less tentative definition, and subject to two amendments, we may say that chemistry is concerned with the composition of matter and with all of the transformations which matter undergoes. The first amendment is restrictive; the chemist is not generally concerned with the detailed composition of the atomic nucleus, but only with its over-all charge and mass. The second amendment enlarges the scope of chemistry considerably by adding that chemistry includes the study of energy in many aspects of its interaction with matter. This definition embraces a very large territory, and we find that there are various "kinds" of chemists, such as the organic chemist or the physical chemist, each pursuing a specialized study within the whole of chemistry.

It will be helpful for us at this beginning of our study of chemistry to examine the structure of science as a whole, and to see how chemistry relates to other aspects of scientific activity. The progress of science is characterized by the application of ideas which are initially quite general to more specific problems. Although it is not always so easy to see that this is true when one considers a particular theory or discovery which finds immediate practical application, the historical record supports such a view.

The most general and abstract ideas (*i.e.*, remote from application to particular problems) are found in mathematics. Although many advances in mathematics, such as the invention of the calculus by Newton and Liebnitz, were no doubt motivated by the need for methods of solving physical problems, the idea content of a mathematical system is independent of any reference to such problems.

A description of the physical world in its most generalized form is found in the laws of physics. Concepts such as mass, energy, temperature, gravitation, radiation, and sound are treated in a general way. No specific statements are made about the details of the chemistry or biology of the objects which obey the physical laws. For example, Newton's law of gravitation states that two bodies of mass m_1 and m_2 separated by a distance r between their centers are mutually attracted by a force F such that

$$F = G\frac{m_1 m_2}{r^2}$$

Just as this law describes the force of attraction between the earth and the moon, or between the earth and an elephant, it describes also the force of attraction between two golf balls. In treating the general relationships which describe matter and energy, physicists make use of mathematical methods which are frequently very sophisticated. It is not too surprising, therefore, to find that the most creative physicists have very often been the source of important advances in mathematics.

The laws and concepts of physics find frequent application in chemistry. It has only been in the last century or so, however, that a deliberate effort has been made to utilize the methods and theories of physics for solving chemical problems. The first scientific journal explicitly devoted to furthering this aim was the Zeitschrift für Physicalische Chemie (Journal of Physical Chemistry), established by the German chemist Ostwald in 1887. Since that time, the borders between physics and chemistry have been open. Advances in physical theory and experimentation are rapidly taken up by physical chemists (or chemical physicists, as they are sometimes called) and applied to problems of essentially chemical interest. We see here again a process of proceeding from a more general theory or development of an experimental method to the more specific work aimed at solving certain kinds of problems.

The ultimate goal of chemistry is to understand how substances react under all possible conditions; alone, and in the presence of other substances.

An historical division of labor, so to speak, has developed during the growth of chemistry. Compounds based largely on carbon (*i.e.*, those containing carbon-carbon and carbon-hydrogen bonds possibly in addition to others) fall within the purview of organic chemistry, while all other chemical substances are classed as inorganic compounds. There is, of course, no hard-and-fast dividing line between inorganic and organic chemistry, as indeed there should not be. The chemistry of living organisms, both plant and animal, is the special concern of biochemists. Since this is a very complex chemistry, biochemists rely heavily on the known behavior of simpler, smaller molecules as a point of departure in understanding the more complex living systems.

Physical chemistry provides a bridge between the generalized results and methods of physics and their more specific application to chemical problems. Physicochemical techniques are used in inorganic, organic, and biochemical work. Analytical chemistry is concerned with methods for determining the compositions of pure substances, mixtures and solutions. The analytical chemist is also concerned with methods for separating the components of mixtures, and with purification of substances. Much use is made of physicochemical techniques in modern analytical chemistry.

The concepts, theories, and practices of chemistry are reflected in the many applications of chemistry to problems in biology, agriculture, geology, ceramics, other natural and physical sciences, and to engineering. The geologist who analyzes the composition of a mineral sample by spectrographic analysis employs a technique which was originally developed by physicists in the nineteenth century, and which has provided physicists and chemists with essential information about the electronic structure of atoms. The agronomist who measures the acidity of a soil sample with a pH meter utilizes a method developed by physical chemists for measurement of the acidic properties of aqueous solutions.

In discussing the structure of science as a flow of idea and practice from the most general to the most specific, we must not overlook the fact that new advances in fundamental science are frequently made possible by technological advances in the applied sciences and in production techniques. For example, the study of the magnetic properties of atomic nuclei, which has been very important for both physics and chemistry in recent years, became possible only within the past twenty years, as a result of advances in electronics. Any number of additional examples might be cited to show that scientific progress is dependent upon the flow of ideas and skills between all areas of scientific activity.

The foundations of chemistry — of all science for that matter — rest on our concepts of matter and energy. For a long time it was thought that these were absolutely conserved *i.e.*, neither gained nor lost. "The energy of the universe is constant," said Clausius, a great physicist of the nineteenth century, and it followed that the mass of the universe is also constant. We know now that this is not true. Einstein concluded in 1905, from the special theory of relativity, that the rest mass of a certain amount of matter m is equivalent to energy E by the simple relationship

$$E = mc^2$$

where c represents the velocity of light, 3×10^{10} cm/sec. Enormous (by ordinary chemical standards) amounts of heat and radiant energy occur during the course of nuclear fission reactions; such as for example, those in atomic bomb explosions and in nuclear reactors. Even larger quantities are evolved in fusion reactions such as in a hydrogen bomb explosion. The quantities of energy evolved represent the loss of measurable amounts of mass from the reacting system. On the other hand, by applying Einstein's equation, it can be seen that the energy changes involved in ordinary chemical reactions are the equivalent of extremely small mass changes, far smaller than can be detected. The combustion of 46 g of ethyl alcohol, for example, represented by the equation

$$C_2H_5OH + 3O_2 \rightarrow 2CO_2 + 3H_2O$$

occurs with a loss in the total mass of the system of only 1.5×10^{-8} g. Unless we are concerned with nuclear reactions, therefore, we can say that the total mass of matter in a closed system (one which does not exchange matter with its surroundings) is essentially constant, regardless of any energy changes which may take place.

1-2 Energy

Energy is commonly defined as the capacity for doing work. The many forms it takes are merely different aspects of the same phenomenon. In the discussion which follows we will consider some of these different forms of energy, with the aim of recognizing their fundamentally similar character.

The first step in understanding and learning to work with a concept in physical science is to learn the dimensions or the units in which it can be expressed. All of the units in which energy can be expressed are reducible to the dimensions of mass, length and time:

$$\text{energy} = \text{mass} \times \text{length}^2/\text{time}^2 \qquad (1\text{-}1)$$

The quantities mass, length and time may be expressed in a number of different units. Length, for example, may be expressed in units of inches, yards, miles, centimeters (cm), furlongs, *etc.* The universally accepted system of units in scientific work is the cgs system, in which length is expressed in centimeters (cm), mass in grams (g) and time in seconds (sec). One g-cm^2/sec^2 is called an *erg*. Of course, in engineering practice, and in applied sciences, one frequently encounters units other than those of the cgs system. It is always true, however, that the units of energy can be related through the definition of the particular unit employed, to the dimensions of Eq. (1-1).

Energy is manifest in so many different ways that it is not easy to construct a neat classification scheme as a basis for discussion. The most familiar form of energy is mechanical, which refers to the energy which material bodies possess by virtue of their motions or positions with respect to one another. An object at rest with respect to some reference object is set into motion by exerting a force on it. The force must be exerted against some resistance; for example, a ball is set into motion by exerting a force against the pull of gravity and friction of the air. When a force has been exerted through a certain distance, work W is accomplished.

$$W = f \times d \tag{1-2}$$

Work has the same dimensions as energy; in fact, work *is* energy. We began our discussion of energy by saying that it is the capacity for doing work.

We can examine Eq. (1-2) from the viewpoint of the dimensions of the quantities involved and determine the dimensions of force. Work of course has the dimensions of energy, mass-length^2/time^2, and distance d adds the dimension of length. We have then

$$\text{mass-length}^2/\text{time}^2 = f \times \text{length} \tag{1-3}$$

Dividing both sides of the equation by the dimension of length, we find that

$$f = \text{mass-length}/\text{time}^2 \tag{1-4}$$

In the cgs system, one gm-cm/sec^2 is called the *dyne*. An erg is then a dyne-cm.

After work has been done on an object, or to use a more scientific term, on a system, the system has acquired energy in the amount of the work done, less any energy lost through heat loss from the system. If we are talking about mechanical energy, the energy possessed by the system may be either potential, or kinetic, or both. To illustrate, consider the case of an archer with a bow and arrow. The archer does work on the bow by pulling the string back into shooting position. The work done by the archer is stored as potential energy in the bow. At the moment that he lets go of the arrow, the bow begins to do work on the arrow by accelerating it. The potential energy of the bow is converted into the kinetic energy of the arrow. If the arrow is shot horizontally, the potential energy of the arrow in the earth's gravitational field decreases as the arrow falls toward the ground; the arrow acquires a kinetic energy in the vertical direction as it loses potential energy. If the arrow is shot vertically into the air, the kinetic energy which it possesses at the instant that it leaves the bow is gradually converted into potential energy as it rises. At the top of its flight, it has lost all kinetic energy, and its potential energy is at a maximum. As it falls back to earth, its potential energy decreases while its kinetic energy increases. It then strikes the earth, and upon impact converts kinetic energy into heat.

It is well to remember that in discussing both kinetic and potential energy there is a certain kind of relativism involved. An object in motion must be in motion *with respect to something.* The arrow in our illustration is in motion with respect to the bow, or the archer. Similarly, potential energy is measured with respect to some reference. In the case of the arrow, we might have arbitrarily said that the zero of potential energy is the surface of the earth. Whatever reference is picked for the potential energy, the convention used in this text is that the more stable potential energy position corresponds to the lower potential energy. Thus the arrow shot vertically reaches a maximum value of potential energy at the top of its flight; its potential energy is lowest at the ground level. If we decided to set the zero of potential energy at a height of four feet, then the arrow would have a negative potential energy when it comes to rest on the ground.

Our discussion of the arrow brings up the question of the distinction between mass and weight. The law of gravity states that the force of attraction of two objects for one another is given by the relation

$$F = G\frac{m_1 M_2}{r^2} \tag{1-5}$$

G is a proportionality constant — the gravitational constant m_1 is the mass of one object, M_2 the mass of the other, and r is the distance separating the centers of mass. For an object at sea level on the surface of the earth, the quantity GM_2/r^2 is a constant, where M_2 is the mass of the earth. We then have $F = gm_1$, where g is a constant representing the quantity GM_2/r^2. Since F is a force, we know its dimensions: mass-length/time^2. Since m has the dimensions of mass, we have for the dimensions of g, length/time^2. But these are the dimensions of acceleration, which is the rate of change of velocity with respect to time. Velocity has the dimensions of length/time (*e.g.*, miles/hour, or cm/sec). Acceleration is then (length/time) /time, or length/time^2 (*e.g.*, miles/hour^2, or cm/sec^2). In the metric system, g has the value 980 cm/sec^2 at sea level. Once g is known, the mass of an object can be determined by measuring the force which the earth's gravitational field exerts on the object. This is always done in a comparative way. A standard mass is chosen, and the masses of other objects are determined by measuring the relative force which the gravitational field exerts on the two objects. This can be done with a balance, such as that used for precise chemical laboratory work. It is important to note, however, that whereas the *mass* of an object is an inherent property, the term *weight* refers to the force exerted by the earth's gravitational field. We have all heard of the weightlessness experienced by astronauts in orbiting space vehicles. There is no net force acting on an orbiting object a few hundred miles out in space, directing it toward the earth, because the orbital motion gives rise to a force which just balances the gravitational force. The weight of the object, which is really a measure of the net force acting on it, is therefore zero. Its mass, however, remains the same, regardless of its location.

Heat energy is the energy which we associate with what we call temperature. As we shall discover in much more detail later, the atoms, molecules or ions of which any substance is composed are in a state of constant motion.

The temperature of a body is a measure of the motional energy possessed by the atoms, molecules or ions. In a gas or liquid, the motional energy may consist of translation, motion through space, and — in the case of molecules, which consist of more than one atom — rotational or vibrational motion. In solids, motion through space is not possible because the individual units of the solid are held in position. There is vibrational motion, however, of each unit about its equilibrium position in the solid. In measuring the temperature of a solid, liquid, or gas, we do not measure the energy possessed by any one unit, but an average value for the atoms, molecules, or ions, of which the substance is composed.

Temperature is measured with respect to any one of a number of scales. In scientific work, the centigrade or Celsius (°C) and Kelvin (°K) scales are most commonly used. The Kelvin scale is an absolute temperature scale, in which zero degrees represents ideally the lowest attainable temperature. The absolute temperature scale is defined, except for a constant multiplier factor which varies from one temperature scale to another, by the second law of thermodynamics. Thus, the Kelvin and Rankine scales, which are both absolute temperature scales, begin at the same point — absolute zero. The Rankine scale, however, has smaller scale divisions than the Kelvin scale (Fig. 1-1). The centigrade scale is defined by the freezing point (0°C) and the boiling point (100°C) of water at one atmosphere pressure. The fixing of these two points defines the size of the centigrade-scale division. The Fahrenheit scale is defined so that the ice point is 32°F and the boiling is 212°F. This leaves 180 divisions between the two refer-

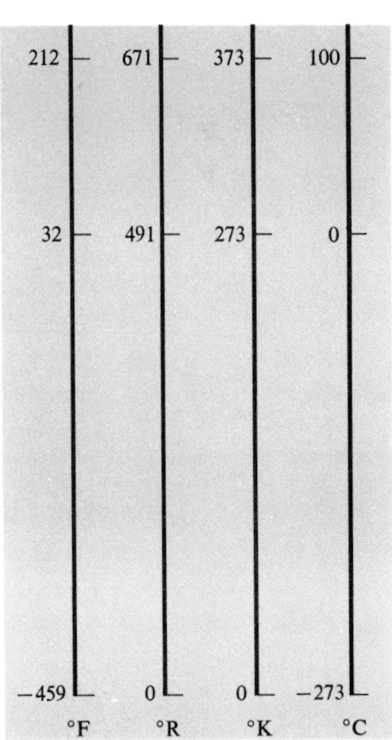

Fig. 1-1 A comparison of various temperature scales.

ence points on the Fahrenheit scale as compared with 100 divisions on the centigrade scale. The Fahrenheit-scale division is therefore only $\frac{5}{9}$ as large as the centigrade scale division. From these facts it is easy to derive the familiar conversion formula,

$$°F = \tfrac{9}{5}°C + 32$$

The Rankine scale is defined so that one degree Rankine is the same as one degree Fahrenheit. Similarly, one degree Kelvin is defined as the same as one degree centigrade. We can summarize by saying that an absolute temperature scale must start at absolute zero, but it could have any size scale division. The Fahrenheit and centigrade scales, on the other hand, not only do not have the same size scale division, they do not have the same zero of reference. Since the use of the centigrade and Kelvin scales in scientific work is universal, it is only necessary to remember the reference points for the centigrade scales, and the fact that the centigrade and Kelvin scales have the same size divisions, with 0°C = 273.15°K.

Chemical energy is a form of potential energy. Chemical reactions involve rearrangements of atoms, molecules, or ions. The heat evolved in a chemical reaction represents a change in the total potential energy of the atoms, molecules, or ions of the products as compared with the initial reactants. When gasoline is combusted in an automobile engine, the gasoline and oxygen molecules are converted into carbon dioxide, carbon monoxide, and water. These products represent a lower potential energy situation for the atoms than for the initial reactants, and heat is evolved. The evolved heat is used, in part, to expand the gaseous products in the cylinder. The work done in this expansion is converted in part to the kinetic energy of motion of the auto.

Electrostatic energy is a form of potential energy resulting from the positions of charges in space. Coulomb's law states that the force between two charges Q and q, separated by a distance r is

$$F = KqQ/r^2 \tag{1-6}$$

The constant K depends on the units chosen for force, charge, and distance. In the cgs system, the charges are expressed in electrostatic units (esu). The esu is defined by saying that two equal charges separated by one cm exert a force of one dyne on one another if each charge is one esu. K is unity in this system of units. For the sake of future reference, we may note here that an electron has a charge of 4.8×10^{-10} esu.

The potential energy of interaction of two charges is given by

$$V = qQ/r \tag{1-7}$$

(We have omitted the proportionality constant K, since we will usually be working with a value of unity for it.)

If q and Q are of opposite sign, the sign of the potential energy V is negative. This corresponds to attractive interaction. When r is infinitely large, V approaches zero. This condition then serves as the reference point for our consideration of potential energy. Since attractive interaction leads to a more stable situation as the distance r is decreased, the potential energy decreases. This is then consistent with the convention we described earlier, that more stable potential energy situations correspond to lower values for the potential energy. If the charges q and Q have the same sign, V is positive and becomes more positive as r decreases. When the charges q and Q are of the same sign, the electrostatic interaction is repulsive, and a less stable configuration results as the charges approach.

Electric fields are frequently encountered in discussing experiments related to charged particles. Suppose that the parallel plates of a condenser are connected to a voltage source of $\mathcal{V}$ volts, as in Fig. 1-2. Current flows

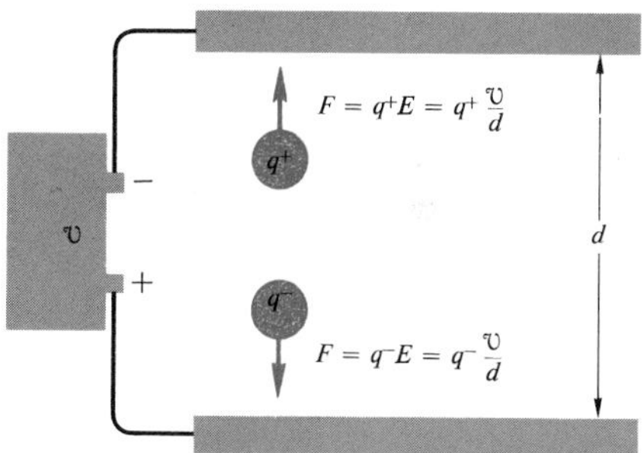

Fig. 1-2 Force on a charged particle in an electrostatic field.

until a certain quantity of charge of opposite sign is present on each of the plates. The amount of charge is proportional to the voltage $\mathcal{V}$ of the source, and to the capacitance C of the condenser. When the current ceases to flow, a voltage $\mathcal{V}$ exists across the capacitor, and an electric force field $E = \mathcal{V}/d$ exists between the plates. Now if a charge of magnitude q is placed in the space between the plates, it experiences a force $F = qE$. The direction of the force depends upon the sign of the charge and the signs of the charges on the plates; a negative charge experiences a force in the direction of the positive plate, and *vice versa.* The charge q also possesses potential energy by virtue of its location in the field between the plates. The electric field may be likened to a gravitational field, as depicted in Fig. 1-3. A positive charge which is close to the positively-charged plate of the capacitor possesses a high potential energy, just as does a mass M which has been lifted in the earth's gravitational field. As the charge moves toward the negatively-charged plate it loses potential energy, just as the mass does in falling. If the charge q is associated with a material object which possesses mass, then the loss in *electrical* potential energy may be seen as a gain in kinetic energy as the charged particle moves in the electric field. Examples of charged particles in electric fields will be dealt with in later material.

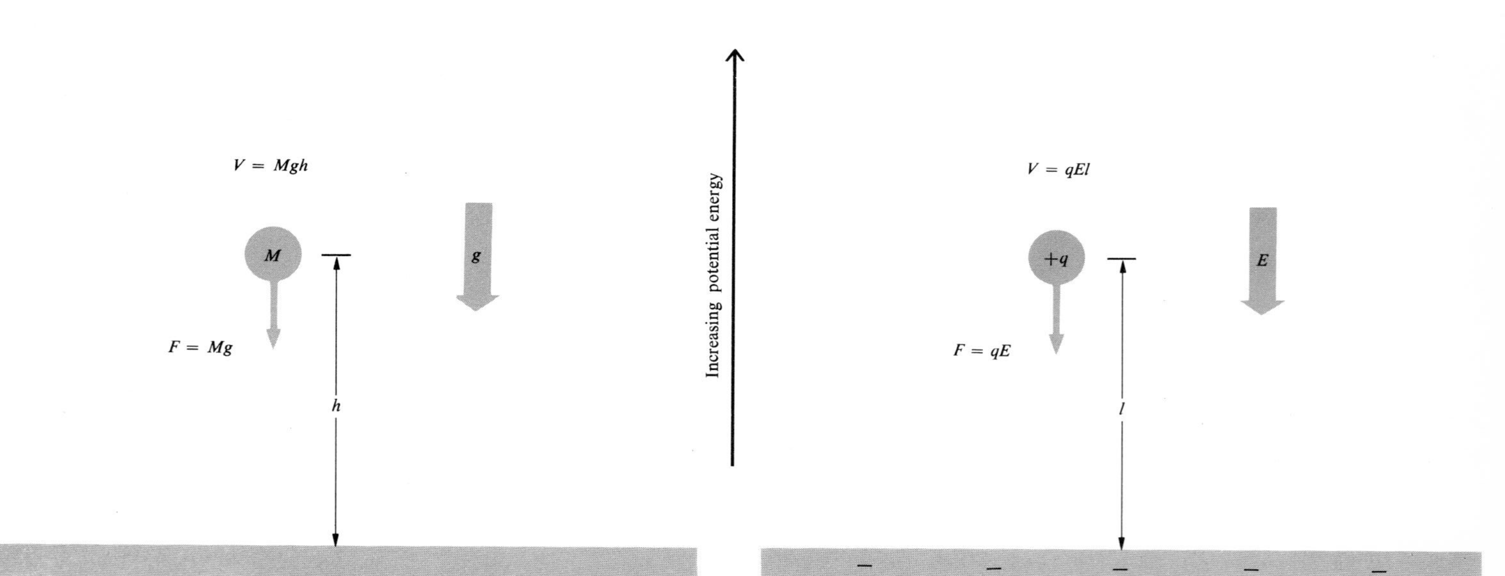

Fig. 1-3 A comparison of electrical and mechanical potential energies.

The energy contained in a permanent magnet, magnetostatic energy, results from the positions of the fundamental units which make up the magnetic material. Magnetostatic energy is thus a form of potential energy.

The motions of charged particles and of magnets give rise to fields and forces which provide the means by which electrical and magnetic energies are converted into other forms of energy. These very important matters are the domain of physics, and we shall not touch upon them in this text except insofar as we have need of them for understanding certain experiments.

Radiant energy is electromagnetic radiation, which has the properties of a periodic wave propagated through space at a characteristic velocity — the velocity of light. These waves are characterized by a frequency (the number of full periods, or cycles, in one second) and by an intensity, which relates to the amplitude of the wave (Fig. 1-4). The product of the frequency

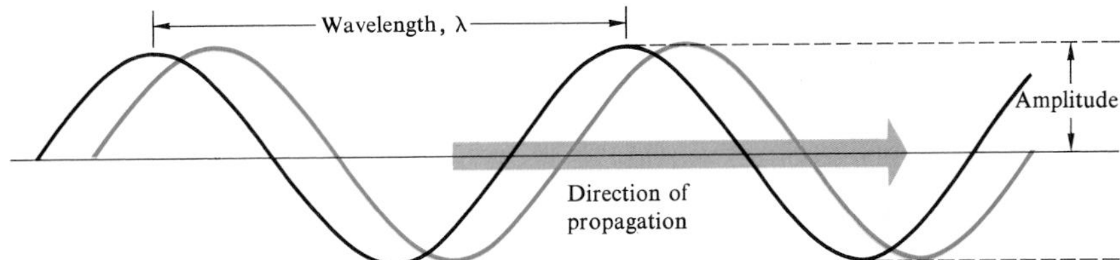

Fig. 1-4 A periodic wave, represented by a sine waveform. The colored line represents the location of the wave a short time interval later, when the direction of propagation is as shown.

ν and the wavelength λ equals the velocity of light: $\nu\lambda = c = 3 \times 10^{10}$ cm/sec. The frequency of a radiant energy may be expressed in cycles/sec; since the unit of cycles is understood, it is common to write just $\sec^{-1}$. The wavelength of the light is similarly expressed in units of length/cycle, or again leaving the dimension of cycle out, just in unit of length. The wavelength may be expressed in meters, centimeters, microns (10^{-6} meters) or Angstroms ($1\text{Å} = 10^{-8}$ cm). Figure 1-5 shows the range of wavelengths which characterize familiar kinds of radiant energy. There is fundamentally, of course, only one kind of electromagnetic radiant energy, but there appear to be many kinds because the interactions of radiant energy with matter are diverse, depending on the wavelength of the radiation. Short wavelengths correspond to high energy radiation, long wavelengths to low energy. Emission and absorption of radiant energy by matter are two closely related phenomena which have been of great interest to physicists and chemists alike. We will now consider one particularly important aspect of this subject, the so-called black body radiation.

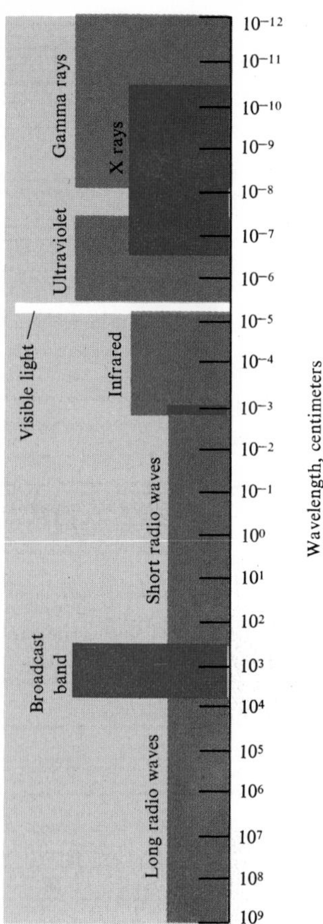

Fig. 1-5 A representation of the electromagnetic spectrum.

1-3 The Quantum Theory

A black body or surface is a good absorber and emitter of radiant energy. The energy emitted from a black body at any particular temperature is not of just one frequency, but consists of a whole range of frequencies, or wavelengths. The emissive power of a black body is experimentally determined by measuring the emission from a small opening in the side of a uniformly heated furnace, as illustrated in Fig. 1-6.

The radiant energy emitted from the furnace is dispersed in a spectrograph (Chap. 4) to separate the radiations of different wavelengths. The total energy in each unit of wavelength is then determined and plotted *vs* the wavelength value for the center of the interval. A curve such as that shown in Fig. 1-7 results. As the furnace temperature is increased, the shape of the curve changes: The total emitted radiation is greater, and the

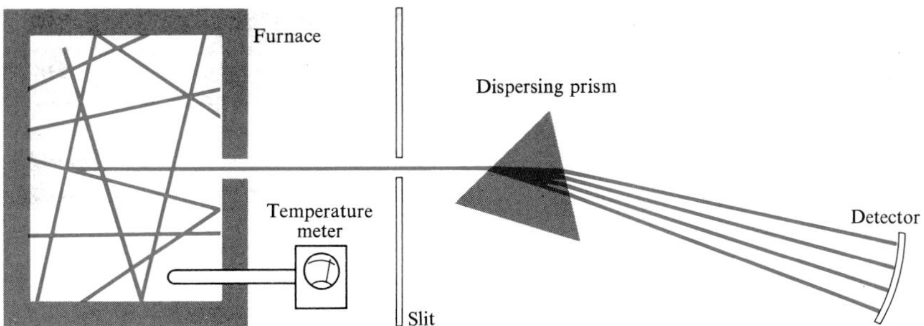

Fig. 1-6 An example of black body radiation. The small hole in the furnace wall serves as a black body surface, with a temperature equal to that in the furnace.

wavelength at which the maximum energy per unit wavelength occurs grows shorter.

The physicists of the nineteenth century attempted to explain these results by assuming that the molecules or atoms of the body were emitting or absorbing energy of all the various wavelengths. It was also assumed that *energy could be emitted or absorbed in arbitrarily small amounts.* The theory based on these assumptions led to the conclusion that the radiation should be distributed among the various wavelengths as shown by the colored line in Fig. 1-7. The disagreement with experiment was quite serious, and could only mean that the theory was based on incorrect assumptions.

Professor Max Planck, a German physicist, discarded the assumption that the energy could be emitted or absorbed in arbitrarily small amounts, and assumed instead that it can be emitted or absorbed only in discrete quantities, in multiples of some smallest unit, called the *quantum.* A crude

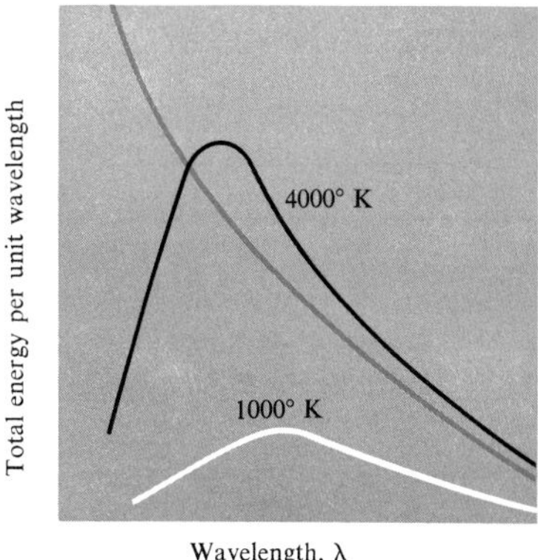

Fig. 1-7 Black body radiation as a function of wavelength.

analogy might be as follows: Suppose that one has an apple tree laden down with a large number of apples. We can say that there is a total amount of "apple" on the tree. This tree loses "apple" as the individual apples become ripe and fall off. Now, it is not possible for the tree to lose "apple" in arbitrarily small amounts, but only in multiples of a single, whole apple. We could say then that the loss of "apple" from the tree is quantized, and the smallest unit or quantum of "apple" is a single, whole apple. In the same way, energy is lost (or gained) by a black body in units of some smallest quantity, or quantum. The magnitude of this quantum of radiant energy is proportional to the frequency ν:

$$E = h\nu \tag{1-8}$$

The proportionality constant h is called Planck's constant. It has the value 6.63×10^{-27} erg-sec. This means that the furnace cannot radiate energy at a frequency ν which corresponds to an energy quantity $\frac{1}{2}\, h\nu$, or $\frac{1}{4}\, h\nu$, or any other quantity than some whole-number multiple of $h\nu$. In the classical model, *any* quantity of energy might be emitted with a frequency ν.

When Planck applied the new quantum assumption to the black-body radiation problem, the theory was in good agreement with experiment. It is difficult to overemphasize the importance of this new development: It required an entirely new and different view of the nature of radiant energy. It had now become possible to think of this energy as a stream of energy bundles much like particles, except for their being massless.

1-4 The Photoelectric Effect

The quantum theory was applied to a number of important problems in the years following 1900, when it was first published. Einstein explained the photoelectric effect, in which electrons are emitted from a metal irradiated with light of sufficiently high frequency, on the basis of the quantum theory. Each quantum of light energy is termed a *photon*. When a photon strikes an electron in the metal, the energy of the photon is transferred to the electron. If the energy of the photon is sufficiently large (*i.e.*, if the product $h\nu$ is large enough), the electron will have acquired enough kinetic energy to overcome the attractive forces holding it in the metal, and will be emitted. Below a certain threshold frequency, therefore, there will be no emission of electrons since the photons do not possess the required energy. If the energy of the photon is greater than that required to remove the electron, the excess energy is imparted to the electron as kinetic energy. We can express the kinetic energy of the electron as

$$E_k = h\nu - h\nu_0$$

where ν_0 represents the threshold frequency and $h\nu_0$ is equal to the energy required to get the electron just barely out of the metal. The difference between this energy and the energy of the photon, $h\nu$, appears as kinetic energy of motion of the electron. Einstein's theory then leads to two conclusions: (a) If the incident radiation is below a certain threshold frequency, there will be no emission of electrons from the metal, regardless of the intensity of the radiation; and (b) if the radiation frequency is above the threshold value, the emitted electrons will exhibit a certain kinetic energy which depends not upon the intensity of the incident radiation, but upon the frequency (Fig. 1-8). These conclusions are quite different from those

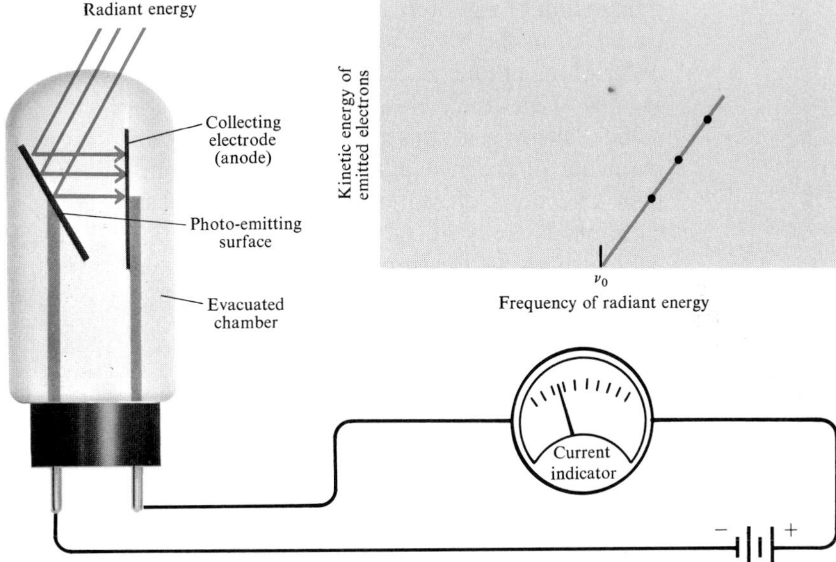

Fig. 1-8 The photoelectric effect.

arrived at from the classical theory. According to the older theory, the metal should be able to soak up energy from the radiation until enough has been absorbed to permit escape of the electron. Thus, even a long wavelength beam of radiation should produce emission of electrons, in proportion to the intensity. With its correctness verified by experimental results, Einstein's theory of the photoelectric effect represented another strong line of evidence for Planck's quantum theory.

As a matter of incidental interest, the energy represented by the threshold frequency is termed the *work function*, and is related to the chemical behavior of the metal.

1-5 The Momentum of the Photon

Einstein's treatment of the photoelectric effect placed emphasis upon the particle-like character of electromagnetic radiation. It is possible, by combining two of the fundamental relationships that we have discussed, to obtain yet another insight into the particulate character of the photon. The energy of the photon, $h\nu$, can be equated to a mass-equivalence, using

Einstein's equation: $E = mc^2 = h\nu$. The equivalent mass of the photon is thus $m = h\nu/c^2$. The momentum of a particle is defined as the product of its mass and its velocity. Since the velocity of the photon is c, the velocity of light, its momentum is then:

$$mc = h\nu/c = h/\lambda \tag{1-9}$$

Insertion of numerical values into these expressions reveals that the momentum of a photon is very slight. Nevertheless, at very high frequencies (small λ), in the X-ray region, the momentum of a photon is of the same order of magnitude as that of an electron which is moving about the outer regions of an atom. Now we learn in mechanics that when two particles collide, there is a conservation of momentum; *i.e.*, that the sum of the momenta of the two particles after collision, with proper allowance for their directions of motion, is equal to the sum of the momenta before collision. There is, of course, also a conservation of energy. When a photon collides with an electron, it may transfer some of its momentum to the electron, as shown in Fig. 1-9. The photon which is observed is seen at some

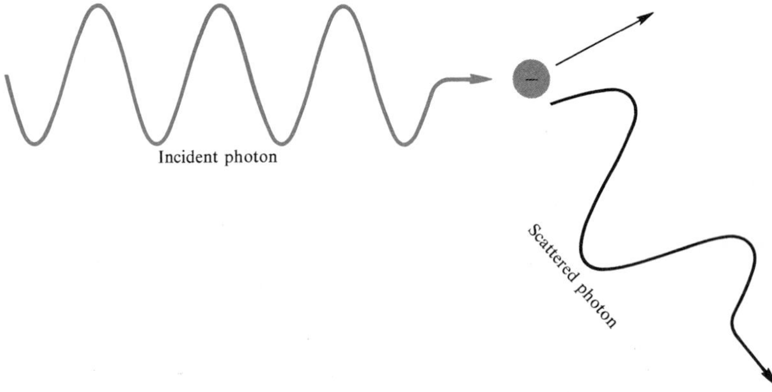

Fig. 1-9 The Compton effect. The wavelength of the photon is changed as a result of transfer of momentum to the electron upon collision.

angle away from the original path of the photon. Most importantly, because the photon has transferred some of its momentum to an electron, it now possesses a lower momentum, and therefore has a lower frequency, or longer wavelength. The effect was discovered and explained by A. H. Compton in 1923, and is termed the Compton effect. It represents still another confirmation of the quantum theory.

If it is assumed that the particle with which the photon collides is at rest, then it can be worked out that the change in wavelength of the scattered photon (in Å) is related to the angle which the path after collision makes with the path before, by the following equation:

$$\Delta\lambda = \frac{2h}{mc}\sin^2\frac{1}{2}\theta \qquad (1\text{-}10)$$

It will be seen from this that the mass of the particle which the photon strikes is in the denominator. The effect is actually quite small even for X rays striking electrons; the coefficient before the angle term in Eq. (1-10) turns out to be about 4.86×10^{-10}. For heavier particles, such as protons or neutrons, it would probably be impossible to observe the effect. We all know from common experience that when two objects collide, the lighter one is the one that rebounds. In a collision, transfer of energy and momentum does not occur to a significant extent unless the two colliding objects possess similar "mass." Compare, for example, the collision of two billiard balls with a collision of one of the balls with the rail of the billiard table.

1-6 Energy Units

We have already seen that the unit of energy in the cgs system is the erg, which is one gm $\times$ cm^2/sec^2. There are, however, a large number of other energy units in use. Many of these are listed in Table 1-1. One of the most common energy units is the calorie, which is used to express heat energy. A calorie was originally defined as the heat required to raise the temperature of one g water from 14.5°C to 15.5°C. It is now defined as equal to 4.184×10^7 ergs. The joule is also commonly employed as a measure of heat energy. One joule is defined as 10^7 ergs. The electron-volt is a unit of energy which is equal to the energy that a single electron acquires by being accelerated through an electric field of one volt. We will refer to this unit of energy in more detail later.

Table 1-1 Some commonly used units of energy.

Unit of energy	Value
erg	1 g-cm^2/sec^2
joule	10^7 ergs
calorie	4.184×10^7 ergs
electron-volt	1.602×10^{-12} ergs
liter-atmosphere	1.01×10^9 ergs

The liter-atmosphere is a unit of energy which results from multiplication of pressure times volume. To see how units are handled in problems of this type, let us see what combination of pressure and volume results in energy units. Pressure (P) is a measure of force per unit area; force is expressed in the cgs (cm-g-sec) system in dynes, which have the dimensions g $\times$ cm/sec^2. Pressure is dynes per cm^2, (g $\times$ cm/sec^2)/cm^2 = g/(cm $\times$

sec^2). Volume (V), of course, has the units cm^3, so that $P \times V$ has units of g $\times$ cm^2/sec^2. These are the units of ergs, as we saw above, so that $P \times V$ has the units of energy. If the pressure is expressed in atmospheres, it is necessary to convert from this unit to dyne/cm^2; similarly, if the volume is expressed in liters, a conversion is necessary. When these conversions are applied, it is found that

$$\begin{aligned} 1 \text{ liter-atmosphere} &= 1.013 \times 10^9 \text{ ergs} \\ &= 24.22 \text{ calories} \end{aligned}$$

Suggested Readings

Darrow, K. K., "The Quantum Theory," *Scientific American*, (March, 1952).

Emmerich, W., M. Gottlieb, C. Helstrom, and W. Stewart, *Energy Does Matter*, New York, N. Y.: Walker & Company, 1964.

Slater, J. C., *Quantum Theory of Atomic Structure*, Vol. 1, Chap. 1, New York, N. Y.: McGraw-Hill Book Company, Inc., 1960.

Excercises

1-1. (a) Name some of the units in which energy can be expressed, and show the factors which must be applied for conversion of these into calories.

(b) Determine the conversion factor from joules to calories.

1-2. How do the terms heat and temperature differ in meaning?

1-3. Carry out each of the following temperature conversions:

(a) 78°F to °C. (d) −89°F to °C.
(b) 478°K to °F. (e) −189° to°K.
(c) 34.5°C to °K. (f) 13.5°R to °F.

1-4. Energy is often readily converted from one form to another. Give an example of each of the following kinds of energy conversions:

kinetic → potential.
chemical → electrical.
radiant → chemical.
radiant → electrical.

1-5. Climbing 20 ft up a ladder, a 160-lb man does about 3.6×10^{10} ergs of work against gravity. If the energy for the work is supplied by the metabolism of sucrose in the man's body, what is the minimum weight of sucrose required, assuming that it is all converted into carbon dioxide and water according to the equation

$$C_{12}H_{22}O_{11} + 12O_2 \rightarrow 12CO_2 + 11H_2O$$

with an evolution of 1349 kcal per 342.3 g of sucrose? In calculating the minimum weight required, assume that the energy released in the conversion of sucrose can be converted completely into work.

1-6. The potential energy of an object near the surface of the earth is given by the product of the force (which we saw is $F = gm$) and the height through which the object is raised relative to the reference point: $V = gmh$. Suppose that a rifle is fired straight upward, and the bullet, weight 18 g, leaves the muzzle with a velocity of 2400 ft/sec. Assuming no air resistance, to what height will the bullet rise? (Recall that at the top of the bullet's flight all of its kinetic energy is converted to potential energy.)

1-7. Assume that the bullet of the previous problem falls into a sand bucket on returning to earth. How many calories of heat will be gained by the sand and bullet as a result of the loss of motional energy?

1-8. Kinetic energy is given by the formula $\frac{1}{2}mv^2$, where m is the mass of an object moving with velocity v. What is the kinetic energy in ergs of a 3550-lb automobile traveling at a speed of 60 miles/hr? Assuming that all of this kinetic energy is converted into heat, how many calories are evolved when the brakes are applied, and the car is brought to a stop?

1-9. Draw an accurate graph of the potential energy of interaction between two charges of one esu, but unlike sign, in the range $r = 1$ cm to $r = 10^{-2}$ cm. Do the same for a pair of 1 esu charges of like sign.

1-10. If 2 cal per cm^2 per min of radiant energy is received at the earth's surface at that point on the earth where the sun is directly overhead, what is the approximate total amount of energy received by the earth in 24 hours? What is the mass-equivalence of this energy?

1-11. A black-body surface can be characterized as a perfect absorber of radiant energy. Explain how a small hole in the side of an otherwise closed box gives a good approximation to a black body.

1-12. What is the frequency of an electromagnetic radiation which possesses a characteristic wavelength of 10 cm?

1-13. The combustion of 92 g of xylene results in the evolution of 1080 kcal of heat. To what loss in mass does this correspond?

1-14. Suppose that the Compton effect experiment is set up, with provision for viewing the scattered photons at an angle of 45° with the path of the unscattered beam. Calculate the change in wavelength to be expected. What fraction of the wavelength is this for a typical X ray (1.5 Å)? For a typical line in the red part of the visible spectrum (7000 Å)?

1-15. Calculate the energy of a quantum of radiant energy if the wavelength of the radiation is 5×10^{-4} cm. [*Hint:* Use the relationship between wavelength and frequency $\nu\lambda = c$.]

1-16. The wavelength limits of human vision are approximately 7000 Å on the low energy side, and 4000 Å on the high-energy side. Calculate the energy of a photon of each of these wavelengths.

1-17. Give the dimensions, in terms of mass, length and time, of each of the following quantities:

(a) velocity.
(b) density.
(c) Planck's constant.
(d) acceleration.

Chapter 2

Sub-Atomic Particles

2-1 Introduction

The idea that matter is built up from some sort of fundamental building blocks has been with man for a long time. The Roman poet and philosopher Lucretius, who lived during the century before Christ, summarized the most advanced natural philosophy of western civilization in his poem, *De Rerum Natura.* He followed the Greek philosophers in asserting that all matter is composed of ultimate particles — indestructible and indivisible — called *atoms.* The atoms of different substances, he argued, differ from one another. The atoms of a substance which has a bitter taste or which burns the flesh must have a rough or barbed surface, whereas the atoms of an inoffensive or bland substance must have a smooth surface.

In 1805, John Dalton advanced the hypothesis that matter is composed of small, indivisible particles called atoms. His theory was much more than a reiteration of Lucretius', however, because Dalton went on to say that, in the reactions of pure elementary substances with one another, it is the atoms which combine to form compounds. With some modifications, this hypothesis withstood the test of time and experiment and became the atomic theory. It was — and is — quite adequate in explaining the weight relationships involved in chemical reactions, but it does not in any way illuminate the question of what causes the atoms to react with one another, or what happens when they react.

Only recently have we had experimental procedures with which to explore the structure of matter at the subatomic level. In this chapter, we will discuss evidence for the existence of the fundamental particles which are of principal concern to chemists. In so doing, we will discuss only a small fraction of the known number, because the majority of these concern only the nuclear physicist, and do not have a direct bearing on chemistry.

We will be concerned in this chapter with the way in which the fundamental charged particles are distributed within the atom. We shall see that the atom is nuclear in character, and that the relatively light, negatively-charged electrons are distributed in a diffuse cloud around the central, positively-charged nucleus. The details of the electron distribution are of vital importance to chemistry, and will be discussed in Chaps. 5 and 6.

2-2 The Electron

The relationship between electrical charge and chemical reaction was demonstrated by Michael Faraday in 1833. Faraday found that when an electrical potential was applied to a pair of electrodes immersed in a salt solution, a current flowed in the circuit. At the same time, chemical reactions occurred at the electrodes. The quantities of substances produced in these chemical reactions varied directly with the total current passed in the circuit. Some years later (1874), the English scientist Stoney inferred from Faraday's laws of electrolysis that discrete units of electricity, which he called electrons, could be associated with atoms. Before one can learn more about electrons, however, they must be detached from the atoms of which they are a part. At about this time, some experiments were being conducted in a seemingly unrelated area. Crookes had discovered that, when a tube containing an anode and cathode connected to a source of about 10,000 volts (Fig. 2-1) was partially evacuated, an electrical discharge

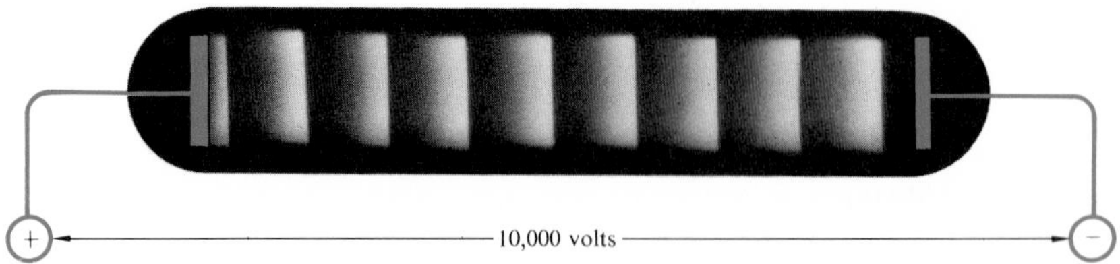

Fig. 2-1 A Crooke's tube. The shadings represent alternating regions of light in the gaseous discharge.

arose. This discharge was observable in the tube as a series of colored regions which changed as the gas pressure was varied. At very low pressures, the light disappeared from the interior of the tube, and the glass walls glowed with a greenish light. If a material object was placed in the tube, a shadow appeared on the otherwise glowing wall; it became clear that the rays which produced the glow were emanating from the negative electrode (cathode). The real significance of these observations was not realized by those who performed these first experiments, but by a later investigator, J. J. Thomson. He is justly famous, not only for the experiments which he performed with the cathode-ray tube in clearing up this problem, but also for the brilliant leadership he exercised as director of the Cavendish Laboratory at Cambridge University in the years 1884–1918.

Thomson concluded that the electron hypothesized by Stoney exists, and he set out to determine some of its properties. From an earlier experiment of Perrin's, it was known that the cathode rays in the Crooke's tube were negatively charged (Fig. 2-2). Thomson deduced that the cathode rays con-

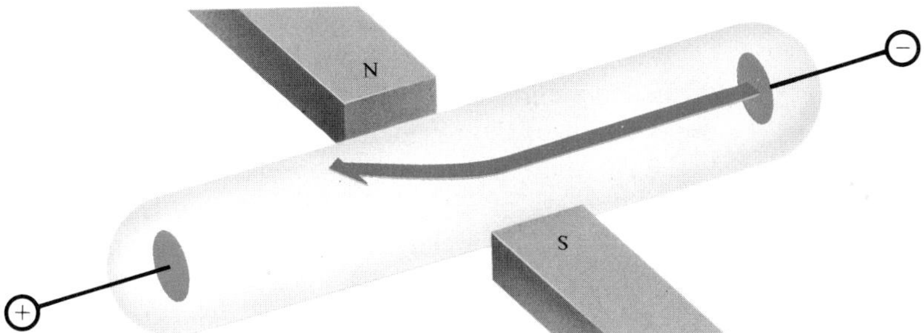

Fig. 2-2 Perrin's experiment. The beam emanating from the cathode is allowed to pass through a magnetic field. From the direction in which the beam is deflected, it is deduced that the particles are negatively charged. The path of the beam is followed with a fluorescent screen, not shown.

sisted of a stream of electrons which were being pulled from the cathode surface. His apparatus for studying the cathode rays is shown in Fig. 2-3. To understand how the experiment was performed and what information could be obtained, it is necessary to consider the effects of an electric field and a magnetic field on a moving electron.

Consider the case in which an electron moving with velocity v_x enters an electric field which is perpendicular to the direction of motion (Fig. 2-4). The electron is acted on by the electric field and is accelerated toward the positive plate. The force which the electron experiences is Ee, the product

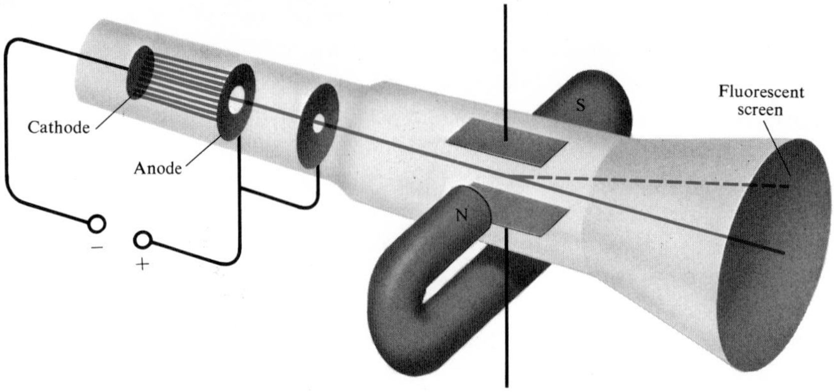

Fig. 2-3 Thomson's experiment to determine e/m for the electron.

of the electric field E and the electronic charge, e. As a result of the deflection in the y-direction, the electron follows a curved path. When the electron has left the vertical electric field, it continues in a straight line.

When an electron enters a magnetic field at right angles to the direction of the magnetic field, it is acted upon by a force Hev_x, where H represents the strength of the magnetic field. This force is at all times perpendicular to the direction of motion of the electron; the path described by the electron is thus the arc of a circle. If the electron leaves the magnetic field before completing a full circle, it continues in a straight line along its path at the time of exit from the field.

Thomson's experiment (Fig. 2-3) consisted of deflecting the stream of electrons by placing a magnetic field across the tube. The direction and magnitude of electric field were then adjusted until the beam of electrons

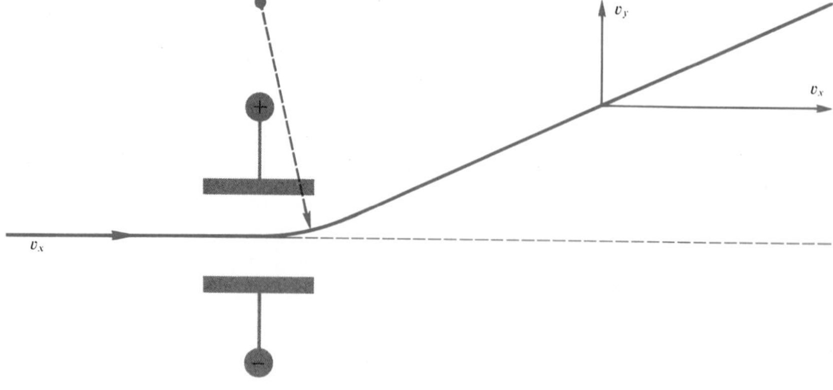

Fig. 2-4 Deflection of a moving electron in an electric field.

returned to its original path. Under these conditions, the two fields were equal and opposite. Then with the magnetic field alone, the deflection of the beam on the screen was determined. From the magnitude of the deflection, and by knowing H and E, it is possible to determine the ratio e/m for the electron.

In succeeding years a number of other experiments were devised for determination of the ratio e/m for the electron. It is significant that the same value was obtained for cathode rays, electrons from a hot cathode, and photoelectrons ejected from a metal by X rays. It thus appears that all of these sources yield the same kind of charged particles. The best value of the ratio e/m is currently $1.7591 \pm 0.0002 \times 10^8$ coulombs/g. (A coulomb is the quantity of charge which flows past a given point in the circuit in one second when the current is one ampere.) This is to be compared with Thomson's original estimate of about 10^8.

The charge on the electron was first measured in 1908 by R. A. Millikan, working at the University of Chicago. His ingenious apparatus is illustrated in Fig. 2-5. A spray of exceedingly fine oil droplets is produced in the upper

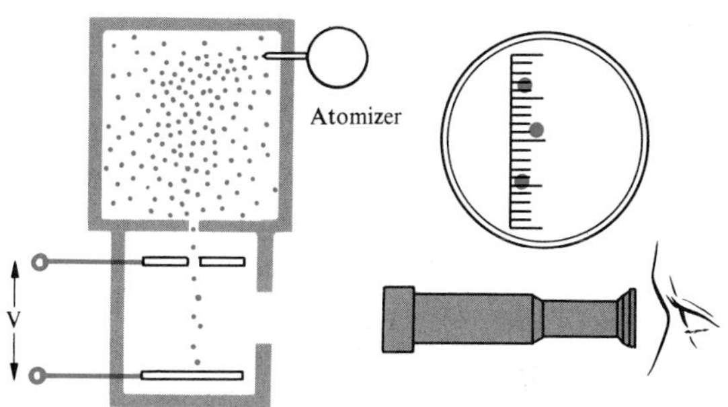

Fig. 2-5 Millikan's apparatus for determination of the electronic charge.

chamber. These settle slowly, and a few pass through the tiny orifice at the bottom and into the field of the telescope. The rate at which a droplet falls through the air (which acts as a viscous medium) is determined by its mass and the force of gravitational attraction. Millikan used his telescope to measure the rate of fall of a particular droplet.

The drops fall because they are acted upon by the force of gravitational attraction. Their rate of fall is limited, however, by the frictional resistance of the air through which they fall. Stokes had shown many years before that for a spherical object the limiting velocity of free fall is given by

$$v = \frac{2}{9}\frac{gr^2(d - d_m)}{\eta}$$

where g is the acceleration due to gravity (see Sec. 1-2), r the radius of the falling sphere, d the density of the sphere, and d_m the density of the medium through which the sphere falls, in this case, air. η is the viscosity of the medium. It takes the falling oil droplets only a very short time to attain the limiting velocity of fall. Since v could be measured, and η, g, d and d_m are known, the radius of a particular sphere could be determined. (In performing his experiments, Millikan found it necessary to add a slight correction term to Stokes' law in order to obtain agreement with measured radii.)

The force acting on the oil droplets is a constant force given by the product of the gravitational constant, g, times the mass of the drop, which in turn is the product of its volume and its density, with a correction for the buoyancy of the air:

$$\text{force downward} = \tfrac{4}{3}\pi r^3 g(d - d_m)$$

Millikan placed a radioactive source close by to produce electric charges on some of the drops. He then applied an electric field E between the plates shown in Fig. 2-5. By varying the potential and using the correct polarity he could cause a charged droplet to move at a certain rate, and even to move upward. The electric field has no effect on neutral oil drops, but it does exert a force of magnitude Eq on those which have acquired an electric charge q. The net force *upward* is then the sum of the gravitational force and the electrostatic force:

$$\text{net force upward} = Eq - \tfrac{4}{3}\pi r^3 g(d - d_m)$$

If it is assumed that the velocity of the drop is proportional to the force acting upon it, then the ratio of the velocity before application of the electric field, v, to the upward velocity with the field on, v_e, is

$$v/v_e = \frac{\tfrac{4}{3}\pi r^3 g(d - d_m)}{Eq - \tfrac{4}{3}\pi r^3 g(d - d_m)}$$

Solving for q we obtain:

$$q = \frac{4\pi r^3 g(d - d_m)}{3E} \times \frac{(v + v_e)}{v}$$

Using the radius r obtained from measurements of velocity of free fall (*i.e.*, with the voltage off), and having measured v and v_e, q can be calculated.

Millikan observed the behavior of many drops, and obtained a great number of different values for q. Millikan noted the very striking fact, however, that he never obtained a value for q less than 4.77×10^{-10} esu, and that all his values were some multiple of this smallest value. He obtained q values which were 1, 2, 3, 4, and so on, up to much higher values of the smallest value. It was clear to Millikan that the electronic charge is a discrete quantity, and that the fundamental unit of charge is about 4.8×10^{-10} esu in magnitude. This charge, he further postulated, is the charge of the electron. Expressed in coulombs, it is 1.6×10^{-19}.

By combining the observed ratio of e/m, 1.76×10^8 coulombs/g, with the value of e, 1.60×10^{-19} coulombs, we obtain the value for m, the mass of the electron:

$$m = \frac{1.60 \times 10^{-19} \text{ coulombs}}{1.76 \times 10^{8} \text{ coulombs/g}} = 9.11 \times 10^{-28} \text{ g}$$

2-3 Positive Rays

It was soon discovered that there is a second kind of ray in addition to the stream of electrons emanating from the cathode in discharge tubes. This is observed by making a small hole in the cathode surface and studying charged particles which move into the space behind the hole. The particles, which constitute this second ray, are evidently accelerated toward the negative electrode, and are thus called positive rays. Some of them pass through the hole in the cathode to the space beyond. Attempts to deflect the particles with magnetic and electric fields showed that they are positively charged, and of a much lower e/m-ratio than the electrons. It was also observed that *the ratio e/m depended upon the gas present* in the tube. The highest ratio was obtained with the gas hydrogen. These positive rays are produced in the tube by collision of the high-speed electrons with the gas atoms. The collision may occur with sufficient force to knock an electron from the gas atom, thus producing a positively charged particle with a mass very little different from that of the original atom. By using the same sort of techniques described above for electron beams, it is possible to measure the ratio e/m for the positive rays with great accuracy. The present-day mass spectrometer, an instrument of great importance in research, is a development from the early instruments built for this purpose (Fig. 2-6). Positive ions are formed in the ion-forming chamber by allowing a stream of energetic electrons to strike a beam of molecules. Except for the thin stream of molecules of the substance under study, the system is under high vacuum. This means that as much gas as possible has been pumped from the system, so that the positive ions formed do not undergo frequent collisions. The positive ions are caused to move down the ion tube by the

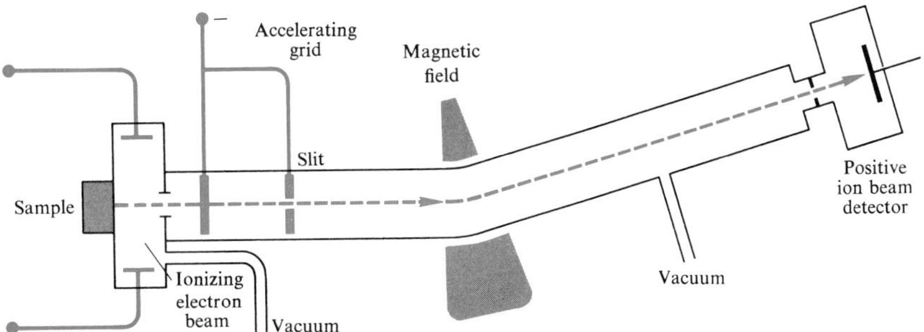

Fig. 2-6 Schematic drawing of a modern mass spectrometer.

presence of a grid at negative potential. After passing through the grid, they move along at a constant velocity. As they enter the region of the magnetic field, they are caused to move in a curved path, as described in the last section. All the ions have a single electronic charge, or some multiple of this, on them. The extent to which an ion of given charge is deflected is therefore determined by its mass and velocity. Thus, for a given voltage on the accelerating grid, and for a given magnitude of magnetic field, only ions of a certain e/m-ratio will pass through the exit slit and reach the ion-beam detector. The spectrometer is swept through a range of e/m-values by steadily changing the magnetic field strength. The type of spectrum which results is shown in Fig. 3-5.

2-4 The Nuclear Atom

The distribution of the positive and negative charges in the atom is a question of great importance. The answer came through the brilliant research work of Ernest Rutherford, who received his training under J. J. Thomson, and who returned in later years as Director of the Cavendish Laboratory. Rutherford's experiment consisted of placing an extremely thin gold foil (5000 Å thickness)* before a source of alpha particles (helium nuclei), as shown in Fig. 2-7. The alpha particles which struck the gold

Fig. 2-7 Rutherford's experiment on deflection of alpha particles by gold foil. The fluorescent screens employed by Rutherford to detect the alpha particles are replaced here by a modern detector which can be set at any angle to count the density of scattered particles.

foil were subsequently detected on screens which luminesce when a particle strikes. It was observed that nearly all of the alpha particles passed through the foil without deflection. A small number (about 1 in 100,000) were deflected. Many of these were deflected at large angles, with the original path as shown in Fig. 2-7.

*One angstrom (Å) equals 10^{-8} cm.

The interpretation of these results is that the gold foil consists largely of space which is devoid of any significant amounts of matter. The alpha particles, while much heavier than an electron, are nevertheless much less massive than the atoms of the metal from which the foil is made. The alpha particles are emitted from the radioactive source with very high energy. The fact that only a small fraction of them is deflected, and some through a large angle, implies that (a) most of the particles encounter no obstacles at all in their passage through more than a thousand layers of metal atoms, and that (b) those which are deflected have encountered a mass which is larger than the alpha particles. The mass of the metal atoms must therefore be concentrated in a small volume. These conclusions are illustrated in Fig. 2-8.

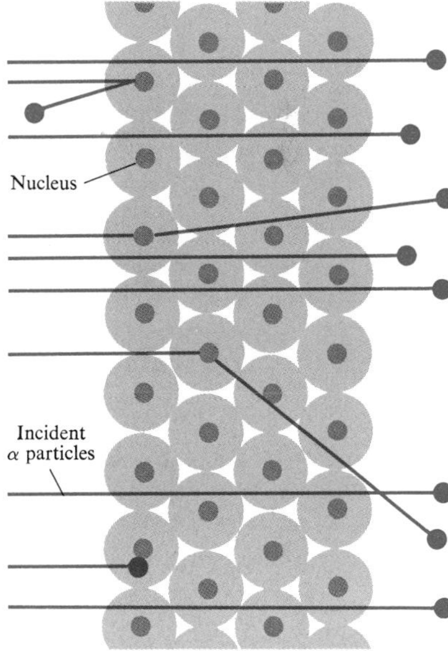

Fig. 2-8 Scattering of alpha particles by the nuclei in a metal foil.

Rutherford reasoned that a massive element of small volume, the nucleus, must contain the positive charge. This reasoning is borne out by detailed analysis of the angular distribution of deflected alpha particles. The much less massive electrons are assumed to occupy the volume around the nucleus, and to impart to the atom its large apparent volume.

2-5 The Proton and the Neutron

The unit of positive charge in the nucleus is termed the *proton*. It possesses a mass some 1847 times that of the electron, and the same magnitude of charge (though of opposite sign) as the electron. The number of protons in the nucleus of an atom is termed the atomic number. All of the atoms of a given element possess the same atomic number; *i.e.*, they all have the same number of protons in the nucleus; and, since the number of electrons must equal the number of protons to maintain electrical neutrality, they all have the same number of electrons in the outer part of the atom.

It was soon evident from studies of positive rays in the mass spectrometer and from simple inspection of the values of atomic weights, however, that atoms which possess equal numbers of protons do not necessarily have the same mass. It became necessary to postulate the existence of another particle, of about the same mass as the proton but possessing no charge, as a constituent of the atomic nucleus. This particle, called the *neutron*, was suggested by Rutherford as early as 1920, on the basis of the preceding arguments. He thought that the neutron might be produced by the union of a proton and an electron. (This view is not accepted at present.) A particle with no net charge is considerably more difficult to detect than one which is charged. It was not until 1932 that the English scientist Chadwick produced unequivocal evidence for its existence as a free particle. The mass of the neutron is determined by the weight relations in nuclear reactions which involve the production of a neutron:

$$^{11}_{5}\text{B} + ^{4}_{2}\text{He} \rightarrow ^{14}_{7}\text{N} + ^{1}_{0}n$$

(The subscript refers to the charge on the nuclear particle, the superscript to the mass number.) The properties and designations of the electron, proton, and neutron are summarized in Table 2-1.

Table 2-1 Properties of the fundamental particles.

Particle	Symbol	Mass (atomic mass units)	Charge
Electron	e^- or β	0.0005486	$-1(1.60 \times 10^{-19}$ coulombs)
Proton	$^{1}_{1}\text{H}$ or p	1.00757	$+1(1.60 \times 10^{-19}$ coulombs)
Neutron	$^{1}_{0}n$	1.00893	0

2-6 Isotopes; Atomic Weight

Atoms which possess the same number of protons may differ in the number of neutrons in the nucleus. They are then called *isotopes* of one another. Every element in the periodic table exhibits isotopic behavior. As an example, the naturally occurring isotopes of tin, element number 50, are shown in Table 2-2.

Each particular combination of protons and neutrons is called a *nuclide*. Not all of the isotopes of an element are necessarily stable. If a given nuclide is not stable, it decomposes in one of a number of possible ways, and is said to be radioactive. Many nuclides which are not found in nature because of their instability have been synthesized in recent years. It is

Table 2-2 Isotopic composition of tin, element 50.

Mass number	Relative abundance (per cent)
112	0.95
114	0.65
115	0.35
116	14.24
117	7.57
118	24.01
119	8.58
120	32.97
122	4.71
124	5.98
Observed atomic weight:	118.70

possible to detect unstable nuclides by observing the products of their decomposition. The use of these nuclides has proven to be a powerful research tool in following the course of chemical reactions and in many other applications (Chap. 23).

The atomic weight of an element is determined by the relative amounts of the various isotopes of the element which are found in nature. The atomic weight of tin, for example, is the weighted average of the individual isotopic weights, as shown in Table 2-1. With very few exceptions, the isotopic composition of a given element is constant throughout the world, insofar as is known. More will be said about the determination of atomic weights in the next chapter.

Exercises

2-1. Summarize the known properties of the three subatomic particles of principal interest to the chemist: the electron, the proton, and the neutron. State the experimental basis for each property that you list.

2-2 Consider Thomson's experiment (Fig. 2-3). Should the glass envelope be evacuated? Explain.

2-3. Suppose that Rutherford's experiment (Fig. 2-7) could be conducted with a foil of lithium metal. In what respects would the results differ from those that he obtained?

2-4. Do you think that, in the deflection of alpha particles by a gold foil, it is necessary that the alpha particles actually make contact before deflection? What forces are operative between the alpha particles and the gold nuclei?

2-5. The electron-volt is a unit of energy. It is the kinetic energy possessed by an electron of charge 4.8×10^{-10} statcoulombs upon falling through a potential gradient of one volt. The energy is

$$4.8 \times 10^{-10} \times \tfrac{1}{300} = 1.6 \times 10^{-12} \text{ erg}$$

(The factor $\frac{1}{300}$ converts to statvolts from volts.)

(a) What is the velocity of an electron that has fallen through a potential of 100 volts?

(b) What is the velocity of a proton that has fallen through a potential of 100 volts?

2-6. Lithium which occurs in nature has an atomic weight of 6.940. It possesses two nuclides, ^{7}Li of mass 7.018 and ^{6}Li of mass 6.017. Calculate the percentages of the two nuclides in the naturally-occurring element.

2-7. Explain the principles on which Millikan's oil-drop experiment is based. How is it possible to tell in this experiment whether the charge on an oil-drop is positive or negative?

Chapter 3

The Gaseous State

3-1 Introduction

A gas is characterized by the property that the particles of which it is composed distribute themselves homogeneously throughout an enclosing volume. The volume of a quantity of gaseous substance is thus the volume of the vessel which encloses it, whereas the volume of a quantity of solid or liquid substance is independent of container volume.

There is no inherent chemical difference between gases, solids, and liquids. Indeed, most substances can exist in any of the three states, depending upon conditions of temperature and pressure. The densities of a substance in the solid and liquid states do not usually differ by more than about 20 per cent, with the liquid state usually being less dense than the solid. (Water is a well-known exception to this generality.) The density of a substance in the gaseous state is much lower than that of the solid and liquid. For example, the density of oxygen gas at 1 atm pressure and 0°C is about 0.0014 g/cm^3, whereas the density of liquid oxygen at -184°C is about 1.48 g/cm^3. The low densities of gases are indicative of a very loose packing of atoms; most of the space enclosing a gas is empty. One of the first observations about a gas which a theory of the gaseous state must explain is this low density. How is it that the molecules of a gas, which are packed relatively tightly in a solid or liquid, should exist in the gaseous state with such large spacings between them? Why does a gas not collapse into a liquid or solid?

In describing the condition, or the state, of a gas, we will find it necessary to specify three quantities: temperature, pressure, and volume. The usual temperature scales are discussed in Sec. 1-2; temperature is also discussed in more detail in this chapter. The dimensions of volume are of course length $\times$ length $\times$ length = length^3. Pressure is a measure of force per unit area, and has the dimensions g/(cm $\times$ sec^2). Rather than express pressure in units which follow directly from these dimensions; *e.g.*, dynes/cm^2 (see Sec. 1-6), it is convenient to express gas pressure in terms of multiples of the pressure which the atmosphere exerts at the earth's surface. One atmosphere of pressure is defined as the force per unit area of a column of mercury 76.0 cm in height, acted upon by the standard gravitational force. The pressure of the earth's atmosphere at any given place and time may not, of course, be precisely 76.0 cm. Its actual pressure can be measured with a mercury barometer (Fig. 3-1). Suppose that a vacuum pump has

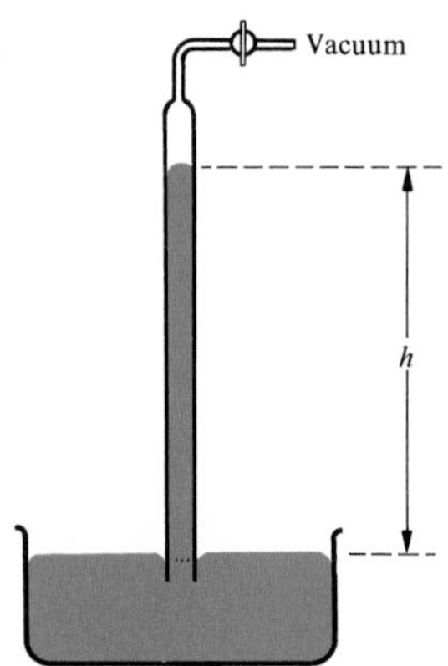

Fig. 3-1 A mercury manometer.

been attached to the outlet at the top of the tube, and the space above the mercury pumped out to a high vacuum; we can assume that there is zero gas pressure above the mercury in the tube. Now consider the forces acting at the surface indicated by the dotted line inside the tube. There is a downward force on this surface from the mass of mercury which is above it. The pressure, or force per unit area, is given by $P = dgh$, where d is the density of mercury, g is the standard earth gravitational constant, and h is the height of the mercury column. There is also an upward force acting on this surface. The earth's atmosphere exerts a force on the surface of the mercury outside the column. The pressure, or force per unit area, acts in an upward direction and is transmitted to the surface which we are considering. When the forces are equal, the height of the mercury in the column remains fixed; a measure of the height h of the column is then a measure of the atmospheric pressure. This is most commonly expressed as mm of mercury,* or in atmospheres, multiples of 760 mm of mercury. To convert the unit of 1 atmosphere of pressure to units of dynes/cm^2 we proceed as follows: The density of mercury is 13.6 g/cm^3. If we have a column of mercury which is 76.0 cm high and 1 cm^2 in cross-sectional area, the total mass of mercury is $76 \times 13.6 = 1033$ g. The pressure is then

$$P = \frac{1033 \text{ g}}{\text{cm}^2} \times 980.6 \text{ cm/sec}^2 = 1.013 \times 10^6 \text{ dynes/cm}^2$$

The earliest form of the Greek atomic theory postulated that the smallest particles of matter (atoms) are in constant motion. Isaac Newton rejected this model and proposed instead a static theory of gases, in which the particles are assumed to be at rest and are held apart from one another by repulsive forces. Robert Boyle had earlier published his pioneer work with gases in which he showed that, at a constant temperature, the product of the pressure and volume of a fixed quantity of gas is constant:

$$P \times V = \text{constant}$$

In 1723, Newton showed that Boyle's law followed from his model, if the repulsive forces which he proposed were inversely proportional to the distance. It was subsequently shown by Bernoulli and others that Boyle's law could be deduced from the model of a gas composed of particles which are in ceaseless motion and between which *no* forces exist. Newton's authority was so great, however, that these workers were more or less ignored.

It was not until 1845 that Joule finally showed conclusively that Newton's theory was incorrect. Joule caused gases to expand from one vessel into a second, evacuated vessel. Only small or zero temperature changes were noted, whereas — according to Newton's theory — the repulsive forces should cause a large temperature change as the average distance between molecules changes.

3-2 Charles' Law

The nearly universal belief in Newton's theory for more than a century did not, however, preclude the performance of some very important experiments. Charles and Gay-Lussac independently studied the effects of temperature on the pressure and volume of gases. A fixed quantity of gas was enclosed in a device illustrated schematically in Fig. 3-2. In one set of experiments, the temperature is varied and the pressure required to hold the volume constant is measured. As the temperature increases, the required

*1 mm mercury pressure is sometimes referred to as 1 Torr.

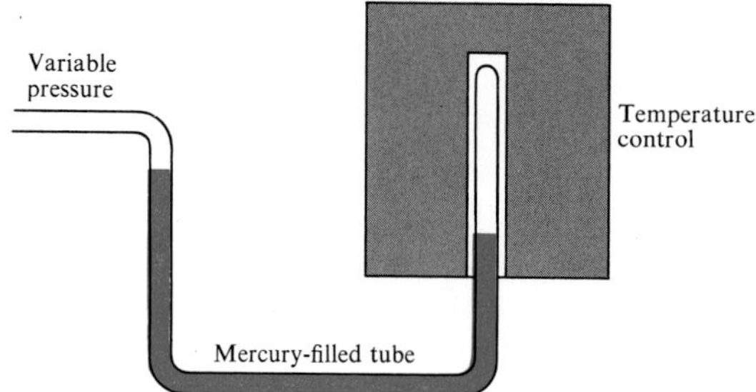

Fig. 3-2 Apparatus for the study of the effects of temperature on gas pressure.

pressure increases. When the data are graphed, the results of such experiments are as shown in Fig. 3-3. The different lines result from the use of different amounts of gas in the cylinder at the start of the experiment. These lines can all be described by the equation

$$P = P_0\alpha t + P_0 = P_0(1 + \alpha t) \tag{3-1}$$

where P_0 is the pressure at zero on the temperature scale and t is the temperature. It is found that, when the centigrade scale is used, α has the value 1/273.1. The slope of the line described by Eq. (3-2) is thus $P_0/273.1$.

A second set of experiments can be performed in which the pressure is kept constant and the volume is allowed to vary with temperature. The volume is found to increase with temperature, and the relationship shown in Fig. 3-3 is also found for volumes. The lines are described by the equation

$$V = V_0\alpha t + V_0 \tag{3-2}$$

We see from Fig. 3-3 that the pressure of a given quantity of gas at constant volume decreases with temperature in a linear fashion. We might ask at what temperature the pressure would become zero if the relationship in Eq. (3-1) remained correct. If Fig. 3-3 is redrawn with an extended centigrade temperature axis, as shown in Fig. 3-4, the pressure extrapolates

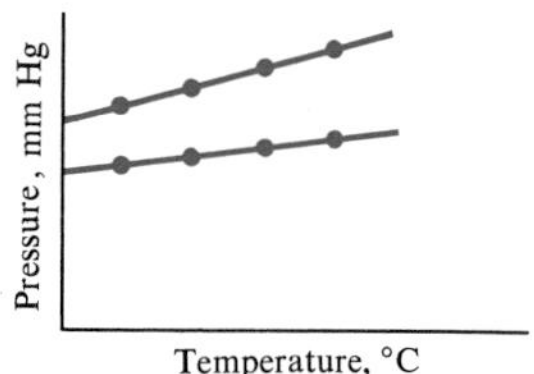

Fig. 3-3 Variation of gas pressure with temperature at constant volume.

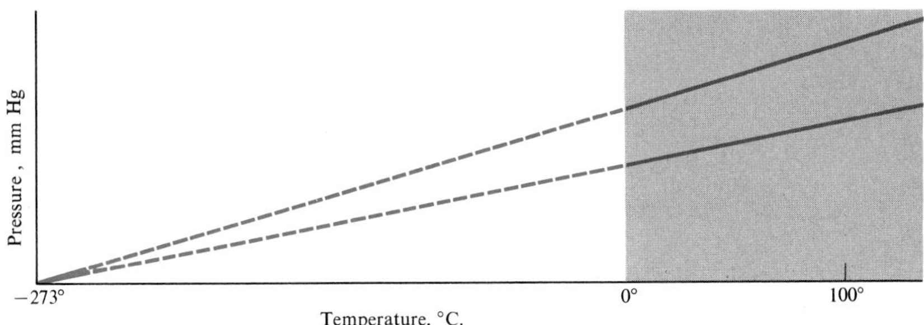

Fig. 3-4 Relocation of temperature origin on the pressure-temperature graph.

to zero at −273.1°C. We can see this algebraically by setting the pressure of the gas, P, in Eq. (3-1) equal to zero. Then

$$0 = P_0\alpha t + P_0 = P_0(\alpha t + 1)$$

Since P_0, which is the pressure of the gas at zero on the temperature scale, is not zero, the quantity in parentheses must be. We have then,

$$\begin{aligned} \alpha t + 1 &= 0 \\ t/273.1 &= -1 \\ t &= -273.1 \end{aligned}$$

Even though a great number of different samples might be studied, each sample giving rise to another line, all lines converge to the same zero pressure. Similarly, graphs show that zero volume occurs at −273.1°C. We define this point as absolute zero. An absolute zero temperature scale, T, can be set up by shifting the origin in Fig. 3-4 to the intersection point. Then:

$$T = t + 273.1 \tag{3-3}$$

where T represents the new absolute temperature scale. Solving for t and substituting in the above equation, we obtain:

$$P = P_0\left(1 + \frac{T - 273.1}{273.1}\right) = \frac{P_0 T}{273.1} \tag{3-4}$$

$$P/T = P_0/273.1$$

The temperature which was zero on the centigrade scale is now 273.1. All the graphs of pressure *vs* temperature now pass through the origin. The relationship between P and T is a straight line, just as before, in Eq. (3-1). The slope of the line is also the same, $P_0\alpha = P_0/273.1$. Only the horizontal location of the origin has been changed. For any given sample we have simply $P = qT$, where q is a proportionality constant, equal to $P_0\alpha$. Similarly, for a gas at constant pressure, $V = qT$. For any given gas at a particular constant pressure, and using some particular absolute temperature scale, q is a constant. Thus, for example, we may have a sample of helium at a pressure of 0.1 mm and a volume of 32.00 cm^3 at $-97.0°C$. Converting to °K, the absolute temperature scale, we have 0.032 liters $= q \times 176.1$; therefore, q for this particular set of conditions is 1.815×10^{-4}. What will be the volume of helium gas, under the same conditions of pressure, if the temperature is raised to 37.5°C? Then:

$$V = 1.815 \times 10^{-3}(310.6) = 56.45 \text{ cm}^3$$

From the relationship $V = qT$, it can be seen that the volume of a quantity of gas held at fixed pressure varies directly with the absolute temperature. This does not mean, however, that at absolute zero the volume of the gas is zero; before this temperature is reached the gas ceases to behave as a gas, since it liquifies or solidifies.

For the so-called "permanent" gases, such as nitrogen, oxygen, and helium, the above relationships are obeyed very well for moderate pressures and temperatures. All real gases deviate from the predicted behavior to some extent, however, and for some gases the deviations are large. We speak of an ideal gas as one that obeys the simple gas laws exactly. The origins of non-ideal behavior of real gases are discussed later.

3-3 The Chemical Reactions of Gases

In his experiments with gases, Gay-Lussac made the discovery that when two different gases react together chemically, the volumes which are consumed (measured at the same temperature and pressure) are in the ratio of small, whole numbers. For example, two liters of hydrogen always react with one liter of oxygen to yield two liters of water vapor. There was at this time (1800–1815) considerable confusion about the nature of gases, as we have seen. Dalton, who accepted Newton's theory of gases, believed that the particles of elementary gases were atoms. This was completely at variance with Gay-Lussac's results, because, for one volume of oxygen gas to give rise to two volumes of water vapor, it would have been necessary for the oxygen atoms to divide in two. Dalton took the view, perhaps understandably, that Gay-Lussac's results were wrong. Gay-Lussac's observations were correctly interpreted by Amadeo Avogadro, who hypothesized that equal volumes of gases at the same temperature and pressure contain equal numbers of *molecules*. A molecule is an entity which consists of a single atom, or of two or more atoms held together by forces which are large in comparison with the motional energy of the atoms. Insofar as the kinetic theory of gases is concerned, each molecule acts as a single particle. Avogadro, assuming that there are two atoms of hydrogen per molecule and two atoms of oxygen per molecule, could then write

$$\text{O—O} + \text{H—H} + \text{H—H} \rightarrow \text{H-O-H} + \text{H—O—H}$$

Avogadro's contribution, unfortunately, was ignored for a long time. In 1858 Cannizzaro revived his idea and applied it to the determination of atomic weights.

By experiment, it is possible to determine the *weight* of one gas that will react with a given weight of a second. If one element is chosen as the standard, the weights of elements that combine with a certain weight of the standard can be determined and a table of *combining weights* can be constructed. Various workers chose different standards in such work (H = 1, O = 1, O = 100, O = 10, O = 16 were among those used), and a great confusion arose in the chemical literature in the years 1810–1860. The difficulties were compounded during this time by two other factors. In the first place, some elements had more than one combining weight; for example, the weight of nitrogen that combines with a unit weight of oxygen depends upon which oxide of nitrogen is being produced. Secondly, Avogadro's hypothesis was almost completely ignored during this period, and there was no clear idea in most chemists' minds of just what sort of atomic or molecular units were reacting. All of the information necessary for a complete understanding of the problem was, in fact, already available to anyone who chose to use it, but — despite all that is said of the objective character of scientific thinking — the leading lights of chemistry in those days were unable to overcome their prejudices in this matter.

3-4 Atomic Weights—Cannizzaro's Method

More than a hundred European chemists met in Karlsruhe in 1860 to discuss the over-all problem of atomic weights. Stanislao Cannizzaro, a young teacher at the Royal University of Genoa, presented each of those attending with a copy of a series of lectures that he had been giving at the University. Cannizzaro had succeeded beautifully in culling, from all of the work and disputation of the previous fifty years, a logical and coherent answer to the problem of atomic weights. His arguments and conclusions were the following:

1. Since a molecule must contain an integral number of atoms, the molecular weight of a compound must contain an integral number of atomic weights. For example, a molecule of a common substance such as propane, used in bottled gas, cannot contain something like four and one-half atoms. There must be a whole number of atoms of both carbon and hydrogen, the two elements of which propane is composed. The weight of a propane molecule must therefore be the sum of a certain number of hydrogen atomic weights plus a certain number of carbon atomic weights. The formula for propane is C_3H_8; each molecule contains three carbon and eight hydrogen atoms.
2. The smallest number of atoms of an element that can occur in any of its compounds is one. Furthermore, it can reasonably be expected that, if a sufficient number of compounds containing a given element are examined, at least one will be found which contains but one atom of that element.
3. One volume of hydrogen (the least dense gas known) reacts with one volume of chlorine to yield two volumes of hydrogen chloride. Accepting Avogadro's hypothesis, this means that the molecule of

hydrogen gas contains an even number of hydrogen atoms. It could conceivably contain four, six, eight, *etc.*, but Cannizzaro chose to assume that it contains two. If the weight of the hydrogen atom is equal to one, the molecular weight of H_2 is two.

4. By use of this idea, one can calculate the weight of another element that reacts with hydrogen to give a gaseous product. For example, consider the reaction of oxygen with hydrogen to yield water vapor. Oxygen gas is, under identical conditions of temperature and pressure, approximately 16 times more dense than hydrogen. From Avogadro's hypothesis, it follows that each molecule of oxygen is 16 times heavier than H_2; its weight is therefore 32. Two volumes of hydrogen and one volume of oxygen are required to produce two volumes of water vapor, which has a density (again under identical conditions of temperature and pressure) just nine times that of hydrogen gas. Therefore, the weight of the molecule of water is 18, and, since each volume of water vapor is produced from one volume of hydrogen, two of the 18 weight units are due to hydrogen, 16 to oxygen. The number of weight units of oxygen in a water molecule is therefore 16, whereas in a molecule of oxygen gas it is 32.

It was tempting to conclude immediately that the atomic weight of oxygen is 16, and that oxygen gas is a diatomic molecule. Cannizzaro, however, determined the number of weight units of oxygen in a number of other compounds. He obtained a value of 16 for carbon monoxide, alcohol, and ether, and a value of 32 for carbon dioxide. From this, he tentatively concluded that the atomic weight of oxygen is 16. Such a conclusion was not certain, because the possibility still existed that oxygen gas is tetratomic, and that there are two atoms of oxygen in compounds in which it has been presumed that there is only one. We know now, of course, that oxygen molecules are diatomic.

It should be noted in passing that, although Cannizzaro made use of gas densities in his determinations of atomic weight, any method for the determination of molecular weights can be used in applying his line of reasoning.

The *combining weight* is defined as the weight of an element that combines with a unit weight of a standard substance. Cannizzaro had chosen hydrogen as a standard with $H_2 = 2$ or $H = 1$. The combining weight of oxygen in forming water is then 8, since 16 units of oxygen are combined with two of hydrogen in this compound. Because oxygen is capable of forming binary compounds easily and with so many elements, it soon became clear that it would be a better standard than hydrogen. Until recently, $O = 16.000$ was the chemical atomic weight standard. On this basis, $H = 1.008$. With the atomic weight scale based on oxygen, the combining weight was defined as the weight of an element that combines with 8.000 g of oxygen. In 1959, the International Union of Pure and Applied Chemistry approved a change in the atomic weight scale which makes the isotope of carbon of mass 12 the atomic weight standard, of mass 12.000.

The limited applicability of Cannizzaro's method for determining atomic weights became apparent when the list of elements which are gaseous or which form gaseous compounds was exhausted. Consider the problems arising with an element such as aluminum: Aluminum can be caused to react with oxygen, and a quantitative study of the reaction shows that the combining weight of aluminum (the weight which combines with 8 g of oxygen) is 9. Since the number of atoms of aluminum per atom of oxygen is not known, this number is not sufficient to determine the atomic weight; all that can be said is that the atomic weight of aluminum is an integral multiple of 9. In the absence of data for the molecular weight of a volatile compound, there is not much more that can be said regarding the atomic weight of this element.

In 1819, Dulong and Petit had announced an empirical rule about the heat capacities of solid elements. The rule states that the heat capacity per gram times the atomic weight of a solid element equals 6.4. This rule is useful in obtaining the atomic weight of aluminum. The heat capacity per gram for this element is about 0.22 cal/°C. Then, $0.22 \times$ atomic weight $= 6.4$, atomic weight $= 29$. This is nearly 3 times the combining weight of aluminum. The factor by which the combining weight must be multiplied to obtain the correct atomic weight is therefore 3, and the atomic weight is 27.

Dulong and Petit's rule does not always work well, especially for elements of low atomic weight, but it was useful in correctly placing the atomic weights for many elements.

Mitscherlich's rule of isomorphism was another of the more useful empirical rules employed about this time as an aid in determining atomic weights. Mitscherlich noted that certain compounds, which one had every reason to expect to be closely related, frequently had the same crystalline shape, and otherwise appeared to be very similar. For example, ferric alum (containing iron) and chrome alum (containing chromium) have the same crystalline shape and general properties. He concluded that the chromium atoms in chrome alum occupy the same relationship to the compound as do the iron atoms in ferric alum; the rest of the crystal remains the same. On this basis, analysis of the alums, which yields the percentage of metal in each case, can also yield the relative atomic weights. For example, there is by analysis 10.9 per cent of iron in ferric alum. There is 10.4 per cent of chromium in chrome alum. If we assume that the two metals bear the same relationship to the rest of the compound in each case, the ratio of the atomic weight of iron to the atomic weight of chromium is therefore determined to be about 1.05. Without invoking the rule of isomorphism, however, one could not rule out the possibility that the ratio of atomic weights is perhaps twice, or half of, this. How do we know that there are equal numbers of iron and chromium atoms relative to the remainder of the alum formula? Mitscherlich's rule was simply that there must be, since the two substances appeared to be entirely similar in appearance and properties. (The term isomorphism has a more exact meaning than we have implied here; it refers to certain features of crystal properties, and will be discussed in Chap. 7.)

Cannizzaro's method, with the aid of these and other empirical rules, provided a sufficient number of atomic weights for Mendeleev and Mayer to begin their formulation of the periodic table. (But more of this in later chapters.)

3-5 Precise Atomic Weights

The development of a particular area of scientific investigation is often closely related to the rate at which accuracy and precision are attained in determination of important quantities. It is certainly true that the development of accurate values for the atomic weights of the elements has been of major importance in chemistry. The traditional method of long standing is the determination of combining weights. In this method, the weight of a given element which combines with another to form a simple compound of known formula is determined as accurately as possible. The method is a "bootstrap" procedure, in the sense that each new atomic weight determined might be used to determine a new atomic weight, or might at least serve in checking an atomic weight determination. Assume, for instance, that the atomic weight of silver is known to be 107.87. A known weight of silver is combined with chlorine to yield a weight of silver chloride of known formula AgCl. The ratio of the weight of silver chloride to the weight of silver is found to be:

$$\frac{\text{weight of AgCl}}{\text{weight of Ag}} = 1.3286$$

Assuming, therefore, that a weight of silver in grams equal to its atomic weight, 107.87 g, were used, the weight of silver chloride formed would be $107.87 \times 1.3286 = 143.32$ g. But the difference between this weight and that of the silver must be the weight of the chlorine, 35.45 g. Having thus determined the atomic weight of chlorine, one might determine the weight of chlorine which combines with a given weight of some other element to form a compound of known formula. As an illustration, suppose that a precisely determined weight of barium chloride is dissolved in water, and that the chloride ion is then completely precipitated as silver chloride. We might have data such as these

$$\text{weight of } BaCl_2 = 67.2843 \text{ g}$$
$$\text{weight of precipitated silver chloride} = 92.6220 \text{ g}$$

From the known atomic weights, we have for the weight of chlorine in silver chloride

$$\frac{35.45}{143.32} \times 92.6220 = 22.911 \text{ g}$$

The weight of barium in the original sample is thus

$$\text{weight of Ba} = 67.2843 - 22.911 = 44.373 \text{ g}$$

Assuming that a weight of chlorine in grams equal to two atomic weights of chlorine combines with one atomic weight of barium, we have

$$\frac{70.90}{22.911} \times 44.373 = 137.3 = \text{atomic weight of Ba}$$

This example has been given to illustrate the principle behind the method of combining weights. In practice, the usefulness of a given compound for determining atomic weight depends upon many experimental considerations. The chemical determination of atomic weights is extremely demanding work, and requires the most exhaustive attention to every detail. Many atomic weights were determined with great accuracy in the laboratory of Professor T. W. Richards at Harvard University in the years 1894–1928. Professor Richards received the Nobel Prize in Chemistry in 1914 as a mark of recognition of the great service that his work rendered to chemistry.

A second method which is of considerable historical importance for determining the atomic weights of some of the lighter elements is the method of gas densities. This method is illustrated in problems at the end of the chapter. The third method, of more recent vintage than the other two, is based on mass spectrometry. The mass spectrum of an element reveals the presence of individual isotopes of that element (Sec. 2-3). If the mass spectrum also reveals with good quantitative accuracy the *mass* of each individual isotope, and its *relative abundance*, the data necessary for an accurate determination of atomic weight are in hand. For example, consider the mass spectrum of mercury vapor, Fig. 3-5. With

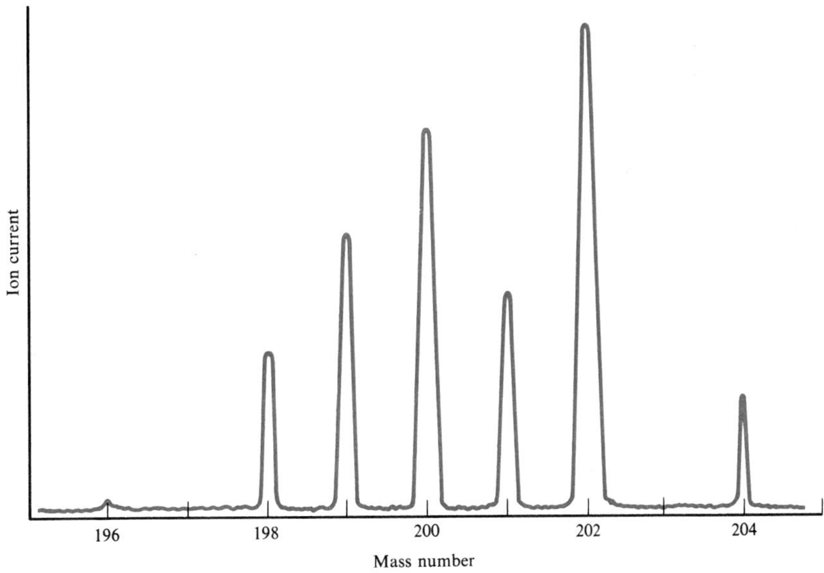

Fig. 3-5 Mass spectrum of Hg^+ ions in mercury vapor.

Table 3-1 Mass spectrum of mercury.

Nuclide	Atomic mass	Fractional abundance
196	195.965	0.0014
198	197.967	0.10039
199	198.967	0.1683
200	199.968	0.2312
201	200.970	0.1323
202	201.970	0.0685
204	203.973	0.2979

Average atomic weight of mercury = 199.596

the use of a precision mass spectrometer, the data of Table 3-1 can be obtained. The mass of each isotope of mercury and the relative abundance are accurately measured. The atomic weight is then given by the sum of each isotopic mass times its fractional abundance.

It might be thought at first that the mass spectrometric method is limited to elements which are quite volatile, but this is actually not the case. By using a high temperature source, it is possible to volatilize an element such as silver, for example, and to obtain a mass spectral determination of atomic weight. Furthermore, a volatile compound of an element might be employed. For example, uranium forms a volatile hexafluoride, UF_6. One might therefore determine the precise masses of ions in the mass spectrum corresponding to UF_6^+. The atomic weight of fluorine is known; very fortunately in this instance, there is only one nuclide of fluorine present in nature, so that the determination of the atomic weight of uranium may be made quite accurately from the mass spectrum of uranium hexafluoride.

3-6 Avogadro's Number

A knowledge of atomic weights is useful to chemists because it provides them (at least indirectly) with a means of counting out atoms, molecules, or ions. For example, from the knowledge that the weights of hydrogen and oxygen are 2 and 32, respectively, it can be said that 2 g of hydrogen gas contains the same number of molecules as 32 g of oxygen gas. Similarly, a certain weight of sodium metal contains a number of sodium atoms which is equal to the number of chlorine molecules contained in (70.9/23) times that weight of molecular chlorine (Cl_2).

The quantity of a substance that is equal in grams to its formula weight is of particular interest, and is given the name *gram-mole*. The formula weights of a number of substances are given as examples:

Formula	Formula weight
Na	23
NaCl	58.5
CH_4	16
Na_2CO_3	106

Since the gram is universally used, as the unit of measure for mass in laboratory work, it is common to refer simply to a *mole* of a substance, with the understanding that the gram-mole is meant.

It is important to realize that a mole of any substance contains a particular number of formula units of that substance, equal to the number of units contained in a mole of any other substance. This particular numerical quantity is called *Avogadro's number*, and has the value 6.023×10^{23}.

For example, 159.8 g, or one mole of molecular bromine, Br_2, contains 6.023×10^{23} Br_2 molecules. A mole of sodium carbonate, Na_2CO_3, which is 106 g of this substance, contains 6.023×10^{23} Na_2CO_3 units. It contains, therefore, 6.023×10^{23} carbonate (CO_3^{-2}) ions, and twice this number of Na^+ ions. In fact, we could say that a mole of sodium carbonate contains one mole of CO_3^{-2} ions and two moles of Na^+ ions.

A distinction is sometimes drawn between substances on the basis of their form; *i.e.*, whether they are ionic or covalent, whether they are represented as atoms or molecules in their formulas, etc. Thus, a mole of a monatomic substance such as argon is referred to as a *gram-atomic weight*, or gram-atom, while a mole of chlorine gas is termed a *gram-molecular weight*, or gram-mole. A mole of an ionic substance, such as NaCl, is called a *gram-formula weight*. It is not necessary to make such distinctions (which are, after all, rather arbitrary) as long as one is careful about stating clearly and completely the nature of the substance in question. Thus, it is not good form to say "a mole of hydrogen," since it is not clear from this whether hydrogen atoms or hydrogen molecules are meant. It is better to say "a mole of molecular hydrogen" if molecular hydrogen (H_2) is meant, or "a mole of atomic hydrogen" if atomic hydrogen is meant. With a substance such as sodium metal this problem does not even arise, since the element is always represented as Na, with no regard to the form in which it actually exists.

Avogadro's number is an important constant in physical science, and much effort has been expended in determining its value, as accurately as possible, by a number of different methods. From the material already dealt with in this text, at least one method is available. The determination of the charge on the electron, 1.60×10^{-19} coulombs, also serves to determine the charge on the proton, since the two must be equal and opposite. The ratio e/m for the proton is determined to be 9.581×10^4 coulombs/g, from which we calculate the mass:

$$\text{mass} = \frac{1.602 \times 10^{-19} \text{ coulombs}}{9.581 \times 10^4 \text{ coulombs/g}} = 1.672 \times 10^{-24} \text{ g}$$

Now, making allowance for the fact that hydrogen gas in nature is a mixture of isotopes, it can easily be calculated that the atomic weight of the nuclide ^1H, which contains just the proton as the nucleus, is 1.0070. Thus 1.0070 g of isotopically pure hydrogen nuclei is equal to one mole of hydrogen nuclei. To find the number of nuclei in this quantity of substance, we divide by the mass per nucleus:

$$\frac{\text{number of nuclei}}{1 \text{ mole}} = \frac{1.0070 \text{ g/mole}}{1.672 \times 10^{-24} \text{ g/nucleus}} = 6.023 \times 10^{23}$$

3-7 The Kinetic Theory of Gases

In discussing the chemical reactions of gases and the problem of atomic weights, we have digressed from the central theme of this chapter, the understanding of the gaseous state. It will be recalled that Joule, in 1845, had performed experiments which showed that Newton's theory of gases was wrong. During the following twelve years a few papers appeared which dealt with a kinetic theory of gases, but it was Clausius who, in 1857, published the most complete and satisfactory form of the theory.

In the kinetic theory it is assumed that:

1. Gases consist of a very large number of perfectly elastic molecules in ceaseless motion.
2. The molecules possess a volume which is negligibly small in comparison with the volume of the container.
3. The duration of molecular collisions is negligibly short in comparison with the average time between collisions.
4. No attractive or repulsive forces exist between the molecules.

A number of deductions about the properties of gases can immediately be made from these assumptions:

1. The pressure of a gas on its container is due to the collisions of the gas molecules with the walls of the container.
2. The average velocity of the molecules remains constant for a particular gas at constant temperature.
3. If the volume of a given quantity of gas is reduced at constant temperature, the number of collisions with the container walls increases, and thus the pressure increases (Boyle's law).
4. When the temperature of a gas is raised, the heat absorbed by the gas increases the kinetic energy of the gas molecules. The increase in pressure attendant upon increase of the temperature at constant volume results from the increased number of collisions per second which each molecule makes with the walls, and from the greater violence of the collisions.

The kinetic energy of motion of a particle of mass m is $\frac{1}{2}mu^2$, where u represents velocity. The total kinetic energy of the molecules which make up a gas, assuming that they all have mass m, is

$$\text{kinetic energy} = \sum_{i=1}^{n} \tfrac{1}{2} m u_i^2 = \tfrac{1}{2} m \sum_{i=1}^{n} u_i^2 \qquad (3\text{-}5)$$

The symbol $\sum_{i=1}^{n}$ represents a summation, to include all of the molecules of the gas. The equation says that the total kinetic energy of the gas is the sum of the kinetic energies of the individual molecules; the sum naturally goes up to n, the total number of molecules in the system. The *average* kinetic energy per molecule is the total divided by the number of molecules, n:

$$\text{average kinetic energy} = \tfrac{1}{2}(m/n)\sum_{i=1}^{n} u_i^2 \tag{3-6}$$

$$= \tfrac{1}{2}\, m\overline{u^2} \tag{3-7}$$

In the second way of writing the equation we have used the symbol $\overline{u^2}$, which is the average squared velocity: $\overline{u^2} = (1/n)\sum_{i=1}^{n} u_i^2$. When we deal with the energy of a gas, therefore, we deal with the *average squared velocity*, which is not the same thing as the *square of the average velocity*. To illustrate this, suppose that we have four molecules, with velocities 4, 6, 10, and 12. The average of these velocities is 8, and its square is 64. On the other hand, the average of the squares of these velocities is $\frac{1}{4}$ (16 + 36 + 100 + 144) = 74. When we use the symbol u for velocity in the discussions which follow, we will generally be talking about the square root of the average squared velocity, which is sometimes called the root-mean-square (rms) velocity:

$$u = \sqrt{\overline{u^2}} \tag{3-8}$$

The molecules of a gas move about at random in their enclosure, colliding with one another and with the walls of the container. At any one instant, some of them are moving fast, others slow. In a later section, we will discuss the way in which the kinetic energy is distributed among the gas molecules; for now, we need only note that the molecules *do* possess an average kinetic energy. We will assume that *all* molecules possess this average kinetic energy, $\frac{1}{2} = m\overline{u^2}$. At any one instant, a molecule might be moving in some arbitrary direction as shown in Fig. 3-6. It is important to note that velocity implies not only speed, but direction. We may say that a car is going 60 miles per hour; when we do, we describe its speed. In order to give its velocity, we must also say *where* it is going. We can think of velocity as a

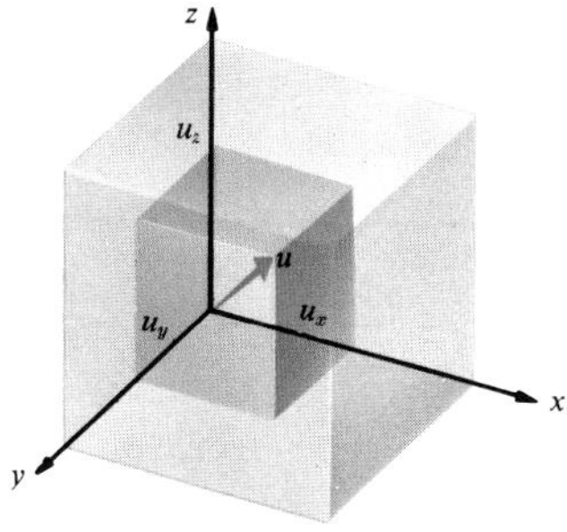

Fig. 3-6 Cartesian coordinate system. The velocity u has components u_x, u_y, and u_z along the x, y, and z-axes, respectively.

vector, or an arrow, which has a definite direction in a coordinate system, such as shown in Fig. 3-6. The length of the vector is a measure of the magnitude of the quantity. The velocity vector can be resolved into components along the three coordinate directions, x, y, and z. For any one molecule that we might single out, the components u_x, u_y, and u_z of its velocity along the three coordinate directions will not, in general, be equal. But if we consider a large number of molecules, their velocities are, on the average, the same in all directions. If this were not so, then there would be a net movement of all of the molecules in one particular direction. This is never observed in a closed container which is in equilibrium with its surroundings. It is obvious from what we have said that the average velocity in the $-x$-direction is the same as that in the $+x$-direction, and similarly for the y- and z-axes. At any one instant, then, we have one-third of the average velocity of the molecules directed along the x-axis, one-third along the y-axis, and one-third along the z-axis. One-half of each third is directed along the $+$, the other half along the $-$ direction.

Now let us consider that we have n molecules, each of mass m, in a cubical container l cm on a side. $n/6$ of the molecules are then moving in the $+x$-direction, with velocity u. (It is immaterial whether we consider that one-sixth of the molecules are moving in the $+x$-direction with velocity u, or that all of the molecules are moving to some extent in the $+x$-direction, with an average component of $u/6$.) We want to determine how many collisions are made with the wall shown by the shaded area in Fig. 3-6. The molecules which are moving in the $+x$-direction will obviously strike the wall, but so will the ones moving in the $-x$-direction, since they rebound from the back wall and are then moving in the $+x$-direction. It is clear, therefore, that we must include all of the molecules whether they are moving along the $+x$- or $-x$-coordinate. Each molecule must travel a distance $2l$ between collisions with the front wall, since after a collision it must travel across the container and rebound off the back wall. Thus, the number of times per second each molecule strikes the front wall is $u/2l$. The total number of collisions with the wall according to our model is then $(n/3)(u/2l)$.

The momentum of a particle of mass m moving with velocity u is mu. When one of the molecules strikes the wall and rebounds, the change in momentum is 2 mu (after collision the momentum is $-mu$ as a result of the reversal of direction, so the total change is $mu - (-mu) = 2mu$). The total change in momentum at the front wall per second is therefore

$$(n/3)(u/2l)2mu = \frac{nm\overline{u^2}}{3l} \tag{3-9}$$

By referring to the discussion of units in Chap. 1, we recall that pressure has the dimensions of force per unit area. This is related to the change in momentum per unit time per unit area. Since the area of the wall is l^2, the pressure at the wall is simply given by

$$p = \frac{nm\overline{u^2}}{3l^3} \tag{3-10}$$

This equation can be written in another way by noting that mn/l^3 has the dimensions of mass/cm³, or density, which we symbolize by ρ. We can, therefore, write

$$p = \frac{1}{3}\rho\overline{u^2} \tag{3-11}$$

Since l^3 represents the volume of the container v, Eq. (3-10) may be written in the form

$$pv = \frac{n}{3}m\overline{u^2} \tag{3-12}$$

The kinetic energy of the gas is given by the product of the average kinetic energy per molecule, Eq. (3-7), and the number of molecules, n.

$$E_k = \text{total kinetic energy} = \frac{1}{2}nm\overline{u^2} \tag{3-13}$$

Combining Eqs. (3-12) and (3-13), we obtain

$$pv = \frac{2}{3}E_k \tag{3-14}$$

Now let it be assumed that the kinetic energy of the gas is directly proportional to temperature. We can then write

$$pv = nkT \tag{3-15}$$

where k is a proportionality constant, called the *Boltzmann constant.* It is one of the most important fundamental constants, and a great deal of effort has been expended in determining it as accurately as possible. The best estimate of its value today is 1.3805×10^{-16} erg/°K-mole.

Now let us suppose that we are dealing with one mole of gas, so that n is N, Avogadro's number. Then, using capital letters for molar quantities,

$$PV = NkT = RT \tag{3-16}$$

R is the molar gas constant, which we shall have occasion to use more frequently than k. Its value in terms of a number of different units is given in Table 3-2. Equation (3-16) is the ideal gas equation. For quantities of gas other than precisely one mole, the right side is multiplied by the number of moles of gas:

$$PV = nRT \tag{3-17}$$

Note that the symbol n is now employed to represent the *number of moles* of gas.

In order to obey the ideal gas equation, a gaseous substance would need to possess all of the properties assumed in the derivation of the kinetic-theory equations. No gases do, but some of them, the so-called permanent gases, come pretty close to obeying Eq. (3-17) in the range of low pressure and at temperatures higher than about 200°K. For much of the less exacting work done with gases, the equation gives satisfactory accuracy. Its major advantage is its simplicity. We now give a number of examples of problems involving gases which can be handled by using Eq. (3-17) or some variant of it.

Table 3-2 Units of, and numerical values for, the molar gas constant *R*.

Units of R	Numerical value
liter-atm/°K-mole	0.08206
calories/°K-mole	1.987
ergs/°K-mole	8.315×10^7
joules/°K-mole	8.315
$\text{ft}^3\text{-}\frac{\text{lb}}{\text{in.}^2}$/°K-mole	10.71

Example 3-1

A quantity of nitrogen gas, confined to a tank of 30-liter volume, pressure 40 atm at 25°C, is to be expanded into a volume of 500 liters and the temperature lowered to 0°C. What will be the final pressure?

This problem is solved most simply by writing

$$\frac{P_1V_1}{T_1} = \frac{P_2V_2}{T_2} = nR$$

since n, the number of moles of gas, is a constant, as is R.

$$P_1 = 40 \qquad V_1 = 30 \qquad T_1 = 298$$
$$P_2 = ? \qquad V_2 = 500 \qquad T_2 = 273$$

Substituting and.solving for P_2, $P_2 = 2.20$ atm.

Example 3-2

What is the density in g/liter of the nitrogen gas after the expansion into the larger volume in Example 3-1?

Let M equal the molecular weight of the gas (g/mole). Then

$$\text{density} = \frac{n \text{ moles} \times M \text{ g/mole}}{V \text{ liters}} \tag{3-18}$$

$$n = \frac{PV}{RT}$$

Substituting in Eq. (3-18),

$$\text{density} = \frac{PM}{RT} \tag{3-19}$$

Replacing the symbols by the data for the final conditions of the nitrogen,

$$\text{density} = \frac{2.20 \text{ atm} \times 28 \text{ g/mole}}{0.082 \text{ liter-atm/°K-mole} \times 273\text{°K}}$$
$$= 2.75 \text{ g/liter}$$

Example 3-3

0.80 g of nitrogen gas is collected in a bottle which is inverted in a water trough. What is the final volume of gas in the bottle when the water level inside is the same as that outside? Barometric pressure = 740 mm Hg, temperature = 25°C.

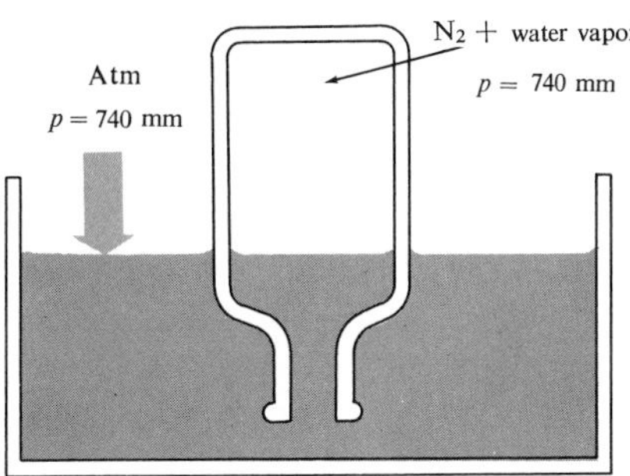

The equality of levels inside and outside of the bottle insures equal pressures. The total pressure in the bottle is due to the impacts of both the nitrogen and water vapor molecules. Since the gas is collected by bubbling through water, it is saturated with water vapor (Chap. 10). From a table of water vapor pressure *vs* temperature, we find that at 25°C the vapor pressure is 24 mm. The pressure of nitrogen in the bottle, therefore, is 740 − 24 = 716 mm. Then

$$(716/760)V = 0.80 \times 0.08206 \times 298/28.0$$
$$V = 0.74 \text{ liter}$$

At this point, a number of students are apt to say, "But you've only accounted for the volume occupied by the nitrogen; what about the volume occupied by the water vapor?" The answer, of course, is that the water vapor is occupying the same volume as the nitrogen gas; its pressure, plus the pressure of the nitrogen, just equals 740 mm.

Example 3-4

A few drops of a liquid of unknown molecular weight with a boiling point below 100°C, are placed in a Dumas bulb. It is placed in a bath of boiling water, with just the tip exposed. The liquid in the bulb boils, and the vapor pushes all of the air out of the bulb. When all of the liquid has just disappeared from the bulb, it is removed from the water, and the outside is cooled quickly to room temperature; the vapor of the unknown substance condenses.

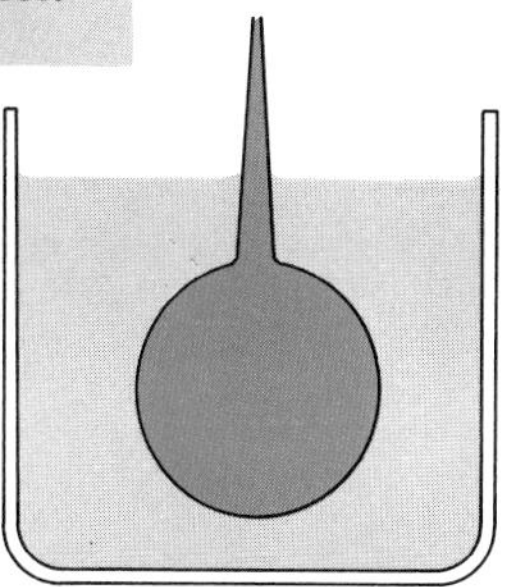

The bulb is weighed, then emptied completely and weighed again. The bulb is then filled with water and weighed. From the following data, calculate the molecular weight of the unknown substances:

Bulb empty	13.025 g
Bulb plus unknown condensate	13.862 g
Bulb filled with water	316.2 g
Atmospheric pressure	735 mm
Temperature of boiling water	98°C

The volume of the bulb (assuming the density of water to be unity) is 316.2 − 13.0 = 303.2 ml. The weight of the unknown vapor which occupies this volume at 98°C and 735 mm pressure is 13.862 − 13.025 = 0.837 g. Using the ideal gas equation, we can solve for the number of moles of gas which occupy the volume 303.2 ml under the stated conditions:

$$n = \frac{PV}{RT}$$

$$= \frac{(735/760) \times 0.303}{0.08206 \times 371} = 0.00964 \text{ mole}$$

molecular weight is in units of g/mole = 0.837 g/0.00964 mole

= 87 g/mole

There are other ways in which this problem can be approached; the student is urged to look for them.

3-8 Avogadro's Hypothesis

The kinetic theory provides a basis for deriving a number of the classical statements about gases. Avogadro's hypothesis, mentioned earlier, was that equal volumes of gases at the same temperature and pressure contain the same number of molecules. The two gases each obey Eq. (3-12):

$$\tfrac{1}{3}\, n_1 m_1 \overline{u_1^2} = pv = \tfrac{1}{3}\, n_2 m_2 \overline{u_2^2} \tag{3-20}$$

Since the gases are at the same temperature, they also possess the same average kinetic energy:

$$\tfrac{1}{2}\, m_1 \overline{u_1^2} = \tfrac{1}{2}\, m_2 \overline{u_2^2} \tag{3-21}$$

By combining these equations, it is simple to see that $n_1 = n_2$, which is Avogadro's hypothesis. (Be sure that you *can* so combine the two equations by carrying out the necessary substitutions.)

3-9 Graham's Law of Diffusion

By rearranging Eq. (3-21) we obtain

$$\frac{\overline{u_1^2}}{\overline{u_2^2}} = \frac{m_2}{m_1} \tag{3-22}$$

Taking the square root of both sides of this equation, we obtain the rms velocities on the left:

$$\frac{u_1}{u_2} = \sqrt{\frac{m_2}{m_1}} \tag{3-23}$$

Suppose that a gas is contained in a vessel which is enclosed except for a tiny pinhole. The molecules of gas which strike the wall surface at the hole leave the vessel. The rate at which the gas leaves is called the *effusion rate.* It is roughly proportional to the rms velocity, u. Graham's law states that the rate at which two gases under identical conditions of temperature and pressure effuse through a hole is inversely proportional to the square roots of their molecular weights, which is just the result shown in Eq. (3-23). Gases of low molecular weight effuse more rapidly than do substances of high molecular weight. The hydrogen fountain is an amusing demonstration of this effect. The apparatus is shown in Fig. 3-7. An inverted beaker con-

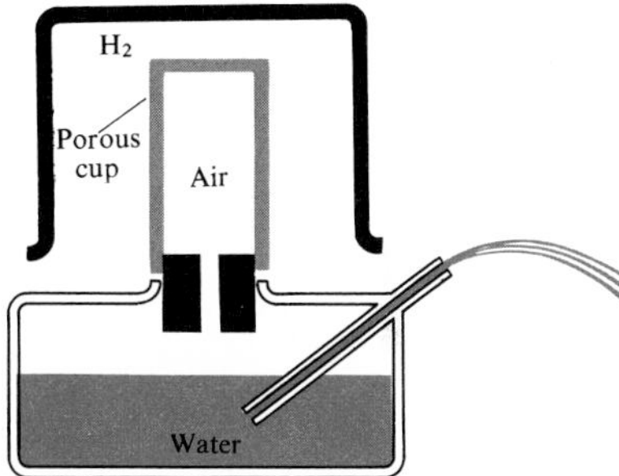

Fig. 3-7 A hydrogen fountain, demonstrating the high diffusion rate of hydrogen gas.

taining hydrogen gas is placed over a porous cup, which is filled with air. Since hydrogen is a gas of lower molecular weight than are the substances which make up the air (average molecular weight 29 g/mole) the hydrogen effuses into the cup through the tiny holes more rapidly than the air effuses outward. The result is a temporary increase of pressure in the gas space in the cup, which forces water out through the tip.

3-10 Dalton's Law of Partial Pressures

Dalton's law of partial pressures, which states that the total pressure that a mixture of gases exerts is equal to the sum of the pressures which the gases would exert if each were present alone, is also easily derived. We can write for each gas present in the volume v, using Eq. (3-14),

$$p_1 v = \tfrac{2}{3}E_1,\ p_2 v = \tfrac{2}{3}E_2,\ etc.$$

The total kinetic energy of all of the gas molecules in the box is equal to the sum of the energies of each of the gases:

$$E_1 = E_1 + E_2 + E_3 + \ldots etc.$$

This conclusion rests on the assumption that, when the gases are mixed, no heat is absorbed or evolved. Hence,

$$p_t v = \tfrac{2}{3}E_t = \tfrac{2}{3}(E_1 + E_2 + E_3 + \ldots) \qquad (3\text{-}24)$$
$$= v(p_1 + p_2 + p_3 + \ldots)$$

We have just shown that a gas which possesses the various properties assumed for an ideal gas satisfies Avogadro's hypothesis, Dalton's law of partial pressures, and Graham's law of diffusion. We might continue in this vein to test some other relationships such as Boyle's law and Charles' law. These follow directly from the ideal-gas equation; they simply correspond to changing certain of the variables while holding others constant. However, we must also ask whether it is *necessary* for a gas to be ideal in the kinetic theory sense to obey these laws. For example, consider Dalton's law of partial pressures. Suppose that we mix two gases A and B, each of which does not obey the ideal gas equation very well in the region of interest. It is still true that the total pressure of a mixture of A and B *might* be equal to the sum of the pressures which A and B would exert if each were present alone in the same volume. Similarly for Avogadro's hypothesis; two gases at the same temperature and pressure *might* contain equal numbers of molecules in the same volume, even though they are far from ideal. We can see in this case, however, that they would have to have the same kind and degree of non-ideality. Real gases are non-ideal in different ways; there is more than one way in which they can fail to conform to the postulates of the kinetic theory enumerated in Sec. 3-7. We will take up the matter of non-ideality, and some ways of accounting for it in the gas equation, in a later section.

3-11 Distribution of Molecular Speeds

On the basis of the kinetic theory of gases, it is not to be expected that all of the gas molecules will possess the same speed. The random collisions of the molecules with one another will cause some of them to have momentarily very high speeds, while at the same time others may be almost motionless. James Clerk Maxwell solved the problem of the distribution of molecular speeds in a gas at constant temperature and pressure. The mathematics required in the derivation of Maxwell's equations is involved and lengthy. His result is that, in a gas that contains a very large number of molecules, the molecules do not all have the same speed. Rather, there is a distribution of speeds among the molecules, and this distribution is of a particular form. It is, in a sense, a problem in statistics. Because there are so many gas molecules in a sample, they behave as a group in a very regular and predictable way. We can show the point involved here by a simple example. Suppose that we cast a number of pennies on the floor, and then calculate the percentage of heads. The most probable distribution between heads and tails is, of course, 50 per cent each. If we perform this experiment with a varying number of pennies, we might get the results shown in Table 3-3.

Table 3-3

Number of pennies thrown	Number of heads up	Per cent heads	Δ
10	6	60	10
100	53	53	3
1000	492	49.2	0.2
1,000,000	500,362	50.04	0.04

It is clear from this table that, as the number of units increases, Δ, the range of the absolute value of deviation from the most probable value, decreases. The molecules of a gas differ from the pennies in that, instead of just two possible states (heads or tails), the molecules can possess a range of energies. The other important respect in which they differ is their number; instead of a mere 10^6 molecules, a mole of gas contains 10^{23} molecules. As a result, the chance is infinitesimally small that the gas will be observed to differ significantly from its most probable behavior as calculated by Maxwell.

A number of useful equations relating to molecular speeds result from Maxwell's treatment. Let n_j be the number of molecules possessing a speed at least as great as u_j. This number is given by the following equation:

$$n_j = n_T\, e^{-\frac{1}{2}mu_j^2/kT} \tag{3-25}$$

The symbol n_T represents the total number of molecules in the system, and k is the Boltzmann constant. Dividing by n_T gives the fraction of molecules possessing a speed equal to or greater than u_j:

$$f_j = \frac{n_j}{n_T} = e^{-\frac{1}{2}mu_j^2/kT} \tag{3-26}$$

The exponential quantity is a negative power; therefore, the larger we choose u_j to be, the smaller the fraction. If u_j is set equal to zero, the fraction f_j is 1, *i.e.*, *all* of the molecules have a speed equal to or greater than zero. If we set u_j equal to infinity, f_j is zero; *none* of the molecules possess a speed equal to or greater than infinity. Note also that if we choose a certain speed u_j, the fraction f_j, grows larger with increasing temperature.

By going through a lengthy and rather difficult derivation, it is possible to use Eq. (3-26) to obtain an equation which gives the number of molecules which possess a speed u as a function of u. A graph of this relationship is shown in Fig. 3-8 for nitrogen gas at two different temperatures. We see that the curve reaches a maximum. The speed corresponding to this maximum, α, is the *most probable speed.* The dotted line corresponds to the square root of the average square velocity, the rms speed, u. We could compute also a simple average speed. All of these different ways of expressing the speeds have some use in one situation or another. For our purposes, it has so far been most useful to use the rms speed. The rms speed for a gas can be calculated most readily by using the ideal-gas equation with Eq. (3-12) for a mole of gas. Then $n = N$ (Avogadro's number), and we have

$$PV = \frac{Nm\overline{u^2}}{3} = RT$$

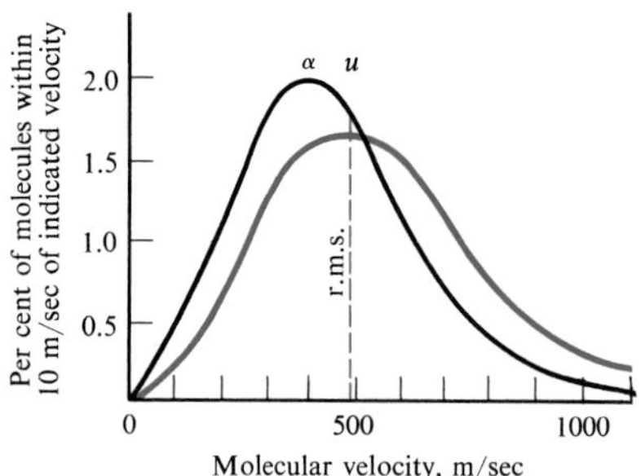

Fig. 3-8 Distribution of molecular speeds for nitrogen at 0°C (273°K). The colored line corresponds to the distribution of speeds at 100°C (373°K). The quantity graphed along the abscissa is the magnitude of the velocity, without regard to the direction in which a particular molecule may be moving.

But Nm is just the molecular weight of the gas, M. Therefore, solving for the rms speed, u:

$$u = \sqrt{\overline{u^2}} = \sqrt{\frac{3RT}{M}} = 1.73\sqrt{\frac{RT}{M}} \tag{3-27}$$

The derivations for the most probable and average speeds are more complex. The results are given here for comparison with u:

$$\alpha = 1.59\sqrt{\frac{RT}{M}} \tag{3-28}$$

$$u_{AV} = 1.41\sqrt{\frac{RT}{M}} \tag{3-29}$$

As a final point in connection with molecular speeds, it should be noted that an increase in temperature causes a general increase in molecular speeds, and a flattening out of the curve. The most probable speed is shifted to a higher value, and more molecules are found in the higher speed range. The effect of temperature upon the velocity distribution is very important in connection with theories of reaction rate, as we shall see later.

3-12 Mean Free Paths and Collision Numbers

If a small sample of a readily-detectable gas such as ammonia is liberated at one end of a long glass tube containing air, an appreciable time elapses before the smell is noticed at the opposite end. Since molecular speeds are very large — perhaps 1000 miles per hour — it might be expected that the odor of ammonia would be noticed immediately. The slow diffusion of gases in a situation such as this is presumed to be due to the collisions of the gas molecules with one another. Obviously, the collisions must be very frequent to so impede the flow of gas down the tube.

The possibility of collision depends upon the finite size of the gas molecules. If they were actually point masses, as we have been content to think of them, the probability of collision would be nil, and diffusion would be rapid. This fact was recognized by Clausius in his treatment of kinetic theory.

A number of approaches to the problem of estimating the sizes of gas molecules were tried during the ninteenth century. Although these methods were all rather crude by our present-day standards, it was nevertheless possible to conclude that molecular diameters were of the order of 10^{-8} cm = 1 Å. The methods used to obtain estimates of atomic and molecular sizes generally agree in a rough way, but they often do not give the same values for a particular substance. This is partly because the notion of atomic size is inherently inexact. It is probably not correct to think of the collision of two atoms as analogous to the coming together of a pair of billiard balls. It is better to replace the billiard balls in this analogy by a pair of balls with hard centers but softer exteriors. In experiments to determine atomic sizes, the extent to which the "softer" outer part of the atom contributes to the apparent size varies.

It is possible to calculate the number of collisions per unit time that a gas molecule makes, if the size is given. Consider a gas composed of molecules — which we shall consider to be spherical — of diameter s. We wish to calculate the number of collisions per second which a particular molecule undergoes. To do this, we neglect the motion of all the other molecules and assume that the molecule of our choice has the rms speed u. Now, a collision will occur whenever the distance between the centers of two molecules is less than s (Fig. 3-9). During one second the molecule moves a distance u. For the purpose of counting collisions, we can say that the effective volume swept out by this molecule is $\pi s^2 u$, since, if the center of another molecule lies within this volume, a collision will occur. If there are L molecules per cm³, there will then be $L\pi s^2 u$ molecules within this volume. The collision number Z, which is number of collisions made by one molecule during one second, is then

$$Z = \pi L s^2 u \tag{3-30}$$

By neglecting the motion of the other molecules, we have introduced an error into the calculations. When this is accounted for in the rigorous version of the kinetic theory, it develops that the correct expression for Z is

$$Z = \sqrt{2}\pi L s^2 u \tag{3-31}$$

The mean free path l is defined as the average distance traversed by a molecule between two collisions; *i.e.*,

$$l = u/Z = \frac{1}{\sqrt{2}\pi L s^2} \tag{3-32}$$

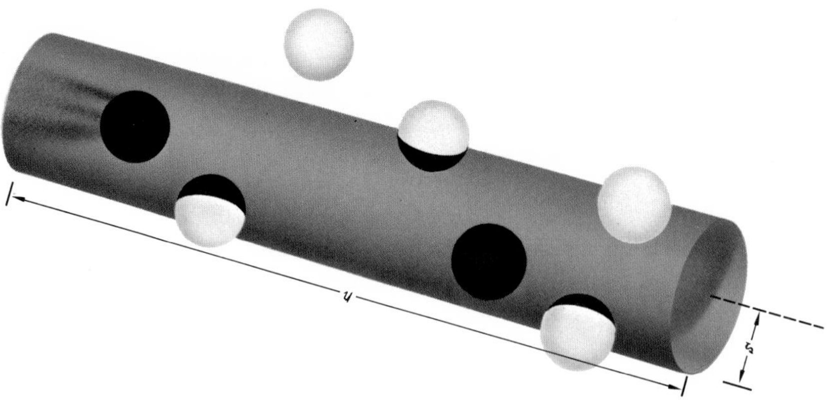

Fig. 3-9 Model for determination of collision number. The cylinder swept out by the molecules is not drawn to scale. It is actually about 10^{12} times longer than it is wide.

Both the collision number and the mean free path are important quantities in considerations of the chemical and physical properties of gases.

As an example of the use of these expressions, we calculate the collision number of oxygen gas at standard temperature and pressure (STP): 0°C, 760 mm pressure. We assume a diameter of 3 Å, molecular weight 32.

$$\begin{aligned} M &= 32, T = 273 \\ u &= 1.73\,(8.315 \times 10^7 \times 273/32)^{1/2} \\ &= 46{,}000 \text{ cm/sec} \end{aligned}$$

We use R in units of ergs here because this gives us u in the desired units of cm/sec. The molar volume at STP is $P/RT = 22.4$ liters. We therefore have 6.023×10^{23} molecules/22.4 liters $\times$ 1 liter/1000 cm^3 $= 2.69 \times 10^{19}$ molecules/cm^3. Then, using Eq. (3-31):

$$\begin{aligned} Z &= 1.414 \times 3.14 \times 2.69 \times 10^{19} \times (3 \times 10^{-8})^2 \times 4.6 \times 10^4 \\ &= 4.95 \times 10^9 \text{ collisions per molecule per sec} \end{aligned}$$

3-13 The van der Waals Equation

So far, our consideration of gases has been restricted to the ideal gas. But we have just seen that there is an inconsistency between the kinetic theory assumption of a point mass and the fact that molecules do collide with one another. The fact that molecules frequently collide is evidence for the finite sizes of gas molecules. In the properties of real gases, there is ample evidence that still another assumption of the kinetic theory, that the molecules exert no forces on one another, is also generally incorrect. We shall now examine the properties of real gases to see just how much they deviate from the behavior of an ideal gas. The state, or condition, of a certain quantity of gas is described completely by specifying its pressure, volume, and temperature. An equation which relates these three variables for a substance is therefore called an *equation of state*. In our discussion of real gases, we shall also develop an equation of state to replace the ideal-gas equation of state, $PV = nRT$.

For a mole of ideal gas ($n = 1$), the quantity PV/RT equals one. If PV/RT is calculated from the properties of a real gas, the extent of deviation from one is a measure of non-ideality for that gas. A graphical presentation of the data can be obtained by plotting PV/RT for the gas at some particular temperature *vs* the pressure P. A few examples of such graphs are shown in Fig. 3-10. It is important to note that, for both N_2 and CO_2, the quantity

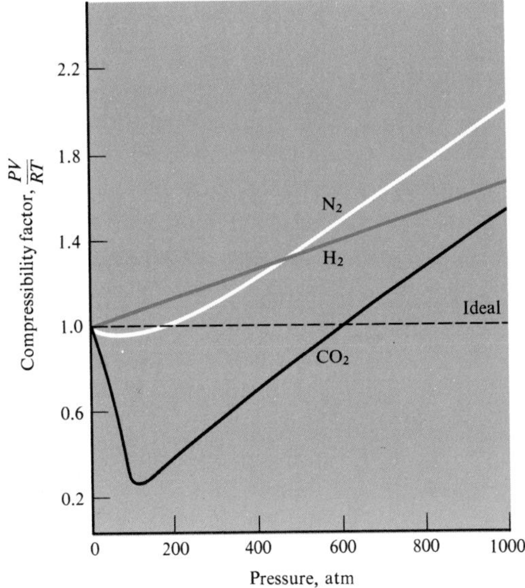

Fig. 3-10 Compressibility factors for several gases (300°K).

PV/RT lies below the value for the ideal gas in certain pressure ranges. PV/RT is frequently referred to as the *compressibility factor*. A value less than one for this quantity indicates that the gas is *more* compressible than an ideal gas. At very high pressures, all gases possess compressibility factors greater than one.

In a general way, it is possible to account for the kind of behavior exemplified in Fig. 3-10 in terms of the two ways in which real gases are expected to deviate from ideal behavior. Consider first a gas of very low molecular weight, such as hydrogen, H_2. While the molecules of such a gas do occupy volume, they may be expected to exert very small attractive forces on one another. The fact that they do occupy volume, however, means that V in the relationship PV/RT is not only the free volume in which the molecules move, it also includes the volume of the molecules. Thus, V will be larger for a real gas than for an ideal gas. The relative importance of the volume of the molecules increases with increasing pressure, since free volume is thereby diminished. Thus, the deviation from ideal behavior increases steadily with increasing pressure.

The ideal gas law can be corrected to account for the finite volume of molecules by subtracting a quantity which corresponds to the volume actually occupied at any one instant by the molecules from the apparent total volume V. This correction term must consist of a constant, which is related to the actual volume occupied by a mole of the molecules, times the number of moles of gas. Thus, for n moles of gas

$$P(V - nb) = nRT \tag{3-33}$$

$V - nb$ may be considered as the "free volume" in which the gas molecules move.

From the fact that gases — even the inert gases such as argon — liquefy at lower temperatures, we infer that there are attractive forces which operate between molecules. At higher temperatures, these forces are not so noticeable because of the relatively much larger kinetic energy of the molecules. At lower temperatures, at which the kinetic energy is smaller, and at higher pressures, at which the average distance between molecules is reduced, the attractive forces become important and represent a significant contribution to the non-ideality of real gases. Of course, the pressures and temperatures at which gases show non-ideality vary over a wide range, and depend on the properties of the particular gas molecule. Attractive forces result in a lower pressure than that which would be observed for ideal behavior; consequently, an additive term is needed in the P-V-T relationship. The

Table 3-4 van der Waals constants.

Gas	a(atm-liter^2/mole^2)	b(liters/mole)
H_2	0.244	0.0266
N_2	1.39	0.0391
Ne	0.211	0.0171
Ar	1.35	0.0322
Kr	2.31	0.0398
Xe	4.07	0.051
CO_2	3.59	0.0427
CO	1.46	0.0392
H_2O	5.47	0.0305
NH_3	4.18	0.0373
CH_4	2.25	0.0427
C_2H_2	4.37	0.0511
C_2H_4	4.46	0.0570
C_2H_6	5.43	0.0641
CCl_4	19.6	0.127

van der Waals equation of state for a non-ideal gas takes account of the two contributions to non-ideality which we have discussed.

$$\left(P + \frac{n^2a}{V^2}\right)(V - nb) = nRT \qquad (3\text{-}34)$$

a and b are constants characteristic of the particular gas, and n is, as usual, the number of moles. Some representative values are listed in Table 3-4.

3-14 Derivation of the van der Waals Constant

The constant b is a measure of the finite volume of the gas molecules. It is possible to calculate the effective radius of the molecules from a knowledge of b. What we want to know is, what is the free volume available to a gas molecule if all the gas molecules have radius s? When two molecules collide, as shown in Fig. 3-11, their centers can get no closer than $2s$. There is a

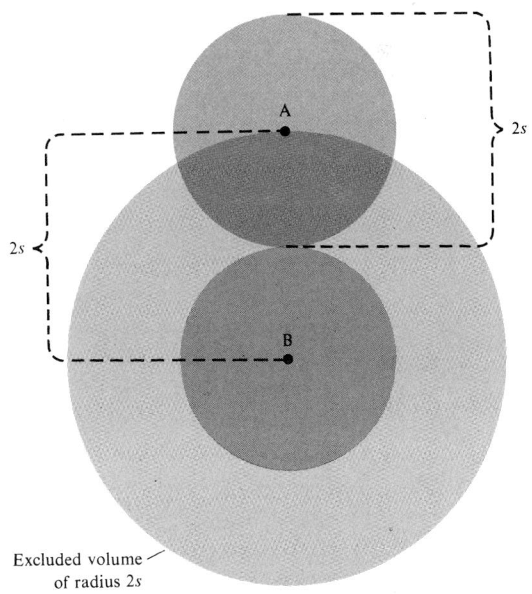

Fig. 3-11 An illustration of the manner in which one sphere of radius s excludes space from another.

sphere of radius $2s$ around molecule B from which the center of molecule A is excluded. The "excluded volume" is then $\frac{4}{3}\pi(2s)^3 = 8(\frac{4}{3}\pi s^3)$. One way to think of this is that molecule B can move wherever it likes, with molecule A moving on its surface. But we could just as well have assigned the excluded volume to molecule B, and let A move wherever it likes. The result is that the excluded volume applies to the pair. The excluded volume per molecule is thus $4(\frac{4}{3}\pi s^3)$, or four times the volume per molecule. The constant b is usually expressed in l/mole. From this, it is a simple matter to calculate the radius s per molecule.

Example 3-5

The van der Waals constant for carbon tetrachloride is reported to be 0.127 liter/mole. Calculate the effective radius of a CCl_4 molecule, assuming it to be spherical.

$$(0.127 \text{ liter/mole}) \times (1 \text{ mole}/6.02 \times 10^{23} \text{ molecules}) \times (1000 \text{ cm}^3/\text{liter}) \times (10^{24} \text{ Å}^3/\text{cm}^3) = 211 \text{ Å}^3/\text{molecule}$$

$$4(\tfrac{4}{3}\pi s^3) = 211 \text{ Å}^3; \qquad s = 2.33 \text{ Å}$$

There are various ways in which the correction to the pressure term in the van der Waals equation can be shown to arise. A common approach is to consider the effect of intermolecular attractive forces on a molecule which is about to collide with a wall, as shown in Fig. 3-12. It is argued that molecule

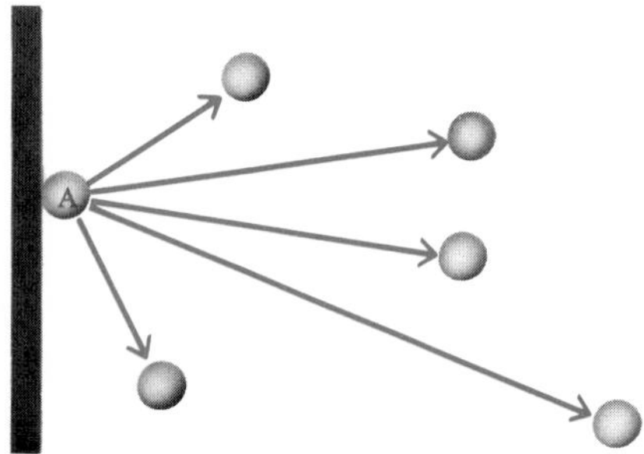

Fig. 3-12 The effect of intermolecular attractive forces on a molecule near the wall of the container. These forces result in a lower pressure than an ideal gas would exhibit.

A experiences a set of attractive forces arising from other nearby molecules, and thus does not strike the wall with the force with which it would if it were an ideal gas molecule. To approach the problem quantitatively, we must have a model for what really goes on. If the molecules are attracted toward one another, we might consider that there is a certain tendency for two molecules which get close together to pair up, forming a dimer:

$$A + A \rightleftharpoons A_2 \tag{3-35}$$

The dimer does not last long; another collision with a different molecule may disrupt it, or it may fly apart by itself. The possibility of building up a trimer, A_3, is very, very small, and may be ignored. Now we can say that at any instant of time, there is a certain small concentration of these dimers. As a result, *the actual number of "molecules" in the gas phase is less than it would be if the gas were ideal.* How does the concentration of dimers depend on pressure, and how can we correct for them? Let us first write down an equilibrium, and an expression for an equilibrium constant. (In order to follow this discussion, you must have some prior acquaintance with how to set up an equilibrium constant. The subject of chemical equilibrium is discussed in Chaps. 14 and 15 of this text.)

$$A + A = A_2$$

$$K = \frac{[A_2]}{[A]^2} \tag{3-36}$$

Since pressure is a measure of the number of molecules per unit volume, it is a suitable unit of concentration. Let P_0 be the pressure which the gas molecules A would have if they did not form dimers. Let P_d be the pressure of the gas dimers. Then the expression for K becomes

$$K = \frac{P_d}{(P_0 - 2P_d)^2} \tag{3-37}$$

If P_d is very small compared to P_0, and we assume that this is the case, then K is approximately given by: $K \approx P_d/P_0^2$. (Prove this for yourself by assuming that P_d is 1 per cent of P_0.) The total pressure in the system, P_T, is given by the sum of the pressures of the A and A_2 molecules. The pressure of the A molecules is $P_0 - 2P_d$, since two A molecules are required to form one A_2 molecule. The pressure of A_2 molecules is P_d. Thus, P_T is given by

$$P_T = P_0 - 2P_d + P_d = P_0 - P_d \tag{3-38}$$

From the approximate expression for K given above, we obtain $P_d = KP_0^2$, and substitute to give

$$P_T = P_0 - KP_0^2 \tag{3-39}$$

P_0 is the pressure which the gas would have *if* it were ideal; *i.e.*, if it did not exhibit attractive interactions between molecules. If the gas were ideal, it would obey the equation

$$P_0 = (n/V)RT \tag{3-40}$$

But P_0 is, from Eq. (3-39), equal to the actual pressure, P_T, plus a small correction term: $P_0 = P_T + KP_0^2$. Using Eq. (3-40) for P_0 on the right side,

$$P_0 = P_T + KR^2T^2(n/V)^2 \tag{3-41}$$

If we assume that KR^2T^2 is a constant, a, characteristic of the gas (remember that K is a characteristic of the particular gas), then $P_0 = P_T + an^2/V^2$. Substituting for P_0 in the ideal gas formula,

$$(P_T + an^2/V^2)V = nRT \tag{3-42}$$

Correcting for excluded volume, as described above, leads to the van der Waals equation. It is not necessary to correct the V^2 term in the denominator, since this would amount to a fairly small correction to what is already a small correction, and the effect would be negligible.

On the basis of our derivation, the van der Waals constant a is equal to KR^2T^2. The units of K, the equilibrium constant, are atm^{-1}, if the concentrations of A and A_2 molecules are expressed in atmospheres. R has the dimensions of liter-atm/°K-mole, and T is, of course, in °K. The result is that a has the dimensions of atm-liter^2/mole^2. From our derivation, it might be expected that a is directly proportional to (temperature)^2. This is not the case, however, since K decreases with increasing temperature. Nevertheless, it is to be expected that the van der Waals constant might vary between two widely different temperatures, and this is indeed found to be the case.

The van der Waals equation is a useful and interesting description of the properties of real gases in terms that are easy to visualize. However, it is not an entirely adequate description of real gases. In order to accurately describe the properties of a typical gas over a wide range of temperature and pressure, it is necessary to introduce more variable parameters than just the a and b constants of the van der Waals equation.

Exercises

3-1. Gallium metal, Ga, which is liquid at temperatures above 30°C, has a density of 5.91 g/cm^3 at 45°C. What would be the height of a column of gallium metal in a gallium manometer at 45°C if the atmospheric pressure were precisely 1 atm?

3-2. A given quantity of gas (assumed to be ideal), confined to a constant volume, exerts a pressure of 330.0 mm at 0°C and a pressure of 375.9 at 38.00°C. From these data, calculate the value of absolute zero on the centigrade scale.

3-3. One liter of oxygen gas, O_2, is collected over water at 20°C and 730 mm pressure. The vapor pressure of water at 20°C is 20 mm. What is the weight of this amount of oxygen? (Assume ideal gas behavior.)

3-4. (a) A balloon initially contains only hydrogen. After 30 minutes, gas from the surrounding atmosphere has diffused into the balloon. How does the ratio of oxygen to nitrogen inside the balloon compare with that outside? (Is it greater, less than, or the same as the ratio of oxygen to nitrogen outside?) Explain.
(b) A balloon is initially filled with carbon dioxide, CO_2. The pressure is 800 mm. After a short time the pressure is 810 mm. Explain why it has risen. (Temperature remains constant.)
(c) A balloon that weighs 29 g has a volume of 125 liters. It is filled with helium at a pressure of 1 atm at 25°C. The average molecular weight of air can be taken at 29.0. What is the maximum weight which the balloon can lift under these conditions, assuming that atmospheric pressure is 1 atm?

3-5. In a measurement of the evolution of gas from a plant, a total of 132 cm^3 was collected at a pressure of 735 mm, 35°C. What would be the volume of the gas at STP? Assuming that the gas is CO_2, how many moles are collected? Assuming that the gas is 50 per cent CO_2, 50 per cent O_2, how many moles (total) are collected?

3-6. A tank of nitrogen gas, N_2, records a pressure of 2300 psi at 30°C. A safety valve on the tank is set to release at 3500 psi. If the tank is heated, at what temperature will the valve release, assuming ideal gas?

3-7. There is a planet X in another galaxy which is identical to our own; the inhabitants, however, employ as their standard of atomic weight, $F = 100.00$. Their "mole" is defined as the number of fluorine atoms in 100 g of fluorine. Complete the following table.

Physical constant	Value on earth	Value on Planet X
1 atm pressure	1.01×10^6 dynes/cm^2	________
Molar gas constant	$\frac{0.082 \text{ liter-atm}}{\text{mole-°K}}$	________
Number of fluorine atoms in 1 mole of fluorine	6.02×10^{23}	________

If we could communicate with Planet X, would it be correct to point out to them that they have made a mistake in the atomic weight scale? Explain.

3-8. Starting from the ideal gas equation, prove that the densities of two gases are proportional to their molecular weights, assuming the same conditions of temperature and pressure.

3-9. Dry air has a composition (in volume per cent) of 78.09 per cent nitrogen, 20.95 per cent oxygen, 0.93 per cent argon, 0.03 per cent carbon dioxide, and traces of other gases. Calculate the density of dry air at 25°C and 1 atm pressure.

3-10. A sample of nitrogen gas is collected over water at 30°C, pressure of 720 mm, volume 660 cm^3. (The vapor pressure of water at 30°C is 35 mm.) Calculate the volume of the N_2 gas when dry at 20°C and 1 atm pressure.

3-11. (a) A large glass bulb, fitted with a stopcock opening, is evacuated and weighed at 32.4652 g. The bulb is then filled with argon at a pressure of 300.0 mm and a temperature of 27.5°C. On reweighing, the bulb weighs 34.0316 g. What is the volume of the bulb? (Assume ideal gas behavior.)

(b) The bulb is re-evacuated and filled with a gas composed of carbon dioxide, CO_2, and oxygen, O_2, unknown proportion. The pressure is again 300.0 mm, and the temperature is 27.5°C. What is the percentage of CO_2 in this gas if the new weighing is 33.9786 g? (Again, assume ideal behavior.)

3-12. The mass spectrum of neon vapor consists of the following masses with the corresponding fractional abundances. From these data, calculate the atomic weight of neon.

Nuclide	Atomic mass	Fractional abundance
20	19.9946	0.9000
21	20.9951	0.0027
22	21.9934	0.0973

3-13. What assumption, in addition to the four listed in Sec. 3-7, is necessary to obtain the ideal gas equation? What evidence can you give that this additional assumption is justified?

3-14. Heliox, a gas mixture consisting of 2.5 per cent oxygen and 97.5 per cent helium, is used in certain underwater craft where the atmosphere is pressurized to about 7 atm so that personnel can enter the water at depths of about 300 ft. Among the problems encountered in using this mixture were rapid failure of vacuum tubes in electronic equipment, and the need to keep the ambient temperature at about 88°F to keep personnel feeling adequately warm. Explain the origin of these difficulties.

3-15. In the calculation of collision number of oxygen gas carried out in the text (p. 58) a diameter of 3 Å was chosen for the molecules, which were assumed to be spherical. Calculate the total surface area of oxygen molecules confined to a cubic container 1 cm on a side, pressure of 1 mm, temperature 0°C, and compare this with the area of the container walls.

3-16. (a) The Dumas bulb experiment described in the text is performed on an unknown liquid, using a mineral oil bath, kept at 200°C, instead of a water bath. From the following data, calculate the molecular weight.

Empty bulb	13.025 g
Bulb plus condensate	13.915 g
Bulb plus water	317.2 g
Barometric pressure	755 mm

(b) What will be the effect of each of the following factors on the calculated molecular weight in part (a)?

1. The bulb expands in the heating from room temperature (at which its volume is measured) to 200°C.
2. Not all of the unknown liquid was evaporated from the bulb before it was removed from the bath.
3. When the unknown vapor condenses, the bulb refills with air, and this is weighed along with the condensed unknown.

3-17. The density of a liquid is to be determined by weighing a known volume. The following data are gathered:

Weight of empty container	32.5163 g
Weight of container with liquid	39.6152 g
Volume of liquid	8.2173 cm^3

The liquid, when placed in the container, displaces air. Therefore, the weights obtained actually represent the weight of the container filled with a very thin fluid (air) in one case, and with a relatively much denser, unknown liquid in the other. By measurement of the humidity, temperature, and pressure, one can calculate the density of displaced air. (These values are also tabulated in handbooks.)

Assuming that the density is 1.173 g/liter in the present problem, calculate the density of the liquid, and compare this value with that obtained when no correction for the displaced air (called the buoyancy correction) is applied. [*Hint:* You will have to carry out some parts of the calculations to 5 significant figures.]

3-18. A diesel engine cylinder has a volume of 0.75 liter at the point in the cycle at which fuel is metered in. Assuming the pressure at this point to be 1.5 atm, and the temperature to be 200°C, what volume of liquid fuel (density 0.85 g/cm^3, molecular weight 80) should be admitted if the vaporized fuel is to have a partial pressure of approximately 5 per cent?

3-19. A bottle is half-filled with an aqueous solution of hydrogen peroxide, H_2O_2, 5 per cent by weight. Assuming that the cap is screwed tightly when the air space in the bottle is saturated with water vapor, the pressure inside is just 1 atm. What will be the pressure in the bottle if all of the hydrogen peroxide decomposes according to the reaction

$$H_2O_2 \rightarrow H_2O + \tfrac{1}{2}O_2$$

Neglect changes in the volume of the liquid, and neglect the solubility of O_2 in water. Assume that the density of the solution and of the water are both 1 g/cm^3, and that the temperature is 27°C.

3-20. Calculate the average molecular speed u, in cm/sec, for the following gases at 25°C: O_2, CO_2, UF_6 (uranium hexafluoride).

3-21. Calculate the value of the molar gas constant R in units of e-v/°K-mole. [*Hint:* Use one of the values of R given in Table 3-1.]

3-22. Calculate the mean free path of N_2 gas molecules at 25°C and at the following pressures: 1 atm, 1 mm, 10^{-6} mm. Assume a spherical molecule of diameter 2.6 Å. What are the units of mean free path as given by Eq. (3-32)?

3-23. Assuming that a gas is kept at constant volume, derive a relationship between collision number and absolute temperature.

3-24. Using Eqs. (3-26) and (3-27), derive an expression for the fraction of molecules which possess a speed equal to or greater than the rms speed, u. [*Hint:* If m in Eq. (3-26) is expressed as molecular weight, then k should be converted to the molar gas constant, R.] Using the resulting expression, calculate the fraction of oxygen molecules which possess a speed greater than u at 300°K. Does this fraction change with temperature?

3-25. It is desired to pass a beam of molecules into an evacuated apparatus, and out through another opening a distance of one meter away without having more than a small fraction of them undergo a collision. This implies that the mean free path should be on the order of 2 meters. Assuming a radius for the molecule of about 1.5 Å, what is the maximum allowable pressure in the apparatus?

3-26. From the data of Table 3-4 it is possible to extract some interesting predictions, and to derive some interesting conclusions. For example: (a) Predict the van der Waals constant b for hydrogen fluoride, HF. (b) Estimate values for the van der Waals constants for radon, Rn. It can be seen from Table 3-4 that the values of the van der Waals constants increase with increasing atomic number among the rare gas elements. What is the significance of this trend?

3-27. Calculate the volume occupied by a mole of Xe at 300°K and 30 atm pressure using the van der Waals equation, and using the ideal gas equation. [*Hint:* Do the ideal gas equation calculation first, then use this result at an approximate value for V in the pressure correction term. After solving the resulting simpler equation for V, compare the two values, decide whether it is necessary to repeat the calculation, using the better approximation for V in the correction term.]

3-28. Repeat Prob. 3-11 (a) using the van der Waals equation rather than the ideal gas law expression (see the *Hint* given in Prob. 3-27). Based on a comparison of the results obtained for the volume using the ideal gas assumption, do you believe that adequate quantitative work could be carried out with the simpler assumption?

3-29. By using the appropriate value for R, calculate the total kinetic energy in calories of a mole of an ideal gas at 25°C.

3-30. Calculate the number of molecules per cm^3 which remain in a volume that has been "evacuated" to a pressure of 10^{-7} mm Hg at room temperature.

3-31. Describe an experiment which might be performed to test the correctness of Maxwell's theory regarding the distribution of molecular velocities.

Chapter 4

Chemical Calculations

4-1 Introduction

Dalton's atomic theory leads directly to two other laws of major importance: The *law of definite proportions* and the *law of multiple proportions*. The first of these states that, when two substances combine to form a particular new substance, the ratio of the weights of the two reacting substances that are consumed in the reaction is always the same. Since the atoms of a particular element are all the same (according to Dalton), it follows that a combination of atoms which gives rise to a particular substance must also always be the same.

The law of multiple proportions is best stated by giving an example. Nitrogen and oxygen combine under one set of conditions to form a compound with the formula NO. The reaction is represented by the equation

$$N_2 + O_2 \rightarrow 2NO \qquad (4\text{-}1)$$

Under different conditions, the reaction of the same two elements may yield a compound of the formula NO_2:

$$N_2 + 2O_2 \rightarrow 2NO_2 \qquad (4\text{-}2)$$

The law of multiple proportions states that the amount of oxygen which combines with a given weight of nitrogen in the second reaction is a simple multiple of that which is required in the first reaction. If this seems obvious already from Eqs. (4-1) and (4-2), it should be recalled that the balanced equations are merely a statement, in shorthand symbols, of what we believe the reaction to be. Therefore, we have already assumed the laws of definite and multiple proportions in the writing of the equations. (We have also assumed that mass is conserved in the reactions.)

In this brief chapter, we will consider some of the applications of these quantitative laws to problems which are encountered daily in ordinary laboratory work.

4-2 Weight-Weight Problems

The weight-weight problem, so-called, is a simple problem encountered frequently in chemical work.

Example 4-1

One wishes to make 50 g of calcium oxide, CaO, by decomposition of calcium carbonate, $CaCO_3$, according to the equation*

$$CaCO_3(s) \rightarrow CaO(s) + CO_2(g)$$

The problem is to calculate the amount of $CaCO_3$ required to produce the desired amount of product, assuming that all of the $CaCO_3$ used will undergo decomposition. We employ a dimensional method in such calculations, one which includes the units of all quantities employed. The advantages of this method are that it is orderly, and that mistakes are minimized.

$$1 \text{ mole } CaCO_3 \rightarrow 1 \text{ mole } CaO + 1 \text{ mole } CO_2$$

We must first determine, using a table of atomic weights, what the formula weight of CaO is, and then calculate how many formula weights (moles) the desired 50 g represent:

$$1 \text{ mole CaO} = 56 \text{ g CaO}$$

$$50 \text{ g CaO} \times \frac{1 \text{ mole CaO}}{56 \text{ g CaO}} = 0.89 \text{ mole CaO}$$

Since 1 mole $CaCO_3$ gives rise to 1 mole CaO, 0.89 mole $CaCO_3$ is required for the production of 0.89 mole CaO.

$$0.89 \text{ mole } CaCO_3 \times \frac{100 \text{ g } CaCO_3}{1 \text{ mole } CaCO_3} = 89 \text{ g } CaCO_3$$

This stepwise calculation can be summarized in one expression:

$$50 \text{ g CaO} \times \frac{1 \text{ mole CaO}}{56 \text{ g CaO}} \times \frac{1 \text{ mole } CaCO_3}{1 \text{ mole CaO}} \times \frac{100 \text{ g } CaCO_3}{1 \text{ mole } CaCO_3} = 89 \text{ g } CaCO_3$$

---

*The letter in parentheses after each formula gives the state [solid (s), liquid (l), or gas (g)] of the compound under the conditions of the experiment.

Two things should be noted about this calculation: (a) The units cancel to leave the desired units, g $CaCO_3$. (b) One term comes from the balanced chemical equation, the ratio of moles of $CaCO_3$ to CaO. The balanced equation is therefore necessary for solving the problem.

Example 4-2

Mercury (II) thiocyanate [$Hg(SCN)_2$], is to be produced from mercury (II) nitrate [$Hg(NO_3)_2$] by the reaction

$$Hg(NO_3)_2 + 2NH_4SCN \rightarrow Hg(SCN)_2 + 2NH_4NO_3$$

How many g of product (theoretically) could be obtained by use of 30 g NH_4SCN, assuming that an excess of $Hg(NO_3)_2$ is present?

$$30 \text{ g } NH_4SCN \times \frac{1 \text{ mole } NH_4SCN}{76.1 \text{ g } NH_4SCN} \times \frac{1 \text{ mole } Hg(SCN)_2}{2 \text{ moles } NH_4SCN} \times \frac{316.8 \text{ g } Hg(SCN)_2}{1 \text{ mole } Hg(SCN)_2} = 62.5 \text{ g } Hg(SCN)_2$$

The weight-weight problem may be complicated in various ways without changing its essential nature. For example, if one of the reactants or products is a gas, the quantity of substance may be expressed in terms of gas volume.

Example 4-3

750 ml of oxygen gas, O_2, is to be collected over water at 25°C. Atmospheric pressure is 740 mm. What quantity of potassium chlorate, $KClO_3$, is required to produce this quantity of gas through decomposition, according to the equation

$$KClO_3(s) \rightarrow KCl(s) + \tfrac{3}{2}\, O_2(g)$$

The experimental arrangement employed in collecting a gas over water is shown in Fig. 4-1. In determining the pressure of oxygen, a correction

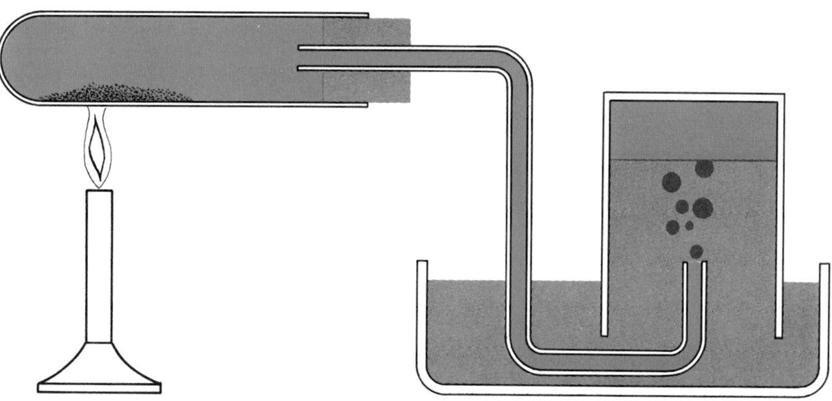

Fig. 4-1 Collection of a gas over water.

must be made for the water vapor pressure in the bottle. At 25°C the vapor pressure of water is 24 mm. The oxygen pressure is, therefore, 740 − 24 = 716 mm. We then calculate the number of moles of oxygen gas:

$$n = \frac{PV}{RT} = \frac{(716/760) \times 0.75}{0.08206 \times 298} = 0.0289 \text{ mole } O_2$$

then,

$$0.0289 \text{ mole } O_2 \times \frac{1 \text{ mole } KClO_3}{\frac{3}{2} \text{ moles } O_2} \times \frac{122.6 \text{ g } KClO_3}{1 \text{ mole } KClO_3} = 2.36 \text{ g } KClO_3$$

4-3 Empirical Formula Problems

The *empirical formula* is one of the first things one wishes to know about a substance being characterized. It is the formula in which the ratios of the numbers of atoms are stated in terms of the lowest possible set of whole numbers. Determination of the empirical formula is dependent upon data acquired in the course of chemical analysis. The problem therefore varies somewhat, depending upon the form in which the data are presented.

Example 4-4

An oxide of manganese, Mn, is found on analysis to contain 69.4 per cent by weight of manganese. What is its empirical formula?

The simplest way to approach a problem of this sort is to assume that we have some convenient quantity of the compound; *e.g.*, 1 g. Then we use the given percentages to calculate the mass of each component in that sample. From these masses, we compute the number of moles of each component; the ratios of the moles present can be obtained by dividing through by the smallest number, and multiplying by an integer, if necessary, to convert to whole numbers.

In 1 g of the oxide, we would have 0.694 g of Mn and 0.306 g of oxygen.

$$0.694 \text{ g Mn} \times \frac{1 \text{ mole Mn}}{54.9 \text{ g Mn}} = 0.0127 \text{ mole Mn}$$

$$0.306 \text{ g O} \times \frac{1 \text{ mole O}}{16 \text{ g O}} = 0.0192 \text{ mole O}$$

Dividing through by the smaller of these, we find that we have 1 mole Mn, 1.51 moles O. Since the analytical method is not entirely free of error and uncertainty, we may take the correct empirical formula to be Mn_2O_3.

Example 4-5

A compound which is known to contain only carbon, hydrogen, and (possibly) oxygen is combusted in the presence of excess oxygen, as shown in Fig. 4-2. The products of the combustion are carbon dioxide, CO_2, and water, H_2O. These are collected and weighed separately.

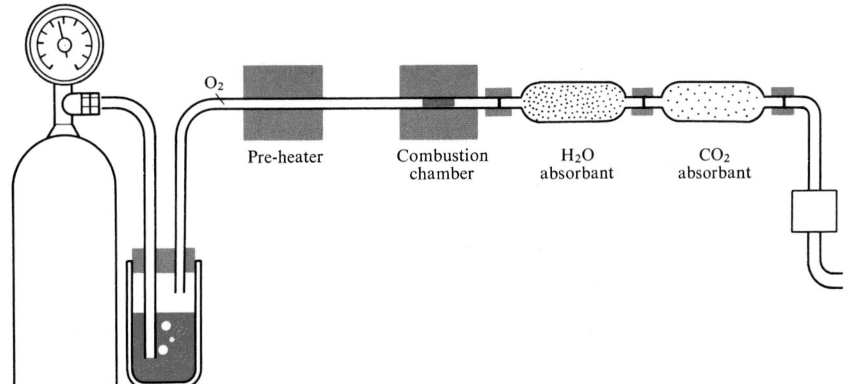

Fig. 4-2 Apparatus for determination of carbon and hydrogen by combustion analysis. The sample is combusted in an excess of oxygen. All of the carbon in the compound is converted to carbon dioxide; all of the hydrogen to water. The carbon dioxide and water are collected in separate absorption tubes which are weighed accurately before and after the combustion.

From the following data, calculate the empirical formula:

Weight of compound combusted:	0.350 g
Weight of CO_2 collected:	0.670
Weight of H_2O collected:	0.411

To determine the empirical formula from these data, we must realize that all of the hydrogen present in the original compound is present as the hydrogen of the water, and that all of the carbon in the original compound is present as the carbon of the carbon dioxide. The first step in treating the data is, therefore, to determine how many grams of hydrogen and carbon are present in the water and carbon dioxide, respectively.

$$0.411 \text{ g } H_2O \times \frac{2 \text{ g H}}{18 \text{ g } H_2O} = 0.0457 \text{ g H}$$

$$0.670 \text{ g } CO_2 \times \frac{12 \text{ g C}}{44 \text{ g } CO_2} = 0.183 \text{ g C}$$

We do not analyze directly for oxygen in this analysis; in fact, the compound is combusted in a large excess of oxygen from the tank. But from the weights of water and carbon dioxide we have calculated the weights of hydrogen and carbon present in the original compound. The difference between their sum and the initial weight of the sample must represent the weight of oxygen.

$$0.350 \text{ g} - 0.229 \text{ g} = 0.121 \text{ g O}$$

The number of moles of each element present is then

$$0.0457 \text{ g H} \times \frac{1 \text{ mole H}}{1 \text{ g H}} = 0.0457 \text{ mole H}$$

$$0.121 \text{ g O} \times \frac{1 \text{ mole O}}{16 \text{ g O}} = 0.00757 \text{ mole O}$$

$$0.183 \text{ g C} \times \frac{1 \text{ mole C}}{12 \text{ g C}} = 0.0152 \text{ mole C}$$

Dividing through by the smallest of these numbers, we have

$$\text{Empirical formula} = C_2H_6O$$

If elements other than carbon, hydrogen, and oxygen are present in the compound; *e.g.*, nitrogen or sulfur, the simple analytical method described must be modified. In general, if a compound contains n different elements, we must analyze for $n - 1$ of these independently; the last can be obtained by difference.

It must be emphasized that the empirical formula is not necessarily the same as the molecular formula. Thus, the molecular formula of the compound in the example above may be C_2H_6O, or $C_4H_{12}O_2$, *etc.* The elemental analysis tells us only the ratio of the numbers of atoms of each of the constituents, expressed as the smallest set of whole numbers.

Exercises

4-1. Carbon and oxygen combine to form more than one compound. Upon analysis, two of the compounds yield the following results:

	Weight carbon	Weight oxygen
Compound A	0.168 g	0.448 g
Compound B	0.515 g	0.686 g

Show how the data illustrate the law of multiple proportions.

4-2. In preparing cyanogen chloride, ClCN, in accordance with the equation

$$NaCN + Cl_2 \rightarrow NaCl + ClCN$$

how many g of ClCN would be obtained by reacting 40 g of Cl_2 with excess NaCN, if the percent yield were 75 per cent?

4-3. It is desired to collect one liter of nitrogen gas over water at 25°C, at barometric pressure 730 mm. What weight of ammonium nitrite is required to generate this amount of nitrogen according to the reaction

$$NH_4NO_2 \rightarrow N_2 + 2H_2O$$

4-4. How many moles of sulfur dioxide, SO_2, can be prepared from 1.5 g of sulfur and 4.5 liter oxygen gas at 1.5 atm, 25°C?

4-5. A sample of methane, CH_4, is collected by bubbling it through water and collecting the gas as in Fig. 4-1. The temperature is 8°C, at which the vapor pressure of water is 8 mm; atmospheric pressure is 735 mm. A volume of 456 cc is collected. How many moles of methane is this?

4-6. A binary compound between chlorine and fluorine is found to be 61.6 per cent by weight in fluorine. What is the empirical formula?

4-7. A compound of iron, chromium, and oxygen is found to have 28.6 per cent oxygen, 25.0 per cent Fe. Calculate the empirical formula.

4-8. When Si_3P_4 is treated with HCl, the products are $SiCl_4$ and PH_3. What is the maximum volume of PH_3 gas at 730 mm and 27°C which can be formed by reaction of 3.0 g Si_3P_4 with 0.75 g HCl? What is the density of PH_3 gas under the conditions described?

4-9. A 1.356-g sample containing tin, Sn, and a number of other elements was dissolved in acid, and the tin recovered as tin (IV) oxide, SnO_2. A weight of 0.685 g SnO_2 was obtained. What is the percentage of tin in the original sample?

4-10. What is the per cent by weight of copper in copper (II) sulfate pentahydrate, $CuSO_4 \cdot 5H_2O$?

4-11. An oxide of vanadium is found to contain 56.0 per cent vanadium. What is its empirical formula?

4-12. What volume of dry carbon dioxide at 725 mm pressure, 30°C, would be collected in the reaction carried out in Example 4-1, p. 70?

4-13. What weight of barium carbonate, $BaCO_3$, is required to produce the same quantity of CO_2 on decomposition as 50 g of $CaCO_3$?

4-14. A compound of carbon, chlorine and hydrogen is found on analysis to consist of 14.2 per cent carbon, 2.36 per cent hydrogen. What is its empirical formula?

4-15. 1.10 g of hydrate of barium chloride, $BaCl_2 \cdot x\,H_2O$, is dried in an oven and reweighed. This procedure is repeated until the compound attains a constant weight, indicating loss of all water. The final weight of the sample is 0.937 g. How many moles of water per mole of the salt are present in the hydrate?

4-16. How many liters of hydrogen at 25°C, 1 atm are required to reduce 120 g of copper (II) oxide, according to the equation

$$CuO(s) + H_2(g) \rightarrow Cu(s) + H_2O(l)$$

4-17. A compound containing carbon, hydrogen and oxygen is to be analyzed by combustion analysis. 0.280 g of the compound is combusted to give 0.507 g carbon dioxide and 0.172 g water. Determine the empirical formula. A sample of the same compound is loaded into a Dumas bulb, which is then immersed in boiling water at 100°C. The following data are obtained: bulb empty, 16.237 g; bulb filled with water, 137.2 g; bulb with condensed vapor, 16.792 g. P = 730 mm. From these data, calculate the molecular weight and the molecular formula of the compound.

4-18. A compound containing only carbon, hydrogen, and sulfur is combusted, using a modified procedure. The sulfur is collected and weighed as SO_2. From the combustion of 0.300 g of compound, 0.220 g of water and 0.260 g of SO_2 are produced. What is the empirical formula of the compound?

4-19. 0.450 g of an anhydrous chloride of tin (Sn) is added to water. A solution of silver nitrate is added until all of the chloride has been removed from the solution as silver chloride, AgCl. The weight of silver chloride recovered is 0.680 g. Determine the empirical formula of the tin compound.

4-20. A large box with a volume of 22 ft^3 is to be used as a dry box. The box atmosphere initially contains a water vapor pressure of 14 mm at room temperature. What is the minimum quantity of barium oxide, BaO, needed to remove this water vapor, assuming that the reaction is

$$BaO(s) + H_2O(g) \rightarrow Ba(OH)_2(s)$$

Assume a total pressure of one atmosphere in the box.

4-21. A mixture of Mn_2O_3 and MnO is treated with hydrogen gas. Hydrogen reacts with the higher oxide, according to the following equation:

$$Mn_2O_3(s) + H_2(g) \rightarrow 2MnO(s) + H_2O(g)$$

2.355 g of the mixture is treated with hydrogen, and the water is collected and weighed at 0.171 g. What is the weight per cent of Mn_2O_3 in the mixture? What is the mole fraction of Mn_2O_3 in the mixture?

4-22. The atomic weight of silver was determined by weighing a certain amount of silver metal, converting this to silver nitrate, $AgNO_3$, and then weighing this. The atomic weight of nitrogen was taken — on the basis of gas density measurements — to be 14.008. The atomic weight of oxygen was set equal to 16.0000. The ratio of the weight of the silver nitrate to the weight of silver was 1.57479. From these data, calculate the atomic weight of silver, to six significant figures.

4-23. Professor A. F. Scott and his students at Reed College reported a chemical determination of the atomic weight of fluorine. [*Journal of the American Chemical Society* **79**, 4253 (1957).] In the method which they employed, the ratio of perfluorobutyryl chloride, C_4F_7OCl, to silver was determined. This ratio was found to be 2.15513. Using the following values for atomic weights, calculate the atomic weight of fluorine: O = 15.999, Cl = 35.453, Ag = 107.870, C = 12.011.

Chapter 5

Atoms

5-1 Introduction

The concept of the atom has been introduced and some atomic properties have been mentioned in previous chapters. These are summarized for review and as an introduction to a more complete discussion of this topic:

1. Atoms are the smallest units of matter which exhibit the ability to enter into chemical reactions (Dalton).
2. When substances react with one another, it is the atoms which combine to form molecules, or atomic aggregates (Dalton).
3. The atom consists of a very small nucleus in which is centered the positive charge and nearly all of the mass, and electrons which are negatively charged and of relatively low mass. The electrons are found in the space comprising most of the volume of the atom (Rutherford).
4. Atomic dimensions are on the order of $1–5 \times 10^{-8}$ cm, or 1–5 Å, in diameter (various workers).

5-2 Atomic Number and the Periodic Table

To trace the origin of the concept of atomic number, one must go back more than a century in the history of chemical science. After a number of the chemical elements had been identified and their chemical behavior characterized, it became clear to some people that certain elements were closely related. A number of attempts were made, therefore, prior to 1850, to classify elements as members of groups. These attempts were largely unsuccessful because they were in advance of a knowledge of the correct atomic weights for many elements. Cannizzaro's contribution, since it provided a method for obtaining the correct atomic weights of the known elements, also paved the way for their systematic classification. It is not surprising then that, only eleven years after Cannizzaro made his work widely known, Mendeleev, a Russian chemist, presented a paper entitled, "On the Relation of the Properties to the Atomic Weights of the Elements." In 1870, Lothar Meyer, a German chemist, published a table of the elements which was nearly identical with Mendeleev's although the two men had worked independently. In these two schemes, the atomic number was assigned to the elements in the order of increasing atomic weight, beginning with the lightest element, hydrogen (atomic number 1). Both Mendeleev and Meyer recognized that the chemical and physical properties of the elements possessed a periodic character. The properties of chlorine, for example, are closely paralleled by those of bromine, an element of greater atomic weight, and also paralleled by those of iodine, which has a still higher atomic weight. By listing these closely related elements in vertical rows, and by placing these vertical rows side by side so that adjacent elements in the horizontal rows were in the order of increasing atomic weight, Mendeleev and Meyer constructed their periodic tables.

Of course, there were gaps in the table because many elements had not as yet been discovered. Mendeleev had painstakingly gathered data on the 63 elements known at the time. He used these data boldly to predict the properties of a few elements which he was convinced were as yet undiscovered, and which would occupy vacant spots in the table. For example, element number 32, which lay between silicon and tin in a vertical row, had not yet been observed. Reasoning that its properties must be intermediate between those of silicon and tin, he described both its physical and chemical properties in detail. When the element germanium was later discovered (the search for it was aided considerably by Mendeleev's predictions) its observed properties were very close to those predicted.

In constructing the periodic table, Mendeleev was guided by his knowledge of the chemical and physical properties of the elements, and by the general rule that the atomic number should increase in the order of increasing atomic weight. There were a number of cases, however, in which this procedure led to difficulties. The atomic weight of tellurium was given at the time as 128, as compared with 127 for iodine. It was clear to Mendeleev that iodine, and not tellurium, belonged under bromine, so he put them in order in which we now find them in the periodic table. He had enough confidence in his method to suggest, in the face of considerable opposition, that the atomic weight of tellurium was wrong. A glance at the modern periodic table shows, however, that the atomic weight given for tellurium was not wrong — but on the other hand, neither was Mendeleev! In a number of other instances in which Mendeleev disputed accepted atomic weights, he proved to be right. There are three notable exceptions to the rule that atomic weight increases regularly with increasing atomic number. These are argon–potassium, cobalt–nickel, and tellurium–iodine.

5-3 Atomic Spectra

When a substance is heated or otherwise made to absorb a large amount of energy, it emits radiation. Spectroscopy is, in part, the study of this emitted energy. Suppose that one wishes to study the emission spectrum of hydrogen. An apparatus similar to that shown schematically in Fig. 5-1 is

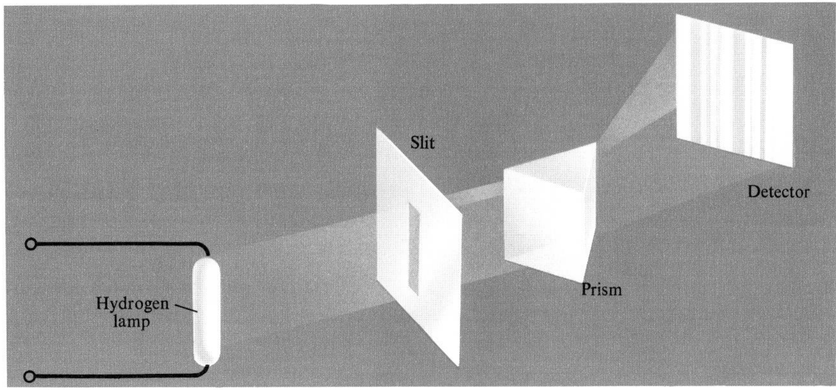

Fig. 5-1 Measurement of the emission spectrum from a hydrogen discharge.

used. The hydrogen is made to absorb energy from an electric discharge, or in some other way, and it then emits radiation. A thin beam of this radiation is passed through a prism, which refracts the rays. The degree to which the rays are refracted, or bent, varies with the wavelength of the radiation. Rays of shorter wavelength are more strongly refracted. The prism thus acts to separate the emitted radiation into different wavelengths. The radiation, after refraction, is detected by allowing it to impinge upon a photographic film.

Now, if the hydrogen were emitting radiation of all wavelengths, there would be a continuous band of rays emerging from the prism, and the developed plate would look like that in Fig. 5-2(a). This is not the case, however; only certain wavelengths of light are emitted, and the film looks

Fig. 5-2 Line spectrum of hydrogen.

like that shown in Fig. 5-2(b). Similar results are obtained for other elementary substances, although the spectra may be much more complex. These are called the *line spectra.* They were obtained long before anyone had a clear idea of the structure of atoms.

In addition to emission, substances may also exhibit absorption of radiation. If white radiation (containing all wavelengths) is passed through hydrogen, the emergent radiation is found to be lacking in certain wavelengths (Fig. 5-4). These have been absorbed by the hydrogen. They correspond precisely to certain of the wavelengths which are emitted by hydrogen when excited.

The interpretation of these results is that, when hydrogen atoms are placed in an electric discharge, or otherwise supplied with energy, the atoms absorb energy and become excited. When the atoms return to their initial condition, this energy is released in the form of radiation. Similarly, in the absorption experiment, the hydrogen atoms absorb certain wavelengths of radiation from the white radiation and become excited. They re-emit this energy, but the emitted energy goes off in all directions. Only the wavelengths which are not absorbed by the hydrogen pass through with high intensity and are detected.

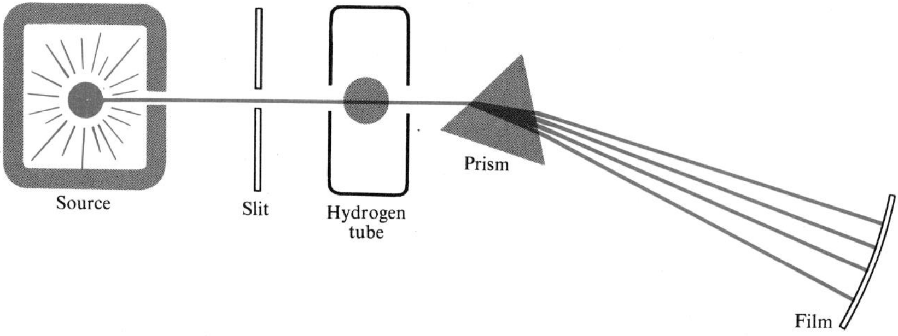

Fig. 5-3 An experiment for determination of the absorption spectrum of atomic hydrogen. Because of numerous practical difficulties, the experiment cannot be conducted as simply as the illustration implies.

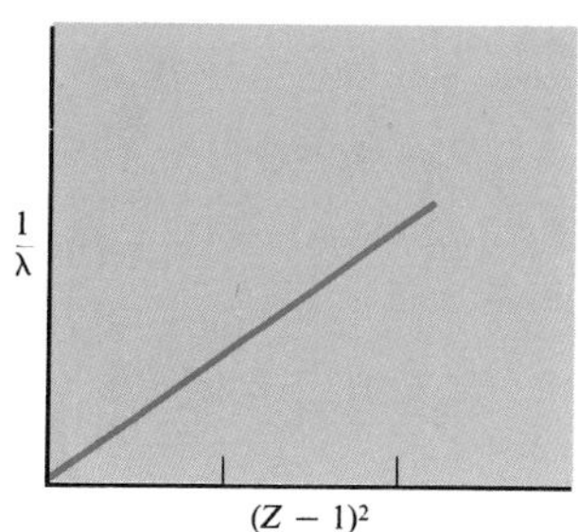

Fig. 5-4 Moseley's relationship between the characteristic X-ray wavelength λ for an element, and the atomic number Z.

The existence of these characteristic line spectra proved to be a stumbling block for Rutherford's theory of atomic structure, in which the atoms were pictured as consisting of a small nucleus containing the positive charge, around which the electrons moved in a circular orbit. Such a model did not explain the line spectra. It also did not fit in well with the requirements of classical electromagnetic theory. The electrons were supposed to have kinetic energy which kept them from falling into the nucleus to which they were attracted. But from electromagnetic theory, an electron moving in a curved path should radiate energy, and should gradually lose its kinetic energy. The classical model was evidently not adequate to deal with the problem. Some missing factor would have to be found before an adequate model could be formulated. Niels Bohr, who worked with Rutherford on the problem, provided this missing factor by incorporating the quantum theory into a model for atomic structure.

5-4 Moseley's Experiments

The particular variety of line spectra discussed in the previous section are termed *optical spectra.* As we shall learn in more detail later, they correspond to changes in the energies of electrons in the outermost parts of atoms. With the exception of hydrogen and the alkali metals, the optical spectra of the elements are exceedingly complex, consisting of many hundreds of lines. Another form of radiant excitation of atoms became known in the period about 1910, when it was discovered that high-energy bombardment caused emission of X-ray spectral lines from substances. By contrast with the optical spectra, the X-ray spectra are quite simple. Certain series of lines were recognized, and it was found that the wavelength of the emitted lines varied in a systematic way as the element undergoing X-ray bombardment was varied.

One very striking difference was noted between the optical and X-ray spectra, aside from the fact that the latter were much simpler. The optical spectra of the elements possess a periodic character; for example, the spectra of alkali metals are similar, as are the spectra of halogens, *etc*. This is what one might expect if the optical spectra are due to changes in energies of electrons in the *outermost* parts of atoms, because the periodic table itself is based on the idea of similar chemical behavior which arises from the character of the outermost electronic region of the atom. The X-ray spectra, on the other hand, showed no such periodicity, but rather just changed in a continuous way from element to element in the table. One might infer that the X-ray spectra, which are much higher in energy than the optical spectra, are due to changes in the energies of the *inner* electrons of the atom. These inner electrons will be influenced mostly by the *magnitude of the nuclear charge*, and not by the details of the electronic arrangement in the outer reaches of the atom. It was this line of reasoning which led Henry Moseley to undertake a systematic study of the way in which the wavelength of a particular series of X-ray spectral lines, the so-called K series, varied in a series of elements.

When Rutherford performed his experiments on the scattering of alpha particles by a gold foil, he measured the angular distribution of the particles which underwent deflection. A major factor in causing deflection, aside from direct collision of the alpha particle with the gold nuclei, is the electrostatic repulsion between the alpha particles (charge $+2$) and the gold nuclei (unknown, but positive, nuclear charge). From these data, he was able to calculate an approximate value for the nuclear charge. It turned out to be (in units of the electronic charge) about half of the atomic weight of gold. Rutherford began to develop the idea that atomic number was related to the charge on the nucleus. This tentative proposal took on a more definite form in the work of Moseley, who was Rutherford's student. Moseley's experiments consisted in bombarding elementary substances with a stream of high-energy electrons. The elements, when excited in this way, emit X rays. Moseley measured the wavelength of the emitted X rays for 38 elements from aluminum to gold.

It was immediately apparent that, as the atomic weight increased, the characteristic wavelength of the X rays decreased. After a careful study of his results, however, Moseley became convinced that it was not the atomic weight which was of fundamental importance, but the *atomic number*. He found a quantitative relationship between the reciprocal of the wavelength of the K X rays emitted by an element and the square of its atomic number:

$$\frac{1}{\lambda} = A(Z - 1)^2 \tag{5-1}$$

where λ is the wavelength, Z is atomic number and A is a constant. The relationship is graphed in Fig. 5-4.

This work provided an insight into atomic structure which had heretofore been missing. Moseley was able, with no difficulty, to establish the atomic numbers of the rare earth elements (58-71) on the basis of their characteristic X-ray wavelengths. Previously, there had been no place for these elements in the periodic table. Moseley's work represents a genuinely inspiring contribution to science. He was faced with problems in the design and construction of his apparatus which were simply staggering. He was only 26 when his results were published in 1913. Three years later, during World War I, he was killed while serving as a signal officer in the British army.

Moseley's work established the relationship between atomic number and nuclear charge, and, at the same time, between atomic number and the number of extranuclear electrons. There still remained the problems of how these charged particles existed in the atom, and what relationship they had to chemical behavior. A number of theories of atomic structure had been proposed, but they were all deficient in one way or another. At this time, the attack on the problem went along two different lines. The physicists were interested in understanding in detail the relationships among the fundamental particles in the atom, and in explaining certain spectral properties of atoms. The chemists were more interested in using Moseley's results to establish relationships between the chemical properties of atoms and the number of extranuclear electrons. We will begin here with the physicists' approach. In a later chapter, we will discuss the chemists' approach, and see that it is essentially an empirical statement of the results which follow from the detailed model for the atom developed by the physicists (and sometimes the chemists).

5-5 The Bohr Model of the Hydrogen Atom

Bohr's contribution consisted in applying Planck's quantum theory (Chap. 1) to an otherwise classical model of the hydrogen atom. His assumptions were:

1. The hydrogen atom consists of a heavy, positively-charged nucleus, around which the single electron moves.
2. Of all of the orbits which are possible according to classical theory, only those which satisfy certain *quantum* conditions are possible in reality. The angular momentum of the electron, mvr (r is the radius of the orbit, v is the linear velocity, and m is the mass of the electron), cannot assume any arbitrary value, but must be a multiple of the quantum, $h/2\pi$.
3. Contrary to classical electromagnetic theory, an electron which is in one of the allowed orbits does not radiate or absorb energy.
4. Radiation is emitted or absorbed when the electron undergoes a transition from one allowed energy state to another. The energy difference between the two orbits is equal to the energy absorbed or emitted.

The model for the Rutherford-Bohr atom is shown in Fig. 5-5. The electron at a distance r from the nucleus is attracted to it (Coulomb's law). The force of attraction is e^2/r^2. Opposing this force, there is a force which tends to move the electron away from the nucleus. Force = mass ×

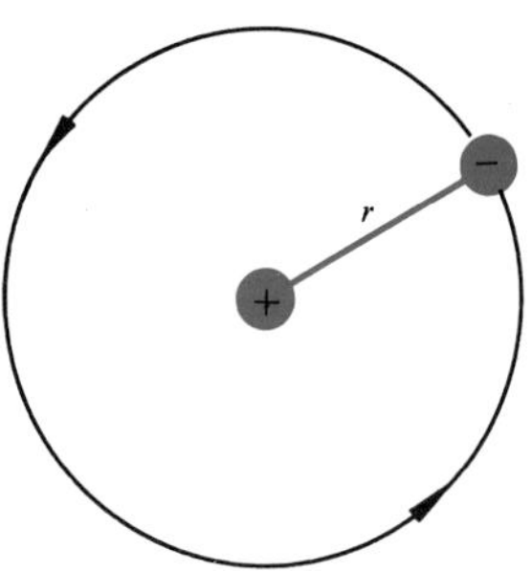

Fig. 5-5 The Bohr-Rutherford model for the hydrogen atom.

acceleration. The acceleration is that which results from the circular motion of the electron, the centrifugal acceleration, v^2/r. Therefore

$$\frac{mv^2}{r} = \frac{e^2}{r^2} \tag{5-2}$$

Bohr's quantum condition is that the angular momentum must be a multiple of $h/2\pi$:

$$mvr = nh/2\pi, \qquad n = 1, 2, 3 \cdots \tag{5-3}$$

Combining Eqs. (5-2) and (5-3), we obtain

$$r = \frac{n^2h^2}{4me^2\pi^2} \tag{5-4}$$

The smallest orbit, with $n = 1$, is then

$$r = \frac{h^2}{4me^2\pi^2} \tag{5-5}$$

On substituting the values for m, e and h in this equation we have

$$\begin{aligned} r &= \frac{(6.63 \times 10^{-27} \text{ erg-sec})^2}{4 \times 9.107 \times 10^{-28} \text{ g} \times (4.8 \times 10^{-10} \text{ esu})^2(3.14)^2} \\ &= 0.530 \times 10^{-8} \text{ cm} \\ &= 0.530 \text{ Å} \end{aligned}$$

The charge of the electron, given in electrostatic units, has the dimensions of $g^{1/2}\,cm^{3/2}\,sec^{-1}$. The student should verify that r is then given in cm. This quantity is often called the Bohr radius, and is given the symbol a_0.

We can write the total energy of the electron as the sum of the kinetic and potential energies.

$$E = mv^2/2 + (-e^2/r) \tag{5-6}$$

From Eq. (5-2), we have $mv^2 = e^2/r$. Substituting into Eq. (5-6),

$$E = -e^2/2r$$

Substituting in the value for r from Eq. (5-4),

$$E = \frac{-2me^4\pi^2}{h^2n^2} \tag{5-7}$$

The energy of the first few orbits is then

$$E = \frac{-2me^4\pi^2}{h^2} = -21.7 \times 10^{-12}\ \text{erg} \qquad n = 1$$

$$E = \frac{-2me^4\pi^2}{4h^2} = -5.42 \times 10^{-12}\ \text{erg} \qquad n = 2$$

$$E = \frac{-2me^4\pi^2}{9h^2} = -2.41 \times 10^{-12}\ \text{erg} \qquad n = 3$$

The values for the energies were obtained by inserting in the appropriate numerical values, as used above in calculating the radius, and as given on page 14 for Planck's constant. It is more usual to express the energy of the electron in another energy unit, the electron-volt (e-v). An *electron-volt* is the kinetic energy gained by an electron after it has been accelerated through an electric field of one volt. One e-v $= 1.6 \times 10^{-12}$ ergs. The energies of the hydrogen electron in e-v are therefore -13.6, -3.39, and -1.51 for $n = 1$, 2, and 3, respectively. The energy levels for hydrogen can be represented on an energy-level diagram, as shown in Fig. 5-6. The zero of energy

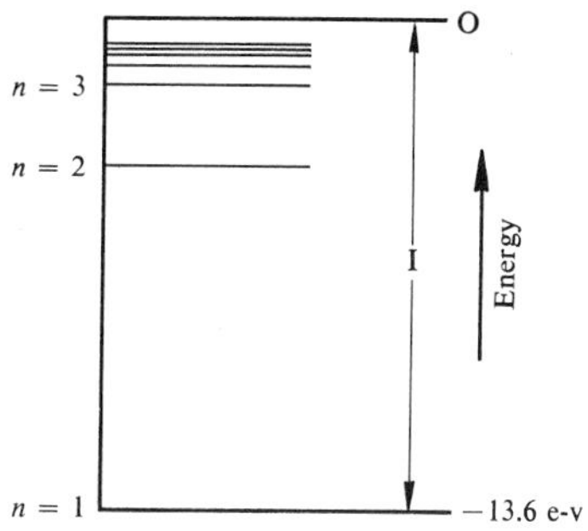

Fig. 5-6 Energy level diagram for hydrogen.

is taken as the energy for the electron when it is completely removed from the atom. The ionization energy for hydrogen in its ground (most stable) state, shown as I in Fig. 5-6, is the energy required to remove an electron from the $n = 1$ level to a very large distance from the nucleus.

According to the Bohr theory, radiant energy is emitted from the atom when the electron undergoes a change in the quantum number from a higher to a lower value. That is,

$$h\nu = \Delta E = E_{n_2} - E_{n_1} \tag{5-8}$$

where ν is the frequency of the emitted radiation and n_2 and n_1 are the quantum numbers of the initial and final states. The energy changes calculated from Eq. (5-8) for various values of n agree very well with the observed line spectrum for hydrogen. The lines are observed to occur in certain series, and each set of lines corresponds to a single value for the quantum number of the final state. For example, there is a series of lines corresponding to the transitions $n_2 \rightarrow n_1$, $n_3 \rightarrow n_1$, $n_4 \rightarrow n_1$, *etc.* To illustrate, for the transition $n = 2$ to $n = 1$ we have

$$h\nu = E_1 - E_2 = -21.7 \times 10^{-12} - (-5.42 \times 10^{-12})$$
$$h\nu = -16.3 \times 10^{-12} \text{ erg}$$

The negative sign indicates that energy is emitted from the atom in the transition. Solving for the frequency, we obtain

$$\nu = \frac{16.3 \times 10^{-12} \text{ erg}}{6.62 \times 10^{-27} \text{ erg-sec}} = 2.46 \times 10^{15} \text{ sec}^{-1}$$

It is understood that ν is in units of *cycles*/sec. The radiant energy absorbed or emitted by atoms is often described in terms of its wavelength rather than its frequency. From Chap. 1, we have

$$\lambda = c/\nu = \frac{3 \times 10^{10} \text{ cm/sec}}{2.46 \times 10^{15} \text{ sec}^{-1}} = 1.22 \times 10^{-5} \text{ cm}$$

Multiplying by 10^8,

$$\lambda = 1{,}220 \text{ Å}$$

Although the Rutherford-Bohr model for the hydrogen atom represented a major step forward, it was not completely satisfactory. The model was extended by other workers to include elliptical orbits, and small relativity corrections were applied. Even with these additions, the model could not be applied satisfactorily to atomic systems with more than one electron, and it was not long before a new quantum-mechanical model was introduced. This new model has produced what we may cautiously hope will be the final answer to the problem of atomic structure, insofar as it concerns chemists. It was developed from a rather formidable mathematical basis; we will limit ourselves here to a brief discussion of the ideas on which it is based, and then to a consideration of some of the results.

5-6 Wave-Particle Duality

It has already been indicated (Chap. 1) that electromagnetic radiation has, in a sense, a dual character. In discussing phenomena such as diffraction or refraction, it is convenient to consider the radiation as wavelike in character. On the other hand, the photoelectric effect is best discussed by considering the radiation as a stream of "particles," called *photons* (Chap. 1). In 1924, the French physicist de Broglie showed that the same duality of behavior should also be observable in particles of very small mass. His conclusion is that a particle of mass m moving with a velocity v should possess a characteristic "wavelength" λ given by

$$\lambda = \frac{h}{mv} \tag{5-9}$$

where h is Planck's constant. To get some idea of the magnitudes of the quantities involved in this equation, imagine the following experiment: Electrons are emitted from a cathode and accelerated toward a plate with a potential of 100 volts (Fig. 5-7). There is a small hole in the plate and some

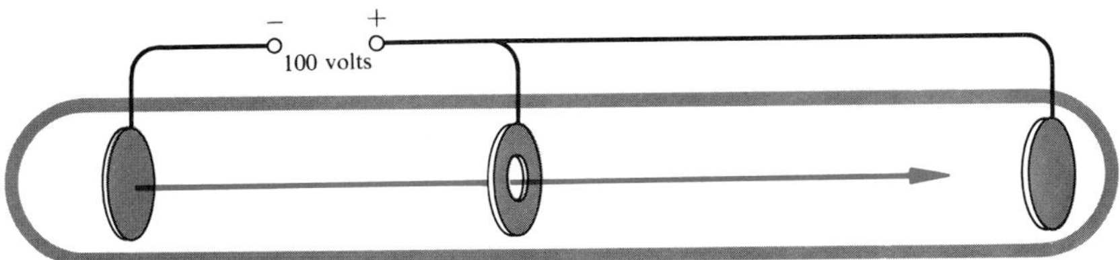

Fig. 5-7 Acceleration of electrons in an electric field.

of the electrons pass through it, continuing beyond at a constant velocity. The momentum of these electrons can be calculated by the methods discussed in Chap. 1. The kinetic energy of the electrons is equal to the energy which they have acquired by "falling" through a potential of 100 volts.

$$mv^2/2 = Ee = \frac{100}{300} \times 4.8 \times 10^{-10}$$

(The factor $\frac{100}{300}$ on the right converts the units of voltage to the necessary cgs units. The charge on the electron, in esu, is already in cgs units, but voltage, in "practical" units, must be multiplied by $\frac{1}{300}$ to convert to the cgs system.) The mass of the electron is 9.107×10^{-28} g. Multiplying both sides by $2m$ gives

$$m^2v^2 = 1.6 \times 10^{-10} \times 2 \times 9.107 \times 10^{-28}$$

and taking the square root:

$$mv = 5.4 \times 10^{-19}\ \text{g-cm/sec}$$
$$\lambda = \frac{h}{mv} = \frac{6.63 \times 10^{-27}\ \text{g-cm}^2\text{/sec}}{5.4 \times 10^{-19}\ \text{g-cm/sec}}$$
$$\lambda = 1.23 \times 10^{-8}\ \text{cm} = 1.23\ \text{Å}$$

That is, the electron moving in the space in back of the plate in Fig. 5-7 can be expected to show — under the right conditions — the properties of a wave phenomenon, with a characteristic wavelength of about 1 Å.

It was only a short while until de Broglie's suggestion was given experimental support. It had been known for a long time that X rays with wavelengths of 1 – 3 Å could be diffracted when the X-ray beam was passed

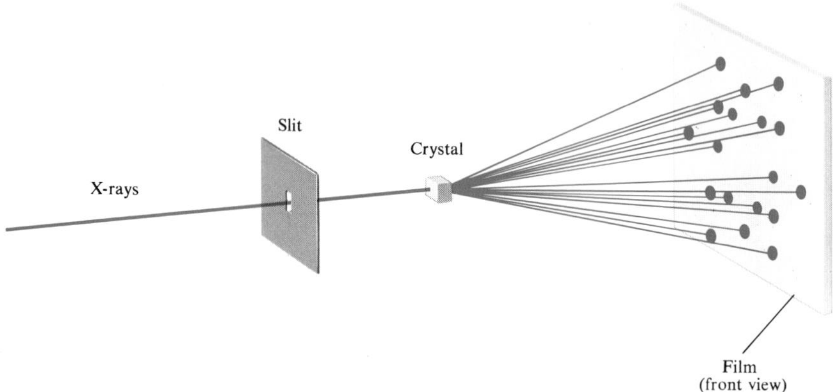

Fig. 5-8 Diffraction of X rays by a crystal.

into a crystal such as sodium chloride (Fig. 5-8). The diffraction, which results from the regular spacing of the atoms in the crystal, shows as a series of spots on a photographic film. In such an experiment, the incident radiation (the X ray) is best viewed as a wave phenomenon. But its characteristic wavelength is of the same magnitude as that which is calculated for electrons which have been accelerated through a potential field of 100 volts. Davisson and Germer in the United States and G. P. Thomson (J. J. Thomson's son) in England used such a beam of electrons to show that diffraction of the electrons did occur in a crystal. Furthermore, it was possible to show that the wavelength of the electrons as calculated from the diffraction pattern is just that which is calculated from de Broglie's equation. Since those first experiments, electron diffraction has grown into a powerful tool for the study of molecular structure and surface films.

According to de Broglie's theory, all particles should exhibit a characteristic wavelength, but it is clear from Eq. (5-9) that as the mass increases the characteristic wavelength decreases. Therefore, for a particle such as a proton or neutron to possess a characteristic wavelength in the region of 1 Å, it must have a much lower velocity than a relatively much lighter electron with that same wavelength. Diffraction of both neutrons and protons has been observed; the work with neutrons in particular has been yielding much valuable information about crystal structures.

5-7 The Uncertainty Principle

At about the time de Broglie was introducing his concept of the wave-particle duality of matter, Werner Heisenberg, a German physicist, introduced the *uncertainty principle*. In order to understand this concept, let us consider the situation illustrated in Fig. 5-9. An observer is following the path of a baseball, and notes its position and momentum at time t_1. The information which the observer receives about the baseball is in the form of

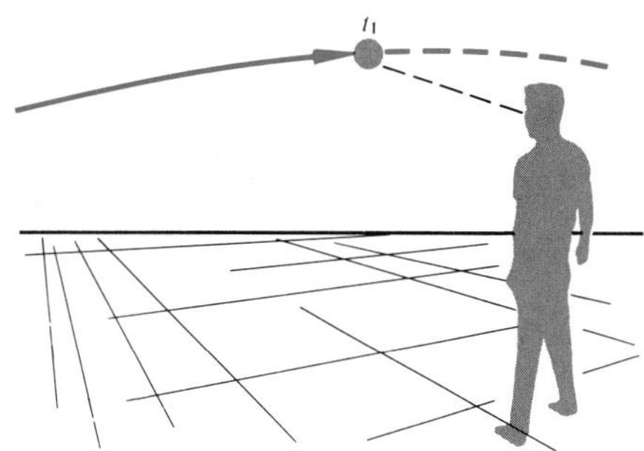

Fig. 5-9 Determination of position and momentum on a macroscopic scale. A baseball in flight is observed by detection of reflected light from the ball.

a stream of photons which have "bounced off" the ball and which strike his eye. We have already seen (Chap. 1) that a photon possesses a certain momentum, given by $p = mc = h/\lambda$. In measuring the position or momentum of a large-scale object such as a baseball, there is no need to consider the possibility that the momentum of the photons which strike it will have an observable effect on its momentum; the baseball is many orders of magnitude too massive for this. Therefore, knowledge of the position and momentum of the ball at time t_1 enables us to predict its position at time t_2. In such an experiment, the sensitivity of the measurements would also be of a macroscopic (large-scale) order. For example, the position of the ball might be measured to within 1 cm.

Now consider the experiment illustrated in Fig. 5-10. The position and momentum of a single electron are to be measured; in principle, this is the

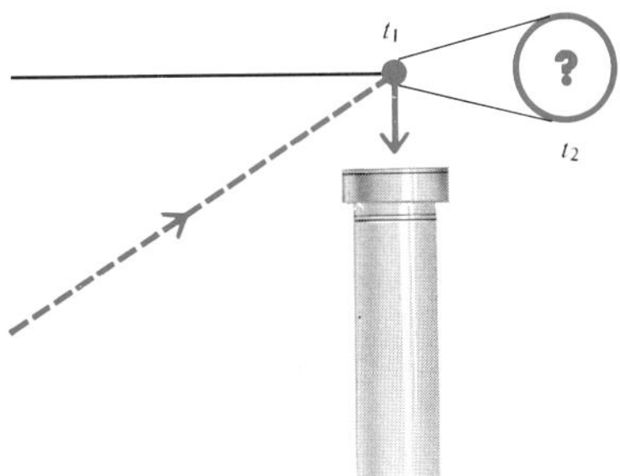

Fig. 5-10 Determination of position and momentum of an electron. An electron in flight is observed by detection of a scattered photon. The momentum and the position of an electron after scattering of the photon are uncertain.

same experiment done with the ball, except that the scale is reduced by many orders of magnitude. It is not necessary to worry, at this point, about the nature of the equipment used for observation. It is true that the electron must be observed in the same way as the baseball — by allowing a photon to impinge upon it, and observing the photon. But because the electron possesses such a small mass, and consequently a small momentum, its momentum is significantly changed as a result of the collision with the photon. As a result, even though the electron's position at time t_1 is known, we cannot predict where it will be at time t_2 because of the uncertainty about its momentum.

We have given only a very qualitative discussion about one of many such problems. General considerations about problems of this kind led Heisenberg to postulate that there is an inherent uncertainty in the knowledge which we can possess, at any one instant, about the position and motion of a single particle. This uncertainty is a part of the nature of things, and does not in any way reflect the quality of our instruments. In short, there is no way to overcome it. This uncertainty exists in all measurements, but it becomes noticeable only when dealing with phenomena on the atomic level.

5-8 A Modern View of the Atom

The modern view of atomic structure is based on a rather complex mathematical model. The fundamental concepts were established in the late 1920's by a number of scientists; Erwin Schrödinger, Werner Heisenberg, and P. A. M. Dirac made particularly illustrious contributions. The formulation which is most generally useful for chemists is the so-called *wave mechanics*. Experimental demonstration of the wave-particle duality of matter, and the limitation set by the uncertainty principle on our ability to measure on the subatomic level, required that the rather mechanical Bohr model of atomic structure be discarded. Furthermore, the notion of a circular or elliptical orbit leads to completely incorrect values for the angular momentum of the lowest energy state of hydrogen. In Schrödinger's wave mechanics, an equation for the total energy of an atomic system is written in such a way that the wave-like character of the particles is implicit in the equation. The result was the now-famous Schrödinger equation, a second-order partial differential equation. Although an equation of this type is very difficult to solve exactly, except in certain simple cases, it possesses certain properties which apply no matter which atomic or molecular system it is applied to. One such property is that the equation holds only when certain functions, called *wave functions*, are inserted into it. The problem of solving the Schrödinger equation for a particular system consists in finding wave functions which are solutions to the equation. Each of these solutions yields a value for the energy of the system. Other properties of the system, besides energy, can be calculated from the wave function and compared with experiment.

The wave functions are algebraic expressions which involve the coordinates; if the problem were expressed in Cartesian coordinates, the wave function would involve x, y, and z. In practice, more complicated coordinate systems must sometimes be used. In dealing with atoms, in which there is just one nucleus which can be placed at the center of the coordinate system, spherical coordinates are frequently employed. In a spherical coordinate system, a position in space can be described in terms of the distance from the origin, r, the angle θ which the vector r makes with the z-axis, and ϕ, the angle which the projection of r into the x-y-plane makes with the x-axis. These are illustrated in Fig. 5-11.

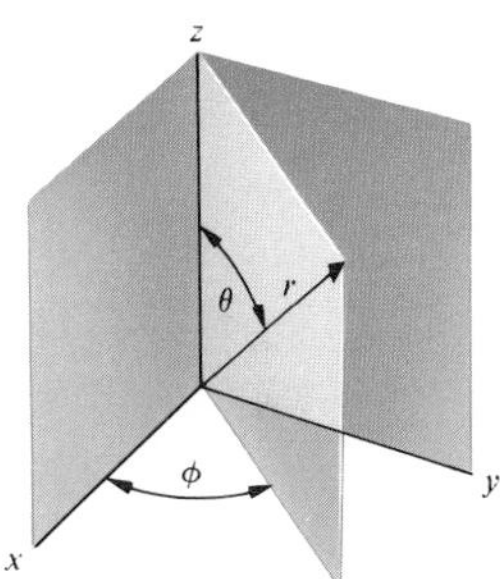

Fig. 5-11 Spherical polar coordinates.

In addition to the coordinates, the wave functions also contain certain constants. The Schrödinger equation is satisfied by the wave function only when the constants have certain values. For example, it may turn out, and it often does, that the equation is satisfied only when a certain constant has the values 0, 1, 2, 3, . . . ; *i.e.*, when the constant has a zero or integral value. These allowed values are analogous to the quantum numbers of the Bohr theory; each different value for a constant in the wave function corresponds to a certain energy value. It may happen, however, that more than one value of a particular constant leads to the same energy. The allowed solutions to the Schrödinger equation thus correspond to certain values of the constants, which we call quantum numbers. The allowed solutions, in turn, lead to certain allowed values for the energy. It is important to realize that the wave-mechanical description of atomic structure is like the Bohr theory in predicting definite allowed energy levels for the electrons. Thus it leads to identically the same result as does the Bohr theory for the allowed energy states of the electron in the hydrogen atom. It is, however, much more general than the Bohr theory, in that it can be applied to any atomic or molecular system. By this, we mean that it is possible to write down the Schrödinger equation for any such system, but it is quite another thing to actually find the wave functions which are solutions to it. The equation may be applied in useful ways, however, without necessarily finding exact solutions. The fundamental principles of wave mechanics have been tested in literally thousands of ways during the past few decades. In its most elaborate form, which includes relativistic corrections, it has been found satisfactory in every respect.

For atoms which contain more than one electron, the wave equation is complicated by the fact that account must be taken of the repulsive interaction which electrons exert upon one another. The difficulties in solving the Schrödinger equation for many-electron atoms are so great that it is not possible to take the electron-electron interactions into account in an exact way if the equation is to be solved at all. Instead, a procedure is used which was jointly developed by an English physicist, Hartree, and two Russians, Fock and Petrashan; the *self-consistent field* (SCF) method. In this method, each electron is assumed to move about the atom in an electrostatic field which is made up of the nucleus and an averaged distribution of the other electrons in the atom. This averaged distribution of the electrons is approximately spherical. The result is that each electron then has a wave function which is like the wave function for hydrogen in that it is based on a spherical coordinate system. If it were not for the assumption that the electron sees only an *averaged* field from the other electrons, each individual electron-electron interaction would need to be accounted for, and there would be no spherical symmetry in the problem. Each electron's behavior would depend on the instantaneous position of every other electron, and so the wave functions would be impossibly complex. The SCF procedure gets around this complexity to a large extent by considering only the average interaction of each electron with all the others. It is, of course, an approximation, and our wave functions for many-electron atoms are not exact as a result. Nevertheless, there is much evidence that the SCF atomic wave functions are quite good approximations, and that they adequately describe many aspects of the electron distributions in atoms.

The term *self-consistent* comes in because it is necessary to use the wave function calculated for a particular electron to obtain the averaged charge distribution seen by one of the others. If a change is made in the wave function for a particular electron, then a corresponding change is produced in the wave functions for all the others, since the wave function obtained for each electron is dependent on the charge distribution assumed for all of the others. The SCF procedure consists in starting with a set of assumed wave functions which are, hopefully, not too far off the mark. The wave function for the first electron is then calculated using the averaged field determined from the assumed set of wave functions for all of the remaining electrons. In general, the wave function so calculated will differ from the one guessed at for that electron. The new wave function is put in place of the assumed one, and the calculation is then carried out for the second electron. The result for that electron is then substituted for the assumed one, and the calculation proceeds for the third electron, *etc.* When this has been done for all of the electrons, a new set of wave functions has been substituted for all of the original set. But now the calculation carried out for the first electron must be repeated, using the new set of wave functions to obtain a second (and presumably better) approximation to the wave function for the first electron. The calculation is thus carried out through cycle after cycle, until the result of a particular cycle is substantially unchanged from that obtained in the previous one. The calculation is thus said to have converged; that is, to have attained *self-consistency*. The procedure may seem to be laborious, and indeed it is. With the aid of high-speed digital computers, however, such calculations have been successfully carried out to a high degree of accuracy for all but the heaviest elements. All descriptions of the electronic structure of atoms commonly used by chemists are based on the self-consistent-field type of approximation in which each electron is assumed to move only under the influence of the *averaged* field of the others.

Our discussion of the Schrödinger equation, and of the methods of wave mechanics, has necessarily been rather general and vague because the complexity of the mathematics does not permit a simple explanation. It is somewhat more rewarding to examine the wave functions which are solutions of the equation. We shall discuss wave functions which are hydrogen-like in that they have the same form as the solutions to the Schrödinger equation for hydrogen. Wave functions of this form are useful, because by a proper choice of certain adjustable parameters they can be made to approximate the more accurate SCF functions. The wave functions are frequently represented by the symbol ψ. They are, as we have mentioned, functions which involve the coordinates of the electron and contain constants which can take on only certain values. There are three such constants; the values which they may take on are termed *quantum numbers*. The three quantum numbers are not entirely independent of one another. Each particular choice of the quantum numbers corresponds to an allowed state of an electron. The wave function which corresponds to this set of quantum numbers is called an *orbital*. Each orbital has associated with it a particular value of the energy. It may happen that more than one orbital will give rise to the same value for the energy. Orbitals which are related to one another in this way are said to be *degenerate*.

Before discussing the quantum numbers in more detail, let us pause to ask what physical significance can be attached to the wave functions. In the Bohr theory, the allowed energy states of the electron correspond to well-defined circular or elliptical paths around the nucleus. In the wave-mechanical model of atoms and molecules, the nature of the electron's motion is not so definite. The wave equation takes account of the wave-like character of the electron. In solving the wave equation, we learn precisely what energies the electron may have, but the notion of precise location is lost. The wave function itself is not directly related to a physically significant quantity. The square of the wave function, ψ^2, on the other hand, does have significance;* it has the meaning of electron probability density. Where ψ^2 is

---

*The wave function ψ may be a complex function (*i.e.*, it may have an imaginary part), depending upon the particular algebraic form which is used. In such an event, the wave function is multiplied by its complex conjugate, and the physically significant quantity is $\psi^* \psi$, where ψ^* is the complex conjugate of ψ.

large, the electron density is large. To see how this interpretation arises, consider the electron as moving about the nucleus in a complicated manner. At any one instant, we cannot tell what its distance from the nucleus will be. It is, however, somewhere in space, so we can say that the probability of finding it somewhere in space is one (*i.e.*, certainty). But what is the probability of finding it at any one instant in a tiny volume element around the point r_1, θ_1, and ϕ_1? This is given by inserting the values r_1, θ_1, and ϕ_1 into the expression for the wave function, and evaluating ψ^2. Where the value for ψ^2 is relatively high, it is probable that the electron will be found there at any one instant. For this reason, ψ^2 is usually referred to as the *probability function.*

There are a number of ways in which the form of the wave function or probability function in space can be represented. The only completely unambiguous way is simply to write the mathematical expression. Unfortunately, most of us cannot readily make the transition from a complicated mathematical expression to a visualization of electron probability density in three-dimensional space. A number of diagramatic representations of either the wave functions themselves, or the probability functions, have been used as visual aids. We will discuss some of these using a simple wave function, that for the lowest energy level of hydrogen, for demonstration. This wave function has the form

$$\psi = \frac{1}{\sqrt{\pi}}\left(\frac{1}{a_0}\right)^{3/2} e^{-r/a_0} \tag{5-10}$$

a_0 is a constant which is equal numerically to 0.529 Å, the radius of the lowest energy state in the Bohr hydrogen atom. It is generally referred to as the *Bohr radius.* We see that this wave function is particularly simple in that there is no dependence on either of the angular variables, but only on r. This means that ψ is spherically symmetrical. The simplest graphical representation is obtained by plotting ψ *vs* r. Since the exponential term is negative, ψ decreases with increasing r, as shown in Fig. 5-12.

Another form of representation is given by the radial distribution function, which is a measure of the probability that the electron will be

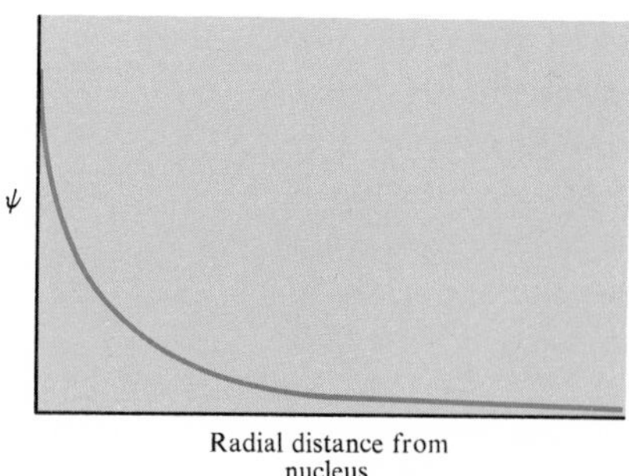

Fig. 5-12 Radial dependence of the hydrogen 1*s* wave function.

found in a spherical shell between r and a slightly larger radius, $r + \delta$, where δ is a small addition to the radius. This function is given by the value for the probability function multiplied by $4\pi r^2$. $4\pi r^2 \psi^2$ is a measure of the total probability of finding the electron at a radial distance r from the nucleus. ψ^2 merely gives the probability density at radius r; to get the *total* probability of finding the electron at this distance, we must multiply ψ^2 by the total volume of the spherical shell at this radius. The volume of a sphere of radius r is $(\frac{4}{3})\pi r^3$. The volume of a slightly larger sphere of radius $r + \delta$ is $(\frac{4}{3})\pi(r + \delta)^3 = (\frac{4}{3})\pi(r^3 + 3r^2\delta + 3r\delta^2 + \delta^3)$. Recall that δ is a *small* increment in radius. The difference in the two volumes is then given by $(\frac{4}{3})\pi(3r^2\delta + 3r\delta^2 + \delta^3)$. Of these terms, only the first needs to be retained since δ^2 and δ^3 are very small compared to δ. The total probability density in the spherical shell between r and $r + \delta$ is then $4\pi r^2\delta\, \psi^2$. Assuming a constant increment δ, we obtain the radial distribution function by graphing $4\pi r^2\psi^2$ *vs* r. The result for the wave function of Eq. (5-10) is shown in Fig. 5-13.

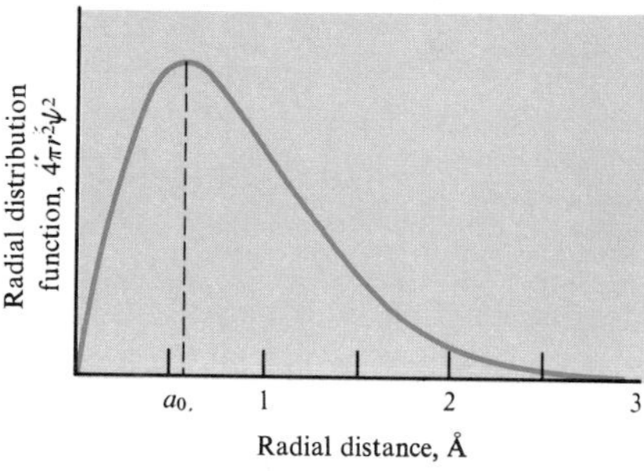

Fig. 5-13 Radial distribution function for a hydrogen 1*s* orbital. Note that it is the probability function, rather than the wave function itself, which is incorporated into the radial distribution function.

The radial distribution function reaches a maximum at precisely 0.529 Å, the Bohr radius, a_0. The radial distribution function is simple to evaluate for a spherically symmetric wave function, but it is not as useful for wave functions which contain angular terms as well.

An additional way to describe the wave function would be to evaluate it at various points in space, and then to connect points for which the wave function has the same value. The procedure is not unlike that of drawing a contour map of terrain. Usually the method is used to give a picture of the electron distribution in two dimensions. An often-used simplification of this procedure is to retain just one of the contour lines, which contains inside its boundaries most (say 90 per cent) of all the value of ψ^2. Examples of these contour representations will be given later, with illustrations in which the contours are pictured in three dimensions.

In making use of these various simplified representations, it must be kept in mind that ψ^2 is quite different from ψ. The latter has the meaning of amplitude, but is not readily interpretable in terms of a simple physical quantity. The wave function may be positive in some regions in space, negative in others. At the boundary between the positive and negative regions, the wave function must equal zero. The result is a so-called *nodal* surface, at which the probability function is also necessarily zero. We will not attempt to describe in detail the characteristics of all the solutions to the wave equation, but rather will restrict our attention to the parts of the probability functions which are of most interest for chemical behavior.

5-9 Atomic Orbitals

In the modern theory of the atom, each possible state of the electron is called an *orbital*. Each orbital is characterized by three quantum numbers, n, ℓ, and m_ℓ. The principal quantum number n corresponds to the quantum number already encountered in the Bohr theory. The azimuthal quantum number ℓ is allowed to take on values from $n - 1$ to zero. Thus for $n = 1$, the lowest energy state, ℓ must be zero. For $n = 2$, ℓ may take on the values 1 and 0. The value of ℓ for an orbital is indicated by a letter, according to the following code:

$\ell = 0$ s
$\ell = 1$ p
$\ell = 2$ d
$\ell = 3$ f
$\ell = 4$ g

Thus if we write $3p$, we refer to the orbital with quantum numbers $n = 3$ and $\ell = 1$. Orbitals with differing values of ℓ, even though n may be the same, correspond to different energy states *except* in the hydrogen atom.

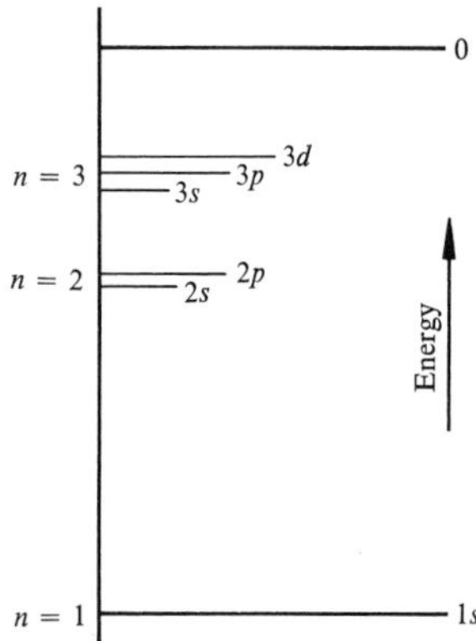

Fig. 5-14 An atomic energy level diagram.

This is shown in Fig. 5-14, where the first few levels are shown. The energy separations between orbitals with differing values of ℓ and equal n are not so large as those for orbitals with equal values of ℓ but differing n.

There is a third quantum number m_ℓ, the magnetic quantum number, which may take on the values $-\ell, \ldots 0, \ldots \ell$. That is, there are $2\ell + 1$ possible values. For example, for a p orbital, $\ell = 1$, m_ℓ may have values $-1, 0, +1$. For a d orbital, $m_\ell = -2, -1, 0, +1, +2$. This means that there are actually three orbitals which have the same symbol $3p$, and five with the symbol $3d$. Orbitals which have the same values of n and ℓ, but differing m_ℓ, have the same energy, and differ from one another only in their orientations in space.

Orbitals which possess different azimuthal numbers ℓ differ in dependence on the angular parts of the wave function. These differences can best be explained by considering the type of nodes which s, p, and d functions may have. An s orbital can have only one type of node, a spherical surface which is concentric about the origin. If we look at a cross-section through a series of s orbitals we observe cross-sections as shown in Fig. 5-15. Note that there is no node in the 1s orbital, one node in the 2s, two nodes in the 3s, *etc.* The radial distribution function for each s orbital is also shown in

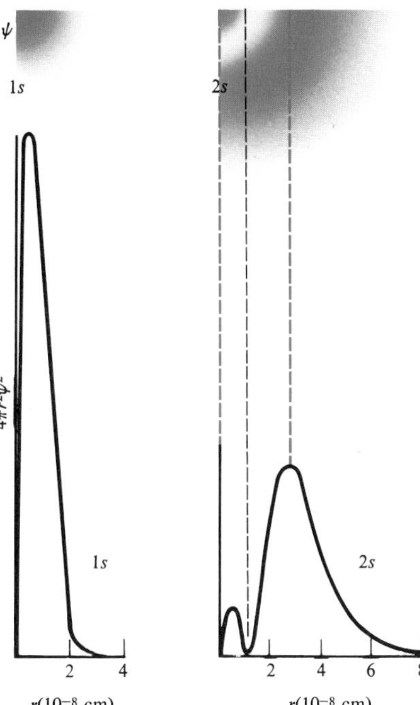

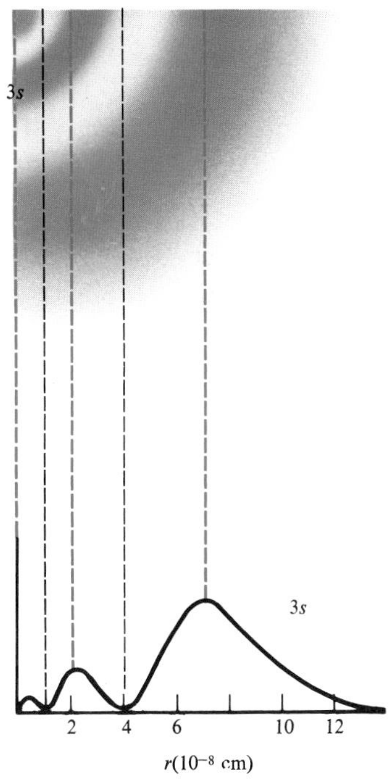

Fig. 5-15 Radial distribution functions and probability functions for hydrogen-like s orbitals of varying major quantum number.

Fig. 5-15. It is interesting to note that the major rise in the radial distribution function occurs at increasing distance from the nucleus with increasing major quantum number, but there is a small portion of the total probability distribution which lies close to the nucleus. In somewhat crude terms, this means that an electron in a 3*s* oribtal, for example, spends a small part of the time close to the nucleus, and most of the time further out. Between these two regions, there is a region in which the electron is seldom present.

The *p* orbitals ($\ell = 1$) possess a nodal *plane* which passes through the origin. There are three equivalent *p* orbitals for each value of the major quantum number, differing only in the orientation of the nodal plane. This is shown in Fig. 5-16, for the 2*p* orbitals. The arrangements of the nodal

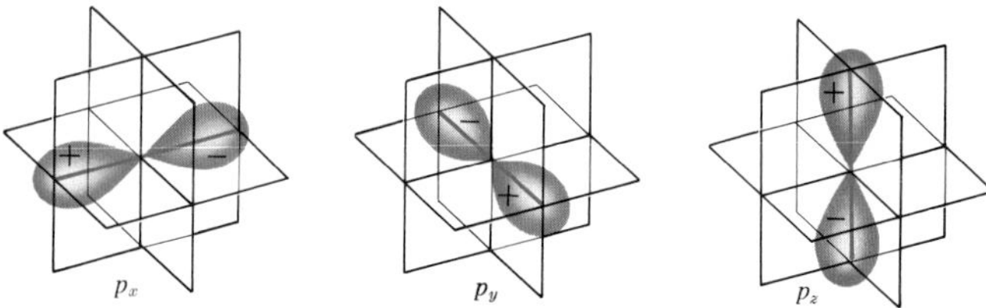

Fig. 5-16 Representations of the wave functions for 2_p atomic orbitals.

planes is precisely the same for the 3*p*, 4*p*, *etc*., orbitals. With an increase in major quantum number an additional set of nodes arise, but it is not necessary for us to concern ourselves with these at this time, because the additional nodes are not of importance in determining chemical properties. The important aspect of the *p* orbitals, whatever value the major quantum number may have, is that there are three equivalent orbitals directed along the three major axes, as shown in Fig. 5-16.

There are five equivalent *d* orbitals for each major quantum number of three and higher. Each *d* orbital possesses two nodal planes, as compared with only one for the *p* orbitals. It is not possible to make a graphical representation of the *d* orbitals so that all five look equivalent, although it is possible to write mathematical expressions for the five in which this is so. The method of representation most commonly employed is shown in Fig. 5-17. It is easy to see for the $d_{x^2-y^2}$ orbital, for example, that there are two vertical nodal planes at right angle to one another, which must separate the regions of opposite sign. We will have occasion, particularly in discussing electronic structures of transition metal complexes, to refer to these pictorial representations of *d* orbitals.

There are seven equivalent *f* orbitals ($\ell = 3$) for each value of the major quantum number of four and above. These are quite difficult to represent pictorially, and we shall not attempt to do so.

To summarize the important aspects of the preceding discussion, we have seen that the orbital designations *s*, *p*, *d*, *etc*., are related to the manner

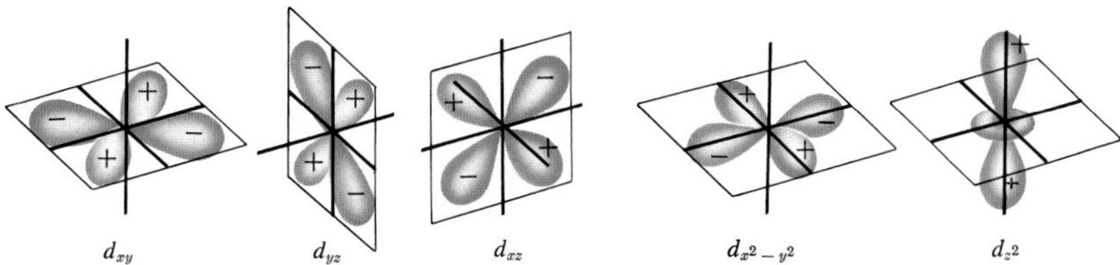

Fig. 5-17 Representations of the $3d$ orbitals.

in which the wave function for the electron changes sign with a change in angular orientation in space. If we begin at some particular distance from the nucleus in a plane which contains the orbital of interest, and make a complete circle about the nucleus, we find that the sign of the function for an *s* orbital is unchanged over the whole path, the sign of the function for a *p* orbital changes, and then changes again as we recross the nodal plane on the other side of the origin. The sign of the function for a *d* orbital changes sign four times, corresponding to the crossing and recrossing of two nodal planes. It must be emphasized again that the sign of the wave function is *not* related to the sign of the electrical charge distribution in space. We always have a negative electronic-charge distribution. It is for this reason, at least in part, that we do not attempt to attach physical significance to the wave function itself but only to the square of the wave function, which is taken as a measure of the probability that the electron will be found at a particular point in space. The wave functions go to zero value on the nodal surfaces, so it follows that the square of the function must also be zero. Regions of high probability density are therefore separated by regions of very low probability density.

In a many-electron atom, the wave functions describing the electrons all involve the same region in space about the nucleus, and there is considerable overlap. It is sometimes a source of concern to students that the functions which apply to different electrons could have substantial values at a single point in space. It must be remembered, however, that the wave functions do not describe the instantaneous behavior of an electron, but are a measure of expectation based on an average over a period of time. The motions of the individual electrons are not known to us, but the fact that the electrons exert a repulsive effect on one another ensures that at any one instant no two electrons are at the same point in space.

It is quite evident from the figures that the electron distribution in an individual *p* or *d* orbital is not spherically symmetric. It is very important to note, however, that if the orbitals in a set have an equal number of electrons, the overall result is a spherically symmetric distribution. For example, if there are three electrons in the $2p$ orbitals, one each in the $2p_x$, $2p_y$, and $2p_z$ orbitals, the resulting distribution is spherically symmetric. Similarly, if there are five $3d$ electrons, one in each of the $3d$ orbitals, or ten $3d$ electrons, with two in each orbital, the resulting distribution is spherically symmetric.

Until now we have been considering the energy levels in atoms, but we have not discussed how the electrons are distributed in these levels. We are in the position of a landlord who owns a many-storied apartment building, but who has relatively few tenants. In hydrogen, the single tenant normally resides in the ground-floor apartment (the 1*s* orbital). Now, we can proceed to the next element of the periodic table, helium, by adding a second proton (and a couple of neutrons) to the nucleus and adding a second electron to the volume element surrounding the nucleus. At this point, it is necessary to ask which orbital this second electron goes into, and also, as we continue this process for the other elements of the periodic table, where the third, fourth, *etc.*, electrons go? The *Pauli exclusion principle* is employed in answering this question. When applied to the problem at hand, it states simply that each orbital can accommodate at most two electrons.

Certain features of atomic spectra can be explained only by assuming that electrons, in addition to their motion about the nucleus, also spin about their own axes. This electron spin gives rise to a fourth quantum number m_s, the electron spin quantum number.

In 1921, the electron was shown by Otto Stern and Walter Gerlach to possess a magnetic dipole moment. In the Stern-Gerlach experiment, a beam of silver atoms issues from a slit in an oven, and is further formed into a narrow ribbon of atoms by a slit at some distance from the oven. The beam then passes through an inhomogeneous magnetic field produced by a magnet with pole faces cut similarly to those shown in Fig. 5-18. Because of the presence of a spin angular momentum in the electron, the one un-

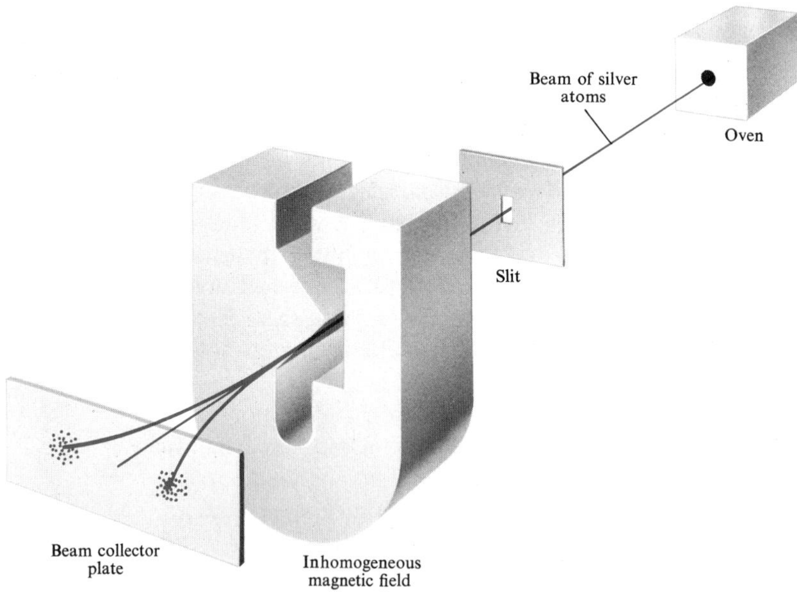

Fig. 5-18 The Stern-Gerlach experiment.

paired electron in a silver atom possesses a magnetic dipole moment. As a result, the silver atoms are acted upon by the magnetic field so as to be pushed toward the left or right, depending upon the orientation of the electron dipole moment. The silver atoms do not all move through the region of magnetic field at the same velocity, so there is a spread in the distribution of silver atoms which are deposited in the collector screen.

In interpreting this experiment, there are a number of steps in the reasoning which leads to the idea that the electron has a "spin". First of all, the experiment demonstrates that the silver atoms possess a magnetic dipole which is affected by the inhomogeneous field. From a knowledge of the properties of the silver nucleus, it can be determined that the magnetic dipole must arise in the *electronic* configuration, and not in the nucleus. It is also necessary to realize that electrons which are paired in an orbital do not exhibit a net magnetic dipole; the magnetic dipole of one electron is cancelled by that of the other. There is but one unpaired electron in the silver atom, and it must be this one electron which gives rise to the observed magnetic dipole. But how does a magnetic dipole arise in the electron? It is not possible to give a totally correct explanation of this in simple terms, but it can be thought of as arising from a spinning of the spherical electron about an axis. The axis is assumed to be oriented; the orientation in a magnetic field is determined by the spin quantum number, m_s, which can take on values of either $+\frac{1}{2}$ or $-\frac{1}{2}$. The deflection of the beam of silver atoms to either the right or left in the Stern-Gerlach experiment corresponds to one or the other of the two possible orientations of the electronic spin in the single unpaired electron.

We can now redraw the energy-level diagram of Fig. 5-14 in such a way that the number of orbitals of each type is shown. Each short line in Fig. 5-19 represents an orbital capable of containing two electrons. Thus, by beginning with the model for the hydrogen atom, for which the energies

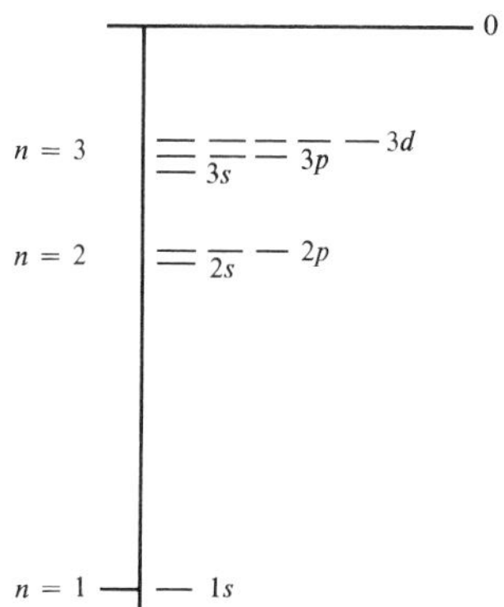

Fig. 5-19 An atomic energy level diagram after provision for the Pauli exclusion principle.

and the numbers of all of the allowed energy levels can be calculated, we may hope to obtain a picture of the electronic structure of atoms containing many electrons. It is well to remember, however, that many complicating factors are of importance in these atoms; *e.g.*, the electrons repel one another, while at the same time they are all attracted to the nucleus. As a result, some of the higher levels are shifted somewhat in relative energy from their values for hydrogen. The electronic structure of the elements is considered in detail in the next chapter.

Exercises

5-1. What is the significance of the fact that the atomic weight of argon is higher than that of potassium, as far as Mendeleev's classification of the elements is concerned?

5-2. By using a handbook and other sources of chemical data, predict as many physical and chemical properties of astatine, element number 85, as you can.

5-3. (a) The characteristic X-ray wavelength emitted from nickel upon bombardment with high energy electrons is 1.656 Å. From this, calculate the value of the constant A in Moseley's equation.
(b) Using the value of A determined in (a), calculate the value of λ to be expected for gold, element number 79.
(c) Explain how Moseley's results could be used to determine whether a metal alloy of unknown composition contains any tin (Sn, element number 50).

5-4. What is the significance of the observation that the emission of radiant energy from excited hydrogen atoms consists only of certain frequencies rather than a continuum of all frequencies? In what way are the line spectra related to Planck's quantum theory?

5-5. If you were required to defend Bohr's theory of the hydrogen atom at about the time it was first published, what would be your arguments in its favor?

5-6. Calculate the *wavelengths* of the light emitted from hydrogen for the electronic transitions: $n_3 \rightarrow n_2$, $n_4 \rightarrow n_2$, $n_5 \rightarrow n_2$.

5-7. Calculate the wavelength of a neutron moving with a velocity of 10^7 cm/sec. (See p. 87 for a similar example.)

5-8. What is a quantum number in terms of the wave mechanical model for atomic structure?

5-9. How many electrons can be fitted into the orbitals which comprize the third quantum shell; *i.e.*, $n = 3$?

5-10. (a) What are the three quantum numbers which characterize an atomic orbital?
(b) Upon which quantum numbers does the energy of an electron in the hydrogen atom depend?
(c) Upon which in atoms other than hydrogen?

5-11. State the Pauli exclusion principle as it applies to electrons in an atom.

5-12. Consider the Li^{+2} ion. This is the same as the hydrogen atom in that there is a single electron and a nucleus, but the nuclear charge is now three instead of one. What can you say about the energy levels for the electron in this system as compared with hydrogen? [*Hint:* In the more general case, the nuclear charge is Ze, and the force of attraction is Ze^2/r^2.]

5-13. Why is it possible, in the Bohr model for hydrogen, to consider the nucleus as fixed and the electron as moving about it? [*Hint:* Consider the relative masses.]

5-14. What is the maximum number of electrons which can possess the following sets of quantum numbers?

(a) $n = 2, \ell = 1$. (c) $n = 3$.
(b) $n = 4, \ell = 3$. (d) $n = 5, \ell = 3, m_\ell = +1$.
(e) $n = 3, \ell = 3$.

5-15. The different orbitals can be classified as to which of the three spherical coordinates shown in Fig. 5-11 are contained in the wave function expression. If a particular coordinate is not present in the wave function expression, then a change in that coordinate does not change the value of the wave function. From examination of the figures in this chapter, determine which of the three spherical coordinates each of the following orbitals depends on: $2s$; $2p_z$; $d_{x^2-y^2}$; $2p_x$.

5-16. The K lines observed by Moseley in his experiments are now known to arise from the ejection of a $1s$-electron from the atom, and then a jump of a $2s$- or $2p$-electron into the vacancy. Is it reasonable that the energy of this last process, which determines the wavelength emitted, should depend primarily on the magnitude of the nuclear charge? Explain. How is the wavelength of the light emitted related to the energy of the electron jump? [*Hint:* See Secs. 1-2 and 1-3.]

5-17. The $2s$ wave function for a hydrogen-like atom of nuclear charge Z is of the form

$$\psi_{2s} = \frac{1}{4}\frac{1}{\sqrt{2\pi}}\left(\frac{Z}{a_0}\right)^{3/2}\left(2 - \frac{Zr}{a_0}\right) e^{-Zr/a_0}$$

As the representation of the $2s$ wave function in Fig. 5-15 shows, the $2s$ orbital possesses a node. Find the value of r in terms of a_0 at which this node occurs.

Chapter 6

The Electronic Structure of Atoms

6-1 Introduction

During the same period that Bohr and Rutherford were developing their theory of the atom, two American chemists, G. N. Lewis and Irving Langmuir were expounding another, quite different, electronic theory. The Lewis-Langmuir theory was based upon Moseley's work, which revealed the number of electrons in the atom for each element, and upon the known chemical behavior of the elements. It was an attempt to classify the electrons in such a way that the position of an element in the periodic table could be associated with a particular electronic arrangement. For example, the elements with atomic numbers 2, 10, 18, 36, 54, and 86 — the inert gases — show an especially low chemical activity. Obviously, then, there is something special about an atom's having these numbers of electrons, something that leads to unusual stability. The periodic recurrence of the inert gases led Lewis and Langmuir to the idea that the electrons are arranged in the atom in groups, or shells. The inert gases are elements which possess a completed shell. Their chemical inertness is to be attributed to the stability of such an arrangement. The chemical behavior of the other elements is explained in terms of the tendency for the atoms to acquire a completed shell by gaining or losing electrons, or by sharing electrons with other atoms.

The Lewis-Langmuir theory is important because it emphasizes the close connection between chemical activity and electronic configuration. Although it does not provide a detailed model of the atom as does the theory discussed in the previous chapter, the two theories are closely related. The relationship is evident from a study of Fig. 6-1. It can be seen that the energy levels occur in groups which correspond to the shells in the Lewis-Langmuir theory.

The electronic structures of the elements can be deduced from a study of their line spectra. As a result of these studies, it is known that the electronic energy levels do not change a great deal, relative to one another, from one atom to the next. For this reason, it is possible to think of the electronic structures of the elements as being "built up" from the hydrogen atom by addition of electrons one at a time (and assuming, of course, that the nucleus always has the same charge as the total number of electrons). Each added electron is placed in the lowest available orbital (keeping in mind the principle that no more than two electrons may occupy the same orbital).

6-2 First Row Elements

We will now begin a consideration of the electronic structures of atoms. Constant reference should be made to Fig. 6-1 as an aid in understanding the following material. We have already seen that for hydrogen, atomic number 1, the single electron is in the $1s$ orbital. In helium, atomic number 2, the second electron is also in the $1s$ orbital. This arrangement is designated by the symbol $1s^2$, where the superscript indicates the number of electrons in the orbital. On the basis of the Pauli exclusion principle, no more electrons can be accommodated in the $n = 1$ level. The next electron (in lithium) goes, therefore, into the lowest energy orbital on the $n = 2$ level, the $2s$ orbital. In beryllium, the fourth electron also goes into this orbital, so the configuration for this element is $1s^2 2s^2$. For boron, the configuration is $1s^2 2s^2 2p$. With the next element, carbon, a new problem arises. The next electron may be placed in the $2p$ orbital which is already singly occupied — in which case it must have the opposite spin of the electron already there; or it may be placed in one of the other $2p$ orbitals — in this event it could have either the same or opposite spin. From atomic spectral studies, it is known that the electron is located in one of the other two orbitals, and that its spin is parallel to that of the other $2p$ electron. We therefore write the configuration for carbon as $1s^2 2s^2 2p 2p$ rather than as $1s^2 2s^2 2p^2$, because the latter description does not make it clear that the electrons are in different p orbitals.

The tendency of the two $2p$ electrons to occupy different p orbitals rather than pair up in the same one is the result of the electrostatic repulsions between the two particles of like charge. From Fig. 5-16 it can be seen that the three p orbitals are differently oriented in space. Two electrons in different orbitals are thus further removed from one another than when they occupy the same orbital, and the electrostatic repulsions between them are

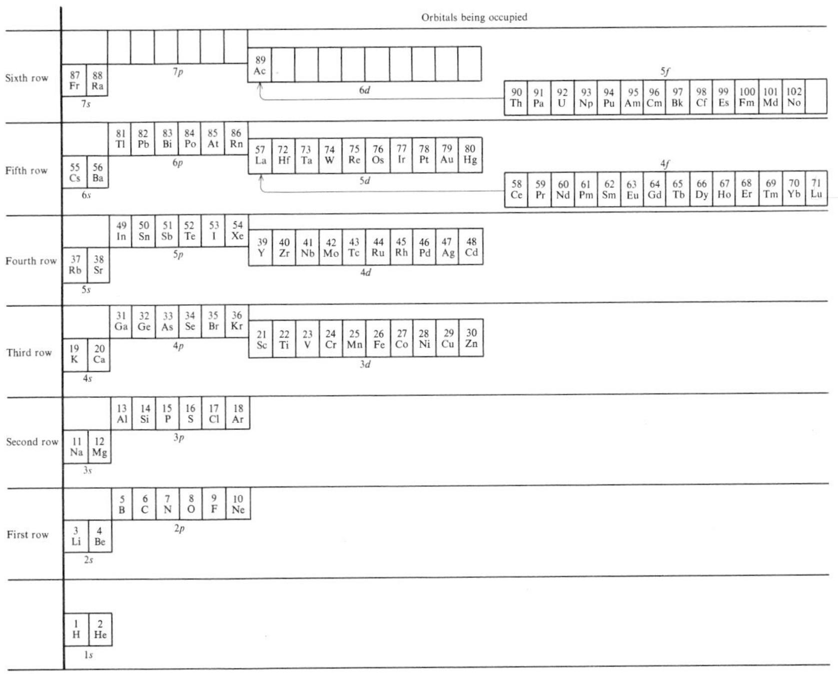

Fig. 6-1 The electronic structures of the elements.

accordingly less. These considerations are embodied in Hund's rule, which states that electrons which occupy equivalent orbitals maintain parallel spins whenever possible.

In nitrogen the same considerations apply; the electronic configuration is $1s^22s^22p2p2p$. In the case of oxygen, the next electron now must go into a $2p$ orbital which is already singly occupied: $1s^22s^22p^22p2p$. The configuration for neon is $1s^22s^22p^22p^22p^2$.

Another method for indicating electronic structures, the box diagram, is convenient for discussion of many chemical problems. The structures of a few of the first ten elements, by this method, are shown:

	$1s$	$2s$	$2p$
H	↑		| |
N	↑↓	↑↓	↑ | ↑ | ↑
F	↑↓	↑↓	↑↓ | ↑↓ | ↑
Ne	↑↓	↑↓	↑↓ | ↑↓ | ↑↓

The major advantage of this representation over that used in the text is that the electron spins are clearly indicated.

6-3 Second Row Elements

The configurations of the elements from sodium through argon are entirely analogous to those in the first row, except that the major quantum number is 3 rather than 2. The electronic configuration for argon, atomic number 18, is $1s^22s^22p^22p^22p^23s^23p^23p^23p^2$. (To shorten the writing, this could be indicated as [Ne]$3s^23p^6$. ([Ne] represents the electronic configuration for neon, the previous rare gas.) Note that the configuration for argon in the $n = 3$ level is the same as that of neon in $n = 2$. This arrangement of filled s and p orbitals in the highest major quantum level occupied is characteristic of the inert gases. Argon does not have a totally filled $n = 3$ level, however; there are no electrons in the $3d$ orbitals.

6-4 Third Row Elements

It might be expected that the highest energy electron in potassium, atomic number 19, would occupy a $3d$ orbital, but this is not the case. In the process of building up the atom from hydrogen to argon by adding electrons and correspondingly increasing the nuclear charge, slight modifications in relative energies occur. As a result, the $4s$ orbital of potassium is lower in energy than the $3d$. Potassium and calcium, therefore, have one and two $4s$ electrons, respectively. The filling of the $3d$ orbitals begins with the element scandium. Since there are five $3d$ orbitals, there are ten elements in the third row which receive electrons into the $3d$ orbitals. These are called the transition elements. Using the box diagram method, we illustrate the configurations for a few of these elements ([Ar] indicates the electronic configuration of argon):

		$4s$	$3d$				
Mn	[Ar]	↑↓	↑	↑	↑	↑	↑
Co	[Ar}	↑↓	↑↓	↑↓	↑	↑	↑
Zn	[Ar]	↑↓	↑↓	↑↓	↑↓	↑↓	↑↓

Following completion of the $3d$ level with zinc, the next six electrons enter the $4p$ orbitals. The configuration for krypton is therefore [Ar]$4s^23d^{10}4p^6$. Note the characteristic inert gas arrangement in the level of highest major quantum number.

6-5 Fourth Row Elements

The electronic structures of the elements from rubidium through xenon are entirely analogous to those of the third row elements; they differ only in that the value of n is increased by one. The elements from yttrium through cadmium form a second transition series, in which the $4d$ orbitals are being occupied. The electronic configuration for xenon is, in the notation we have adopted, $[Kr]5s^2 4d^{10} 5p^6$.

6-6 Fifth Row Elements

Cesium and barium possess one and two $6s$ electrons, respectively. The next electron, in lanthanum, goes into a $5d$ orbital, just as in the previous row (yttrium) it had gone into a $4d$ orbital. With element 58 (cerium), however, a new set of orbitals enters the picture; these are the $4f$ orbitals. Beginning with cerium, these seven orbitals are successively filled in the next fourteen elements (see Fig. 6-1). In order not to make the periodic table unduly long, these elements are listed at the bottom of the table. They are called the *inner transition elements*, or the lanthanum series, or rare earth elements. Following the filling of the $4f$ orbitals there follows, beginning with hafnium, the remainder of the transition series of ten elements in which the $5d$ orbitals are being occupied. Then the $6p$ orbitals are occupied, so that the configuration for radon is $[Xe]6s^2 4f^{14} 5d^{10} 6p^6$.

6-7 Sixth Row Elements

Radium possesses two $7s$ electrons. Following actinium, which has a single $6d$ electron, there begins the filling of the $5f$ orbitals which presumably goes on through the new synthetic elements. A second inner transition series is thus formed.

6-8 Summary

We have now completed a survey of the electronic structures of all of the elements. It is at first sight a very complicated matter, but with the aid of the periodic table it can be mastered quickly. Fig. 6-2 shows a block diagram of the periodic table in which the symbol for each element has been omitted. Instead, for each group of elements, the symbol for the set of orbitals being occupied is shown. A study of this figure shows that it is possible to group the elements according to the orbital type which is being filled. We have already named three of these groups: the transition metals, with d orbitals being filled; the inner transition elements, with f orbitals being filled; and the noble gases, with the configuration s^2p^6 in the highest occupied level

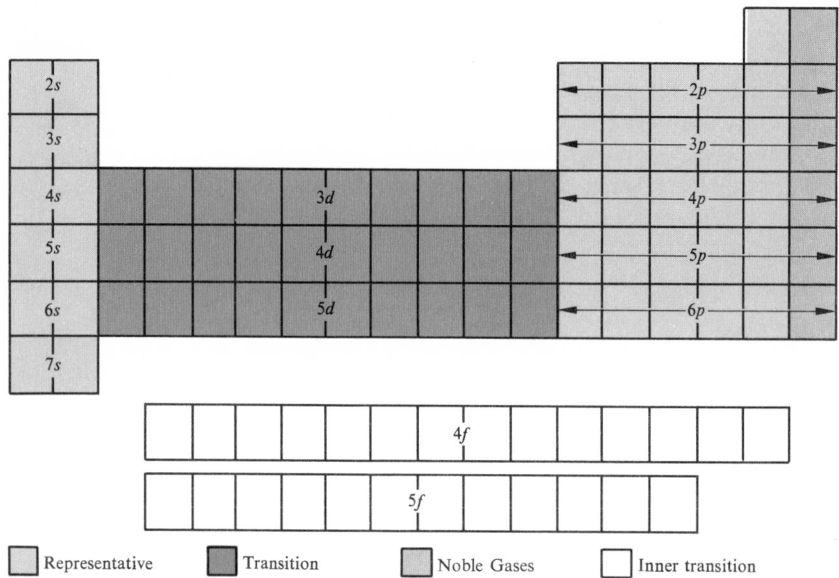

Fig. 6-2 Classification of the elements according to orbitals being filled.

(except for helium, with $1s^2$). The other elements, for which electrons are being added to either an *s* or *p* orbital, are called the *representative* elements.

To complete the discussion of electronic configurations, we must introduce one slightly complicating factor which applies to a few of the elements. Consider the configurations which one might expect, on the basis of our discussion so far, for the elements nickel, copper, and zinc:

		4*s*	3*d*
Ni	Ar	↑↓	↑↓ ↑↓ ↑↓ ↑ ↑
Cu	Ar	↑↓	↑↓ ↑↓ ↑↓ ↑↓ ↑
Zn	Ar	↑↓	↑↓ ↑↓ ↑↓ ↑↓ ↑↓

The electronic structure shown above is not found for copper. Instead, it is found that the 3*d* subshell contains ten electrons and the 4*s* orbital contains only one:

		4*s*	3*d*
Cu	Ar	↑	↑↓ ↑↓ ↑↓ ↑↓ ↑↓

The element chromium provides another example of this effect; in this instance the 3*d* subshell contains five electrons, and the 4*s* orbital only one:

		4*s*	3*d*
Cr	Ar	1	1 | 1 | 1 | 1 | 1

In these cases, the energy gained by the complete filling, or half-filling, of the 3*d* subshell is sufficient to cause the electron which would normally occupy the 4*s* orbital to move into the 3*d* level. The 4*s* and 3*d* orbital energy levels are not very much different in energy, so the energy involved in producing the effect is probably not very great, and no pronounced chemical consequences result from it. Similar slight variations in electron configuration are found for a few other elements. The electron configurations of all of the elements are shown in Appendix D.

6-9 Ionization Potentials

It is apparent from Fig. 6-2 that the electronic configurations of the elements are periodic in character. It is reasonable to expect, then, that atomic properties which are related to electronic configuration will also exhibit periodicity. The most important of these properties is the *ionization potential*, which is defined as the energy required to remove an electron from a gaseous atom to a distance of infinity. It is, therefore, the energy required for the process

$$M(g) \rightarrow M^{+}(g) + e^{-}$$

where M is any gaseous atom. It is not necessary to restrict the definition to the removal of only one electron; the second ionization potential is defined as the energy required to remove an electron from M^{+} to a distance of infinity:

$$M^{+}(g) \rightarrow M^{+2}(g) + e^{-}$$

Third and higher ionization potentials have also been defined and measured.

The electron removed from the atom or ion in the ionization process is always the highest energy electron. It is instructive to note how the energy of this electron varies with atomic number. Fig. 6-3 shows the first ionization potential I_1 graphed *vs* the atomic number. The over-all periodicity is quite evident. Ignoring for the moment the lesser reversals in trend, there is a gradual increase in I_1 with increasing atomic number in any horizontal row. But for the elements in any one group, there is a gradual decrease in I_1 with increasing atomic number. Note, for example, the gradual decrease in the alkali metal series lithium through cesium, or in the inert gas series helium through radon.

It is possible to explain the observed variations in I_1 in terms of a few simple considerations. The first of these, the *effective nuclear charge*, is

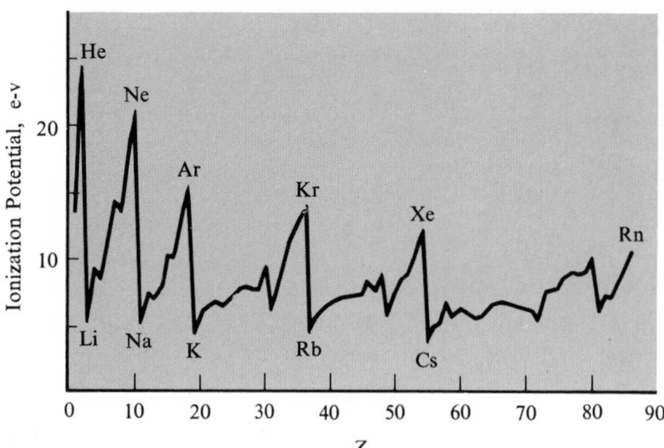

Fig. 6-3 Ionization potential of the elements versus atomic number.

illustrated by the comparison of hydrogen with helium. In hydrogen, there is a single $1s$ electron and a singly charged nucleus. The attractive potential energy operating between these two is just

$$V = \frac{-e^2}{r_H}$$

where e is the electron charge and r_H is the average distance of the electron from the nucleus. In helium, the nucleus is doubly charged, and there are two $1s$ electrons. The electrons repel one another, and this repulsive interaction between the electrons reduces the effective nuclear charge to a value intermediate between two and one. The potential may therefore be expressed as

$$V = \frac{-Ze^2}{r}$$

where Z has a value between 2 and 1, and the average value for r is somewhat smaller than in the hydrogen atom. Another way of stating the same conclusion is to say that each electron acts to partially screen the second electron from the nucleus, so that it is attracted by less than the full nuclear charge.

In lithium, the third electron goes into a $2s$ orbital; its most probable distance from the nucleus is therefore larger than for the two $1s$ electrons. Since the nuclear charge has increased from two to three, the two $1s$ electrons are held more tightly than they are in helium, and they effectively shield the nucleus from the more distant $2s$ electron. The effective nuclear charge for the $2s$ electron is therefore about one. Since the value for r is larger, the ionization potential is correspondingly less than for hydrogen.

The comparison of beryllium with lithium follows the same argument as that used in comparing helium with hydrogen. There is a slight drop in I_1 for boron as compared with beryllium, resulting from the higher energy of the $2p$ as compared with the $2s$ orbital. As each of the other two $2p$ orbitals becomes occupied, in carbon and then nitrogen, there is a steady increase in the value of I_1. This represents another example of increasing effective nuclear charge, an effect which seems general throughout the table. It is summarized by the statement that *electrons with the same values of n and ℓ are relatively ineffective in screening one another from the nuclear charge.* Therefore, in a series of elements in which electrons are being ionized from a set of orbitals with the same values of n and ℓ, there is a gradual increase in ionization potential with atomic number.

There are a number of exceptions to the above rule, but they are not overly important. The first exception is the slight drop in I_1 in going from nitrogen to oxygen (it also occurs at the same place in the next two periods). This results from the requirement that the fourth p electron go into an orbital which already contains one electron:

	$1s$	$2s$	$2p$
N	↑↓	↑↓	↑ | ↑ | ↑
O	↑↓	↑↓	↑↓ | ↑ | ↑
F	↑↓	↑↓	↑↓ | ↑↓ | ↑

This is a relatively minor effect, however, and is not reflected in the values of I_1 for fluorine or neon.

The highest energy electron in sodium is in the $3s$ orbital. The other ten electrons are in levels with lower major quantum number, and form the inert gas configuration of neon. They are very effective in shielding the nucleus from the $3s$ electron, so that the effective nuclear charge for this electron is only about one. Since it is further removed from the nucleus than the $2s$ electron in lithium, its ionization potential is lower.

One further point in connection with Fig. 6-3 needs mentioning. Note that, in the first transition series extending from atomic number 21 to 30, there is a more or less continual increase in the ionization potential. This is another application of the rule mentioned above. One of the consequences of this increase is that the values of I_1 for the B subgroup elements are larger than for their counterparts in the A subgroups. For example, the ionization potential for copper is higher than for potassium. The higher value of I_1 for the B subgroup elements is an important factor in their lower reactivities, as compared with the elements of the A subgroups.

The successive ionization potentials for the elements from sodium through argon are listed in Table 6-1. The ionization potentials are expressed in terms of electron-volts. (One electron-volt per mole is equal to 23,060 calories.) When the rare gas configuration of neon is reached (the dotted line) an enormous increase in potential is required for removal of the next electron. These data are of considerable interest in connection with the ionic bond, and will be referred to again in the next chapter.

Table 6-1 Successive values of ionization potential.

Element	Ionization potential (electron-volts)						
	I_1	I_2	I_3	I_4	I_5	I_6	I_7
Na	5.1	47.3					
Mg	7.6	15.0	80.1				
Al	6.0	18.8	28.4	120			
Si	8.1	16.3	33.4	45.1	167		
P	11.0	19.6	30.1	51.3	65	220	
S	10.4	23.4	35.0	47.3	72	88	280
Cl	13.0	23.8	39.9	53.5	68	97	114
Ar	15.8	27.6	40.9	59.8	75	91	124

6-10 Radial Charge Distribution

The most probable distance from the nucleus for the single 1*s* electron in hydrogen is 0.53 Å, as we saw in the previous chapter (Fig. 5-13). The most probable distance for the two 1*s* electrons in helium is 0.30 Å, less than that for hydrogen because the effective nuclear charge is greater than one. As the nuclear charge increases with increasing atomic number, the 1*s* electrons are acted upon by an ever greater attractive force. Their most probable distance from the nucleus therefore decreases continuously with increasing nuclear charge. While it is not possible to calculate exactly the electronic charge distribution in many-electron atoms, rather accurate approximate calculations can be made. They show that in krypton (atomic number 36) the most probable distance of the 1*s* electrons from the nucleus is only

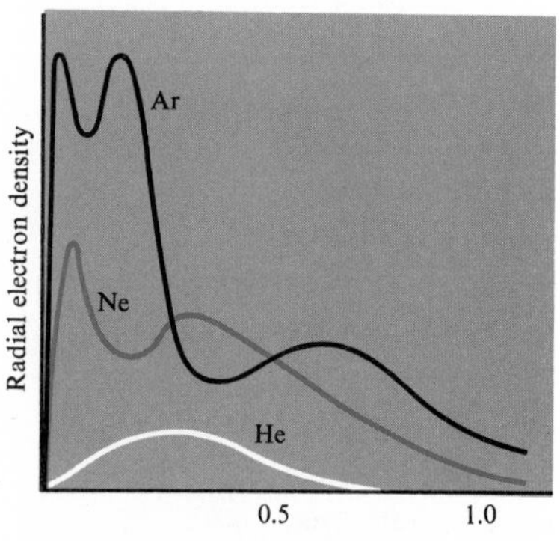

Fig. 6-4 Radial electron distributions in helium, argon, and neon. Note that the maxima in the electron densities, corresponding roughly to the various values for the principal quantum number, are clearly seen.

0.015 Å! The most probable distance for the 2*s* electrons is 0.090 Å; for the 3*s* electrons it is 0.25 Å; and for the 4*s* electrons, 0.74 Å.

It is evident, then, that the electrons in the heavier atoms are more closely bunched together than in the lighter atoms. Fig. 6-4 shows the radial charge distribution for a few inert gas atoms which have a spherical electron distribution. Note in particular how the electron density builds up around the nucleus in the heavier atoms. A second point worth noting is that the charge distribution does not have just one maximum in the heavier atoms, but a number of maxima, corresponding to differing values for the principal quantum number. Thus, the idea expressed in the Lewis-Langmuir theory, that the electrons are arranged in the atom in shells, finds expression also in the results of detailed quantum mechanical calculations.

6-11 Atomic Radii

The ionization potential is one of a very few properties of interest which can actually be measured for single atoms. Many other so-called atomic properties are measured for the atoms as parts of molecules. They are not, therefore, properties only of the atom, but depend to some degree on the nature of the molecule containing the atom, and on other factors. Provided that this limitation is kept in mind, it is possible to observe interesting and useful relationships between these quantities and atomic number.

The distances which separate atoms in molecules can be determined by a number of modern techniques, including X-ray and electron diffraction, infrared and microwave spectroscopy. From the results of such studies, a set of atomic, or covalent, radii for the elements has been established. A graph of atomic radii *vs* atomic number is shown in Fig. 6-5.

It is evident, on a comparison of Figs. 6-3 and 6-5 that an increase of ionization potential roughly parallels a decrease in radius. This is because a higher effective nuclear charge, which leads to a high ionization potential,

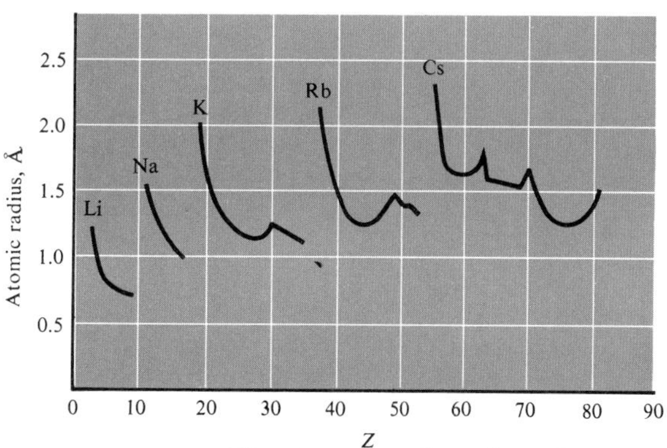

Fig. 6-5 Atomic radius of the elements versus the atomic number.

also causes the electrons in the outer shell of the atom to be pulled in closer to the nucleus, so that the atom has a smaller radius.

6-12 Electron Affinity

The tendency of an isolated ion or neutral atom to acquire an additional electron is known as the *electron affinity*. Actually, the definition of electron affinity can be extended to the acquisition of more than one electron, to give dinegative ions, *etc.* The electron affinity is a difficult quantity to measure experimentally, and only a few data are available as a result of direct measurement. There are various ways, however, in which other data can be handled to give at least rough estimates of electron affinities. Table 6-2 lists the electron affinities of some elements. In the case of the halogens, the process

$$X(g) + e^- \rightarrow X^-(g)$$

proceeds with evolution of heat; *i.e.*, the ion is more stable than the separated atom and electron. The sign attached to the electron affinity is therefore negative. For other ions, however, particularly the oxide ion, O^{-2}, the net result of adding two electrons is a *less* stable species, so that the sign of the electron affinity is positive.

Table 6-2 The electron affinities of some elements.

Element	Ion formed	*E* kcal/mole
F	F^-	−79.5
Cl	Cl^-	−83.3
Br	Br^-	−77.6
I	I^-	−70.6
O	O^-	−32
O	O^{-2}	+170
S	S^{-2}	+90

It is clear from an inspection of the values for the halogens that electron affinity does not vary to any great extent with vertical position in the group. At first sight this seems anomalous, since we think of fluorine as a strongly electron-attracting element, which should exhibit a very great tendency to acquire an electron. There is an opposing effect, however, which arises from the electron-electron repulsions which are introduced by the presence of the additional electron. The $2p$ orbital into which the electron must go in forming F^- is much smaller in spatial extent than is the $3p$ orbital of the added electron in Cl^-. The average distance of the electron from all the other electrons is therefore smaller in F^-, thus largely balancing the greater nuclear-electron attraction. The electron affinity is positive in the case of the dinegative ions as a result of the added electron-electron repulsions brought on by adding two additional electrons.

The electron affinity for a particular atom or ion is just the reverse of the ionization potential for the species of one greater negative charge. Thus, the electron affinity of F is just the reverse of the ionization potential of F^-. One possible use of this idea is for the purpose of estimating electron affinities. The ionization potentials of neutral atoms and positive ions are well known. An isoelectronic series (*i.e.*, a series in which all members have the same electron configuration but differ in nuclear charge) might be extrapolated to yield at least approximate values for electron affinities. By way of example, the isoelectronic series F, Ne^+, Na^{+2}, Mg^{+3}, Al^{+4}, Si^{+5} is shown in Fig. 6-6. The graph is non-linear, but by a process of curve fitting, the data

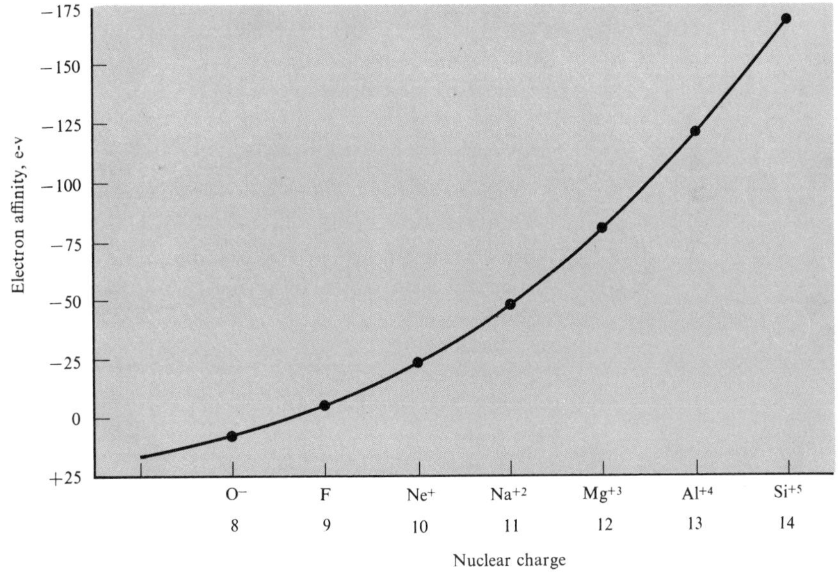

Fig. 6-6 Electron affinity versus nuclear charge in an isoelectronic series.

for the points to the right of F might have been extrapolated to give the value for F. By extrapolation one step further, using the known value for F, one can estimate that the electron affinity for the O^- species, which is isoelectronic with the other members of the series, is about +9 e-v. The value for the electron affinity of O^- can be obtained from the data listed in Table 6-2. We have

$$O + e^- = O^- \qquad -32 \text{ kcal/mole}$$
$$O + 2e^- = O^{-2} \qquad +170 \text{ kcal/mole}$$

By subtracting the first from the second (changing the sign of the energy term by the subtraction) we have:

$$O^- + e^- = O^{-2} \qquad +202 \text{ kcal/mole}$$

Since there are 23 kcal/e-v., we have about +8.8 e-v., in agreement with the extrapolated value.

The variation in electron affinity with nuclear charge in an isoelectronic series such as that shown in Fig. 6-6 is readily understandable. For a given number of electrons, the attraction of the atom or ion for an added electron is greater for a greater nuclear charge.

Exercises

6-1. List all four quantum numbers for each electron in the fluorine atom, $Z = 9$, in its lowest energy state.

6-2. Draw the box diagram representation of the electronic structure of each of the following elements: S, Ti, Br.

6-3. Describe the characteristic electron configurations of the following:
(a) the transition elements.
(b) the inert gases.
(c) the representative elements.

6-4. Although rubidium and silver each possesses one electron in the outermost (5*s*) shell, they differ greatly in chemical properties. Explain.

6-5. Within a single group of the periodic table; *e.g.*, Group VI, how do you expect each of the following to change with increasing atomic number?
(a) atomic radius.
(b) ionization potential.
(c) electronic configuration.
(d) density.

6-6. Which of the following atoms has no unpaired electrons in its ground state (lowest energy) configuration? Ca, P, Be, Cr, Xe.

6-7. Which of the following atoms should have a spherically symmetric electron distribution? Li, Cu, N, F, Ne, Fe.

6-8. By making use of Figs. 6-3 and 6-5, make a graph of ionization potential *vs* atomic radius for the elements from lithium through chlorine. What general conclusions and statements can you draw from this graph?

6-9. Explain the difference in ionization potential between
(a) sodium and magnesium.
(b) copper and potassium.
(c) neon and argon.

6-10. With reference to Table 6-1, explain the large difference between the first and second ionization potentials of sodium. Why is this difference so much smaller in magnesium?

6-11. Indicate how the Pauli exclusion principle enters into determining the most stable electronic configuration for:
(a) carbon.
(b) oxygen.
(c) nitrogen.

6-12. Consider a helium atom in which one of the electrons has been excited to the 4*s* level; *i.e.*, an atom with the electronic configuration 1*s*4*s*. What do you estimate would be the effective nuclear charge seen by each of the two electrons? Explain.

6-13. Write the electron configurations of the following elements: Pb, As, Co.

6-14. How do you explain the fact that the ionization potential increases so greatly for removal of electrons after the inert gas configuration has been reached? (Refer to Table 6-1.)

6-15. Considering the wave-mechanical model for atomic structure, explain the significance of the concept of atomic size.

6-16. The ionization potential of hydrogen is 13.6 e-v, the second ionization potential of helium is four times this, or 54.4 e-v. If the inner core of electrons in sodium and magnesium could be assumed to completely shield the 3*s* electrons from ten units of nuclear charge, what ratio would you expect between the first ionization potential of Na and the second of Mg? Compare this value with the observed ratio (Table 6-1) and comment on the source of the discrepancy.

6-17. Which of the following species do you expect to have the higher electron affinity?
(a) K or Cl.
(b) Ca^{+2} or Ca^{+}.
In each case explain your answer.

6-18. From the data of Table 6-1, and using the procedure discussed in this chapter, estimate the electron affinity of Na.

6-19. In light of the positive value obtained for the electron affinity of O^{-2} do you think that a negative value is likely for the electron affinity of any multiply charged negative ion?

Chapter 7

The Chemical Bond I. Ionic Bonds

It was shown in the preceding chapter that atomic properties such as ionization potential could be explained in terms of the electronic structures of atoms. We shall now apply this knowledge of atomic properties and electronic configurations to gain an understanding of the chemical activity of the elements and their compounds.

7-1 Valence

Valence is defined as the capacity of an atom to enter into chemical combination with other atoms. It is possible to assign a value to the valence displayed by an atom in a particular compound. For example, hydrogen is found to form a single chemical bond to one other atom in nearly all of its compounds. The valence of hydrogen therefore is generally taken to be one. Oxygen, on the other hand, exhibits a valence of two in water and in its other compounds. The valences of the other elements can be defined in terms of the combining weight. It will be recalled that the combining weight of an element is the quantity of that element which will combine with 8 grams of oxygen. In compounds which are not oxides, the combining weight

Table 7-1 Oxide and hydride formulas of second row elements.

Element	Na	Mg	Al	Si	P	S	Cl	Ar
Empirical formula of oxide	Na_2O	MgO	Al_2O_3	SiO_2	P_2O_5	SO_3	Cl_2O_7	None
Apparent valence	1	2	3	4	5	6	7	0
Empirical formula of hydride	NaH	MgH_2	AlH_3	SiH_4	PH_3	SH_2	HCl	None
Apparent valence	1	2	3	4	3	2	1	0

usually can be calculated indirectly. The valence displayed by an element in a particular compound is equal to the atomic weight divided by the combining weight.

The characteristic valences of the elements are closely related to their positions in the periodic table. An illustration of this is found in Table 7-1, in which the formulas of the hydrides and certain of the oxides of the elements from sodium through argon are listed. It is to be noted, first of all, that the valences of phosphorus, sulfur, and chlorine in their oxides are different from those in their hydrides. Furthermore, it is just these elements which also form other oxides, in which their apparent valences differ from either of those shown in Table 7-1. The particular oxides shown in the table were chosen because they correspond to the maximum valence which these elements show. It is clear, then, that the valence is not a fixed property of the atom, but may vary with the chemical environment.

Secondly, it should be noted that there is a regular variation in valence with atomic number for both the oxides and hydrides. Such regularities as these provide the basis for the Lewis-Langmuir electronic theory of valence. According to this theory, the chemical activity of atoms is associated with the greater stability of the inert gas electronic arrangements. Atoms which do not have such an electronic configuration are found to enter into chemical reactions with other atoms in such a way that, by sharing, gaining, or losing electrons, they acquire the inert gas configuration.

The ionic compounds, in which the atoms involved have completely lost or gained one or more electrons, represent the simplest example of chemical activity. We shall treat the ionic bond in detail in this chapter and then proceed to a discussion of other types of bonding in later chapters.

7-2 The Energetics of Ion Formation

The driving force for all types of chemical activity is energetic in character. Just as a ball rolls downhill to come to a new position of rest at a lower energy, so a particular arrangement of atoms or molecules undergoes spontaneous chemical reaction to come to a new arrangement with a lower energy. The loss in energy is reflected in heat or work. Consider the reaction of single, gaseous sodium atoms with single chlorine atoms to produce sodium and chloride ion pairs separated by 2.8 Å:

$$Na(g) + Cl(g) \rightarrow Na^{+} \cdot Cl^{-}(g) \qquad (7\text{-}1)$$

This process can be thought of as the resultant of three separate processes. The first of these is the ionization of the sodium atom:

$$Na(g) \rightarrow Na^{+}(g) + e^{-} \qquad E = 118 \text{ kcal/mole} \qquad (A)$$

This process requires 118 kcal of energy per mole of sodium atoms. The second step is the formation of the chloride ion from the chlorine atom and an electron:

$$Cl(g) + e^{-} \rightarrow Cl^{-}(g) \qquad E = -83 \text{ kcal} \qquad (B)$$

83 kcal of energy per mole of chlorine atoms is evolved in this step. This energy is called the *electron affinity* of chlorine. The third step involves the coming together of the two oppositely-charged particles from a distance of infinity to their equilibrium distance:

$$Na^{+}(g) + Cl^{-}(g) \rightarrow Na^{+} \cdot Cl^{-}(g) \qquad E = -119 \text{ kcal} \qquad (C)$$

By adding A, B and C we obtain the desired reaction, that of Eq. (7-1). The energy involved in this reaction is just the sum of the energies of A, B, and C,

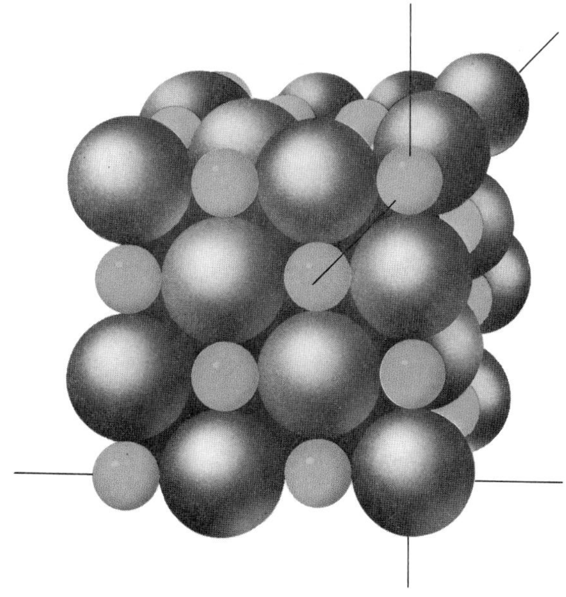

Fig. 7-1 A drawing of the sodium chloride lattice. The smaller spheres represent Na^{+} ions, the larger represent Cl^{-} ions.

$E_t = -84$ kcal. In other words, the ion pair $Na^+ \cdot Cl^-$ separated by 2.8 Å is stable relative to the isolated, neutral atoms by 84 kcal/mole.

We are accustomed to seeing and working with ionic compounds, not as ion pairs in the gas state, but as solid, crystalline materials. The crystalline substance, a three-dimensional array of positively- and negatively-charged particles (Fig. 7-1), is merely an arrangement which is built up by putting a large number of ion pairs together in a regular way. The stability of this array of charged particles is measured by the *lattice energy*. The lattice energy is defined as the energy evolved when the individual ions are brought together from infinite separation to form one mole of the ionic solid. For sodium chloride, for example, the lattice energy is -184 kcal/mole. By comparing this with the -119 kcal in reaction C, we see that a mole of sodium chloride in the solid lattice is more stable by 65 kcal than the same quantity of sodium chloride existing in the gas phase as ion pairs. This is quite a large energy difference; in the light of this quantity, it is not surprising to find that many ionic materials have high melting and boiling points. Table 7-2 lists these quantities for a number of ionic substances. Those boiling points which have been observed are all above 1300°C. By way of comparison, boiling points in the vicinity of 100°C are usual for non-ionic compounds with molecular weights in the range of 100-200 g/mole.

Table 7-2 Melting and boiling points (in °C) of some ionic substances.

Compound	Melting point	Boiling point
LiCl	613	1353
$LiNO_3$	255	—
NaCl	801	1413
$NaNO_3$	307	—
KCl	776	1500
RbCl	715	1390
MgO	2500-2800	—
BaO	1923	2000
SrO	2430	—
$MgCl_2$	708	1412
$CaCl_2$	772	1600
$SrCl_2$	873	—
$BaCl_2$	962	1560
Al_2O_3	2050	2250

7-3 The Lattice Energy

Direct experimental determination of the lattice energy is quite difficult, and has been accomplished only for a few of the alkali halides. Born and Haber developed a method for calculating the lattice energy from other,

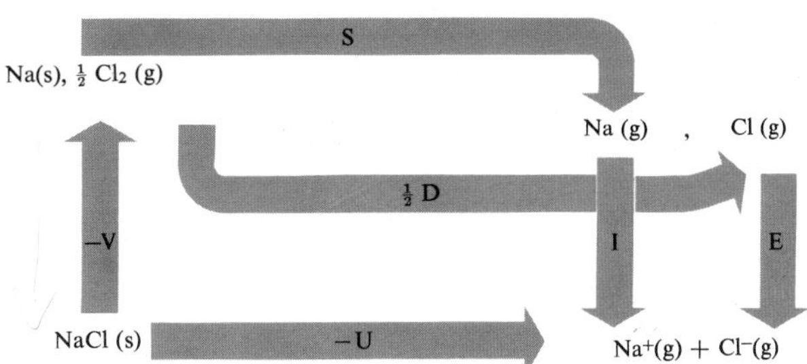

Fig. 7-2 The Born-Haber cycle.

more readily accessible experimental data. Their cycle makes use of a thermodynamic principle that the total net heat evolved in going from one state to another is the same, regardless of the path taken. The cycle is shown in Fig. 7-2. The lattice energy U is defined as the heat evolved in the process Na^+ (g) + Cl^- (g) → NaCl (s); the heat for the reverse process is thus the reverse of this, $-U$. But it is possible to conceive of another route from solid NaCl to the separated gaseous ions. We make use of the following processes:

(a) Heat of formation [Na(s) + $\frac{1}{2}Cl_2$(g) → NaCl(s)]
$$V = -98.2 \text{ kcal}$$

(b) Heat of sublimation of sodium [Na(s) → Na(g)]
$$S = +26.0 \text{ kcal}$$

(c) Heat of dissociation of chlorine [$\frac{1}{2}Cl_2$(g) → Cl(g)]
$$\tfrac{1}{2}D = +28.6 \text{ kcal}$$

(d) Electron affinity of chlorine [Cl(g) + e^- → Cl^-(g)]
$$E = -83.3 \text{ kcal}$$

(e) Ionization potential of sodium [Na(g) → Na^+(g) + e^-]
$$I = +118.0 \text{ kcal}$$

The first of these is directly measurable experimentally; it is simply the heat evolved when the elements in their normal states react to form a mole of solid NaCl. The heat of formation is not to be confused with the lattice energy, which refers to formation of solid NaCl from separated, gaseous ions, and not to the elements in their normal states. The other quantities are self-explanatory, or have been met with before in Chap. 6. Note that when a process, as written, proceeds with evolution of heat, a negative sign is attached to the numerical quantity. A great deal of difficulty with sign conventions can be avoided by thinking in terms of whether the process, as written, represents an evolution or absorption of heat. This is usually determinable from qualitative considerations. Thus, it clearly will require energy to cause the evaporation of a mole of solid sodium metal to the vapor, or to rupture a Cl_2 bond to form atomic chlorine, or to ionize a mole of sodium metal. The sign attached to the energy quantities for these processes is therefore positive. On the other hand, heat is evolved in the formation of solid sodium chloride from the elements, and in formation of Cl^- from gaseous chlorine atom, so the signs of V and E are negative. If any of these processes are reversed in forming the cycle, then the sign of the energy term is also reversed. If heat is evolved for a process in one direction, it must be absorbed for the reverse process.

If we reverse the sign of Eq. (a) above, and then add all five equations together, we obtain

$$NaCl(s) \rightarrow Na^+(g) + Cl^-(g) \quad -U$$

In terms of the energy quantities listed we have

$$-U = -V + S + \tfrac{1}{2}D + E + I \qquad (7\text{-}2)$$
$$-U = -(-98.2) + 26.0 + 28.6 - 83.3 + 118.0 \text{ kcal}$$
$$-U = 187.5 \text{ kcal}$$
$$U = -187.5 \text{ kcal}$$

The values needed for calculation of the lattice energy of a number of ionic substances are listed in Table 7-3.

Table 7-3 Values (kcal/mole) of electron affinity, dissociation energy, sublimation energy, and ionization potential for some ions.

Element	Ion	E	$\frac{1}{2}D$	Element	S	I	Ion
F	F^-	−80	19	Li	38	124	Li^+
Cl	Cl^-	−83	28	Na	26	118	Na^+
Br	Br^-	−78	23	K	22	100	K^+
I	I^-	−71	18	Cs	19	89	Cs^+
O	O^{-2}	+170*	26	Ag	69	174	Ag^+
				Cu	82	177	Cu^+
				Ni	95	592	Ni^{+2}
				Mn	57	530	Mn^{+2}

It is possible to treat the properties of ionic lattices in theoretical terms with a fairly high degree of accuracy by using just a few simple ideas. Let us consider first the ion pair, $Na^+ \cdot Cl^-$, isolated in the gas phase. A graph of the potential energy of this ion pair as a function of the distance r between the ions is shown in Fig. 7-3. The behavior of the curve at large r is readily understandable; the ions are attracted to one another by Coulomb's law, $V = Z_+Z_-e^2/r = -e^2/r$. If the ions were point charges, the curve would continue on down to negative infinity at $r = 0$. But the ions *do* have size, arising from the electron distribution around the nucleus. As the two ions come close together, the electron distributions begin to overlap. This intro-

*Note that the net electron affinity for addition of both electrons to the oxygen atom is positive; *i.e.*, energy must be put into the system to form O^{-2} from the separated oxygen atom and two electrons.

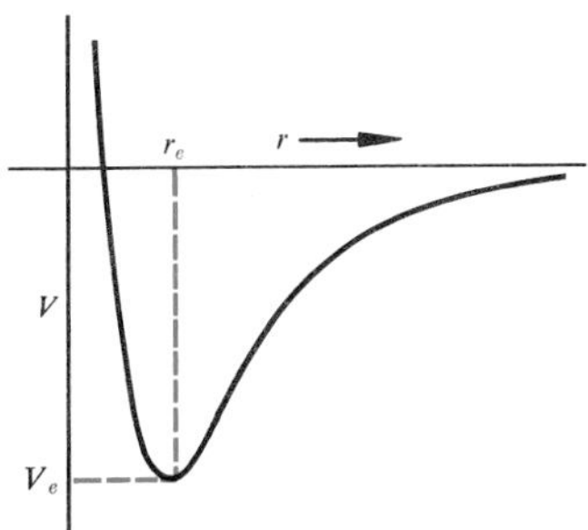

Fig. 7-3 The potential energy curve for an ion pair.

duces a very sharp repulsive energy. Both Na^+ and Cl^- have an inert gas configuration, and there are no vacant orbitals in the valency shell of either ion. As the electron distribution from one ion intrudes upon that of the other, there is, according to the Pauli principle, no possibility for an effective sharing of atomic orbitals. Thus the only interaction which occurs is the repulsive force between the electron shells, coupled with a certain amount of nuclear-nuclear repulsion as well.

There is no simple way to arrive theoretically at a formula for the dependence of the potential energy on r in the range of small r. On the basis of experiment, however, it is known that the dependence is very steep, as shown in Fig. 7-3. The overall dependence of the potential energy on r can be written in a single equation of the form

$$V = \frac{-e^2}{r} + \frac{be^2}{r^n} \tag{7-3}$$

The second term represents the repulsive interaction which occurs at small values for r. The value for n varies somewhat with the particular ions, but would be perhaps eight for the Na^+Cl^- ion pair. The constant b determines the point at which the repulsive term begins to predominate; *i.e.*, it helps to fix the value of r at which the minimum in V occurs. At large values of r the second term is unimportant; in the range of small r it predominates.

The electrostatic potential energy of the ions in a three-dimensional sodium chloride crystal is of the same form as Eq. (7-3), but things are a bit more complicated. Each ion in the lattice sees an entire array of ions of unlike charge, and another array of ions of like charge. Consider one of the positive ions in the lattice (Figs. 7-1 and 7-4). It is surrounded by six chloride ions at a distance d, giving rise to a potential energy term $-6e^2/d$. There are twelve sodium ions at a distance $\sqrt{2}d$, giving rise to a term $+12e^2/\sqrt{2}d$. Eight chloride ions at a distance $\sqrt{3}d$ give rise to a term $-8e^2/\sqrt{3}d$, *etc.* If this process of summing all of the potential energy terms over the whole lattice is continued, a series of the form

$$V = Z_+Z_-(e^2/d)(6 - 12/\sqrt{2} + 8/\sqrt{3} - 6/2 + 24/\sqrt{5} - 24/\sqrt{6} + \ldots)$$

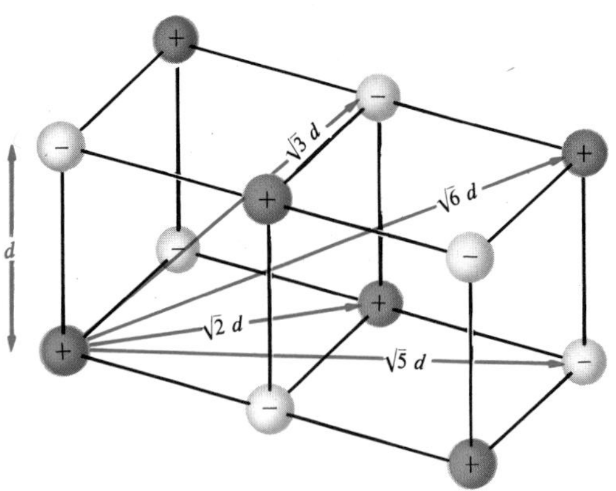

Fig. 7-4 Distances to neighboring ions in the sodium chloride lattice.

results. The expression in parentheses is an infinite series, which converges to the value 1.7476. From the manner in which it is derived, we see that it is characteristic of the lattice type and would be equally applicable to *any* ionic AB compound which possesses the sodium chloride type of lattice. If the charges on the ions were $+2$ and -2 instead of $+1$ and -1, this would merely place a $Z_+Z_- = -4$ in the numerator outside the series. Other lattice types give rise to similar series which, however, converge to different values. These constants are referred to as *Madelung constants*, represented by the symbol A.

The lattice energy can be represented by the formula

$$U = -\frac{Ne^2\alpha^2 A}{r}\left(1 - \frac{1}{n}\right) \tag{7-4}$$

N is Avogadro's number, A is the Madelung constant, and α is equal to the highest factor common to the charges of the ions which make up the lattice. The constant n in the term in parentheses is the exponent of r in the repulsion term in Eq. (7-3). The repulsion term does not make a large contribution to the lattice energy of a three-dimensional array because it is operative only between ions which are in immediate contact, whereas the electrostatic repulsion terms, having only a $1/r$ dependence, extend over a much larger range in the lattice. If, for example, n is 8, as it would be for NaCl, then $(1 - 1/n)$ would be 0.88, so the repulsion term represents only about a 12 per cent contribution to the lattice energy.

Examination of Eq. (7-4) tells us that the lattice energy depends on the following factors:

1. *The changes on the ions.* Highly-charged ions give rise to ionic crystals with large lattice energies. Thus it is clear, for instance, why MgO should have a much higher melting temperature and be a much harder crystal than LiCl.

2. *The distance between ions.* A small value for r leads to a larger lattice energy. This aspect of ionic structures is dealt with in the following section.

3. *The lattice type.* The Madelung constant has a different value for each type of lattice. Values of A for a number of the more common lattice types, along with the appropriate values for the largest common factor α, are listed in Table 7-4.

Table 7-4 Values of the Madelung constant for various lattice types.

Structure type	A	α
Sodium chloride	1.74756	1
Cesium chloride	1.7627	1
Zinc blende (ZnS)	1.638	2
Fluorite (CaF_2)	5.0388	1
Rutile (TiO_2)	4.816	2
Corundum (Al_2O_3)	25.0312	1

7-4 Ionic Radii

The distances separating the ions in the lattice in the equilibrium, or most stable state, are determined by two opposing factors. The attractive potential between ions of opposite charge varies as $1/r$, and thus increases as the distance separating the ions decreases. At short distances, however, the electron distributions in the two ions make contact, and very strong repulsive forces become important. The equilibrium distance represents a balance between these attractive and repulsive forces.

From a knowledge of the distances separating ions in a large number of ionic lattices involving various positive and negative ions, it is possible to obtain a set of ionic radii. The radii for a number of ions with inert gas configurations are given in Table 7-5. Two interesting and important trends are illustrated by the data in this table. Note that, in the series of *isoelectronic* (same electronic configuration) *ions*, O^{-2}, F^-, Na^+, Mg^{+2}, Al^{+3}, there is a steady decrease in the radius. This decrease is caused by the increase in nuclear charge in the same series. With the number of electrons remaining constant, the electrons are attracted more strongly to the nucleus as the nuclear charge increases. The apparent size of the ion therefore decreases.

Table 7-5 Radii (in Å) of ions with inert gas configurations.

Groups									
I		II		III		IV		VII	
Li^+	0.68	Be^{+2}	0.30			O^{-2}	1.45	F^-	1.33
Na^+	0.98	Mg^{+2}	0.65	Al^{+3}	0.45	S^{-2}	1.90	Cl^-	1.81
K^+	1.33	Ca^{+2}	0.94	Sc^{+3}	0.68	Se^{-2}	2.02	Br^-	1.96
Rb^+	1.48	Sr^{+2}	1.10	Y^{+3}	0.90	Te^{-2}	2.22	I^-	2.19
Cs^+	1.67	Ba^{+2}	1.31						

The variation in radius in any one family of elements, provides a second interesting trend. There is a general increase in radius with increasing period; *i.e.*, as the highest occupied principal quantum number increases.

The radii of both positive and negative ions are important in determining the lattice energy. For a given charge on the ions, the lattice energy decreases as the distance separating the ions at equilibrium increases. This is illustrated by the data in Table 7-6. In the series of sodium halides, the lattice energy is largest for the smallest negative ion.

Table 7-6 Comparison of lattice energy (in kcal/mole) with ionic radii (in Å).

Salt	Radius of negative ion	Lattice energy	Salt	Radius of positive ion	Lattice energy
NaF	1.33	219	LiCl	0.68	203
NaCl	1.81	188	NaCl	0.98	188
NaBr	1.96	175	KCl	1.33	171
NaI	2.19	150	RbCl	1.48	163
			CsCl	1.67	160

Similarly, in the second series, in which the size of the positive ions is varied, the lattice energy decreases with increasing size of the positive ion.

7-5 X-Ray Diffraction

The arrangements of atoms and ions in solid substances is determined from diffraction data, principally X-ray diffraction. The X-rays observed and measured by Moseley (Sec. 5-4) are emitted from a metal upon bombardment of the metal surface by high-energy electrons. Although the X rays that come off are distributed in wavelength, by far the most intense rays are the *K* rays. A typical setup for use of the X rays for diffraction is shown in Fig. 7-5. It happens that the X rays come off the metal at a rather small angle with the metal surface. Most of the X ray energy which is not of the K_a wavelength is removed by a suitable filter. Then the X rays, which have been collimated (formed into a narrow parallel beam), by the slit system are allowed to fall on the crystal of interest. Copper metal is a commonly used target material for generation of X rays. The wavelength of the K_a line for this metal* is 1.542 Å. But in a typical crystal, the distances separating the atoms or ions are on the order of 1-3 Å. In this situation, the lattice of the crystal acts as a diffraction grating, and causes diffraction, or scattering of the radiation. The precise manner in which the X-ray photons interact with the lattice is rather complex, but the following simple picture serves to give some idea of what is involved.

---

*There are actually two K_a lines, the K_{a_1} and K_{a_2}, which are close in wavelength, 1.540 Å and 1.544 Å. These behave as a single wavelength source for all but the most exacting X-ray work, and are seldom distinguished in practice.

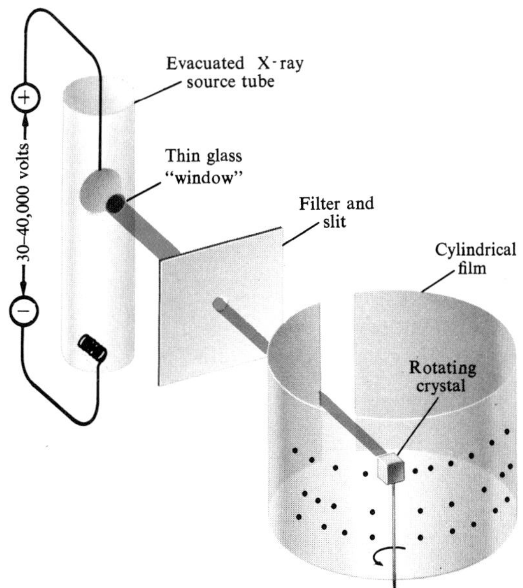

Fig. 7-5 Diffraction of X rays by a crystal.

When a photon impinges on an atom or ion, it interacts with the electrons and is scattered off at some angle. The scattered photons have a different wavelength because of the Compton effect (Sec. 1-5), but the effect is very small, and the wavelength is essentially unchanged. The probability of scattering by a particular atom or ion, which determines the intensity of the scattered radiation, is proportional to the square of the number of electrons in the atom or ion. From this, we see immediately that the heavy atomic weight elements in a lattice will make the largest contributions to the X-ray diffraction. There is a relationship between the angle which the path of the scattered photon makes with the original path and the intensity of the scattering. This is a well-known function which we need not go into. The more important factor which we must consider is the effect of having a three-dimensional array of atoms or ions, all of which are potentially able to scatter an impinging photon. To see what effect a regular arrangement of scattering points might have, let us consider a one-dimensional array of such points, as shown in Fig. 7-6. All of the points lie on a line, and are equally spaced with interval d. Let the incoming beam of X rays make an angle θ with the perpendicular to this line. Now consider a line of scattered photons which leave at an angle ϕ to the perpendicular. The incoming radiant energy possesses a wavefront, indicated by the dotted line, which is perpendicular to the direction of propagation. Along this line, the photons are all in phase. To think of the X rays for the moment as a wave phenomenon (Fig. 1-4), the crests and troughs of all the waves at the points connected by the dotted line are the same. Now, after the X-ray

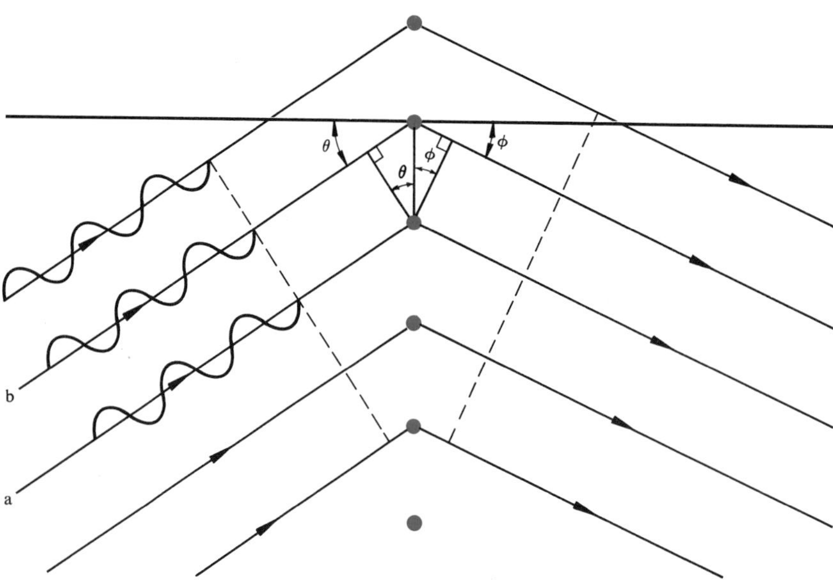

Fig. 7-6 Scattering of photons by a row of lattice points.

beam moves on to the line of atoms and is scattered at the angle ϕ, the wavefront is lined up along the dashed line. In order for the light to have appreciable intensity in this direction, all of the components of the wavefront must again be in phase. But consider two of these components, *a* and *b*. We see that the *b* component actually moves a greater distance to arrive at its point along the wavefront after scattering. Unless this greater distance is precisely a complete wavelength or some multiple of it, the *b* component will not be in phase with the *a*. It requires only a simple application of trigonometry to see that the extra distance that the *b* component moves is $d \sin \theta + d \sin \phi$. This distance must be qual to $n\lambda$ where n is an integer, 1, 2, 3 . . . and λ is the wavelength of the incident X ray. If we fix the angle θ between the line of atoms and the incident beam, then for a given value of the wavelength, and a given distance d between atoms, there are only certain angles ϕ at which the scattered rays will be in phase, and will give a measurable intensity. The problem is much more complex for a three-dimensional array of atoms, since the condition

$$n\lambda = d(\sin \theta + \sin \phi) \tag{7-5}$$

must be met for every regular set of atoms, in each dimension, before the scattered rays will be in phase.

Using an experimental setup such as that shown in Fig. 7-5, the angles at which scattering occurs in phase can be determined by measuring the locations of the spots on the photographic film. In recent years, more elaborate counting devices which can be swept through various angles have come into use. In practice, the crystal is moved through a repetitive turning or rocking motion to expose the various planes of atoms to the beam. The photographic plate is then analyzed to determine not only the angles at which intensity is found, but also to determine the relative intensities of the scatterings. As already mentioned, the scattering power of atoms is related to atomic number; determination of the relative intensities of the spots is necessary to determine the precise locations in the lattice of the different atoms or ions which make it up.

W. H. and W. L. Bragg in England were pioneers in developing the use of X rays for crystal-structure determinations. They conceived of a simple and useful way to envision the fulfillment of a special angle condition for diffraction. Imagine that the X-ray beam impinges upon a crystal, which consists of planes of atoms separated by a distance d. The X rays are assumed to be reflected from these planes, as shown in Fig. 7-7. If the incoming rays make an angle θ with the planes, then the beam components, which are reflected from adjacent layers, all are reflected off at the same angle. The

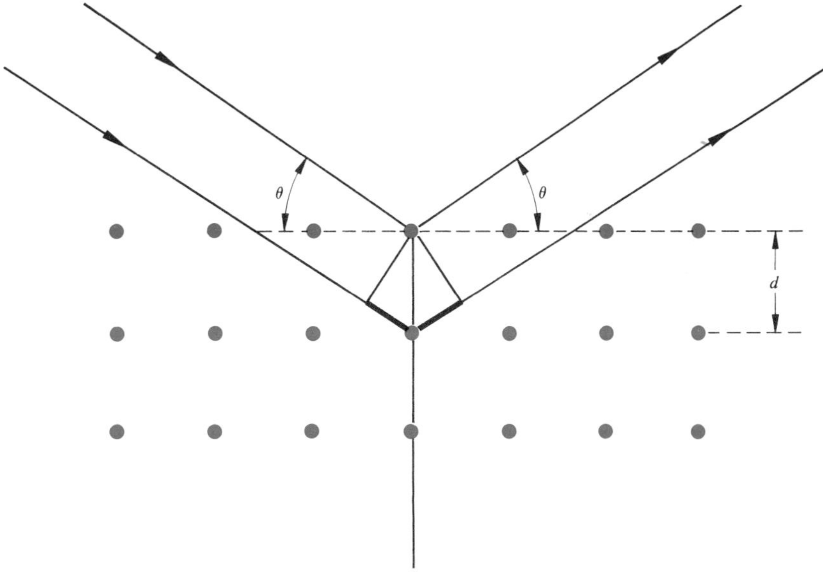

Fig. 7-7 Diffraction of incident light by a set of lattice planes. The Bragg condition is satisfied when the difference in path lengths for light reflected from adjacent layers is a multiple of the wavelength: $2d \sin \theta = n\lambda$.

difference in path length travelled by components reflected from adjacent planes is then $2d \sin \theta$. (Can you show this?) Again, if intensity is to be observed these components must differ in path length travelled by a multiple of the X-ray wavelength, so

$$n\lambda = 2d \sin \theta \tag{7-6}$$

This simple expression is called the *Bragg equation*. It is possible to show, by proceeding with the analysis given above for a one-dimensional line of atoms, extended to three dimensions, that the Bragg equation does result from the more realistic picture in which each atom is thought of as a scattering point. When the distances between planes of atoms are not the same along the three different axis directions in a crystal, or when the angles between the planes of atoms are not 90°, the equations for the diffraction condition become much more complex.

X-ray diffraction techniques have led to the elucidation of the structures of many thousands of solid substances. From a knowledge of the arrangements of atoms and ions in solids, chemists are able to formulate theoretical models relating the nature of the units which make up the solid and the geometrical arrangements which are found, as well as other physical and chemical properties of solids.

7-6 The Structures of Solid Substances

As a prelude to discussion of the structures of ionic substances, we must first consider some general aspects of solid-state structures. Although there is a tremendous difference in the properties of solid argon (melting point 84°K) and a substance such as MgO (melting point over 2500°K), there is,

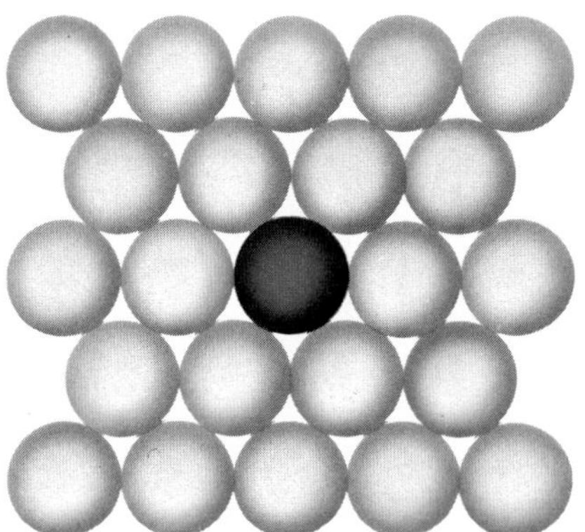

Fig. 7-8 Close-packing of spheres in a layer.

nevertheless, a common characteristic in the manner in which the atoms or ions are packed into the solid. It is profitable, therefore, to consider the question of packing, whether of atoms or ions, in the most general way possible.

Consider a three-dimensional arrangement of spherical objects. We wish to know the most efficient mode of packing of the spheres, in terms of space utilization. Within a single plane, the most efficient array is as shown in Fig. 7-8, in which each sphere has six neighbors in the plane. In constructing a three-dimensional array, we can put another such layer on top of the first. The second layer is not directly over the first, but located so that each sphere of the second layer is in contact with three atoms of the first layer, as in Fig. 7-9. In adding a third layer, a choice arises. The third layer

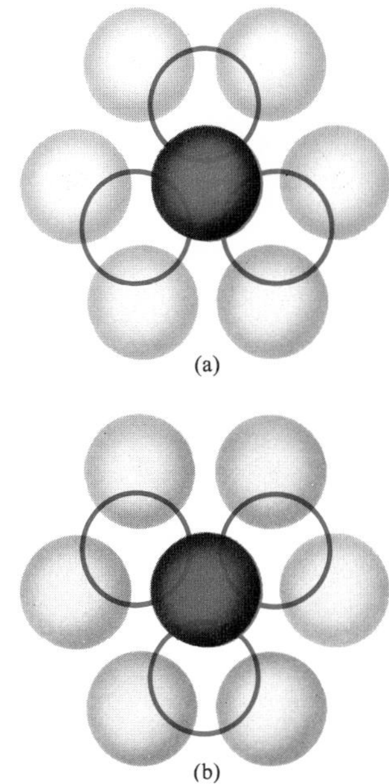

Fig. 7-9 Relative positions of two successive layers of close-packed spheres.

can be oriented so that it lies directly over the first, or it can be situated in a set of triangular depressions which are offset from the first layer. The efficiency of space utilization is the same in each case. When the first alternative is chosen, the successive layers might be labelled *ababab* . . . , in terms of their orientation. This arrangement is called *hexagonal-close-packing*. When the second alternative is chosen, the successive layers can be labelled *abcabcabc*. . . . This arrangement is termed *cubic-close-packing*. The two alternative arrangements are depicted in Figs. 7-10 and 7-11.

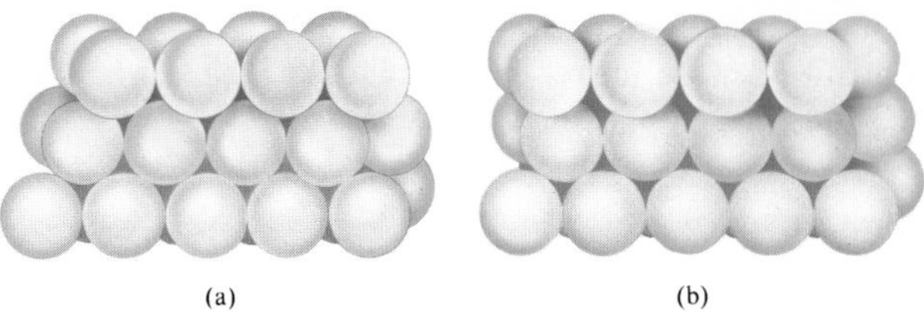
(a) (b)

Fig. 7-10 Models depicting hexagonal and cubic-close-packing arrangements. The difference in the two forms lies in the orientation of the third layer with respect to the first. In hexagonal-close-packing, (a), the atoms in the third layer lie directly over those in the first. In cubic close-packing, (b), the third layer is displaced slightly from the first. In both structures, however, each sphere possesses twelve nearest-neighbors.

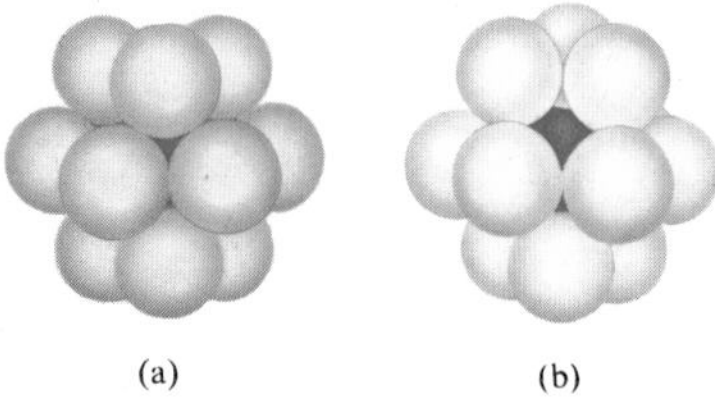
(a) (b)

Fig. 7-11 A view of the packing of nearest neighbors around an individual atom in a close-packing arrangement. Note that in cubic-close-packing, (b), the triangle formed by the three top atoms is rotated 60° relative to that formed by the lower three atoms. In hexagonal-close-packing, (a), the top three atoms are directly over the bottom three.

The close-packing arrangements are the most compact manner of arranging spheres. In either close-packing arrangement, the spheres occupy 74 per cent of the space. The remainder is in the form of interstices, or crevices, which may be of great importance, as we shall see. Each sphere in a close-packed arrangement is in contact with twelve other atoms, as shown in Fig. 7-11.

The plane of atoms shown in Fig. 7-8 is not the only type of plane to be found in a close-packed structure. There are other, similarly dense planes, and other planes which contain the atoms in less dense arrangements. For example, there are planes in which the atoms are arranged in a simple rectangular array, with each sphere in direct contact with only four others. By looking at the cubic-close-packing arrangement in such a plane, we find that it can be viewed as a simple rectangular lattice with spheres on the faces of the simple cube, as shown in Fig. 7-12. The cubic-close-packing arrangement is sometimes referred to as face-centered cubic.

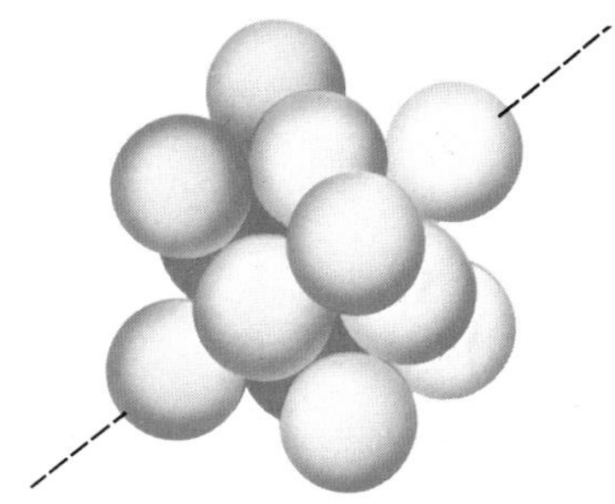

Fig. 7-12 Cubic-close-packing, seen as a face-centered cubic lattice. The dotted axis, passing diagonally through the cube, is perpendicular to the close-packing planes as shown in Figs. 7-10 and 7-11.

The rare gases all crystallize (at low temperatures, of course) into close-packing arrangement. All of the metallic elements also crystallize in one of the close-packing arrangements, or in a body-centered cubic arrangement which is almost as efficient as close-packing. The rare gases and the metals consist of a single kind of atom, so that it is not surprising that they might form close-packed arrangements. But what of ionic substances, which consist of two or more different ions? It is possible to consider ionic substances in terms of close-packed arrangements because it frequently happens that the negative ions of the lattice have a much larger radius than the positive ions. The negative ions, therefore, can be considered to form a close-packed lattice, with the positive ions fitting into the interstices, or holes, between the larger spheres. Within the close-packed arrangement, it is possible to distinguish the interstices according to the number of surrounding close-packed spheres. There are *trigonal* sites, with three surrounding spheres, *tetrahedral* with four, *octahedral* with six, and *cubic* with eight. The trigonal sites are easy to visualize, Fig. 7-13(a). The tetrahedral site is formed from three spheres with a fourth situated over them, as shown in Fig. 7-13(b); the foui spheres form a tetrahedral arrangement. In the space between three spheres of one layer and three of the next layer which are situated over it as shown in Fig. 7-13(c), there is an octahedral hole, surrounded equivalently by the six spheres. It is sometimes easier to visualize

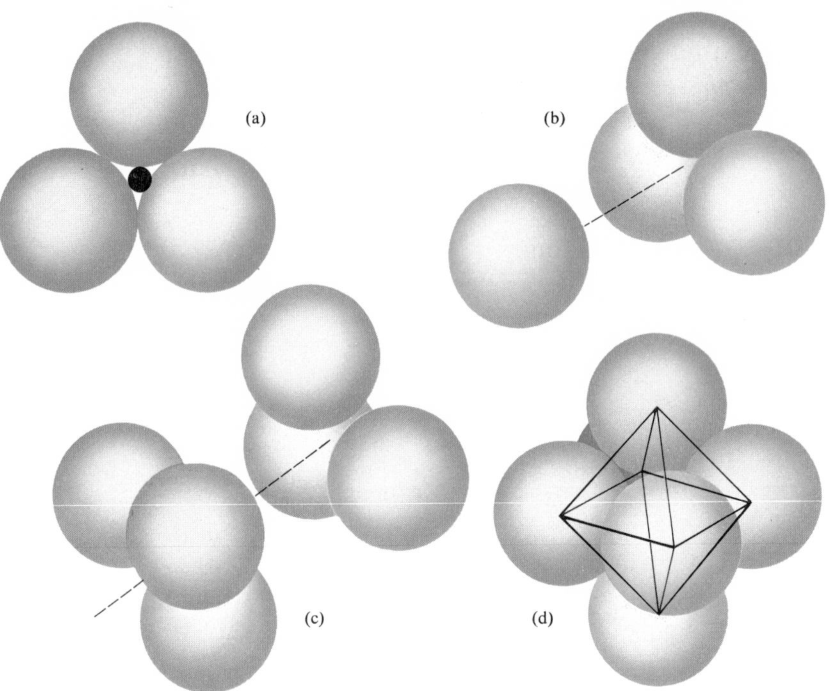

Fig. 7-13 Interstices within the close-packed structure. (a), trigonal, (b), tetrahedral, (c), (d), octahedral.

the octahedral hole by turning the arrangement so that four of the surrounding spheres lie in the horizontal plane. The six then appear as shown in Fig. 7-13(d). By looking for the plane in which the spheres seem to be arranged in a simple rectangular arrangement, it is possible to locate the cubic hole, about which there are eight equivalent spheres. How many holes of each type are there in a close-packed arrangement? It turns out that there are four trigonal sites per close-packed sphere, two tetrahedral sites, one octahedral site, and $\frac{1}{2}$ cubic site.

We have said that ionic substances may frequently be viewed as close-packed arrangements of the negative ions with the positive ions fitted into the interstices, or holes. But which type of hole does the positive ion go into? The answer is to be found in two considerations. In the first place, the positive ion should be in contact with the maximum number of negative ions, in order to maximize the attractive electrostatic interactions. This consideration would then always argue for occupancy of the hole with the largest number of surrounding spheres: cubic, octahedral, tetrahedral, trigonal sites, in order of decreasing preference. There is an additional consideration, however, in the repulsions which the negative ions exert on one another. As a general rule, if the anions are in direct contact, unfavorable repulsions are maximized. Direct contact of the negative ions can be avoided by putting a positive ion into a hole which would not be quite large enough for it if the negative ions touch, so that they are forced apart slightly by the presence of the positive ion. It is easy to calculate the size of positive ion which will just fill each of the types of holes discussed above. The calculation for the trigonal hole is shown in Fig. 7-14. The size of the

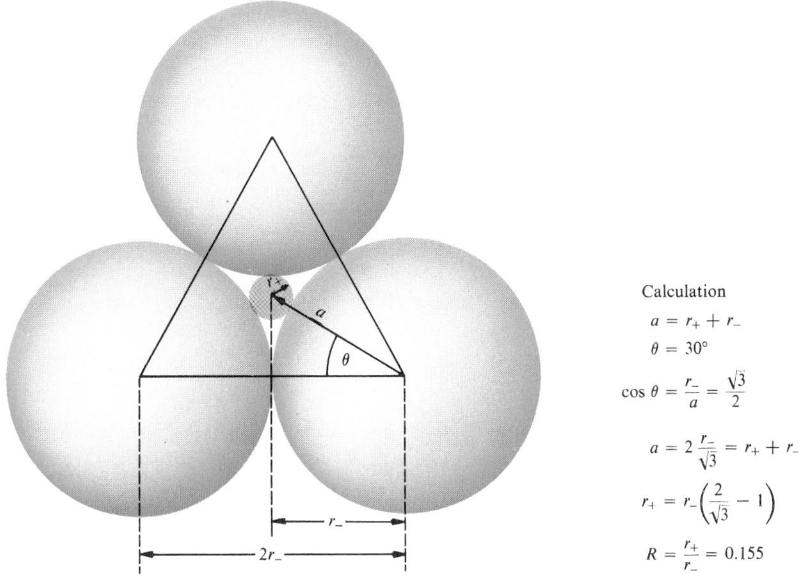

r_+
a
θ
r_-
$2r_-$
Calculation
$a = r_+ + r_-$
$\theta = 30°$
$\cos\theta = \frac{r_-}{a} = \frac{\sqrt{3}}{2}$
$a = 2\frac{r_-}{\sqrt{3}} = r_+ + r_-$
$r_+ = r_-\left(\frac{2}{\sqrt{3}} - 1\right)$
$R = \frac{r_+}{r_-} = 0.155$

Fig. 7-14 Calculation of the minimum radius ratio to prevent negative ion contact in a trigonal hole.

positive ion at which the larger spheres just come into contact is given in terms of the radius ratio, the ratio of the radius of the positive ion to that of the negative ion: $R = r^+/r^-$. Table 7-7 lists each type of hole, the coordination number (the number of bordering spheres) and the smallest radius ratio which still prevents contact of the larger spheres.

As a rough rule, the positive ion occupies the hole of largest coordination number which will still prevent direct contact of the negative ions. The

Table 7-7 Radius ratio requirements for various sites.

Site	Coordination number	Radius ratio requirement
Trigonal	3	0.155 or larger
Tetrahedral	4	0.225 or larger
Octahedral	6	0.414 or larger
Cubic	8	0.73 or larger

rule is not always followed, however, because there are frequently other factors which contribute to the energetics, notably covalent bonding. In Table 7-8 are listed a number of one-to-one ionic compounds, with their radius ratios and structure types. Note that the rule is not always obeyed.

Table 7-8 Radius ratio and structure type for a number of ionic compounds.

Compound	Radius ratio	Coordination number of positive ion	Predicted from radius ratio
CsBr	0.86	8	8
BaS	0.74	6	6—8
AgCl	0.70	6	6
KBr	0.68	6	6
AgI	0.58	4	6
CdS	0.53	4	6
MgO	0.46	6	6
LiCl	0.33	6	4
BeO	0.22	4	3—4

A crystalline solid consists of a regular array of atoms, or ions, in three dimensions. Every crystal is characterized by a repeating unit; when the locations of all of the constituents of this unit are known, the structure of the entire crystal is known. This smallest repeating unit from which the entire crystal structure may be built up is called the *unit cell.* The dimensions of the unit cell, and the locations of all atoms and ions within it, are determined from X-ray diffraction data, or — less often — from other types of diffraction data. We shall now consider some of the more important types of unit cells which characterize simple ionic structures (Fig. 7-15). In these representations, the ions are not shown in their true sizes, since if this were done, the spheres would occupy most of the space and it would not be possible to envision their relative locations. One should keep in mind, however, that the principles discussed above apply to these structures. That is, the negative ions are generally larger in radius, and the positive ions occupy interstitial positions in what is nearly a close-packed lattice of negative ions. The coordination numbers of the ions vary from one structure to the next; the negative ions do not usually have twelve equivalent negative ions for near-neighbors as they would if they were strictly in a close-packed lattice.

The drawings of Fig. 7-15 show only a single unit cell of each lattice. It is evident that the number of atoms necessary to completely describe all of the arrangements varies with the cell type. At the same time, one should not be deceived by appearances. An ion which is on the face of a unit cell is half in one cell and half in the next one, and so makes only a one-half contribution to each. An ion on an edge is shared by four unit cells, and so makes only a one-quarter contribution to each. An ion on a corner is shared

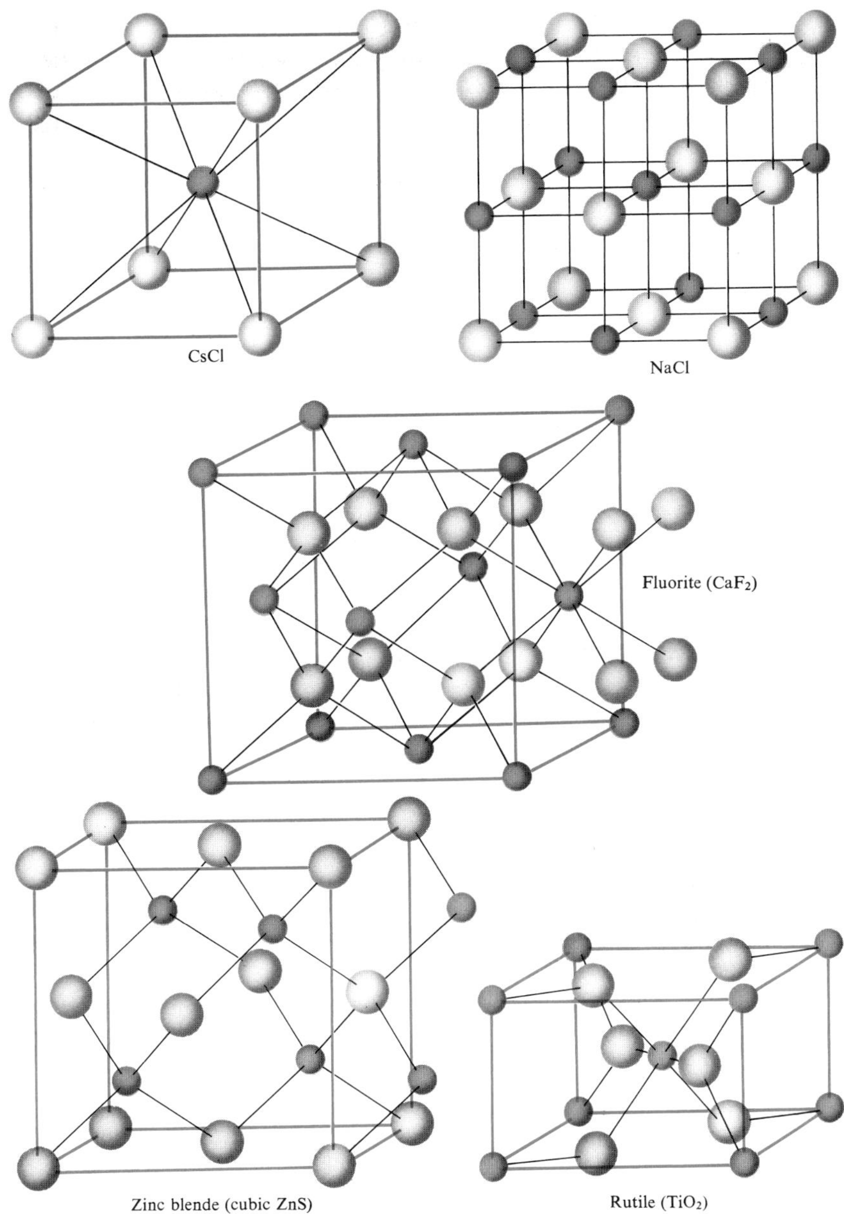

Fig. 7-15 Unit cells for a number of important crystal structure types.

by eight unit cells, and so makes only a one-eighth contribution to each. For example, in the CsCl structure, the Cs^+ ion is totally within one unit cell, but there are eight Cl^- ions on the eight corners, each of which makes only a one-eighth contribution to the unit cell, for a total Cl^- contribution of one.

The point is that the contents of the unit cell must possess the correct empirical formula of the compound. The student should verify that this is the case in the other structures shown, and should be able to ascertain how many formula units are contained within the unit cell in each case.

The sodium chloride structure can be thought of in either of two ways. In one view, it is a pair of interpenetrating lattices, each of which is the same. It is evident from the figure that the positive ions alone form a lattice which is identical with that formed by the negative ions alone. Each lattice is a face-centered cubic. Since the Cl^- ions are larger than Na^+ ions, it is useful to visualize the lattice as a closely-packed face-centered lattice of Cl^- ions with the Na^+ ions in octahedral holes. This is consistent with radius ratio considerations; the Na^+ ions force the Cl^- ions slightly apart, and prevent their contact.

The CsCl lattice can similarly be thought of as interpenetrating lattices of positive and negative ions, or the Cs^+ can be thought of as occupying the cubic holes of a Cl^- lattice.

The Zinc blende structure is an excellent example of the utility of thinking of the lattice in terms of the larger negative ions. The S^{-2} ions (radius 1.90 Å) are much larger than Zn^{+2} (radius 0.74 Å). The S^{-2} ions alone form a lattice which is face-centered cubic. Zn^{+2} ions occupy tetrahedral holes in the lattice, as we expect from the fact that the radius ratio is 0.39. Since this value is considerably larger than the minimum value of 0.225 for tetrahedral sites, the Zn^{+2} ions force the S^{-2} ions apart, so that strictly speaking they are not close-packed. Since there are two tetrahedral holes per S^{-2} ion, only half are used by the Zn^{+2} ions. There is another form of ZnS, called *wurtzite*, in which the S^{-2} ions are in a hexagonal close-packed arrangement rather than in the cubic-close-packed arrangement as in zinc blende. In the wurtzite structure, the zinc ions also occupy tetrahedral sites.

The radius ratio for fluorite, CaF_2, is $R = 0.71$. Although this is slightly lower than the minimum value for cubic coordination of the cation, each Ca^{+2} ion is surrounded by eight equivalent F^- ions in the fluorite structure. The other important structure for the AB_2 type of compound is rutile, TiO_2. The radius of Ti^{+4} is 0.68 Å, so that the radius ratio for rutile is $0.68/1.45 = 0.47$. Our rule suggests that Ti^{+4} ions will be found in octahedral sites. This is in fact what is found, as is evident from examining the environment of the central Ti^{+4} ion in Fig. 7-15. Since there are but half as many titanium ions as oxide ions in the formula, only half of the available octahedral holes in the oxide lattice are occupied by Ti^{+4} ions. Table 7-9 lists a number of ionic substances which crystallize in each of the aforementioned types of crystal structures.

Two ionic substances that exhibit the same crystal structure are said to be *isomorphous*. All of the compounds in Table 7-9, for example, which possess the NaCl arrangement are isomorphous.

Substitution of one ion for another in an ionic lattice is frequently observed in minerals. The most important requirement for substitution to

Table 7-9 Crystal structure types for ionic solids.

Sodium chloride		Cesium chloride	Zinc blende or wurtzite	fluorite	rutile
LiF	CaTe	CsCl	AgI	CaF_2	MgF_2
LiCl	KF	CsBr	ZnO	CdF_2	NiF_2
LiBr	KCl	CsI	ZnS	SrF_2	MnF_2
LiI	KBr		ZnSe	BaF_2	FeF_2
NaF	KI		ZnTe	$SrCl_2$	ZnF_2
NaCl	RbF		CdS	$BaCl_2$	MnO_2
NaBr	RbCl		CdSe	ZrO_2	TiO_2
NaI	RbBr		CdTe	ThO_2	SnO_2
AgF	RbI		HgS	UO_2	
AgCl	MgO		HgSe		
AgBr	MgS		HgTe		
CaO	MgSe		CuBr		
CaS	CdO				
CaSe					

occur is that the radius of the substituting ion be the same, or nearly the same, as that of the ion it is replacing. As an illustration, Al^{+3}, Cr^{+3}, and Fe^{+3} ions are mutually replaceable in many ionic structures. It is not even necessary that the two ions involved have the same charge, provided that electroneutrality can be maintained by another substitution elsewhere in the structure. Thus in some minerals, Ca^{+2} replaces Na^{+}, and at the same time Al^{+3} replaces Si^{+4}.

As a general rule, substitution of ions with equal charge can be expected to occur if the radii of the ions do not differ by more than about 15 per cent (Goldschmidt's rule).

If two ionic compounds have the same crystal structure, if the radii of the positive ions are about the same, and if the radii of the negative ions are about the same, they may form crystals — *solid solutions* — in which both compounds are present. If one of the ions (*e.g.*, the negative ion) is common to both compounds, then solid solution and substitution are synonymous. For example, NaCl and AgCl form a wide range of solid solutions in which Na^{+} may be thought of as substituting for Ag^{+}, or *vice versa*.

Solid solution formation may occur, at least to a limited extent, even between compounds which differ in their type formulas; *e.g.*, between KCl and $CaCl_2$. Consider the possibility of a dilute solution of $CaCl_2$ in KCl. A little $CaCl_2$ is added to molten KCl and thoroughly mixed, and then the melt is allowed to solidify. How can we determine whether solid solution formation has occurred? If no solid solution formation occurs, then the solid is merely an intimate mixture of essentially pure KCl and $CaCl_2$ crystals. The density of a quantity of the solid material is then an average of the densities of the two solid materials, weighted according to the relative amounts present. $CaCl_2$ is more dense than KCl (2.15 g/cm^3 for $CaCl_2$ *vs*

No solid solution formation
Observed
Density of solid
Per cent $CaCl_2$ in KCl

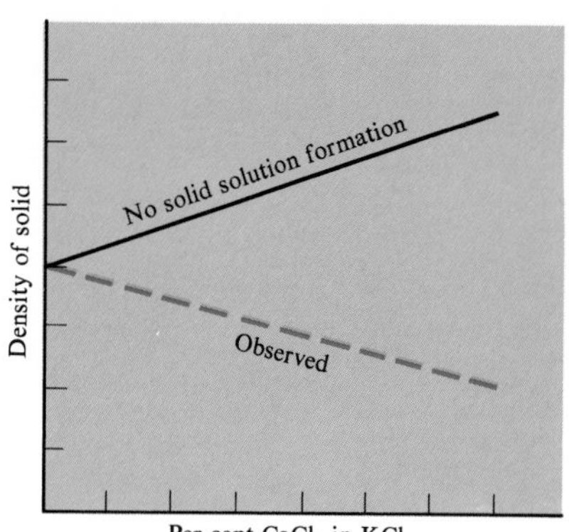

Fig. 7-16 Effect of added calcium chloride on the density of solid potassium chloride.

1.98 g/cm³ for KCl). The apparent density of the mixture as a function of the concentration of $CaCl_2$ added would then vary as the solid line in Fig. 7-16. But when the experiment is carried out, it develops that the density actually *decreases* as a function of added $CaCl_2$, as given by the dotted line of Fig. 7-16. This result is understandable if there has been solid solution formation. Since we have a very little bit of $CaCl_2$ in the presence of a large amount of KCl, the $CaCl_2$ must fit into the KCl lattice. However, there are two chloride ions per positive ion in $CaCl_2$, so there must be a few more negative than positive ions in the lattice of the solid solution. There are,

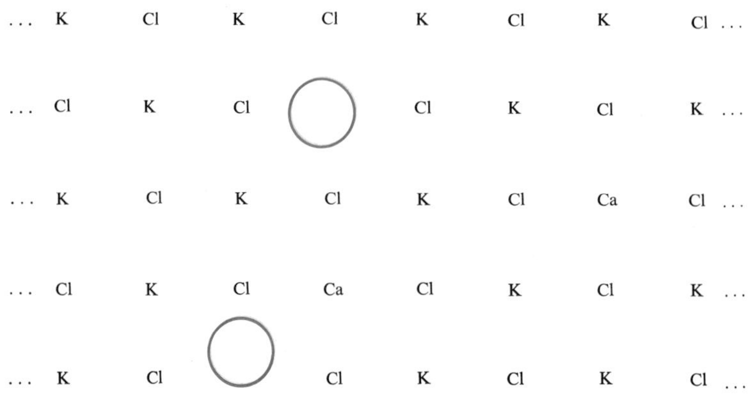

... K Cl K Cl K Cl K Cl ...
... Cl K Cl Cl K Cl K ...
... K Cl K Cl K Cl Ca Cl ...
... Cl K Cl Ca Cl K Cl K ...
... K Cl Cl K Cl K Cl ...

Fig. 7-17 Schematic representation of potassium chloride lattice containing calcium chloride as solid solution. Note that the number of positive ion vacancies is equal to the number of Ca^{2+} ions.

therefore, a few vacancies in the lattice, where there would otherwise be positive ions, as a result of introducing the $CaCl_2$. Since a Ca^{+2} ion has about the same mass as a K^+ ion, the substitution of Ca^{+2} for K^+ does not change the density of the crystal significantly. It is the vacancies, which have no matter in them, which result in a decrease in density for the crystal as a whole.

A schematic drawing of a lattice which possesses the type of vacancy just described is shown in Fig. 7-17. It must be kept in mind that the lattice is always electrically neutral. The vacancies can exist in the lattice only because of the presence of doubly-charged Ca^{+2} ions in numbers equal to the number of vacancies. A simple way of viewing the mixed crystal is to think of a close-packed lattice of Cl^- ions with K^+ ions in octahedral holes. Two K^+ ions can be removed and replaced by one Ca^{+2} ion, with maintenance of electrical neutrality.

7-7 Electrolysis of Ionic Melts

Solid ionic substances are not conductors of electricity. Although they are made up of charged particles, the forces which hold each particle in place in the lattice are so great that they do not move under the influence of an applied electrical potential. In the molten state, however, such substances are excellent conductors. The ions, no longer held in fixed positions, are free to migrate when a potential is applied. Consider the experiment shown in Fig. 7-18. A pair of inert electrodes is immersed in molten sodium chloride, and an electrical potential is applied. The ions migrate under the influence of this potential, the Na^+ ions moving toward the negative electrode and the Cl^- ions toward the positive electrode. If nothing more than this occurred, there would soon be a stop to the flow of charge because of the electrostatic forces which would result from a build-up of negative charge in one part of the bath and of positive charge in another part. We observe, however, that current does continue to flow, and that chemical reactions occur at both electrodes. Sodium metal is deposited at the negative electrode and chlorine gas is evolved at the positive electrode. We can write *half-reactions* for each of these processes as follows:

$$Na^+ + e^- \rightarrow Na$$
$$Cl^- \rightarrow \tfrac{1}{2}Cl_2 + e^-$$
$$\overline{Na^+ + Cl^- \rightarrow Na + \tfrac{1}{2}Cl_2}$$

The sum of these two half-reactions, the over-all cell reaction, represents the electrolysis of sodium chloride. The distinguishing feature of an electrolysis reaction is that electrical energy is employed to produce a chemical reaction.

Electrolysis reactions such as that employed in the preparation of chlorine and sodium involve both oxidation and reduction of chemical substances. *Oxidation* is defined in these systems as the *loss of electrons*, and the electrode at which oxidation occurs is called the *anode*. *Reduction*, conversely, is the *gain of electrons*, and the electrode at which reduction occurs is called the *cathode*. Because positive ions generally undergo reduction at the cathode they are called *cations*, while negative ions are called *anions*. Oxidation-reduction reactions occur in many types of chemical systems, and are not at all restricted to electrolysis cells.

Other chemical reactions, which proceed spontaneously without the need for electrical energy, may also involve oxidation and reduction. In these reactions, it is convenient to label the reactants as either oxidized or reduced. To illustrate, consider the laboratory preparation of chlorine, by reaction of manganese (IV) oxide with hydrochloric acid solution:

$$4\,HCl(aq) + MnO_2(s) \rightarrow MnCl_2(aq) + Cl_2(g) + 2H_2O(1)$$

In this reaction, the chloride ion is oxidized to produce chlorine gas, just as in the electrolysis cell. However, the electrons released in this step do not have an external circuit to pass through; they go directly to the chemical acceptor of electrons, in this case, MnO_2. Manganese (IV), in acquiring the two electrons, is converted into manganese (II). In accordance with the definitions given earlier, one would say that the chloride is oxidized and that the manganese is reduced. Another, and quite common, way of putting it is to say that the chloride acts as a reducing agent, and the manganese dioxide acts as an oxidizing agent. Other examples of oxidation-reduction reactions are to be found in this chapter and throughout the remainder of the text. As a preparation for working with and understanding these reactions, Appendix A should be read.

To maintain a balance of charge in the system shown in Fig. 7-18, the

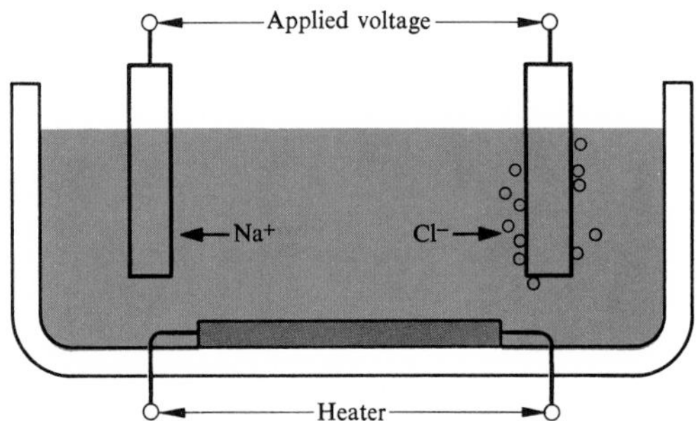

Fig. 7-18 Electrolysis of molten sodium chloride.

number of sodium ions reacting at the cathode must equal the number of chloride ions reacting at the anode. Furthermore, the number of electrons which flow from the source of potential into the cathode must equal the number of sodium ions reduced there. Similarly, the number of electrons which flow from the anode into the source must equal the number of chloride ions which are oxidized. These conclusions are summarized in Faraday's law of electrolysis, based on observations made by Michael Faraday about 1834:

1. The quantity of a substance produced at an electrode varies directly with the amount of electricity passed through the cell.
2. The number of electrons taken up at the cathode is equal to the number of electrons evolved at the anode.

The quantity of electrons required to produce one mole of sodium at the cathode is termed the *faraday*. Since one mole of sodium contains Avogadro's number of sodium atoms, 6.02×10^{23}, one faraday is a quantity of charge equal to 6.02×10^{23} electrons. In terms of commonly used units, it is equal to 96,500 coulombs, or ampere-seconds. These ideas are best illustrated with a problem. Suppose that we wish to determine what quantities of magnesium and chlorine are produced from the electrolysis of molten magnesium chloride with a current of 10 amperes flowing for a period of 3 hours. The cell reactions are as follows:

$$\begin{array}{r} Mg^{+2} + 2e^- \rightarrow Mg \\ 2Cl^- \rightarrow Cl_2 + 2e^- \\ \hline Mg^{+2} + 2Cl^- \rightarrow Mg + Cl_2 \end{array}$$

$$30 \text{ ampere-hours} \times \frac{3600 \text{ seconds}}{1 \text{ hour}} \times \frac{1 \text{ coulomb}}{1 \text{ ampere-second}} \times \frac{1 \text{ faraday}}{96{,}500 \text{ coulombs}} = 1.12 \text{ faradays}$$

Now, from the fact that magnesium is divalent, we know that it will require two faradays of electricity to produce one mole of magnesium. Thus it is said that there are two *equivalents* per mole of magnesium, or that the equivalent weight of magnesium is one-half of the atomic weight.

$$1.12 \text{ faradays} \times \frac{1 \text{ mole Mg}}{2 \text{ faradays}} \times \frac{24.12 \text{ g Mg}}{1 \text{ mole Mg}} = 13.5 \text{ g Mg}$$

Similarly,

$$1.12 \text{ faradays} \times \frac{1 \text{ mole Cl atoms}}{1 \text{ faraday}} \times \frac{35.46 \text{ g Cl}}{1 \text{ mole Cl}} = 39.7 \text{ g Cl}$$

or we could have written

$$1.12 \text{ faradays} \times \frac{1 \text{ mole } Cl_2}{2 \text{ faradays}} \times \frac{70.92 \text{ g } Cl_2}{1 \text{ mole } Cl_2} = 39.7 \text{ g } Cl_2$$

7-8 Electronic Structures of Ions; Nomenclature

Although the formation of ionic bonds leads, in many cases, to ionic structures which are in agreement with the expectations of the Lewis-Langmuir theory, it is also true that many ions do not possess an inert gas configuration. Examples are found among the transition and inner-transition elements. It usually is true, however, that the orbitals of highest principal

Table 7-10 Electronic configurations of some transition metal ions.

Element	Electronic configuration	Ion	Electronic configuration
Fe	[Ar] $4s^2$ $3d^6$	Fe^{+3}	[Ar] $3d^5$
Cr	[Ar] $4s$ $3d^5$	Cr^{+3}	[Ar] $3d^3$
Mn	[Ar] $4s^2$ $3d^5$	Mn^{+2}	[Ar] $3d^5$
Cu	[Ar] $4s$ $3d^{10}$	Cu^{+2}	[Ar] $3d^9$
Au	[Xe] $6s$ $5d^{10}$ $4f^{14}$	Au^{+3}	[Xe] $5d^8$ $4f^{14}$

quantum number are vacated in forming these ions; the deviations from the inert gas structure lie in the presence of electrons in the lower *d* or *f* orbitals. Some examples of electronic configurations of transition metal ions are given in Table 7-10. The attainment of a particular charge always involves the loss of the 4*s* electrons first, and *may* involve the loss of some of the lower level *d* electrons; the transition metals often exhibit more than one valence, corresponding to the loss of a varying number of *d* electrons in addition to the *s* electrons. Some examples of this are Cr^{+2} and Cr^{+3}, Fe^{+2} and Fe^{+3}, Cu^{+1} and Cu^{+2}. Generally, under a given set of conditions, one of the valence states of an ion is more stable than the others.

Ions which consist of more than one atom are very common. The atoms which make up such an ion are covalently bonded to one another, and the over-all charge is best thought of as being distributed throughout the entire system. In considering the interaction of a polyatomic ion with other charged species, however, the charge may be regarded as located at the center of the ion. Table 7-11 lists a number of polyatomic ions and their names.

The naming of chemical compounds is, like all of science, in a continual state of change. Articles which were published in the literature a hundred years ago are sometimes quite difficult to read because of now-unfamiliar names attached to compounds. Common names which are specific to a particular compound are generally undesirable, since they are not systematic and reveal little or nothing about the constitution of the material.

The name of the positive ion comes first in the naming of ionic compounds. If the cation is derived from an element which exhibits only one valence, the name of the element alone is sufficient — as in sodium chloride, magnesium sulfate, *etc.* If the element commonly shows two valences, it is usual to signify the lower valence by adding the suffix *-ous* to a term desig-

Table 7-11 Polyatomic ions.

Formula	Name	Formula	Name
BO_3^{-3}	Borate	ClO_4^-	Perchlorate
SO_4^{-2}	Sulfate	$SnCl_6^{-2}$	Hexachlorostannate
PO_4^{-3}	Phosphate	NO^+	Nitrosyl
ClO^-	Hypochlorite	VO^{+2}	Vanadyl
ClO_2^-	Chlorite	NH_4^+	Ammonium
ClO_3^-	Chlorate	H_3O^+	Hydronium
CrO_4^{-2}	Chromate	MnO_4^-	Permanganate

nating the element — as stannous, ferrous, *etc.* The higher valence state is indicated by the suffix *-ic* — as stannic, ferric, *etc.*

The name of the anion, if it is monatomic, ends in *-ide* — as sulfide, chloride, *etc.* Polyatomic anions have characteristic names depending upon their structures and their relationship to other anions containing the same elements. Discussion of these species, some of which are shown in Table 7-11, is deferred until the chapter on non-metals.

A system of naming (devised by Albert Stock) which has come into wide usage avoids many of the disadvantages of the method described above. The anions are named as before, but the cations, when they are monatomic metal ions, are given the name of the element followed by its valency in Roman numerals. Examples of this system are tin (IV) chloride, $SnCl_4$; iron (II) sulfate, $FeSO_4$; manganese (IV) oxide, MnO_2. The last compound is alternatively called manganese dioxide, the prefix *di-* indicating the number of oxide ions.

Exercises

7-1. Provide a brief definition of the following terms:
(a) valence.
(b) isomorphous.
(c) cathode.
(d) radius ratio.
(e) electrolysis.
(f) coordination number.
(g) solid solution.
(h) latice energy.

7-2. A distinction is sometimes made between ionic and covalent valence. The ionic valence of element in a monatomic ion is equal to the charge on the ion. Is this definition consistent with the definition of valence in terms of combining weight given in Sec. 7-1? Explain.

7-3. Which ionic compound in each of the following pairs would you expect to have the higher lattice energy? Explain.
(a) BaO or CaO.
(b) CaO or KCl.
(c) Na_2S or K_2Se.

7-4. The heat of formation of solid CsCl is 107 kcal. Calculate the lattice energy for this salt.

7-5. The lattice energies of lithium chloride and lithium iodide are calculated theoretically to be 202 and 177 kcal/mole, respectively. Using the values in the Born-Haber cycle, calculate the heats of formation of these two salts. Discuss your results in terms of the concepts introduced in this chapter.

7-6. The lattice energy of AgBr is calculated to be 197 kcal/mole. From this, calculate the heat of formation of AgBr. Compare this value with the heat of formation of NaBr, for which the lattice energy is calculated to be 170 kcal/mole. Of what significance is this for the use of AgBr in photographic film emulsions?

7-7. The heats of formation of solid NiO and MnO are -57.8 and -92.0 kcal/mole, respectively. Calculate the lattice energies of these two compounds. On the basis of this calculation, which cation do you think has the larger radius? It is interesting that nickel (II) ion and lithium ion have the same radius, 0.68 Å. Furthermore, the radii of the oxide and fluoride ions are not much different (Table 7-5). Compare the lattice energies of LiF and NiO, and account for the difference. (The standard heat of formation of LiF is -146 kcal/mole.)

7-8. Calculate the lattice energy of calcium fluoride, assuming a value of 8 for n, and using a Ca-F distance of 2.34 Å, observed from X-ray diffraction studies.

7-9. Lithium fluoride has the sodium chloride structure, as shown in Fig. 7-1. From X-ray diffraction data, it is known that the length of the unit cell as shown is 4.016×10^{-8} cm. The density of LiF has been measured to be 2.662 g/cm^3. From these data, it is possible to calculate Avogadro's number. (To do this you must know the number of molecules per unit cell, which you should be able to deduce from a study of Fig. 7-15, with the aid of the comments in the text.)

7-10. A crystal consisting of planes of atoms separated by 1.976 Å is subjected to X-ray diffraction, using copper K_α radiation, of wavelength 1.542 Å. At what angles of incident radiation with the crystal will diffraction intensity be observed?

7-11. Assuming that KCl is capable of forming a solid solution in small concentration in NaCl, would you expect the density of the crystal to be greater or less than that of pure NaCl? Assuming that KCl forms a solid solution in low concentration in $CaCl_2$, would you expect the density of the crystal to be less than, greater than, or equal to, a weighted average of the densities of KCl and $CaCl_2$? Explain.

7-12. Write the electronic structures for each of the following ions:
(a) Co^{+3}.
(b) Ga^{+3}.
(c) Fe^{+2}.

7-13. From the data in Table 7-5, predict the coordination number of the cation in each of the following compounds:
(a) SrS.
(b) BeS.
(c) MgSe.
(d) AlF_3.
(e) Na_2S.

7-14. The radius of the Mn^{+2} ion is 0.91 Å. Name a compound which might be isomorphous with MnO. Name another, different compound which might exhibit solid solution behavior with MnO.

7-15. 30 g of potassium is deposited from a bath of molten KCl in a period of 3 hours. What is the average current during the period? Write the electrode reactions occurring during electrolysis.

7-16. A current of 12.5 amps is employed to electrolyze molten $MgCl_2$. How long will it take to obtain a deposit of metal which is 1 mm thick on the surface of a cathode which has a surface area of 5 cm^2? (You must look up the density of Mg metal.)

7-17. The radius of the hydride ion H^- is given as 2.08 Å. Assuming that NaH is an ionic substance, how would its lattice energy compare with that of the sodium halides?

7-18. Show from trigonometric considerations that the anions come into contact for coordination number 6 at a radius ratio of 0.414.

7-19. The ionic radii of Ni^{+2}, Co^{+2}, and Fe^{+2} are 0.68 Å, 0.78 Å, and 0.75 Å, respectively. Predict the structures of the oxides and sulfides of these elements.

7-20. Predict the radii of the nitride anion N^{-3}, and of the silicon (IV) cation, Si^{+4}.

7-21. Compare the following set of Group IB and IIB ionic radii with the corresponding values for Group IA and IIA elements. What is the explanation for the difference? Is the general character of the comparison consistent with other data so far considered for the two groups of elements?

Cu^+	0.96	Zn^{+2}	0.74
Ag^+	1.26	Cd^{+2}	0.97
Au^+	1.37	Hg^{+2}	1.10

Chapter 8

The Chemical Bond II. Covalent Bonding

8-1 Introduction

The ionic bonds considered in the previous chapter constitute one extreme in the spectrum of chemical bonding in which the bonding forces between atoms consist of electrostatic terms arising from the complete transfer of one or more electrons between atoms. At the other extreme, we find bonding forces between atoms which are the same, as in H_2, so that there can be no net transfer of electronic charge. In this chapter we will consider the second type of bonding, which comes under the classification of covalent bonding.

Suppose that we attempt to formulate a general view of bonding in covalent molecules without any preconceptions about how it should be described. Let us say that we know the nuclear configuration; *i.e.*, we know the locations of the nuclei relative to one another in the equilibrium, or lowest energy, state of the molecule. This is the sort of information which might be obtained from an X-ray diffraction study of a crystalline form of the compound of interest. We now wish to arrive at an understanding of the *electronic* structure of the molecule.

An electron which is added to a system consisting only of nuclei will, of course, be attracted to all of the nuclei. Just as in the case of the atom, the Schrödinger wave equation is appropriate to describe the states of the electron. The Schrödinger equation for the molecule is much more complex, however, because there are a number of nuclei, located at different points in space, which exert attractive forces on the electron. Addition of the other electrons, one by one, to attain the neutral molecule adds the further complication of electron-electron repulsion. The problem, even for the simplest molecules, is too complex to be solved exactly. There are, however, a number of useful observations and generalizations which might be made in the interests of simplifying the problem somewhat.

Since the electrons must move in accordance with a wave equation in the molecule, just as in the atomic case, the wave functions which are solutions to the equation for the molecule represent allowed energy states for the electrons, just as in the atomic case. These allowed states in the molecules are termed *molecular orbitals.* Our goal is to attain some feeling, perhaps some semi-quantitative understanding, of these molecular orbitals which is akin to the understanding we have of the atomic orbitals. It is useful to note that the chemical behavior of the elements is periodic in character. That is, the types of compounds formed by an element in one row of the periodic table are frequently very similar to those formed by the element below it. This tells us that it is the valence shell electrons, those occupying orbitals in the atom which are above the next lowest number inert gas orbitals, which are largely responsible for chemical activity. Secondly, we should note that the valences exhibited by the atoms frequently correspond to the number of electrons required to fulfill the nearest rare gas electronic configuration. This suggests that there is a condition for molecules, just as for atoms, in which the attainment of a certain number of electrons results in stability. By analogy with the atoms, this condition must correspond to the complete filling of a set of molecular orbital sublevels. But the filling of sublevels and levels in atoms arises because of the restrictions on the quantum numbers which electrons may have. In the same way, the Pauli exclusion principle operates in the molecule to restrict each molecular orbital to occupancy by two electrons of opposite spin.

8-2 Molecular Orbitals in Diatomic Molecules

The most useful approach so far to the electronic structure of molecules has been to describe the structures of the molecules in terms of quantities which are fairly well known, the atomic orbitals. The idea is that the individual atoms in a molecule do not completely lose their identity; that electrons which are around a chlorine atom in a molecule, for example, experience forces which are not unlike those experienced by an electron in a chlorine atom. Atomic orbitals are therefore combined to form the molecular orbitals; the method is called the *linear-combination-of-atomic-orbitals molecular-orbital* (LCAO MO) *method.* To see how it works, consider the simplest diatomic molecule, H_2. We know that in the hydrogen atom there is one electron which occupies the 1*s* orbital. Now consider two hydrogen atoms situated only 0.75 Å apart, the equilibrium distance in H_2. There are two electrons in this H_2 molecule; each experiences the attractive interaction with the two nuclei, and each is repelled by the other. (The total energy of the system contains also the nuclear-nuclear repulsion energy.) Let us suppose that the molecular orbital in which the electron moves can be described as a linear combination of the two hydrogen 1*s* functions ψ_{1s_1} and ψ_{1s_2}. Then

$$\psi_{\text{MO}} = a\psi_{1s_1} + b\psi_{1s_2} \tag{8-1}$$

The constants a and b are chosen to give the best value for the energy. In this example we can guess that a is numerically the same as b. Since the two atoms in the molecule are the same, there is no reason why one of the wave functions making up the combination should be preferred over the other. A second condition on the constants a and b is that they must lead to a normalized wave function, so that when $\psi^*_{\text{MO}}\,\psi_{\text{MO}}$ is summed over all space, the result equals one. The reason for this requirement is that $\psi^*_{\text{MO}}\,\psi_{\text{MO}}$ is a probability function. If there is one electron in the orbital described by ψ_{MO}, the probability of finding an electron in the molecular orbital somewhere in space must be equal to one. The two conditions which have just been stated for a and b result in two different solutions. In one solution, a and b have the same sign; in the other, they have opposite signs. The first leads to an energy for the system which is lower than the sum of the energies of two isolated hydrogen atoms, the second leads to a higher energy. The first corresponds to a bonding molecular orbital, the second to an anti-bonding molecular orbital.

The results can be diagrammed as shown in Fig. 8-1. As a general rule, when a certain number of atomic orbitals are combined to form molecular orbitals, an equal number of molecular orbitals must result. In this first example, the two molecular orbitals which are formed have quite different energies. The bonding molecular orbital is stabilized relative to the energies of the starting atomic orbitals; the anti-bonding molecular orbital is destabi-

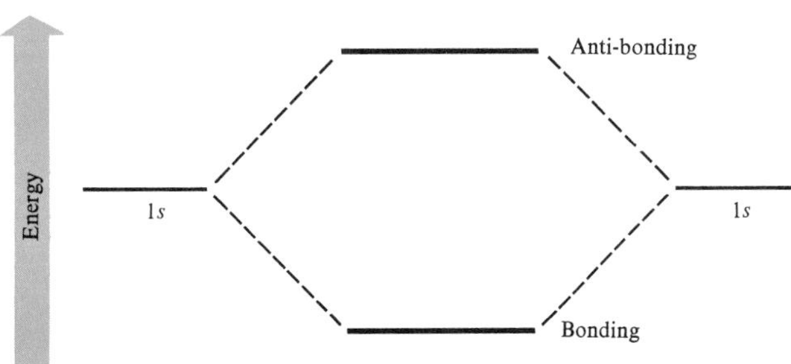

Fig. 8-1 Molecular orbital diagram for hydrogen.

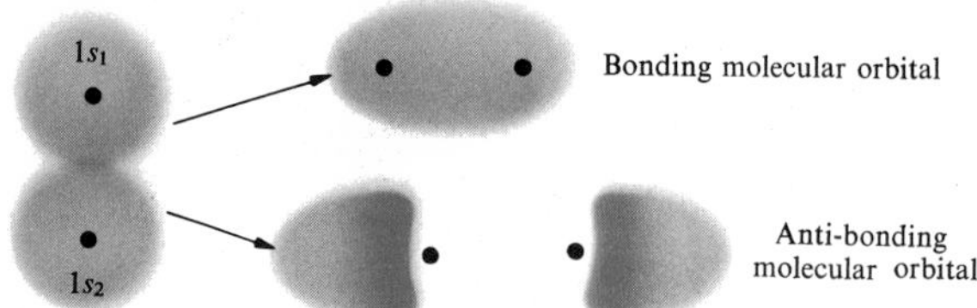

Fig. 8-2 Schematic representation of the combination of two hydrogen 1*s* orbitals to form bonding and anti-bonding molecular orbitals.

lized by an equal amount. The two electrons which must go into H_2 are placed in the lowest energy orbital, which just fills it. The binding energy of H_2 is the result of the lower energy of the bonding molecular orbital over the energies of the hydrogen 1*s* orbitals in the separated atoms.

We must now inquire into the nature of this bonding molecular orbital. Why is there a stabilization of energy over that which characterizes the atomic orbitals of the separated atoms? Fig. 8-2 shows schematically how the molecular orbital is formed from the atomic orbitals. In the bonding orbital, the electron density is concentrated in the region between the nuclei. In simple terms, the electrons are stabilized by interacting with not one, but two, nuclei. This stabilization is attained when the electrons are in the region between the nuclei. In the anti-bonding orbital, the electrons are required to be in the regions *outside* the internuclear space. They do not interact with more than one nucleus at a time, and are actually in a less stable state than in the separated atoms.

Students sometimes find it difficult to see why there should be an anti-bonding orbital at all, or why electrons should ever be found in it. The answer lies largely in the Pauli exclusion principle. The number of molecular orbitals formed from combination of a given number of atomic orbitals must equal the number of atomic orbitals with which we start. If we are to have a bonding molecular orbital by combining two atomic orbitals in a constructive way, then another combination of the same two atomic orbitals must yield yet another orbital. This second orbital must be formed so that there is no interaction with the first, in order that the Pauli exclusion principle be obeyed. What this means in terms of the probability function is that the second orbital must have relatively high probability density in regions of space where the bonding orbital does not, and *vice versa*. The precise form of the orbitals is dictated by the mathematics of the problem.

We can summarize our conclusions regarding the formation of the H_2 molecule by approaching the problem in another way. Suppose that we begin with two hydrogen atoms separated by a great distance, and examine the energy change as they are brought closer together. Figure 8-3 shows the potential energy of this pair of atoms as the distance between them is varied. It is seen that, as the distance between them decreases, there is an increasing

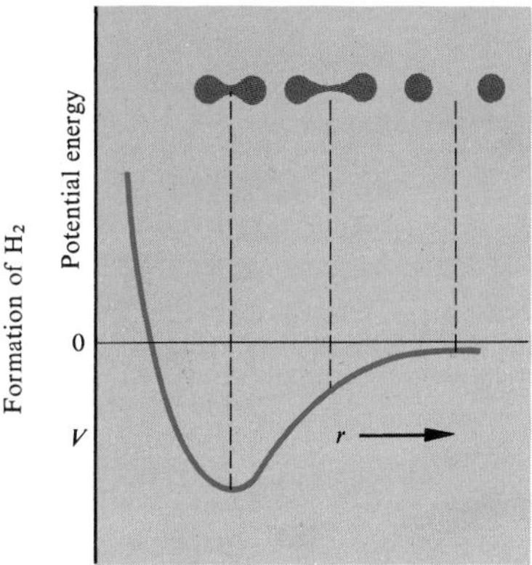

Fig. 8-3 Potential energy diagram for hydrogen.

attractive potential energy. This attractive interaction results from the overlapping in space of the orbital of one hydrogen atom with that of the other. If the electrons in these two atoms have opposite spins, it is possible for them to share the region of space which is common to both orbitals. As the distance between the nuclei decreases, this region of overlap becomes larger, and the potential energy continues to decrease. At very short distances, however, the nuclei repel one another with a force which counterbalances the attractive forces, and the potential energy reaches a minimum value.

The energy diagram of Fig. 8-1 helps to explain why two helium atoms do not combine to form a stable molecule. The electronic structure of each helium atom consists of two electrons in a 1*s* orbital. To form the molecular orbitals in a hypothetical He_2 molecule we would form a linear combination of the two 1*s* orbitals of the He atoms. The result would be similar to that in Fig. 8-1, except that the energy scale would be slightly different because of the greater nuclear charge of He. But the important point about He_2 is that there are *four* electrons which must be accommodated. After placing two electrons into the bonding orbital, there are still two more, which must be placed into the anti-bonding orbital. The loss in stability which results from putting the two electrons into the anti-bonding orbital just cancels the stabilization gained from the pair in the bonding orbital. There is, therefore, no net tendency for two He atoms to enter into chemical combination. What about an ionic species such as He_2^+? The electron removed from He_2 to form the He_2^+ ion comes out of the highest energy (least stable) orbital in the molecule. This would leave two electrons in the bonding molecular orbital and only one in the anti-bonding molecular orbital. The result would be a slight net bonding effect, so that the species He_2^+ should have some stability relative to a separated He atom plus an He^+ ion. The He_2^+ species has been observed spectroscopically in high pressure helium which has been subjected to ionizing electron bombardment.

Having discussed the possibilities of diatomic molecule formation by the first two elements of the periodic table, let us now consider the third element, lithium. We have in this instance a new situation to confront, namely that there is more than one atomic orbital involved in the electronic configuration of the atom itself. The electronic configuration of Li is $1s^22s$. In forming linear combinations of the lithium atomic orbitals, which atomic orbitals must be combined with which? Can we consider a combination of a $1s$ orbital on one of the lithium atoms with a $2s$ orbital on the other? The answer is that there are certain restrictions on which atomic orbitals can or should be combined. The most strict of these involves symmetry, and will be discussed shortly in connection with another example. Another important restriction is that the atomic orbitals which are combined together should have about the same energies in the separated atoms. Thus, combination of a $1s$ orbital of one lithium with the $2s$ orbital of the second lithium does not lead to much gain in bonding stability, because the two atomic orbitals differ considerably in energy in the first place. We can make the simplifying approximation that when equivalent orbitals are present, only the combination of these with one another needs to be considered. Later, in considering diatomic molecules formed from two different atoms, we will see that some modification of this simple assumption is necessary. With this simplifying assumption, the bonding energy level diagram for the Li_2 molecule appears as in Fig. 8-4. Note that combination of the $1s$ orbitals leads to a bonding and anti-bonding combination, and the same sort of thing happens for the $2s$ orbitals. The positions of the starting atomic orbitals on the energy scale

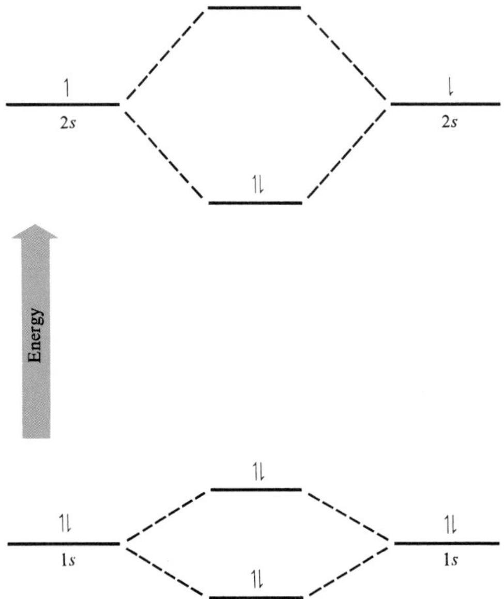

Fig. 8-4 Molecular orbital energy diagram for lithium.

correspond to the known relative atomic energies of the orbitals in the Li atom. It is necessary to place six electrons in the molecular orbitals; they are placed as shown in Fig. 8-4. It is important to note that, just as in He, the four electrons which go into the molecular orbitals formed from the 1*s* functions essentially result in no bonding. It is clear that for all the elements from helium onward, this same thing will occur. Thus we can see that there is no net contribution to the bonding from the inner shell electrons. Only the 2*s* electrons of Li make a contribution to the bonding in Li_2. The stabilizing interaction of the 2*s* orbitals in Li_2 is smaller than the analogous interaction of the 1*s* orbitals in H_2. This is in part because of the greater internuclear distance in Li_2; the dissociation energy of Li_2 is considerably less than that of H_2 (see Table 8-1).

Consideration of the bonding tendency in beryllium, Be_2, does not present any new problems; the reader should be able to see that there is essentially no tendency toward molecule formation, as a result of the necessity of adding two electrons to the upper anti-bonding molecular orbital of Fig. 8-4. The next element, boron, presents a new problem. The electronic configuration of B is $1s^2 2s^2 2p$. We must now consider the manner, or rather, manners, in which 2*p* orbitals may be combined to form molecular orbitals. In an isolated atom, the three *p* orbitals of a given major quantum number are degenerate. Formation of a diatomic molecule, however, introduces a unique direction, the internuclear axis, into the system. Let us establish the convention that the molecular axis lies along the *x*-axis. This makes the p_x orbital different from the p_y and p_z orbitals; the latter two remain degenerate. Formation of the diatomic molecule thus results in a partial removal of degeneracy. The orbitals are aligned with respect to the axis system as shown in Fig. 8-5.

σ and π Orbitals — We must now consider which *p* orbitals can be combined with which to form molecular orbitals. Until now, we have considered the question of what factors determine the degree of energy

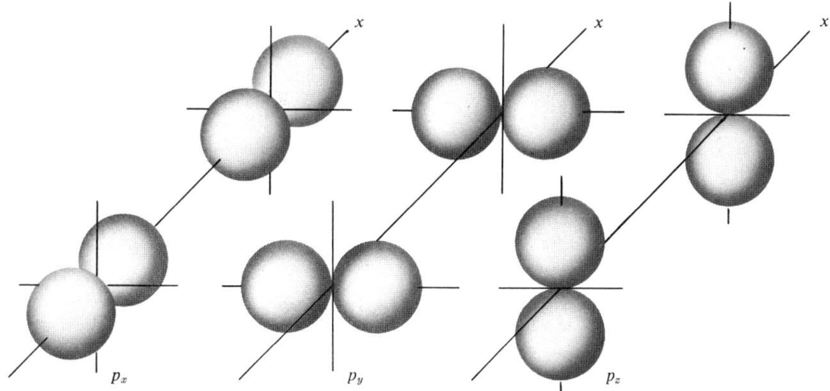

Fig. 8-5 Combination of *p* orbitals in a diatomic molecule.

lowering which results from the combination of two atomic orbitals to form a molecular orbital. Aside from the matter of the relative energies of the two orbitals, discussed above, the most important other factor is *orbital overlap*. The atomic orbitals which are being combined share space together, although they are not centered on the same location. The overlap is a measure of the degree to which the wave functions of the two atomic orbitals lead to a finite probability density for the electron over the space. We have learned that the quantity $\psi\psi$ for a wave function at some point in space is a measure of the probability density at the point in space. If we now have two different wave functions, ψ_a and ψ_b, the quantity $\psi_a \psi_b$ has a similar connotation. The most important quantity, however, is what we have when this is summed over all space. The overlap is the sum of the function $\psi_a\psi_b$ taken over all space. When this is large, the degree of interaction between two orbitals of comparable energy is large. When this is small, or zero, the degree of interaction is small, or zero. We can frequently tell immediately, merely by inspection, that the overlap between certain atomic orbitals is zero. For example, consider the combinations in Fig. 8-6. A p_y orbital must

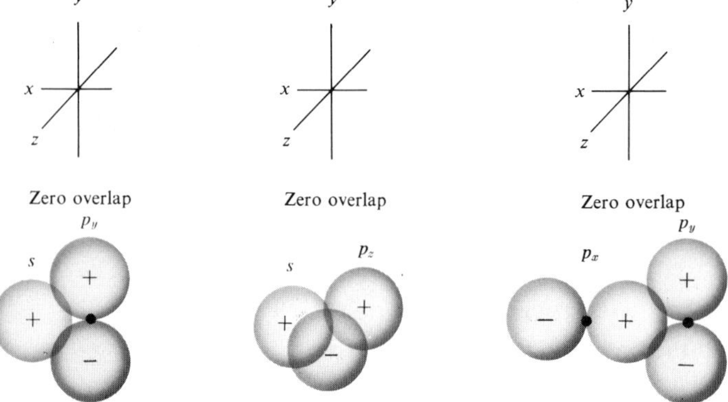

Fig. 8-6 Effect of symmetry on the overlap of atomic orbitals in a diatomic molecule.

have zero overlap with an *s* orbital on the other center, because for every point in space where the product $\psi_{1s}\,\psi_{2p_y}$ is positive, there is a point on the opposite side of the nodal plane where the function is negative. The sum over all space must therefore be zero. The same thing is true of a $2p_x$ orbital on one center and a $2p_y$ or $2p_z$ on another. It is clear, then, that the combinations of atomic orbitals in diatomic molecules is quite restricted.

It is useful to have a classification system for molecular orbitals which is based on symmetry. In diatomic molecules, the internuclear axis furnishes the reference. If one looks along the internuclear axis, the orbitals which form the molecular orbitals can be classified according to how many times the wave function changes sign as a function of rotation about the axis. This

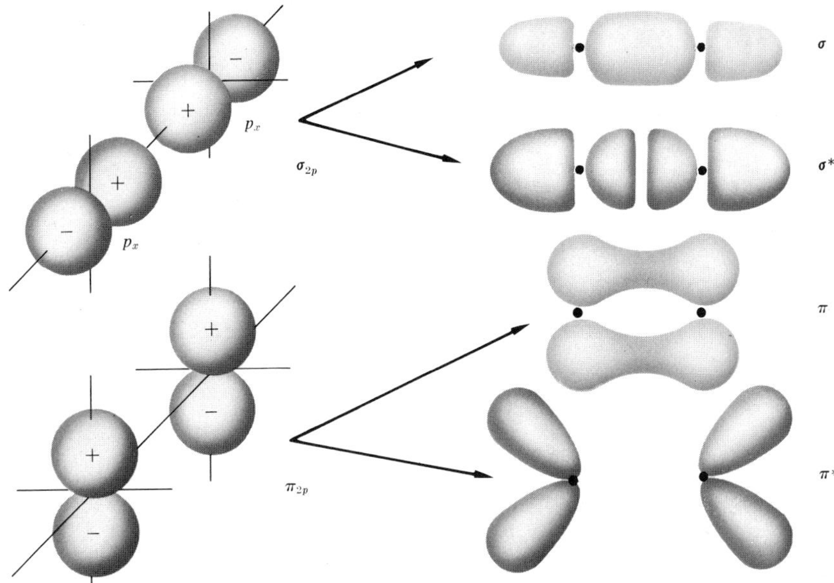

Fig. 8-7 Formation of σ and π molecular orbitals from p orbitals.

is shown in Fig. 8-7. Those orbitals for which there is no sign change are labelled σ (sigma); those for which there is one sign change are labelled π (pi).

The overlap of the $2p_x$, or σ_{2p}, orbitals in diatomic molecules is greater than the overlap of the $2p_y$, or $2p_z$, or π_{2p}, orbitals. The energy level scheme for diatomic molecules from the elements through neon is shown in Fig. 8-8. Of course, the actual splitting in energy between bonding and anti-bonding, and the relative energies of the starting atomic orbitals, varies with each element, but the general features of the energy scheme are sufficiently similar throughout so that we can consider a single bonding scheme. Note that the $2p_y$ and $2p_z$ orbitals are shown as degenerate, as is required by the symmetry of the system. Note also that the lowering of the bonding molecular orbital resulting from combination of the $2p_x$ orbitals is greater than that resulting from combination of the $2p_y$ or $2p_z$ orbitals. This is a consequence of the greater overlap attained by the $2p_x$ orbitals, which are directed toward one another, as opposed to the $2p_y$ and $2p_z$ orbitals, which overlap in a sideways fashion to form the π molecular orbitals. The notation described earlier for designation of the molecular orbitals is used, with the additional feature that the anti-bonding molecular orbitals are distinguished by the * symbol. The electronic configuration of any homonuclear diatomic (*i.e.*, A = B) can be deduced from the diagram of Fig. 8-8, by simply placing electrons in the orbitals in the order of increasing energy, taking note of the Pauli principle, so that there is a maximum of two electrons per MO, with paired spins. It needs only be noted further that Hund's rule (Sec. 6-2) is

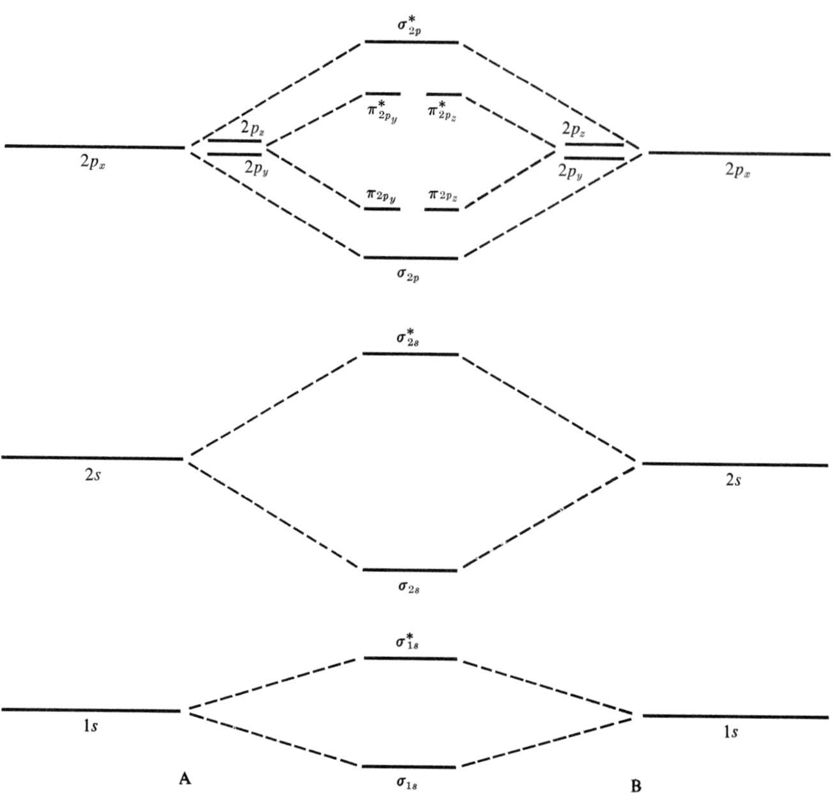

Fig. 8-8 Generalized molecular orbital diagram for homonuclear diatomic molecules. The only known substance for which the ordering of the molecular orbitals is different from that shown is boron. For this molecule, it appears that the π_{2p} orbitals are slightly lower in energy than the σ_{2p}.

obeyed for molecules as well as for atoms. The electronic configurations of the diatomic molecules can be written in a manner similar to that for atoms, as in Table 8-1. This table lists the electronic configurations for all of the neutral diatomic molecules of the first ten elements. Also listed is the number of net bonding electrons (that is, the number of electrons in bonding orbitals less the number placed in anti-bonding orbitals) and the observed dissociation energies and bond lengths. Note that in the N_2 molecule there are three bonding electron pairs, one of σ symmetry and two of π symmetry. The bond between the nitrogen atoms is thus a triple bond. The N_2

Table 8-1 Electronic configurations of the homonuclear diatomic molecules of the elements hydrogen through neon. Bond lengths and dissociation energies are obtained from spectroscopic studies of gaseous species.

Diatomic molecule	Number of electrons	Electron configuration	Net number of bonding electrons	Dissociation energy (e-v)	Bond length (Å)
H_2	2	σ^2_{1s}	2	4.476	0.75
He_2	4	$\sigma^2_{1s}\ \sigma^{*2}_{1s}$	0	0	—
Li_2	6	$\sigma^2_{1s}\ \sigma^{*2}_{1s}\ \sigma^2_{2s}$	2	1.05	2.67
Be_2	8	$\sigma^2_{1s}\ \sigma^{*2}_{1s}\ \sigma^2_{2s}\ \sigma^{*2}_{2s}$	0	0	—
B_2*	10	$\sigma^2_{1s}\ \sigma^{*2}_{1s}\ \sigma^2_{2s}\ \sigma^{*2}_{2s}\ \sigma^2_{2p}$	2	3.0	1.59
C_2	12	$\sigma^2_{1s}\ \sigma^{*2}_{1s}\ \sigma^2_{2s}\ \sigma^{*2}_{2s}\ \sigma^2_{2p}\ \pi^2_{2p}$	4	6.4	1.31
N_2	14	$\sigma^2_{1s}\ \sigma^{*2}_{1s}\ \sigma^2_{2s}\ \sigma^{*2}_{2s}\ \sigma^2_{2p}\ \pi^4_{2p}$	6	9.76	1.09
O_2	16	$\sigma^2_{1s}\ \sigma^{*2}_{1s}\ \sigma^2_{2s}\ \sigma^{*2}_{2s}\ \sigma^2_{2p}\ \pi^4_{2p}\ \pi^{*2}_{2p}$	4	5.11	1.21
F_2	18	$\sigma^2_{1s}\ \sigma^{*2}_{1s}\ \sigma^2_{2s}\ \sigma^{*2}_{2s}\ \sigma^2_{2p}\ \pi^4_{2p}\ \pi^{*4}_{2p}$	2	1.6	1.43
Ne_2	20	$\sigma^2_{1s}\ \sigma^{*2}_{1s}\ \sigma^2_{2s}\ \sigma^{*2}_{2s}\ \sigma^2_{2p}\ \pi^4_{2p}\ \pi^{*4}_{2p}\ \sigma^{*2}_{2p}$	0	0	—

*The ordering of molecular energy levels for this molecule is slightly different than shown in Figure 8-8. See caption of Figure 8-8.

molecule is an exceedingly stable diatomic molecule, as evidenced by its high dissociation energy.

The O_2 molecule furnishes an excellent example of the operation of Hund's rule. There are a total of sixteen electrons in the O_2 molecule. The energy levels σ_{1s}, σ_{1s}, σ^*_{1s}, σ^*_{2s}, σ_{2p} are occupied in that order, and take a total of ten electrons. The π_{2p} orbitals which are next in energy and degenerate; they accept a total of four electrons, leaving just two more. The next energy level encountered is the anti-bonding π_{2p} pair. Because these are degenerate, after one electron has been placed in one of them, the second goes into the other member of the pair, with parallel spin, in accordance with Hund's rule. The energy level diagram is shown in Fig. 8-9. The result, which is really quite remarkable, is that the O_2 molecule contains two unpaired

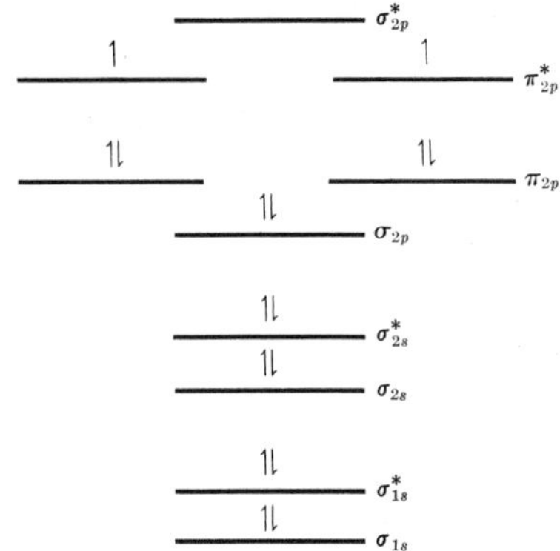

Fig. 8-9 Electron configuration in the oxygen molecule.

electrons. Molecules which possess unpaired electrons behave in a distinctive way in the presence of a magnetic field, and are termed *paramagnetic*. It is one of the major triumphs of the molecular orbital model that it so neatly accounts for the presence of these unpaired electrons, which have been observed in O_2.

Finally, it should be noted that the net number of bonding electrons for the hypothetical Ne_2 molecule is zero, which is consistent with the low chemical activity of neon, and the non-existence of Ne_2.

Heteronuclear diatomics — The bonding scheme we have been discussing can be extended to heteronuclear diatomic molecules as well.

The simplest example of a heteronuclear diatomic molecule is helium hydride ion, HeH^+. This species, isoelectronic with H_2, has been observed; it proves to be remarkably stable, with a dissociation energy of about 45 kcal/mole. In formulating a molecular orbital diagram for HeH^+, we have to contend with the fact that the starting atomic orbitals do not have the same energies. Figure 8-10 shows an energy diagram in which the orbital

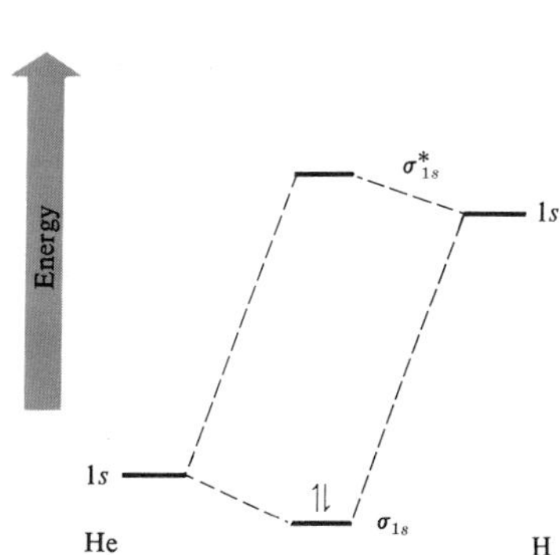

Fig. 8-10 Molecular orbital energy level diagram for HeH^+, the simplest heteronuclear diatomic molecule.

energies are shown in correct relationship. We might begin as before, and assume that molecular orbitals can be formed as a linear combination of the hydrogen and helium 1*s* orbitals:

$$\psi_{MO} = a\psi_{1s_{He}} + b\psi_{1s_H}$$

Since the nuclear charge of helium is twice that of hydrogen, the energy of the helium orbital is lower than that of hydrogen. It is evident that our earlier assumption regarding the equality of a and b is not satisfactory. We do not have the necessary background at this point to go into details about how one decides on the relative values to be assigned to a and b. It is intuitively clear, however, that the bonding electrons will, on the average, spend the greater portion of time in the vicinity of the helium nucleus. This is equivalent to stating that in the bonding molecular orbital, a is greater than b. The helium-hydrogen bond is thus polar to some extent, in that the electrons are "shared" unequally by the two nuclear centers. In considering the great variety of bonding situations in all of chemistry, the notion of bond polarity will be found to arise again and again. The general idea is simple enough, but assignment of quantitative estimates to the polarity has proven to be difficult.

8-3 Extension to Larger Molecules

The model for bonding in diatomic molecules which we have just presented was formulated by Hund in Germany and Mulliken in the U. S., in the early 1930's. The application to molecules with more than two atoms becomes increasingly difficult. Nevertheless, many of the ideas involved in understanding the bonds in diatomic molecules can be carried over to more complex systems. One of the most useful of these is the notion of symmetry as a means of classifying bonds. We will find that it is possible to distinguish σ and π bonds in many molecules. The idea of essentially inert inner shells of electrons is another extremely useful idea, applicable to molecular structure in general. As a good approximation, in discussing bond formation, we can neglect the electrons of the atom which correspond to the next lower inert gas configuration. This approximation would not be a good one if we were intent on calculating molecular energies to a high degree of accuracy, but it suffices very well for qualitative and even for semi-quantitative considerations. Finally, the formulation of the molecular orbitals in terms of the atomic orbitals of the starting atoms is a workable framework in which to consider molecular structure.

Even before the advent of wave mechanics, and before development of molecular orbital theory, a theory of chemical bonding was developed by the American chemists G. N. Lewis and Irving Langmuir. Their model was essentially empirical, in that it offered no detailed explanation for the stabilities of covalent substances. Instead, Lewis and Langmuir, working more or less independently, concentrated on associating chemical activity with the electronic structure of the atom. They noted that every rare gas element after helium is characterized by an octet of electrons in the outermost electron shell. The low chemical activity of these elements points to an unusual stability for this octet arrangement. The valences of the other elements could be explained in terms of a tendency to attain such an arrangement, by sharing, gaining, or losing electrons. Thus chlorine, which possesses seven electrons in the outermost shell, $n = 3$, acquires an electron from sodium to become Cl^-. But in forming a covalent bond with hydrogen in HCl, the chlorine also acquires an added electron, in effect. If we wished to be quite precise in constructing a molecular orbital diagram for HCl we would need to consider mixing of the hydrogen $1s$ orbital with a number of atomic orbitals on chlorine. On the other hand, we have already seen that orbitals which do not have the correct symmetry, *or* which do not have at least comparable energies, do not mix to a significant extent. This tells us that it is sufficient to picture the bonding in HCl as involving the $1s$ orbital of hydrogen, and the $3s$ and $3p$ orbitals of chlorine. All of the electrons which occupy the $1s$, $2s$, and $2p$ orbitals in the chlorine atom can be considered to occupy these same orbitals in the molecule. Now, because of symmetry restrictions (see Fig. 8-6), the hydrogen $1s$ orbital can interact with either the $3s$ or $3p_x$ orbital of chlorine, but not with the $3p_y$ or $3p_z$ orbital. The preference for $3s$ or $3p_x$ orbital would depend on the overlap, and on energy matching. It appears that the $3p_x$ orbital is most favorable, so that we can describe the hydrogen-chlorine bond approximately as occurring between the hydrogen $1s$ and chlorine $3p_x$ orbitals.

Now, chlorine has seventeen electrons, and hydrogen has one. Ten of these we have decided to assign to the $1s$, $2s$, and $2p$ orbitals of the chlorine; the other eight must be placed in orbitals involving the valence shell atomic orbitals or molecular orbitals. Two electrons are placed in the bonding orbital between the hydrogen $1s$ and chlorine $3p_x$ orbital, leaving six. These are assigned in groups of two to the chlorine $3s$, $3p_y$, and $3p_z$ orbitals. In terms of molecular orbital theory, this fills all of the orbitals which are fairly low in energy, thus leading to a chemically stable molecule. In terms of the Lewis-Langmuir theory, we have achieved a rare gas configuration about hydrogen and chlorine, by virtue of the sharing of electrons in the bond between two atoms. We can thus see that molecular orbital theory is a quantitative version of the more qualitative and empirical Lewis-Langmuir theory, in which the number of bonds which the elements form is dictated by the requirement that a rare gas configuration of eight electrons be achieved. Thus, the compounds listed in Table 8-2 are typical covalent compounds of the elements in Groups IV to VII.

Table 8-2 Typical covalent compounds. of elements of Groups IV to VII.

Group IV	Group V	Group VI	Group VII
CCl_4	NF_3	$(CH_3)_2O$	HBr
SiH_4	PCl_3	H_2S	CH_3Cl
$PbCl_4$	AsH_3	CH_3SeH	Cl_2O
$Sn(CH_3)_4$			ICl

Localized Bonds — The molecular orbitals of a large molecule extend in principle over the entire molecular framework. At the same time, it is also clear that in many instances the bonding orbitals consist of a high concentration of charge in the region immediately between two atoms. This fact makes it possible to speak in terms of *localized bonds.* For example, in the molecule methyl chloride,

```
 ..
:Cl:
H:C:H
 ..
 H
```

after neglecting the $1s$, $2s$, and $2p$ electrons on chlorine, and the $1s$ electrons on carbon, we still have fourteen electrons to worry about in a description of the bonding. If we tried to describe the molecular orbitals of this molecule, it would be necessary to consider a mixing together of orbitals from all the nuclear centers. Experience shows, however, that in this case the bonding can be adequately described in terms of localized bonds, pairs of electrons localized between each pair of atoms, and pairs of electrons on the chlorine atom in the $3s$, $3p_y$, and $3p_z$ orbitals. These seven pairs of electrons are indicated in the structure above by pairs of dots.

In drawing the electronic structures of molecules, it is convenient to employ an abbreviated notation in which only the valence shell electrons are indicated.

:Cl:
H—C—Cl:
:Cl:

H—S—H

H—O—O—H

:F—N—F:
:F:

Unshared electron pairs are indicated by a pair of dots, and shared electron pairs are indicated by dashes. The examples illustrate the method. These representations, commonly referred to as Lewis structures, are *not* intended to convey information about the geometries of molecules.

8-4 Partial Ionic Character and Electronegativity

Although the covalent bond is conceived of as a sharing of a pair of electrons by two atoms, it does not follow that the electrons are shared equally. If the two atoms are the same, as in H_2, Cl_2, *etc.*, each atom must have an equal share, but if the two atoms are different, one atom may exert a greater influence on the electrons than the other. We have seen a simple example of this in HeH^+. In the limiting case, one atom exerts such a relatively larger influence than the other, that the bond is said to be ionic. Most covalent bonds lie somewhere between these two extremes of a completely ionic or a completely non-polar bond.

The relative tendency of an atom to attract electrons to itself in a bond with another atom is measured by its *electronegativity.*

On the basis of a number of experimental tests, it is possible to assign values to the electronegativities of the elements. These quantities are presumed to be essentially constant for an element from one compound to another. The elements with the largest values of electronegativity attract bonding electrons most strongly. This quantity is related to two other properties of atoms which we have already considered, the ionization potential and the electron affinity. Atoms which show a high value of ionization potential also show a strong attraction for the electron pair in a bond. This effect is further enhanced if the atom also shows a tendency to acquire an added electron, as measured by the electron affinity. On this basis, we should expect that the metallic elements (to the left in the periodic table), with low ionization potentials and no electron affinity, will have low electronegativities. On the other hand, the non-metallic elements (on the right side of the periodic table), with higher ionization potentials and an observable tendency to acquire an added electron, should have high values of electronegativity.

We have already observed that the ionization potential decreases with increasing size of the atom in any one vertical column, and therefore we expect to find that electronegativity decreases with increasing atomic weight.

These expectations are borne out by the data shown in Table 8-3. It is seen that fluorine has the highest value of electronegativity. The electronegativity decreases in going to the left from this element toward the metallic elements, and in proceeding downward in the Group VII elements.

The electronegativity difference between the atoms X and Y in the bond X-Y is a measure of the polarity of the bond. For example, the bond in H-F (electronegativity difference 1.8) is more polar than the bond in H-I (electronegativity difference 0.4).

Table 8-3 Electronegativities of the non-metallic elements.

				H 2.1
B 1.9	C 2.5	N 3.0	O 3.5	F 3.9
	Si 1.8	P 2.1	S 2.6	Cl 3.0
		As 2.0	Se 2.4	Br 2.8
		Sb 1.9	Te 2.1	I 2.5

The concept of electronegativity is important because so many of the chemical properties of the elements can be systematized in terms of this quantity. For example, the strength of the bond between two atoms increases with the difference in their electronegativities. To illustrate this, the strengths of the bonds between hydrogen and the halogens are listed in Table 8-4, with the values of the electronegativity difference.

Table 8-4 Bond strength *vs* electronegativity difference for the hydrogen halides.

Bond	Bond strength (kcal/mole)	Electronegativity difference
H-F	135	1.8
H-Cl	103	0.9
H-Br	87	0.7
H-I	71	0.4

8-5 Hybridization

Carbon compounds — Covalent bonds differ from ionic bonds in that electrons are *shared* between atoms, rather than transferred from one atom to another. They also differ in having directional character. Ionic bonds depend entirely upon the electrostatic forces between unlike charges and, of themselves, have no directional characteristics. On the other hand, we find that, in covalent molecules, the atoms assume positions not only with characteristic internuclear distances but also with characteristic angles between the bonds. It is important to recognize that a satisfactory theory of chemical bonding must also provide an explanation of these bond angles. The problem is well illustrated by the compounds of carbon. It is known that in methane (CH_4), carbon tetrachloride (CCl_4), and similar molecules, the four bonds from the carbon are all equivalent, and that the molecules possess tetrahedral symmetry.* It is necessary to reconcile this fact with the chemical behavior that we expect of carbon based on its electron configuration:

	1*s*	2*s*	2*p*
C	⇅	⇅	↑ ↑

First of all, it is clear that, if carbon were to form compounds while in this electronic configuration, it should only be divalent, since there are only two unshared electrons available for bond formation. Furthermore, it would not then have a rare gas configuration. On the other hand, if one of the two 2*s* electrons were promoted to the 2*p* orbital, there would be four electrons available for bond formation:

	1*s*	2*s*	2*p*
C	⇅	↑	↑ ↑ ↑

Since the 2*p* orbital is a higher energy orbital than the 2*s*, energy is required to accomplish this. However, this energy is more than recovered by the increased bond-forming power of carbon.

There remains a further difficulty. Although the promotion of the 2*s* electron has provided a means by which the carbon can exhibit a valence of four, the four bonds are not all equivalent. One bond involves a 2*s* orbital on the carbon, while the other three involve 2*p* orbitals. The equivalence of these four bonds is explained in terms of the concept of *hybridization*. From

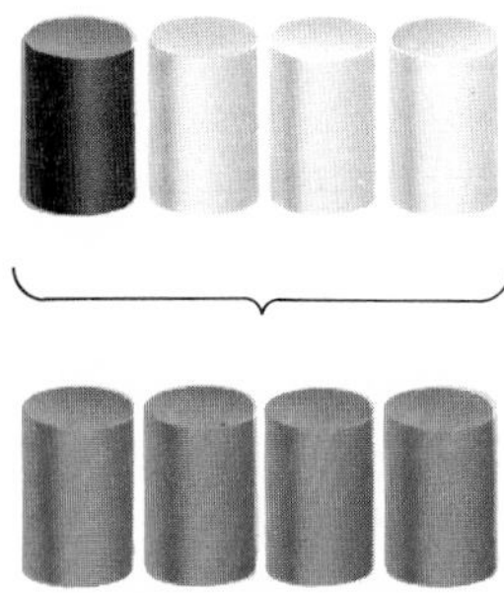

Fig. 8-11 Hybridization analogy.

*The tetrahedral symmetry of carbon in compounds as this was inferred many years ago by van't Hoff and Le Bel. It has been verified many times by a variety of modern experimental techniques.

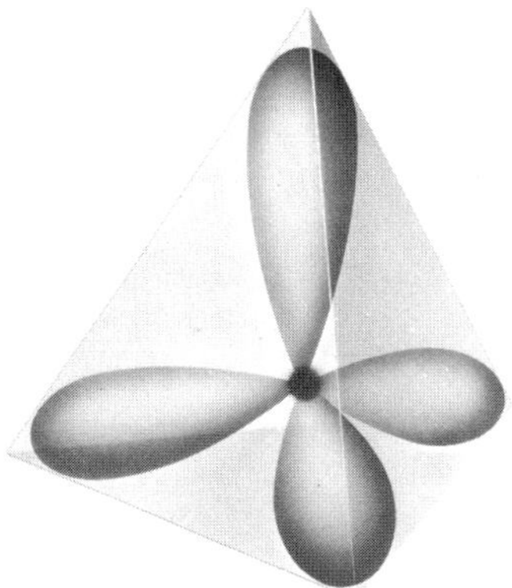

Fig. 8-12 Approximate electron distribution for tetrahedral hybrid orbitals.

the drawings of the orbitals given in Chap. 5, it is clear that neither the s nor the p orbitals are directed toward the corners of a tetrahedron. But it is found that if one s and three p orbitals are "mixed," a set of four equivalent orbitals is obtained. Each of these orbitals is 75 per cent p orbital and 25 percent s orbital. A rather crude analogy to the process of hybridization is shown in Fig. 8-11, in which one can of black and three cans of white paint are mixed to give four equivalent cans of gray paint. It should be remembered that the atomic orbitals can be expressed mathematically. The idea of hybridization is derived from the mixing of mathematical expressions.

It develops that the four equivalent orbitals which result from the hybridization are directed toward the four corners of a tetrahedron. These four tetrahedrally directed hybrid orbitals are designated as sp^3. The orbitals are represented in Fig. 8-12.

We have proceeded in a stepwise fashion in showing the hybridization of the carbon atom. The steps are shown in Fig. 8-13. It must not be imagined, however, that the atom goes through such a stepwise procedure to reach the

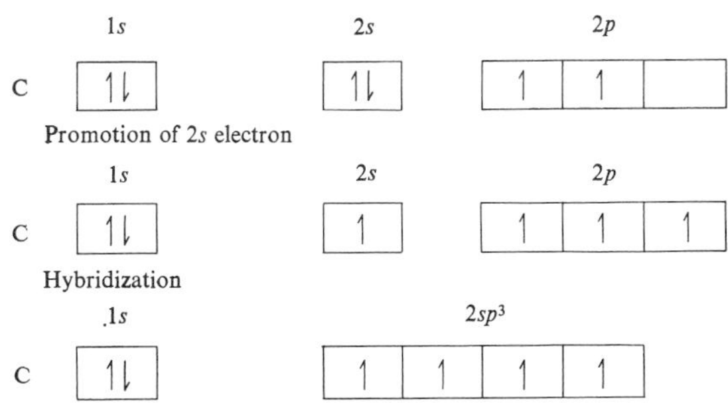

Fig. 8-13 Stepwise formation of the tetrahedral valency state of carbon.

condition in which it is found in a compound. The fact is that carbon is tetrahedrally surrounded by bonding atoms in many of its covalent compounds. We are attempting to describe the bonding in the compounds in terms of combinations of orbitals which are appropriate to the isolated atoms. While the results of doing this are generally quite encouraging, it must not be forgotten that this is an approximation. Hybridization is a means of adjusting the basic set of atomic orbitals which we use so as to obtain the most highly localized bond description possible. By means of hybridization, it is possible to obtain sets of hybrid orbitals which point in the directions of bonding atoms, so that the bonding electron pair may be thought of as localized in the region between the atoms.

The other elements of Group IV, silicon, germanium, tin and lead, also exhibit tetrahedral symmetry in their four-covalent compounds. It may be concluded that the bonds in these atoms are also sp^3 hybrids. In silicon, these are formed from the 3*s* and 3*p* orbitals, and in germanium, from the 4*s* and 4*p* oribtals, *etc.*

Group V elements — The Group V elements are normally three-covalent. In the nitrogen atom, which contains five electrons in the valence shell, the electronic configuration is

	1*s*	2*s*	2*p*
N	↑↓	↑↓	↑ | ↑ | ↑

It is clear that if nitrogen forms three bonds while in this electronic configuration, the three bonds will be equivalent. Since the *p* orbitals are directed along axes at right angles, the three bonds in ammonia, NH_3, should make 90° angles with one another (Fig. 8-14). On the other hand, we can imagine the 2*s* and 2*p* orbitals in nitrogen to be hybridized. This process is most easily conceived of by temporarily removing the five electrons in the valence

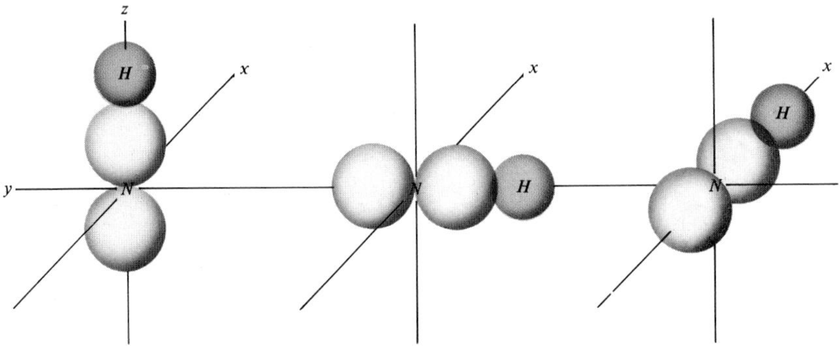

Fig. 8-14 Formation of N-H bonds in ammonia from 2_p atomic orbitals on N.

shell, allowing the orbitals to hybridize, and then replacing the electrons one at a time:

	$1s$	$2s$	$2p$
N	↑↓	↑↓	↑ | ↑ | ↑

	$1s$	$2sp^3$
N	↑↓	| | |

	$1s$	$2sp^3$
N	↑↓	↑↓ | ↑ | ↑ | ↑

The electrons are placed in the orbitals singly until all of the orbitals contain at least one electron, and then are placed in the orbitals with paired spins. The result is that one of the hybrid orbitals contains a pair of unshared electrons, while the other three are singly occupied and available for formation of bonds. It is clear that, on this basis, the nitrogen is still three-covalent. The difference between having the orbitals hybridized or not hybridized lies in the bond angles. The nybrid orbitals are directed toward the corners of a tetrahedron, and thus make angles of about 110° with one another, in contrast with the 90° angles expected for pure p orbitals. A second difference lies in the disposition of the unshared pair of electrons. If the orbitals are hybridized, this unshared pair is located in an sp^3 hybrid orbital which is directed in space toward one of the corners of the tetrahedron. If the orbitals are not hybridized, the unshared pair occupies the $2s$ orbital, which is spherically symmetrical about the nucleus. These alternative models for the ammonia molecule are illustrated in Fig. 8-15. There is no way at present of directly determining the location of non-bonding electrons such as the unshared pair in ammonia. The HNH bond angle in this compound has been determined, however. It has the value 106°45′, which is close to that which would be expected for a perfect tetrahedron in ammonia,

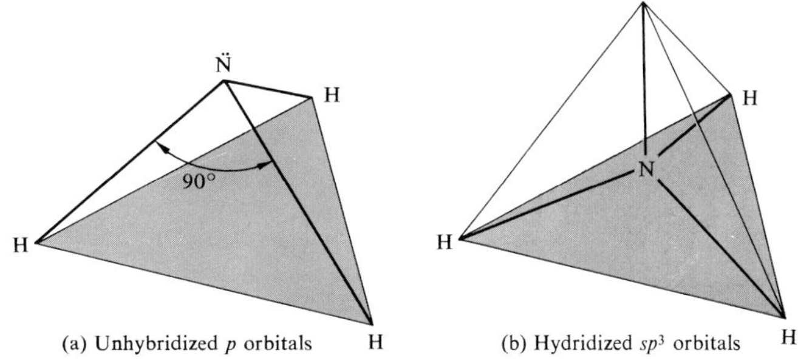

Fig. 8-15 Alternative models for ammonia: (a) unhybridized p orbitals; (b) hybridized sp^3 orbitals.

even if the orbitals are hybridized, since the groups located at the four corners are not all the same.

One further bit of evidence supports the notion that the nitrogen orbitals are hybridized in ammonia. It is possible to form the ammonium (NH_4^+) ion (we shall see later under what conditions it forms) by the addition of a proton to the ammonia molecule. This proton can be pictured as attaching itself to the unshared pair of electrons on the ammonia molecule as follows:

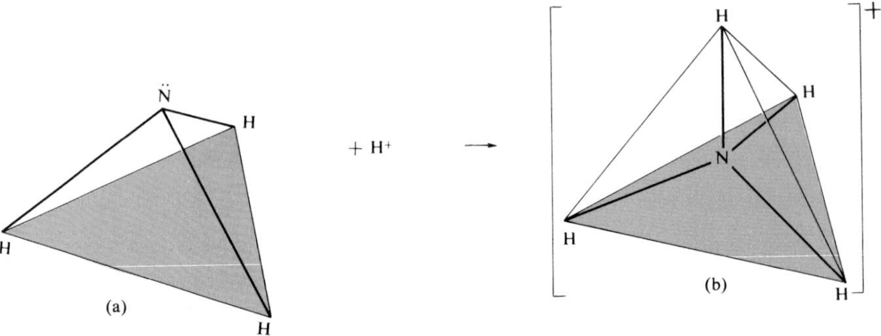

The ammonium ion is isoelectronic with methane; the only difference between NH_4^+ and CH_4 is the added nuclear charge of one on the central nucleus in NH_4^+. It is not surprising, therefore, to find that the ammonium ion is tetrahedral, with four hydrogens equivalent. This is not proof that the orbitals are hybridized in NH_3; they may *become* hybridized in the process of forming the NH_4^+ ion. When the behavior of nitrogen is compared with that of the other Group V elements, however, it is good evidence for hybridization.

The bond angles in PH_3 are 93°, considerably closer to the 90° we expect for pure *p* orbitals than we observed in ammonia. Furthermore, PH_3 is a very poor proton acceptor; *i.e.*, the PH_4^+ ion is highly unstable relative to NH_4^+. This is in agreement with the idea that the bonds in phosphorus are not hybridized, since the unshared electron pair in an *s* orbital would not be expected to attract a proton very strongly. Similar considerations apply for the other Group V elements.

Group VI elements — Turning to the Group VI elements, we find it convenient to consider the 2*s* and 2*p* orbitals in oxygen as being hybridized:

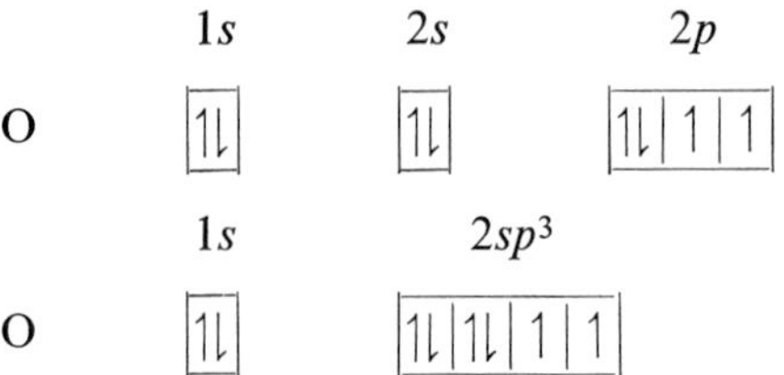

We see that on this basis the water molecule, H_2O, should exhibit an HOH bond angle near $109\frac{1}{2}°$, the tetrahedral angle; it is found to be $104\frac{1}{2}°$. Furthermore, the two unshared electron pairs would then be in sp^3 hybrid orbitals.

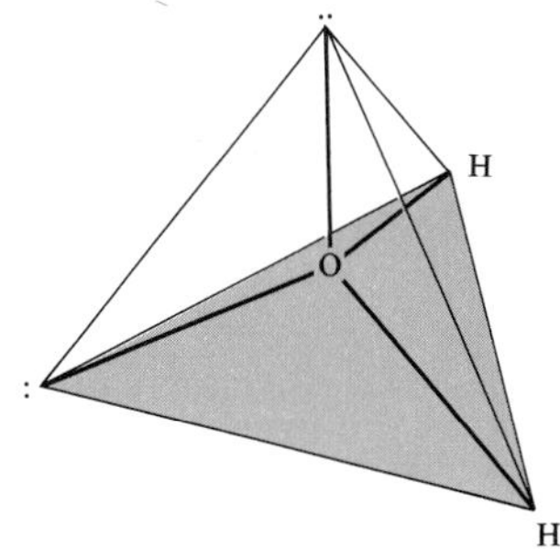

Fig. 8-16 Structure of the water molecule, assuming approximately sp^3 hybridization of the oxygen orbitals.

As in ammonia, we might expect that a proton could become attached to one of these to form the hydronium (H_3O^+) ion. Under the proper conditions, this ion does indeed form very readily. It is reasonable to consider, therefore, that the water molecule has the structure shown in Fig. 8-16.

The H—S—H bond angle in H_2S is found to be 95°, close to the angle expected for pure *p* orbitals. Furthermore, the H_2S molecule shows essentially no tendency to attach a proton to form H_3S^+ under any conditions. The other Group VI elements behave in the same way as does sulfur in these respects.

Since the Group VII elements form only a single covalent bond, there is no way to determine the probable state of hybridization from bond angles. It can be noted, however, that HF readily accepts a proton to form the H_2F^+ ion, a property of covalent hydrides which we have so far identified with the presence of hybridized orbitals. None of the other hydrogen halides (Group VII hydrides) exhibit this property.

While the concept of hybridization is well established, its application in many instances is still a matter for disagreement among chemists. A number of other factors are of importance in determining bond angles (repulsions between atoms not bonded to one another, repulsions between electron pairs, *etc.*), and the interpretation of molecular structures in terms of hybridization alone is not adequate.

In the methane molecule, the carbon atom has a pair of electrons in each of the valence shell orbitals. Each of these pairs is shared with a hydrogen nucleus, and all four pairs are equivalent. It follows from the Pauli exclusion principle that each electron pair should exert a repulsive influence on the other pairs. They are therefore oriented as far from one another in space as possible; *i.e.*, they assume the tetrahedral arrangement. There are four electron pairs about the nitrogen atom in ammonia, just as there are four pairs about the carbon atom in methane. The difference between the two cases is that only three of the electron pairs in ammonia are shared with hydrogen nuclei. The repulsive effect exerted by the unshared pair is greater than that exerted by one of the shared pairs, because the electronic charges of the latter are partially compensated for by the hydrogen nuclei. The result is that the shared pairs are pushed together somewhat. The H—N—H bond angle is therefore slightly less than the tetrahedral angle. A similar explanation applies to the H—O—H bond angle in water.

The magnitude of the change in bond angle which results from the greater repulsion of the unshared pair (or pairs) is determined by how much difference exists between the repulsive effects of unshared pairs, as compared with shared pairs, of electrons. In nitrogen and oxygen compounds the difference is usually small, because the central atom is strongly electronegative, and pulls the shared pairs in closely. The compensating effect of the atoms bonded to the central atom is, therefore, not very strong. On the other hand, in compounds of the less electronegative elements such as phosphorus and sulfur, the shared electron pairs are further from the central atom. They are, therefore, more easily pushed in toward one another by the unshared pair (or pairs) of electrons, and a smaller bond angle results.

Considerations of this kind can be applied to molecules in which the central atom contains in its valence shell more than eight electrons; *e.g.*, to the bond angles in BrF_5 or ClF_3 (Chap. 19). The rule in all cases is simply that an unshared electron pair exerts a greater repulsion than does a shared electron pair on the other electron pairs in the valence shell. It follows from this rule that the most symmetrical structures will result when all of the electron pairs about the central atom are equivalent. Thus, we find tetrahedral angles in all of the Group IV compounds in which the central atom exhibits a valence of four. Species such as the PO_4^{-3}, SO_4^{-2}, ClO_4^-, and NH_4^+ ions are also tetrahedral in symmetry.

8-6 Multiple Bonds

The electronic structures of many covalent compounds can be formulated adequately only if it is assumed that there are multiple bonds between certain atoms. Consider, for example, ethylene, C_2H_4. The number of valence shell electrons which must be placed in formulating the Lewis structure for this substance is twelve (four from each carbon, one from each hydrogen). It is evident that the four hydrogens must be bonded to the carbons, and that there is a pair of electrons to be shared in each of these bonds. This leaves us with four electrons which must be shared between the two carbons. If all four hydrogens were bonded to one carbon, methane, and not ethylene, would be the result. If three were bonded to one carbon, and one to another, the following Lewis structure

```
   H  H
   |  |
H—C—C:
   |
   H
```

would result. This is an unsatisfactory structure, since the octet rule is not obeyed for one of the carbons. We are left then with

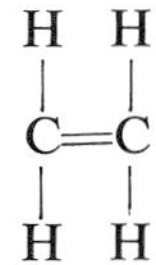

as the only reasonable Lewis structure. This does have a double bond between the two carbons.

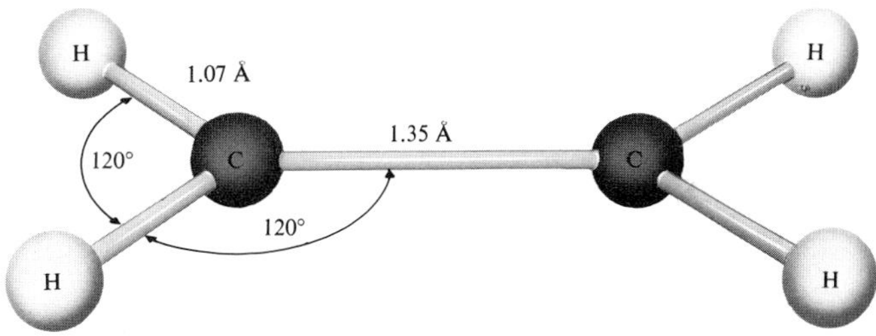

Fig. 8-17 The geometrical structure of ethylene.

The geometrical structure of ethylene is shown in Fig. 8-17. The nuclei lie in a plane. The H—C—H bond angle is 120°. These details can be explained in terms of a model for the bonding which is similar to that employed in describing the multiple bonds between atoms in homonuclear diatomic molecules. In the present case, however, we must also take account of the bonds from carbon to hydrogen. If we consider just the arrangement of bonding atoms around each carbon, we see that there are three atoms lying in a plane, with 120° angles between the internuclear axes. We now ask what hybridization scheme we would need to obtain three orbitals on a carbon atom which are directed in this manner. The mixing together of an s and two p orbitals provides three hybrid orbitals of this type. We may begin our picture of the ethylene structure, then, by assuming that the bonds between the carbon atoms are formed from the sp^2 hybrid orbitals. Using the box diagram representation, we have

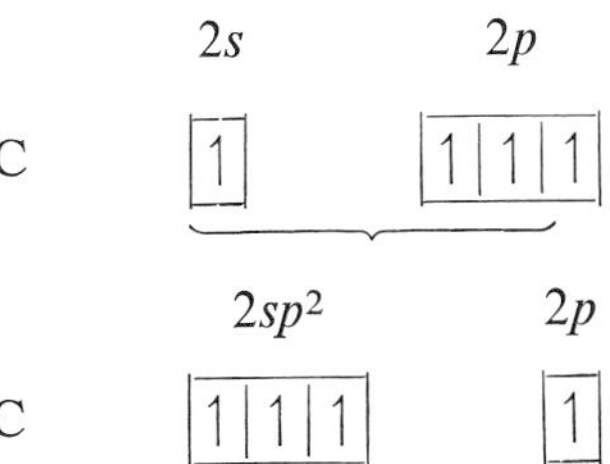

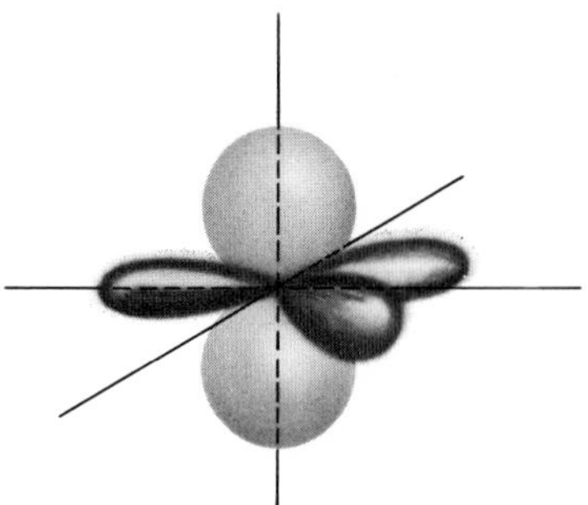

Fig. 8-18 Spatial representation of $sp^2 + p$ orbital set.

Each carbon atom then employs three of its valence shell electrons in forming three single electron pair bonds to the three neighboring atoms. Each carbon has in addition one valence shell orbital, the $2p$ orbital, which contains the remaining valence electron. The $2p$ orbital is perpendicular to

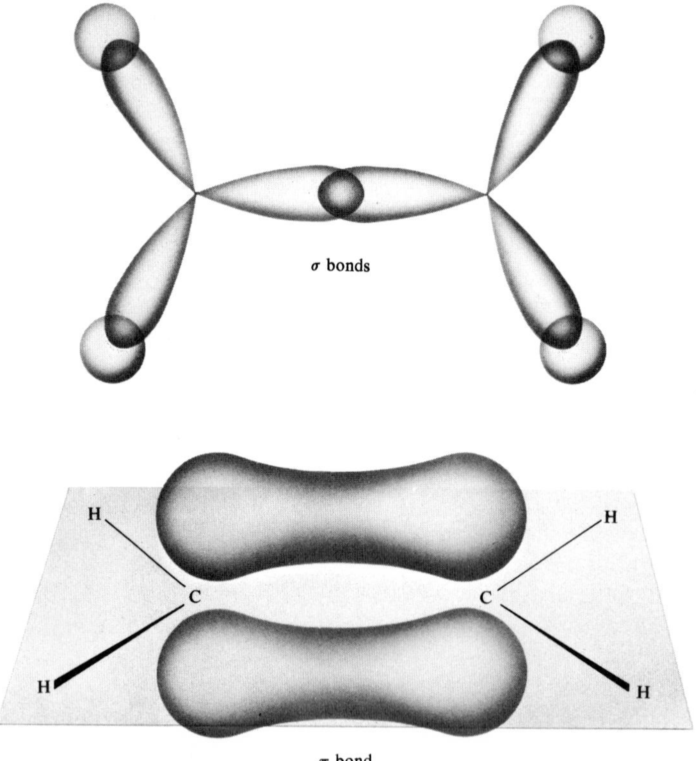

Fig. 8-19 The σ and π bonding structure in ethylene.

the plane formed by the three sp^2 hybrid orbitals, as shown in Fig. 8-18. The $2p$ orbitals on the carbon atoms overlap with one another to form a bond of π symmetry. The double bond between the carbon atoms in ethylene is thus composed of a σ type bond, formed by an sp^2 hybrid orbital of each carbon, and a π type, formed from a $2p$ orbital of each carbon. The bonding structure is shown schematically in Fig. 8-19.

Acetylene, C_2H_2, furnishes yet another example of a covalent compound in which it is necessary to postulate multiple bonding. In this case, a triple bond is required between the carbons, as indicated by the following Lewis structure:

$$H{-}C{\equiv}C{-}H$$

Acetylene is a linear molecule. Each carbon is therefore bonded to two other atoms, with a 180° angle between the internuclear axes. Formation of the sp hybrid leads to a pair of orbitals which are directed along an axis at 180° to one another, with two p orbitals perpendicular to this axis. The triple bond in acetylene, shown in Fig. 8-20, is thus very similar to that in N_2. The carbon-carbon bond of acetylene is quite a strong bond, as is the bond in N_2.

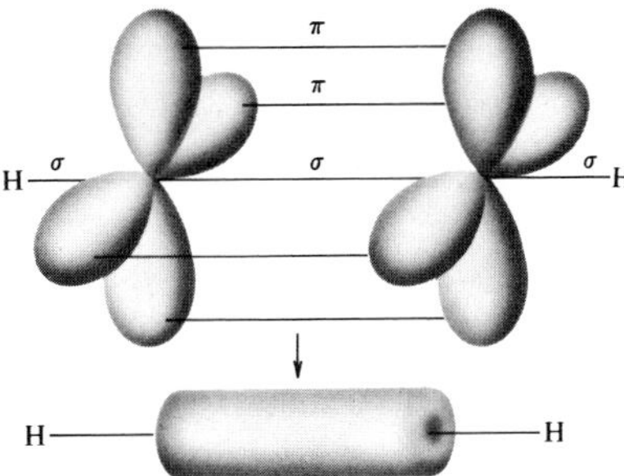

Fig. 8-20 Schematic representation of acetylene orbitals, showing the manner in which the π bond is formed.

The formation of sp hybrid orbitals is a particularly simple example of hybridization, since only two orbtials are involved. It is possible to see how the two hybrid orbitals arise by considering the ways in which we might combine an s and a p orbital. (We will avoid talking specifically about $2s$ and $2p$ orbitals, since the principle is the same regardless of the major quantum number.) Let us denote the hybrid orbital by the symbol ψ_{HY}. There are just two ways in which the s and p orbitals can be combined; with the same sign, or with opposite sign. Thus, we have

$$\psi_{HY_1} = \psi_s + \psi_p; \qquad \psi_{HY_2} = \psi_s - \psi_p$$

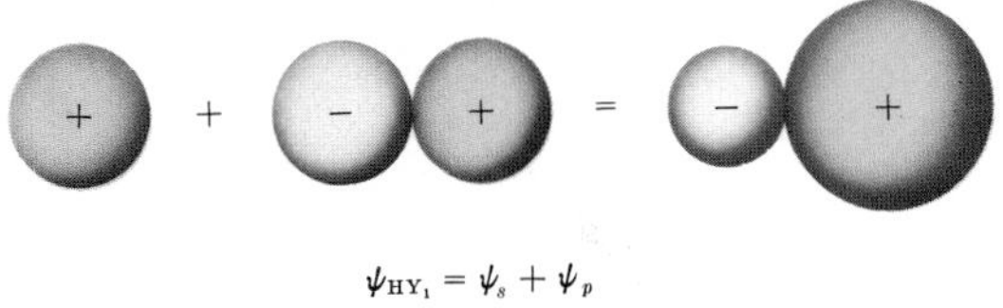

$$\psi_{HY_1} = \psi_s + \psi_p$$

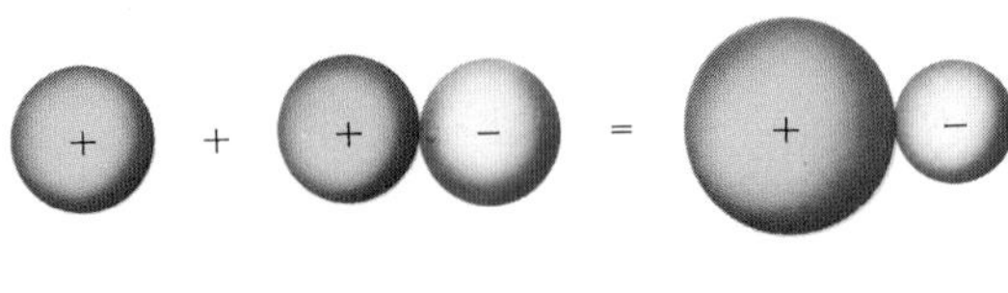

$$\psi_{HY_2} = \psi_s - \psi_p$$

Fig. 8-21 Mixing of *s* and *p* orbital to form two *sp* hybrid orbitals.

The result of carrying out this mixing is illustrated in Fig. 8-21. The hybrid orbital has a large value for the wave function in that region of space where the two orbitals being mixed have the same sign, and a low value in the region where they have an opposite sign. The *s* orbital has the same sign everywhere, but the *p* orbital has a nodal plane, with opposite signs on each side. The result of the two possible combinations is thus a pair of hybrid orbitals that have regions of high electron density directed in opposite directions.

Similar considerations apply to formulation of other types of hybrids. It is not quite as easy to see just how the mixing occurs when more than two orbitals are involved. In addition to the hybrids formed between *s* and *p* orbitals, others involving *d* orbitals are also possible, and are frequently considered in explaining certain bonding situations. We will discuss some of these others shortly.

Multiple bonding may occur also between unlike atoms, as in the examples shown in Fig. 8-22. In all of these examples, it is possible to formulate

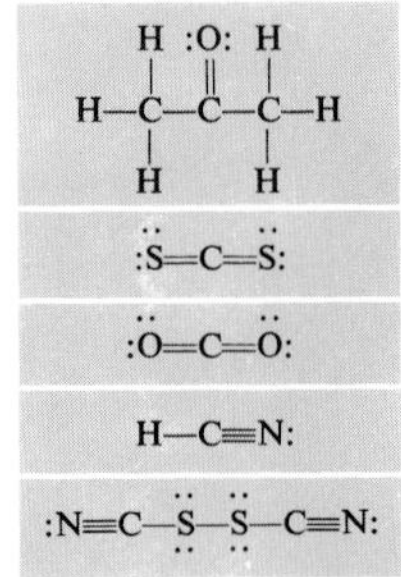

Fig. 8-22 Examples of Lewis structures of molecules containing multiple bonds.

the multiple bond in terms of a σ component and a π component, as discussed above. The bonding between sulfur and carbon represents an example of multiple bonding between atoms from different rows as well as different groups of the periodic table. The π bonding is formed from overlap of a carbon $2p$ orbital with a sulfur $3p$ orbital. The fact that the major quantum numbers of the orbitals are different results in a decrease in the strength of the π bond, because of a lower degree of overlap of the orbitals and because the energy matching is not as good. The symmetry properties are the same, however, as for a C═O or C═C bond.

Because more than one electron pair is shared between the two nuclei, multiple bonds are characterized by higher bond energies than single bonds. Furthermore, there is a shortening of the distance between atoms in multiple bonds as compared with single bonds. Table 8-5 lists some typical bond distances and bond energies for carbon and nitrogen bonds. The values listed are simply typical values, since the quantities involved vary over a narrow range with the particular molecular environment in which the bond is formed (except for N_2, which is unique).

Table 8-5 Values of bond length and thermochemical bond energies. (After T. L. Cottrell, *Strengths of Chemical Bonds*, Second Edition, Butterworths, London, 1958.)

Bond	Bond length (Å)	Thermochemical bond energy (kcal/mole)
C—C	1.54	83
C═C	1.35	146
C≡C	1.21	200
N—N	1.47	40
N═N	1.24	100
N≡N	1.09	226
C—N	1.47	73
C═N	—	147
C≡N	1.15	213

8-7 Delocalized Orbitals

In the discussion of covalent bonding so far, we have assumed that the bonding can be adequately described in terms of orbitals localized between two atoms. There are many molecules and ions, however, for which this assumption is not satisfactory for at least a part of the bonding. The classic example of delocalized bonding orbitals is benzene, C_6H_6. The benzene molecule is planar, with the carbon atoms arranged at the corners of a hexagon. Each carbon is bonded to two other carbons, and to one hydrogen. The three bonding atoms thus surround each carbon in a plane, with 120° angles between the internuclear axes. This bonding situation requires sp^2 hybrid

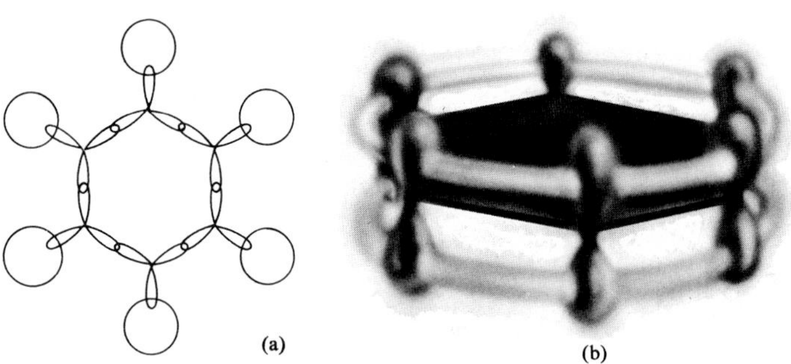

Fig. 8-23 Two views of the bonding in benzene. The view in (a), looking down on the plane of the molecule, shows the σ bonding. The view in (b) shows the π bonding. The hydrogen atoms have been omitted from (b).

orbitals. The bonding in the molecular plane is therefore as shown in Fig. 8-23(a). In addition to this σ bonding framework, there is a π bonding system, formed from the $2p$ atomic orbital perpendicular to the molecular plane possessed by each carbon atom, with a single electron still to be employed in bonding. The π bond which is formed from these $2p$ orbitals differs from the bond in ethylene, however, in a very important respect: Each carbon orbital interacts with *two* other carbon orbitals, not just one. The result is that a ring of π orbitals is formed, as illustrated in Fig. 8-23(b). The overlap of each carbon $2p$ orbital with two neighbors results in a delocalization of the electrons which occupy the π molecular orbitals. In the case of benzene, the electrons are essentially free to move about the entire circle of orbitals. Since each carbon contributes one electron to the π system,

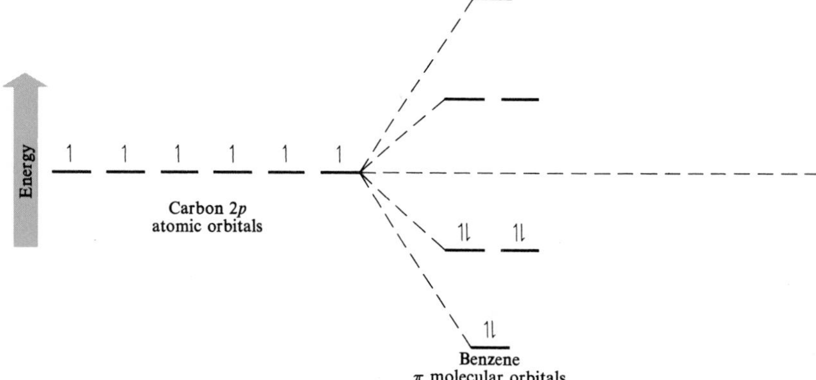

Fig. 8-24 Energy level diagram for the π orbitals of benzene.

there are three electron pairs in the π molecular orbitals. On the average, then, there is one-half of an electron pair between any two carbon atoms. Because of the electron-electron repulsions, the six electrons tend to be equally distributed over the perimeter formed by the $2p$ orbitals.

The bringing together of the six carbon $2p$ orbitals to form the delocalized π molecular orbitals is merely a somewhat more complex version of the formation of molecular orbitals between just two orbitals, as discussed in Sec. 8-2. An energy level diagram for the benzene π orbitals has been worked out; it is shown in Fig. 8-24. The six carbon $2p$ orbitals, all of the same energy, are linearly combined to give six π molecular orbitals. Note that in two cases, the molecular orbitals formed are degenerate; *i.e.*, they have the same energy. Since there are six electrons involved, there are three electron pairs to be placed in the π molecular orbitals. These three pairs just fill the bonding orbitals, leaving the anti-bonding orbitals vacant.

In summary, then, the molecular orbital model for the bonding in the benzene molecule consists of a σ bonding framework which lies in the molecular plane, and a π bonding system which is made up of carbon $2p$ orbitals perpendicular to the molecular plane. The σ bonds are all localized, in the sense that each bonding electron pair is concentrated in the region between the nuclei. The π bonds, however, are delocalized; each π bonding electron pair is distributed over the entire system of six carbon $2p$ orbitals. On the average, this distribution is equal among all of the bonds, so that all carbon-carbon bonds are equivalent. Each carbon-carbon bond is essentially a full σ bond, and one-half of a π bond. The observed C—C bond length in benzene is 1.40 Å, intermediate in length between a C—C single bond, 1.54 Å, and a C=C double bond, 1.35 Å. Benzene is an extraordinarily stable molecule, in part because of the extra stability obtained through the delocalization of the π electrons. Delocalization leads to an extra stability because it provides a means for lessening electron-electron repulsions.

Delocalization of π electrons is seen in many other molecules and ions. The same sort of bonding scheme as employed for benzene serves to explain the structures of these species as well. It is, however, not always so easy to visualize the π bonding. There is an alternative approach to the structures of such species, called *resonance theory*, which is quite easy to visualize, and which serves as a qualitative indication of stabilization through delocalization.

8-8 Resonance

The Lewis structures which we write for molecules must always be in accord with the facts of chemistry. Thus, we found that to represent the methane molecule it was necessary to write the four C—H bonds as being equivalent. We are able to understand the structure in terms of the concept of hybridization, which involves an extension of our previous notions about atomic orbitals. In a similar way, the notion of a localized electron pair bond must be broadened to include the delocalized π bond in benzene, in order to satisfactorily account for the known geometry. In attempting to write Lewis structures for benzene, we find that there are two very obvious ones which are equivalent except for placement of the double bonds:

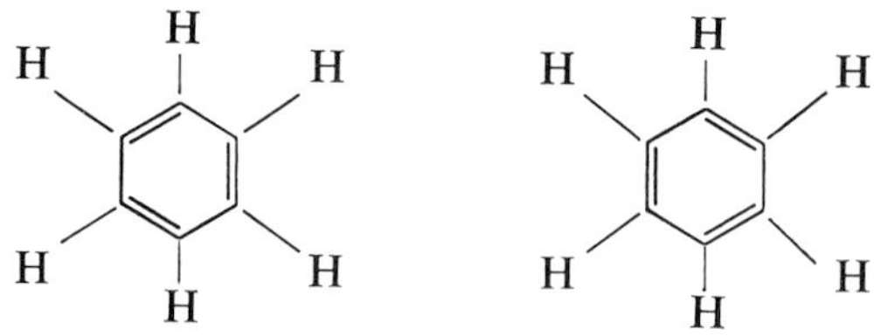

Neither of these alone can be a correct representation for benzene, because there would have to be alternating single and double bonds, with consequent variation in carbon-carbon bond distances and bond energies, whereas benzene is known to have perfect hexagonal symmetry. We can, however, describe the benzene molecule by saying that it is an average of the two equivalent structures shown. The two forms shown are called *canonical structures*. The real molecule is a hybrid mixture of these two. It is as though the molecule raced back and forth, or resonated, between the two extreme forms with great rapidity, so that all we observe is the average of the two.

It is important to realize at the outset that it is not correct to think of the molecule as continually changing back and forth among all of the possible forms as represented by the canonical structures. There is no experimental evidence that this is the case. Rather, the concept of canonical forms (called *resonance theory*) is an artifice that we use to talk about a molecule for which we cannot adequately write one single structure. We can only say that the molecule behaves *as though* it were an average of two or more ordinary Lewis structures.

The two canonical structures shown for benzene are equivalent, differing only in the placement of the double bonds. It is possible, however, to write Lewis structures representing other sets of equivalent canonical forms, such as the following three equivalent forms for benzene, which are sometimes referred to as the Dewar structures:

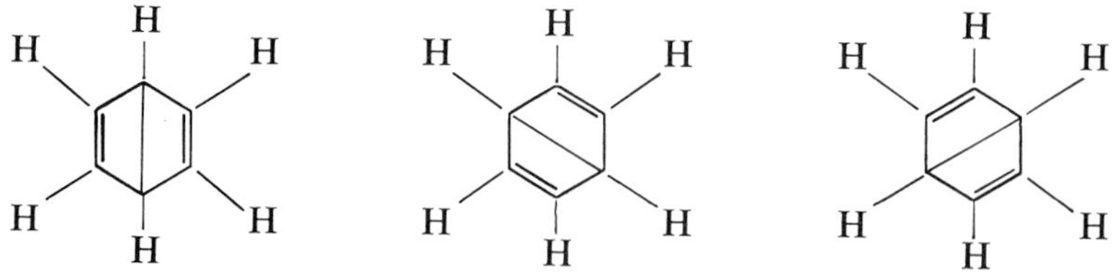

These structures are not as stable as the two already discussed, since they involve a very long covalent bond. They therefore would not be expected to make a very important contribution to the actual structure of the molecule. In general, the more stable a particular structure is, the greater is its contribution to the observed properties of the compound. For this reason, we need usually concern ourselves only with those canonical structures which

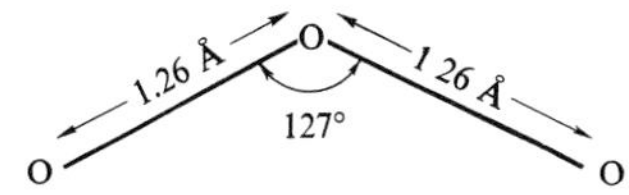

Fig. 8-25 Structure of ozone, O_3.

are energetically most stable, and which therefore make the major contributions to the observed structure.

As a second example of resonance, consider the ozone molecule, O_3. It is known that this molecule has the geometry shown in Fig. 8-25. From the distances separating the atoms it is clear that the two terminal oxygens are not bonded to one another. The other interesting point is that the two O—O distances are equal. Now, when one attempts to write Lewis structures for this molecule, the following two are obtained:

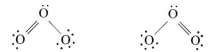

Neither one of these by itself can be the correct structure, since in each case one of the O—O bonds is double, the other single. This would require that the two O—O distances be different, which is contrary to what is found. It is possible to describe the O_3 molecule by saying that its structure is what one would expect for the average of the two shown. It is interesting that the O—O distance observed is about midway between the values expected for single and double O—O bonds. This is consistent with the requirement that each O—O bond is a double bond in one of the structures and a single bond in the other. As further examples of resonance, the important canonical forms for sulfur dioxide, acetate ion, and nitrate ion are shown in Fig. 8-26.

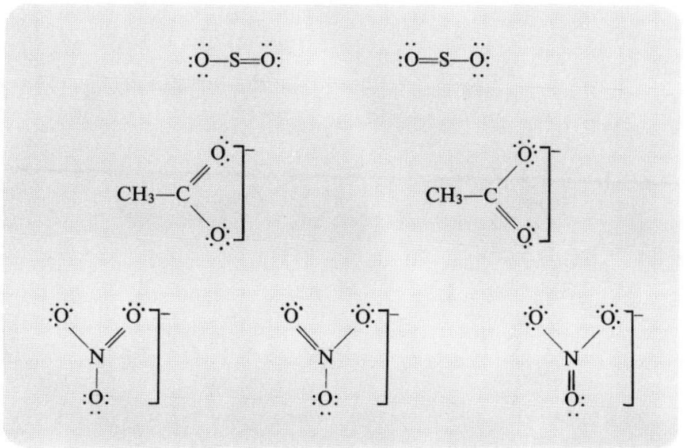

Fig. 8-26 Canonical forms of some molecules and ions.

The idea that a molecule can be described in terms of more than one ordinary Lewis structure is generally referred to as resonance theory. The theoretical basis of resonance theory is fundamentally the same as that of molecular orbital theory; resonance leads to a stabilization of the molecule by providing a means for delocalization of electronic charge over a larger number of atoms. For qualitative discussions of molecular structure, it is frequently very helpful to use the language of resonance theory. Whenever it is possible to describe the structure of a molecule or ion in terms of two or more equivalent (or nearly equivalent) Lewis structures, the conditions for resonance stabilization are present, and an extra stability can be anticipated. In order to arrive at this same conclusion from molecular orbital theory, one might have to employ a considerably more sophisticated approach.

8-9 Expanded Valence Shells

There are a number of compounds in which it is quite clear that one of the atoms possesses more than eight electrons in the valence shell. Sulfur hexafluoride, SF_6, and phosphorus pentachloride, PCl_5, are examples. These cases are exceptions to the valence rules given in the Lewis-Langmuir theory.

It is possible to understand the valences of the central atom, and also the geometries of these compounds, by once again invoking the concept of hybridization. Consider the electronic structure of sulfur:

	$3s$	$3p$	$3d$	$4s$
S [Ne]	⇅	⇅ | ↑ | ↑	| | | |	

We have already learned that the $3d$ orbital is unoccupied in this atom and, indeed, that in potassium the $4s$ orbital is occupied before the $3d$ orbital. Nevertheless, to increase the bond-forming ability of the atom under certain

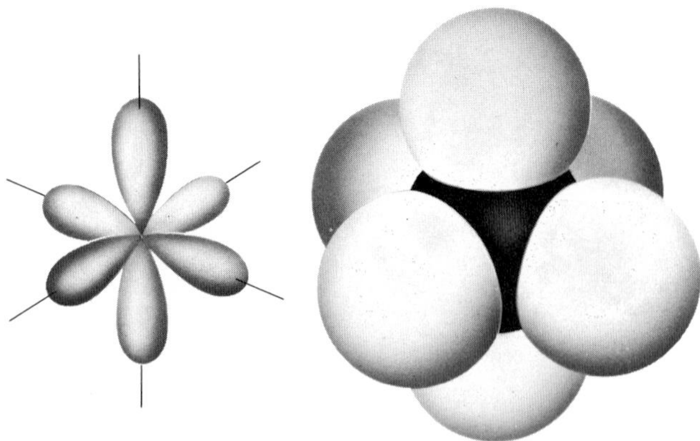

Fig. 8-27 sp^3d^2 hybrid orbitals, and the structure of SF_6.

conditions, two of the $3d$ orbitals may form a set of hybrid orbitals with the $3s$ and $3p$ orbitals:

$3sp^3d^2$

S [Ne] | 1 | 1 | 1 | 1 | 1 | 1 |

The result of this hybridization is a set of six equivalent orbitals which are directed towards the corners of an octahedron. It is these six orbitals which are used in forming the six bonds to fluorine in SF_6 (Fig. 8-27). This kind of hybridization does not always occur. (There is, for example, no SCl_6.) Generally speaking, it occurs when the bond to be formed is particularly strong, as are the sulfur-fluorine bonds in SF_6. Other examples of molecules or ions in which such bonds are formed are SeF_6, TiF_6^{-2}, and $SnCl_6^{-2}$.

The five bonds used by phosphorus in the compound PCl_5 are also hybridized, as follows:

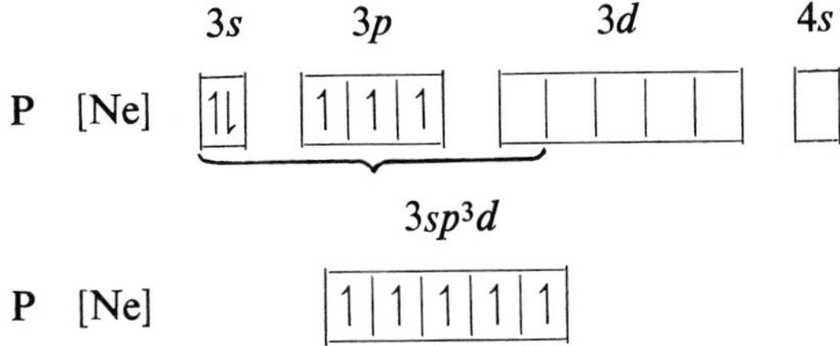

This particular set of orbitals possesses the directional characteristics shown in Fig. 8-28. Note that in sp^3d bonding the central atom has ten electrons in the valence shell, whereas in sp^3d^2 bonding there are twelve.

8-10 The Dative Bond

In the single covalent bonds which have been discussed up to this point, it has been possible to think of the bond as being formed from a sharing of a

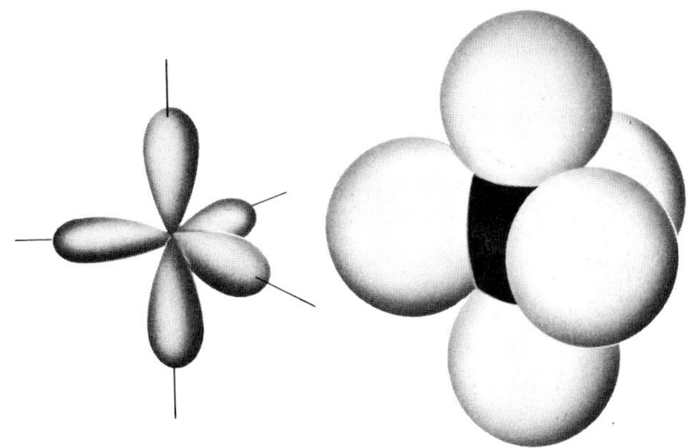

Fig. 8-28 sp^3d orbitals, and the structure of PCl_5.

pair of electrons, with each atom contributing one electron. There is no particular reason, however, why the two electrons involved could not have been associated originally with only one of the atoms. To illustrate the point, consider the compound $POCl_3$, phosphorus oxytrichloride, and compare its structure with that of PCl_3.

```
                      ..
                     :O:
       ..             |
       P              P
      /|\            /|\
    Cl Cl Cl       Cl Cl Cl
```

In PCl_3, the phosphorus atom possesses a pair of unshared electrons; in $POCl_3$, these electrons have been "donated" to the oxygen atom. The result is that the phosphorus atom still has eight electrons in its valence shell, and oxygen — by acquiring the two electrons from phosphorus — also has eight. This type of bond, in which one of the atoms is considered to have supplied both of the shared electrons, is termed the *dative*, or *coordinate covalent*, bond. There is actually no particular reason to give this bond a special name, although the formation of such bonds is treated formally as a part of general acid-base theory (Chap. 12). The structures for covalent compounds can be written from a knowledge of the total number of electrons in the valence shells of the atoms involved, and a knowledge of how many electrons should be located in the valence shell of each atom. This number is ordinarily eight, the exceptions being hydrogen with two, and occasionally atoms with expanded valence shells, Further examples of compounds containing one or more dative bonds are afforded by the oxychlorine acids.

```
  ..                        ..
 :O:                       :O:
  |   ..               ..   |   ..
 :Cl—O—H              :O—Cl—O—H
  ..  ..               ..  ..  ..
Chlorous acid          Chloric acid

               ..
              :O:
          ..   |   ..
         :O—Cl—O—H
          ..   |   ..
              :O:
               ..
         Perchloric acid
```

The anions derived from these acids results from the removal of the proton (see Table 7-8).

8-11 Covalent Atomic Radii

A number of problems are associated with the concept of a covalent atomic radius. To illustrate the most elementary difficulty encountered, consider the molecule iodine monochloride, ICl. What are the radii of the iodine and chlorine atoms in this molecule? By one or another experimental method, the internuclear distance has been found to be 2.30 Å. The internuclear distance in Cl_2 is 1.98 Å, and in I_2 it is 2.667 Å. One-half of each of these values might be taken as the radii of the chlorine and iodine atoms, 0.99 Å and 1.33 Å, respectively. The sum of iodine and chlorine radii is then 2.32 Å, which is in fairly good agreement with the observed bond distance in ICl.

We must not immediately conclude, however, that we have chosen the correct radii for chlorine and iodine in ICl. The iodine-chlorine bond is somewhat polar, since chlorine is more electronegative than iodine. The chlorine atom therefore assumes a greater share of the bonding electron pair in ICl than it does in Cl_2. Because of this, it may have a somewhat larger radius than in Cl_2, since the added electron density around the atom — through repulsive interaction with the other electrons — would cause an increase in atomic size. Correspondingly, iodine may have a somewhat smaller radius in ICl than in I_2. If the two effects just counterbalance one another, as they well might, the observed bond distance in ICl would still be as predicted.

A second, and perhaps more serious, difficulty arises in discussing the radius of an element in a series of compounds in which it is bonded to different numbers of atoms. What of the radius of chlorine, for example, in the series of acids, HClO, $HClO_2$, $HClO_3$, and $HClO_4$? As the number of oxygen atoms about the chlorine atom increases, and electron density is drained from it, the atomic radius decreases. It is obvious, therefore, that there is no single number which can represent the radius of chlorine in all of its compounds.

Provided that the above limitations are kept in mind, the single-bond, covalent radii of elements are useful in discussing the general shapes of molecules, and in predicting bond distances in favorable cases. They are shown graphed *vs* atomic number in Fig. 6-5; Table 8-6 lists values for some of the more familiar elements.

Table 8-6 Single-bond, covalent radii.

Element	Radius (Å)	Element	Radius (Å)
Hydrogen	0.30	Arsenic	1.21
Boron	0.88	Antimony	1.45
Aluminum	1.22	Oxygen	0.66
Gallium	1.30	Sulfur	1.04
Carbon	0.77	Selenium	1.17
Silicon	1.15	Tellurium	1.37
Germanium	1.22	Fluorine	0.64
Tin	1.4	Chlorine	0.99
Nitrogen	0.70	Bromine	1.14
Phosphorus	1.10	Iodine	1.33

Professor J. C. Slater has recently made an interesting observation regarding the significance of the covalent atomic radii. Detailed calculations of the electron distributions in atoms are now becoming available, and the radial distribution of electrons in atoms is better understood. Graphs such as those in Fig. 6-4 are available for many of the elements. All of these radial distributions exhibit the shell structure shown in Fig. 6-4. It is rather remarkable that the covalent radii based on the observed single covalent bond distances are very close to the distances of maximum radial charge density in the outermost shell in the atoms. This finding is consistent with the notion that the covalent bond results from overlap of atomic orbitals. When the atoms have come together to such a distance that the outer charge density maxima in the atoms coincide, the overlap of bonding orbitals should be at a maximum.

8-12 Forces Between Covalent Molecules

The properties of covalent compounds depend to a large extent upon the magnitudes and kinds of forces which the molecules exert upon one another. Since the molecules are neutral, these attractive forces are not the powerful charge-charge interactions operative in ionic substances. They are, however, electrostatic in character.

In a diatomic molecule formed between two different atoms, the bonding electron pair is displaced toward one atom or the other. This causes the center of negative charge for the molecule to lie at a different point than does the center of positive charge, and the molecule is said to be polar. Examples of such molecules are HCl, BrCl, *etc.* The polarity of diatomic molecules increases with increasing electronegativity difference until in the limit, ionic "molecules" such as Na^+Cl^-, are formed. The polarity of a molecule which is made up of many covalent bonds is determined by the sum of all of the individual bond polarities. Often the symmetry of the molecule is such that the net sum of all bond polarities is zero, and the molecule as a whole is non-polar. CCl_4 is an example of such a molecule. Each C—Cl bond is polar, but because of the symmetrical, tetrahedral structure, the resultant of all four bond polarities is zero. On the other hand, H_2O is a polar molecule, as is NH_3 or PCl_3.

Table 8-7 Melting and boiling points of some polar and non-polar compounds.

Polar compound	Dipole moment (Debye)	Melting point (°C)	Boiling point (°C)	Non-polar compound	Melting point (°C)	Boiling point (°C)
HCl	1.08	−112	−84	CH_4	−184	−162
HBr	0.79	−88	−67	CO_2	—	−78
HI	0.38	−51	−36	Cl_2	−102	−34
H_2O	1.84	0	100	CS_2	−108	46
NH_3	1.45	−78	−33	C_6H_6	5	80
$CHCl_3$	1.86	−63	61	CCl_4	−23	76
PCl_3	0.78	−91	75	Br_2	−73	59
H_2S	0.89	−83	−62	I_2	114	183
CH_3OCH_3	1.29	−138	−24	CBr_4	48	189
ICl	1.2	27	97			
CH_3CN	3.9	−43	82			

The polarity of a molecule is expressed by the *dipole moment*, which is given by the magnitude of the charges times the distance separating them. It is symbolized by $\longrightarrow$, in which the head of the arrow points toward the negative end of the dipole. Table 8-7 lists the dipole moments of a number of common substances. The unit employed is the Debye, 10^{-18} esu-cm. An electron and a proton separated by a distance of 1 Å represent a dipole of 4.8 Debye (4.8×10^{-10} esu $\times$ 10^{-8} cm).

Dipoles are capable of exerting attractive or repulsive forces upon one another, depending upon the distance separating them and the orientation of one dipole with respect to the other. A system of dipolar molecules tends to assume an arrangement which represents the greatest attractive forces between the molecules. This tendency to assume an ordered arrangement is countered by the kinetic energy of motion of the molecules; at low temperatures the electrostatic forces prevail, and the substance is a crystalline solid. As the temperature increases, the motional kinetic energy increases; when it is great enough to overcome the electrostatic forces, the substance melts and, at a higher temperature, eventually boils.

In addition to polarity, the molecular weight is also important in determining the melting and boiling points of a compound. The list of non-polar compounds in Table 8-7 is in the order of increasing molecular weight; a corresponding, general increase in melting and boiling points is indicated. The reasons for this dependence on molecular weight have to do with a very general kind of attractive force which atoms and molecules exert on one another, called *London*, or *dispersion, forces*. These arise as a result of small, momentary polarizations of part of one molecule by part of another, nearby molecule. Interactions of this kind give rise to small dipoles of fleeting existence, which, by their actions on one another, produce the attractive forces. The magnitude of these forces increases as the number of electrons in the molecule increases. Another important property of the London forces is that they are very short range. The molecules must be close together for them to be operative.

One of the most important applications of the idea of London forces is to the noble gas elements. These elements exist as single atoms which have neither polarity nor any great tendency to react chemically. Yet it is true that these substances can be both liquefied and solidified; thus, by implication, some kind of attractive force exists between the atoms. From Table 8-8,

Table 8-8 Melting and boiling points of the rare gas elements.

Element	Atomic number	Atomic weight	Melting point (°K)	Boiling point (°K)
He	2	4	0.9	4.2
Ne	10	20	24.4	27
Ar	18	40	83.9	87
Kr	36	84	104	121
Xe	54	131	133	164
Rn	86	222	202	211

it can be seen that the melting and boiling points increase with increasing atomic weight. Theoretical calculations show that these facts can be accounted for by supposing that the attractive forces between the atoms are London forces.

The dispersion forces are seen also in non-ideal gas behavior. The magnitude of the van der Waals *a* constant, Table 3-4, can be associated with intermolecular attractive forces which, for many of the substances listed, are largely dispersion forces. The increase in *a* with atomic weight in the noble gas series provides a good example of this.

8-13 The Hydrogen Bond

Hydrogen is unique among the elements in that its valence electron is the only electron in the atom. When hydrogen is bonded to a highly electronegative atom such as nitrogen, oxygen, or fluorine which exerts a strong attraction for the bonding electron pair, the effect is very much like that of a bare proton at the end of the bond. This proton may be attracted to the unshared electrons of an atom in another molecule (Fig. 8-29); this attractive interaction is termed the *hydrogen bond*, and is represented as

$$X\text{—}H\cdots Y$$

where X is the atom to which the hydrogen is normally bonded and Y is the atom to which it is hydrogen bonded. It has been found experimentally that hydrogen bonds of significant strength are formed when the atoms X and Y are either nitrogen, oxygen, or fluorine; if either X or Y is another element (*e.g.*, chlorine), much weaker hydrogen bonds are formed.

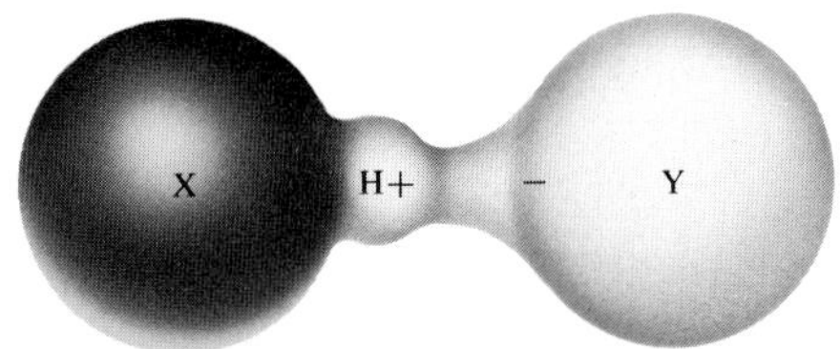

Fig. 8-29 A representation of the hydrogen bond.

The energy of the hydrogen bond is generally around 5 kcal/mole, about 5–10 per cent of the energy of an ordinary single covalent bond. For this reason, the hydrogen bond is really a secondary kind of bonding and must be considered separately. The bond energy is large enough that it is not easily broken by molecular collisions at room temperature; at higher temperatures it may be disrupted more easily as a result of a high-energy collision.

The most obvious examples of substances in which hydrogen bonding is present are the hydrides of nitrogen, oxygen, and fluorine. A comparison

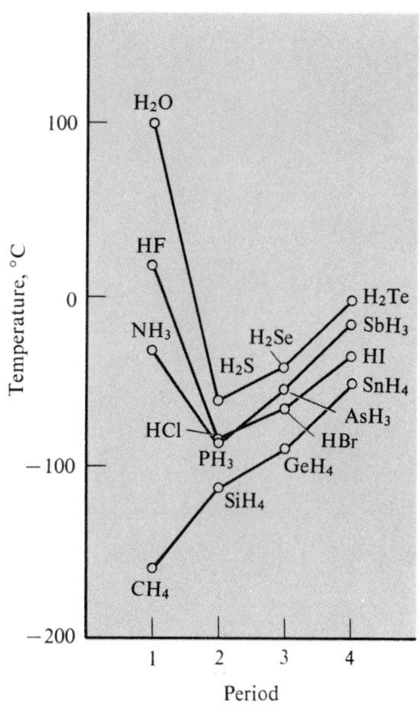

Fig. 8-30 Boiling points of Group IV, V, VI and VII hydrides.

of the melting and boiling points of the hydrides of the Group V, VI, and VII elements shows immediately that these quantities are abnormally high for H_2O, NH_3, and HF (Fig. 8-30). To understand the reasons for this, consider ice as an example. The structure of ice is known to consist of a regular three-dimensional arrangement of water molecules, with each oxygen surrounded tetrahedrally by hydrogen atoms (Fig. 8-31). Two of these are bonded to the oxygen by a normal oxygen-hydrogen bond; two are held by hydrogen bonds. To melt ice, it is necessary to increase the motional energy of the molecules sufficiently to break a large number of the hydrogen bonds; a higher than ordinary temperature is needed to do this.

In liquid water just above the freezing point, a very large number of hydrogen bonds still remain. In fact, the water has a structure which closely resembles that of ice, except for being somewhat looser. As the temperature is raised, this rather open structure continues to break down because of the rupture of more and more hydrogen bonds. It is rather interesting that in the temperature range from 0 to 4.2°C, the density *increases* with increasing temperature because of this effect. This is in contrast to the decrease in density with increasing temperature which is usually observed.

Throughout the entire range of liquid water, there is a continuous breakdown of hydrogen bonding with increasing temperature. The boiling point for water is also abnormally high, therefore, in comparison with the hydrides of the other Group VI elements. A great many other properties

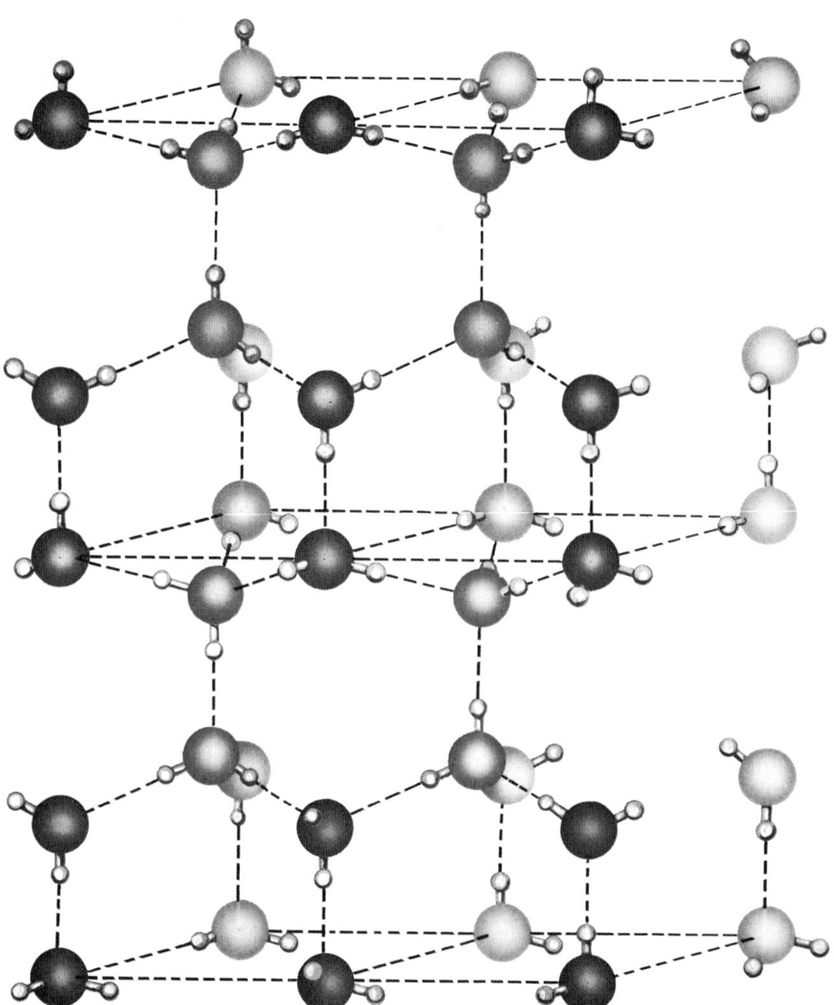

Fig. 8-31 The structure of ice.

of water are the result, directly or indirectly, of the presence of hydrogen bonds. Since water is, in a sense, "the" chemical of life as we know it on earth, it is not hard to appreciate why hydrogen bonding is viewed by chemists and biologists as being just as important as any other kind of bonding.

The "abnormal" behavior of water is paralleled by ammonia and hydrogen fluoride. These compounds have been referred to as "water-like" solvents because they are similar to water in the types of substances that they dissolve, and in their general physical behavior. Hydrogen bonding increases in this series of compounds in the following order: Ammonia, water, hydrogen fluoride.

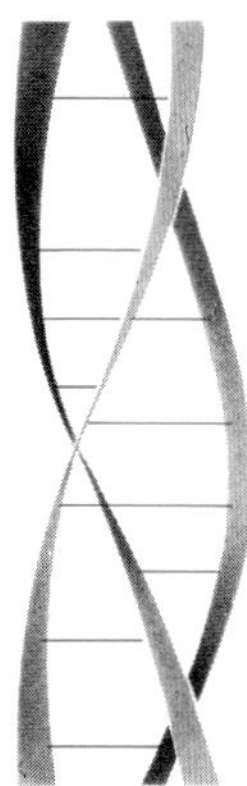

Fig. 8-32 A schematic representation of collagen, a protein molecule. The helical bands represent strands of molecules; the dotted lines represent hydrogen bonds which connect one strand with another. The molecular weight of the entire molecule is around 360,000. It is about 2800 Å in length, and 10–15 Å in width.

Many molecules found in living systems owe some of their important properties to the presence of hydrogen bonds. For example, proteins in connective tissue (collagen) are found to be very high-molecular-weight molecules (about 360,000) which are long and rather thin. They consist of strands of molecules wound around one another in a helical fashion, the strands being held together by hydrogen bonds (Fig. 8-32). This is but one example among many in which hydrogen bonding is a vital part of the chemical and physical constitution of a biochemical molecule.

At this point, it must seem that the many aspects of covalent-bond formation provide a bewilderingly large number of possibilities for compound formation. This is indeed the case. It is by no means possible for even the most experienced chemist to predict the stabilities or chemical and physical properties of most hypothetical compounds from theoretical principles. It is absolutely necessary that each chemist build up his knowledge of the properties of *known* substances by careful attention to these matters in his reading and in the laboratory. Once a thorough understanding of a large body of this descriptive matter has been attained, the chemist can, with the aid of theoretical principles, reason out the properties of hypothetical substances which he recognizes to be analogous to one or more known compounds. Many of the later chapters of this book are devoted to discussion of the properties of known compounds. There, the principles already learned will be applied.

Suggested Readings

Davidson, H. R., "Dipole Moments and Molecular Structure," *J. Chem. Educ.*, Vol. 27, 1950, p. 598.

Fowles, G. W. A., "Lone Pair Electrons," *J. Chem. Educ.*, Vol. 34, 1957, p. 187.

Huggins, M. L., "Hydrogen Bonding in High Polymers and Inclusion Compounds," *J. Chem. Educ.*, Vol. 34, 1957, p. 480.

Lewis, G. N., *Valence and the Structure of Atoms and Molecules*, New York: Chemical Catalog Co., 1923.

Palmer, W. G., *Valency* (2nd ed.), New York: Cambridge University Press, 1959.

Sienko, M. J., R. A. Plane, and R. E. Hester, *Inorganic Chemistry*, New York: W. A. Benjamin, Inc., 1965. (Chapters 1 and 2, Part I.)

Exercises

8-1. Given that a bond is to be formed between two atoms, X and Y, indicate what factors you would consider in attempting to predict whether the bond will be ionic or covalent.

8-2. Write the Lewis structures for the following compounds:

(a) SCl_2. (d) H_2CO. (g) C_2H_4.
(b) CH_3OH. (e) BiI_3. (h) H_2O_2.
(c) SiH_4. (f) H_2SO_4.

8-3. Explain why the molecule Ne_2 is not stable.

8-4. Draw an energy level diagram for the species Be_2^{+2}. What neutral diatomic molecule is this isoelectronic with?

8-5. Assume that nitric oxide, NO, can be treated as a homonuclear diatomic. This is not a bad assumption, since the nuclear charges of N and O are nearly the same. Draw an energy level diagram for NO. From the data given in Table 8-1 estimate the NO bond dissociation energy.

8-6. Draw a graph of bond dissociation energy *vs* bond distance for the diatomics listed in Table 8-1, excluding H_2. What conclusion can you draw regarding the relationship between bond energy and bond length?

8-7. The equilibrium bond distance in H_2 is 0.75 Å. Suppose that the nuclei could be held at a distance of 1.00 Å. Would the energy separation between the bonding and anti-bonding orbitals (Fig. 8-1) be larger or smaller than for the equilibrium distance? Explain.

8-8. Draw an energy level diagram for lithium hydride, LiH. The energies of the hydrogen 1*s* and lithium 2*s* orbitals can be assumed to be given by the negatives of ionization potentials for the atoms, Fig. 6-3. Which atom acquires the larger share of the bonding electron pair? Does the molecule have a dipole moment?

8-9. Write the important canonical structures for each of the following:

(a) CO_2. (c) NO_3^-.
(b) SO_3. (d) SO_2.

8-10. Predict the order of covalent bond strength in the following series of bonds.

(a) N—Cl, N—Br, N—F.
(b) O—H, S—H, Se—H.

Explain your answer.

8-11. What is the significance of the concept of localized bonds?

8-12. The Mulliken definition of electronegativity states that this quantity is proportional to half the energy difference between the negative and positive ions. This energy can be ascertained from a knowledge of the ionization potential and electron affinity: Electronegativity $= K(I - E)/2$. K is a constant which scales the quantities calculated so that they fall in the same range as electronegativities calculated by another method. The negative sign in the formula is necessary because we wish the absolute value of the *difference* in energies of the two charged species. From the data given in Table 6-2 and Appendix D, calculate the relative electronegativities of the halogens and oxygen. (Be careful of units!) Find the value of K necessary to scale the chlorine value to 3.0, and use this value of K for the other elements. Compare the results with the values listed in Table 8-3.

8-13. Give an example of a compound in which one of the atoms possesses more than eight electrons in the valency shell. Explain the structure of this compound in terms of hybridization of atomic orbitals.

8-14. The perchlorate ion, ClO_4^-, is known to be tetrahedral in symmetry. What kind of hybrid orbitals are being used by chlorine in this ion? How does this compare with chlorine in HCl?

8-15. Describe the concept of hybridization. Give an example of a compound in which this is presumed to occur. What experimental evidence is this presumption based upon?

8-16. What orbital set, hybridized or non-hybridized, is appropriate to each of the following bond sets?

(a) BiH_3, HBiH bond angle = 91°.
(b) BeH_2, linear molecule.
(c) The carbon-oxygen σ bonds from carbon in CO_2, a linear molecule.
(d) SiH_4, a tetrahedral molecule.
(e) BF_3, a planar molecule.

8-17. On the basis of the data listed in Table 8-5, predict a value for the C═N bond length.

8-18. List two conditions which are necessary in order that the interaction of two atomic orbitals result in a bonding orbital of appreciable stability. Give examples in which each of these conditions is met, and examples in which these conditions would not be met.

8-19. Indicate briefly the origin of the unusually high melting and boiling points of ammonia as compared with the hydrides of the other Group V elements. Draw structural formulas to illustrate the argument.

8-20. Hydrogen fluoride is known to associate in the gas phase to give species such as $(HF)_2$, $(HF)_3$, . . ., $(HF)_n$. Draw the structural formulas for a few of these.

8-21. How would you expect the magnitude of the bond dipole to vary in the following series of bonds: Al—Cl, Si—Cl, P—Cl, S—Cl? How do you account for the fact that $SiCl_4$ is a non-polar molecule, whereas PCl_3 is definitely polar?

8-22. When tin tetrachloride is added to an aqueous solution containing concentrated chloride ion (*e.g.*, concentrated HCl) the hexachlorostannate ion is formed. Write the equation for this reaction. Illustrate it by drawing structural representations for the reactants and products. Show the geometries of both reactants and products, and indicate the hybridizations of the tin atoms. Would you expect to observe similar behavior for carbon tetrachloride? Explain.

8-23. Many proteins undergo a sudden change in properties at a particular temperature when they are subjected to heat. Provide one possible reason for this change.

8-24. In terms of one or more of the concepts discussed in this chapter, present an interpretation of the dipole moments observed in each of the following series of compounds:

(a) PH_3	0.55 Debye,		(b) SiF_4	0 Debye,
PCl_3	0.80 Debye,		PF_3	1.02 Debye,
PF_3	1.02 Debye.		PF_5	0 Debye.

8-25. It is frequently necessary to know some of the facts about the geometrical structure of a molecule before a Lewis structure can be written. Assuming that the structures are symmetrical, write Lewis structures for the three oxides of nitrogen, N_2O_3, N_2O_4, and N_2O_5.

8-26. In one of its forms, white phosphorus, the element phosphorus exists as P_4 molecule. This molecule has tetrahedral symmetry. Draw a Lewis structure for the molecule. What must be the angles between bonds at each phosphorus atom?

Chapter 9

Chemical Thermodynamics

9-1 Definitions and Concepts

Thermodynamics is primarily a study of the relationship between heat and other forms of energy, particularly the transformation of heat into work. Every event of chemical interest involves the transfer of energy from one form to another and/or the interconversion of heat and other forms of energy. The laws of thermodynamics provide the basis for a quantitative understanding of the energy relationships among chemical compounds and chemical reactions. Before discussing the laws themselves and their applications to chemistry, it is necessary to define some terms and concepts.

Temperature — It is a part of everyone's experience that material bodies have a characteristic property which we call temperature. Furthermore, we know that a body which has a high temperature loses heat to its cooler surroundings, with a resultant lowering of its temperature. Our bodily senses tell us something about the temperatures of objects, provided that they are not too hot or too cold. It is easy to tell, for example, that a section of concrete pavement lying in the shade is cooler than an adjacent section which is exposed to sunlight. But a more quantitative and versatile measure of temperature is needed in scientific work.

We might proceed to define a temperature scale in the following way: We note that a particular object possesses a certain set of measured properties, such as volume, color, electrical resistance, *etc.*, which are subject to change as the temperature of the body is varied. We choose one of these, *e.g.*, the electrical resistance, to employ as our temperature-dependent, measured variable. A pair of easily reproduced constant-temperatures, such as the boiling and freezing points of water, are chosen as reference temperatures. The electrical resistivity of the object is measured as these two temperatures by bringing the object into thermal equilibrium with the reference systems. We assign a number to each of the two reference temperatures — 0 for the freezing point, and 100 for the boiling point. Then the interval between the freezing and boiling points is 100; a change in electrical resistivity of

$$\frac{R_{100} - R_0}{100}$$

where R_{100} and R_0 are the resistivities at the two reference temperatures, might be employed as a unit change in temperature on this scale. This last step is equivalent to making the assumption that the resistance varies linearly in the interval from the freezing to the boiling points of water.

This probably would not be a correct assumption. The electrical resistivity might actually obey a relationship represented by the black line in Fig. 9-1. Then, when the resistivity is exactly intermediate between the two reference values, the temperature — instead of being 50°, as the assumption of a linear relationship would have it — would actually be 56°. Of course, this would not be obvious to the person conducting such an experiment, since he would have no external reference with which to compare his results. Any other physical property of the object might be utilized similarly in establishing a temperature scale, but, in general, these scales would not coincide with one another at temperatures between 0° and 100°.

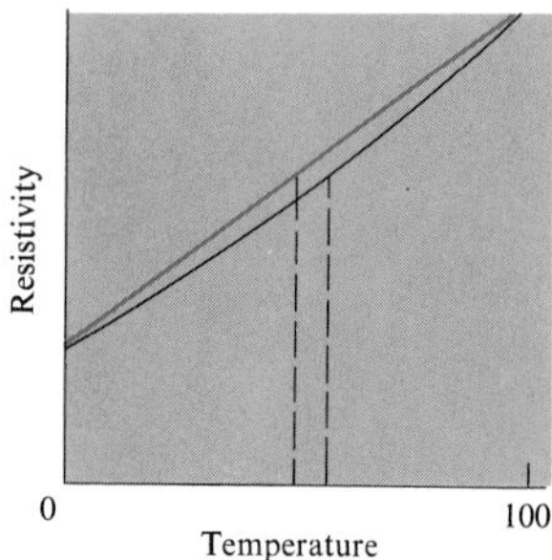

Fig. 9-1 A comparison of assumed and actual change of resistance with temperature between two reference points.

It is important, therefore, that the property of the object chosen for measurement of the temperature be one that varies linearly with temperature, or, at any rate, varies in a way which is precisely known from thermodynamic arguments. Once a temperature scale has been obtained by some such means, the electrical resistivity of the object could be used very nicely, since the curve of resistance *vs* temperature would be accurately determined.

It is possible to establish a temperature scale on the basis of the fundamental principles of thermodynamics. The volume of a given quantity of an ideal gas is proportional to this thermodynamic scale; many real gases behave so nearly ideally that they can be used for temperature measurement. Gas thermometers provide the precise temperature readings which are utilized in calibrating more convenient devices, such as mercury and alcohol thermometers.

Energy — According to Ostwald, the founder of physical chemistry, "Energy is work or anything which can be produced from or converted into work." In thermodynamics, we are concerned with the presence of energy in one or another of its forms, as described in Chap. 1, and with its appearance as heat or work in the course of a chemical event. Under a given set of conditions, an isolated body of matter possesses a fixed and definite amount of energy. It is not generally possible to say what this amount of energy actually is; the important thing, however, is to know what *changes* in energy the body undergoes during a process such as a chemical reaction. Frequently, the energy of the body is caused to change in some way and then to return to its initial state. It is important to recognize that we must have at our command some *experimental* way of verifying that the body has or has not returned to its initial condition. To do this, we must measure a number of properties the body possesses: A sufficient number so that when all of these properties are at certain values the body can have only one energy. Consider as an example a block of iron: Suppose that we measure its temperature and its position. Are these two properties enough to determine its energy? The iron could be heated to red-heat, cooled, and returned to its initial temperature and position; it would then have the same energy as it had initially. But it could also be placed in a strong magnetic field and then returned to its initial temperature and position. It would now have a different energy, despite the fact that the temperature and position were the same as they had been initially. Obviously, the magnetization is another property which should be included among those necessary to determine the energy of the iron. There is no simple way to tell when a sufficient number of properties have been measured in any given instance. It depends upon the substance being studied and upon the experiment being performed.

Heat — It was at one time thought that heat was a fluid, called *caloric*, which existed within bodies and passed from one body to another without loss. The best scientific minds of the 18th century — men such as Davy, Cavendish, and Count Rumford — disputed this notion when the connection between heat and the motions of bodies became evident. Thus Davy wrote, "It seems possible to account for all of the phenomena of heat if it be supposed that in solids the particles are in a constant state of vibratory motion, the particles of the hottest bodies moving with the greatest velocity." Another scientist, Sadi Carnot, who made very important contributions to thermodynamics, had this to say (about 1830): "Heat is simply motive power, or rather motion which has changed form. Whenever there is destruction of motive power there is at the same time production of heat in quantity proportional to the quantity of motive power destroyed."

James Prescott Joule, a brewer from the town of Salford, England, and a pupil of Dalton's, performed a series of experiments during the years 1843—1878 which were of great importance. He showed that one could measure the appearance of heat in a system, and relate this to the work done on the system. He further established that there is a constant ratio between the heat produced and the work done which is dependent only on the units chosen for the two quantities. This ratio is called the *mechanical equivalent of heat*. The currently accepted value is 4.184×10^7 ergs/gram-calorie. Joule's experiments provided an unequivocal basis for the rejection of the caloric theory.

Systems — In dealing with the energy changes that matter undergoes, we must confine our attention to definite bodies or quantities of matter. A *system* is any space or any material that we wish to consider in this way. It is separated by an imaginary boundary from all of the rest of the universe, which we call the *surroundings*. The system may exchange energy or mass with the surroundings, or it may be isolated in whole or in part. The important thing is that we should be able to observe and control the exchanges that do occur. The most commonly considered systems are *closed* systems, those that do not exchange matter with the surroundings. If the system also does not exchange energy, it is said to be *isolated*. If the system does not exchange heat with the surroundings, it is said to be *adiabatic*. Note that an adiabatic system may exchange work. An example of such a system is shown in Fig. 9-2. A certain quantity of gas is confined in a cylinder which

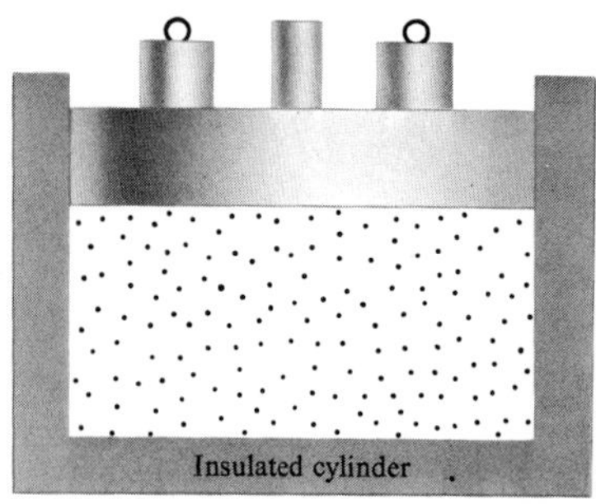

Fig. 9-2 Cylinder for carrying out an adiabatic expansion.

is constructed of a perfect insulating material, under a weight of 20 lbs. One of the weights is removed and the gas pressure in the cylinder causes the piston to rise to a new equilibrium position. The confined gas (the system) has not exchanged heat with the surroundings, but it has done work in raising the piston: It has undergone an adiabatic expansion.

The terms *adiabatic* and *isothermal* are often confused. An isothermal process is one which proceeds at a constant temperature. In the example of an adiabatic expansion just given, the temperature of the gas in the system would not remain constant, so it would not be an isothermal process.

9-2 The First Law of Thermodynamics

The laws of any science are simply statements of experience. The first law of thermodynamics is no exception: It states that neither work nor energy can be derived from nothing. The energy of an *isolated* system is, then, a constant as long as it remains isolated. If we withdraw energy from the system by causing it to do work, the energy of the system must then decrease by an amount equal to the work done. It is often said that the first law was the result of the failure of all attempts to construct a perpetual motion machine capable of doing work without loss of energy from the system.

Consider a system which possesses a certain quantity E_1 of energy. This system is caused to undergo a change such that the resultant energy is E_2. The first law tells us that the change in the energy of the system, $\Delta E = E_2 - E_1$, must be exactly equal to the total of all the heat and work which pass into the system during the change

$$\Delta E = q(\text{absorbed}) - W(\text{done})$$

The heat q is always regarded as positive when it is absorbed by the system. W is taken to be work done *by* the system on the surroundings. Since work done by the system results in a lowering of energy, the sign before W is negative. Work done by the surroundings on the system is negative, so that the two minus signs cancel, leading to a positive contribution to the energy. Confusion about the signs can be avoided by remembering simply that any change that causes removal of energy from the system makes a negative contribution to ΔE, whereas any change that results in addition of energy causes a positive contribution.

From the nature of the system's energy, it is clear that ΔE depends only upon the initial and final states, and not upon how the system was treated in going from one state to the other. The values for q and W, however, do depend upon the manner in which the change was carried out. But, although q and W individually are dependent upon the manner of execution of the change, the difference $(q - W)$ cannot be, since it is equal to ΔE.

Any property of a system that has a particular value for each state of the system is called a *state function*. A state function is not determined by the manner in which the system happened to get to a particular state. For example, the pressure, temperature, and volume are state functions of an ideal gas. The changes in these functions when the system undergoes a change are determined by specifying the initial and final states; the manner in which the system is made to pass from one state to the other does not affect the change in the state functions. For any system, the state variables are related to one another through an equation of state. For example, the ideal gas equation, $PV = nRT$ is the equation of state for an ideal gas. Because the state functions P, V, and T are related through this equation, it is sufficient to specify any two of them for a given quantity of gas n, to determine the third. Similarly, the energy of an ideal gas can be determined by specifying the temperature. In general, then, specification of only certain of the state functions for a system suffices to determine the values for the others. We frequently do not know the equation of state for a real system, and much experimental work is simply an effort to learn just what the relationships between the state functions of a system are.

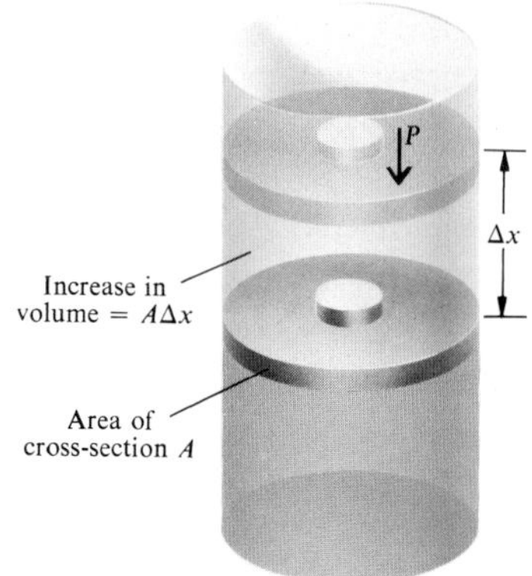

Fig. 9-3 The expansion of a gas.

Work of expansion — In illustrating the application of the laws of thermodynamics, it is convenient to talk about the expansion of gases as an example of work. Consider the cylinder in Fig. 9-3, fitted with a frictionless piston, and containing a gas at pressure P. The total force acting on the piston is the product of the pressure and the area of the piston face, $P \times A$. If the piston is moved through a small distance ΔX, the work (force $\times$ distance) is $P \times A \times \Delta X$. But $A \times \Delta X$ is a volume element, so we can write $W = P \times \Delta V$. Note that, if the gas in the cylinder is expanding, it is doing work on the surroundings; and, since ΔV is positive (the final volume is larger than the initial volume), $P \times \Delta V$ is positive. On the other hand, if the gas is being compressed, work is done on the gas, ΔV is negative and $P \times \Delta V$ is negative. This is in accordance with the convention that, when work is done by the system, W will be positive.

The work done by the gas in expanding from volume V_1 to volume V_2 is written

$$W = \int_{V_1}^{V_2} P\,dV \tag{9-1}$$

This integral is exact for any pressure-volume work. We will not find it necessary to solve problems in integral calculus in order to discuss the work done by expansion under various conditions. The following special cases will be important in subsequent discussion:

1. The gas expands into a vacuum. This means that P, the opposing pressure, is zero, $P \times \Delta V$ is zero, and no work is done.
2. The gas expands against a constant pressure. Then, with P constant, $W = P(V_2 - V_1) = P\Delta V$.
3. The gas expands reversibly. The concept of reversibility is very important in thermodynamics, because reversible processes represent limiting cases in which many thermodynamic relationships can be solved exactly. Consider the system in Fig. 9-3, under conditions such that the pressure exerted on the piston from outside is exactly equal to the pressure exerted by the gas from within. The piston will remain stationary and no work will be done. Now imagine that the outside pressure is decreased infinitesimally; the very slightly higher pressure inside the cylinder will cause the piston to rise a little. But, as the piston moves, the volume inside the cylinder will increase and the pressure will simultaneously decrease; so that, after the piston has moved an infinitesimal distance, equilibrium will again be established at a new volume $(V + \Delta V)$ and new pressure $(P - \Delta P)$. By proceeding in such a series of steps, always decreasing the pressure by infinitesimal amounts, we would eventually achieve a finite change in volume $(V_2 - V_1)$. But, if the changes in the opposing pressure are made infinitesimally small, such an operation would require an infinity of time! This limiting case represents the truly reversible expansion of a gas. In fact, it is easily seen that we cannot ever carry out such an operation, but the concept of a reversible process is, nevertheless, very important.

A little reflection, with the aid of the ideas introduced in the paragraph on systems, will make it clear that there is more than one way in which an expansion might be carried out, insofar as temperature is concerned.

Adiabatic expansion — In this process, the walls of the cylinder are impervious to the passage of heat, so that the heat in the system (the gas) remains constant. Remember that this does not mean that the temperature remains constant. From the first law, since $q = 0$,

$$\Delta E = -W$$

If the gas does work, W is positive, so ΔE is negative; *i.e.*, the work which the gas does is done at the expense of the energy of the system.

Isothermal expansion — In this process, the temperature of the gas remains constant. We allow heat to pass through the walls of the cylinder as necessary to maintain a constant temperature. We illustrate this type of process by first considering an ideal gas, for which the equation relating pressure, volume, and temperature is $PV = nRT$. The number of moles of gas is represented by n, R is the gas constant, and T is the absolute temperature. Note that PV, which has the dimensions of energy, is dependent *only* upon temperature. It follows that, at constant temperature, the energy of an ideal gas is constant During an isothermal expansion of an ideal gas, then, $\Delta E = 0$ and $q = W$; *i.e.*, the system acquires heat equal in quantity to the work done. For real gases, the energy depends upon other variables in addition to temperature. An isothermal expansion of a real gas would not result in $\Delta E = 0$. For example, the energy of a gas which obeys the van der Waals equation of state (page 61), involves, in part, the attractive forces between the molecules which lead to the a/V^2 correction term. Expansion of the gas causes a change in the degree to which the molecules exert attractive forces upon one another, and thus leads to a change in the energy of the gas, even though the temperature remains constant. The amount of heat which must be absorbed in this case is then $q = \Delta E + W$; *i.e.*, it is equal to the energy dispelled as work of expansion, plus the quantity necessary to overcome the attractive forces, given by ΔE.

To illustrate these ideas, let us consider a mole of ideal gas in a cylinder with a movable piston, at a pressure of 2 atm, 0°C. From the ideal gas equation, the volume is calculated to be 11.2 1. Now, consider an isothermal expansion to a volume of 22.4 liters. At this volume, the final pressure of the gas is 1 atm. It will be recalled that the expression for the work done in any expansion of a gas is

$$W = \int_{V_1}^{V_2} P\,dV$$

It should first of all be noted (see Fig. 9-3) that P in this expression is the *external* pressure on the piston. If the process is to be carried out reversibly, P must be different by only an infinitesimal amount from the pressure of the gas inside; we may say that it is the same. The pressure inside the cylinder is, for an ideal gas, $P = nRT/V$. Substituting this for P above, we have

$$W = nRT \int_{V_1}^{V_2} \frac{dV}{V}$$

The term outside the integral sign is a constant for an isothermal expansion. The integral of dV/V is $\log_e V$, so we have

$$W = nRT(\log_e V_2 - \log_e V_1)$$

$$W = nRT \log_e \frac{V_2}{V_1}$$

It should be noted that this value for the work is the *maximum* work which can be obtained from the isothermal expansion of a gas from volume V_2 to V_1. It is a general principle worth remembering that *the maximum work is always obtained from a process carried out under a given set of conditions when the process is carried out reversibly*. Inserting the appropriate values we have

$$W = 2.303 \times 1.987 \times 273 \times \log_{10} \frac{22.4}{11.2}$$

$$= 375 \text{ cal}$$

Since we are considering an isothermal process for an ideal gas, $\Delta E = O$, and $q = W = 375$ cal. The work of expansion is exactly balanced by a flow of heat into the system.

Enthalpy — If a process is carried out at constant volume, and no electrical or other special kinds of work are involved, W will be equal to zero and $\Delta E = q_v$. The subscript v is used to denote that the heat quantity is that for a constant volume process. For a constant pressure process, on the other hand, there may be pressure-volume work $P\Delta V$. Then

$$\Delta E = q_p - P\Delta V$$

or

$$q_p = \Delta E + P\Delta V$$

Constant pressure processes are important, since most chemical changes are observed under these conditions.

There is another thermodynamic function, termed the *enthalpy*, which is defined by the expression

$$H = E + PV \tag{9-2}$$

Since enthalpy is related to quantities which are all state functions, it follows that H is also a state function. If the system undergoes a change in state at constant pressure,

$$\Delta H = \Delta E + P\Delta V = q_p \tag{9-3}$$

We see that ΔH is then equal to the heat absorbed by the system for a constant pressure process q_p. It is a very useful function because it is readily observed experimentally. When ΔH is negative, corresponding to *loss* of heat from the system, the process is said to be *exothermic;* when ΔH is positive, the process is *endothermic*, and the system *absorbs* heat from the surroundings.

Most chemical reactions and changes of state are studied under conditions of constant pressure (1 atm). In some special instances (*e.g.*, heats of combustion), the reactions are carried out at constant volume. We may assume, without any serious error, that $\Delta E = q_v$, the heat of the process at constant volume. Then

$$q_p = q_v + P\Delta V = q_v + PV_2 - PV_1$$

where V_2 and V_1 are the volumes of the system at final and initial conditions, respectively. If only solids and liquids are involved in a reaction, the $P\Delta V$ term is so small in comparison with the heats usually observed that it can be neglected. Then

$$q_p = q_v \quad \text{and} \quad \Delta H \approx \Delta E$$

When gases are involved, we can write

$$PV_2 = n_2RT \quad \text{and} \quad PV_1 = n_1RT$$

where n_2 and n_1 are the number of moles of gas in the products and starting materials, respectively. Then

$$q_p = q_v + \Delta nRT$$

For example, in the reaction of H_2 and O_2 at 298°K, 1 atm pressure, to yield water

$$H_2(g) + \tfrac{1}{2}O_2(g) = H_2O(l)$$
$$\Delta H = -68{,}318 \text{ cal}$$

Here, since the heat at constant pressure is given for liquid water as the product, $\Delta n = -\frac{3}{2}$. Then

$$\Delta nRT = -\tfrac{3}{2} \times 1.987 \times 298 = -888 \text{ cal}$$

Therefore,

$$\Delta E = \Delta H - \Delta nRT = -67{,}430 \text{ cal}$$

9-3 Heat Capacity

Consider a system in which the only possible kind of work is pressure-volume work. If heat is added to such a system at constant volume, the system does no work and all of the heat is directed to an increase in the energy. Under these conditions, we may write

$$\Delta E = C_v\Delta T = q_v$$

C_v represents the heat required to raise the temperature 1°C at constant volume; it is termed the *heat capacity at constant volume.* On the other hand, when heat is added to the system at constant pressure,

$$\Delta E + P\Delta V = C_p\Delta T = q_p$$

The heat added under these conditions is applied both to increasing the energy of the system and to doing pressure-volume work (expansion). C_p is termed the *heat capacity at constant pressure.* We could also have written

$$\Delta(E + PV)_p = C_p\Delta T$$

where the subscript p reminds us that we are dealing with a constant pressure process. Then, using the definition of enthalpy, $H = E + PV$,

$$\Delta H = C_p\Delta T$$

In summary,

$$C_v = \Delta E/\Delta T \tag{9-4}$$

$$C_p = \Delta H/\Delta T \tag{9-5}$$

In general, C_p is larger than C_v because, in addition to increasing the energy of the system by an amount ΔE, there is always the additional work of expansion for the constant pressure conditions. For gases, this term is appreciable, but for solids and liquids, which have only a small coefficient of expansion, C_p and C_v are very nearly the same.

On the basis of the kinetic theory, it can be shown (see Chap. 3) that the pressure-volume product for a mole of an ideal gas equals $Nm\overline{u^2}/3$, where N is Avogadro's number, m is the mass of a molecule, and $\overline{u^2}$ is the mean square velocity of the molecule. The translational energy (*i.e.*, the kinetic energy) can be written

$$E_t = \tfrac{1}{2}Nm\overline{u^2}$$

so that we then have

$$PV = \tfrac{2}{3}E_t = RT$$
$$E_t = \tfrac{3}{2}RT$$

All of the thermodynamic energy of an ideal gas can be identified with E_t. Then, for a change in the system at constant volume,

$$C_v = \frac{\Delta E}{\Delta T} = \frac{3}{2}R$$

For a change at constant pressure, we have

$$C_p = \frac{\Delta H}{\Delta T} = \frac{\Delta E + P\Delta V}{\Delta T}$$

But $P\Delta V/\Delta T = R$, from $PV = RT$, so that

$$C_p = \tfrac{3}{2}R + R = \tfrac{5}{2}R$$

R has the value 1.987 cal/°K-mole, so that values of C_v and C_p for a mole of ideal gas would be about 3 and 5 cal/°K-mole, respectively. These are the values observed for monatomic gases such as helium, argon, *etc.* For diatomic and more complex molecules, these values are not observed. The reason for this is that it is no longer true that all of the energy of the molecules is translational; *e.g.*, diatomic molecules possess two axes about which rotation may occur [Fig. 9-4(a)]. (Rotation about the x-axis is not included, because the moment of inertia about this axis is very close to zero.) Rotational motion contributes an energy RT to the molecules, so that $E = \tfrac{5}{2}RT$. C_v is then $\tfrac{5}{2}R$, and $C_p = \tfrac{7}{2}R$.

There is yet another mode of motion which a diatomic molecule may exhibit, the vibration along the molecular axis [Fig. 9-4(b)]. This motion is not found to any appreciable extent at room temperature, but at higher temperatures, it is. One finds, therefore, that the heat capacities of diatomic

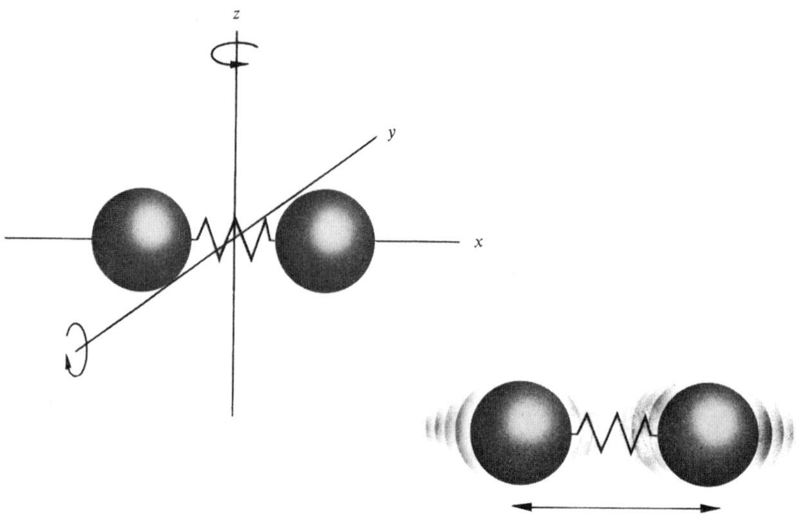

Fig. 9-4 Rotational and vibrational modes for a diatomic molecule.

and all other polyatomic gases vary with temperature until a sufficiently high temperature is reached for all of the possible modes of molecular motion to be activated. The temperature dependence of the heat capacity of a diatomic molecule is shown schematically in Fig. 9-5. Each form of motion which a molecule may possess is termed a *degree of freedom.* For a diatomic molecule, there are three degrees of translational freedom (corresponding to motion along the *x*-, *y*-, and *z*-axes), two degrees of rotational freedom, and one degree of vibrational freedom. Each degree of freedom is capable of contributing to the heat capacity of the gas. At very

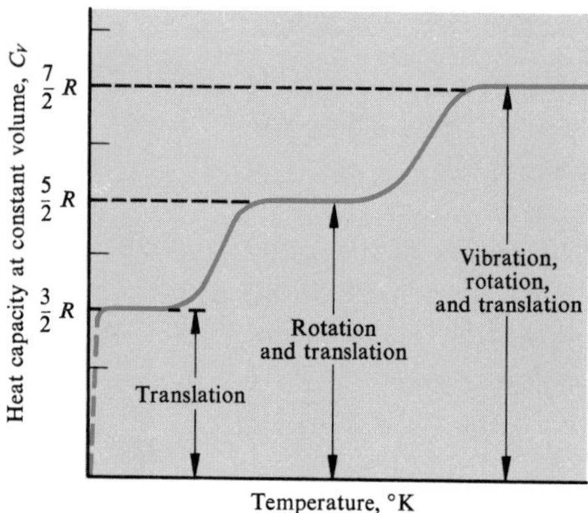

Fig. 9-5 The variation of heat capacity at constant volume, C_v, for a diatomic molecule.

high temperatures, the energy of a system of diatomic molecules is distributed between all of these degrees of freedom, and the maximum heat capacity is observed. At lower temperatures, certain of the degrees of freedom are not fully "activated," so to speak, and the observed heat capacity is less than the maximum possible.

According to Planck's quantum theory, all forms of energy change are quantized; in other words, molecules always change energy in quantum

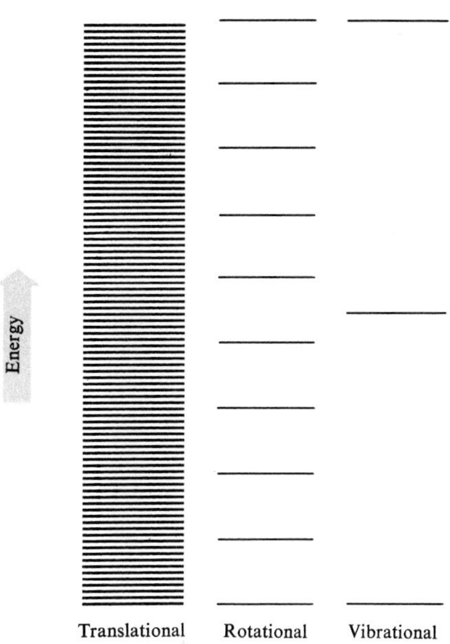

Fig. 9-6 Quantization of energy among the degrees of freedom possible for a diatomic molecule. The lowest possible energy associated with each form of motion has been arbitrarily set at the same level, to emphasize the energy gaps between adjacent levels. This is not strictly correct, because of a phenomenon known as the zero-point energy, but this is not of importance for the purposes of the present discussion.

jumps, not continuously. The quantization of energy for the degrees of freedom of a diatomic molecule are shown schematically in Fig. 9-6. The quantum jumps between different quantities of translational energy are so small that for all practical purposes, a molecule can have any value of translational energy. The jumps between the rotational energy levels are larger. At low temperatures, most of the molecules in a diatomic gas are in the lowest possible rotational energy; very few acquire the necessary

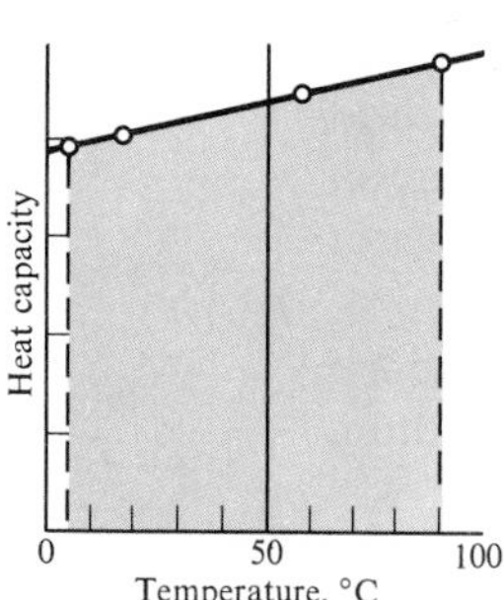

Fig. 9-7 Heat capacity versus temperature for liquid benzene.

energy through collisions to make the jump to the next higher rotational state. As the temperature is raised, however, more energetic collisions result, and more molecules may acquire the energy to make the quantum jump to higher rotational energy levels. Once this conversion of translational energy into rotational energy becomes frequent for a large fraction of the molecules, the rotational degrees of freedom become a "storage place" for energy, and contribute to the heat capacity.

The vibrational energy levels are, of course, also quantized. They are much more widely separated in energy than are the rotational energy levels. The transfer of energy from translational and rotational into vibrational does not become effective, therefore, until high temperatures are attained. Heat capacities are generally expressed as cal/°K-mole. Sometimes the gram heat capacity, cal/°K-g, is employed. The *specific heat* of a substance is the ratio of its heat capacity to that of water at 15°C. It is the same numerically as the gram heat capacity. Some representative gram heat capacities are listed in Table 9-1.

Examination of the data for benzene shows that the heat capacity for this substance is not constant. A different amount of heat is required to

Table 9-1 Heat capacities of various substances.

Substance	Temperature (°C)	C_p (cal/°K-g)
Aluminum	20	0.214
Bromine	20	0.107
Mercury	20	0.0332
Lead bromide ($PbBr_2$)	0	0.0502
Ice	0	0.492
Water	15	1.000
Carbon tetrachloride	20	0.201
Benzene	5	0.389
	20	0.406
	60	0.444
	90	0.473

raise the temperature of a given amount by 1°C at 20° than is needed at 60°. This sort of behavior is the rule, rather than the exception. For a number of reasons, it is desirable to have available a mathematical expression relating heat capacity and temperature. Such a relationship is obtained by experimentally evaluating the heat capacity as a function of temperature, and graphing the results *vs* temperature. This has been done for benzene in Fig. 9-7.

Over the temperature range involved, the graph is essentially linear; the equation for a straight line $C_p = 0.385 + 9.5 \times 10^{-4}\, T$ (where T is in °C).

Suppose that one wishes to know how many calories of heat are required to raise the temperature of 1 g of benzene from 5°C to 85°C. Since the heat capacity is not constant in this temperature range, we cannot simply multiply a constant C_p by ΔT. The amount of heat required is represented by the shaded area in Fig. 9-7.

9-4 Thermochemistry

Whenever a closed system undergoes chemical reaction or a change in state, there is, in general, an absorption or evolution of heat. A knowledge of the heats involved in such processes is valuable to the chemist from both the theoretical and practical points of view. Thermochemistry encompasses the experimental determination of the heats involved in chemical reactions and in changes of state, and includes the use of these quantities to determine other heat quantities, many of which are not directly measurable.

In tabulating the results of thermochemical studies, it has become the practice to list the results obtained at 25°C and 1 atm of pressure — these are standard conditions. There often is ambiguity in the choice of physical states of reactants or products. For example, in the reaction of hydrogen and oxygen to yield water, the reaction could be written as above, with liquid water as the product, or it could be written

$$H_2(g) + \tfrac{1}{2}O_2(g) \rightarrow H_2O(g)$$

The heats of these two reactions are not the same, since the initial and final states are not identical. The convention in use is to list the states of all substances as those states in which they would normally exist at 25°C and 1 atm pressure. These are the so-called standard states. For gases, the standard state is 1 atm of pressure. Thus H_2O should be written as $H_2O(l)$. For reactions run at 1 atm pressure, and with all substances in their standard states, the thermodynamic symbol is written with a superscript degree, as $\Delta H°$.

To understand the uses to which thermochemical data can be put, we must recall that enthalpy is a state function; *i.e.*, ΔH for a process depends only upon the initial and final states, and not upon the path taken. We make use of this in the following way: Suppose that we wish to know the heat of a reaction A $\rightarrow$ B which it is not possible to carry out in the laboratory. Often, we can still obtain the desired information if it is possible to convert B into some other substance, C, and if A, in turn, can be converted into C:

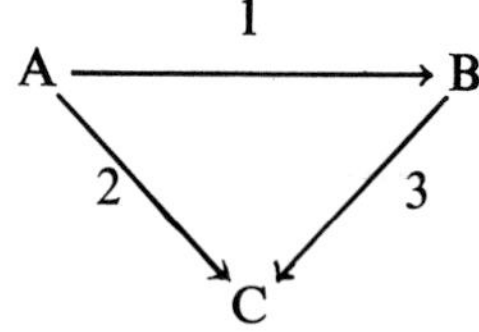

Since the heat for a reaction such as A → B depends only upon the initial and final states, and since A → C plus −(B → C) equals the same final and initial states as does A → B, it follows that

$$\Delta H_1 = \Delta H_2 - \Delta H_3$$

As a specific example, suppose that we wish to know the heat of formation of methyl alcohol.

$$C(s) + 2H_2(g) + \tfrac{1}{2}O_2(g) \rightarrow CH_3OH(g) \qquad \Delta H^\circ = ?$$

We cannot measure the heat of this reaction by reacting hydrogen and oxygen gases with carbon directly, but we can proceed in the following way:

First of all, we measure the heat of combustion of methyl alcohol

$$CH_3OH(g) + \tfrac{3}{2}O_2(g) \rightarrow CO_2(g) + 2H_2O(l) \qquad \Delta H^\circ = -182.58 \text{ kcal}$$

or, writing the reverse reaction

$$\text{(A)} \quad CO_2(g) + 2H_2O(l) \rightarrow \tfrac{3}{2}O_2(g) + CH_3OH(g) \quad \Delta H^\circ = +182.58 \text{ kcal}$$

We also need to know the heats of formation of CO_2 and H_2O:

$$\text{(B)} \qquad C(s) + O_2(g) \rightarrow CO_2(g) \qquad \Delta H^\circ = -94.06 \text{ kcal}$$
$$\text{(C)} \qquad 2H_2(g) + O_2(g) \rightarrow 2H_2O(l) \quad \Delta H^\circ = -136.64 \text{ kcal}$$

[Note that ΔH for reaction (C) is double that given previously; we have doubled the quantities in the equation.]

Adding these last two reactions together and adding the total to (A), we have for the over-all material balance and ΔH°:

$$C(s) + \tfrac{1}{2}O_2(g) + 2H_2(g) \rightarrow CH_3OH(g) \quad \Delta H^\circ = -48.12 \text{ kcal}$$

The "law" which is usually quoted in connection with the procedure we have just used was enunciated by Hess in 1840: "If a chemical reaction occurs in stages, the algebraic sum of the amounts of heat evolved is equal to the total evolution of heat when the reaction occurs directly". The law is a consequence of the fact that ΔH is a state function; it must be presumed when applying the law that the reactions under consideration are all run under the same conditions of constant pressure or constant volume.

Whenever a chemical reaction takes place isothermally, the entire heat of the reaction can be attributed to changes in chemical energy and the other (usually small) volume, surface, or phase changes. There are some important classes of reaction which deserve separate consideration.

Heat of formation — The molar heat of formation of a compound is the heat evolved or absorbed when one mole of a compound is formed from the elements, all of the substances being in their standard states. The temperature is usually specified as 25°C. *As an arbitrary convention, the enthalpies of all elements in their standard states at 25°C are taken to be zero.* The heat of formation of $H_2O(l)$, as seen from the data given above, is −68,318 cal. The negative sign indicates that the reaction is exothermic; *i.e.*, heat is lost by the system. Standard enthalpies of formation for a number of substances are tabulated in Appendix E.

Heat of combustion — When a compound containing (usually) carbon, hydrogen, and perhaps other elements (oxygen, sulfur, nitrogen, *etc.*) is combusted at 25°C with oxygen to yield CO_2, H_2O, and oxidation products of the other elements, the heat evolved is termed the heat of combustion. For example,

$$CH_4(g) + 2O_2(g) \rightarrow CO_2(g) + 2H_2O(l) \quad \Delta H^\circ = -212.8 \text{ kcal}$$

Some representative heats of combustion are listed in Table 9-2. In practice, these reactions are often studied under constant volume conditions. The substance is combusted in a small bomb which is immersed in a calorimeter; the escape of the gaseous products is thus prevented. The correction necessary to convert the constant volume heat to ΔH has already been discussed.

Table 9-2 Some standard heats of combustion at 25°C (kcal/mole).

Substance (gaseous)		ΔH°_{298}
Methane	CH_4	−212.79
Ethane	C_2H_6	−372.81
Propane	C_3H_8	−526.3
Ethylene	C_2H_4	−337.28
Methyl alcohol	CH_3OH	−182.58

We can generalize on the above considerations and say that in a reaction

$$A + B \rightarrow C + D + \ldots$$

the heat of the reaction is equal to the sum of the heats of formation of each of the products, minus the sum of the heats of formation of each of the reactants.

$$\Delta H = \Delta H_f(\text{products}) - \Delta H_f(\text{reactants})$$

The following example is illustrative:

$$2FeO(s) + \tfrac{1}{2}O_2(g) \rightarrow Fe_2O_3(s) \quad \Delta H^\circ = ?$$
$$\Delta H^\circ = \Delta H_f^\circ(Fe_2O_3) - 2\Delta H_f^\circ(FeO) - \tfrac{1}{2}H_f^\circ(O_2)$$

We have

(A)	$2Fe(s) + \frac{3}{2}O_2(g) \rightarrow Fe_2O_3(s)$	$\Delta H_f° = -196.5$ kcal
(B)	$2Fe(s) + O_2(g) \rightarrow 2FeO(s)$	$2\Delta H_f° = -127.4$ kcal

The heat of formation of O_2 is, by definition, zero. The heat for the reaction in question is then (A) − (B) = −69.1 kcal.

Sources of thermochemical data — The most extensive tabulation of chemical thermodynamic properties is found in Circular No. 500 of the U.S. National Bureau of Standards. The American Petroleum Institute Research Project 44 contains data for many hydrocarbon compounds. These two volumes can be found in almost any chemical library.

Thermochemical bond energies — It is important for chemists to have some quantitative data regarding the energies of chemical bonds. Such information is necessary in establishing relationships between bond-forming power and other properties of atoms, in predicting chemical behavior, and in evaluating the conclusions drawn from chemical theory.

There are a variety of ways in which bond energies can be evaluated. We are concerned here with only the thermochemical method. The energy of a bond in a diatomic molecule is measured by the heat which is absorbed in the reaction:

$$X—X \rightarrow 2X$$

For a polyatomic molecule, the thermochemical data relate to the heat of dissociation of the molecule into atoms, but not to the individual bonds. Consider H_2O as an example. We have the following data available:

(A)	$H_2(g) + \frac{1}{2}O_2(g) \rightarrow H_2O(g)$	$\Delta H° = -57.80$ kcal
(B)	$H_2(g) \rightarrow 2H(g)$	$\Delta H° = 104.18$ kcal
(C)	$O_2(g) \rightarrow 2O(g)$	$\Delta H° = 118.32$ kcal
(A)	$-(B) - \frac{1}{2}(C)$ gives	
(D)	$2H(g) + O(g) \rightarrow H_2O(g)$	$\Delta H° = -221.14$ kcal

This value of ΔH (with opposite sign) is the sum of the energies required to break first one O—H bond and then the other. These two values are almost certainly not the same; but we can say that, as a good approximation, $\frac{1}{2}\Delta H$ for reaction (D) is the average value of the O—H bond energy, 110.6 kcal.

By using the values for heats of formation listed in NBS Circular No. 500, it is possible to compute average thermochemical bond energies quite easily. As an illustration, we will evaluate the average energy of the P—H bond in PH_3. The reaction we are interested in is

$$P(g) + 3H(g) \rightarrow PH_3(g) \qquad \Delta H = ?$$

The values for the heats of formation, in kcal of each of the three species are listed in NBS Circular No. 500 as ΔH_f°. They are: (1) P(g), 75.18; (2) H(g), 52.09; (3) PH_3(g), 2.21. The required ΔH is then

$$(3) - 3 \times (2) - (1) = -229.24 \text{ kcal}$$

Dividing this by three gives the average value of the P—H bond energy, 76 kcal.

Change of state — Because chemical reactions may be carried out with the reactants and products in a variety of physical states, it is important to have data available for the heat evolved or absorbed in the transitions of substances from one state to another. These data are also of interest in themselves, in that they often provide some insight into the chemical nature of the substances.

1. *Heat of vaporization.* The molar heat of vaporization ΔH_v is the heat required to convert 1 mole of a liquid substance to the gaseous state. Its negative is the *heat of condensation*. The heat of vaporization is dependent upon temperature, as the data listed in Table 9-3 show.

Table 9-3 Heats of vaporization.

Substance	Temperature (°C)	ΔH_v (cal/mole)
Water	0	10,767
Water	20	10,565
Water	40	10,360
Water	100	9,717
Methyl alcohol	0	9,145
Carbon tetrachloride	0	8,000
Benzene	80.1	7,350

2. *Heat of fusion.* The molar heat of fusion ΔH_c is the heat absorbed when 1 mole of a solid substance is converted to the liquid state at 1 atm of pressure. Some representative values are listed in Table 9-4.

Table 9-4 Heats of fusion.

Substance	Temperature (°K)	ΔH_c (cal/mole)
Copper	1356	2662
Lithium nitrate	523	6100
Carbon tetrachloride	250	600
Sulfur	392	424
Water	273	1437

3. *Heat of sublimation.* The molar heat of sublimation ΔH_s is the heat required to convert 1 mole of a solid substance to the gaseous state. By applying Hess' law, it can be seen that $\Delta H_s = \Delta H_v + \Delta H_c$, where the two latter quantities are known for the same temperature.

9-5 Calorimetry

A calorimeter is a device used for measurement of the heat of a chemical reaction or a physical transition. There are many different designs of calorimeters; we will discuss the more common types and some considerations involved in their use.

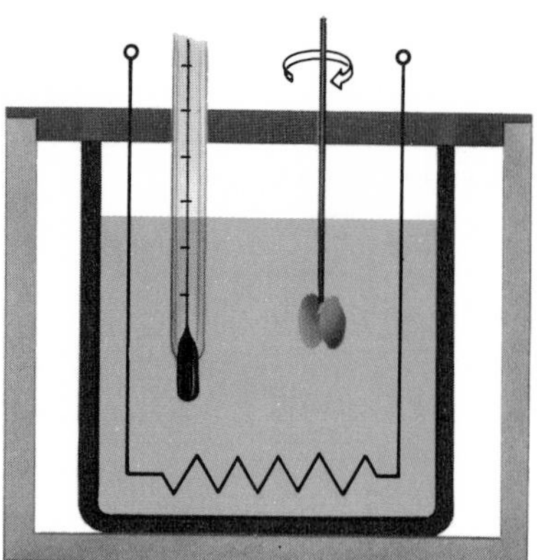

Fig. 9-8 Schematic illustration of a simple calorimeter.

Figure 9-8 shows a schematic drawing of a simple calorimeter. It is operated at constant pressure (atmospheric), and is essentially an adiabatic, non-isothermal device. The calorimeter proper is a Dewar flask, a double-walled vessel with an evacuated, sealed double-wall space. The inside walls of the evacuated space are silvered to further reduce heat loss.

A heating coil, thermometer and stirrer are led in through a block of insulating material which forms the top of the calorimeter. Another opening (not shown) is provided for introduction of samples. The steps involved in measuring the heat of solution of a salt in water will be outlined in order to illustrate the procedures employed. First, an amount of water equal to that to be used in the experiment is introduced into the calorimeter. After allowing the system to come to temperature equilibrium, a current of known amperage is passed through a heating coil for a period of time t. The total heat evolved in the calorimeter from the passage of i amperes of current flowing for t seconds through a resistance R is

$$q = i^2Rt$$

(The units of q are watt-seconds, or joules; 1 calorie is equal to 4.185 joules.) This amount of heat will cause a rise of ΔT degrees in the temperature of the system. The heat capacity of the entire system is then

$$C = q/\Delta T$$

The system is again allowed to come to temperature equilibrium, the weighed amount of solute is added, and the temperature rise is again noted. From the observed temperature rise, $\Delta T'$, and the known value for C (the small heat capacity of the added solute is often neglected),

$$q' = C\Delta T'$$

One of the principal sources of uncertainty in calorimetry is the loss of heat from the calorimeter by conduction, convection, radiation, *etc.* If

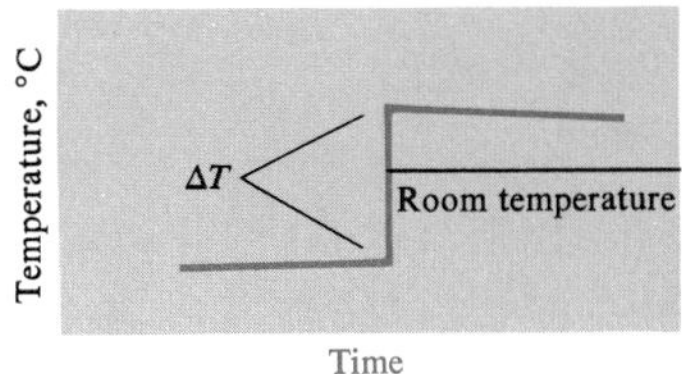

Fig. 9-9 Temperature change resulting from an exothermic process in a calorimeter.

the reaction is essentially instantaneous, the correction can be made fairly easily. It is best done by beginning with the system at a temperature slightly below room temperature. The temperature of the calorimeter system is measured as a function of time, and then — at a known time — the reaction is carried out. After a few seconds, the temperature is again read as a function of time. At this time, it should be above room temperature. All of the data are graphed as shown in Fig. 9-9. The time *vs* temperature lines are extrapolated to the time of reaction. The difference here is the correct value of ΔT. Other, more elaborate methods are involved in correcting for heat loss during slow reactions.

One method of avoiding the problem of correcting for heat loss is shown schematically in Fig. 9-10. A heater is constructed to fit around the calorimeter. A temperature-sensing device is led from the calorimeter to a control unit. The control unit regulates the current passing through the heater so that the difference between the temperature inside and outside the calorimeter is zero. Under these conditions, no heat transfer occurs between the calorimeter and its surroundings.

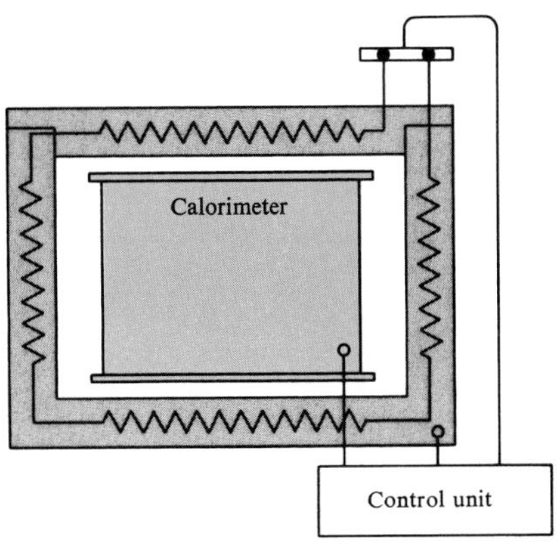

Fig. 9-10 A schematic drawing of a device for maintaining adiabatic conditions in a calorimeter.

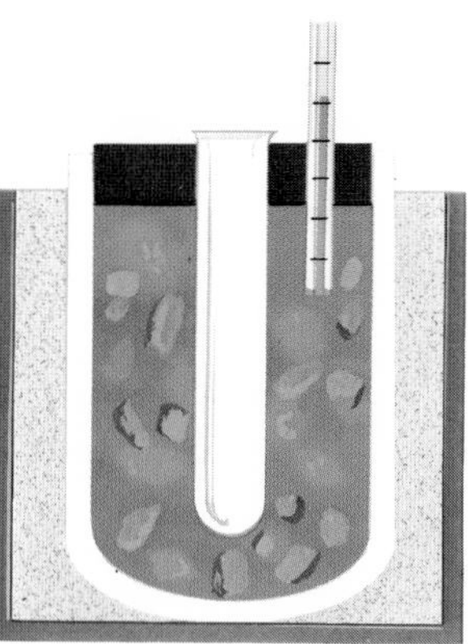

Fig. 9-11 An isothermal, or Bunsen, calorimeter. The reaction tube is immersed in an ice-water mixture contained in an insulated flask. The heat evolved in the reaction tube causes a certain amount of ice to melt. The volume change which results from the melting is measured in the capillary tube which projects from the flask.

A bomb calorimeter is an adiabatic device, as are the other calorimeters, but it differs in being operated under conditions of constant volume. It is generally used to determine heats of combustion. The sample to be burned is placed in a small metal bomb with an oxidant such as sodium peroxide. When the bomb and its contents have come to thermal equilibrium in the calorimeter, the mixture inside is ignited electrically. The calorimeter-plus-bomb assembly is calibrated by burning weighed samples of material for which the heat of combustion is accurately known.

In the calorimeters mentioned thus far, the heat evolved in the process under study is absorbed by the calorimeter and its contents with a corresponding rise in temperature. The Bunsen calorimeter in Fig. 9-11 is an isothermal device; the heat evolved is absorbed by a water-ice mixture. The quantity of ice which melts is a measure of the total heat evolved. As long as some ice remains, the temperature of the mixture remains fixed at zero degrees. The quantity of ice which melts as a result of a change taking place in the calorimeter can be determined by measuring the change in volume. The density of ice is less than that of water, so the melting of a certain quantity of ice is accompanied by a decrease in total volume.

9-6 Entropy

The first law of thermodynamics is a statement of our experience that no machine is capable of spontaneously creating the energy for doing work. By means of the first law, we express the relationship between the work done on or by the system and the heat gained or lost by the system in any given change. It is incomplete, however, in one very important respect: It does not tell us whether any given change is really feasible. For example, it tells us that *if* we drop a brick onto a hard surface, a certain amount of heat — equal to the change in potential energy — will be evolved. It also tells us that *if* a brick rises from the floor to our hand, a certain amount of work will be done, and that this amount of work will be equal to the change in heat energy of the brick plus the heat supplied from outside. Now, we know that if we let go of a brick it will spontaneously drop to the floor, and we know also that it will *not* spontaneously rise from the floor; the first law does not include this distinction.

Changes occurring in nature proceed spontaneously in the direction of a stable state, or equilibrium. A state function which would indicate whether a particular process might occur spontaneously, and in which direction, would therefore be a measure of the stability of the system. The thermodynamic state functions which are defined in terms of the first law of thermodynamics do not satisfy this new requirement. Let us consider, for example, the enthalpy change, ΔH, accompanying a process. It is true that most processes which proceed spontaneously also are exothermic; *e.g.*, the burning of a match, the mixing of water and sulfuric acid, the reaction of iron filings with sulfur. We are tempted to say that spontaneous processes are those which proceed with evolution of heat. A little searching, however, reveals many other processes which proceed spontaneously with either zero or positive ΔH. Among these are changes in state, such as the melting of an ice cube at zero degrees and above, and vaporization of any liquid at its boiling point. Heat must be supplied for the process to proceed, and yet it occurs spontaneously. The dissolving of potassium chloride in water is another example of a spontaneous, endothermic process; as the salt dissolves in water, the temperature of the solution decreases. The expansion of an ideal gas, as shown in Fig. 9-12, provides an example of a process which proceeds spontaneously with no change in enthalpy. When the stopcock connecting the two vessels is opened, the gas confined to flask A expands spontaneously into the evacuated flask, B.

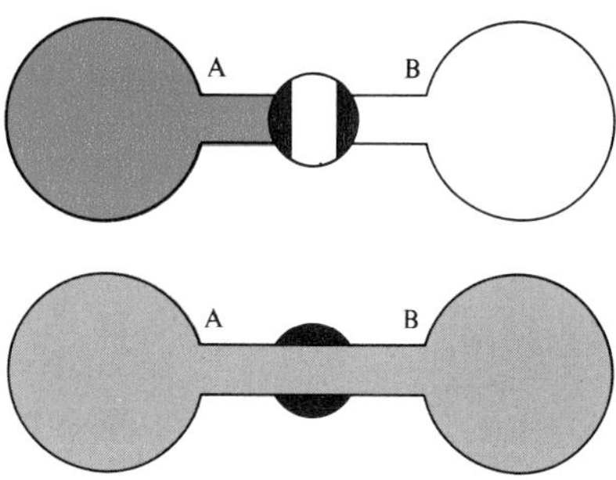

Fig. 9-12 Expansion of a gas into an evacuated cylinder.

These examples of changes which occur spontaneously in spite of a positive or zero enthalpy change have one common characteristic, which is that the final state is more disordered or random than was the initial state. In a solid lattice the molecules are ordered with respect to one another, whereas in a liquid, the molecules are free to move about, and there is not a large degree of ordering of one molecule with respect to another. To describe a crystalline solid, we must be very specific about the location of each molecule with respect to the others; in a liquid, on the other hand, the locations of the molecules with respect to one another are not so specific, since there is no regular lattice. Many different arrangements are essentially equivalent from the standpoint of potential energy. Because of the less stringent requirements on the locations of individual molecules, we say that the liquid is more disordered, or random, than the solid; melting results in an increase in randomness. The vapor state for a substance is more random than the liquid state for a slightly different reason; namely, the large increase in volume. While the molecules of a liquid are not highly ordered with respect to one another, they are constrained to occupy a volume which is not much larger, if at all, than in the solid. In the vapor state, however, the volume is very much larger. The vapor is more random in character than the liquid because of the larger volume in which each molecule is free to move.

The dissolving of potassium chloride in water represents a composite of the two effects just mentioned. The order present in the potassium chloride lattice is lost upon dissolving, and the volume in which the ions are free to move is enlarged to include the entire solution. There are, as we shall see in discussing ionic solutions, other effects accompanying solution which may actually lead to *more* ordering, but in most cases, the solution of a solid in a liquid leads to an increase in randomness of the system.

In these examples, the enthalpy change is positive; *i.e.*, heat must be supplied to overcome potential energy effects which favor the original state. The spontaneous character is present even though the system must extract heat energy from the surroundings. Heat flows, in effect, uphill in all of these examples. The last example given, expansion of an ideal gas, involves no enthalpy change, and is spontaneous because of the increase in randomness associated with an increase in volume of the system.

Clausius first proposed a new thermodynamic state function, called *entropy*, as a measure of the stability of the system. Entropy is most easily understood in terms of probabilities. One way to state the matter is this: The most stable state of an isolated system (*i.e.*, one which does not exchange work, heat or matter with the surroundings) is the state of maximum probability. But what is the state of maximum probability in any particular case? In order to consider this question, we must look at the system from a microscopic point of view; *i.e.*, in terms of the individual molecules. Certain properties of a system which we ordinarily observe, such as temperature or pressure, are *macro*scopic properties; they reflect the average behavior of a very large number of particles, and are not characteristic of the molecules individually. In a gas, for example, the molecular velocities are distributed over a wide range. An individual molecule might at any instant have any velocity from zero to very large. The pressure exerted by the gas on its container, however, is the result of an average impact of the molecules on the walls. The pressure observed is a measure of an average effect; the likelihood of experimentally detecting deviations from this for gases under normal conditions is vanishingly small.

The probability of a particular state can be calculated from examination of the microstates. The microstates are determined by considering the locations and energies of the individual molecules, assuming that we can attach labels to them. In effect, we wish to know how many possible, energetically equivalent ways there are of having a particular distribution of molecules. To illustrate, consider the system shown in Fig. 9-13. We have a gas consisting of six molecules free to move in the spaces A and B. Let us assume that the molecules have the same energy everywhere in the

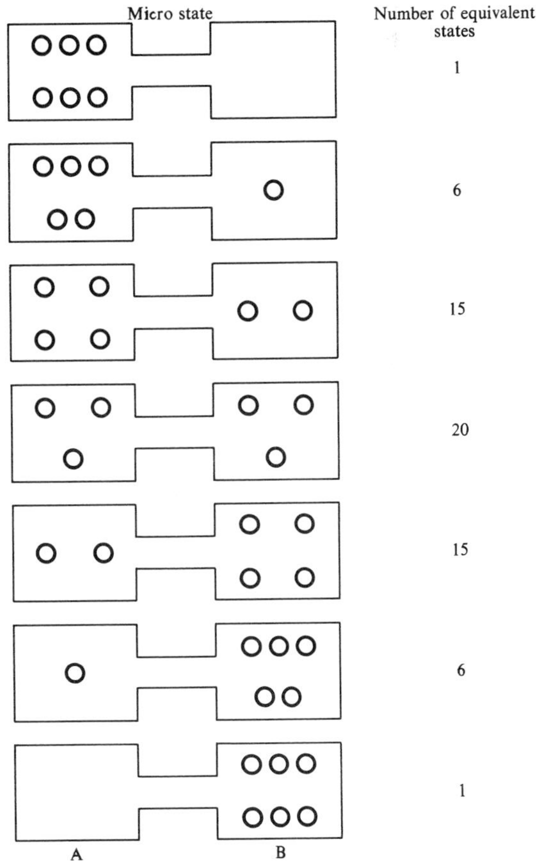

Fig. 9-13 Relative probabilities of distribution of six particles between equal volumes A and B.

volume, so that there is no energetic preference for one location over another. We now ask what are the probabilities of achieving distributions in which all six molecules are in A, in which five are in A and one is in B, *etc.* This is a problem in elementary probability theory, and is discussed in most college algebra texts. The probability of attaining each distribution is simply proportional to the number of ways in which each distribution can be attained. We see that the most probable distribution is one in which the six molecules are equally distributed in the two volume elements. There is a fairly large probability, however, that a non-symmetrical distribution; *e.g.*, four-two, will occur. This would be detectable in this case, since the number of particles is so small. If we have on the order of 10^{23} molecules, however, the probability of achieving a large enough difference in the number of molecules in the two volumes so that it could be detected is essentially zero. The result is that a gas is observed to occupy volumes A and B equally. No one microstate of the system is more probable than any other; it is simply that there are more microstates corresponding to the symmetrical distribution. The expansion illustrated in Fig. 9-12 is thus seen as a process in which the system moves spontaneously toward the most probable state.

Entropy is denoted by the symbol S. It can be defined in terms of the equation

$$S = k \log_e W \tag{9-6}$$

where k is the Boltzmann constant (see page 49), and W is the number of microscopic states available to a system which is under a given set of conditions. When we are discussing quantities of matter which contain on the order of 10^{23} molecules, the number W is fantastically large; except for certain ideal cases, we do not know what it is. It is usually not possible to calculate the entropy for a system simply by inserting an appropriate value for W into the equation. Equation (9-6) is therefore useful to us at this point as a guide to understanding the nature of the entropy function.

Once we have specified a particular macroscopic state for a system (by specifying the pressure, temperature, volume, *etc.*), the number of microstates W is, at least in principle, determined. This means that entropy is a state function — a particularly important point. Since the entropy is a state function, the entropy *change* for a system is determined only by specification of the initial and final states, and is not determined by whether the change occurs reversibly or irreversibly.

The concept of entropy was first considered by the 19th century physicist, Clausius. He noted that it was possible to use some earlier work of Carnot's, dealing with the concept of reversibility, to define a new thermodynamic function. We will not try to repeat all of the steps in Clausius' reasoning, but rather try to understand the nature of his result. Clausius was able to show that the entropy change for an isothermal process occurring at temperature T is given by the expression

$$\Delta S = q_{\text{rev}}/T \tag{9-7}$$

The entropy change is equal to the heat which would be absorbed or evolved *if* the process were carried out reversibly, divided by the absolute temperature. It is very important to note that we must always use the heat change for the reversible process in calculating the entropy change of a system, regardless of whether the actual change from state 1 to state 2 is carried out reversibly. Since entropy is defined as a state function, ΔS cannot depend upon the path taken in going from state 1 to state 2. For an isothermal change, q_{rev} is a state function; there is only *one* way to carry out the change reversibly and isothermally, whereas there are infinitely many ways in which to do it irreversibly. What can we say about a process that is not isothermal? Every such process can be thought of as a succession of little steps in which a small quantity of heat is absorbed reversibly at constant temperature, followed by an adiabatic step ($q = 0$) in which the temperature changes, then another reversible absorption of heat, *etc.* The entropy change is the sum of q_{rev}/T for all of the little isothermal steps in the process. The steps can be made as little as we like, until in the limit we have the integral

$$\Delta S = \sum_{1}^{2} \frac{q_{rev}}{T} = \int_{1}^{2} \Delta \frac{q_{rev}}{T} \tag{9-8}$$

where Σ denotes a sum over all the little steps in the path from 1 to 2. Figure 9-14 shows this relationship schematically. In proceeding from

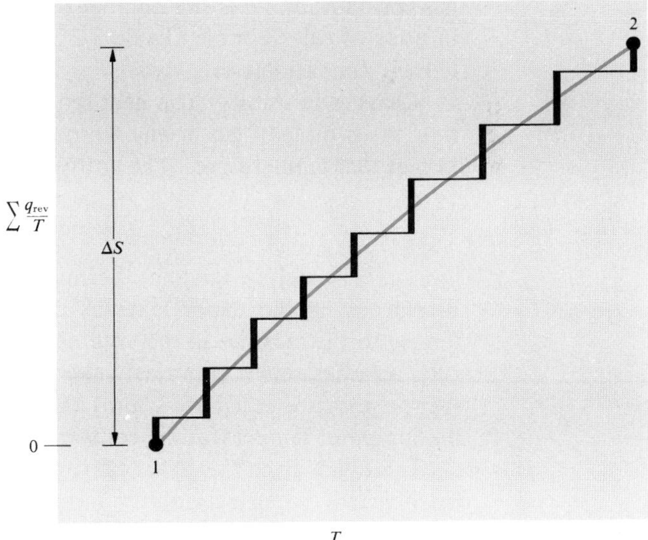

Fig. 9-14 Entropy change in a non-isothermal process, considered as a set of successive adiabatic (horizontal segments) and isothermal (vertical segments) processes.

state 1 to 2, the system acquires added heat in a succession of isothermal steps (represented by the vertical lines), and thus increases in entropy. The smooth line results from making the steps infinitesimally small. It is possible to visualize more than one reversible pathway for a non-isothermal process, with different values of q_{rev}. How, then, can Eq. (9-8) define the entropy change? The answer is that although q_{rev} may not be a unique quantity for a change, $\Sigma\, q_{rev}/T$ *is* unique, no matter which reversible path is taken.

Looked at from the standpoint of probabilities, the entropy change for a process is given by the following:

$$\Delta S = S_2 - S_1 = k \log_e W_2 - k \log_e W_1 = k \log_e \frac{W_2}{W_1} \tag{9-9}$$

If the final state is more probable than the initial state; *i.e.*, has a greater number of microstates W, ΔS is positive. From the viewpoint of the connection with macroscopic properties of the system, the entropy change can be evaluated by starting with Eq. (9-7). At this point, it is worthwhile for us to consider a number of typical processes, to deduce from general considerations whether the entropy change should be positive or negative, and perhaps to obtain formulas for calculating the change.

Volume changes — Entropy increases with an increase in the volume of the system, for the reasons given above. For quantities of matter on a molar scale, Eq. (9-9) leads to the formula $\Delta S = nR \log_e (V_2/V_1)$, where V_1 and V_2 are the initial and final volumes, respectively, and R is the molar gas constant, and n is the number of moles of substance. If R is expressed in units of cal/°K-mole (Table 3-2), S is in these same units, since the log term is dimensionless.

Changes in state — The heat required to reversibly convert a mole of a liquid to a mole of gas at any temperature is given by the heat of vaporization at that temperature. The entropy change on vaporization is, therefore,

$$\Delta S_{vap} = \frac{\Delta H_{vap}}{T} \tag{9-10}$$

Since ΔH is positive (an endothermic process), ΔS is positive. The increase in entropy in the vapor state as compared with the liquid state is due largely to the increase in the volume occupied by the substance. Trouton's rule, a rather rough empirical rule, states that at the normal boiling point (vapor pressure equal to 1 atm) the heat of vaporization divided by the boiling point temperature is constant, and equal to about 21.

Table 9-5 gives the heats of vaporization, normal boiling points, and entropies of vaporization for a number of substances. The values for many substances are quite close to 21, but there are exceptions. The most notable are the unusually high values for liquids which are known to be rather strongly hydrogen bonded. Thus, Trouton's constant for water is 26, much higher than the "normal" 21. The process of vaporization for a hydrogen-bonded liquid involves rupture of hydrogen bonds between molecules. The loss of these bonds results in an additional increase in disorder in passage from the liquid to vapor state.

Table 9-5 Trouton's constants for some liquid substances.

Substance	Boiling point (°K)	Heat of vaporization (kcal/mole)	ΔS_{vap}
Benzene	353	7.35	20.8
Cyclohexane	354	7.19	20.3
Carbon tetrachloride	350	7.17	20.5
Dimethyl ether	248	5.14	20.7
Sulfur hexafluoride	222	4.08	18.3
Tin (IV) chloride ($SnCl_4$)	386	8.3	21.5
Mercury	630	13.89	22.1
Lead	2023	43.0	21.3
Water	373	9.720	26.0
Ammonia	240	5.58	23.3
Phosphine (PH_3)	185	3.49	18.8
Ethyl alcohol	352	9.22	26.2

In a similar way, the entropy of fusion is given by the heat of fusion divided by the melting temperature. The major contribution to the entropy change in this case comes from loss of order in the solid. Table 9-4 lists the data required to calculate ΔS_c for a few substances. It is apparent that ΔS_c is small compared with ΔS_{vap}; for example, ΔS_c for water is 5.26 cal/°K-mole, as compared with $\Delta S_{\text{vap}} = 26$ cal/°K-mole.

Freezing and condensation, processes which are the reverse of fusion and vaporization, proceed with entropy changes which are the same numerically, but opposite in sign, to those described above. This is necessarily so, since temperature is the same, but ΔH is changed in sign. Negative values for ΔS correspond to an increase in order.

Change in temperature — Equation (9-8) furnishes the basis for calculating changes in entropy due to temperature changes. The process can be divided (in principle) into a number of very small steps. The heat absorbed reversibly is equal to the heat capacity times the temperature change. At constant pressure,

$$q_{\text{rev}} = C_p \, \Delta T$$

and

$$\Delta S = \sum_{T_1}^{T_2} \frac{C_p \Delta T}{T} \approx C_p \log_e \frac{T_2}{T_1} \tag{9-11}$$

The approximate equality is applicable only when the heat capacity is essentially constant over the temperature range. In practice, these expressions are handled as integrals in doing numerical evaluations. It is clear that entropy increases with increase in temperature; it increases more rapidly with temperature at constant pressure than at constant volume ($C_p > C_v$) because of the extra contribution from an increase in volume on warming at constant pressure.

Sample Problem

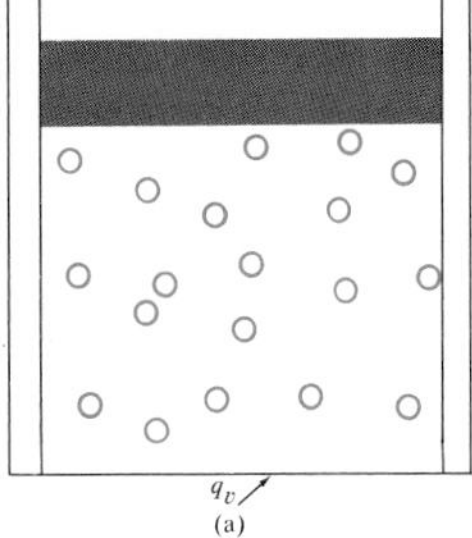

(a)

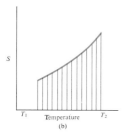

Consider one mole of an ideal gas confined in a cylinder at 300°K, under a pressure of 3 atm. What is the entropy change upon heating the gas to 400°K at constant volume? At constant pressure?

The entropy change is given by the formula $\Delta S = q_{\text{rev}}/T$. The heat required to change the temperature by a small amount ΔT is given by the product of the heat capacity and the temperature change, $q_v = C_v \Delta T$. This is the same as the reversible heat. When we insert this into the expression for ΔS we have

$$\Delta S_v = \sum_{T_1}^{T_2} C_v \Delta T / T_i$$

Each step in the summation corresponds to a tiny change in the temperature from temperature T_i to temperature $T_i + \Delta T$. If the number of steps is made very large (*i.e.*, if ΔT is made very small), the summation becomes an integral, and a formula similar to that in Eq. (9-11) results:

$$\Delta S_v = C_v \log_e \frac{T_2}{T_1}$$

S

T_1 Temperature T_2

(b)

In our case C_v is $\frac{3}{2} \times 1.987$ cal/degree-mole (see Chap. 3), $T_2 = 400$, and $T_1 = 300$. We have

$$\Delta S_v = 2.303 \times 2.980 \times \log_{10} 1.333$$
$$= 0.858 \text{ cal/degree}$$

This increase in entropy results from the flow of heat into the gas during the heating. If the heating is done at constant pressure, the factor which changes is the heat capacity term. At constant pressure, $q_p = C_p \Delta T$. C_p is larger than C_v (Sec. 9-3), and the entropy change in the constant pressure process is correspondingly larger than in the constant volume process. $C_p = \frac{5}{2} \times 1.987$ cal/degree-mole.

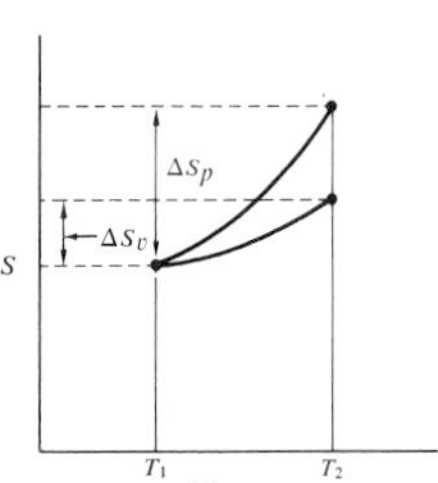

(c)

$$\Delta S_p = 2.303 \times 4.975 \times \log_{10} 1.333$$
$$= 1.43 \text{ cal/degree}$$

The entropy change in the constant pressure process is larger because the volume occupied by the gas increases in the process. In order for the heating to occur at constant pressure, a volume increase must occur.

Note that the actual pressure under which the gas is confined does not enter into the entropy change. Only the volume change (in the constant pressure process) is important. What would the entropy change have been if the cylinder contained two moles of ideal gas rather than one? ΔS would have been twice as great in each case, since the heat capacity of the system is the heat capacity per mole times the number of moles of substance involved.

A final "food for thought" question: Can you calculate ΔS from a two-step process in which the gas is first heated to 400°K at constant volume and then expanded isothermally to give the same pressure as originally? Carry this calculation through for the mole of ideal gas in the above example.

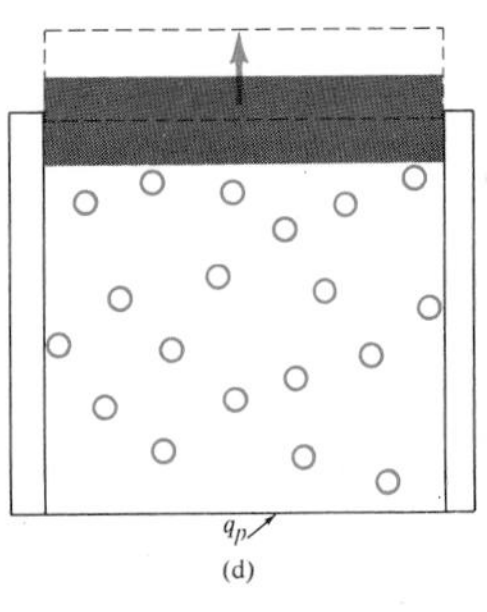

(d)

Equation (9-11) suggests a very interesting possibility: If the heat capacity of a substance over the entire temperature range from absolute zero to, say, room temperature were known, and if the entropy at absolute zero is known, the *absolute* value of the entropy could be determined. There is no way at present of directly measuring the entropy of a substance at absolute zero. It is possible, however, from considering the statistical properties of solid crystalline substances, to state a third law of thermodynamics: *The entropy of a perfectly ordered crystalline solid tends toward zero as the absolute zero of temperature is approached.* The measurement of heat capacities at extremely low temperatures is very difficult, but, fortunately, the experimental approach can be supplemented by theoretical values. The absolute values of entropy are known for a number of substances.

How does the idea of an increasing entropy with increasing temperature fit in with the idea of an increase in randomness? Even at constant volume, the entropy increases quite markedly with increasing temperature. The answer lies in the idea of quantized energy levels discussed in Sec. 9-3. The heat capacity for a diatomic gas increases with temperature, as various degrees of freedom for energy are brought into play. The availability of degrees of freedom for location of energy leads to increased randomness in a manner analogous to the increased randomness resulting from an increase in volume. Instead of the possibility that the molecules of the system might be located at more points in space, there is the possibility that the molecule might be found in more different conditions of internal energy. As more degrees of freedom become activated with increasing temperature, the heat capacity increases. Figure 9-5 shows a typical graph of C_v *vs* temperature for a diatomic molecule. Figure 9-15 shows the corresponding graph of C_v/T *vs* temperature. The absolute entropy is

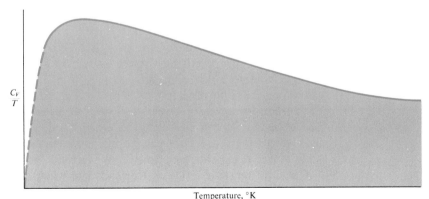

Fig. 9-15 Absolute entropy of a diatomic molecule. The absolute entropy is the area under the curve, measured from absolute zero to the temperature of interest.

given by the total area under this curve, from absolute zero to the temperature of interest. A graph of C_p/T *vs* temperature is similar, but the area under the curve is larger, because there is an added entropy contribution due to expansion.

Mixing — The mixing of two or more substances to form a solution is generally accompanied by increase in entropy. Mixing leads to an increase in randomness because each of the components is free to move in a larger volume than before. For example, consider the mixing of ideal gases A and B, which can be brought about by opening the stopcock separating them, as shown in Fig. 9-16. We can calculate the entropy change for the process by focusing attention on each gas separately. Before mixing, A is confined to the left-hand flask. After mixing it has, in effect, expanded into the second flask as well. The entropy change for this expansion of A is given by $\Delta S_A = n_A R \log_e V_2/V_1 = n_A R \log_e (V_A + V_B)/V_A$. Similarly, the

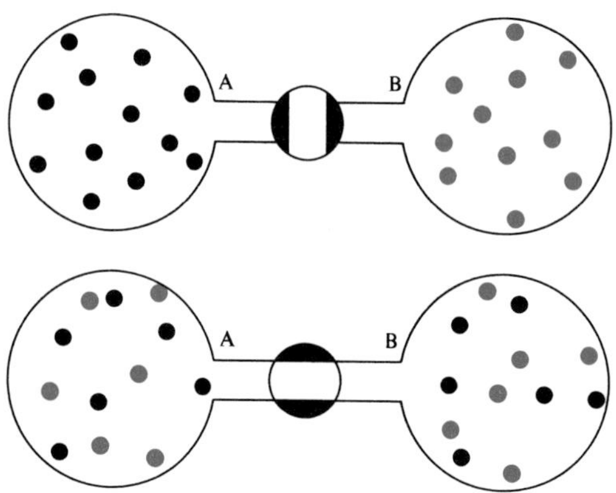

Fig. 9-16 Spontaneous mixing of two gases.

entropy change for expansion of gas B is $\Delta S_B = n_B R \log_e (V_A + V_B)/V_B$. The total entropy of mixing for the system is $\Delta S = \Delta S_A + \Delta S_B$. Since the total volume which each gas occupies after mixing is greater than it was before, ΔS is positive.

The entropy of mixing is not so easy to calculate if there are interactions between the components, as when a salt dissolves in water. In cases where there are not such strong interactions, however, it can be safely said that the entropy of mixing is positive.

Chemical reaction — The entropy change when substances react may be positive or negative, depending upon the nature of the reaction. In reactions where a gas is produced from solid or liquid reactants, a positive entropy of reaction is likely. For example, in the reaction

$$CaCO_3(s) \rightarrow CaO(s) + CO_2(g) \qquad \Delta S^\circ_{25^\circ C} = +38.4 \text{ cal/°K-mole}$$

the entropy change is positive, mainly because of the formation of gaseous product. When a reaction results in a greater number of particles, as in the dissociation of a molecule, the entropy change is again positive:

$$H_2O(g) \rightarrow H_2(g) + \tfrac{1}{2}O_2(g) \qquad \Delta S^\circ_{25^\circ C} = +10.6 \text{ cal/}^\circ\text{K-mole}$$

It should be evident that dissociation leads to an increased randomness, since the hydrogen and oxygen molecules can move independently in the available volume. Further examples of the entropy changes accompanying chemical processes are to be found throughout the text. It must be kept in mind that when the entropy change accompanying a chemical reaction, or any other process, is discussed, the entropy change of the system only is usually meant. An entropy change also occurs in the surroundings, of such sign and magnitude that the net entropy change for the universe as a whole is zero (for a reversible process) or greater than zero (for an irreversible process). This last statement is a consequence of the second law of thermodynamics, which is discussed in the next section.

9-7 The Second Law of Thermodynamics

We have stated that entropy is a measure of the stability of a system. Now consider a completely isolated system. If it is not in a state of maximum stability, it may spontaneously undergo a change in the direction of greater entropy; *i.e.*, greater stability. Once arrived at the state of maximum stability, however, it cannot spontaneously change. A spontaneous change in an isolated system, therefore, is one which results in an increase in entropy of the system. Since the surroundings cannot undergo any change at all if the system is isolated, the net effect is an increase in the total entropy of the universe. One definition of the second law is simply that a spontaneous (and therefore irreversible) process results in an increase in the entropy of the universe. The term *spontaneous* as used in thermodynamics does not imply anything about the rapidity with which a change occurs. The change may be slow or fast; the system is responding, however, to a finite "push" or driving force. For example, the reversible expansion of a gas, discussed in Sec. 9-2, involves expansion against a vanishingly small push, in the form of a tiny pressure differential. An irreversible expansion, on the other hand, would occur if the difference between the external pressure and the internal pressure exerted by the gas is finite. This pressure difference might vary from quite small to the limit, where the external pressure is zero. The degree of irreversibility is, therefore, variable.

The system of interest may not be isolated, but may exchange heat or work with the surroundings. A change in the system may then be reflected in a corresponding change of some kind in the surroundings. If the system is adiabatic, so that no heat is exchanged with the surroundings, it is still true that any spontaneous change must be in the direction of increasing entropy. The entropy of the surroundings does not change in these circumstances. If heat is exchanged with the surroundings, the entropy of the system may increase or decrease. For example, in the reversible vaporization of a mole of water at 100°C and 1 atm pressure, the entropy of the system increases by 26 cal/°K-mole. This increase in entropy is accompanied by the absorption of heat from the surroundings, with a consequent loss of 26 cal/°K-mole of entropy there. The net entropy change for this process is thus zero. Now consider the direct reaction of a mole of hydrogen and one-half mole of oxygen at 25°C to form gaseous water. The entropy change for this process is −10.6 cal/°K-mole, the opposite of the value given on page 231, since the reverse reaction is involved. In the course of this reaction, therefore, the system undergoes a decrease in entropy. How can this be, since the reaction is spontaneous? The answer lies in the fact that this system is not isolated or adiabatic. In order for the reaction to occur isothermally, the heat evolved in the reaction, 57,800 cal, is passed to the surroundings. The surroundings, therefore, undergo an entropy change of 57,800/298 = +194 cal/°K-mole. Clearly, the net entropy change for the universe as a whole has increased. The second law of thermodynamics states, in one form, that *the entropy of the universe increases in any spontaneous process*; *it remains constant for reversible processes.* Since all processes which occur in nature proceed under a finite driving force, it follows that the entropy of the universe is constantly increasing. Although individual systems may be highly ordered, and therefore have relatively low entropy, the order has been achieved at the expense of an even greater degree of disorder in some other part of the universe.

The push, or driving force, which occasions spontaneous change results from energy differences on the one hand, and the tendency toward maximum randomness on the other. Equilibrium results when the energy within a system has been minimized, and the randomness has been maximized. The tendency toward maximum randomness places a limitation on the convertability of heat into other forms of energy. In order to obtain work from a system by utilizing its heat content, it is necessary to set up an engine of some sort which makes use of the spontaneous flow of heat. For example, the engine might be an internal combustion engine, a turbogenerator or a rocket engine. But spontaneous heat flow occurs only if the heat sink is at a lower temperature than is the heat source. Thus, the

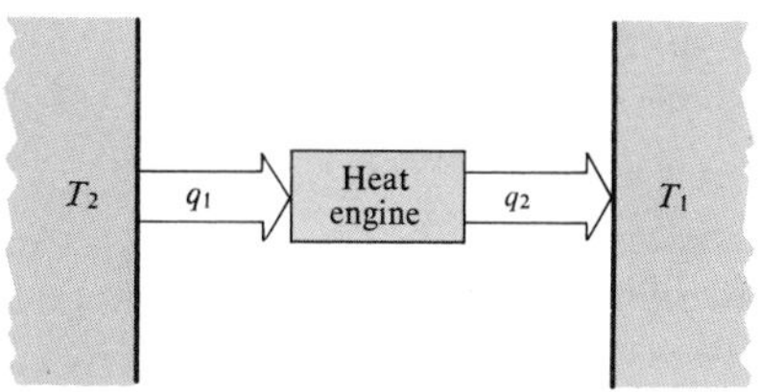

Fig. 9-17 Heat flow in a generalized heat engine.

heat engine (indicated schematically in Fig. 9-17) can convert heat from the heat source at temperature T_2 into useful work only if $T_2 > T_1$. Furthermore, an analysis of this system using thermodynamic principles shows that, even if the engine operates reversibly, so that it has the maximum possible efficiency, it can convert only a portion of the heat received from the source into useful work. The ratio of the maximum work to the quantity of heat received is given by the relationship

$$\frac{W_{\text{max}}}{q_1} = (T_2 - T_1)/T_2 \tag{9-12}$$

Any process which utilizes the energy in a material by first converting that energy into heat is restricted in efficiency, regardless of the efficiency of the engine itself. The performance of a real engine can be gauged by comparing its efficiency with that which a reversible engine would have under the same conditions. Equation (9-12), therefore, is useful as a guide in estimating the efficiency of real engines. For example, consider a steam engine operating between the boiling point of water and 35°C. If the pressure in the system is 1 atm, the boiling point is 100°C, and the theoretical efficiency is given by

$$\frac{373 - 308}{373} = 0.174$$

If the system is pressurized, the temperature of the steam can be increased. At 50 atm the boiling point of water is 265°C. The theoretical efficiency under these conditions is

$$\frac{538 - 308}{538} = 0.428$$

If the temperature of the discharge is made lower, the efficiency of the engine also increases. In the limit — when the discharge temperature is absolute zero — the efficiency of the engine is unity. Of course, this condition can never be obtained in practice.

Another interesting special case arises when the temperatures of the two heat reservoirs are the same ($T_2 = T_1$). Under these conditions, the efficiency is zero. From this result it can be stated that *it is not possible to convert heat into work by means of an isothermal cycle.*

These observations upon the heat engine provide the basis for another statement of the second law*: *It is impossible to construct a machine, operating in cycles, which will produce no effect other than the absorption of heat from a reservoir and its conversion into an equivalent amount of work.* The key phrase is "equivalent amount of work". The work obtained is, as we have seen, always less than the heat absorbed.

---

*S. Glasstone, *Thermodynamics for Chemists*, D. Van Nostrand Co., Inc., New York, 1947.

9-8 Free Energy

For isolated systems, the condition that the system be at equilibrium is very simple; the entropy is at a maximum. Chemists, however, are most often interested in isothermal systems at constant pressure. A new thermodynamic state function, the free energy G, sometimes called the Gibbs free energy, embodies the condition for equilibrium. For an isothermal system at constant pressure, the condition of equilibrium corresponds to a minimum in the Gibbs free energy. G is defined by the equation

$$G = H - TS$$

For a change at constant pressure and temperature,

$$\Delta G = \Delta H - T\Delta S \tag{9-13}$$

The free energy is a measure of the maximum net work available in the system. Any change which proceeds spontaneously must occur because of the finite push or driving force which we have spoken of before. The existence of this driving force can be put to use in doing work. If the process were carried out reversibly, so that the maximum work were extracted from the change, the amount of useful work realized would be ΔG. (Useful work, in this case, is work other than the work of expansion of gases in the system.) The sign of ΔG must be negative for a process which proceeds spontaneously; *i.e.*, there must be a loss in power to do work. Equilibrium corresponds to a minimum in the free energy function. A system which is at equilibrium cannot change spontaneously; but if it cannot change spontaneously, the free energy change with respect to some possible change in the system must be zero. This is true because ΔG is a measure of the maximum work extractable from a change; if the change is between two states which can exist in equilibrium with one another, no useful work can be derived. Free energy is a state function, since it is defined entirely in terms of quantities which are state functions. This means that ΔG for any process is determined only by the initial and final states, and not by the path taken in carrying out the change. We note that ΔG corresponds to the *maximum* amount of net useful work obtainable from a given change. We may not get this amount of work out, and indeed, we seldom come very close in practical situations. ΔG is a measure of the work which *might* be gotten out if the process were conducted reversibly.

The significance of the free energy function can be further appreciated by examining Eq. (9-13). A negative ΔG is required for the reaction to proceed spontaneously. If it were not for entropy effects, all exothermic reactions (ΔH negative) would be spontaneous. The entropy contribution, $-T\Delta S$, represents an additional source of driving force for reaction. If ΔS is positive, meaning that the final state is more disordered than the initial, $-T\Delta S$ yields a negative contribution to ΔG; *i.e.*, it favors a spontaneous reaction. We can distinguish various types of changes according to the signs of ΔS and ΔH. This is done in Table 9-6. When ΔH and ΔS work in opposite directions on ΔG, temperature becomes the overriding factor in determining the direction in which a spontaneous change occurs. It is important to emphasize once again that the term spontaneous as used in thermodynamics does not imply anything about the rapidity with which a process occurs. Thus, when ΔH and ΔS are both negative for a process, it occurs spontaneously at low temperatures. But this means only that *if* the process occurs, it will occur in the direction indicated by negative ΔG. It may be that the rate of the process is extremely slow at low temperatures. The rate might be changed by adding a catalyst, by initiating the reaction with light, or by some other means, but the value for ΔG remains the same.

It is difficult to give hard and fast rules for predicting the course of chemical reactions at a function of temperature, because both ΔH and ΔS may vary with temperature. The categories of Table 9-6 are, therefore, to be taken only as a general guide.

Just as it is possible to define and tabulate a standard enthalpy of formation for elements and compounds (Sec. 9-4), standard free energies of formation can also be tabulated. The convention is employed that the standard state refers to the pure substance at 1 atm pressure, and (usually) 25°C (298°K). The standard free energies of formation of the elements in their stable form under these conditions is defined to be zero, just as for the standard enthalpies of formation. We do not, of course, know the absolute values of these thermodynamic quantities for substances. Our interest, however, is in evaluating *changes* in them during some process such as a chemical reaction. Beginning with the definition of standard values for the elements, the free energies of formation of compounds can be evaluated from experimental data. Through the relationship $\Delta G^\circ = \Delta H^\circ - T\Delta S^\circ$, the standard entropies of formation can be calculated from the values given for the other two thermodynamic functions. One must be cautious here, however, because *absolute* entropy values are known and tabulated for many substances. The absolute and standard entropies are not the same. For example, the absolute entropy of nitrogen gas at 298°K is 45.77 cal/°K-mole. The standard entropy on the other hand, is zero, since $T\Delta S_f^\circ = \Delta H_f^\circ - \Delta G_f^\circ = 0$, by definition. We will use the symbol ΔS_f° for the standard entropy values, and S° for the absolute entropy values. In calculating the change in entropy in a chemical reaction, it does not matter whether the standard entropies or absolute entropies are used, since only the difference matters. One must be consistent, however, in using the same scale for all of the quantities employed in the calculation. Some values of standard enthalpy and free energy of formation, and some absolute entropy values are listed in Appendix E.

As a summarizing example of how the laws of thermodynamics may be applied to a chemical system, let us consider the reaction of a solution containing Cu^{+2} ions with zinc metal:

$$Cu^{+2}(aq) + Zn(s) = Cu(s) + Zn^{+2}(aq) \tag{9-14}$$

Table 9-6 Behavior of the free energy function for various types of systems.

ΔH	ΔS	Behavior of ΔG	Examples
$-$ (or O)	$+$ (or O)	ΔG is negative at all temperatures. The process proceeds spontaneously at all temperatures.	$O_3(g) \rightarrow \frac{3}{2}O_2(g)$ $[\Delta H\,(-), \Delta S\,(+)]$ $H_2(g) + Cl_2(g) \rightarrow 2HCl(g)$ $[\Delta H\,(-), \Delta S\,(\sim O)]$
$+$ (or O)	$-$ (or O)	ΔG is positive at all temperatures. The process proceeds spontaneously in the reverse direction at all temperatures.	$\frac{3}{2}O_2 \rightarrow O_3(g)$ $2NO_2(g) \rightarrow N_2O_4(g)$ $[\Delta H\,(\sim O), \Delta S\,(-)]$
$-$	$-$	ΔG is negative at low temperatures, and positive at high temperatures. The process proceeds spontaneously at low temperatures, but tends to reverse at high temperatures.	$Cl(g) + Cl(g) \rightarrow Cl_2(g)$ $NO(g) + \frac{1}{2}O_2(g) \rightarrow NO_2(g)$ $Mn(s) + O_2(g) \rightarrow MnO_2(s)$
$+$	$+$	ΔG is positive at low temperatures, and negative at high temperatures. The process proceeds spontaneously in the reverse direction at low temperatures, but tends to proceed in the forward direction at high temperatures.	$Cl_2(g) \rightarrow Cl(g) + Cl(g)$ $NO_2(g) \rightarrow NO(g) + \frac{1}{2}O_2(g)$ $MnO_2(s) \rightarrow Mn(s) + O_2(g)$

This reaction might be carried out in a variety of ways. If the materials are mixed together in an isothermal calorimeter at 25°C, the reaction proceeds with evolution of heat, and with no useful work being accomplished. Since there are no gases involved, pressure-volume work is negligible. Therefore, according to the first law, $\Delta E = q$. If one mole of zinc is employed with a solution containing a mole of Cu^{+2} approximately 1 molar concentration, the heat evolved is 51,400 cal. Thus, we know the value for q and ΔE, $-51{,}400$ cal. We now wish to learn the values for ΔH, ΔS, and ΔG. The enthalpy change is easy to determine, since at constant pressure, $\Delta H = \Delta E + P\Delta V$. But since the last term is essentially zero, $\Delta H = \Delta E = -51{,}400$ cal. In order to learn the values for ΔS and ΔG, we must obtain information about the reaction when it is carried out reversibly. It would be incorrect, for example, to set $\Delta S = -51{,}400/298$, since ΔS is related to the heat evolved when the process is conducted *reversibly*. The value 51,400 cal is the heat evolved when the process is conducted completely *irreversibly*.

Our system can be made the basis for an electric cell, as shown in Fig. 9-18. We will go into the matter of electric cells in some detail in a later chapter; for now it is sufficient to note that by using the cell under conditions where the cell reaction is conducted reversibly, the free energy change for the reaction can be determined. For the reaction to proceed reversibly, it is necessary that it proceed at an infinitely slow rate. If the voltage is measured under conditions such that an infinitesimal current is flowing in the external circuit, the conditions approximate those for a reversible reaction. When this is done, the voltage $\mathcal{E}$ of the cell is 1.107 v. The energy represented by this voltage is given by $n\mathcal{F}\mathcal{E}$, where n is the number of equivalents of charge per mole, 2, and $\mathcal{F}$ is the faraday, which converts voltage to units of energy. It has the value of 23,060 cal/volt-equiv. The free energy change is then

$$\Delta G = -n\mathcal{F}\mathcal{E} = -2 \times 23{,}060 \times 1.107 = -51{,}070 \text{ cal}$$

We can be sure that this is the free energy change for the reaction because the free energy change is defined as the maximum useful work obtainable

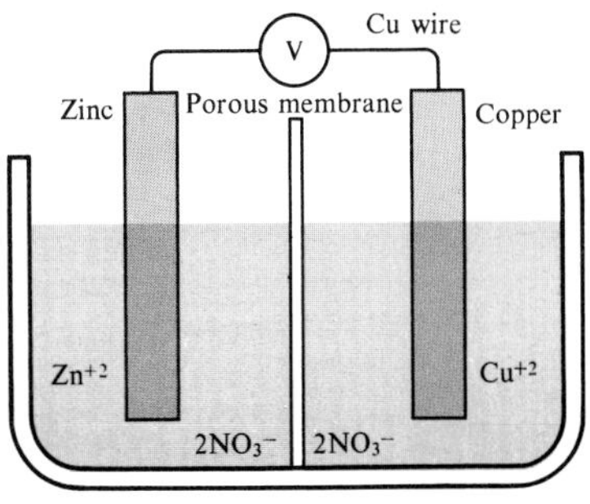

Fig. 9-18 Electric cell for copper-zinc system.

from the reaction. By measuring the voltage under conditions in which the reaction is proceeding reversibly, we have measured the maximum useful work obtainable.

Thus, $-51{,}070$ represents w_r, the reversible work for the isothermal, process. Let us now return to the first law; we have already determined that ΔE for the isothermal process is $-51{,}400$ cal. Since ΔE is a state function, this must be the value of ΔE for the process no matter how it is carried out. But if we proceed by the reversible path, $\Delta E = q_r - w_r$. We have then

$$-51{,}400 = q_r - (-51{,}070)$$
$$q_r = -330 \text{ cal}$$

There is no direct way of measuring the heat evolved when the reaction is carried out reversibly (since it takes forever to complete). By making use of the state function property, however, q_r is readily calculated.

Using this value, it is now possible to calculate ΔS.

$$\Delta S = -330/298 = -1.13 \text{ cal/deg}$$

It might be pointed out once more that the entropy change *for the system* is the same for both of the methods of carrying out the reaction. Since the initial and final states of the system are the same, it follows of necessity that the entropy change must be the same. The entropy change *in the surroundings* is, however, quite different in the two cases. When the reaction is carried out reversibly, the heat evolved is 330 cal. The entropy change for the surroundings is then $330/298 = 1.13$ cal/deg. The over-all entropy change for the universe is the sum of the entropy change for the system plus that for the surroundings: $1.13 + (-1.13) = 0$.

When the reaction is carried out irreversibly (in the calorimeter), the heat evolved is 51,400 cal. The entropy change for the surroundings is then $51{,}400/298 = 176$ cal/deg. The over-all entropy change in this instance is $-1.13 + 176 = 174.9$ cal/deg. The over-all entropy change is positive, as is always the case when a process proceeds irreversibly.

Suggested Readings

Bridgeman, P. W., "Reflections on Thermodynamics," *American Scientist*, Vol. 41 (1953), p. 549.

Mahan, B. H., *Elementary Chemical Thermodynamics*, New York: W. A. Benjamin, Inc., 1963.

Nash, L. K., *Elements of Chemical Thermodynamics*, Reading, Massachusetts: Addison-Wesley Publishing Company, Inc., 1962.

Waser, J., *Basic Chemical Thermodynamics*, New York: W. A. Benjamin, Inc., 1966.

Exercises

9-1. Using the assumption that the volume of a gas is proportional to the absolute temperature, calculate a value for absolute zero from each of the following sets of data:

(a) At 1 atm of pressure, the volume of 1 mole of nitrogen gas is 22.401 liters at 0°C, and 30.627 liters at 100°C.
(b) At 0.1 atm of pressure, the volume of 1 mole of nitrogen gas is 224.13 liters at 0°C, and 306.20 liters at 100°C. [*Hint:* Use five- or six-place log tables.]

9-2. Devise an experiment which would permit the evaluation of the mechanical equivalent of heat. What is needed here is a method by which a known amount of mechanical energy (kinetic, potential, *etc.*) can be converted into heat.

9-3. What is meant by the term *state function*? Why is work not, in general, a state function?

9-4. Calculate the work done by the system in each of the following processes:
(a) A mole of ideal gas expands isothermally from a volume of 5.5 liters to a volume of 10.5 liters against a zero external pressure at 25°C.
(b) Two moles of an ideal gas expand isothermally from a volume of 5.5 liters to a volume of 10.5 liters against an external pressure of 1 atm at 25°C.

9-5. A mole of helium gas is expanded adiabatically from a volume of 11.2 liters to a volume of 16.4 liters, against an external pressure of 0.5 atm, beginning from an initial temperature of 25.00°C. The same experiment is carried out with xenon. In each case, the final temperature is measured accurately. In which system will the temperature be lower? Explain your answer. [*Hint:* Consider the effects of non-ideality.]

9-6. Find the work (in calories) performed in each of the following processes:
(a) Two moles of an ideal gas is expended isothermally from 25 to 60 liters against a constant pressure of 0.5 atmosphere at 25°C.
(b) Two moles of an ideal gas is expanded isothermally and reversibly from 25 to 60 liters at 25°C.
This problem illustrates the relationship between the work performed in a reversible process as compared with that performed in a process which is not reversible. How do the values of q and ΔE compare in these two cases?

9-7. One mole of an ideal gas is expanded isothermally and reversibly from a volume of 10 liters to a volume of 20 liters at 27°C. It is then cooled reversibly at constant pressure until its volume is again 10 liters. What are net values of ΔE, q and W for the over-all process?

9-8. The combustion of methane in a bomb calorimeter (*i.e.*, at constant volume) at 25°C according to the reaction

$$CH_4(g) + 2O_2\ (g) \rightarrow CO_2(g) + 2H_2O(1)$$

yields 211.6 kcal of heat per mole of methane. What is ΔH for the reaction?

9-9. At constant pressure, the heat capacities of gaseous chlorine and gaseous iodine at 298°K are 8.11 and 13.1 cal/deg-mole, respectively.

Since both gases consist of diatomic molecules, why are the two heat capacity values not the same? (In other words, what can you deduce about the spacings of the vibrational energy levels for I_2 *vs* Cl_2?)

9-10. At constant pressure, the heat capacity of nitrogen gas, in the temperature interval 300 to 1000°K, is expressed by the empirical equation

$$C_p = a + bT$$

where the values for a and b are as follows:
$a = 6.524$ cal/deg-mole
$b = 1.25 \times 10^{-3}$ cal/deg^2-mole
Some *constant volume* heat capacities are listed below for nitrogen gas:

Temp (°K)	500	700	900
C_v (cal/deg-mole)	5.08	5.36	5.68

(a) Using the empirical equation given above, compare the calculated values of heat capacity with the observed values.
(b) Determine graphically the heat energy required to heat 1 mole of nitrogen gas from 400°K to 800°K at constant pressure. (If you are familiar with calculus you can do the problem numerically, by integrating C_p between the temperature limits.)
(c) Why does the heat capacity of nitrogen vary with temperature?

9-11. The heat of vaporization of methyl alcohol at 25°C is 8,950 cal/mole. From the data on p. 214 regarding the heat of formation of $CH_3OH(g)$, calculate the heat of formation of $CH_3OH(l)$.

9-12. The heat of combustion of benzene, $C_6H_6(l)$, at 25°C, is -781.0 kcal/mole. Calculate the heat of formation of benzene.

9-13. Discuss the temperature variation in the molar heat of vaporization of water (Table 9-3) from a molecular point of view. Why does it decrease with increasing temperature? Also compare ΔH_v for water, methyl alcohol and carbon tetrachloride at 0°C and indicate the source of the variation.

9-14. Using Appendix E, calculate the heats of each of the following reactions:
(a) $\underset{\text{ethylene}}{C_2H_4(g)} + H_2(g) \rightarrow \underset{\text{ethane}}{C_2H_6(g)}$.
(b) $3H_2(g) + N_2(g) \rightarrow 2NH_3(g)$.
(c) $H_2(g) + F_2(g) \rightarrow 2HF(g)$.
(d) $3H(g) + N(g) \rightarrow NH_3(g)$.
(e) $CaCO_3(s) \rightarrow CaO(s) + CO_2(g)$.
(f) $3Cl_2(g) + 2Sb(s) \rightarrow 2SbCl_3(s)$.
(g) $2Fe(s) + \frac{3}{2}O_2(g) \rightarrow Fe_2O_3(s)$.
(h) $Fe_2O_3(s) + 2Al(s) \rightarrow 2Fe(s) + Al_2O_3(s)$.
(i) $Ag^+(aq) + Cl^-(aq) \rightarrow AgCl(s)$.
(j) $CaCO_3(s) + 2H_3O^+(aq) \rightarrow Ca^{+2}(aq) + 3H_2O(l) + CO_2(g)$.

9-15. From the following heats of formation, taken from NBS Circular No. 500, calculate the average thermochemical energy of the bond to chlorine for each of the Group V elements.

P(g)	74.18 kcal/mole	PCl_3(g)	−73.22 kcal/mole
As(g)	60.64 kcal/mole	$AsCl_3$(g)	−71.5 kcal/mole
Sb(g)	60.8 kcal/mole	$SbCl_3$(g)	−75.2 kcal/mole
Bi(g)	49.7 kcal/mole	$BiCl_3$(g)	−64.7 kcal/mole
Cl(g)	29.01 kcal/mole		

(a) What can you say about the way in which this quantity varies as a function of atomic number in this group?
(b) By extrapolation (or inspired guess), estimate a value for the nitrogen-chlorine bond energy. Compare with the value based on experimental evidence, given in class.

9-16. In addition to pressure-volume work, systems of chemical interest may also perform other kinds of work. Calculate the work in calories required at 25°C to change 1 mole of water from a single spherical drop to a spray with particle size of 10^{-4} cm diameter. The surface tension of water at 25°C is 72 dyne/cm. [*Hint:* This problem can be solved easily by merely considering units. Work has the dimensions of energy, dyne-cm. Surface tension, which is a measure of the force acting to decrease the surface area, is given in units of dyne/cm.]

9-17. The "law" of Dulong and Petit — really an empirical rule announced by them in 1819 — says that the molar heat capacities of all solid elements are the same, the value being 6.4 cal/deg-mole. This means, then, that the heat capacity per gram times the atomic weight should equal 6.4. The rule was thus useful in obtaining some idea of the value for the atomic weight of a newly-discovered element. Use a handbook of chemical data to test the rule on the elements with the following atomic numbers: 22, 24, 26, 27, 32, 42, 47, 50, 53, 56, 74, 78, 79, 82. Compute the molar heat capacity in each case, compare the average with the empirical value of 6.4, and calculate the average deviation of the results.

9-18. The standard molar enthalpies of formation of gaseous acetylene (C_2H_2), ethylene (C_2H_4), and ethane (C_2H_6), are +54.19, +12.50, and −20.24 kcal/mole, respectively. Calculate the heat per *mole* of each substance evolved on combustion. Calculate the heat evolved per *kilogram* of each substance. Assuming that the heats of vaporization are about the same, which substance would make the most efficient fuel in terms of heat units per unit weight of liquid fuel?

9-19. 4.00 g of KCl is added to 200 ml water in a calorimeter of the type shown in Fig. 9-8. The temperature is observed to change from 25.818 to 24.886°C. The heater in the calorimeter has a resistance of 6.00 Ω. A current of 4.00 amps is passed for a period of 20.0 sec. The temperature in the calorimeter is observed to increase by 2.290°C. From these data, calculate the molar heat of solution of KCl in water at 25°C.

9-20. Which has the higher entropy of fusion, lithium nitrate or sulfur? (See Table 9-4) What explanation can you give to account for the difference?

9-21. Liquid sodium and potassium are miscible with one another (*i.e.*, they form a solution) over the entire concentration range. Since the two elements are very similar, what do you predict for ΔH for this process? What is the direction of free energy change for solution formation? Can this be accounted for in terms of the expected entropy change? Explain.

9-22. Assuming that no changes other than volume are involved, calculate the entropy change on mixing one mole of liquid sodium (density 0.95 g/cm^3) with one mole of liquid potassium (density 0.85 g/cm^3).

9-23. What do you predict for the entropy change of the system for each of the following processes carried out isothermally and at constant pressure? Indicate your reasoning.
(a) $3H_2(g) + N_2(g) \rightarrow 2NH_3(g)$.
(b) 50 g benzene are mixed with 25 g carbon tetrachloride to form a solution.
(c) $BaCl_2(s) + 2H_2O(g) \rightarrow BaCl_2 \cdot 2H_2O(s)$.
(d) $NH_4Cl(s) \rightarrow NH_3(g) + HCl(g)$.

9-24. The enthalpy changes for the four processes in Prob. 9-23 are as follows: (a) small negative; (b) about zero; (c) fairly large negative; (d) large positive. From this information, indicate whether the processes should proceed spontaneously in the direction as written at high temperatures; at low temperatures.

9-25. Indicate whether each of the statements is true or false. Where the statement is false as written, indicate why it is false, or under what conditions it might be false.
(a) Heat and work are entirely interchangeable.
(b) The work done by a system is equal to the heat absorbed by a system.
(c) At constant temperature, the energy of a system is constant.
(d) The energy of an isolated system is constant.
(e) The energy of a system which undergoes a spontaneous process must decrease.
(f) A system which undergoes a spontaneous change must necessarily become more disordered.
(g) The enthalpy of a system does not change during an adiabatic process.
(h) The free energy of a system decreases in a spontaneous process.
(i) Spontaneous processes are those which proceed rapidly under a given set of conditions.

9-26. From the data in Table 9-3, calculate the entropy change for vaporization of a mole of benzene at 80.1°C and a mole of water at 100°C. What structural argument can be given for the difference in the two quantities?

9-27. Calculate the quantities, $\Delta G°$, $\Delta H°$, and $\Delta S°$ from the thermodynamic data listed below for the isothermal decomposition of carbonyl chloride, $COCl_2$, according to the reaction

$$COCl_2(l) \rightarrow CO(g) + Cl_2(g)$$

The decomposition is to be conducted at 25°C, and under such conditions that the product gases are in their standard states; *i.e.*, at approximately one atmosphere of pressure

	ΔG_f°(kcal)	ΔH_f°(kcal)
Carbonyl chloride	−48.96	−53.5
Carbon monoxide	−32.81	−26.42

ΔG_f° and ΔH_f° refer to the standard free energy and enthalpy of formation, respectively.

9-28. Indicate in some detail how the second law of thermodynamics may be thought of as
(a) a definition of temperature.
(b) a definition of heat.
(c) a statement about entropy change and its relation to heat flow.

9-29. One very short and easily-remembered version of the first and second laws of thermodynamics goes as follows — First law: You can't win; Second law: You can't even break even. Elaborate on these two statements, using a flashlight battery as your example.

9-30. The heat capacity of liquid water in the temperature range from 0 to −30°C is about 1.0 cal/g. Now suppose that one mole of liquid water which has been supercooled to −30°C is placed in an adiabatic container which has no heat capacity. Assume that crystallization begins, and that the system moves to equilibrium. The heat of fusion of ice at 0°C is 80 cal/g. The heat capacity of ice in this temperature range is 0.49 cal/g. What is the temperature of the system at equilibrium? How much ice is present? What change in the entropy of the system has occurred? What change in free energy? [*Hint:* Use the state function property, devise a path for the change which is easy from a computational standpoint. You actually don't need quite all of the information given.]

9-31. Using the data of Appendix E, calculate ΔG°, ΔH°, and ΔS° for each of the following processes:
(a) $2HI(g) + Cl_2(g) \rightarrow 2HCl(g) + I_2(g)$.
(b) $C_2H_2(g) + H_2(g) \rightarrow C_2H_4(g)$.
(c) $C_2H_2(g) + 2H_2(g) \rightarrow C_2H_6(g)$.
(d) $3CaO(s) + 2Fe(s) \rightarrow Fe_2O_3(s) + 3Ca(s)$.

9-32. Utilizing the data from (b) and (c) of Prob. 9-31, calculate ΔG°, ΔH°, and ΔS° for the process

$$C_2H_4(g) + H_2(g) \rightarrow C_2H_6(g)$$

9-33. In NBS Circular No. 500, the vaporization of gallium metal is reported as follows:

$$Ga(l) \rightarrow Ga(g) \quad (p = 0.0006 \text{ mm Hg}) \qquad T = 1210°K$$
$$\Delta H = 63.8 \text{ kcal/mole}$$

Calculation of ΔS_{vap} from these data leads to a much larger value than those reported in Table 9-5. Calculate a correction to the value obtained from the above data which will provide a more useful comparison with the data of Table 9-5.

9-34. The following questions pertain to the copper-zinc system which is discussed in the text.
(a) Does the free energy change for the reaction represented by Eq. (9-14) depend upon the manner in which it is carried out?
(b) Does the entropy change for the system depend upon the manner in which it is carried out?
(c) Suppose that the reaction is carried out by mixing the zinc metal with the copper solution in an isothermal calorimeter at 25°C, as described. The heat which is liberated is then used to drive a small heat engine, which has a Dewar vessel filled with a dry ice-acetone bath at −78°C for its cold reservoir. What is the maximum amount of work that can be realized from the reaction (per mole of reactants) under these conditions?
(d) The answer for part (c) is not the same amount of work as that which can be obtained by using the reaction in an electric cell, since the work obtainable from the system in this instance is 51,070 cal. How does this fact fit in with the answer to part (a)?

9-35. Suppose that proposals have been advanced to carry out research leading to the large scale industrial development of each of the following reactions:
(a) Synthesis of acetic acid from methane,

$$CH_4 + CO_2 \rightarrow CH_3CO_2H$$

(b) Synthesis of methanol from methane,

$$CH_4 + CO_2 \rightarrow CH_3OH + CO$$

(c) Recovery of sulfur from flue gases,

$$2CO + SO_2 \rightarrow 2CO_2 + S$$

On the basis of thermodynamic considerations, how would you evaluate the chances of these proposals for ultimate success? Which, if any, would you recommend for further study?

9-36. Nitrogen trioxide is an unstable species which seems to be present in certain reactions of nitrogen compounds. The thermodynamic functions for it are: $\Delta G_f^\circ = 27.7$ kcal/mole, $\Delta H_f^\circ = 17.0$ kcal/mole, at 25°C. Calculate the free energy and enthalpy changes for each of the following reactions:
(a) $NO_3(g) + NO_2(g) \rightarrow N_2O_5(g)$.
(b) $NO_3(g) \rightarrow \frac{1}{2}N_2(g) + \frac{3}{2}O_2(g)$.
(c) $NO_3(g) + SO_2(g) \rightarrow NO_2(g) + SO_3(g)$.

Chapter 10

Liquids and Solutions

10-1 The Liquid State

The liquid state is the most difficult of the three states of matter to treat from a theoretical point of view. In a gas, the molecules are sufficiently far removed from one another so that intermolecular interactions are relatively weak in comparison with the kinetic energy. In a solid, the intermolecular (or interionic) interactions are relatively much stronger. A theoretical approach to the solid is made somewhat more feasible, however, by the highly ordered character of the solid lattice; at any rate, we know where the units of the lattice are with respect to one another. In a liquid, the intermolecular interactions are of about the same magnitude as those in the solid. The liquid state seems to differ from the solid mostly by the loss of long range order; *i.e.*, the existence of a regular lattice arrangement of units which extends for some distance throughout the substance.

When solids melt, there is usually an increase in volume on the order of 10 to 20 per cent. This increase in volume is associated with the loss of long range order in the solid.

The molecules of a solid possess kinetic energy in the form of vibrational motions about their equilibrium positions. As the temperature increases, the average vibrational energy also increases. Assuming that the vibrational energies in a solid are distributed in the same way that velocities are distributed in a gas, it is easy to imagine that some of the surface molecules will possess sufficient energy to escape from the attractive forces of the surrounding molecules; consequently, the solid should possess a vapor pressure. Solids do indeed exhibit vapor pressure, which varies in the same general way with temperature as does the vapor pressure of a liquid.

The melting of a solid involves more, however, than just the loss of molecules from the surface. Just as some of the molecules on the surface acquire the energy required to escape from their equilibrium positions into the vapor phase, so also some of the molecules in the interior of the solid acquire enough vibrational energy to escape from their equilibrium positions, causing a disruption of the lattice at that point. These interior molecules generally move into relatively unstable positions between other molecules, and eventually return to a stable position. There is, then, an equilibrium in the interior of the solid between molecules moving into unstable positions and those returning to stable positions. The lattice depends for its stability upon the regular arrangement of molecules which characterizes the solid. A few disruptions at scattered points will not cause it to collapse, but when these reach a certain number per unit volume it will do so. The key point about melting is just this: The melting of the solid — composed of a very large number of molecules — is determined by the behavior of a much smaller number of molecules. Transitions of this kind are termed *cooperative*, since the molecules seem to act together in undergoing the phase change at a single, characteristic temperature.

Crystallization, the reverse of melting, occurs when a number of molecules of sufficiently low energy are grouped together to form the nucleus of a crystal. Once such a crystal nucleus has formed, further solidification may occur by growth on its surfaces. It is often possible to lower the temperature of a liquid below the point at which crystallization should occur. The liquid is then said to be *supercooled.* Supercooling occurs because of the lack of a nucleus on which crystal growth may proceed, and represents a metastable state of the system. Any kind of disturbance of the system may initiate nucleation, with the result that rapid growth of crystals ensues. This disturbance may consist of introduction of a dust particle, scratching the inside wall of the container or perhaps merely tapping it. Addition of a tiny seed crystal of the substance is one of the best ways of initiating crystallization.

It should not be forgotten that, even though melting or crystallization occurs at a single temperature, heat must still be absorbed or evolved in proportion to the amount of material present in order for all of the material present to change its state. For example, 18 grams of water is converted from liquid to solid at 0°C by the evolution at this temperature of

1,430 calories of heat. Similarly, absorption of the same quantity of heat is required to convert 18 grams of ice to liquid water at 0°C.

It might appear at first glance that the gas kinetic theory, which was designed to treat an assembly of non-interacting molecules separated by large distances, could not be applied profitably to the liquid state. The distances between molecules in the liquid state are small, and the intermolecular forces are comparatively large. One can imagine a gas to be continuously compressed until the average distance between molecules

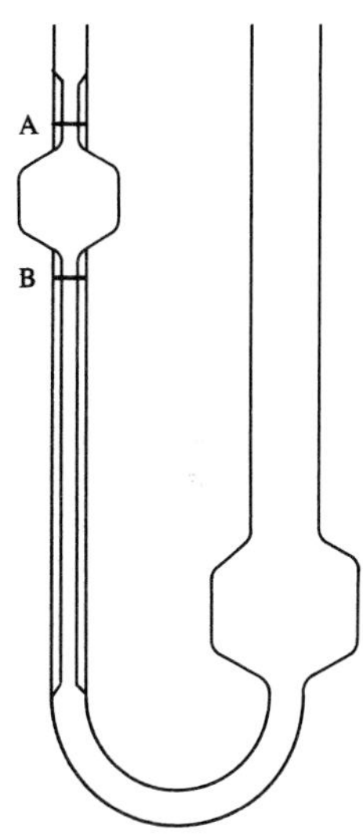

Fig. 10-1 An Ostwald viscosimeter, used in determining the viscosities of liquids. The instrument is employed in determining *relative* viscosities, and must be calibrated with a liquid of known viscosity. A known volume of liquid is introduced into the viscosimeter and drawn up into the region between the marks A and B. At an initial time t_0, the bottom level of the liquid is at B. The liquid is then allowed to fall through the capillary, and the time required for all of it to pass B is noted. The equation which relates the viscosity to the time of flow and to the density ρ of the liquid is $\eta = k\rho t$, where η is viscosity. The constant k is characteristic of the viscosimeter, and is determined by making measurements with a liquid of known viscosity.

approaches the distance characteristic of the liquid state, but this change by itself does not make a liquid. It is necessary to make allowance for the attractive forces acting between molecules. When this is done, the model system which results bears many qualitative resemblances to the gaseous state, but at the same time possesses the properties characteristic of liquids.

The intermolecular forces which operate between molecules in a liquid are evidenced by the quantity of heat required to separate the molecules from one another in the process of vaporization. There are, in addition, other physical properties associated with liquids which are also of value in providing insight into the character of the intermolecular forces. *Viscosity* refers to the resistance to flow of a bulk quantity of liquid. One method of measuring the viscosity involves determining the time required for a given quantity of liquid to flow through a narrow capillary tube under a given pressure (Fig. 10-1). The viscosities of liquids are generally expressed in *millipoises.* A *poise* is a dyne-sec/cm^2. Table 10-1 lists the viscosities of a number of familiar liquid substances. In fluid flow, it is necessary for molecules or groups of molecules to move with respect to one another. The ease with which such movement can take place is dependent upon the magnitude of the forces operating between the molecules, as well as upon the molecular weight and chemical structure of the liquid. As the average kinetic energy of the molecules increases with increasing temperature, the influence of the intermolecular forces becomes relatively less important, and viscosity decreases.

Surface tension refers to the forces which act to minimize the surface area of a given volume of liquid. It is responsible for the tendency of water droplets to assume a spherical shape when falling through the air, or when placed on a greasy surface. The molecules on the surface of a liquid are attracted by the intermolecular forces down into the body of the liquid. The pull which is thus exerted on the surface molecules tends to decrease the surface area. As the temperature of the liquid is raised, the intermolecular forces become less important in comparison with the average kinetic energy, so that surface tension decreases with increasing temperature.

Water is an excellent example of a liquid in which the intermolecular forces of attraction are particularly large. The major source of the attrac-

Table 10-1 Viscosities in millipoises of some liquid substances at 20°C.

Substance	Viscosity
Benzene	6.5
Carbon tetrachloride	9.7
Water	10.019
Ethyl alcohol	11.9
Acetic acid	12.2
Ethylene glycol	199
Glycerol	15,000

tive potential is the intermolecular hydrogen bond. The surface tension and viscosity of water are both high in comparison with the values for other substances of comparable molecular weight.

The molecules of a liquid are free to move about with respect to one another. The random motions of molecules in a liquid are evidenced in the phenomenon of diffusion. Diffusion in liquids is much slower than in gases, because the mean free paths of the molecules in the liquid state are very short. The volume in which the molecules of the liquid move is occupied to a large extent by other molecules, so that there is very little "free volume". The situation is quite different in a gas, in which most of the volume of the system is free volume, and only a small fraction of the space is occupied by molecules. The correction for the finite volume of molecules in the van der Waals equation of state, in the expression $(V - nb)$, is thus very small for gases, and very large for liquids.

It is possible to gain some appreciation of the size of the free volume available to molecules in a liquid by considering the entropy change on vaporization of a liquid. Let us choose krypton for consideration, since it is a simple substance composed only of spherical atoms, and the intermolecular forces are small. At the boiling point, 120°K, the liquid is in equilibrium with the vapor at a pressure of 1 atm. The molar volume of the gas phase is thus $V = 0.08206 \times 120/1 = 9.85$ l. The entropy change is given by the heat of vaporization, 2.158 kcal/mole, over the temperature: $\Delta S = 2{,}158/120 = 18.0$ cal/°K-mole. If this entropy change were due entirely to volume change (and it must be very nearly all due to that) we would have, from Eq. (9-9)

$$\Delta S = 2.303\, R \log_{10} \frac{V_2}{V_1}$$

where V_2 is the free volume in the gas phase, and V_1 is the free volume in the liquid phase. Then,

$$18.0 = 2.303 \times 1.987 \log_{10} \frac{V_2}{V_1}$$

$$\log_{10} \frac{V_2}{V_1} = 3.920$$

$$\frac{V_2}{V_1} = 8{,}310$$

But we know that the volume of the gas phase, V_2, is 9.85 l. V_1, therefore, is 1.18 ml. Now the density of krypton at the boiling point is 2.413 g/cm^3, so the volume of a mole of liquid krypton at 120°K is 34.8 ml. Thus, we see that V_1, the *free* volume in liquid krypton at the boiling point, is about 3 per cent of the total volume of the liquid. This much free volume is sufficient to allow the atoms to move with respect to one another, so the liquid has the familiar properties of flow, *etc.* At the same time, the average distance between molecules in liquids is not much different from those in the close-packed solid. It is not surprising, then, that liquids resist compression to about the same extent as do solids, or that the average force between molecules is about the same as in the solid.

In liquids which are composed of more complex polar molecules, the interactions between molecules may lead to some short-range order in the liquid. At any one instant there may be a distribution of clusters of molecules of varying sizes, in which the molecules are held in fixed positions with respect to one another. The number and sizes of such clusters is temperature dependent. Just above the melting point they may exist in large numbers, and be quite large; but at higher temperatures, as the molecules gain kinetic energy, the clusters tend to break up. The clusters do not persist as such for long; rather, they are continually forming and breaking up, as the molecules of the liquid collide and transfer energy from one molecule to another. Part of the heat capacity of a liquid is due to the energy needed to break up the clusters. The clusters formed through hydrogen-bonding interactions are quite important in determining the properties of water (Sec. 8-13).

10-2 Vapor Pressure

Suppose that a liquid is introduced into the bottom of a closed cylinder, as shown in Fig. 10-2, and that the space above the liquid is initially free of any gas. At any given time, a certain fraction of the molecules in the liquid possess sufficient energy to escape from the attractive forces of the surrounding molecules. During a certain interval of time, some of these, located at or near the surface, will acquire the necessary direction of motion to escape into the vapor phase. We should find, therefore, that a vapor pressure rapidly develops in the gas phase. But as the number of molecules present in the gas phase becomes considerable, the chance increases that a gas phase molecule will strike the liquid surface and remain. In time, the rate at which molecules return to the liquid phase equals the rate at which others are escaping into the gas phase. This situation repre-

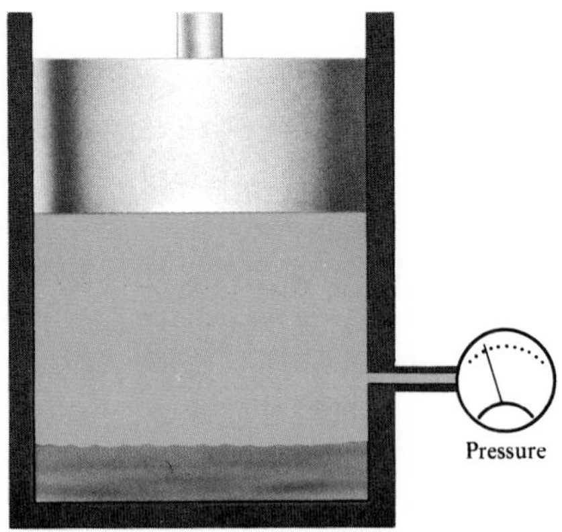

Fig. 10-2 Vapor pressure of a liquid.

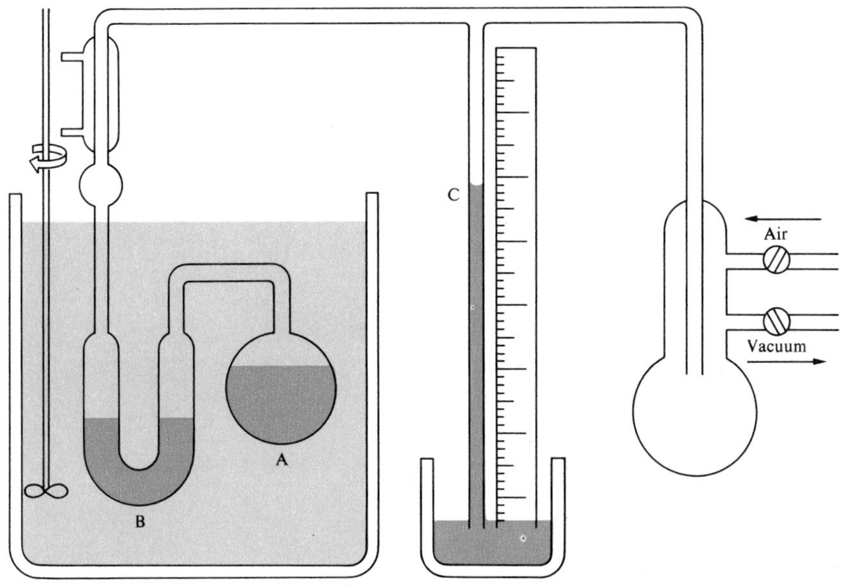

Fig. 10-3 Isoteniscope for determination of the vapor pressure of a liquid as a function of temperature.

sents a dynamic equilibrium; the number of molecules in the gas phase is constant as a function of time because the rates of the two opposing processes are equal. This type of system must be distinguished from a static one, in which the system is constant as a function of time because all of the processes are going on at zero rate.

If the piston is lowered after the system is at equilibrium, there occurs a momentary increase in the pressure of the vapor. But as soon as this occurs, the rate at which molecules pass from the gas phase to the liquid increases, whereas there is no change in the rate at which molecules pass from the liquid phase to the gas. As a result, the pressure decreases from the momentary high to the value it had before the compression. In the same way, if the volume available to the gas is increased by raising the piston, the pressure in the gas phase decreases and again returns to the same value. We see, then, that the vapor pressure of a liquid is a function of temperature only. The determination of vapor pressure as a function of temperature is shown in Fig. 10-3. The apparatus is called an *isoteniscope.* The liquid to be studied is placed in the bulb A and in the U-tube portion of the apparatus at B. The portion of the apparatus which contains the liquid of interest is thermostated at the desired temperature. Now the apparatus is opened to a vacuum pump. The liquid in A and B begins to boil vigorously when the pressure decreases to the vapor pressure of the liquid. In the course of this boiling, all air is expelled from the space between A and B. Now air is admitted to the system slowly until the liquid

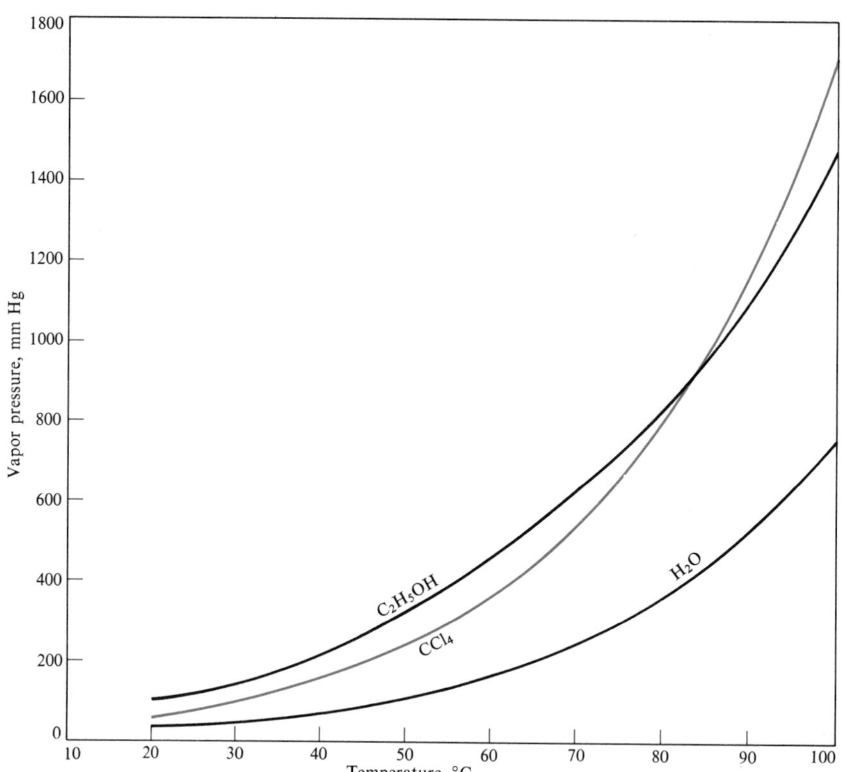

Fig. 10-4 Vapor pressure-temperature curves for a few liquids.

levels in the two arms of the U-tube are the same. At this point, the pressures acting on the two arms must be the same. But the only pressure exerted on the right hand side of the U-tube is from the vapor pressure of the liquid. Hence, by reading the height of the mercury column at C, and knowing the barometric pressure, the vapor pressure of the liquid can be obtained. The experiment can be preformed at various temperatures to obtain the vapor pressure *vs* temperature relationship.

The variation of vapor pressure with temperature for three substances in the range from 20 to 100°C is shown in Fig. 10-4 and Table 10-2. Note that the vapor pressure increases with temperature in a very non-linear fashion. The vapor pressure increases with temperature ever more rapidly until the critical temperature is reached. The *critical temperature* is the highest temperature at which it is possible for a given substance to exist as liquid. Above the critical temperature, no amount of pressure can cause the gas to pass through a distinct phase change into the liquid state. The nature of the vapor pressure *vs* temperature curves can be understood if we recall the way in which the distribution of molecular speeds varies

Table 10-2 Vapor pressure (mm Hg) *vs* temperature.

Temperature (°C)	CCl_4	C_2H_5OH	H_2O
20	91.0	44	17.5
30	142	78	31.8
40	215	134	55.3
50	314	220	92
60	447	350	149
70	621	541	234
80	843	813	355
90	1122	1187	526
100	1463	1693	760

with temperature, Sec. 3-11. We saw there that the fraction of molecules which possess a velocity at least as great as some arbitrary value, say, u_j is

$$f_j = n_j/n_T = e^{-\frac{1}{2}mu_j^2/kT} \tag{10-1}$$

Suppose that a molecule must have at least velocity u_j to escape from the liquid into the gas phase. Assume also that the distribution of energies in the liquid state follows the same pattern as for the gas state. Then, the vapor pressure, which is determined by the rate at which molecules escape from the liquid into the vapor, is proportional to f_j. Now let us take the $\log_e$ of Eq. (10-1)

$$\log_e f_j = -\tfrac{1}{2}mu_j^2/kT \tag{10-2}$$

Since f_j is proportional to vapor pressure P, it follows that $\log_e P$ will also depend upon a term which involves $1/T$ on the right-hand side. The equation which one obtains from thermodynamic considerations is known as the *Clausius-Clapeyron equation*

$$\log_e P = \frac{-\Delta H_v}{RT} + C \tag{10-3}$$

This equation is actually very much like Eq. (10-2). Instead of the energy which each individual molecule must have to "evaporate", $\frac{1}{2}mu_j^2$, we have the *molar* heat of vaporization, ΔH_v. In the denominator we have R, the molar gas constant, which is just Avogadro's number times the Boltzmann constant, k. The constant C is added to give the correct vapor pressure for some particular temperature, but the first term is the important one, since it gives the temperature dependence.

Expressed in base-10 logs, the Clausius-Clapeyron equation is

$$\log_{10} P = \frac{-\Delta H_v}{2.303 \times 1.987 \times T} + \mathrm{C} \tag{10-4}$$

If we wish to know the ratio vapor pressure at two different temperatures, T_1 and T_2, we obtain

$$\log P_2 - \log P_1 = \frac{-\Delta H_v}{4.58\,T_2} + \frac{\Delta H_v}{4.58\,T_1}$$

$$\log \frac{P_2}{P_1} = \frac{-\Delta H_v}{4.58}\left(\frac{1}{T_2} - \frac{1}{T_1}\right) \tag{10-5}$$

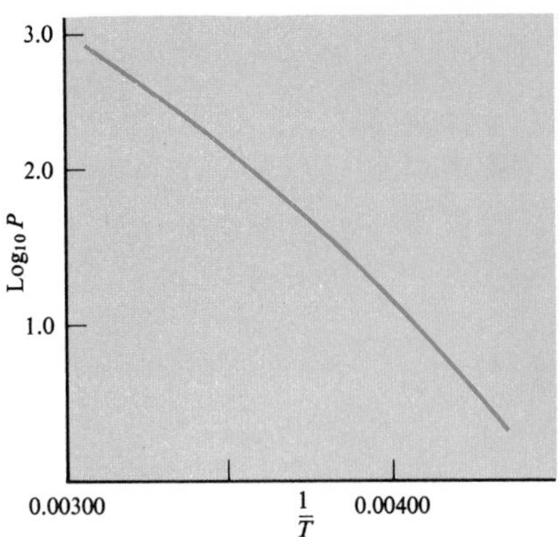

Fig. 10-5 Log_{10} vapor pressure *vs* $1/T$ for liquid bromine. The relationship is not strictly linear over the entire temperature covered, but the data over a narrow temperature range can be well approximated by a straight line.

Equation (10-3) predicts that the log of the vapor pressure of a liquid will vary linearly with $1/T$, and that the slope of the relationship is $-\Delta H_v/4.58$. This means that a study of the temperature variation of the vapor pressure can yield the heat of vaporization of a liquid. The linearity of the relationship between $\log_{10} P$ and $1/T$ is, however, dependent upon ΔH_v remaining constant with temperature. This is generally found to be the case over modest temperature ranges. As an illustration, the vapor pressure over liquid bromine varies with temperature as shown in Fig. 10-5. The student is encouraged to construct graphs of $\log_{10} P$ *vs* $1/T$ from the data given in Table 10-2, and to compare the ΔH_v values obtained from these graphs with those listed in Table 9-3.

The *boiling point* of a liquid is the temperature at which the liquid vapor pressure becomes equal to the pressure acting upon the surface of the liquid. The normal boiling point is defined as the temperature at which the vapor pressure equals one atmosphere. Boiling is associated with the formation of vapor within the body of the liquid, giving rise to bubbles. This occurs only when the vapor is in contact with an essentially infinite gas space (for example, a beaker of liquid being heated on the lab bench), or when the liquid is being pumped on, so that gas is continuously withdrawn.

Boiling is a very curious phenomenon which is not, as yet, very well understood. In distillations, it is important that the liquid pass into the vapor state by a smooth, steady formation of gas bubbles. It often happens, however, that bubbles do not form readily. The liquid acquires more heat than it should, and becomes superheated. There is then a sudden, almost explosive conversion of liquid into the vapor state, which thoroughly disrupts the carefully established conditions in the distillation column. In an effort to avoid this, various objects such as marble chips, may be added to the liquid to enhance bubble formation.

10-3 Solutions

A *solution* is generally defined as a honogeneous system of variable composition. The term *homogeneous* means, in this instance, that the composition is the same everywhere throughout the system, and that no part is separated by a physical boundary from any other. The variable composition of the solution classes it as a mixture, as opposed to a pure substance. A pure substance has a definite, fixed composition.

Although the chemist considers many of the substances he works with in the laboratory to be pure, it is a fact that they all contain impurities to some degree. For the most part, these are dissolved in the parent substance, so that the "pure" substances are really solutions. Trace quantities of impurities are often of critical importance. For example, semiconductor materials such as silicon and germanium, used in the construction of transistors, contain controlled amounts of certain dissolved "impurities." The study of solutions is important for an understanding of these matters as well as for the more common uses of solutions in everyday chemical practice.

Solutions may be classified according to the phase (solid, liquid, or gas) of the system.

Gaseous solutions — All gas mixtures are solutions, since gases always mix to produce a homogeneous system. The atmosphere, consisting of nitrogen (78.09 per cent), oxygen (20.95 per cent), argon (0.93 per cent), carbon dioxide (0.03 per cent) and other gases in trace quantities, is an example of a gaseous mixture.

Liquid solutions — A liquid solution may be obtained as a mixture of two substances which are themselves liquids, such as ethyl alcohol and water mixtures, or it may result from addition of a solid or gas to a liquid. Occasionally, two solid substances will form a liquid solution; an equimolar mixture of sodium and potassium is liquid at room temperature, although both elements individually are solid.

In describing solutions, it is common to speak of the substance present in greatest quantity as the *solvent*, and of other substances as the *solutes*. A solution is termed dilute if it contains a relatively small amount of solute. A *binary solution* is constituted from two substances, one of which is designated the solvent, the other the solute. Of course, if the two substances are present in about equal concentrations there is no point, other than convenience, in calling either one the solvent.

The composition of a solution, though variable, is often restricted within certain limits. If addition of more solute to a system after its concentration has reached a certain value does not result in an increase in the amount dissolved, such a solution is said to be *saturated* with respect to that solute. As an operational definition, a solution can be said to be saturated with respect to a solute when there is a quantity of undissolved solute in contact with the solution at equilibrium.

Ethyl alcohol and water form a single, homogeneous phase when mixed in any proportions. Liquids for which this is true are said to be *miscible* with one another. By contrast, carbon tetrachloride and water do not form solutions in one another to any appreciable extent, and are termed *immiscible*. The ability of two substances to form solutions over a limited range of composition is termed *partial miscibility*. In general terms, it may be said that substances which are alike chemically are likely to be miscible, while substances which differ widely in chemical properties tend to be immiscible.

Solid solutions — Although solid solutions are not as commonplace as liquid solutions, they are nevertheless important. Many alloys are solid solutions of two or more elements. As another example, many solid elements take up gases in large quantities to form solutions of the gas in the metal. Palladium, when properly prepared, is capable of taking up to 0.25 mole of molecular hydrogen per mole of metal at room temperature, to form a solid solution of atomic hydrogen in the metal.

Concentrations of solutions — Solution concentrations may be expressed in a number of ways:

1. *Weight fraction or weight percentage.* The weight of solute per unit weight of solution. For example, a 10 per cent aqueous solution of sugar contains 10 g of sugar in 90 g of water, or 10 g of sugar per 100 g of solution. The weight fraction of sugar is 0.1.

2. *Mole fraction.* The number of moles of solute per unit total moles of solution. For example, a solution containing 20 g of ethanol (C_2H_5OH) and 80 g of water contains $20/46 = 0.43$ moles of ethanol and $80/18 = 4.4$ moles of water. The mole fraction f of ethanol is then

$$f = \frac{n_2}{n_1 + n_2} = \frac{0.43}{0.43 + 4.4} = 0.090$$

3. *Molarity.* The number of moles of solute per liter of solution. The symbol for molarity is M. One liter of a $1M$ solution of sodium chloride (NaCl) would be prepared by placing one formula weight (one mole) of sodium chloride, 58.5 g, in a vessel and adding approximately one-half liter of water, stirring until it had all dissolved. Additional water would then be added with stirring until the total volume was just one liter. Since the concentration is expressed in terms of the *volume* of the solution, it is always true that solutions of the same molarity contain the same number

of moles of solute in a given volume of solution, even though the solutes may be different.

Molarity is a convenient measure of concentration, and is widely used in laboratory work. One of its disadvantages is that the molarity of a solution changes with temperature because of expansion or contraction of the liquid.

4. *Molality.* The number of moles of solute per 1000 g of solvent. The symbol for molality is *m*. A 1*m* solution of sodium chloride (NaCl) in water is made up by adding 58.5 g of sodium chloride to 1000 g of water. Note that molality and molarity are not the same, although for dilute aqueous solutions they are not much different in magnitude. The major advantage of the molality unit is its temperature independence. The relative inconvenience of making up solutions by weight is, however, something of a disadvantage.

5. *Formality.* The number of formula weights of solute per liter of solution. The symbol for formality is *F*. With the term "mole" used in the looser sense which we prefer, formality is not distinguished from molarity.

6. *Normality.* The number of equivalents of solute per liter of solution. The essential difference between normality and molarity rests in the focusing of attention not on the solute as a whole, but only on that part of the solute which is of importance in a particular instance. If 98 g (one mole) of sulfuric acid (H_2SO_4) is dissolved in water to make a liter of solution, its concentration is 1*M*. But this solution contains two moles of acidic, or replaceable, hydrogen; it may be said that it contains two equivalents ($2 \times 6 \cdot 10^{23}$) of hydrogen ion, or that it is 2 normal (2*N*). The number of equivalents per mole for any acid is just equal to the number of replaceable hydrogens per molecule. For a base, the number of equivalents per mole is equal to the number of moles of hydrogen ion with which one mole of the base will react in an acid-base type reaction.

Normality is also used to express the concentrations of solutes in oxidation-reduction reactions. In these reactions, interest centers on the number of electrons gained or lost, and normality is defined in terms of the number of equivalents of electrons gained or lost per mole of solute. Because normality refers directly to the chemical behavior of the solute, it is best treated with a discussion of the type of reaction involved (see Appendix A).

One simple fact is worth remembering in connection with normality: It is always some multiple (usually a whole number) — 1, 2, 3, *etc.* — of the molarity. For sulfuric acid, the multiple is two; for aluminum hydroxide, Al $(OH)_3$, it is three. The *equivalent weight* of a solute is equal to the formula weight divided by the number of equivalents per mole. Thus, the equivalent weight of sulfuric acid is

$$98 \text{ g/mole} \times 1 \text{ mole/2 equiv} = 49 \text{ g/equiv}$$

Example 10-1

To illustrate the calculation of the various concentration units, consider an aqueous solution made up from 300 g of $MnBr_2$ per liter of solution. The density of this solution at room temperature is 1.252 g/cm^3.*

1. *Weight fraction or weight percentage.* The weight of a liter of solution is 1252 g. Since there are 300 g of manganese (II) bromide, $MnBr_2$, in this total weight, the weight fraction is $w_2 = 300/1252 = 0.239$.

2. *Mole fraction.* The formula weight of $MnBr_2$ is 214.8 g. There is, therefore, $300/214.8 = 1.40$ moles of $MnBr_2$ in one liter of the solution. The weight of water in this quantity of solution is $1252 - 300 = 952$, which is $952/18 = 52.8$ moles of water. The mole fraction of $MnBr_2$ is

$$f_2 = \frac{1.40}{1.40 + 52.8} = 0.0259$$

3. *Molarity.* The molarity is just the number obtained above in (2) for the number of moles of $MnBr_2$ in a liter of the solution: 1.40*M*.

4. *Molality.* To compute molality, we need to know the number of grams of $MnBr_2$ per 1000 g of water. We know that there are 300 g in 952 g of water; in 1000 g there are then $1000/952 \times 300 = 316$ g. The molality is

$$m = \frac{316/214.8}{1000 \text{ g solvent}} = \frac{1.47 \text{ moles}}{1000 \text{ g solvent}}$$

5. *Formality.* The formality is the same numerically as the molarity.

6. *Normality.* To calculate the normality of the solution, we would need to know the chemical context in which the concentration is needed. We will, therefore, defer any consideration of this concentration unit to later chapters.

10-4 Raoult's Law

A solute, on being dissolved, is distributed uniformly throughout the solution. If the mole fraction of solute is f_2, the fraction of solute molecules on the surface of the solution is also f_2. Assuming that the energy distribution of solvent molecules is not changed by introduction of the solute (a good assumption), and assuming further that the energy required of a solvent molecule for escape from the liquid is unchanged, the number of solvent molecules escaping per unit time into the vapor phase is reduced to f_1 times that for the pure solvent. If the solute is non-volatile, it does not contribute to the vapor pressure; essentially none of the solute particles escape from the solution. The result is, then, that the total number of molecules escaping from the liquid phase per unit time is reduced. At the same time, the possibilities for return of gas phase molecules to the liquid phase remain essentially unchanged. The equilibrium conditions have therefore been changed in the direction of fewer molecules in the vapor

*The densities of solutions are usually reported as specific gravities. The specific gravity is the ratio of the solution density to the density of water at 4°C. Since the latter is essentially 1.000, specific gravity is the same numerically as density, but is dimensionless.

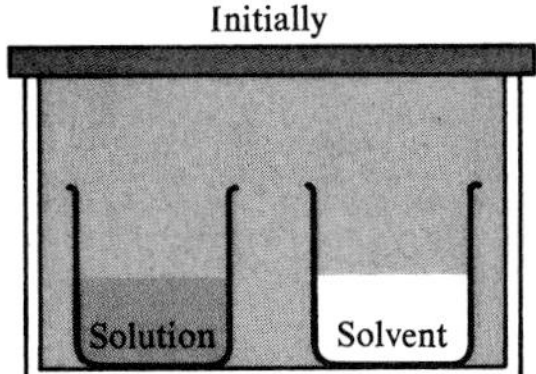

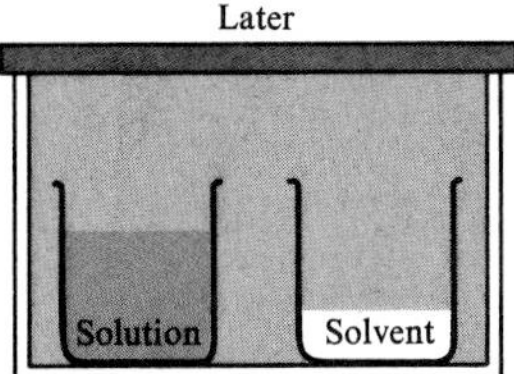

Fig. 10-6 Comparison of the vapor pressures of water over the pure solvent and over a concentrated solution. The vapor pressure in the container tends to attain the value characteristic of the pure solvent. Since this is a greater vapor pressure than that which would exist at equilibrium over the solution, water condenses into the solution. Thus, there is effectively a transfer of water from the solvent container to the solution.

phase, and the vapor pressure is reduced. Furthermore, the vapor pressure reduction is proportional to the mole fraction of solute. A demonstration of the lower vapor pressure of solvent molecules over the solution is shown in Fig. 10-6. Solvent molecules are observed to transfer from a beaker containing pure solvent to one containing solution.

We can describe this situation mathematically by expressing the vapor pressure of the solvent as the product of the vapor pressure of pure solvent and the mole fraction of solvent in the solution: $P_1 = f_1 P_1^0$. If the solute also has an appreciable vapor pressure in its pure form, the above argument can be extended to it as well.

An *ideal solution* may be defined as one whose properties are precisely the average of the properties of each of its constituents, weighted according to the mole fraction of each constituent. Thus, an ideal solution consisting of an equal number of moles of A and B would have a boiling point exactly midway between the boiling points of A and B. Although ideal solutions are rarely, if ever, encountered, the concept is important for two reasons: Many real systems approach ideal behavior and can be treated as approximately ideal; and deviations from ideality, which result from unequal forces acting between different species, are useful clues in the study of solutions. Figure 10-7 shows a diagram of boiling point *vs* composition for benzene-toluene mixtures. This particular system deviates slightly from ideal behavior in that the observed boiling points lie along the solid line, whereas the boiling points for ideal solutions would follow the dotted line.

Raoult's law states that *the vapor pressure of each substance in a solution is equal to the vapor pressure of the pure substance times the mole fraction of*

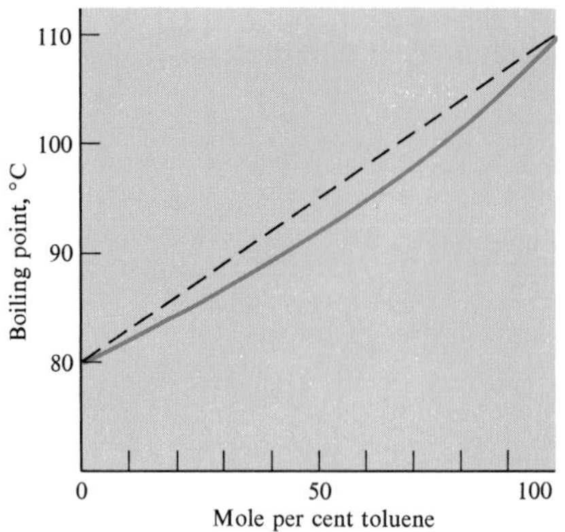

Fig. 10-7 Boiling point *vs* composition diagram for benzene-toluene mixtures.

the substance. From this statement, it follows that the total vapor pressure over a solution is equal to the vapor pressure of each pure constituent times its mole fraction, summed over all of the components

$$P_t = P_1^0 f_1 + P_2^0 f_2 + \ldots + P_n^0 f_n \tag{10-6}$$

A consideration of the vapor pressure of the solvent over a solution is one of the most important applications of Raoult's law. When the vapor pressure of the solute is negligibly small (*e.g.*, as it is with an ionic solute), the total vapor pressure over the solution is essentially just that of the solvent

$$P_t = P_1^0 f_1 = P_1$$

where P_1 is the vapor pressure of the solvent over the solution. Since f_1 is less than 1, P_1 is less than P_1^0, the vapor pressure of pure solvent. The depression of the vapor pressure of the solvent is demonstrated by the experiment shown in Fig. 10-6.

In connection with Raoult's law, it is important to understand that the lowering of the vapor pressure of the solvent depends only on the *number* of entities present as solute, and not on their nature. Thus, 1000 g of water contains 1000/18.02 = 55.53 moles of water. If one mole of sugar is dissolved in 1000 g of water to form a 1-*molal* solution, the mole fraction of solvent is

$$f_1 = \frac{55.53}{1 + 55.53} = 0.9822$$

The vapor pressure of water at 25°C is 23.756 mm. The vapor pressure of the 1-molal sugar solution at the same temperature is then 0.9822 × 23.756 mm = 23.335 mm, *assuming ideal behavior*. This represents a lowering of the vapor pressure by 0.421 mm. The same conditions would apply to a

1-molal solution of any other non-electrolyte. But consider now a 1-molal solution of sodium chloride. The dissolving of one mole of sodium chloride in water results in the formation of two moles of particles in solution: 1 mole of Na^+ ions and 1 mole of Cl^- ions. The mole fraction of water in a 1-molal NaCl solution is thus

$$f_1 = \frac{55.53}{2 + 55.53} = 0.9652$$

The vapor pressure of water over the solution is $0.9652 \times 23.756 = 22.930$ mm. The vapor pressure lowering is 0.826 mm, about twice that for the 1-molal solution of a non-electrolyte. By a similar argument, it can be seen that the vapor pressure lowering for a 1-molal $CaCl_2$ solution should be about three times that for the 1-molal non-electrolyte solution.

The fact that aqueous solutions of electrolytes show vapor pressure lowerings which are some multiple of that found for non-electrolyte solutions of the same molality was of great importance in the development of an adequate theory of these solutions. It represented strong evidence in favor of Arrhenius' concept of dissociation of ionic substances in solution. The vapor pressure lowering itself is quite difficult to measure accurately, but its effect on other properties of the solution — notably the melting and boiling points — is easily observed with good quantitative accuracy.

10-5 Henry's Law

The gaseous space over a solution contains (at least in principle) molecules of all of the components of the solution. A mixture of ethyl alcohol and water, for example, exerts a vapor pressure due to both the alcohol and the water. On the other hand, if the solute is quite non-volatile, its vapor pressure may be negligible, as in the case of ionic substances dissolved in water.

In studying the vapor pressures of solutions, one may focus attention on either the solute or the solvent. In the former case, Henry's law is of major interest; in the latter case, Raoult's law is particularly useful.

The English chemist William Henry (1774–1836) studied the effect of pressure upon the solubility of gases in water. He discovered that, for a given gas at constant temperature, the solubility is proportional to the pressure of the gas over the water. His findings can be generalized to the statement that *for any substance dissolved in a given solvent, the pressure of that substance over the solution is proportional to its concentration in the solution.*

$$P_s = K[S]$$

The symbol $[S]$ indicates the concentration in solution and P_s represents the pressure. K is called the Henry's law constant. It is dependent upon the solute, the solvent, and the temperature. The law is really a limiting

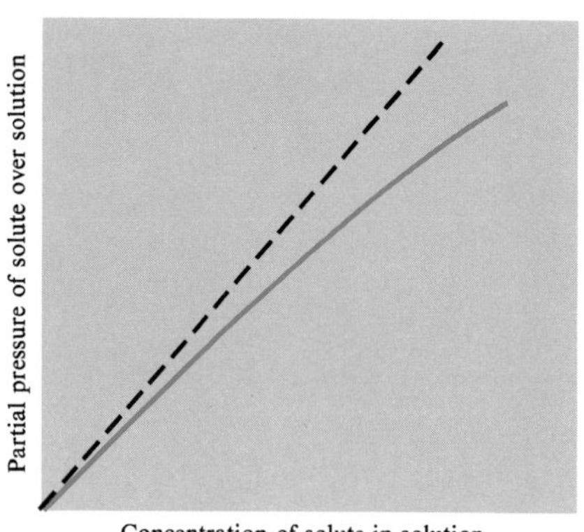

Fig. 10-8 Graphical illustration of Henry's law. The behavior of a system obeying Henry's law is represented by the dotted line.

relationship which is obeyed only in dilute solutions. This is shown in Fig. 10-8, where the limiting slope of the vapor pressure *vs* solubility curve represents the Henry's law constant K.

The Henry's law constants for a given solute in a variety of solvents are a crude index of interactions between solute and solvent. Attractive interactions would lead to increased solubility of the solute, and to a smaller Henry's law constant. The constant is smaller because the solubility of the gas is greater for a given pressure, so that the ratio $P_s/[S] = K$ decreases. Table 10-3 lists the values of K for a variety of gases in water.

Table 10-3 Henry's law constants for some gases in water.

Gas	K (mm Hg/mole fraction) at 25°C
Argon	3.0×10^7
Nitrogen	6.3×10^7
Bromine	5.2×10^4
Hydrogen sulfide	4.1×10^6
Sulfur dioxide	3.3×10^4
Carbon dioxide	1.2×10^6

Henry's law is not the equivalent of Raoult's law. Raoult's law would require that the vapor pressure of the solute be precisely equal to the product of the vapor pressure of pure solute times the mole fraction of solute

$$P_2 = f_2 P_2^0 \tag{10-7}$$

If Henry's law is written in this form, using mole fraction for the concentration unit,

$$P_2 = f_2 K \tag{10-8}$$

By comparing these two equations, we see that the Henry's law constant for a substance which produced a solution obeying Raoult's law would be just the vapor pressure of pure solute. But many gases are dissolved in water at temperatures which are above the critical temperature for the substance, so that it is not even possible to speak of a vapor pressure for the "pure solute". Interactions between the solute and solvent may lead to a Henry's law constant which is quite different from that predicted by Raoult's law.

For example, the pressure of HCl vapor over pure, liquid HCl at 25°C is 46.6 atm. The Henry's law constant for HCl, if the solution obeyed Raoult's law, would therefore be 46.6 atm. We can write Eq. (10-7) in the form $P_{HCl} = 46.6\ f_{HCl}$, where P_{HCl} is the pressure of HCl vapor over the solution, and f is the mole fraction of HCl in the solution. If we set P_{HCl} equal to 1 atm, f_{HCl} is $1/46.6 = 0.022$ mole HCl/mole solvent. That is, the solubility of HCl in an ideal solution at 25°C, with one atm HCl vapor over the solution, would be 0.022 in mole fraction units. In practice, the solubility of HCl (or any other solute) in a solvent is measured with a known vapor pressure of the solute over the solution. From the solubility data, the Henry's law constant is calculated. Table 10-4 lists a number of solubility data for HCl in various solvents at 25°C, all in units of moles HCl/mole solvent, with 1 atm HCl in the vapor phase. It is evident that the solutions are far from ideal.

Table 10-4 Solubility of HCl in various solvents at 25°C, 1 atm HCl vapor.

Solvent	Solubility (moles HCl/moles solvent)
Ideal	0.022
n-Decane ($C_{12}H_{26}$)	0.048
Benzene	0.055
Chlorobenzene (C_6H_5Cl)	0.030
Dibutyl ether ($C_4H_9OC_4H_9$)	0.62
Acetic acid	0.12
Ethanol (C_2H_5OH)	0.82

Henry's law claims only a direct proportionality between the vapor pressure of solute and the concentration of solute in the solution. Even this weaker requirement is often not met in solutions. As the concentration increases, there may be a tendency for the solute molecules to group together, whereas this does not occur in more dilute solutions. As the environment experienced by the solute changes with concentration, so the tendency for solute molecules to escape to the vapor phase changes. Consequently, a non-linear variation of solute vapor pressure with concentration is seen, as indicated by the solid line in Fig. 10-8.

10-6 Boiling Point Elevation of Solutions

Figure 10-9 shows a typical variation of vapor pressure with temperature in the region of the boiling point for a liquid. Although the variation of vapor pressure with temperature is non-linear as we have seen (Fig. 10-4), the curvature is very slight if we consider only a very small temperature interval, as we do in Fig. 10-9. The normal boiling point of the liquid is the temperature at which the vapor pressure just equals the atmospheric pressure. Addition of a non-volatile solute causes the solvent vapor pressure to decrease in accordance with Raoult's law. The vapor pressure *vs* temperature curve for the solution is represented by the colored line in Fig. 10-9. The lowering of the vapor pressure over the solution as compared with that over pure solvent is proportional to the concentration of solute, so the colored line in the figure represents the situation for just one particular concentration. It can be seen that the addition of the non-volatile solute causes an increase in the normal boiling point, since a higher temperature is required to bring the vapor pressure over the solution to 1 atm. For any given solvent, a 1-molal solution of a non-volatile non-electrolyte will cause a raising of the boiling point which is a constant for that solvent. Some boiling point constants K_b are listed in Table 10-5. K_b is measured experimentally for a solvent by measuring its boiling point at 1 atm, then adding a known weight of a non-volatile solute to a known weight of the solvent and determining the boiling point of the resulting solution. From

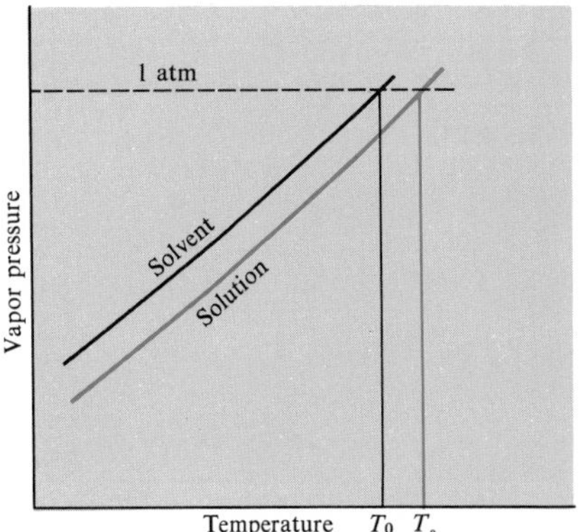

Fig. 10-9 Vapor pressure curves for pure solvent and for solution at 1 atm pressure in the vicinity of the boiling point. T_o represents the boiling point of pure solvent, T_s the boiling point of solution.

a knowledge of the molality of the solution, the boiling point elevation constant is calculated from the expression $\Delta T = K_b m$.

Table 10-5 Boiling point elevation constants.

Solvent	K_b (°C)*
Water	0.512
Benzene	2.53
Carbon tetrachloride	5.03
Chloroform	3.63
Cyclohexane	2.79

Example 10-2

1.20 g of an organic compound of molecular weight 180 is dissolved in 40.0 g benzene. The boiling point at 1 atm pressure is elevated 0.422°C relative to the pure solvent. Calculate the boiling point elevation constant for benzene. Since the concentration unit appropriate for this problem is molality, we must first calculate the number of moles of solute per 1000 g of solvent:

$$\frac{1.20 \text{ g solute}}{40.0 \text{ g solvent}} \times \frac{1 \text{ mole solute}}{180 \text{ g solute}} \times \frac{1000 \text{ g solvent}}{1000 \text{ g solvent}} = \frac{0.167 \text{ moles solute}}{1000 \text{ g solvent}}$$

The boiling point elevation constant is given by the observed temperature rise divided by the molality:

$$K_b = \frac{0.422°\text{C}}{\left(\dfrac{0.167 \text{ moles solute}}{1000 \text{ g solvent}}\right)} = 2.53°\text{C}/1 \text{ mole solute}/1000 \text{ g solvent}$$

Once the boiling point elevation constant for a solvent has been established, the measurement of the boiling point elevation of a solution is useful in determining the molecular weight of a solute.

Example 10-3

0.860 g of a solid of unknown molecular weight is dissolved in 36.2 g CCl_4. The boiling point of the solution is elevated 0.615°C relative to the pure solvent. Calculate the molecular weight of the solute, assuming that it is a non-volatile non-electrolyte. K_b for CCl_4 is 5.03°C/1 mole/1000 g solvent.

$$\frac{0.86 \text{ g solute}}{36.3 \text{ g solvent}} = \frac{23.8 \text{ g solute}}{1000 \text{ g solvent}}$$

$$0.615°\text{C elevation} \times \frac{1 \text{ molal solution}}{5.03°\text{C elevation}} = 0.122 \text{ molal solution}$$

$$\frac{23.8 \text{ g solute}}{1000 \text{ g solvent}} = \frac{0.122 \text{ moles solute}}{1000 \text{ g solvent}}$$

$$1 \text{ mole solute} = 195 \text{ g solute}$$

*For one mole of non-volatile non-electrolyte per 1000 g solvent.

It should be noted that the solute is assumed to be non-volatile. If it is not, the vapor pressure of the solute contributes to the total vapor pressure. As a result, the boiling point is elevated to a lesser extent, and may even be lowered if the solute is more volatile than the solvent.

If the solute is an electrolyte and dissociates in the solvent, the boiling point elevation is greater than it would be for a non-electrolyte solution of the same concentration. This results from the fact that the decrease in solvent vapor pressure, which is responsible for the boiling point elevation, depends only upon the number of particles in solution, and not on their kind. For example, the boiling point of an aqueous 0.1-molal sodium chloride solution is elevated about twice as much as that of a 0.1-molal sugar solution.

A Quantitative Treatment of Boiling Point Elevation

We can derive an expression which relates the boiling point elevation to other properties of the solution by making use of the Clausius-Clapeyron formula, Eq. (10-3). From Raoult's law, we have for the vapor pressure of the solvent over the solution,

$$P_1 = f_1 P_1^0$$

where, as before, P_1^0 is the vapor pressure of the pure solvent, and f_1 is the mole fraction of solvent. Taking the log of both sides,

$$\log_{10} P_1 = \log_{10} f_1 + \log_{10} P_1^0$$

Substituting for $\log_{10} P_1^0$ from Eq. (10-3),

$$\log_{10} P_1 = \frac{-\Delta H_v}{2.3\,R}\left(\frac{1}{T}\right) + C + \log_{10} f_1$$

We are interested in comparing the pure solvent with the solution when both are at their boiling points. Calling the temperature of the pure solvent at the boiling point T_0 and that of the solution T_s, we have

$$(\log_{10} P_1^0)_{T_0} = (\log_{10} P_1)_{T_s}$$

$$\frac{-\Delta H_v}{2.3\,R}\left(\frac{1}{T_0}\right) + C = \frac{-\Delta H_v}{2.3\,R}\left(\frac{1}{T_s}\right) + C + \log_{10} f_1$$

Rearranging, dropping C from both sides,

$$\frac{-\Delta H_v}{2.3\,R} T_s - \left(\frac{-\Delta H_v}{2.3\,R} T_0\right) = T_0 T_s \log_{10} f_1$$

$$T_s - T_0 = \Delta T = 2.3\,R \frac{T_0 T_s \log_{10} f_1}{-\Delta H_v}$$

Since T_0 and T_s do not differ by very much, we can write $T_0 T_s = T_0^2$. Furthermore, since $f_1 + f_2 = 1$, $f_1 = 1 - f_2$.

$$\Delta T = \frac{2.3\,R\,T_0^2}{-\Delta H_v} \log_{10}(1 - f_2)$$

If f_2 is small, as it ordinarily is in the boiling point elevation experiment, we can use the approximation

$$2.303 \log_{10}(1 - f_2) \approx -f_2$$

[*e.g.*, $2.303 \log_{10}(1 - 0.0150) = 2.303 \log_{10}(0.985) = -0.0151$]
Using this approximation, we therefore write

$$\Delta T = \frac{R T_0^2}{\Delta H_v} f_2 \tag{10-9}$$

From this expression, it is clear that the change in the boiling point ΔT is proportional to the mole fraction of solute f_2, and inversely proportional to the heat of vaporization of solvent. The quantity f_2 can be converted to molality if it can be assumed that the quantity of solute is small in comparison with the quantity of solvent.

$$f_2 = \frac{\text{moles solute}}{\text{moles solute} + \text{moles solvent}} \approx \frac{\text{moles solute}}{\text{moles solvent}}$$

To convert these units to molality, *i.e.*, moles solute per 1000 g of solvent, we must find out how many moles of solute there are in one g of solvent, and multiply by 1000. The number of moles of solute per one g of solvent is f_2 divided by the molecular weight of solvent. The molality m is, thus,

$$m = \frac{1000}{\text{molecular weight of solvent}} f_2$$

As an example, consider a benzene solution in which the mole fraction of solute is 0.01. Since the mole fraction of solute is small, we may use the preceding approximation:

$$f_2 = \frac{0.01 \text{ moles solute}}{1 \text{ mole (solvent + solute)}} \approx \frac{0.01 \text{ mole solute}}{1 \text{ mole solvent}}$$

The molecular weight of benzene is 78 g/mole. Using the preceding method, we have

$$\frac{0.01 \text{ mole solute}}{1 \text{ mole solvent}} \times \frac{1 \text{ mole benzene}}{78 \text{ g benzene}} \times 1000 \text{ g benzene}$$

$$\left(\frac{10}{78}\right) \frac{\text{moles solute}}{1000 \text{ g benzene}} = 0.128\ m$$

We have, finally,

$$\Delta T = R T_0^2 \frac{Mm}{1000\ \Delta H_v} \tag{10-10}$$

where M is the molecular weight of solvent.

On comparing Eq. (10-10) with the expression employed in the previous section for working boiling point elevation problems, it can be seen that the boiling point elevation constant K_b is equal to $RT_0^2M/1000\ \Delta H_v$. Thus, by making use of the vapor pressure *vs* temperature data for a liquid, and from a knowledge of its molecular weight and boiling point, it is possible to calculate the boiling point elevation constant.

10-7 Freezing Point Lowering of Solutions

The *freezing point* of a liquid can be defined as the temperature at which the vapor pressure over the liquid becomes equal to its vapor pressure over the solid (Fig. 10-9). Addition of a non-volatile solute lowers the solution vapor pressure in accordance with Raoult's law. This is shown by the colored line in Fig. 10-9. The solute has no effect on the vapor pressure of the solid phase, however, provided that the solute does not form a solid solution with the solvent. The result of adding solute, therefore, is that the intersection of the vapor pressure curves for the two phases is shifted to a lower temperature. It is possible to assign a freezing point lowering constant to a solvent which is analogous to the boiling point elevation constant. Some examples of these freezing point lowering constants (sometimes called *cryoscopic constants*) are given in Table 10-6.

The freezing point lowering experiment is very similar to that in which boiling point elevations are observed. The same considerations obtain with regard to the effect of electrolytes as compared with non-electrolytes. A 1-molal aqueous sodium chloride solution depresses the freezing point of water about twice as much as a 1-molal solution of a non-electrolyte such as sugar, or glycerin. Molecular weights are conveniently determined by freezing point lowering measurements in solvents for which the cryoscopic constant is known. Benzene and cyclohexane are frequently used as solvents because their freezing points occur at convenient temperatures.

Table 10-6 Freezing point lowering constants and freezing points for liquids.

Compound	K_f*	Freezing point, °C
Acetic acid	3.90	16.6
Benzene	4.90	5.48
Nitrobenzene	7.00	5.7
Water	1.86	0
Cyclohexane	20.5	6.5
Camphor	37.7	180

Example 10-4

0.644 g of a compound of unknown molecular weight is dissolved in 34.7 g nitrobenzene. The freezing point is depressed 0.895°C. What is the molecular weight of the unknown substance?

$$\frac{0.644 \text{ g solute}}{34.7 \text{ g solvent}} \times \frac{1000 \text{ g solvent}}{1000 \text{ g solvent}} = \frac{18.55 \text{ g solute}}{1000 \text{ g solvent}}$$

$$0.895°\text{C} \times \frac{\left(\dfrac{1 \text{ mole solute}}{1000 \text{ g solvent}}\right)}{7.00°\text{C}} = 0.128 \frac{\text{moles solute}}{1000 \text{ g solvent}}$$

$$0.128 \text{ moles solute} = 18.55 \text{ g solute}$$

$$1 \text{ mole solute} = 145 \text{ g solute}$$

*In units of °C/1 mole solute/1000 g solvent.

10-8 Colloids

Ordinary solutions are classed as homogeneous because there are no means ordinarily available for detecting a distinct boundary between the solute molecules and the solvent phase. Typically, the solute molecules or ions are on the order of up to 1,000 in molecular weight, and perhaps up to 12–15 Å across. Larger particles, consisting of single molecules of high molecular weight, or particles consisting of agglomerations of ions or molecules, may, however, become homogeneously distributed in a solvent. When one dimension of the particles is on the order of about 10^4 Å, with the others about 10 Å, they may form a colloidal suspension in a solvent. In a colloidal suspension, the boundary between solute and solvent can be detected. Colloidal systems are classified as *lyophobic* or *lyophilic*, according to whether the solvent and solute particles are strongly attracted to one another (lyophilic) or not (lyophobic). Lyophilic colloids include starches, proteins and other large molecular systems which occur in living systems. The very large DNA (deoxyribonucleic acid) molecule which carries the genetic information of the cell is an example of the upper range of molecular weights attained by molecular colloids, with a molecular weight in the range of 1,000,000. There are many functional groups on such molecules which are able to enter into hydrogen bonding with the water of the solvent. Furthermore, many of these molecules are polyelectrolytes, which means that they contain sites of positive and/or negative charge at frequent intervals along the molecule. As a result, there is a strong interaction with solvent, resulting in a lyophilic colloid. The attractive forces between solvent and solute in a lyophilic colloid help to keep the colloidal particles separated from one another. On the other

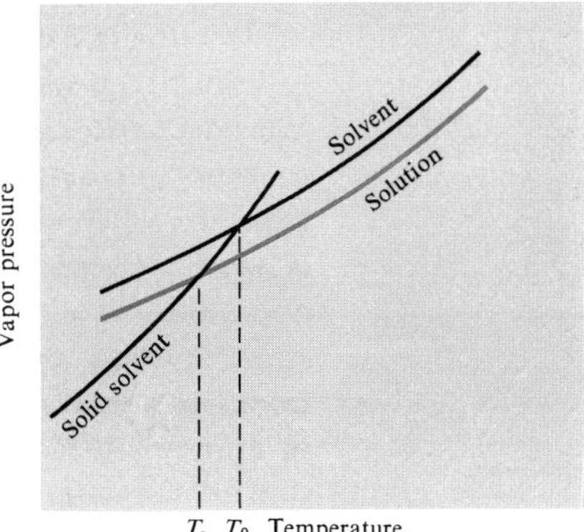

Fig. 10-10 Vapor pressure curves of pure solvent, pure solid solvent and solution near the freezing temperatures. T_o represents the freezing point of the pure solvent, T_s the freezing point of the solution.

hand, lyophobic colloidal suspensions are not generally stable. The particles may be prevented from condensing and precipitating out, however, by two effects. The first is the kinetic energy of motion of the particles. If the weight of the individual particles is low enough, collisions with solvent molecules can cause energy transfer. The colloidal particles can be seen under an ultramicroscope to undergo a random motion, called *Brownian motion.* This motional energy may be sufficient to prevent aggregation of the colloid. An additional factor may be the presence of charges on the surface of the colloid. For example, if extremely small gold particles are formed in water containing sodium chloride, the chloride ions are adsorbed on the gold particles. Thus, all of the colloidal particles acquire the same charge, and electrostatic repulsions prevent aggregation. Although lyophobic colloids are inherently unstable, they can be prepared in such a way that they will remain indefinitely in the form of suspensions. Colloidal suspensions of sulfur and of gold in water have been maintained for over fifty years.

Colloidal suspensions exhibit many properties which distinguish them from what we commonly think of as true solutions. Some of these properties are extremely important for practical reasons. For example, *gelation* may occur when a lyophilic colloid absorbs so much of the solvent that the quantity of free solvent is materially reduced. *Thixotropy,* another property of certain colloids which is of some importance, refers to the ability of a gel to appear as very viscous, or even solid, but at the same time to flow easily under the influence of a shearing force. Many other properties of colloids are of importance in the study of living systems. The interested reader is referred to the reading list at the end of the chapter.

10-9 Phase Diagrams

In our discussion so far of liquids and solutions, we have had occasion to consider a number of cases in which two phases are in equilibrium. We have, for example, considered the equilibrium between a liquid and its vapor, or between a solid and a solution. It is possible to portray systems in which phases are in equilibrium with one another on a *phase diagram.* Phase diagrams range from very simple to extremely complex, depending upon the number of variables that the system possesses. The variables that we need to consider are the concentrations (or mole fractions), temperature and pressure. The number of *components* is the number of different substances which must be specified to determine the composition of the system. For example, a system which consists of pure water in equilibrium with pure ice is a one-component system, with two phases. On the other hand, the system which is described by the diagram in Fig. 10-7 is a two-component system, since we must know that both benzene and toluene are present to know what the system is all about. In this particular diagram one of the degrees of freedom, the pressure, is constant at one atmosphere. The phase diagram tells us about the temperature at

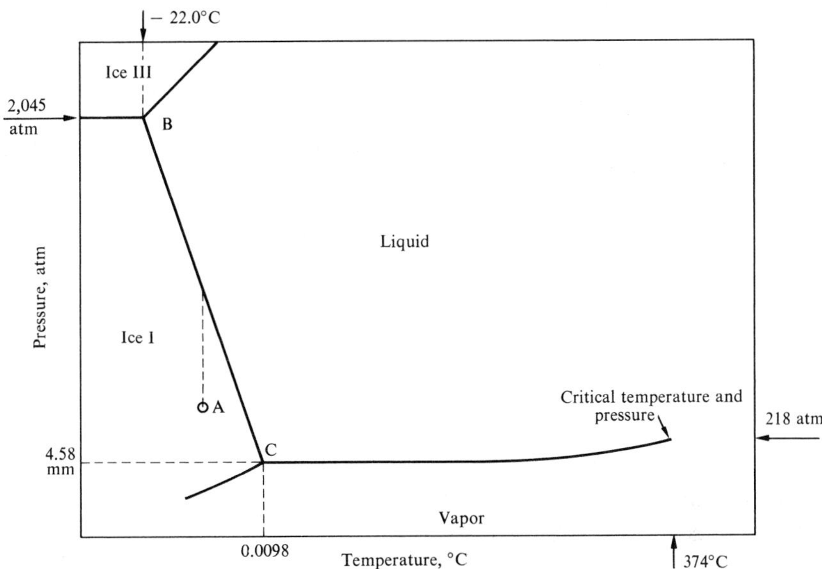

Fig. 10-11 Phase diagram for water. The temperature and pressure scales are somewhat distorted, to show a wider range of these variables.

which the liquid and gas phases are in equilibrium at one atmosphere, as a function of the composition.

We will not pursue the subject of phase diagrams in great detail, but it is of interest to examine thoroughly a phase diagram for a one-component system, such as pure water. Since we have a one-component system, the composition cannot vary. The only two variables remaining are temperature and pressure. We can thus see that a two-dimensional plot which has temperature as one coordinate and pressure as the other should be adequate to completely describe the equilibria that might exist between the three phases, solid, liquid, and gas.

The phase diagram for water is shown in Fig. 10-11. Each phase is indicated by an area. The phases are in equilibrium at the temperatures and pressures indicated by the lines that separate the phases. For example, at a temperature of −12°C and a pressure of 300 atm, the water at equilibrium is entirely in the form of ice. (See point A in Fig. 10-11.) If the temperature is maintained constant, however, and the pressure increased to 1150 atm, an equilibrium between ice and liquid water is reached. Perhaps the most interesting point on the graph is the *triple point*, the single combination of pressure and temperature at which all three phases can exist in equilibrium, indicated by point *C*. The triple point occurs at 0.0098°C and 4.58 mm water vapor. We are accustomed to thinking of the freezing point of water as 0.0000°C; this, however, is the freezing point of water in the presence of 1 atm of air. Since air is slightly soluble in water, the freezing point is slightly depressed, from 0.0098 to 0.0000°C.

Figure 10-11 is incomplete in that only a limited range of pressure and temperature is shown. Nothing of much interest occurs at higher temperatures, but at higher pressures new solid phases (*i.e.*, different forms of ice) are found. The various forms of ice differ in the manner in which the molecules are packed within the lattice. The structure for ice I, ordinary ice, is shown in Fig. 8-31. It is a rather open structure, with a rather low density. Under the influence of pressure, the structure is modified in various ways to permit more efficient packing, thus leading to denser forms of ice. Ice II is in equilibrium with ordinary ice (ice I) at about −31°C and 2100 atm. The ice I-ice III-liquid triple point is shown on Fig. 10-11, point B. Seven different forms of ice are known. Ice VII is in equilibrium with ice VI and liquid water at 81.6°C and 21,680 atm. This is again a triple point; a single combination of pressure and temperature at which three distinct phases exist in equilibrium. We see that if we consider just two phases in equilibrium, we can be anywhere on a line on the phase diagram. Another way of saying this is that there is one degree of freedom. Within limits, we can choose a temperature *or* a pressure; once we have chosen one, the other is determined. A triple point, where three phases are in equilibrium, is represented by a point on the phase diagram; there are no degrees of freedom. If there is just one phase present, *e.g.*, the vapor phase, we can be anywhere within an area on the diagram. That is, we can, again within limits, choose a certain temperature *and* pressure, and still be within the area that represents the gas phase alone. Thus there are two degrees of freedom.

For the one-component system, then, there is a relationship between the number of phases present, ϕ (pronounced "fee"), and the number of degrees of freedom, Δ (delta): $\Delta = 3 - \phi$. This little equation is a special case of the *phase rule*, which is more general in that it also takes into account systems of more than one component. The phase rule states simply that the number of degrees of freedom which a system possesses is equal to the number of components, c, minus the number of phases, ϕ, plus two:

$$\Delta = c - \phi + 2$$

The factor two represents the two variables, pressure and temperature. It is clear that this is a general form of the equation which we found for the one-component system ($c = 1$). The interested student is referred to the list of suggested readings for further material on the phase rule.

Suggested Readings

Bernal, J. D., "The Structure of Liquids," *Scientific American*, August, 1960.

Daniels, F. and R. A. Alberty, *Physical Chemistry*, 2nd Ed., New York: John Wiley & Sons, Inc., 1961.

Hildebrand, J. H., *Introduction to Kinetic Molecular Theory*, New York, N.Y.: Reinhold Publishing Corporation, 1966.

Vold, M. J. and R. D. Vold, *Colloid Chemistry*, New York, N. Y.: Reinhold Publishing Corporation, 1964.

Exercises

10-1. Calculate the molarity and molality of each of the following solutions:
(a) 88 g water and 12 g sodium sulfite, Na_2SO_3; specific gravity 1.115 at 20°C.
(b) 74 g water and 26 g tin (II) chloride; specific gravity 1.225 at 25°C.

10-2. A solution is made by dissolving 235 g $CaCl_2 \cdot 2H_2O$ in 645 g water. The density of the resulting solution at 25°C is 1.195 g/cm^3. Express the concentration of calcium chloride in the solution in terms of mole fraction, molarity, and molality. Which of these concentration units is temperature dependent?

10-3. Concentrated nitric acid is 70 per cent HNO_3 by weight and has a specific gravity of 1.42 at room temperature. Calculate the molarity. What is the normality as an acid?

10-4. 1.50 g of acetic acid, CH_3COOH, is dissolved in water to make 1 liter of solution. Another 1.50 g is dissolved in benzene to make a liter of solution. The densities of pure water and benzene at 25°C are 0.997 and 0.879, respectively. Calculate the concentration of acetic acid in each solution in units of mole fraction, molarity, and molality. (Since the solutions are quite dilute, the densities of the solutions may be approximated by the densities of the pure solvents.)

10-5. According to Trouton's rule, the entropy of vaporization of a liquid is about 21 cal/°K-mole. The rule seems to apply quite well to lead, with a boiling point of 2023°C, as well as to liquids with a boiling point of approximately 350°C. What does this imply about the temperature dependence of the free volume in the liquid state? Explain.

10-6. Which contains the greater mole fraction of solute, a $1M$ or a $1m$ aqueous solution? Explain. What property of the solvent determines whether the molarity or molality represents the greater mole fraction of solute?

10-7. Assuming that the solutions behave ideally, calculate the molality and molarity of ethanol in a solution made up from 32.5 g ethanol (C_2H_5OH) and 186 g water. The density of ethanol may be taken to be 0.787 g/ml; that of water, 1.00 g/ml.

10-8. A student performing an experiment to determine the vapor pressure of carbon tetrachloride at various temperatures, using the apparatus shown in Fig. 10-3, failed to achieve removal of all of the air from the space between A and B. What effect does this have on his measured values for vapor pressure? What effect would it have on his estimate of the heat of vaporization of CCl_4?

10-9. Calculate the heat of vaporization of CCl_4 at 60°C from the data in Table 10-2.

10-10. Using the Henry's law constant given in Table 10-3, calculate the *molar* concentration of bromine in water when the pressure of bromine vapor over the solution at equilibrium is 6 mm Hg.

10-11. The vapor pressure of water at 50°C is 92.5 mm Hg, while that of ethyl alcohol, C_2H_5OH, at the same temperature is 222 mm Hg. Assuming an ideal solution, calculate the total vapor pressure at 50°C over a solution made up from 100 g water and 46 g alcohol. Again assuming ideal behavior, what is the total vapor pressure over a solution made up from 100 g water and 42.4 g lithium chloride (LiCl)?

10-12. 1.350 g of an organic compound is dissolved in 46.52 g of chloroform. The boiling point of the solvent is raised 0.730°C as a result. Calculate the molecular weight of the unknown substance.

10-13. 0.300 g of acetic acid, CH_3COOH, is dissolved in 36.0 g benzene. The freezing point of the solution is 0.340°C lower than that of the pure solvent. What is the molecular formula for acetic acid in this solvent?

10-14. 0.400 g naphthalene, $C_{10}H_8$, molecular weight 128.2, is dissolved in some carbon tetrachloride; the boiling point of the solution increases by 0.458°C. Now 0.538 g of a substance of unknown molecular weight is added to the naphthalene solution. The boiling point increases by an additional 0.430°C. What is the molecular weight of the unknown substance?

10-15. Consider the following vapor pressure data for iso-amyl alcohol, $C_5H_{12}O$:

Temperature (°C)	Pressure (mm Hg)	Temperature (°C)	Pressure (mm Hg)
100	238.6	130	743.2
110	358.6	140	1033
120	523.3		

From these data, determine the boiling point elevation constant by the method described in the text. (Remember to convert temperature to °K.)

10-16. Calculate a Henry's law constant, in units of mm Hg, for HCl in benzene from the data listed in Table 10-4. What explanation can you offer to account for the very large differences in solubility for HCl in the various solvents listed in the Table?

10-17. The isopiestic method for determining molecular weights is as given in the following example: 0.365 g of biphenyl ($C_{12}H_{10}$), molecular weight 154.2, is placed in arm A of the apparatus shown. 0.273 g of an unknown substance is placed in arm B. Some benzene is poured into each arm, and the apparatus is pumped out to remove air and sealed off. It is then placed in a constant temperature bath for a few days, at the end of which time equilibrium in the system is established.

The quantity of benzene in each arm is then measured and found to be 36.20 g in arm A and 42.65 g in arm B. From these data, calculate the molecular weight of the unknown substance. (We must assume, of course, that both the known and unknown substances dissolve in benzene. The only reminder which should be necessary is that, if equilibrium is established, the vapor pressure of benzene over the solutions must be the same in each arm. From this point on, Raoult's law is all that is required.)

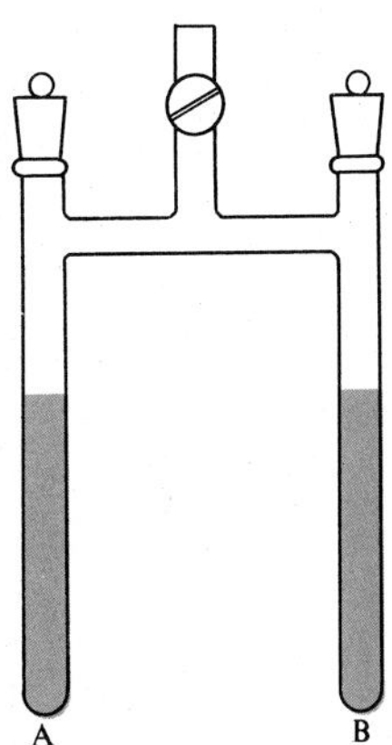

10-18. Suppose that you have a boiling point elevation apparatus with which you can determine temperature differences of ± 0.01°C with reasonable confidence. The apparatus has a capacity of 15 ml of solution. If you are required to determine the molecular weights of substances which have molecular weights of approximately 250 g/mole to an accuracy of 2 per cent using CCl_4 as solvent (Table 10-5), what is the minimum weight of substance you require for a single determination?

10-19. From the knowledge that ice is less dense than water, and that melting is an endothermic process, explain why the melting point of ice decreases with increasing pressure (Fig. 10-11).

10-20. Assuming an average density of 1.2 g/cm^3, calculate the weight of a cylindrical particle which is 15 Å in diameter and 2,000 Å in length. What would be the molecular weight of the substance if this corresponds to the dimensions of a single molecule?

10-21. Indicate whether each of the following is a lyophilic or lyophobic colloid:
(a) Cigarette smoke in the air.
(b) The butterfat in homogenized milk.
(c) Jello.
(d) An extremely fine silver iodide precipitate in water.
(e) Eggnog.

Chapter 11

Solutions of Electrolytes

11-1 Solute-Solvent Interactions

Water which has been carefully purified is not a good conductor of electricity. (It does conduct to a very slight extent; the reason for this will be discussed in Chap. 12, "Acids and Bases.") If one dissolves some sodium chloride, or any other ionic compound, in the water, the resulting solution is a good conductor. The explanation for this behavior was given by Svante Arrhenius in 1887, when he suggested that the current was being carried by the separated ions of the salt. Today, there is so much evidence in favor of this theory of electrolytic conduction that it can no longer be questioned.

When an ionic crystal is placed into water, the surface of the crystal becomes covered with water molecules which interact strongly with the ionic charges, as shown in Fig. 11-1. This interaction arises because the ions which are on the surface of the crystal are not in a symmetrical environment, as are those in the bulk of the solid. The water molecules, which are dipolar (p. 190), are attracted to the surface through charge-dipole forces. As ions leave the surface, they become completely surrounded

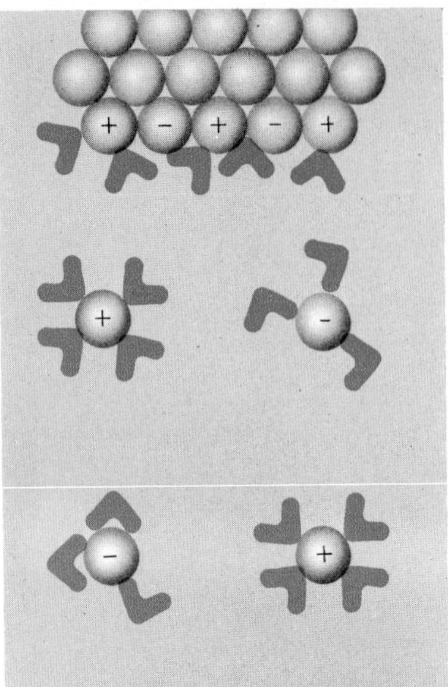

Fig. 11-1 Hydration of ions in aqueous medium.

by water molecules. This sheath of solvent around the ion acts to prevent it from interacting with an ion of opposite charge, or with the solid surface. Furthermore, the attraction which ions of opposite charge have for one another is much lower in the solution than it would be in the gas phase because of another property of the solvent — its dielectric constant. Two charges of opposite sign, which are separated by a distance r in a vacuum, are attracted to one another with a force given by

$$F = -e^2/r^2 \qquad (11\text{-}1)$$

where e is the magnitude of the charge. If the same two charges, separated by the same distance r, are now placed in a medium such as water (or any other solid, liquid, or gas), the force operating between them is

$$F = -e^2/\epsilon r^2 \qquad (11\text{-}2)$$

where ϵ is the dielectric constant of the medium. It can be seen that when the dielectric constant is high, the attracting force between the ions is low. We may expect, then, that solvents which are capable of maintaining a separation of charges will possess a high dielectric constant. Table 11-1 lists the dielectric constants of a number of media. Water, it will be noticed, possesses an unusually high dielectric constant.

Any substance which, when dissolved in a polar solvent such as water, gives rise to an electrically conducting solution is referred to as an *electrolyte*. This designation includes salts such as sodium chloride, potassium hydroxide, barium chlorate, *etc.*, all of which are ionic in the solid state.

Table 11-1 The dielectric constants of some liquids.

Compound	Temperature (°C)	ϵ
Benzene	25	2.25
Chloroform	20	4.81
Carbon disulfide	20	2.64
Ammonia	−33	22.4
Methyl alcohol	25	32.6
Phosphoryl chloride ($POCl_3$)	22	13.3
Hydrogen fluoride	0	84.0
Ethyl alcohol	25	24.3
Water	25	78.5
Dimethyl ether	25	5.02

Therefore, the production of ions in the solution involves only a separation of the ions. Hydrogen chloride, ammonia, acetic acid, and a number of other similar substances are also electrolytes. They differ from the ionic substances, however, in that they are covalent molecules in the absence of water. The presence of ions in solutions of these substances results from a chemical reaction of the dissolving substance with the solvent. These reactions are, in part, the subject of the chapter "Acids and Bases," and will not be elaborated on at this point. Suffice it to say that, whatever the mechanism by which the ions are produced, the properties of the electrolytic solutions are qualitatively the same.

It might be thought that, if the ions are indeed well separated from one another in solution, they would behave in some respects as independent entities. Subject to the restriction that the charges must be uniformly distributed throughout the solution as a whole, this is, in fact, the case. Thus, in a solution of (*e.g.*) sodium nitrate and potassium chloride, the four ions, Na^+, K^+, Cl^-, and NO_3^- are uniformly distributed throughout the solution. In writing chemical equations involving the reactions of ions in solutions, therefore, it is correct to write the ions as separate entities. An example will serve to illustrate this point. When a solution of sodium chloride is added to a solution of silver nitrate, a silver chloride precipitate forms. The equation representing this reaction may be written as

$$NaCl(aq) + AgNO_3(aq) \rightarrow NaNO_3(aq) + AgCl(s) \qquad (11\text{-}3)$$

On the other hand, if we write the ions as separated, the equation is of the form

$$Na^+(aq) + Cl^-(aq) + Ag^+(aq) + NO_3^-(aq) \rightarrow Na^+(aq) + NO_3^-(aq) + AgCl(s) \qquad (11\text{-}4)$$

From this it can be seen that the sodium and the nitrate ions are essentially unchanged in the chemical reaction. If we delete these from the balanced equation, we have what is called the *net ionic equation*:

$$Ag^+(aq) + Cl^-(aq) \rightarrow AgCl(s) \qquad (11\text{-}5)$$

This equation represents the essence of the change which is taking place. If potassium chloride and silver perchlorate had been used instead of sodium chloride and silver nitrate, the net ionic equation would be the same, since the essentials of the reaction — formation of solid silver chloride — would remain the same.*

The nature of the equilibrium in a saturated solution is exemplified by the equation representing a saturated sodium chloride solution

$$NaCl(s) \underset{}{\overset{H_2O}{\rightleftharpoons}} Na^+(aq) + Cl^-(aq)$$

This equation expresses the dynamic character of the system: Two processes are continuously occurring in a saturated solution which is in equilibrium with a quantity of solid solute. The solute — sodium chloride, in this instance — is dissolving; *i.e.*, sodium and chloride ions are leaving the surface of the crystals and dispersing in the water phase. At the same time, sodium and chloride ions are leaving the water phase and attaching to the crystal surfaces. Since the concentration of the solute in a saturated solution does not change as a function of time, it must be true that the rate at which the first process occurs in a given area of crystal surface is just equal to that at which the second process occurs over the same area. The dynamic character of the equilibrium is evidenced in a number of ways. For example, if one places a number of small crystals of sodium chloride in contact with a quantity of saturated solution for a period of time, the small crystals gradually disappear and are replaced by a smaller number of larger crystals. One could easily verify, however, that at all times the quantity of solid in contact with the solution remains constant. Though we may not know just why the larger crystals tend to grow at the expense of the smaller ones, it is still likely that such a change could only occur through the intermediacy of the dynamic processes described above.

A second experiment which demonstrates the character of the equilibrium between the solid and the dissolved solute is performed by adding to a saturated sodium chloride solution some solid sodium chloride in which a radioactive isotope of chlorine is used. In a short time, this radioactivity is present in the solution as well as in the solid.

11-2 Effect of Temperature on Solubility

When sodium chloride is dissolved in water, a certain quantity of heat is *absorbed*. We can make an explicit account of this by writing the solution process as

$$NaCl(s) \xrightarrow{H_2O} Na^+(aq) + Cl^-(aq) + \text{heat } \textit{absorbed}$$

*The notation (aq) to designate species in aqueous solution will be employed intermittently throughout the remainder of the text, particularly when there is some possibility that the reaction conditions may not be clearly understood. In the absence of any notation, it should be understood that species reacting in water solution are hydrated by the solvent.

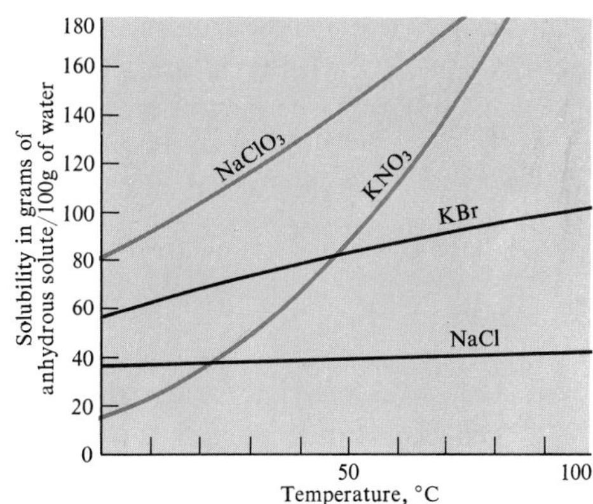

Fig. 11-2 The effect of temperature on solubility of salts in water.

From thermochemical considerations, it follows that the reverse reaction proceeds with the *evolution* of the same quantity of heat

$$Na^+(aq) + Cl^-(aq) \rightarrow NaCl(s) + \text{heat } \textit{evolved}$$

In a system at equilibrium (in this case, a saturated solution in contact with sodium chloride) the rates of the two reactions proceed at equal rates, and the rate at which heat is absorbed in the forward reaction just equals the rate at which heat is evolved in the reverse step. If, however, the temperature is increased the equilibrium will no longer exist; the reaction which proceeds with the absorption of heat will speed up relative to the opposing reaction, until a new set of equilibrium conditions is reached. The concentration of dissolved sodium chloride will therefore increase with increasing temperature. In general, if the solution process proceeds with the absorption of heat, the solubility increases with increasing temperature; if, however, the solution process proceeds with the evolution of heat, the opposite is true. Furthermore, the *degree* of temperature dependence is directly related to the *amount* of heat evolved or absorbed in the solution process. For example, the heat absorbed when sodium chloride dissolves in water is quite small; the solubility of this salt is correspondingly not very dependent upon temperature (Fig. 11-2). On the other hand, the solubility of potassium nitrate increases very rapidly with temperature, because a considerable quantity of heat is absorbed when this salt dissolves in water (Fig. 11-2).

The solution of most ionic substances in water is an *endothermic* (heat absorbing) process, whereas the solution of most gases in water is an *exothermic* (heat evolving) process.

11-3 Thermochemistry of Electrolytic Solutions

The *heat of solution* is the heat evolved when one mole of a solute is dissolved in a given quantity of a liquid solution. The heat of solution is dependent upon the quantity of solvent employed; one therefore finds that the concentration of the solution is specified, as, for example, one-molar, or one part of solute per x parts of solvent. If the amount of solvent used is sufficiently large, however, the heat of solution *per mole of solute* is not noticeably dependent upon the amount of solvent. Thus, the process written

$$n\mathrm{H_2O} + \mathrm{CaCl_2(s)} \rightarrow \mathrm{CaCl_2(aq)} \qquad \Delta H = -19{,}600 \text{ cal}$$

refers to the dissolving of a mole of calcium chloride in a large amount of water to form a dilute solution (strictly speaking, an infinitely dilute solution).

One can calculate the heat of formation of an ionic substance in aqueous solution from a knowledge of the heat of formation of the solid substance and the heat of solution of the solid substance:

$$(A)\quad \mathrm{Ca(s)} + \mathrm{Cl_2(g)} \rightarrow \mathrm{CaCl_2(s)} \qquad \Delta H_f^\circ = -190.2 \text{ kcal}$$
$$(B)\quad \mathrm{CaCl_2(s)} + n\mathrm{H_2O} \rightarrow \mathrm{CaCl_2(aq)} \qquad \Delta H^\circ = -19.6 \text{ kcal}$$
$$(A) + (B)\quad \mathrm{Ca(s)} + \mathrm{Cl_2(g)} + n\mathrm{H_2O} \rightarrow \mathrm{CaCl_2(aq)} \qquad \Delta H^\circ = -209.8 \text{ kcal}$$

Since, to a good approximation, the ions in a solution of an electrolyte subsist independently of one another, we can think of the heat of formation of a strong electrolyte in water as the sum of the heats of formation of the individual ions. But, because we can never hope to actually observe the properties of a single kind of ion alone in water, we cannot separate the total heat of formation for two ions into individual values unless we arbitrarily set a value for the heat of formation of some one ion. This has been done: The heat of formation of hydronium ion H_3O^+ in an aqueous solution has been set equal to $-68{,}300$ cal/mole. (The nature of the hydronium ion, H_3O^+, and the reason for its use, will become clear in the next chapter, "Acids and Bases". This convention is not followed in some texts, and only the hydrogen ion, H^+, is written. The hydronium ion consists of the hydrogen ion plus a water molecule. The heat of formation of H^+ in water is defined to be zero. Then the heat of formation of the hydronium ion is just the heat of formation of the extra water molecule which is needed to form H_3O^+, or $-68{,}300$ cal/mole. The important point in all this is that *the same species is being considered in all cases; namely, the hydrogen ion in water*. It is simply a matter of whether it is to be represented as H^+ or H_3O^+.)

The heat evolved when one equivalent of a strong acid is neutralized with one equivalent of a strong base is independent of the particular acid or base employed. From this, it can be inferred that the essential reaction is

$$\mathrm{H_3O^+(aq)} + \mathrm{OH^-(aq)} \rightarrow 2\mathrm{H_2O(l)} \qquad \Delta H^\circ = -13{,}800 \text{ cal}$$

If we combine this with the reaction representing the heat of formation of water

$$\mathrm{H_2(g)} + \tfrac{1}{2}\mathrm{O_2(g)} \rightarrow \mathrm{H_2O(l)} \qquad \Delta H^\circ = -68{,}300 \text{ cal}$$

we obtain

$$H_2O(l) + H_2(g) + \tfrac{1}{2}O_2(g) \rightarrow H_3O^+(aq) + OH^-(aq) \quad \Delta H^\circ = -54{,}500 \text{ cal}$$

Since the heat of formation of $H_3O^+(aq)$ has been set equal to $-68{,}300$ cal/mole, the same as that of $H_2O(1)$, the heat of formation of $OH^-(aq)$ is $-54{,}500$ cal/mole.

Once the heats for formation of H_3O^+ and OH^- have been established, it is a simple matter to obtain the values for other cations and anions from a knowledge of the heats of formation of strong acids and bases in aqueous medium. Some values for heats of formation of ions in aqueous medium at 25°C are listed in Table 11-2.

Table 11-2 Heats of formation of ions in water at 25°C.

Ion	ΔH° (kcal/mole)	Ion	ΔH° (kcal/mole)
H_3O^+	-68.3	Co^{+2}	-16.1
Li^+	-66.6	Ni^{+2}	-15.3
Na^+	-57.3	Cu^{+2}	-15.4
K^+	-60.0	Ag^+	-25.3
NH_4^+	-31.7	Al^{+3}	-125.4
Mg^{+2}	-110.4	Fe^{+3}	-11.4
Ca^{+2}	-129.8	Cl^-	-40.0
Sr^{+2}	-130.4	Br^-	-28.9
Ba^{+2}	-128.7	I^-	-13.4
Mn^{+2}	-52.3	NO_3^-	-49.4
Fe^{+2}	-21.0	NO_2^-	-25.4

The uses to which such data can be put are illustrated by the calculation of the heat of the reaction

$$FeO(s) + 2H_3O^+(aq) \rightarrow 3H_2O(l) + Fe^{+2}(aq)$$

The heat of the reaction is equal to the sum of the heats of formation of the products, minus the sum of the heats of formation of the reactants. The required data are

(1) $Fe^{+2}(aq)$ $\quad \Delta H_f^\circ = -21.0$ kcal
(2) $H_2O(l)$ $\quad \Delta H_f^\circ = -68.3$ kcal
(3) $FeO(s)$ $\quad \Delta H_f^\circ = -127.4$ kcal
(4) $H_3O^+(aq)$ $\quad \Delta H_f^\circ = -68.3$ kcal

$(1) + 3(2) - (3) - 2(4) = 38.1$ kcal; the reaction is endothermic.

11-4 Electrolysis of Solutions

Electrochemistry deals with those electrical properties of liquids and solutions which derive from their ability to conduct electric current. Solutions of electrolytes in water are the most important single area within the subject.

In solutions of electrolytes in water, the individual ions are (in dilute solution, at any rate) independent of one another, the only requirement being that they are distributed uniformly throughout the entire solution. Since the solvent molecules and solute ions are in constant thermal motion at ordinary temperatures, the ions diffuse at random in solution. If a pair of electrodes is inserted into the solution and a voltage is applied, the ions no longer diffuse completely at random; under the influence of the potential gradient, they move more in one direction than in the other. The positively-charged ions migrate toward the negative electrode, while the negatively-charged ions move in the opposite direction. The situation is much like that encountered in electrolysis of fused ionic substances, as discussed in Sec. 7-7. The major difference between the behavior of aqueous solutions and that of fused salts is in the character of the electrode reactions, since water may be oxidized or reduced in preference to the solute ions at either or both electrodes. Faraday's laws are nevertheless obeyed; in fact, these laws were the outcome of Faraday's work with aqueous solutions.

Let us consider the electrolysis of a sodium chloride solution. The reaction at the anode — by definition, an oxidation reaction — may be oxidation of either chloride ion or water. The two half-reactions are

$$2Cl^- \rightarrow Cl_2(g) + 2e^-$$
$$6H_2O \rightarrow O_2(g) + 4H_3O^+ + 4e^-$$

We find that in dilute solutions of sodium chloride both reactions occur; the relative amounts of chlorine and oxygen produced depend upon the applied voltage, the current density (amperes per cm^2 of electrode surface), and the temperature. In concentrated solutions, the product is essentially all chlorine.

The product appearing at the cathode cannot be sodium for the obvious reason that, if sodium were produced, it would react immediately with water. The reaction sequence would be

$$2Na^+ + 2e^- \rightarrow 2Na \qquad (11\text{-}6)$$
$$2Na + 2H_2O \rightarrow 2Na^+ + 2OH^- + H_2(g) \qquad (11\text{-}7)$$

The second reaction is simply an oxidation-reduction reaction in which sodium metal reduces water. The over-all effect, embodied in the sum of the two reactions, is the reduction of water

$$2H_2O + 2e^- \rightarrow 2OH^- + H_2 \qquad (11\text{-}8)$$

Merely on the basis of observing the evolution of hydrogen gas at the cathode, we cannot say whether sodium ion is reduced and subsequently reduces water, or whether water is reduced directly. The end result is the reduction of water, in any case. In concentrated brine solution, therefore, the two half-reactions and over-all cell reaction are

$$2Cl^- \rightarrow Cl_2(g) + 2e^- \qquad (11\text{-}9)$$
$$2H_2O + 2e^- \rightarrow 2OH^- + H_2(g) \qquad (11\text{-}10)$$
$$\overline{2H_2O + 2Cl^- \rightarrow Cl_2(g) + 2OH^- + H_2(g)} \qquad (11\text{-}11)$$

This cell reaction is employed commercially in the production of chlorine gas, Chapter 17.

In the electrolysis of a sodium sulfate solution, the solvent enters into both electrode reactions to the exclusion of both sodium and sulfate ions. The sulfate ion is more stable than the chloride ion, and unless special circumstances prevail, water is preferentially oxidized. Since the solute in this case does not enter into either electrode reaction, it serves only as a carrier of charge. It is instructive to examine the migration of ions in such a solution to see what changes in the solution composition occur as the reaction proceeds. Initially, as shown in Fig. 11-3a, the composition of

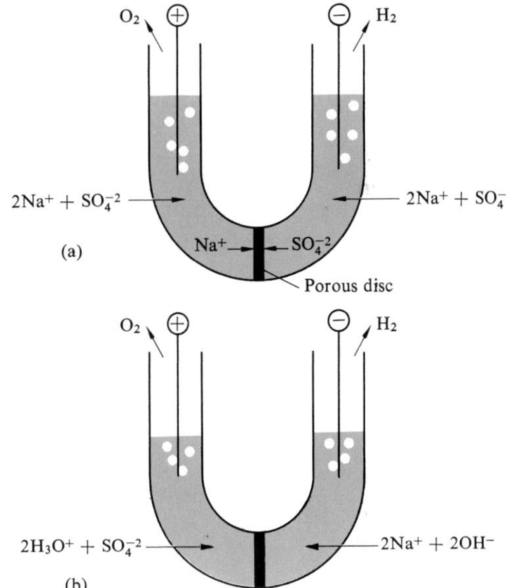

Fig. 11-3 Electrolysis of aqueous sodium sulfate solution. As electrolysis proceeds, the solution in the anode compartment becomes acidic, that in the cathode compartment becomes basic.

the solution is homogeneous throughout. When the potential is applied, the cations migrate toward the cathode, the anions toward the anode. The reactions occurring at the two electrodes permit the migration to continue by generating charged particles as reaction products. At the anode, the product is hydronium ion, Eq. (11-12); at the cathode, it is hydroxide ion, Eq. (11-13).

$$6H_2O \rightarrow O_2(g) + 4H_3O^+ + 4e^- \qquad (11\text{-}12)$$
$$4H_2O + 4e^- \rightarrow 2H_2(g) + 4OH^- \qquad (11\text{-}13)$$

Despite the ion migration, or really, because of it, electroneutrality is maintained throughout the solution. After the electrolysis has proceeded for some time, the ion concentrations are as shown in Fig. 11-3(b). The cathode compartment has become richer in hydroxide ion, OH^-, and the anode compartment has become richer in hydronium ion, H_3O^+.

Electrolysis is an important chemical tool: It is used in electroplating, in refining metals such as copper, lead and zinc, and in producing chlorine and a number of chemical compounds by electrochemical oxidation. Many of these examples are discussed in the appropriate places in the text, but we will summarize here the types of electrode reaction which may be encountered.

Reduction of a metal ion — Metals which are not too high in chemical activity may be reduced at a cathode in aqueous solution. The reduction of copper (II) ion is typical:

$$Cu^{+2}(aq) + 2e^- \rightarrow Cu \qquad (11\text{-}14)$$

Sodium, magnesium, and aluminum are among the elements which are obtained commercially by electrolytic reduction of fused melts.

Oxidation of a non-metallic anion — This type of reaction is typified by the oxidation of chloride ion, Eq. (11-9). In other similar reactions, the products are not the gaseous free elements. Electrolysis of a potassium chloride solution, for example, leads to the production of chlorate ion if carried out under suitable conditions:

$$Cl^- + 9H_2O \rightarrow ClO_3^- + 6H_3O^+ + 6e^- \qquad (11\text{-}15)$$

Reduction of anions — Because of the charge relationship, anions tend to migrate away from the cathode during electrolysis. There is, however, no reason why they cannot be reduced. In fact, there are a number of important electrochemical procedures in which this occurs. Chrome plating is done from baths containing CrO_3 in acid solution. Under these conditions, the chromium is present in the solution as $Cr_2O_7^{-2}$, or some similar anionic species. The migration of anion away from the cathode is counteracted by keeping the solution well stirred. The reduction reaction is represented by the equation

$$12e^- + Cr_2O_7^{-2} + 14H_3O^+ \rightarrow 2Cr + 21H_2O \qquad (11\text{-}16)$$

In addition to the aforementioned applications of electrolysis, there are a number of analytical procedures which are based on electrolytic methods. To give just one example, a metal alloy which contains copper, zinc, and aluminum could be analyzed for copper by dissolving the alloy in acid and then electrolyzing the solution. By careful choice of voltage, electrode material, and other variables, copper could be selectively electroplated from the solution, leaving the other, more active metals in solution.

11-5 Conductance of Electrolytic Solutions

The fact that electrolytic solutions are conductors of electric current forms the basis for conductance studies. The interest now is not on the electrode reactions (in fact everything possible is done to avoid them), but on the quantitative measure of the ability of the solution to carry current.* Electrode reaction is minimized by employing alternating current rather than direct current potential. A typical conductance-measuring experiment is diagrammed in Fig. 11-4. The cell containing the electrolytic

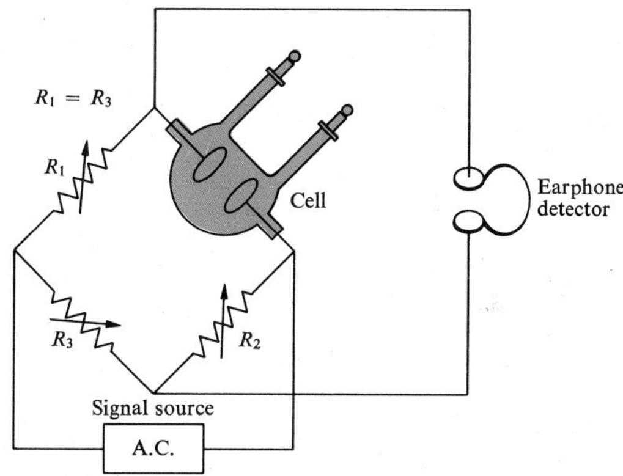

Fig. 11-4 Experimental arrangement for conductance studies.

solution is made part of a resistance bridge. When the bridge is balanced (when R_2 is equal in value to the cell resistance), no signal is heard in the headphones. From the values obtained for the cell resistance, the specific conductance C of the solution can be determined. It is a familiar fact that the resistance of an electrical conductor increases with the length of the conductor, and decreases with an increase in cross-sectional area:

$$R = \rho(l/A) \tag{11-17}$$

R = measured resistance l = length of conductor
A = cross-sectional area ρ = specific resistance

The specific resistance is a constant which is characteristic of the conducting medium. The specific conductance C is $1/\rho$. It can be determined from the measured cell resistance if the dimensions of the electrodes and their distance of separation are known. Because of difficulties in making such measurements, however, it is usual to "calibrate" the cell by measuring its resistance when it contains a substance of known specific conductance.

*Conductance is the logical inverse of resistance, and is in fact just equal to $1/R$, where R is the resistance.

Since the specific conductance is a characteristic of the conducting medium, it is a more general quantity than the measured resistance in that it is independent of the particular cell in use. The specific conductance of a $0.1N$ sodium nitrate solution, for example, is the same for any given temperature regardless of the type of cell in which it is placed. However, the specific conductance of this solution *is* different from that of a sodium nitrate solution of another concentration. It would be roughly twice as great as that of a $0.05N$ solution of the same salt.

To take account of the fact that the concentration of a solution may vary, the equivalent conductance Λ is employed. This is equal to the specific conductance divided by the normality, and multiplied by 1000; in effect, the specific conductance is divided by the number of moles of solute per cm^3 of solution:

$$\Lambda = \frac{1000C}{N} \tag{11-18}$$

Example 11-1

The specific conductance of a $0.0200N$ potassium chloride solution at 25°C has been determined to be 0.002768/ohm-cm. Calculate the equivalent conductance of this salt.

This particular solution is one of the standards used to calibrate a cell which is to be employed in conductance studies. Note the units attached to the specific conductance. You should be able to verify that these are correct from the discussion given above. The equivalent conductance is given by

$$\Lambda = \frac{1000 \times 0.002768}{0.0200} = 138.4 \text{ cm}^2/\text{ohm-equivalent}$$

Example 11-2

When the $0.0200N$ KCl solution is added to a particular cell, its resistance is found to be 284 ohms. An unknown solution has a resistance of 164 ohms in the same cell. Calculate the specific conductance of the unknown solution.

The specific resistance of the KCl solution is given by $1/0.002768 = 361.3$ ohm-cm. We need to have a value for the ratio l/A in Eq. (11-17); this ratio is characteristic of the cell, and is termed the *cell constant*. Inserting the known quantities,

$$284 = 361.3 \times (l/A)$$
$$(l/A) = 0.786 \text{ cm}$$

Since this ratio is a constant characteristic of the cell, it applies to the unknown solution as well, so that

$$164 = \rho \times 0.786$$
$$\rho = 206 \text{ ohm-cm}$$

The *specific conductance* of the solution is the reciprocal of this:

$$C = 1/206 = 0.00485/\text{ohm-cm}$$

The ability of an electrolytic solution to conduct current depends upon the ability of the ions to move through the body of the solution under the influence of the applied field. The nearest thing to this movement on a macroscopic scale is the fall of a spherical particle through a viscous medium under the influence of gravitational potential; *e.g.*, the fall of a marble through water. This situation was discussed in connection with Millikan's oil drop experiment, Sec. 2-2. An analysis of this problem shows that the rate of fall is dependent upon the viscosity of the medium (the marble would fall much more slowly through a thick oil), the magnitude of the gravitational force, and the radius of the sphere. (A larger radius, with no change in mass, would result in a lower rate of fall.) When we carry this over to the problem of the movement of ions under the electrical potential in water, the variables involved are the charge on the ion (since this determines the force) and its radius. By means which we need not consider here, it is possible to calculate from the experimental data the mobility of an ion in solution; *i.e.*, the measure of how rapidly the ion moves under the influence of a certain potential. Table 11-3 is a short list of some ionic mobilities, based on a standard value of 100 for the hydronium ion.

Table 11-3 The relative mobilities of ions in aqueous solution, based on a value of 100 for the hydronium ion.

Cation	Mobility	Anion	Mobility
H_3O^+	100.0	Cl^-	21.8
Li^+	11.05	Br^-	22.4
Na^+	14.3	I^-	21.9
K^+	21.0	OH^-	56.6
Ca^{+2}	34.0		
La^{+3}	59.7	SO_4^{-2}	45.6

Since the solvent is the same in all cases, and the value of the applied potential can be set, the variables which remain are the magnitude of the charge (but not its sign), and the radius of the ion. From this point of view, the data in Table 11-3 are very interesting indeed. In the series of alkali metal ions — lithium, sodium, and potassium — we would certainly expect the ionic size to increase and the mobility to decrease, if no other factors were involved. Instead, we find that the mobility increases quite markedly in the series. We must conclude that the lithium ion is larger in aqueous solution than the sodium ion, and that the sodium ion in turn is larger than the potassium ion. The reason for this apparent reversal in size lies in the hydration of the ions. The small lithium ion is very strongly hydrated; water molecules are held so tightly that they move with the ion through the solution. The effective size of the ion is therefore determined by the number of water molecules held tightly in hydration. The lithium ion, with its sheath of hydrating water molecules, represents a larger body moving through the solvent than the potassium ion with its hydration sheath.

The effect of hydration of cations is also evident from a comparison of cation and anion mobilities. The radius of the chloride ion, for example, is much greater than that of the potassium ion, and yet the mobilities of the two ions in solution are about the same. It is known that anions are not as strongly hydrated as cations, partly because of their larger size and partly because water molecules cannot cluster about a negative charge as easily as about a positive charge. The potassium ion with its hydration sphere thus becomes about equal in radius to the hydrated chloride ion.

Besides furnishing the kind of interesting information just discussed, conductance experiments have been useful in a number of other ways. Studies of the conductance of solutions of a strong electrolyte as a function of concentration show that equivalent conductance decreases with increasing concentration. This very probably results from a decrease in ion mobilities with increasing concentration. At one time, this decrease in equivalent conductance was ascribed to incomplete dissociation of ionic solutes. The explanation of this effect which is now generally accepted — at least for concentrations up to perhaps $0.01M$ — is due to Debye and Hückel, and is generally referred to as the Debye-Hückel interionic attraction theory. Although it is true that ions are able to move about in the solution quite freely, it is also true — as Debye and Hückel pointed out — that, because of the attractive forces which exist between unlike charges, each positive ion will on the average be in a surrounding in which there are more negative than positive charges in the near vicinity. When a potential is applied across the solution, each ion tends to migrate toward the electrode of opposite sign. The positive ion begins to drift toward the negative electrode, but the negative ions, which predominate in the surroundings of the positive ion, tend to go the other way. The result of this effect is a "drag" on the ion which decreases its mobility. What we have said for positive ions applies as well, of course, for anions. The effect of the drag is small in very dilute solutions because the average distance between ions is large. As concentration increases, the effect increases in importance; hence, the decrease in ion mobilities with concentration.

Exercises

11-1. From a consideration of the data in Table 11-1, arrange the following solvents in the order of their ability to dissolve ionic substances: Ethyl alcohol, chloroform, hydrogen fluoride, dimethyl ether. Explain your choice of order.

11-2. Using Eq. (11-2), calculate the ratio of the distances separating two opposite charges for the same value of attractive force in benzene and water. (*I.e.*, we want (r in benzene)/(r in water) for the same value of F in both solvents.) Is this ratio of any significance insofar as the solubilities of ionic substances in the two solvents are concerned? Explain.

11-3. Write complete ionic and net ionic equations analogous to Eqs. (11-4) and (11-5) for each of the following:
(a) Barium sulfate is precipitated by adding a sodium sulfate, Na_2SO_4, solution to a barium chloride, $BaCl_2$, solution.
(b) Barium sulfate is precipitated by addition of a potassium sulfate, K_2SO_4, solution to a barium perchlorate, $Ba(ClO_4)_2$, solution.

11-4. Describe an experiment which might be performed at a single temperature to determine whether the solubility of sodium perchlorate in water increases or decreases with increasing temperature.

11-5. Write half-reactions and a balanced chemical equation for electrolysis of a dilute aqueous solution of sodium fluoride (NaF). State which ionic species tend to concentrate in the anode and cathode compartments (Fig. 11-3) as electrolysis proceeds.

11-6. How many grams of lead would be deposited at the cathode from an aqueous solution of lead (II) nitrate, $Pb(NO_3)_2$, by a current of 2.5 amperes flowing for a period of 3 hours?

11-7. Explain briefly the origin of the interaction which leads to hydration of ions in water.

11-8. Would you expect the ionic mobility of the cation to decrease with concentration more rapidly for $La(NO_3)_3$ or $NaNO_3$? Explain.

11-9. In electrolyzing a potassium chloride solution to produce chlorate anion, Eq. (11-15), it is found that 37.5 g of potassium chlorate is produced in a period of 4 hours. What is the average current flow during this period?

11-10. Using the data in Table 11-3, arrange the following 0.1M solutions in the order of increasing conductivity (assuming that the same cell is employed for all): $CaCl_2$, NaBr, LiCl, KOH, $LaCl_3$, HCl.

11-11. By making use of Hess' law, and using the data in Tables 7-3 and 11-2, calculate the heat evolved in the process, $M^+(g) \rightarrow M^+ (aq)$ for the alkali metals, lithium, sodium, and potassium. Explain the variation in this number in terms of concept developed in this chapter.

11-12. The ionic radii of K^+ and Ca^{+2} ions in crystals are 1.33 and 1.06 Å, respectively. Which ion do you think would have the larger effective size in aqueous solution? Is your answer consistent with the data in Table 11-3? Explain.

11-13. You are given a 0.0100M solution of potassium chloride and told that the specific conductivity is 0.00141 mhos/cm at 25°C (a mho is a conductivity unit, ohm^{-1}). You are then given a conductivity apparatus and a 0.0156M solution of sodium nitrate. With the KCl in the cell, its resistance in the bridge is 573 ohms. With the $NaNO_3$ solution in the cell, its resistance is 425 ohms. From these data calculate the specific and equivalent conductances of the sodium nitrate solution.

11-14. The equivalent conductances of sodium chloride solutions depend on concentration, as shown by the following data:

Normality	0.001	0.01	0.05	0.10
Equivalent conductance	123.7	118.5	111.1	106.7

Explain the concentration dependence of the equivalent conductance. From these data it should be possible to estimate the limiting equivalent conductance; *i.e.*, the equivalent conductance for an infinitely dilute solution of sodium chloride, in which interionic effects are absent. Graph the values of Λ *vs* concentration and, in a second plot, against $\sqrt{\text{concentration}}$. In which case is it easier to extrapolate the data to infinite dilution?

11-15. (a) From the following data for the heats of formation of aqueous salts, calculate average values for the heats of formation of $SO_4^{-2}(aq)$ and $CrO_4^{-2}(aq)$. (Use Table 11-2)

$MgSO_4$	−327.3	$SrSO_4$	−347.4
$CaSO_4$	−346.7	$MgCrO_4$	−321.4
$BaSO_4$	−345.6	$CaCrO_4$	−336.0

(b) Optional: Discuss these two ions with regard to electronic structures, geometry, *etc.*

11-16. Solid potassium nitrate is added to water in an adiabatic calorimeter, and it dissolves. What change has occurred in the entropy of the system? (Refer to Fig. 11-2). What change would occur in the entropy of the system if the dissolving process were carried out isothermally? Can you account for these changes in entropy in a general way in terms of the ideas developed in Chap. 9?

Chapter 12

Acids and Bases

12-1 Introduction

The concept of acids and bases has been important in chemistry for a long time. In the seventeenth century, Robert Boyle wrote that an acid could be defined as a compound which possessed a certain set of observable properties, among which were the following:

1. Dissolves many substances.
2. Precipitates sulfur from its solutions in alkalies.
3. Causes blue vegetable dyes to turn red.
4. Loses its properties on contact with alkalies.

To this list we might add sour taste, reaction with active metals to yield hydrogen, and many others. It should be noted that a compound must satisfy *all* of these simple criteria to qualify as an acid. A further important point is that these requirements are all applicable to aqueous solutions of the acid, although they apply in varying degrees to other solvents as well.

Acid-base theory began when chemists attempted to make statements about the nature of compounds which possess the properties associated with acids. Lavoisier, for example, stated that all acids must contain oxygen. When this was shown to be false it was replaced by another "one-element" theory: All acids contain an easily replaceable hydrogen atom. Many other acid-base theories have been advanced over the years. While most of these are valid within narrow limits of applicability, they are not based upon a clear understanding of the principles of chemical valence. Not until the Lewis-Langmuir theory had been established could a satisfactory *general* theory of acid-base behavior be formulated. It is not surprising, therefore, that the most useful general acid-base theory is called the Lewis theory. It is applicable to a great variety of chemical systems, including many which do not contain either water, oxygen atoms, or hydrogen atoms.

The Brönsted theory of acids and bases was developed at about the same time as the Lewis theory. It is applicable to water and water-like solvents, and is widely used. We shall, therefore, begin with a discussion of acid-base behavior from the Brönsted point of view.

12-2 Acid-Base Behavior in Water

In developing his theory of the dissociation of electrolytes in water, Arrhenius defined an acid as a substance which produces hydrogen ions in water, and a base as a substance which produces hydroxyl ions in water. Thus, HCl, HNO_3, and other common acids were presumed to dissolve in water to produce H^+ ions, while NaOH, $Ba(OH)_2$, and other alkalies dissolve in water to produce OH^- ions. Arrhenius assumed that all of these compounds simply dissociated in water to yield their constituent ions, as

$$HCl \rightarrow H^+ + Cl^- \qquad (12\text{-}1)$$
$$NaOH \rightarrow Na^+ + OH^-$$

But the pure acids are covalent compounds, so something more than simple dissociation is involved when they are dissolved in water. Furthermore, compounds such as ammonia dissolve in water to produce hydroxyl ions. Again, since ammonia does not contain an OH^- group, this argues against simple dissociation as the origin of acid-base behavior.

An acid is defined in the Brönsted theory as a substance which is capable of donating, or giving up, a proton. Hydrogen chloride, a covalent compound, gives up a proton upon dissolving in water, not by a simple dissociation, but by *reaction* with water:

$$HCl + H_2O \rightarrow H_3O^+ + Cl^- \qquad (12\text{-}2)$$

A base is defined as a proton acceptor. In the reaction above, *water is acting as a base* because it accepts a proton from the HCl molecule. This reaction with water is typical of acids, which are all covalent substances. Consider the reaction of a weak acid such as acetic acid with water:

$$CH_3COOH + H_2O \rightarrow H_3O^+ + CH_3COO^- \qquad (12\text{-}3)$$

This reaction does not proceed to an appreciable extent; in a 0.1*M* solution of acetic acid only about 1.3 per cent of the acetic acid molecules are in the dissociated form, whereas with HCl the reaction is essentially complete. The acetate ion, CH_3COO^-, is a base; *i.e.*, it is capable of accepting a proton to form the neutral acetic acid molecule. Therefore, when acetic acid is dissolved in water, there is a competition between acetate ions and water molecules for the protons. From the fact that relatively few of the protons become attached to water molecules to form the H_3O^+ ion, it can be concluded that acetate ion is the stronger base. The chloride ion, by contrast, is a very weak base, since it does not compete effectively with water molecules for the proton.

The basic character of the acetate ion can also be appreciated by considering the reverse reaction in Eq. (12-3). Hydronium ion reacts with acetate ion to produce water and acetic acid. The hydronium ion acts as an acid to donate a proton to the base, the acetate ion. In the Brönsted theory, the hydronium ion is referred to as the *conjugate acid* of the base, water; acetate ion is the *conjugate base* of the acid, acetic acid. In general, any acid-base reaction is described as

$$\text{acid} + \text{base} \rightarrow \text{conjugate acid} + \text{conjugate base} \qquad (12\text{-}4)$$

This scheme is also applicable to the reaction of ammonia with water to yield a basic solution:

$$NH_3 + H_2O \rightarrow NH_4^+ + OH^- \qquad (12\text{-}5)$$

In this reaction *water acts as an acid*, and the ammonia molecule as a base. The ammonium ion is the conjugate acid of ammonia, while the hydroxyl ion is the conjugate base of water. Ammonia is not strongly basic in water solution; the reaction represented by Eq. (12-5) occurs only to a slight extent. In a 0.1*M* ammonia solution, only about 1.3 per cent of the ammonia molecules are in the form of the conjugate acid. Another way of stating this is that the ammonium ion is a stronger acid than the water molecule. The tendency for the ammonium ion to donate a proton to the hydroxyl ion to form water and ammonia is greater than the tendency of the water molecule to donate a proton to ammonia.

Many hydroxyl ion-producing substances, such as NaOH and $Ba(OH)_2$, are ionic, and already contain the hydroxyl ion before dissolving in water. When these are dissolved, there is no reaction with the solvent except in the sense that hydration of the ions is considered a reaction. The process can be represented as

$$NaOH(s) \xrightarrow{H_2O} Na^+(aq) + OH^-(aq)$$

It should be noted that the Brönsted theory is different from the Arrhenius theory in that the characteristic species in an aqueous acid solution is the hydronium ion, H_3O^+, not the proton, H^+. The characteristic species in basic solution remains the hydroxyl ion.

Although it is sometimes argued that there is no conclusive evidence for the existence of the hydronium ion, the formation of this species follows logically from the Lewis-Langmuir theory of valence. The proton in water will therefore be represented throughout as H_3O^+. In recent years, a considerable body of evidence has accumulated to support this convention. (See Chap. 3 of R. P. Bell, *The Proton in Chemistry*, Cornell University Press, 1960.)

12-3 Hydrolysis

Autohydrolysis of water — Water may act as either an acid or base, depending upon the substance with which it is caused to react. This idea can be extended to include the reaction of water molecules with one another in a general acid-base reaction:

$$H_2O + H_2O \rightarrow \underset{\text{conjugate acid}}{H_3O^+} + \underset{\text{conjugate base}}{OH^-} \qquad (12\text{-}6)$$

This reaction does occur; in pure water at room temperature there is a 10^{-7} molar concentration of both H_3O^+ and OH^-. The reaction of one water molecule with another in this way is termed *autohydrolysis, autoprotolysis*, or *autoionization.* It is important to understand that this reaction, as well as all of those already discussed [Eqs. (12-2), (12-3), and (12-5)], leads to a dynamic equilibrium. Water molecules are continuously reacting with one another to form the product ions, and these are continuously combining to form water molecules. We observe a constant concentration of H_3O^+ and OH^- ions, but this is an average concentration. No individual ion exists for long before it combines with its opposite number and is replaced by another ion.

Equation (12-6) shows that the reaction of water molecules with one another always produces equal numbers of hydronium and hydroxyl ions. Pure water, or any aqueous solution, which contains equal numbers of hydronium and hydroxyl ions, is said to be *neutral.* If there are more hydroxyl than hydronium ions, as in solutions of NH_3 or NaOH, the solution is said to be *basic.* If hydronium ions are in excess, as in solutions of HCl, the solution is *acidic.*

Hydrolysis of anions — When sodium chloride, an ionic salt, is dissolved in water the solution remains neutral. A solution of sodium acetate, on the other hand, is definitely basic. There is evidently an excess of hydroxyl ions in the latter solution. In attempting to understand this, we may note first of all that neither sodium nor chloride ion exhibits any tendency to react with water molecules in such a way as to produce either hydronium or hydroxyl ions. By contrast, the acetate ion does show fairly strong base character. This is evidenced in the reverse reaction of Eq. (12-3), in which the acetate ion accepts a proton from the hydronium ion to form acetic acid and water. In the same way, the acetate ion might be expected to accept a proton from a water molecule (which is capable of acting as an acid) to form acetic acid and the hydroxyl ion:

$$\underset{\text{base}}{CH_3COO^-} + \underset{\text{acid}}{H_2O} \rightarrow \underset{\text{conjugate acid}}{CH_3COOH} + \underset{\text{conjugate base}}{OH^-} \tag{12-7}$$

Of course, H_3O^+ is a much stronger acid than H_2O, but the forward reaction in Eq. (12-7) might nevertheless be expected to proceed to some extent. That it does so is evidenced by the basicity of sodium acetate solutions.

The reaction of acetate ion with water to form hydroxyl ions is typical of the action of many anions. The general reaction, which involves a splitting of the water molecule, is termed *hydrolysis.* For the reaction to proceed, the anion must be a fairly strong base. But if it is a strong base, the acid from which it is derived — its conjugate acid — must be a weak acid. Table 12-1 lists a number of anions which undergo hydrolysis in water.

Table 12-1 Anions and their conjugate acids.

Anion	Name	Conjugate acid
HSO_3^-	Bisulfite	H_2SO_3
NO_2^-	Nitrite	HNO_2
CH_3COO^-	Acetate	CH_3COOH
SO_3^{-2}	Sulfite	HSO_3^-
ClO^-	Hypochlorite	$HClO$
CN^-	Cyanide	HCN
CO_3^{-2}	Carbonate	HCO_3^-
S^{-2}	Sulfide	HS^-

The more extensively the anion is hydrolyzed, the more basic its solution will be for a given concentration. A 0.1*M* sodium acetate solution is not as basic as, for example, a 0.1*M* sodium cyanide solution. Note that a number of multiple-charge ions are listed in Table 12-1. These are capable of reacting with more than one mole of water, as exemplified by the behavior of the carbonate ion:

$$CO_3^{-2} + H_2O \rightarrow HCO_3^- + OH^-$$
$$HCO_3^- + H_2O \rightarrow H_2CO_3 + OH^-$$

The first reaction occurs much more extensively than the second, however, so that if one begins with the carbonate ion, the amount of hydroxyl ion produced by the first reaction is many times that produced from the second reaction of HCO_3^-. For example, in a 0.1*M* solution of sodium carbonate about 4 per cent of the total carbonate is in the form of HCO_3^- ions, and only about 0.00001 per cent is in the form of H_2CO_3. It is clear from this that the second reaction is much less extensive than the first.

Hydrolysis of cations — A solution of ammonium chloride in water shows acidic properties. By analogy with the behavior of solutions in which hydrolysis of anions occurs, it may be reasoned that a proton is transferred from ammonium ion to a water molecule:

$$NH_4^+ + H_2O \rightarrow NH_3 + H_3O^+ \quad (12\text{-}8)$$

The reverse reaction of Eq. (12-5) occurs to a large extent, showing that NH_4^+ is a fairly strong acid. The forward reaction in Eq. (12-8) might therefore be expected to occur. It is not really a hydrolysis reaction in the sense in which the term was used earlier, since no splitting of the water molecule is involved, but it is nevertheless classed as one.

While the acidity of solutions of ammonium chloride and other like salts is easily understood, it is not so immediately clear why a solution of salts such as iron (III) chloride or zinc (II) nitrate and many others should be acidic. Acidic behavior is exhibited, however, by salts consisting of an anion derived from a strong acid (chloride, nitrate, *etc.*) and a trivalent or one of many divalent cations. In Chaps. 10 and 11 it is pointed out that all cations in aqueous solution are hydrated. *The extent of hydration increases with increasing charge and decreasing radius of the cation.* The water molecules which hydrate the cation are in a different environment from those which make up the bulk of the solvent. The unshared electron pairs on the oxygen are attracted to the positive charge residing on the cation, and a general displacement of charge in the entire molecule results, as illustrated in Fig. 12-1. As a result of the charge displacement, the protons in hydrating water molecules are more easily removed; *i.e.*, these molecules are stronger acids than those in the bulk of the solvent. The number of water molecules participating in the hydration of a particular

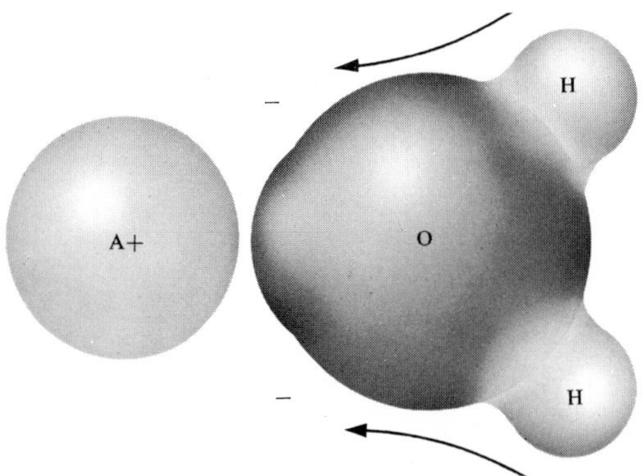

Fig. 12-1 Displacement of electrical charge in a hydrating water molecule. The displacement, which results in increased acidity of the water molecule, is greatest for highly charged ions of small radius.

cation is generally not known. For some metal ions which form crystalline compounds containing a definite number of water molecules about the cation, such as $Cu(H_2O)_4SO_4(H_2O)$ or $Cr(H_2O)_6Cl_3$, the ion in solution is often written with the same hydration number. The equation representing hydrolysis of a hydrated metal ion is written, in general as

$$M(H_2O)_n^{+m} + H_2O \rightarrow M(H_2O)_{n-1}(OH)^{+m-1} + H_3O^+ \qquad (12\text{-}9)$$

In this reaction, one of the hydrating water molecules loses a proton to a solvent molecule, leaving a hydroxyl ion in the hydration sphere. The reaction occurs to an appreciable extent only when M is highly charged and quite small.

It is possible for the water molecules in the hydration sphere to lose collectively more than one proton, as

$$M(OH)(H_2O)_{n-1}^{+m-1} + H_2O \rightarrow M(OH)_2(H_2O)_{n-2}^{+m-2} + H_3O^+ \quad (12\text{-}10)$$

The loss of each successive proton becomes progressively more difficult. In the presence of a strong proton acceptor such as OH^- ion, however, a number of protons may be lost, as exemplified by the behavior of the aluminum (III) ion. This ion is usually written as having a hydration number of six.

$$Al(H_2O)_6^{+3} + OH^- \rightarrow Al(OH)(H_2O)_5^{+2} + H_2O$$
$$Al(OH)(H_2O)_5^{+2} + OH^- \rightarrow Al(OH)_2(H_2O)_4^{+} + H_2O$$
$$Al(OH)_2(H_2O)_4^{+} + OH^- \rightarrow Al(OH)_3(H_2O)_3(s) + H_2O$$

Note that addition of three equivalents of OH^- leads to precipitation of the neutral species $Al(OH)_3(H_2O)_3$. If this reaction were to be described by an equation in which the acid character of the hydrated metal ion is not considered, it would be written as

$$Al^{+3} + 3OH^- \rightarrow Al(OH)_3(s)$$

An excess of OH^- over that required to precipitate the neutral species causes the precipitate to dissolve, forming a hydroxy-anion:

$$Al(OH)_3(H_2O)_3(s) + OH^- \rightarrow Al(OH)_4(H_2O)_2^-$$

This behavior is known as *amphoterism*, which characterizes a substance that undergoes reaction with both acid and base. The amphoteric substance is $Al(OH)_3(H_2O)_3$, which dissolves on treatment with either acid or base.

The description just given of metal ion hydrolysis is actually somewhat oversimplified. With many metal ions the hydroxy-anions formed in aqueous solution are larger units, containing two or more metal ions. The metal ions are joined together by hydroxide ion bridges, to form species which are usually referred to as *polyanions*. Despite the formation of polyanions in some cases, the simple description of hydrolysis given above is adequate for the lighter metals.

12-4 Acid-Base Behavior in Non-aqueous Solvents

The Brönsted theory is easily extended to other solvents in which proton transfer can occur. For example, hydrogen chloride gas can be dissolved in anhydrous liquid ammonia (boiling point −33°C):

$$HCl + NH_3 \rightarrow NH_4^+ + Cl^- \qquad (12\text{-}11)$$

It is apparent that, in this reaction, the ammonia molecule acts as a base. Ammonia is, in fact, much more basic than water. As a result, many acids which are only weakly ionized in water — such as acetic acid — are just as strongly ionized in ammonia as in hydrogen chloride:

$$CH_3COOH + NH_3 \rightarrow NH_4^+ + CH_3COO^-$$

The equation representing the autoionization of ammonia is entirely analogous to that for water:

$$NH_3 + NH_3 \rightarrow \underset{\text{conjugate acid}}{NH_4^+} + \underset{\text{conjugate base}}{NH_2^-} \qquad (12\text{-}12)$$

From this reaction, it is seen that the characteristic base species in ammonia is the amide ion, NH_2^-. Thus potassium amide, KNH_2, is the analogue in liquid ammonia of potassium hydroxide, KOH, in water.

100 per cent sulfuric acid is another solvent in which the Brönsted theory is applicable. This solvent, in contrast to liquid ammonia, is very strongly acidic; it exhibits a strong tendency to transfer a proton to a solute species. When sodium acetate is dissolved in sulfuric acid there is an essentially complete transfer of protons to the acetate ions:

$$CH_3COO^- + H_2SO_4 \rightarrow CH_3COOH + HSO_4^- \qquad (12\text{-}13)$$

By contrast, the analogous equation in water, Eq. (12-7), occurs to only a slight extent.

The realization that many liquids other than water could be treated in terms of a common acid-base theory has been important in the development of new applications for these substances. By taking advantage of variation in chemical and physical properties, it has often been possible to conduct a particular chemical reaction in one solvent when it could not be carried out at all in the others.

12-5 Lewis Theory of Acids and Bases

All of the molecules or ions which are bases in the Brönsted scheme possess at least one pair of unshared electrons. In fact, one could define a Brönsted base as a substance which is capable of electron-pair donation to a proton. But molecules or ions of this category may also donate an electron pair to systems other than just the proton. Lewis defined acid-base properties in terms of this general donor-acceptor action: "A base is a base because it can donate an electron pair . . . an acid is an acid because it can accept an electron pair . . . ." Note that the focus of interest has been shifted from the proton to the electron pair. As an example of an acid-base reaction which is encompassed by this definition, but which is outside the realm of the Brönsted theory, consider the reaction of trimethylamine with boron trifluoride:

$$(CH_3)_3N\!: + BF_3 \rightarrow (CH_3)_3N\!:\!BF_3 \qquad (12\text{-}14)$$

A stable compound is formed by the action of the Lewis base, trimethylamine, on the Lewis acid, boron trifluoride. To understand why BF_3 should act as an acid, it is necessary to study the electronic structure of this molecule

```
 ..    ..
:F : B : F:
 ..    ..
    ..
   :F:
    ..
```

Note that there are but three electrons in the valence shell of the boron atom; a fourth orbital in the valence shell is vacant.

	1s	2s	2p
B	[↑↓]	[↑↓]	[↑ | |]

	1s	$2sp^2$	2p
B	[↑↓]	[↑ | ↑ | ↑]	[ ]

The three hybrid sp^2 orbitals are directed toward the corners of an equilateral triangle, and are employed in bonding the three fluorine atoms. The remaining $2p$ orbital is normal to the plane of the three sp^2 bonds (Fig. 8-18).

The interaction of $(CH_3)_3N$ with BF_3 consists in the donation of the unshared electron pair on the nitrogen atom to the vacant orbital on boron. In the process, the boron orbitals undergo a reorganization, so that the geometry in the complex is one in which the three B—F bonds are bent back (Fig. 12-2).

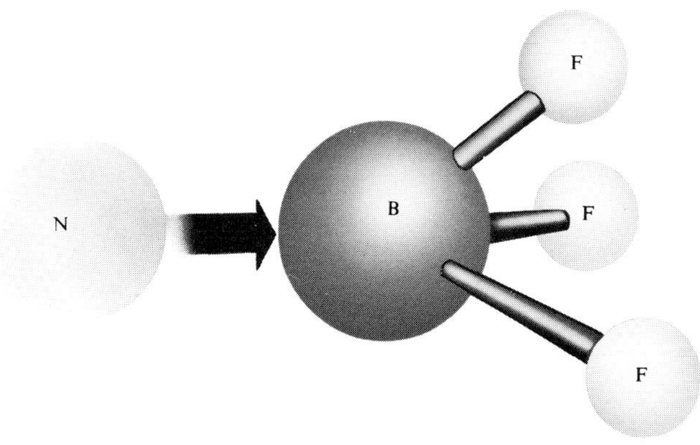

Fig. 12-2 Geometry about boron in the NH_3BF_3 addition compound.

A number of other Group III element compounds, including $AlCl_3$, BCl_3, $GaCl_3$, and $Al(CH_3)_3$, are also capable of acting as Lewis acids toward molecules with unshared electron pairs. Elements such as tin, arsenic, *etc.*, which are capable of expanding their valence shells are also Lewis acids. The reaction of $SnCl_4$ with chloride ion in concentrated hydrochloric acid solution is a typical Lewis acid-base reaction. A pair of electrons on each chloride ion is donated to the acid molecule, $SnCl_4$ (Chap. 8):

$$SnCl_4 + 2Cl^- \rightarrow SnCl_6^{-2} \qquad (12\text{-}15)$$

The reaction of carbon dioxide with water to form carbonic acid, H_2CO_3, furnishes another example of a Lewis acid-base interaction. Here the CO_2 molecule accepts the electron pair on the oxygen atom of water through rupture of a double bond:

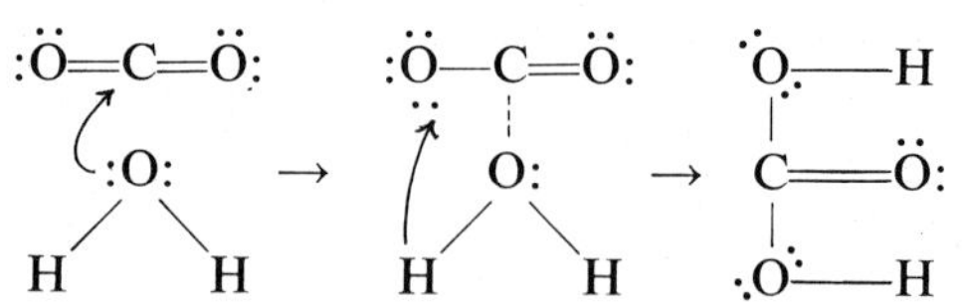

$$:\ddot{O}{=}C{=}\ddot{O}: + H{-}\ddot{O}{-}H \rightarrow :\ddot{O}{-}C{=}\ddot{O}: \cdots \ddot{O}(H)_2 \rightarrow (HO)_2C{=}\ddot{O}: \qquad (12\text{-}16)$$

The addition of sulfur dioxide to pyridine is an example of a similar reaction:

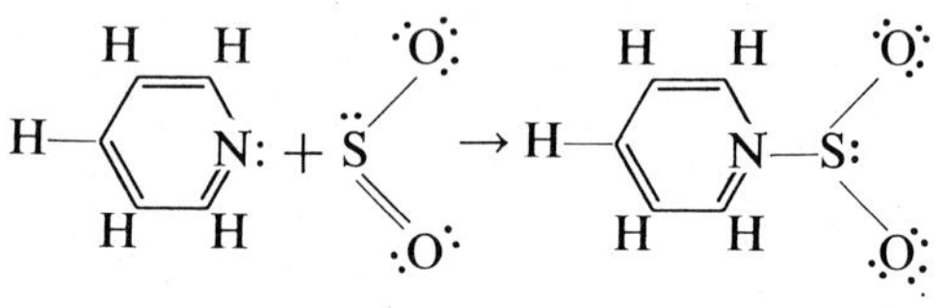

$$C_5H_5N: + SO_2 \rightarrow C_5H_5N{-}SO_2 \qquad (12\text{-}17)$$

(Pyridine is a six-membered ring structure. Each corner of the hexagon represents a C—H group, as in benzene, with the exception of one corner, which is a nitrogen atom.)

The Lewis theory is applicable to a great variety of chemical reactions. It has been of considerable aid in understanding the nature of many reactions of importance in both inorganic and organic chemistry. The Brönsted theory is a special case of Lewis acid-base interaction, in which the acid is always a proton donor. In the reaction of ammonia with water, the hydrogen atom attached to oxygen acts as the acid:

$$H_3N: + H{-}O{-}H \rightarrow H_3N \cdots\cdot H{-}O{-}H \rightarrow NH_4^+ + OH^- \qquad (12\text{-}18)$$

By viewing the reaction in this way, it can be seen that the transfer of the proton from water to ammonia is dependent upon the hydrogen-bonding property of the OH bond in the water molecule.

12-6 Neutralization

It is an essential of all acid-base theories that the acid species react rapidly and quantitatively with the base species to yield a product which lacks the distinctive properties of either the acid or base. A reaction of this type is termed *neutralization*. For neutralization to be complete, the total acid added to the system must be equal to the total base initially present, or *vice versa*. For example, 100 ml of a 0.1*M* NaOH solution would require 100 ml of a 0.1*M* — or 50 ml of a 0.2*M* — HCl solution for neutralization. In general,

$$M_A \cdot V_A = M_B \cdot V_B$$

where M_A and M_B represent the molarities of the acid and base solutions, respectively, and *V*'s refer to the volumes of the solutions. This equation is applicable only when the acid and base contain the same number of acid or base functions per molecule. It applies, for example, to the reaction of HCl with NaOH or of H_2SO_4 with $Ba(OH)_2$. It would not be applicable, however, to the reaction of H_2SO_4 with NaOH.

In a solution of H_2SO_4 there are two replaceable (*i.e.*, acidic) hydrogens per mole of solute. The molarity must therefore be doubled to obtain a measure of the total replaceable hydrogen per unit volume of solution. The concentration of acidic hydrogen is given by the normality, defined as the number of equivalents per liter (Chap. 10). Then,

$$N_A \cdot V_A = N_B \cdot V_B$$

where the *N*'s refer to normalities.

Indicators — The addition of an acid solution to a base solution (or of a base solution to an acid solution) to determine the point at which complete neutralization occurs is termed *titration*. The point of complete neutralization, the *endpoint*, is detected by means of an indicator.

Indicators are colored substances similar to dyes, obtained from vegetable matter or from chemical synthesis. They are generally weak acids or bases, and the acid form is colored differently than is its conjugate base. As an example of how indicators are used, consider methyl orange, a yellow dye which is a weak base. When this substance, which we will represent by the symbol IN^-, is added to a basic solution in water it imparts a yellow color to the solution. As acid is added to the basic solution, it reacts with the OH^- ion so that the concentration of this species decreases. So long as there is an excess of OH^- the indicator remains in the base form, which is yellow. When sufficient acid has been added to neutralize all of the base initially present, there is no longer an excess of OH^-. Any further addition of acid will result in reaction with the methyl orange, converting it to the acid form:

$$\underset{\text{yellow}}{IN^-} + H_3O^+ \rightarrow \underset{\text{orange}}{HIN} + H_2O \tag{12-19}$$

The endpoint is thus detected by the change in color. Acid indicators, of which phenolphthalein is an example, undergo similar equilibria:

$$HIN + OH^- \rightarrow IN^- + H_2O \tag{12-20}$$

The actual concentration of H_3O^+ at which conversion of the indicator from one form to the other occurs is dependent upon the strength of the acid or base character of the indicator. For this reason, care must be taken to select for a titration an indicator which will undergo color change in the desired range of H_3O^+ concentration. The quantitative aspects of this problem are treated in more detail in Chap. 18.

Indicators are added to the system to be titrated in very small amounts, so as not to produce a significant change in the properties of the system. An additional requirement is that they must not react with the components involved in the titration in any way other than by an acid-base reaction. The progress of an acid-base titration can also be followed by other methods, including some instrumental methods which are much more convenient and precise than the visual use of indicators.

Suggested Readings

Duke, F. R., "Acid-Base Reactions in Fused Alkali Nitrates," *J. Chem. Educ.*, Vol. 39, 1962.

Gould, E. S., *Inorganic Reactions and Structure* (2d ed.), New York, N.Y.: Henry Holt & Company, 1962.

Sisler, H. H., *Chemistry in Non-Aqueous Solvents*, New York, N. Y.: Reinhold Publishing Corporation, 1961.

Van der Werf, C. A., *Acids, Bases and the Chemistry of the Covalent Bond*, New York, N. Y.: Reinhold Publishing Corporation, 1961.

Exercises

12-1. Define an acid and a base according to the Brönsted theory; according to the Lewis system. Give an example in each case.

12-2. Which is the stronger acid in water, the hydonium ion or the ammonium ion? Support your answer by choosing particular reactions in which the two ions may be compared. Draw the structural models for these two ions, showing all covalent bonds and unshared electron pairs.

12-3. Anhydrous hydrogen fluoride is a strongly acidic medium. Write the equation for autoionization of this solvent. Write the equation representing what happens when each of the following compounds is dissolved in HF:

(a) Sodium trifluoroacetate, CF_3COONa. (c) Ammonia.
(b) Water. (d) NaOH.

12-4. Write the equation for the reaction of hydrogen cyanide (HCN) with water, labeling the acid, base, conjugate acid, and conjugate base. From the fact that a solution of sodium cyanide in water is basic, what can be said about the acid strength of HCN in water?

12-5. The ionization of phosphoric acid according to the reaction

$$H_3PO_4 + H_2O \rightarrow H_2PO_4^- + H_3O^+$$

proceeds to an appreciable extent in dilute solutions. Write the equations representing further ionization of phosphoric acid. Would you expect these to occur to a greater or less extent than the first ionization? Explain.

12-6. State whether each of the following is an acid, a base, or both, in water according to the Brönsted theory: HI, $HSeO_4^-$, PO_4^{-3}, $Fe(NO_3)_3$.

12-7. Assuming that cadmium (Cd) is bound to four water molecules as the divalent ion in water, write equations to demonstrate its amphoteric character.

12-8. How might sulfide ion, S^{-2}, participate in a hydrolysis reaction? Write an equation representing hydrolysis of this species.

12-9. Approximately how many ml of concentrated sulfuric acid (98 per cent H_2SO_4, density 1.84 g/cm^3) would you need to prepare 500 ml of a 0.175*M* solution? After the solution is prepared, a 25.00 ml sample requires 83.65 ml of 0.1000*N* sodium hydroxide for neutralization. What is the normality of the H_2SO_4 solution?

12-10. Arrange the following compounds in order of increasingly acidic solution upon dissolving in water, assuming the concentrations to be the same: KCl, $FeCl_3$, $NiCl_2$, $BaCl_2$. Explain your choice of order.

12-11. Discuss the reaction of SO_3 with water in terms of the Lewis acid-base theory. Does this reaction come within the scope of the Brönsted theory? Explain.

12-12. How many ml of 0.126*N* NaOH are required for neutralization of 72.0 ml of 0.200*N* H_2SO_4?

12-13. 0.652 g of $Ba(OH)_2$ is added to water and titrated with 0.215*N* HCl solution. How many ml are required for neutralization?

12-14. 0.400 g of a compound containing nitrogen is decomposed with sulfuric acid. The solution is made basic, and ammonia — which is the decomposition product of the nitrogen — is distilled off. This is collected in water and titrated with acid. 81.0 ml of 0.0510*N* HCl is required. What is the percentage by weight of nitrogen in the compound?

12-15. 0.2070 g of an unknown organic acid is dissolved in 100.0 ml of 0.02560*N* sodium hydroxide. 36.50 ml of 0.02400*N* hydrochloric acid solution is required to titrate the resulting solution. From these data, calculate the equivalent weight of the unknown acid.

Chapter 13

Chemical Kinetics

13-1 Introduction

Kinetics is a branch of chemistry which deals with the rates at which chemical reactions proceed. This aspect of chemistry is important from both the practical and theoretical viewpoints. In a practical sense, a chemical reaction which is useful for synthesis of needed substances must proceed at a rate which is neither too rapid nor too slow. The study of chemical kinetics provides basic knowledge necessary for proper control of the variables which affect the reaction rate.

The ultimate goal of chemistry is a complete understanding of the chemical behavior of all substances. This understanding would include a detailed knowledge of the manner in which substances combine to form reaction products. The detailed manner in which a reaction proceeds under given conditions is called the *mechanism.* We are a long way at present from having this kind of knowledge, but good progress has been and is being made in determining the mechanisms of many important reactions. One of the most useful tools in this endeavor is the study of reaction rates. In this chapter, the factors which affect the rate of a chemical reaction, current theories of reaction rates, and some of the tools employed in studying rates will be considered.

13-2 Reaction Rate and the Rate Constant

The rate of a chemical reaction is defined in terms of the rate of disappearance of one of the reactants from the system. Alternatively, it may be defined in terms of the rate of appearance of one of the products. For example, if the equation for a reaction were

$$2A + B \rightarrow C \quad (13\text{-}1)$$

the rate could equally well be expressed as $-\delta[A]/\delta t$, $-\delta[B]/\delta t$, or as $\delta[C]/\delta t$, where $\delta[\;]$ means a small change in concentration and δt a correspondingly small change in time. The negative sign indicates that the concentrations of A and B decrease with time. One must always state which particular substance is to be used as the reference in discussing the rate. In the example above, $-\delta[A]/\delta t$ is twice $-\delta[B]/\delta t$. Depending upon the particular reaction, a variety of techniques can be used to observe a rate of appearance or disappearance. For example, in the reaction of an organic chloride RCl (where R represents an organic group) with water according to the following equation:

$$RCl + H_2O \rightarrow ROH + HCl \quad (13\text{-}2)$$

The reaction can be followed by observing the rate at which hydrogen chloride, HCl, appears in the system. This could be done by titrating samples taken from the system with a standard sodium hydroxide solution at regular intervals. The results of such an experiment might be as shown in Fig. 13-1.

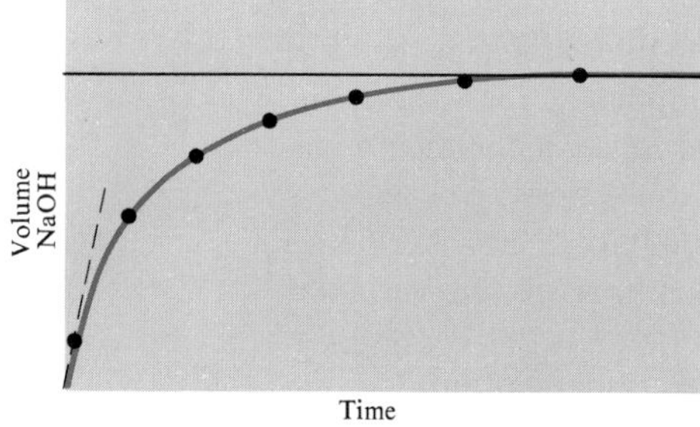

Fig. 13-1 Titration of hydrogen chloride evolved in hydrolysis of an organic chloride. The volume of sodium hydroxide solution is proportional to the quantity of alkyl halide which has undergone reaction at each time of measurement.

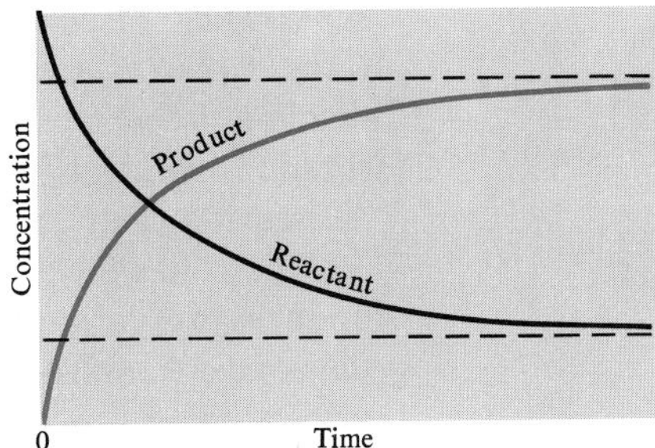

Fig. 13-2 Variation of the reactant and product concentations with time. The dotted lines represent the concentrations of materials when the system has arrived at equilibrium, and no further changes of concentrations with time are occurring. If the reaction were to proceed essentially to completion, as many reactions do, the reactant concentration would approach zero.

The rate is measured by how rapidly the concentrations of reactants or products change with time. It is therefore represented by the slope of the graph shown in Fig. 13-1. The *initial rate* is the limiting value of the slope at zero time; it is shown by the dotted line in the figure. The rate decreases with time, and approaches zero (horizontal slope) as time increases. This condition corresponds to completion of the reaction, or attainment of equilibrium.

As a second example, the rate of the reaction shown in Eq. (13-3) might be followed by observing the rate of appearance of oxygen gas, using a gas burette.

$$H_2O_2 \rightarrow H_2O + \tfrac{1}{2}O_2 \tag{13-3}$$

A graph of the concentration or quantity of a reactant would generally appear as shown by the line labelled as "reactant" in Fig. 13-2. If, on the other hand, the concentration or quantity of a product is being followed, as in the above examples, the graph of product concentration or quantity *vs* time is as shown in Fig. 13-2.

It is not always a simple matter to so arrange experiments that the concentration of the desired substance can be measured at various times. Occasionally, some over-all physical property of the system can be measured and evaluated in terms of the extent to which the reaction has proceeded. If the reaction shown in Eq. (13-1) were a gas phase reaction, for example, and were being conducted at constant volume and temperature, the total pressure in the system would be a good measure of the extent of the reaction. This is true because there is but one mole of gas in the product for three moles of reactant, so the pressure would decrease steadily as the reaction proceeds. Various other properties such as density, volume, optical rotation, spectral behavior, *etc.*, can be used in special cases to yield the desired information.

The rate of a reaction at a particular temperature may be dependent upon many variables. Among these, the most obviously important are the concentrations of the reactants. It is generally not possible to tell from inspection of the balanced equation precisely how the reaction rate will depend upon reactant concentrations. In the reaction of hydrogen peroxide in water used as an example above, for instance, it *might* turn out that the rate is in accord with one of the following expressions:

$$\text{rate} = \delta[O_2]/\delta t = -\tfrac{1}{2}\,\delta[H_2O_2]\,/\delta t = k\,[H_2O_2] \qquad (13\text{-}4)$$

or

$$\text{rate} = \delta[O_2]/\delta t = -\tfrac{1}{2}\,\delta[H_2O_2]\,/\delta t = k\,[H_2O_2]^2 \qquad (13\text{-}5)$$

These expressions state that the rate of the reaction, as measured by the change in concentration of one of the reactants, is proportional to some function of the concentration of the reactant H_2O_2. The proportionality constant k is named the *specific rate constant.* An experimental study of the system is required in order to discover the correct rate expression.

For a great many reactions, it is possible to write the rate as proportional to the product of a number of concentrations raised to some power:

$$\text{rate} = k(c_1^{n_1})(c_2^{n_2})(c_3^{n_3})\ldots \qquad (13\text{-}6)$$

Reaction rate expressions in Eqs. (13-4) and (13-5) are of this type. When the rate can be defined in terms of such an expression, the *order* of the reaction is defined as the sum of the exponents n_1, n_2, *etc.* If reaction (13-3) follows rate expression (13-4), for example, it would be termed a first-order reaction, since the sum of the exponents is one. Similarly, if it followed Eq. (13-5) it would be called a second-order reaction.

The reaction order with respect to a particular reactant is revealed by the manner in which the initial rate of the reaction varies as a function of the initial concentration of that reactant, all other variables being held constant. Table 13-1 shows data of this variety for the reaction of ammonium and nitrite ions at 30°C in aqueous solution:

$$NH_4^+ + NO_2^- \rightarrow N_2 + 2H_2O \qquad (13\text{-}7)$$

The concentration of ammonium ion is held relatively constant in this series of data, while nitrite ion concentration is varied over a wide range. Assuming that the rate expression for this reaction is of the general type shown in Eq. (13-6), we can write

$$\text{initial rate} = k[NH_4^+]^m[NO_2^-]^n \qquad (13\text{-}8)$$

Substances other than ammonium and nitrite ions may enter into the rate expression, but the data in Table 13-1 are obtained under conditions in

Table 13-1 Rate data for the ammonium-nitrite ion reaction, Eq. (13-7).

Total nitrite concentration (moles/l)	Total ammonium concentration (moles/l)	Initial rate (moles/l/sec)
0.0940	0.196	643×10^{-8}
0.0507	0.196	338×10^{-8}
0.0249	0.196	156×10^{-8}
0.0100	0.198	65×10^{-8}
0.0049	0.198	33×10^{-8}
0.0024	0.198	18×10^{-8}

which everything but the initial nitrite ion concentration is constant. The change in the nitrite ion concentration is, therefore, the only variable which can affect the initial rate. Thus, Eq. (13-8) can be simplified to the form

$$\text{intial rate} = \text{constant} \times [NO_2^-]^n \qquad (13\text{-}9)$$

Taking the base-ten log of both sides of this equation, we have

$$\log_{10} (\text{initial rate}) = \log_{10} (\text{constant}) + n \log_{10} [NO_2^-]$$

A graph of the logs of the initial rates *vs* the logs of the nitrite ion concentrations for the data in Table 13-1 should then give a straight line with slope n, the order of the reaction with respect to the nitrite ion. The graph is shown in Fig. 13-3; the slope is one within the expected experimental uncertainty, so we may conclude that the rate is first order with respect to nitrite ion.

Once the over-all rate of the chemical reaction is known — and also the order with respect to all of the rate-determining substances [*i.e.*, the

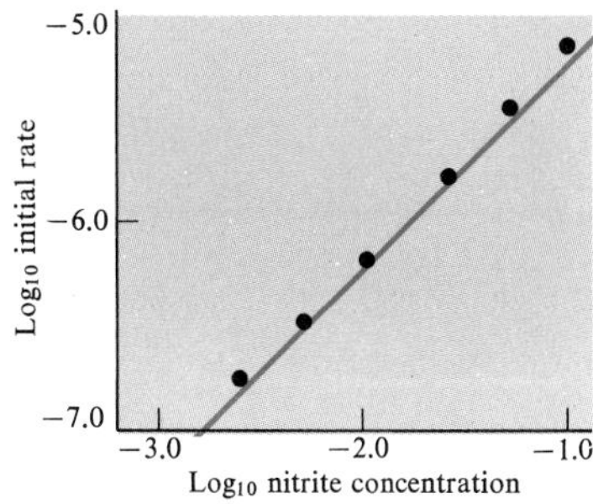

Fig. 13-3 $\log_{10}$ initial rate *versus* $\log_{10}[NO_2]^-$ for the ammonium nitrite reaction. The slope, 0.97, is the apparent order of the reaction with respect to nitrite ion.

values of n_1, n_2, *etc.* in Eq. (13-6) — the rate constant k can be calculated. The evaluation of k is most easily performed by putting the rate expression in a different form. To illustrate, consider a first-order reaction of the form

$$-\delta[A]/\delta t = k[A] \qquad (13\text{-}10)$$

This expression simply states that the rate at which A reacts is proportional only to the concentration of A. If the left-hand term is written as a differential we have, after some rearranging,

$$\frac{-d[A]}{[A]} = k\,dt \qquad (13\text{-}11)$$

Integration of this expression leads to

$$-\log_e\,[A] = kt + \log_e C$$

$$\text{or}\quad -\log_{10}[A] = \frac{kt}{2.303} + \log_{10} C \qquad (13\text{-}12)$$

or

$$[A] = C\,e^{-kt} \qquad (13\text{-}13)$$

Equation (13-12) is most useful for evaluating k. A graph of $\log_{10}$ [A] *vs* t yields a straight line with slope $-k/2.303$ and intercept $-\log_{10} C$. Setting $t = 0$ yields $\log_{10} A_0 = -\log_{10} C$, where A_0 is the initial concentration of A. Equation (13-13) is therefore

$$[A] = [A_0]e^{-kt} \qquad (13\text{-}14)$$

To illustrate the application of the preceding discussion to a particular problem, consider the decomposition of dimethyl ether in the gas phase, a reaction which proceeds according to the equation

$$(CH_3)_2O \rightarrow CH_4 + H_2 + CO$$

The rate of the reaction is to be followed manometrically; *i.e.*, by observing the over-all pressure in the system. The following data are obtained at 504°C:

Time (sec)	Total pressure (mm)	Increase in pressure $P - P_o$ (mm)
0	312	0
390	408	96
777	498	186
1195	562	250
3175	779	467
Infinity	931	619

(The last entry refers to the pressure after a very long time has elapsed and no further reaction appears to be occurring.)

To evaluate k, we must find the rate at which the concentration of dimethyl ether changes with time. From the equation for the reaction, it can be seen that three molecules of gas are produced for each molecule that decomposes. At any time t, then, the pressure of each of the substances in the system is

$$\begin{array}{cccc} (CH_3)_2O \rightarrow & CH_4 + & H_2 + & CO \\ P_0 - P_r & P_r & P_r & P_r \end{array}$$

where P_r is the number of mm of pressure *loss* of the dimethyl ether due to reaction. The total pressure P in the system at time t is then

$$P = (P_0 - P_r) + 3P_r = P_0 + 2P_r$$

$$P_r = \frac{P - P_0}{2}$$

P_r is calculated by dividing the quantity in the last column in the table by two. The initial pressure P_0 minus pressure loss P_r is the pressure of dimethyl ether in the system. We then have the following data:

Time (sec)	P_r (mm)	Pressure of ether (mm)
0	0	312
390	48	264
777	88	224
1195	125	187
3175	234	78
Infinity	314	0

It is interesting to note that the reaction proceeds essentially to completion, since after a very long time has elapsed the ether pressure is zero. The graph of the logarithm (base 10) of the ether pressure *vs* time is shown in Fig. 13-4. The linearity of the relationship shows that the reaction is first order. The slope of the line is $-k/2.303$; the value for k so obtained is 4.31×10^{-4}. The dimensions of k are sec^{-1}, so that it actually is not dependent upon the units in which the concentration of the ether is expressed.

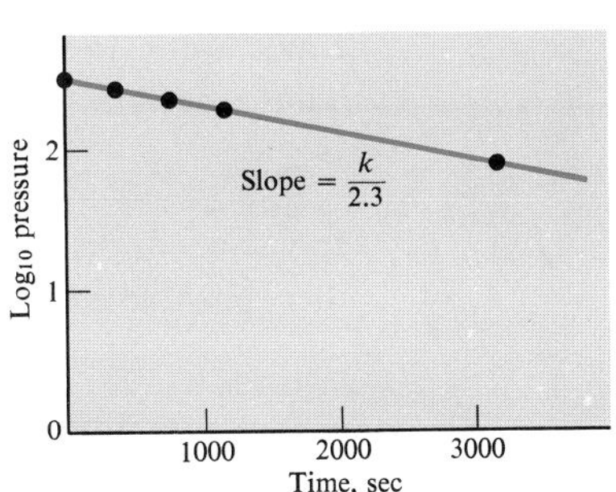

Fig. 13-4 Log_{10} dimethyl ether pressure *versus* time.

It is very important to note that this example involves two distinct aspects of the kinetics problem. In the first place, we have shown how one proceeds to treat the raw data (in this case, total pressure) by working with the balanced equation for the over-all reaction to obtain the concentration of the reactant. Secondly, these data must be graphed in their log form *vs* time to obtain a linear plot. From the slope of this line, we obtain the rate constant.

The expressions for the rate constant of reactions of higher order are rather more involved. By proceeding in a similar manner, it is possible to evaluate these constants from experimental data. The expression for a second-order reaction which is first order with respect to two different substances is

$$\text{rate} = k[\text{A}][\text{B}]$$

If the equation is rearranged to yield an expression for k, one obtains

$$k = \frac{\text{rate}}{[\text{A}][\text{B}]}$$

Rate has the dimensions of concentration unit divided by time, since it is a measure of the rate of change of concentration as a function of time. The over-all dimensions are therefore

$$k = \frac{\text{concentration/time}}{(\text{concentration})^2} = \frac{1}{\text{concentration-time}}$$

When concentration is expressed in moles/liter, the resulting units are liter/mole-sec. In these units, a typical bimolecular rate constant for a gas phase reaction might be 10^{-5} to 10^{-6} liter/mole-sec.

Often a reaction which is actually bimolecular appears to be first order. This situation arises when one of the reactants is present in large excess. The hydrolysis of an organic iodide in water provides an example:

$$\text{RI} + \text{H}_2\text{O} \rightarrow \text{ROH} + \text{HI}$$

If the reaction is conducted in water as a solvent, the concentration of water does not change appreciably during the course of the reaction. Thus, though the rate expression may be

$$-\delta[\text{RI}]/\delta t = k[\text{RI}][\text{H}_2\text{O}]$$

the concentration of water is effectively constant. It may therefore be incorporated into the rate constant, so that

$$-\delta[\text{RI}]/\delta t = k'[\text{RI}]$$

The reaction is termed pseudo-first-order, and k' is called the pseudo-first-order rate constant.

The concept of *reaction half-life* is quite useful, particularly in connection with first-order reactions. The half-life is the time required for the concentration of the reactant to decrease from some initial value at starting time t_0 to one-half of the initial concentration. It is quite easy to express the half-life for a first-order reaction in terms of the first-order rate constant. Let $[A_0]$ be the concentration of reactant at time $t_0 = 0$. Then the concentration of A at a later time t is given by Eq. (13-14). Now let $[A] = \frac{1}{2}[A_0]$. Then $e^{-kt_{1/2}} = \frac{1}{2}$. Taking logs of both sides, $kt_{1/2} = -\log_e(\frac{1}{2}) = 0.693$. Thus, at a time which is given by $0.693/k$ from the starting time the concentration of reactant will have decreased to one-half of the initial value. In another time $0.693/k$ from the first starting time it will have decreased to $\frac{1}{2} \times \frac{1}{2}$ times the initial value, *etc.*

The half-life of a reaction which is not first order is not so simple to evaluate. Whereas $t_{1/2}$ for a first-order reaction is independent of concentration, the same is not true for second-order or other, more complex reactions.

13-3 Gas Kinetic Theory of Reaction Rates

The formula for the number of collisions undergone by one molecule in a gas mixture in one second was derived in Chap. 3. Consider a system composed of two kinds of molecules, A and B, in which A and B react with one another to form a product C:

$$A + B \rightarrow C$$

For A and B to react, they must come together in a collision in the gas phase. In asking about the rate at which A reacts with B, therefore, the first question of interest is the number of collisions of A molecules with B molecules. The problem is the same as the collision number problem treated in Chap. 3, with one exception: If A and B differ in molecular weight, some kind of mean value must be used for the average velocity. We will assume that the effective average velocity in collisions of A and B molecules is the geometric average, $u_{Av} = (u_A u_B)^{1/2}$. We assume also that the effective collision diameter is the average of the collision diameters of A and B: $s_{AB} = (s_A + s_B)/2$. Thus, if there are n_B molecules of B per cm^3, the number of collisions of a single A molecule with B molecules in one second is given by

$$Z = 2\pi s_{AB}^2 u_{Av} n_B \tag{13-15}$$

Multiplying Eq. (13-15) by n_A (the number of A molecules per cm^3) gives the total number of collisions occurring between A and B molecules in one cm^3 in one second:

$$Z_{AB} = 2\pi s_{AB}^2 u_{Av} n_A n_B \tag{13-16}$$

To illustrate the magnitude of the collision number which this equation yields, suppose that the molecular weights of A and B are 32 and 46, respectively, and that the partial pressure of each gas in the mixture is 0.5 atm at a temperature of 25°C. Let s_{AB} be 3 Å.

1. We first calculate n_A and n_B from the ideal gas law:

$$n_A = n_B = \frac{\text{molecules}}{\text{cm}^3} = \frac{\text{molecules}}{\text{mole}} \times \frac{\text{moles}}{\text{liter}} \times \frac{1\text{ liter}}{1000\text{ cm}^3}$$

$$\frac{\text{moles}}{\text{liter}} = \frac{n}{V} = \frac{P}{RT} = \frac{0.5\text{ atm}}{0.0821\,\dfrac{\text{l—atm}}{\text{mole—°K}} \times 298\text{°K}} = 0.0204$$

$$n_A = n_B = \frac{6.023 \times 10^{23}\text{ molecules}}{1\text{ mole}} \times \frac{(0.0204\text{ moles})}{1\text{ liter}} \times \frac{1\text{ liter}}{1000\text{ cm}^3}$$

$$= 1.23 \times 10^{19}\,\frac{\text{molecules}}{\text{cm}^3}$$

2. The average velocity is given by combining the expressions for the rms velocities of A and B [Eq. (3-27)].

$$u_{Av} = 1.73\left[\frac{RT}{M_A^{1/2}\,M_B^{1/2}}\right]^{1/2} \tag{13-17}$$

$$= 1.73\left[\frac{8.315 \times 10^7 \times 298}{5.65 \times 6.78}\right]^{1/2}$$

$$= 44{,}000\text{ cm/sec}$$

3. Using Eq. (13-16)

$$Z_{AB} = 1.41 \times 3.14 \times 9 \times 10^{-16}\text{ cm}^2 \times 4.40 \times 10^4\text{ cm/sec}$$
$$\times (1.23 \times 10^{19}\text{ molecules/cm}^3)^2$$
$$= 2.60 \times 10^{28}\text{ collisions/cm}^3\text{-sec}$$

Z_{AB} represents the rate at which A and B molecules collide. If this rate of collision is assumed to be equal to the reaction rate, then

$$Z_{AB} = k[\text{A}][\text{B}] \tag{13-18}$$

If the concentrations of A and B are expressed in units of molecules per cm³, then

$$k = Z_{AB}/n_A n_B \tag{13-19}$$

$$= \sqrt{2}\pi\, s_{AB}^2\, u_{Av} \tag{13-20}$$

Inserting the values for s_{AB} and u_{Av}, we have

$$k = 1.414 \times 3.14 \times 9 \times 10^{-16} \times 4.36 \times 10^4$$
$$= 1.74 \times 10^{-10}\text{ cm}^3\text{/molecule-sec}$$

To obtain k in the more commonly employed units of liter/mole-sec, we must multiply this value by Avogadro's number and divide by 10^3. The resulting value for the bimolecular rate constant is then

$$k = 1.05 \times 10^{11}\text{ liter/mole-sec}$$

The first point to be noted is that this number is many orders of magnitude larger than the rate constants which are ordinarily observed; it was noted earlier that values of k of about 10^{-5} liter/mole-sec were commonly found for bimolecular reactions. Secondly, if the collision number actually were to represent the rate at which reaction occurs, then the reaction rate would vary with the half-power of the temperature. This follows from the fact that the only quantity in the formula for Z [Eq. (13-16)] which is temperature sensitive is u_{Av}, which varies in proportion to $T^{1/2}$, as we see from Eq. (3-27). Now a dependence upon the half-power of temperature would not represent a very strong temperature dependence. For example, in order for a reaction which had a certain rate at room temperature to double in rate, the temperature would need to be raised to four times room temperature, about 1200°K! We know that reaction rates are very much more sensitive to temperature than this. One of the rough rules of thumb which students are taught in their first introduction to chemistry is that the rate of a reaction doubles for each 10°C rise in temperature. It is clearly necessary to invoke a new theoretical model to account for the strong temperature dependence of most reactions.

13-4 Activation Energy

Since the number of reactions of A with B molecules is so much smaller than the number of collisions occurring between these two species in the same period of time, it is clear that only a fraction of the collisions leads to chemical reaction. The velocities of both A and B molecules are distributed in magnitude in accordance with the Maxwell distribution (p. 55), and the molecules may be moving in any direction relative to one another at the moment of collision. For these reasons, the relative kinetic energy of collision (that is, the relative kinetic energy along the line of centers) may vary from zero to a very large value (Fig. 13-5). The simplest ex-

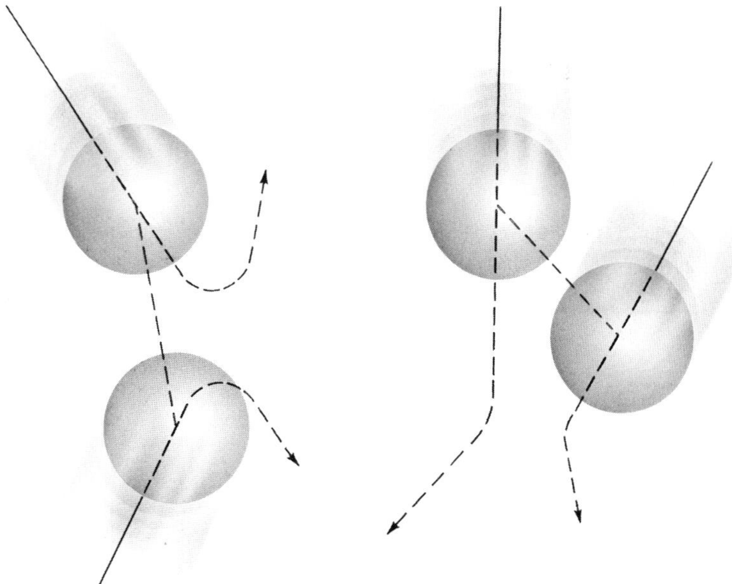

Fig. 13-5 Example of collisions occurring with differing relative kinetic energies of collision.

planation for the fact that only a fraction of the collisions results in reaction is that some minimal value of the relative kinetic energy of collision is required for chemical reaction to occur. This minimal energy is termed the *activation energy*. The requirement of an activation energy is an indication that the reactants do not simply pass directly into products on collision, but must first form an intermediate species of some type, a species higher in energy than the sum of the energies of the colliding reactants. The excess of energy contained in the intermediate form is derived from the kinetic energy of the collision. If the relative kinetic energy of collision of the reactants is insufficient to form the intermediate species, the reactants simply rebound from the collision without reacting. If it is sufficient, the intermediate is formed and subsequently decomposes into the products. The formation of the intermediate state, or *activated complex*, is shown in the energy diagram in Fig. 13-6.

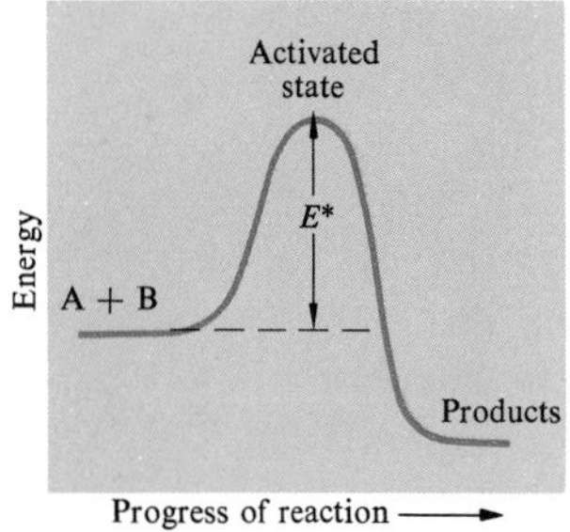

Fig. 13-6 Energy diagram for the reaction A + B → Product.

However, one can think of other reasons why the fraction of collisions which leads to reaction is small. It is reasonable to suppose, for example, that the orientation of the colliding molecules with respect to one another might be important. If relatively inactive parts of the colliding molecules come into contact upon collision, the reactants might very well separate without reacting. It is difficult to imagine, though, that this factor could be responsible for the very large discrepancy — of the order of 10^{16} — between theory and experiment. Another serious shortcoming is the fact that this *steric factor*, as it is called, does not account for the observed temperature dependence of most reactions. On the other hand, a temperature dependence which is correct for most reactions follows in a logical way from the concept of an activation energy. The steric factor thus is not an alternative to the idea of an activation energy, but rather an additional factor which plays a secondary role in affecting reaction rates.

In the case of a reaction such as that discussed above — one which involves simply the collision of two reactants — the rate of reaction varies with temperature in the manner shown in Fig. 13-7. Svante Arrhenius

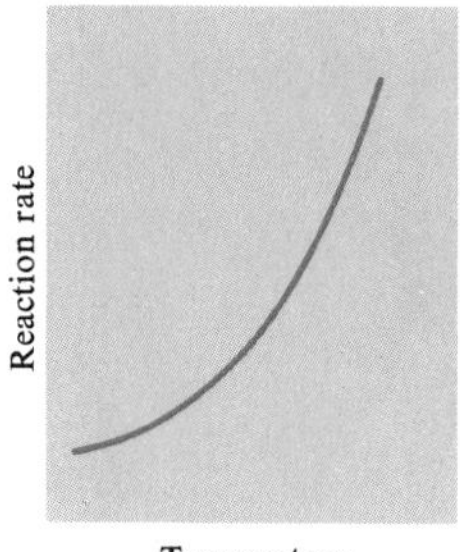

Fig. 13-7 Typical variation of reaction rate with temperature.

suggested on the basis of empirical observation that the rate constant k could be expressed as

$$k = Ae^{-E^*/RT} \tag{13-21}$$

where E^* is the activation energy and A is a constant which differs from one reaction to another. It is of interest to see whether this result can be obtained by using the gas kinetic theory coupled with the concept of the activation energy as the minimum relative kinetic energy of collision. We can write the rate constant k as equal to the expression given in Eq. (13-20) times a factor q which represents the fraction of these collisions which results in formation of the activated state:

$$k = \sqrt{2}\pi s_{\mathrm{AB}}^2\, u_{\mathrm{Av}} \times q \tag{13-22}$$

We must now deduce the manner in which the fraction q varies with temperature. From Eq. (3-26) we obtain the fraction of molecules in the system which possess an energy at least as great as some pre-determined value. If we let this energy be the activation energy, E^*, we obtain

$$f = e^{-E^*/RT} \tag{13-23}$$

The fraction q we are after is proportional to f. The proportionality constant contains, among other things, the steric factor for the particular reaction. We need not know what this is for the moment. Let us assume that the steric factor, along with molecular diameters, s_{AB}, average velocity, u_{Av}, and any other constants can be put into one constant. The velocity does have a slight temperature dependence, as we have already seen, but this can be ignored in comparison with the exponential term. The rate constant can then be written

$$k = A\, e^{-E^*/RT}$$

This, of course, is just the Arrhenius equation. We have merely shown that it can be obtained by combining the gas kinetic theory of collision rates with the concept of an activation energy, and then using the Maxwell-

Boltzmann distribution of molecular energies. In principle, all of the quantities could be calculated for any particular system. In practice, this proves to be beyond the present range of chemistry, except for exceedingly simple reactions such as

$$H + H_2 \rightarrow [H\cdots H\cdots H] \rightarrow H_2 + H \qquad (13\text{-}24)$$

In the first place, we cannot calculate the energetics of the activated state with any reliability. In addition, even assuming that we know the activation energy from experiment, we still cannot calculate the rate constant with reliability because we do not know the appropriate values for all of the quantities which go to make up A. The importance of the kinetic theory model for reaction is not that it is a complete and detailed picture of how and at what rate the reaction occurs, but that it provides a basis on which to view and understand the reaction process in a general way. A large majority of the reactions which the chemist encounters for substances in solution behave in the manner expected from the Arrhenius equation. By taking the log of both sides of Eq. (13-21) we obtain a form of the equation which is useful in evaluating the activation energy:

$$\log_{10} k = \log_{10} A - \frac{E^*}{2.3RT} \qquad (13\text{-}25)$$

We again have the familiar linear relationship between the log of a variable quantity and $1/T$. For systems which obey the Arrhenius equation, a graph of $\log_{10} k$ *vs* $1/T$ leads to a linear relationship with slope $-E^*/2.3R$. When R is expressed as 1.987 cal/°K-mole, E^* is obtained in units of calories.

13.5 Kinetics and Reaction Mechanisms; Catalysis

The details of the paths by which substances pass from one form to another are of interest to the chemist. From a knowledge of the manner in which a certain reaction occurs, it may be possible to devise other chemical systems that will undergo reaction in a desirable way. Kinetics is a tool in the attempt to learn about the mechanism, which is a description of the reaction path. It is important to realize at the outset that we seldom have more than very scanty information on which to discuss the mechanism. The information we obtain from kinetics studies is in the form of the rate law (the dependence of the rate on concentrations of various substances), and perhaps, information regarding the temperature dependence. But in studying the rate, we are examining the characteristics of only the slow, rate-determining step. In a more complex process which proceeds through a series of consecutive steps, we may thus really only learn about one of these steps, the slowest. There is the further difficulty that we can never be sure that we have hit upon the really unique mechanism, or pathway, by which the reaction proceeds. We may devise a model for the mechanism which is consistent with all that is known about a reaction, but this is not in itself a guarantee that we have hit upon the one and only correct picture. Perhaps, because of our limited knowledge of chemistry, we have overlooked an approach to the problem which would lead to another, quite different picture of how the reaction proceeds. In spite of these limitations, however, there are many chemical reactions for which the mechanism is well established on the basis of data gathered from a number of different experiments.

Let us first consider a reaction which involves a simple transfer of an atom from one species to another in a bimolecular collision. Such a reaction is exemplified by the reaction of nitric oxide and ozone:

$$NO + O_3 \rightarrow NO_2 + O_2 \qquad (13\text{-}26)$$

This is a very exothermic reaction which evolves 48 kcal/mole of oxygen produced. The activation energy is quite small, only about 2.5 kcal/mole, so that a large fraction of collisions result in reaction. The rate expression is given by

$$\frac{d(O_2)}{dt} = k(NO)(O_3)$$

The rate constant k is equal to about 5×10^8 liter/mole-sec. The free energy surface for this reaction appears as shown in Fig. 13-8. The reaction coordinate, which measures the progress of the reaction, is a complicated function of all of the bond distances and angles. We do not know what this is like in any detail except for the very simplest of reactions, as for example, in Eq. (13-24). It is of interest, however, to inquire about the nature of the activated complex, or *transition state*, as it is sometimes referred to. From the experimental result that the reaction is first order with respect to each reactant, we assume that the rate-determining step involves just one molecule of each reactant. The reaction consists in merely the passing of an oxygen atom from O_3 to NO. Now O_3 is a bent molecule

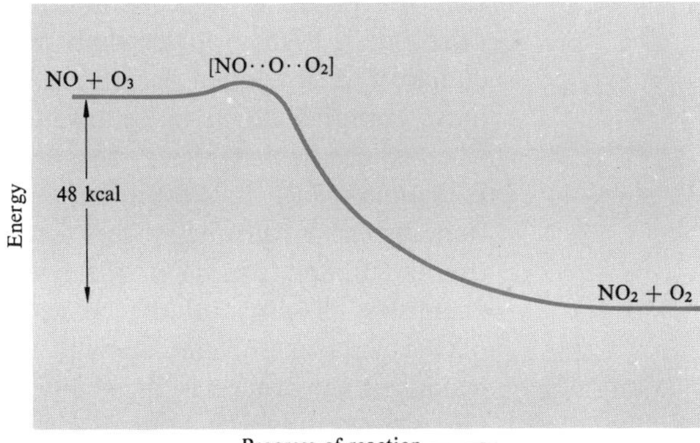

Fig. 13-8 Energy diagram for the $NO + O_3$ reaction.

(Fig. 8-25), and so is NO_2. We might guess that in the transition state, the arrangement is such that both NO_2 and O_3 are recognizable; for example,

$$\left[\begin{array}{ccccc} & N\cdots & O & \cdots O & \\ / & & & & \backslash \\ O & & & & O \end{array}\right]$$

The fact that the reaction is first order with respect to each reactant permits us to discard a number of other possible pathways. For example, assume that

$$\begin{array}{ll} NO + NO \rightarrow N_2O_2 & \text{(slow)} \\ N_2O_2 + O_3 \rightarrow N_2O_3 + O_2 & \text{(fast)} \\ N_2O_3 \rightarrow NO + NO_2 & \text{(fast)} \end{array}$$

If this mechanism were correct, the reaction would be second order with respect to NO, and zero order with respect to O_3, contrary to what is observed. As a more likely alternative, consider this one:

$$\begin{array}{ll} O_3 + M \rightarrow O_3^* + M & \text{(slow)} \\ O_3^* \rightarrow O_2 + O & \\ O + NO \rightarrow NO_2 & \end{array}$$

In this type of mechanism, which depends upon O_3 picking up the energy needed for dissociation from collision with any other molecule M, the rate would be found to depend upon the total pressure, and upon the concentration of O_3. This is again in contradiction to what is found. The rate law thus serves to exclude incorrect mechanisms. It does not necessarily lead us to the correct one, however, since there may be more than one pathway which leads to essentially the same kinetic behavior.

Catalysis — In addition to temperature and the concentrations of reactant species, other factors frequently are important in affecting the rate of a reaction. If a substance not included as one of the reactants in the balanced equation exerts an effect upon the rate, or if one of the reactants or products appears in the rate expression with its concentration raised to a higher power than that to which it appears in the balanced equation, it is termed a *catalyst*. This definition of a catalyst, although more complicated, is preferable to the one often given (that a catalyst is a substance which affects the rate of a chemical reaction without itself being changed). It is often impossible to recover the catalyst from a reaction system in an unchanged form. From a practical point of view, the use and understanding of catalysts is tremendously important to modern chemistry. Nearly all of the new products of the chemical industry — plastics such as polyethylene and Dacron, high octane gasolines, *etc.* — are produced by processes in which catalysis plays a vital role. Despite its importance, however, chemists still do not fully understand all forms of catalysis.

The general function of a gas phase catalyst can be deduced from the kinetic molecular theory of reaction rates. Since the number of collisions occurring between reactants at a fixed temperature and concentration is a constant, the only way in which a catalyst could operate to increase the rate would be to increase, by one means or another, the fraction of collisions which are effective in causing reaction. If the reaction occurs in the gas phase, and is a simple bimolecular reaction of the type

$$A + B \rightarrow C$$

a catalyst — let us call it X — could operate only in a particular way. It might at first be thought that it could intervene in the collision of A and B at the moment of collision, and somehow lower the energy necessary for the reactants to form the product. An analysis of the frequency with which three bodies come together in the gas phase in collision, however, shows that the number of such collisions per unit time is quite small. Therefore, the catalyst must be effective by first undergoing collision with *one* of the reactants — say, A — and forming some sort of intermediate which is stable enough to stay together for an extended period of time, long enough for the intermediate to collide with a B molecule. The relative kinetic energy of collision of the intermediate with B, which is required for reaction, must be small so that a reaction leading to formation of products and freeing of the catalyst occurs. The sequence of steps is illustrated in the following scheme:

$$A + X \rightarrow A \cdot X \text{ (intermediate)}$$
$$A \cdot X + B \rightarrow C + X$$

1. *Homogeneous catalysis.* Homogeneous catalysis by a chemical agent (*i.e.*, catalysis in which all of the species are in the same phase) generally proceeds according to such a scheme. Another, usually less important role that the catalyst might play is in reducing the steric requirement on collision of B with the intermediate.

The introduction of catalyst into a homogeneous reaction system implies the substitution of a more rapid step, involving the catalyst, for the spontaneous step. This, of course, results in a change in order of the reaction with respect to the reactants as well as a change in rate. Consider as an example the reaction of nitrous oxide:

$$2N_2O \xrightarrow{k_1} 2N_2 + O_2 \tag{13-27}$$

In the absence of a catalyst, this reaction is second order in N_2O. Nitric oxide, NO, acts as a catalyst, probably through the following sequence of steps:

$$NO + N_2O \xrightarrow{k_2} N_2 + NO_2 \tag{13-28}$$
$$2NO_2 \rightarrow 2NO + O_2 \tag{13-29}$$

The first of these is the slower, and therefore the rate-determining reaction. The rate expression for disappearance of N_2O in the presence of a catalyst is therefore

$$\frac{-d(N_2O)}{dt} = k_2(NO)(N_2O)$$

whereas in the uncatalyzed reaction it is

$$\frac{-d(N_2O)}{dt} = k_1(N_2O)^2$$

As it happens, the activation energies for the catalyzed and uncatalyzed reactions are about the same. The specific rate constant for the catalyzed reaction, k_2, is about ten times that for the uncatalyzed reaction. This example illustrates the very important point that, in general, the concentration of catalyst appears in the rate expression for a catalyzed reaction. In the present example, the rate is first order in catalyst, but the order might vary, depending upon the detailed mechanism of the catalyzed process.

As another example of homogeneous catalysis, consider the oxidation of thallium (I) ion to thallium (III) ion by cerium (IV) ion:

$$2Ce^{+4} + Tl^{+} \rightarrow 2Ce^{+3} + Tl^{+3} \tag{13-30}$$

This reaction proceeds slowly in aqueous solutions because the loss of two electrons from the thallium (I) ion requires that two cerium (IV) ions be present simultaneously; this means that a trimolecular collision is required, and these are relatively infrequent. Addition of manganese (II) ion causes a great increase in the reaction rate. The key to the catalytic effect of this ion lies in its ability to exist in both the (III) and (IV) oxidation states, in addition to the (II) oxidation state. The mechanism of the catalyzed reaction is

$$\begin{aligned} Ce^{+4} + Mn^{+2} &\rightarrow Ce^{+3} + Mn^{+3} \\ Ce^{+4} + Mn^{+3} &\rightarrow Ce^{+3} + Mn^{+4} \\ Mn^{+4} + Tl^{+} &\rightarrow Mn^{+2} + Tl^{+3} \end{aligned} \tag{13-31}$$

All of the steps in the catalyzed process are now bimolecular, and the reaction rate is in the range that one would expect for a bimolecular process.

The reaction of chlorine with benzene affords a third, and very important, example of homogeneous catalysis. The reaction leads to formation of chlorobenzene and hydrogen chloride:

$$C_6H_6 + Cl_2 \rightarrow C_6H_5Cl + HCl \tag{13-32}$$

It is catalyzed by iron (III) chloride, which functions to disrupt the chlorine-chlorine bond:

$$FeCl_3 + Cl_2 \rightleftharpoons FeCl_4^- + Cl^+ \tag{13-33}$$

The reaction of Eq. (13-33) proceeds to only a very slight extent, but it is kinetically important because the Cl^+ ion is extremely reactive. The steps which follow in rapid succession are

$$C_6H_6 + Cl^+ \rightarrow C_6H_6Cl^+ \tag{13-34}$$

$$C_6H_6Cl^+ + FeCl_4^- \rightarrow C_6H_5Cl + FeCl_3 + HCl$$

2. *Heterogeneous catalysis.* This occurs when a substance in another phase catalyzes the reaction; *e.g.*, when a solid catalyzes a gas phase reaction. In this type of catalysis, the catalyst acts by providing a surface upon which the reactants can come together and react more readily than they do in the homogeneous phase. The addition of hydrogen to ethylene is an example of a gas phase reaction which is catalyzed by certain specially prepared metal surfaces:

$$H_2 + H_2C{=}CH_2 \xrightarrow{Ni} H{-}CH_2{-}CH_2{-}H \tag{13-35}$$

The metal surfaces act as a catalyst by initial adsorption of hydrogen gas on the metal surface; the hydrogen-hydrogen bond is weakened as a result. When an ethylene molecule strikes the metal surface near a point at which a hydrogen molecule has been previously adsorbed, addition of hydrogen to ethylene occurs readily.

Heterogeneous catalysts generally have very large surface areas per unit weight. Special treatment of one kind or another is usually necessary in preparing the catalyst so that the surface will be highly active. For example, an effective chromium (III) oxide catalyst is prepared by precipitating hydrous chromium (III) oxide, $Cr(OH)_3 \cdot XH_2O$ from aqueous solution, and then heating the precipitate slowly to a temperature of about 400°C under vacuum. The solid product, a partially hydrated chromium (III) oxide, is crushed and sieved to obtain a uniformly small particle size. When prepared in this manner, the catalyst has an effective surface area of about 300 square meters per gram. It is effective in catalyzing the addition of hydrogen to ethylene at 0°C. One of the interesting studies made using this catalyst involves the addition of deuterium, D_2, to ethylene, $CH_2{=}CH_2$. The product of the reaction was found to be ethane-1, 2-d_2,

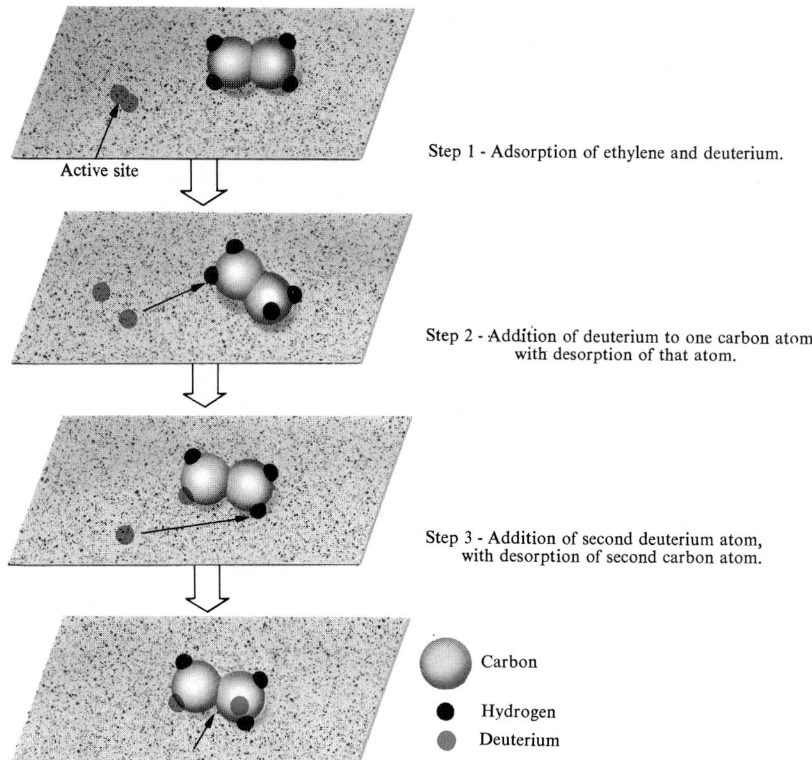
Active site
Step 1 - Adsorption of ethylene and deuterium.
Step 2 - Addition of deuterium to one carbon atom, with desorption of that atom.
Step 3 - Addition of second deuterium atom, with desorption of second carbon atom.
Carbon
Hydrogen
Deuterium

Fig. 13-9 Schematic illustration of the hydrogenation of ethylene on a heterogeneous catalyst. By using deuterium, it can be determined that the two atoms which are added to the ethylene molecule are fixed one on each carbon atom.

CH_2D—CH_2D. This is a strong indication that the mechanism of the reaction proceeds as shown in Fig. 13-9. The significant point in this experiment is that no products of the form CHD_2—CH_2D, or any others containing more than one deuterium on each carbon atom, were found. If one or more of the original carbon-hydrogen bonds of the ethylene were broken during the time in which the molecule is adsorbed on the surface, a replacement by deuterium would be very probable, since it is present in large amounts. It is unlikely, therefore, that the original carbon-hydrogen bonds are disrupted during the period of adsorption.

Studies of this type provide important information on how catalysts operate, but a great deal more work is needed before the catalysts themselves are understood. As an illustration of the state of our knowledge in this area, the nature of active sites on the surface of the chromium (III) oxide is not known, nor is it known just why one method of catalyst preparation is superior to another.

Suggested Readings

Eyring, H. and E. M. Eyring, *Modern Chemical Kinetics*, New York, N. Y.: Reinhold Publishing Corporation, 1963.

King, E. L., *How Chemical Reactions Occur*, New York, N. Y.: W. A. Benjamin, Inc., 1963.

Exercises

13-1. Give a brief definition of each of the following:
(a) Reaction rate.
(b) Pseudo-first-order reaction.
(c) Activation energy.
(d) Rate constant.
(e) Heterogeneous catalysis.

13-2. Indicate the ways in which the rate of the reaction

$$2H_2(g) + Se_2(g) \rightarrow 2H_2Se(g)$$

might be expressed.

13-3. The reaction $SO_2(g) + 2H_2(g) \rightarrow S + 2H_2O$ is first order in both SO_2 and H_2. Write the rate expression for the reaction. What units would the rate constant have if the concentration of the reactants were expressed in atmospheres of pressure?

13-4. Explain why the rate of a simple bimolecular reaction between two gases A and B changes with the pressures of the gases. Explain why the rate increases with increasing temperature.

13-5. Indicate an experimental procedure by which the rate of the reaction described in Eq. (13-27) might be determined.

13-6. The reaction of iodide with hypochlorite ion

$$I^- + OCl^- = Cl^- + OI^-$$

occurs in fairly strong base solution (NaOH about $1M$), $T = 25°C$. The reaction was followed by observing the concentrations of the reactants at various times after initial mixing, starting with equal concentrations of the reactants

Time (sec)	$[I^-] = [OCl^-]$, M
0	0.00200
3	0.00158
5	0.00135
8	0.00112
14	0.00073

From these data, graphically estimate the rate $-d[I^-]/dt$ at $t = 4$ and 8 sec. What is the order of the reaction in the two reactants?

13-7. The rate of a hetergeneous reaction between a solid and a liquid is found to depend on the state of the solid surface. Explain how this factor might influence the reaction rate.

13-8. Consider the bimolecular reaction of hydrogen and iodine in the gas phase:

$$H_2 + I_2 \rightarrow 2HI$$

It is to be studied in a cylinder fitted with a piston. Indicate the effect which each of the following would have on the reaction rate, and on the rate constant:
(a) Increase in temperature, volume constant.
(b) Increase in volume, temperature constant.
(c) Add an inert gas (for example, helium), volume and temperature constant.
(d) Add a gas X which acts as a catalyst, volume and temperature constant.

13-9. Sulfur monoxide, SO, reacts with O_3 to give SO_2 and O_2. SO is a quite unstable molecule, which must be specially produced for a kinetics study. What do you predict for the order of this reaction with respect to the reactants? Suggest a mechanism, and indicate whether the reaction should have a high or low activation energy.

13-10. The dissolution of a solid in a liquid is a process occurring at a surface. Assuming a well-stirred solution which is far from saturated, how would the rate of solution of a solid consisting of spherical granules 0.10 mm in radius compare with the rate if the granules were 2.5 mm in radius, assuming that the total quantity of solid is constant?

13-11. Consider Eq. (13-33) as an acid-base reaction, and identify the acid, base, conjugate acid, and conjugate base. (Note that although this is not a protonic system, and so is not within the Brönsted system, it is nevertheless possible to extend the acid-base concept to this type of system.) What does this suggest insofar as the possible use of substances other than $FeCl_3$ as catalysts is concerned?

13-12. Assume that Eq. (13-33) is the slow step in the catalyzed chlorination of benzene, with over-all Eq. (13-32). What is the order of the chlorination reaction in benzene? In catalyst?

13-13. The decomposition of dimethyl ether, the reaction used as an example on p. 312, was also studied at 552°C. The following data were obtained.

Time (sec)	Pressure (mm)
0	420
114	743
219	954
299	1054
564	1198
Infinity	1258

From these data, calculate the rate constant at 552°C. Using your value for k at 552°C and the value for k at 504°C given on p. 313, calculate the activation energy for the reaction.

13-14. The action of a catalyst in a bimolecular reaction system such as that described on p. 323 provides some interesting kinetic problems. There are actually three reactions to consider:

$$A + X \xrightarrow{k_1} AX$$
$$AX \xrightarrow{k_2} A + X$$
$$AX + B \xrightarrow{k_3} \text{products} + X$$

Assuming that A, B, and X are present in comparable concentrations, derive the form for the rate expression for the following relative magnitudes of the rate constants:

(a) k_1 much greater than k_2; k_3 much greater than k_1.
(b) k_1, k_2, and k_3 are of comparable magnitudes.

[*Hint:* After the reaction has gotten started, the concentration of X will become effectively a constant, since it is re-formed in the last two reactions as rapidly as it is used up in the first reaction. Write a rate expression for the over-all rate of disappearance of X, therefore, and equate it to zero.]

13-15. Assuming that the rate of the reaction expressed by Eq. (13-7) is first order in both ammonium ion and nitrite ion, calculate an average value for the specific rate constant from the data in Table 13-1. (Assume that the concentrations given are actually the concentrations of nitrite and ammonium ions in solution at the start of the reaction.)

13-16. Would you expect the entropy change on adsorption of ethylene as shown in Fig. 13-8 to be positive or negative? Explain.

13-17. The Friedel-Crafts reaction, shown in the following equation, is catalyzed by aluminum chloride, $AlCl_3$, or by iron (III) chloride:

$$CH_3\overset{\overset{O}{\|}}{C}{-}Cl + C_6H_6 \xrightarrow{AlCl_3} C_6H_5{-}\overset{\overset{CH_3}{|}}{C}{=}O + HCl$$

Describe a possible mechanism for the catalysis.

13-18. What is the half-life of the dimethyl ether decomposition reaction described in Fig. 13-4 and the accompanying table?

13-19. The first order rate constant for the hydrolysis of *t*-butyl chloride in an isopropyl alcohol-water mixture at room temperature is 0.535 hr^{-1}. What fraction of the original concentration of the chloride remains after 30 minutes from the starting time? What is the half-life of the reaction in minutes?

13-20. Using the same kind of approach as in the discussion of half-life in the text, calculate the ratio A/A_0 which corresponds to a time which is just $1/k$, the first-order rate constant. (If we had a simple way of measuring A/A_0, we could determine the first-order rate constant to a rough approximation by measuring the time required to achieve the ratio for which you are asked to calculate the correct value.)

Chapter 14

Chemical Equilibrium I. Principles

14-1 Law of Mass Action

When hydrogen and iodine are brought into contact in the gas phase at some convenient temperature, a chemical reaction ensues, leading to production of hydrogen iodide:

$$H_2 + I_2 \rightarrow 2HI$$

The rate of the reaction, measured by the rate of disappearance of H_2 or I_2, or by the rate of appearance of HI, steadily decreases with time from the initial value. The rate eventually becomes zero, although both H_2 and I_2 still remain in the system. Under these conditions, the system is in a state of chemical equilibrium.

On the basis of the kinetic-theory model for reaction rate, it is clear that H_2 and I_2 molecules are still reacting to form product, since collisions between these two species continue to occur. The static condition of the system is then only an apparent one; the concentrations of the reactants remain constant because H_2 and I_2 molecules are being produced at a rate which just equals the rate of their disappearance.

The reaction of H_2 and I_2 to form HI is a simple bimolecular gas phase reaction.* The reaction rate decreases steadily from the initial value as the reactants are consumed, because fewer collisions between H_2 and I_2 molecules occur per unit time. The reverse reaction, the bimolecular collision of two HI molecules to form H_2 and I_2, does not occur initially because the concentration of HI is zero. As product is formed, however, the rate of this reverse reaction steadily increases. Equilibrium is attained when the rates of the two opposing reactions are equal. The rate of the forward and reverse reactions can be written

$$\text{rate of forward reaction} = k_f[I_2][H_2]$$
$$\text{rate of reverse reaction} = k_r[HI]^2$$

where k_f and k_r are the rate constants for the forward and reverse reactions, respectively. At equilibrium, the two rates are equal, and

$$k_f[I_2][H_2] = k_r[HI]^2$$
$$\frac{k_f}{k_r} = K = \frac{[HI]^2}{[H_2][I_2]} \qquad (14\text{-}1)$$

where, as before, the square brackets indicate concentrations.

From Eq. (14-1) it can be seen that the concentrations of product and reactant species at equilibrium are determined by the ratio of the rate constant of the forward reaction to that for the reverse reaction. This ratio is called the equilibrium constant K.

Guldberg and Waage, in 1867, proposed that this type of treatment applies to all chemical reactions. For a general reaction of the form

$$nA + mB + \ldots \rightarrow sP + rQ + \ldots$$

they proposed that the equilibrium constant could be written as

$$K = \frac{[P]^s[Q]^r \ldots}{[A]^n[B]^m \ldots} \qquad (14\text{-}2)$$

that is, the product of the concentrations of the products, each raised to the power of its coefficient in the balanced equation, divided by a similar product of concentrations for the reactants. The Guldberg and Waage treatment of chemical equilibrium is known as the *law of mass action*, and

---

*Very recent evidence has been published which indicates that the hydrogen-iodine reaction may not be a simple bimolecular process as chemists have assumed for so long. The alternative mechanism is not distinguishable from the simple bimolecular process so far as ordinary kinetics studies are concerned. It is possible that the reaction proceeds by a prior dissociation of I_2 molecules, in an equilibrium step:

$$I_2 \rightleftarrows 2I$$

There is then a reaction of iodine atoms with H_2:

$$H_2 + I \rightarrow H_2I$$
$$H_2I + I \rightleftarrows [H_2I_2] \rightleftarrows 2HI$$

is based on the assumption that the rates of the forward and reverse reactions could be written as

$$\text{rate of forward reaction} = k_f[\text{A}]^n[\text{B}]^m \ldots$$
$$\text{rate of reverse reaction} = k_r[\text{P}]^s[\text{Q}]^r \ldots$$

It is now well established that the rate of a chemical reaction need not be proportional to concentrations of reactants raised to the power of their coefficients in the balanced equation. The original formulation of the law of mass action was therefore based on an incorrect premise. Nevertheless, the expression for the equilibrium constant obtained by application of the law of mass action *is* correct. It is possible to show on the basis of other arguments that the equilibrium constant must have the form dictated by the over-all equation, as illustrated in Eq. (14-2). It is therefore not necessary to know anything about the mechanism of a reaction, or of the rate law expression which applies, in order to write the expression for the equilibrium constant. To illustrate, consider the hydrogen-bromine-hydrogen bromide system:

$$H_2 + Br_2 \rightleftharpoons 2HBr$$

The equilibrium constant for this reaction is

$$K = \frac{[HBr]^2}{[H_2][Br_2]} \tag{14-3}$$

This is precisely the same form of expression as that obtained for the hydrogen-iodine-hydrogen iodide system, as it must be if the form of K is determined only by the over-all balanced equation. But the mechanisms by which the two reactions proceed are altogether different, and the two reactions follow quite different rate laws. The reaction of hydrogen and bromine to form HBr does not proceed by a simple bimolecular process, but by a series of chain reactions, as shown in the following equations:

$$
\begin{aligned}
Br_2 &\rightleftharpoons 2Br \\
Br + H_2 &\rightarrow HBr + H \qquad \text{(slow)} \\
H + Br_2 &\rightarrow HBr + Br \qquad \text{(fast)} \\
H + HBr &\rightarrow H_2 + Br \qquad \text{(fast)}
\end{aligned}
$$

From this example, it is clear that the form of the equilibrium constant is determined only by the over-all balanced equation and not by the path by which the reaction proceeds.

14-2 Thermodynamic Approach to Equilibrium

Let us consider the following gas phase equilibrium system:

$$nA + mB \rightleftharpoons rC + sD \tag{14-4}$$

That is, n moles of A react with m moles of B to produce r moles of C and s moles of D. We might say, alternatively, that r moles of C and s moles of D react to produce n moles of A and m moles of B. The numbers n,

m, r, and s do not, of course, refer to the actual number of moles of each species present, but only to the ratios of molecules which react. The reactants and products might be present in any possible mixture of relative proportions. The only requirement is that at equilibrium, some amount, however small, of each substance must be present.

The equilibrium constant for the system is, by the method just described,

$$K = \frac{[C]^r[D]^s}{[A]^n[B]^m} \tag{14-5}$$

Let us now consider this system from a thermodynamic point of view. To simplify the discussion for the moment, we consider that the substances A, B, C, and D are all ideal gases. We shall first consider the free energy for each substance in the system, and show that, at constant temperature, the free energy of each substance depends upon pressure in a way that we can describe with a simple equation. Then we will consider the *total* free energy change in going from reactants to products. We know that at equilibrium the free energy change must be zero. This is, in fact, the definition of equilibrium. If the free energy change is not zero, there is a net tendency for spontaneous change in the direction which leads to lower free energy, as discussed in Sec. 9-8. From the definition of equilibrium, we can derive an expression for the equilibrium constant, which we shall show to be closely related to the expression in Eq. (14-5). There is a new element introduced by the thermodynamic approach which has to do with use of concentration or pressure in the equilibrium constant expression. We will consider this in some detail.

Consider one of the gaseous substances involved in the equilibrium of Eq. (14-4). If the gas is ideal, we can define the standard state for it as one atmosphere pressure at some temperature T. In addition to the standard molar free energy, there is also a corresponding standard molar enthalpy and entropy. We have thus,

$$G^\circ_A = H^\circ_A - TS^\circ_A \tag{14-6}$$

where the superscripts indicate the standard state. We will not label each quantity with the temperature T, because in this discussion, temperature is assumed to be constant. The substance A is at some standard pressure P° (1 atm) and some standard volume V°. Now, suppose that a change in both pressure and volume occurs, to give a new pressure P and volume V. We want to know the value for the corresponding change in free energy of A, ΔG_A. To know this, we must evaluate ΔH_A and ΔS_A.

$$\Delta H_A = H_A - H^\circ_A = E_A - E^\circ_A + PV - P^\circ V^\circ$$

We are dealing with an ideal gas, for which the internal energy varies only with temperature. Therefore, $E^\circ_A = E_A$. Furthermore, since $P^\circ V^\circ = RT$, and $PV = RT$, we see that $\Delta H_A = 0$. The entropy change in A is given by the expression obtained earlier (page 226):

$$S - S^\circ = \Delta S = R \log_e \frac{V}{V^\circ} \tag{14-7}$$

We can make a useful substitution here. From the ideal gas law, $V^\circ = RT/P^\circ$ and $V = RT/P$. Substituting into Eq. (14-7) we have, after, cancelling out the constant quantities,

$$\Delta S = R \log_e \frac{P^\circ}{P} \tag{14-8}$$

Putting these results into the equation for ΔG,

$$\begin{aligned} \Delta G_A &= G_A - G_A^\circ = H_A - H_A^\circ - T(S_A - S_A^\circ) \\ &= \Delta G = RT \log_e \frac{P}{P^\circ} \end{aligned} \tag{14-9}$$

[Go through all the steps needed to obtain Eq. (14-9).] We can write this equation another way by doing a little rearranging:

$$G = G^\circ + RT \log_e P \tag{14-10}$$

The last term in this formula arises because we define P° as one atmosphere. This is a very useful expression. It says that the free energy of one mole of A is given by the free energy in the standard state G°, plus a correction term to convert to the actual pressure, P. Similar expressions apply, of course, for the other substances involved in the equilibrium. Now let us consider the free energy change involved in the forward reaction of Eq. (14-4). Proceeding as before (Chap. 9), we write,

$$\Delta G = rG_C + sG_D - nG_A - mG_B \tag{14-11}$$

Now let us substitute for each of the G's in this equation the corresponding expression, Eq. (14-10). Then

$$\begin{aligned} \Delta G = {} & rG_C^\circ + rRT \log_e P_C + sG_D^\circ + sRT \log_e P_D \\ & -nG_A^\circ - nRT \log_e P_A - mG_B^\circ - mRT \log_e P_B \end{aligned}$$

Recalling that $ab \log_e Y = b \log_e Y^a$, we have

$$\begin{aligned} \Delta G &= rG_C^\circ + sG_D^\circ - nG_A^\circ - mG_B^\circ + RT \log_e \frac{P_C^r P_D^s}{P_A^n P_B^m} \\ &= \Delta G^\circ + RT \log_e \frac{P_C^r P_D^s}{P_A^n P_B^m} \end{aligned} \tag{14-12}$$

Now it is very important to recognize what the terms in this expression mean. ΔG° is the free energy change that would occur if n moles of A and m moles of B were to be converted into r moles of C and s moles of D, *with all of these substances present in their standard states.* One way to imagine this is to suppose that we have a huge container with all of these substances present, each at one atm of pressure. Then this conversion would not significantly change the pressures of any of the gases if there were many more moles than these of all four present. Alternatively, imagine that we have a system in which there is one mole total of A, B, C, and D, each present at one atm. We now allow a very small fraction of reaction to occur, say 10^{-4} times n moles of A, *etc.* The free energy change ΔG° expressed in Eq. (14-12) would then be 10^4 times the free energy change for the slight reaction allowed. Since only trifling quantities of the substances would have reacted, the pressures of each would still be essentially 1 atm, so that the conditions required for evaluation of $\Delta G°$ would still be present. The second term in Eq. (14-12) is a correction to the free energy change for the non-standard condition of the reactants A, B, C, and D. ΔG is not, in general, equal to $\Delta G°$, because the pressures of all of the substances involved are not, in general, equal to one atmosphere.

Now let us go on to the next step in the analysis by supposing that the system is at equilibrium. This means that the particular combination of pressures of A, B, C, and D is such that there is no net tendency for the reaction to proceed in either the forward or reverse direction. This in turn means that ΔG is zero. It is very important to see that at equilibrium it is the value of ΔG, and not $\Delta G°$, which is zero. But if ΔG is zero, then

$$0 = \Delta G° + RT \log_e \left(\frac{P^r_{\mathrm{C}} P^s_{\mathrm{D}}}{P^n_{\mathrm{A}} P^m_{\mathrm{B}}} \right)_{\mathrm{eq}}$$

$$G° = -RT \log_e \left(\frac{P^r_{\mathrm{C}} P^s_{\mathrm{D}}}{P^n_{\mathrm{A}} P^m_{\mathrm{B}}} \right)_{\mathrm{eq}} = -RT \log_e K_{\mathrm{eq}} \qquad (14\text{-}13)$$

The ratio of pressures in the log term is a true constant, since, for a particular temperature,

$$\log_e \left(\frac{P^r_{\mathrm{C}} P^s_{\mathrm{D}}}{P^n_{\mathrm{A}} P^m_{\mathrm{B}}} \right)_{\mathrm{eq}} = \frac{-\Delta G°}{RT}$$

There is only one possible value for $\Delta G°$ at any particular temperature, so the term on the right-hand side is a constant.

To this point, we have considered an equilibrium involving a mixture of ideal gases. If the gases are non-ideal, or if substances in solution, or solids, *etc.* are involved, an analogous treatment can be applied in a general way. No matter what state the substance may be in, it is possible to define a standard state with corresponding standard free energy for the substance. For any state of the substance other than the standard one, we write

$$G = G° + RT \log_e a \qquad (14\text{-}14)$$

where a is a quantity called *activity*. The activity of a substance represents an effective concentration. For the standard state, $a = 1$, so the last term drops out, and $G = G°$, as it must. If substances were ideal, activity and concentration would be synonymous; in the case of an ideal gas, the activity would be just the pressure. When the substance is non-ideal, some correction must be applied to the concentration to allow for the non-ideality. The activity can be expressed as a product of the concentration times an *activity coefficient*, γ:

$$a = \gamma c \qquad (14\text{-}15)$$

When the substance is ideal, $\gamma = 1$, but this is not generally the case for real systems.

We see from this treatment that the thermodynamically correct equilibrium constant is written in the form

$$K_{eq} = \frac{a_C^r \times a_D^s}{a_A^n \times a_B^m} \tag{14-16}$$

The difference between this equation and Eq. (14-5) is that here we have made no assumptions regarding the ideality or non-ideality of the components of the equilibrium. The difficulty with Eq. (14-16) is that we usually do not have information about activity. As a result, there is nothing to be done but assume that the activity coefficient γ is unity, and that activity and concentration are numerically the same. When the system is non-ideal, as is often the case, this approximation is not a very good one. The so-called equilibrium constant, which is obtained from using concentrations directly, is then not really independent of the equilibrium concentrations as it should be if activities were used.

As an illustration of this thermodynamic concept of equilibrium, let us consider the simple system described by Henry's law (p. 261). The Henry's law constant K is the equilibrium constant for the equilibrium of a substance A between the liquid and gaseous phases:

$$\text{A (solution)} \rightleftharpoons \text{A (gas)}$$

At equilibrium, the free energy of A is the same in each phase:

$$G_{A(sol)} = G_{A(gas)}$$

The free energy of A in each phase can be expressed as

$$G_{A(sol)} = G^\circ_{A(sol)} + RT \log_e a_{A(sol)}$$
$$G_{A(gas)} = G^\circ_{A(gas)} + RT \log_e a_{A(gas)}$$

Then, since

$$G_{A(sol)} = G_{A(gas)}$$

$$\Delta G^\circ = G^\circ_{A(gas)} - G^\circ_{A(sol)} = -RT \log_e \frac{a_{A(gas)}}{a_{A(sol)}} \tag{14-17}$$

By assuming that the activity of A in the gas phase is given by the pressure, and in solution by the concentration, we obtain Henry's law:

$$a_{A(gas)} = P_A; \qquad a_{A(sol)} = [A]_{sol}$$

$$\Delta G^\circ = -RT \log_e \frac{P_A}{[A]_{sol}} = -RT \log_e K \tag{14-18}$$

$$K = \frac{P_A}{[A]_{sol}} \tag{14-19}$$

The approximation that the activity in the gas phase is proportional to the pressure is, for low pressures, a good one. But the proportionality between concentration and activity in solution is generally not followed, except in dilute solution. The relationship between concentration and activity in the solution can be expressed by the equation

$$\mathscr{a}_{\mathrm{A}} = \gamma[\mathrm{A}]_{\mathrm{sol}} \tag{14-20}$$

In dilute solutions, γ approaches one. The equilibrium we are studying would follow the dotted line in Fig. 10-3 if γ were to remain equal to one in the more concentrated solutions. It is just because it does not that the equilibrium falls off along the solid line. The *true* equilibrium constant is given to a good approximation by the expression

$$K_{\mathrm{eq}} = \frac{P_{\mathrm{A}}}{\gamma[\mathrm{A}]_{\mathrm{sol}}} \tag{14-21}$$

The ratio of this equilibrium constant to the approximate one given by Eq. (14-19) is

$$\frac{K_{\mathrm{eq}}}{K} = \frac{1}{\gamma} \tag{14-22}$$

In dilute solutions, where $\gamma \rightarrow 1$, the two K's are equal. In the more concentrated solutions, the value for the activity coefficient is less than one, and the true equilibrium constant is greater than the approximate K. If the value for K_{eq} is obtained in very dilute solutions, where $\gamma \rightarrow 1$, then the values for γ at higher concentrations can be obtained from Eq. (14-22) by determining the apparent equilibrium constant and comparing it with the value for the dilute solutions.

The most important applications of activities to equilibrium problems are those dealing with solutions of electrolytes. The variation in activity coefficient with concentration for ionic solutes in water can be understood in terms of the Debye-Hückel theory (p. 290). In very dilute solutions, the ionic surroundings of any particular ion are not heavily populated; *i.e.*, the average distance between ions is large. Under these conditions, the ions behave independently and the activity coefficients are unity. With an increase in concentration, the surroundings of each ion become more heavily populated with other ions, and the "drag" effect mentioned earlier operates to reduce the activity. Since this effect becomes steadily more important as the concentration increases, the activity coefficients steadily decrease. The rate of decrease depends upon the size and charge of the ions, the charge being the more important factor. It is not possible to define (at least in most cases) separate activity coefficients for the cation and anion; instead, the mean activity coefficients for the two kinds of ions are obtained. The mean activity coefficients of some solutions are listed in Table 14-1. Note that the mean activity coefficient decreases much more rapidly with concentration for the 2:2 electrolytes ($ZnSO_4$, $CuSO_4$) than for the 1:1 electrolytes (NaOH, HCl, KCl).

The Debye-Hückel theory applies only to rather dilute solutions. In more concentrated solutions — starting at perhaps $0.1M$ for singly-charged ions (1:1 electrolytes) — other effects which are not included in the theory

Table 14-1 Mean activity coefficients of electrolytes at 25°C.

Molality	HCl	NaOH	KCl	$ZnSO_4$	$CuSO_4$	$CdCl_2$
0.0000	1.000	1.000	1.000	1.000	1.000	1.000
0.0001	0.9891	—	—	—	—	—
0.0002	0.9842	—	—	—	—	—
0.0005	0.9752	—	—	0.780	—	0.880
0.001	0.9656	—	0.965	0.700	0.762	0.819
0.002	0.9521	—	—	0.608	—	0.743
0.005	0.9285	—	0.927	0.477	—	0.623
0.01	0.9048	0.899	0.902	0.387	0.404	0.524
0.02	0.8755	0.860	0.869	0.298	0.320	0.456
0.05	0.8340	0.805	0.817	0.202	0.216	0.304
0.10	0.798	0.759	0.770	0.150	0.150	0.228
0.20	0.768	0.719	0.719	0.104	0.110	0.163
0.50	0.769	0.681	0.652	0.0630	0.067	0.100
1.00	0.811	0.667	0.607	0.0434	—	0.066
2.00	1.011	0.685	0.578	0.035	—	—

begin to operate. As the concentration increases, the activity coefficients begin to increase and eventually become greater than one.

It must be remembered when working equilibrium constant problems that *the substitution of concentrations for activities is an approximation.* In electrolyte solutions, especially, the error which results may be serious. It is not an easy matter to make activity coefficient corrections, especially in solutions containing a number of different electrolytes. For this reason, we usually content ourselves with using concentrations, keeping in mind at all times that if critical work were to be done, the corrections would have to be made.

14-3 Effect of External Variables on Chemical Equilibrium

The equilibrium constant expresses a mathematical relationship which must hold true between the activities of substances in a chemical reaction when the system is at equilibrium. In general, there is more than one set of activities which will satisfy the relationship. For example, in the reaction of $A \rightleftharpoons B$, for which $K = (A)/(B)$, all that is required is that the ratio of activities of A and B be equal to K. The values which (A) and (B) take on might be anything. If K were equal to 10, say, then (20)/(2), (30)/(3), or (73)/(7.3) would all satisfy the equilibrium constant expression. In special cases, because of the nature of the system, a unique set of activities is required to satisfy K. For example, if calcium carbonate is heated in a closed container, the following equilibrium is established.

$$CaCO_3(s) \rightleftharpoons CaO(s) + CO_2(g) \tag{14-23}$$

The equilibrium constant expression is

$$K = \frac{(CaO)(CO_2)}{(CaCO_3)} \tag{14-24}$$

where the quantities in parentheses indicate activities. But we have defined the activity of a pure solid substance to be one. If there are any $CaCO_3(s)$ and $CaO(s)$ present at all, therefore, the activity for each substance is one, and $K = CO_2(g) \approx P_{CO_2}$. Since there is only one activity on the right, this activity must be equal to K. For any given temperature the value of K is determined, and so is the pressure of CO_2 that can exist in equilibrium with CaO and $CaCO_3$. If we were to somehow slip another crystal of $CaCO_3$ or of CaO into a system at equilibrium without otherwise disturbing the system, there would be no change in the pressure of CO_2. If we tried to increase the CO_2 pressure by adding more of the gas, the excess would merely combine with CaO to give again the correct pressure of CO_2 at equilibrium. The equilibrium between liquid and gaseous states of a pure substance is another example of this type. The vapor pressure in equilibrium with liquid is a constant for a given temperature so long as the liquid is pure (Sec. 10-2). If the activity of the liquid is lowered by dissolving some solute, then the equilibrium vapor pressure is also lowered, so that K remains constant. Vapor pressure lowering was discussed in Chap. 10.

The parameters which can be varied to change the conditions under which a reaction system attains equilibrium are temperature, pressure, volume, and activities. The approach usually taken in analyzing the effect of a change in one of these variables is due to Le Chatelier. *Le Chatelier's principle* states that if one of these variables is changed for a system at equilibrium, the equilibrium shifts in such a way as to counteract the change, if this is possible. For example, if the pressure is increased, the equilibrium shifts so as to lead to a lower pressure, if a shift in one direction or the other can do this.

Let us consider the variables mentioned above in turn. What is the effect of change in temperature? If the temperature of the system is changed, the equilibrium constant itself is changed. If there is a net evolution of heat for the reaction in the forward direction (exothermic reaction), the equilibrium constant decreases in magnitude with increase in temperature. If the reaction is exothermic in the forward reaction, it is endothermic (heat-absorbing) in the reverse direction. Upon addition of heat to the system, the reaction proceeds more rapidly in the reverse direction, the direction corresponding to the absorption of heat. The equilibrium constant therefore decreases. It is possible to express the temperature dependence of the equilibrium constant in a more quantitative way, as shown in the following section. The above statements do not really explain the effect of temperature, but merely serve as a useful guide for qualitative prediction.

We have already encountered one example of the relationship between heat change in a reaction and the effect of temperature on the equilibrium constant, in the temperature dependence of solubility, Sec. 11-2. The equilibrium constant between a solid substance A and its solution can be written as

$$A(solid) \rightleftharpoons A(solution) \tag{14-25}$$

$$K = \frac{A(solution)}{A(solid)} \tag{14-26}$$

According to the rule just given, if the forward reaction in Eq. (14-25) is exothermic, the magnitude of the equilibrium constant will decrease with increasing temperature. But since the activity of solid A is a constant, this can only mean that the activity of A in solution decreases; *i.e.*, A becomes less soluble, as temperature increases. For most ionic substances dissolving in water, the forward reaction is endothermic, so that the equilibrium constant *increases* with increasing temperature, as shown in Fig. 11-2.

Consider now the effect of a change in pressure. According to Le Chatelier's principle, if the pressure is increased, the equilibrium shifts in the direction which leads to smaller volume, since this would counteract the increased pressure. For systems entirely in the liquid or solid state, the effects of pressure on equilibrium are very small for ordinarily attained pressures. If the pressure is made very high, however (*e.g.*, thousands of atmospheres), rather substantial changes may occur. For example, most liquids solidify on application of high pressure. This happens because most substances have a smaller molar volume in the solid than in the liquid state. When pressure is applied, the solid state then becomes relatively more stable.

When gases are involved, the equilibrium shifts in the direction of the smaller number of moles of gas when pressure is increased. For example, in the reaction

$$N_2 + 3H_2 \rightleftharpoons 2NH_3 \tag{14-27}$$

application of high pressure shifts the equilibrium toward ammonia. Another way of looking at the effect of pressure in such a system is to examine the equilibrium constant expression itself:

$$K = \frac{(NH_3)^2}{(N_2)(H_2)^3} = \frac{P^2_{NH_3}}{P_{N_2} \times P^3_{H_2}} \tag{14-28}$$

In this expression, we have approximated the activities by pressures. At any given temperature, pressure is a unit of concentration, as we can see by writing the ideal gas law in the form $P = (n/V)RT$. Pressure is merely moles per liter times a proportionality constant. Now if the pressure in the ammonia system were suddenly doubled with no shift in equilibrium, all of the individual pressures would suddenly be doubled. There would thus be a new factor 2^2 in the numerator, and $2^3 \times 2$ in the denominator. Thus, K cannot be a constant if there is no shift in equilibrium. In order that K remain constant, the ammonia pressure must change by more than two and the nitrogen and hydrogen pressures by less than two. There must, in other words, be a shift toward relatively more ammonia upon increase in pressure.

It is important to keep in mind that a change in pressure does not change the value for the equilibrium constant. The conditions of equilibrium may be shifted by pressure change, but the value for K remains the same.

Changes in the volume of a system may produce significant shifts in equilibrium only when gases are involved. Consider the ammonia system, Eq. (14-27). If the system is at equilibrium and the volume suddenly halved, the effect would be precisely the same as if the pressure of each reagent were suddenly doubled. Changes in volume can be handled by simply asking what effect the volume change will have on the pressures, and then proceeding as just described for pressure changes.

According to Le Chatelier's principle, a change in activity of one of the substances in a chemical equilibrium causes a shift in equilibrium in such a way that the activity change is counteracted. If the activity of one of the reactants is increased, for example, the equilibrium shifts in the direction of more product, to use up some of the increased activity of reactant. Thus, if hydrogen gas is added to the nitrogen-hydrogen-ammonia system at equilibrium, and there were no shift in equilibrium, the only effect would be an increase in the denominator of Eq. (14-28). Thus K would grow smaller, which is in contradiction to the requirement that K change only with temperature. In order that K not be changed, therefore, there must be a shift in equilibrium toward more ammonia. The magnitude of the shift in equilibrium would be a function of the activities of the various substances at equilibrium, and the amount of added hydrogen. Methods of solving such problems numerically are treated in more detail in the following chapter. We are concerned here only with the general notion of shift in equilibrium.

It is important to distinguish a change in activity from merely a change in the quantity of substance present. In the equilibrium system produced by the decomposition of calcium carbonate, Eq. (14-23), there are two solid phases, each possessing unit activity. The solid phase possesses unit activity as long as there is any quantity at all present. Addition of more solid does not change this. Remember that activity is a thermodynamic property akin to concentration. What we require is the "concentration" of calcium carbonate in calcium carbonate. This, of course, is a constant which is not changed by merely adding more calcium carbonate. From these considerations, we see that it should be possible to arrive at the equilibrium condition by beginning from widely different starting conditions. We might, for example, begin with calcium oxide and excess carbon dioxide gas. The gas would combine with calcium oxide until the carbon dioxide activity reached the equilibrium value. On the other hand, one might begin with pure calcium carbonate, or with a mixture of calcium carbonate and calcium oxide. It is usually possible, in any reaction system, to begin the approach to equilibrium from a variety of starting conditions. The relative quantities of substances remaining at equilibrium will vary, depending upon what there was to begin with. The equilibrium constant expression must be satisfied, however, in every case, since K is independent of the starting conditions.

Units — The equilibrium constant given by Eq. (14-1) is a dimensionless quantity. The concentration units in the numerator cancel those in the denominator, provided that all concentrations are expressed in the same units, so that the magnitude of K is independent of the concentration units employed. A cancellation of the concentration units occurs only when the total of the exponents in the numerator equals that in the denominator. In all other instances, the equilibrium constant possesses dimensions, and its numerical value depends upon the concentration units employed. For example, in the ammonia system, Eq. (14-28), if the concentrations are expressed in units of atmospheres of pressure, K has the units of $1/\text{atm}^2$. If concentrations are expressed in units of moles/liter, K has the units of $\text{liter}^2/\text{mole}^2$. Of course, the numerical value would be different in the two cases. It is therefore important that the units in which K is expressed be stated.

Example 14-1

The equilibrium constant of Eq. (14-5) is $1.84 \cdot \text{liter}^2/\text{mole}^2$ at 350°C. What is the magnitude of K in units of $1/\text{atm}^2$?

The factor necessary to convert from moles/liter to atmospheres of pressure is easily obtained from the ideal gas law.

$$P = \left(\frac{n}{V}\right)RT$$

n/V has the dimensions of moles/liter. Therefore, to convert from moles/liter to atm one needs to multiply by RT. Since we wish, in this example, to convert from $(\text{liters/mole})^2$ to $1/\text{atm}^2$ we need to multiply K by $1/(RT)^2$.

$$RT = 0.08206 \times 623 = 51.1; \qquad 1/RT = 0.0196$$

$$\begin{aligned} K_p &= K_c \times 1/(RT)^2 \\ &= 1.84 \times (0.0196)^2 \\ &= 7.09 \times 10^{-4}/\text{atm}^2 \end{aligned}$$

Effect of catalysts — A catalyst might affect the rate at which a system approaches chemical equilibrium; it cannot, however, affect the position of equilibrium. It is easy to see that this must be so from thermodynamic considerations. A catalyst affects the reactants in the course of their reaction, but does not alter the reactants or products themselves to a significant degree. It therefore cannot alter the activities of these substances. Since the activities of the substances entering into the equilibrium are not affected, the equilibrium condition itself remains unchanged. From a kinetic point of view, the catalyst operates to lower the activation energy for reaction, or to increase the probability of reaction in some other way, as by changing the steric factor (Chap. 13). The effect of the catalyst on the forward reaction must, however, be matched by its effect on the reverse reaction, since we require that the path taken by the reactants in forming products be reversible. In other words, whatever a catalyst can do for the forward reaction it must also do for the reverse reaction, which is like a movie film run backwards.

14-4 Effect of Temperature on the Equilibrium Constant

To understand how temperature affects the value of the equilibrium constant, let us consider the hydrogen iodide system discussed at the beginning of the chapter. Assuming that this is a simple bimolecular process, the equilibrium constant is given by the ratio of the forward to reverse reaction rates, Eq. (14-1). Figure 14-1 illustrates the energy changes which occur as the reactants form the activated complex, and then the products.

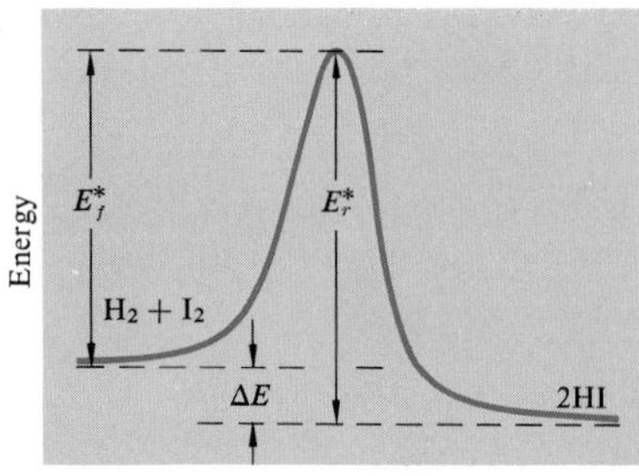

Fig. 14-1 Energy diagram for the hydrogen-iodine-hydrogen iodide system. Note that the activation energy for the reverse reaction is greater than that for the forward reaction.

It will be noted that the energy of the products is lower than that of the reactants. $\Delta E = E_f^* - E_r^*$ is equal to the energy evolved in the forward reaction. (This is in accord with the usual convention in that, if energy is evolved, ΔE is negative in sign.) The reverse reaction — the formation of H_2 and I_2 from HI — is endothermic by an amount ΔE.

The rate constant for the forward reaction can be written as in Eq. (13-21).

$$k_f = A_f e^{-E_f^*/RT}$$

Similarly, the reverse reaction rate constant is written

$$k_r = A_r e^{-E_r^*/RT}$$

A_f and A_r have different values for the two processes, but we do not need to know these. At equilibrium, K is equal to k_f/k_r. Lumping the constants together,

$$k_f/k_r = K = Be^{-(E_f^*-E_r^*)/RT}$$
$$K = Be^{-\Delta E/RT} \qquad (14\text{-}29)$$

where B is a constant.

It is clear from Eq. (14-29) that the equilibrium constant depends upon the value of ΔE, which is essentially independent of temperature. If the forward reaction is exothermic (ΔE is negative) the value for K will in general be large, whereas if the opposite is true (Fig. 14-2) K will tend to be small. It must be remembered that K is determined also by the factors which go into making up the pre-exponential term, B, but we are concerned here with the temperature dependence of K rather than with its magnitude. B is essentially temperature independent over small temperature ranges.

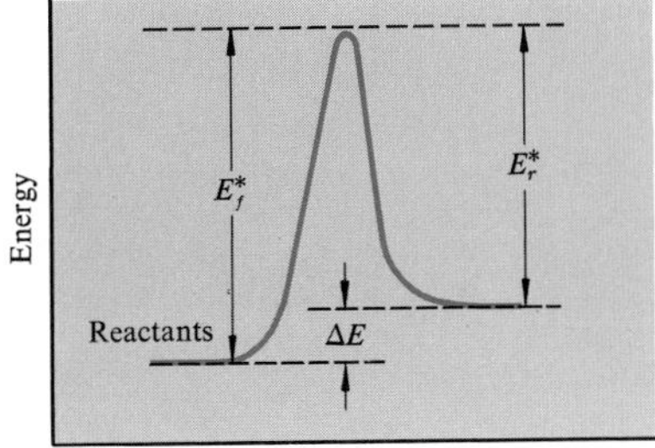

Fig. 14-2 Energy diagram for a reaction in which the activation energy for the reverse reaction is less than that for the forward reaction.

If the reaction is being studied under conditions of constant pressure rather than constant volume, ΔE in Eq. (14-29) should be replaced by ΔH. We then have

$$K = Be^{-\Delta H/RT}$$

Taking $\log_{10}$ of both sides,

$$\log_{10} K = \frac{-\Delta H}{2.3\ RT} + \log_{10} B \qquad (14\text{-}30)$$

From this relationship, it is possible to determine ΔH for a reaction system by graphing $\log_{10} K$ *vs* $1/T$. The result should be a line with slope $-\Delta H/2.3\ R$. We see also from this equation that if ΔH is negative, log K must decrease as T increases. This is in accord with the rule discussed earlier, that the equilibrium constant for an exothermic reaction decreases with increasing temperature. Conversely, if ΔH is positive, then log K increases with increasing temperature.

Example 14-2

The equilibrium constant

$$K = \frac{[HI]^2}{[H_2][I_2]}$$

has the following values at various temperatures:

Temp (°K)	667	731	764
K	59.8	48.8	45.4

what is the value of ΔH?

From Eq. (14-30),

$$\log_{10} K = -\Delta H/2.3RT + \log_{10} B$$

This is the equation of a straight line with $\log_{10} K$ as one variable and $1/T$ as the other. The slope of the line is $-\Delta H/2.3R$. Graphing the above data in this way,

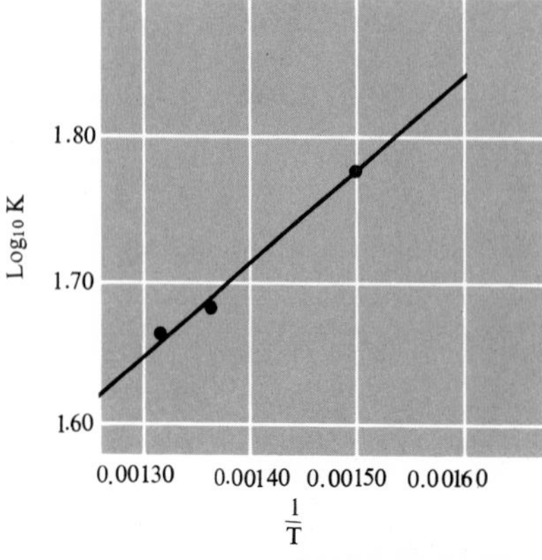

we find

$\log_{10} K$	1.777	1.688	1.657
$1/T$	1.50×10^{-3}	1.37×10^{-3}	1.31×10^{-3}

$$\text{Slope} = 700$$

$$-\Delta H = 2.3R \times 700 = 2.3 \times 1.987 \times 700$$

$$\Delta H = -3210 \text{ cal/mole}$$

The result is, then, that the products are formed from hydrogen and iodine with the evolution of about 3 kcal/mole. The reaction is thus exothermic. The value of the equilibrium constant decreases with increasing temperature in accordance with the rule given earlier.

Exercises

14-1. What relationship exists between the equilibrium constant for a given reaction and the mechanism by which the reaction proceeds?

14-2. (a) Write the equilibrium constant expression for each of the following gas-phase reactions:

1. $SO_3 + H_2 \rightleftharpoons SO_2 + H_2O$.
2. $4NH_3 + 5O_2 \rightleftharpoons 4NO + 6H_2O$.

(b) State the units in which each of the equilibrium constants is expressed if the gas concentrations are given in units of atmospheres of pressure.

14-3. The reaction $3H_2 + N_2 \rightarrow 2NH_3$ is exothermic. How does the equilibrium constant for this reaction vary with temperature?

14-4. Write equilibrium constant expressions for each of the following reactions:
(a) $Fe_2(SO_4)_3(s) \rightleftharpoons Fe_2O_3(s) + 3SO_3(g)$
(b) $Cl_2O(g) + S(s) \rightleftharpoons SOCl_2(l)$
(c) $2H_2S(g) + SO_2(g) \rightleftharpoons 2H_2O(g) + 3S(s)$
(d) $H_2S + I_2 \rightleftharpoons 2HI + S$ (all reactants in solution)

14-5. A gaseous mixture of boron trifluoride, ammonia, and boron trifluoride-ammonia addition compound is at equilibrium with respect to the reaction

$$BF_3(g) + NH_3(g) \rightleftharpoons BF_3NH_3(g)$$

Complete the table below by inserting "increase," "decrease," or "no change" for each pressure listed across the top, and for each of the changes listed in the vertical row, all made at constant temperature.

Change	P_{BF_3}	$P_{BF_3NH_3}$	$(P_{BF_3NH_3}/P_{BF_3})$
Decrease in total volume of system			
Addition of argon at constant volume			
Addition of ammonia at constant volume			

14-6. The reaction $3PCl_5 + P_2O_5 = 5POCl_3$ can be caused to proceed to equilibrium by heating the substances together in a sealed tube. Describe two different compositions, not involving the same substances, which might be placed in the tube before sealing.

14-7. Derive the expression for the equilibrium constant for the reaction of Eq. (14-23) from thermodynamic principles. You must use Eq. (14-14) and proceed through the development as outlined in the text.

14-8. Write the equilibrium constant for the reaction

$$SO_2(g) + \tfrac{1}{2}O_2(g) \rightleftharpoons SO_3(g)$$

Write the equilibrium constant for the reaction

$$2SO_2(g) + O_2(g) \rightleftharpoons 2SO_3(g)$$

Will these two expressions yield the same values for the concentrations of the various gases involved in the equilibrium? Explain.

14-9. The solubility of gaseous sulfur dioxide in water at various temperatures when the SO_2 pressure is one atmosphere is as follows:

Temperature	Solubility*
10°	15.4
25°	8.96
30°	7.7
50°	4.2

*$gSO_2/100gH_2O$

From these data calculate the heat of solution of SO_2 in water.

14-10. Draw a diagram similar to that in Fig. 14-1 for a system which is uncatalyzed, as compared with one which contains a catalyst which lowers the activation energy. Is the equilibrium constant changed by the catalyst?

14-11. Consider the following equilibrium system:

$$PCl_3(g) + Cl_2(g) \rightleftharpoons PCl_5(g)$$

The reaction is endothermic as written. What is the effect on the equilibrium (shift to the right, to the left, or no change) of each of the following changes in the equilibrium system?
(a) Increase in total pressure.
(b) Addition of equal molar quantities of PCl_3 and PCl_5 at constant volume.
(c) Addition of chlorine gas at constant volume.
(d) Doubling of the volume.
(e) Temperature increase.

14-12. From the data given in Sec. 14-4, predict the equilibrium constant for the hydrogen iodide system at 800°K.

14-13. A given solute A is dissolved in water, and its vapor pressure over the solvent measured at a number of concentrations. From the following data, determine the activity coefficient of A at 0.45, 0.55, and 0.60*M* concentrations.

Concentration (*M*)	Pressure (mm)
0.05	0.77
0.10	1.52
0.20	2.92
0.30	4.10
0.40	5.00
0.50	5.77
0.60	6.35

14-14. Experiments are performed to determine the distribution of bromine between water and carbon tetrachloride

$$Br_2\ (water) \rightleftharpoons Br_2\ (CCl_4)$$

when the two immiscible liquids are in contact with one another. It is found that the value obtained for K varies slightly with the quantity of bromine employed. What is the probable reason for this variation? Under what conditions will the best approximation to the thermodynamic value for K be obtained?

14-15. Sulfur dioxide reacts with oxygen gas to give sulfur trioxide, as expressed by the equations of Prob. 14-8. At 1000°K, the equilibrium concentrations of the three substances in one mixture were found to be 0.562 atm of SO_2, 0.101 atm O_2, and 0.332 atm SO_3. Calculate the value for K for each of the balanced equations in Prob. 14-8. What relation exists between the numerical values of the two constants?

14-16. The Debye-Hückel limiting law, an outcome of the Debye-Hückel theory of interionic attraction, states that the log of the mean activity coefficient γ for an electrolyte should vary with the ionic strength according to the equation

$$\log_{10} \gamma = -\text{constant}\ z_+ z_- \mu^{1/2}$$

where z_- and z_+ represent the charges on the anion and cation, respectively. For example, for NaCl, $z_- = 1$, $z_+ = 1$. The ionic strength μ is given by the expression

$$\mu = \tfrac{1}{2}(c_1 z_1^2 + c_2 z_2^2 + c_3 z_3^2 + \ldots)$$

where the sum is taken over *all* of the ions in the solution. c represents the molarity of each ionic species, and z the charge on that species. Test the Debye-Hückel limiting law, using the data in Table 14-1 for KCl, $ZnSO_4$, and $CdCl_2$ solutions to 0.02 m. Since the solutions are quite dilute, the molality may be substituted for molarity in computing the ionic strength. Two distinct tests may be applied:
(a) Does $\log_{10} \gamma$ vary linearly with $\mu^{1/2}$?
(b) If test (a) is satisfactory, the slopes of the lines obtained should vary with the magnitude of the $z_+ z_-$ term, since the constant in the expression is supposed to be the same for all electrolytes.

Chapter 15

Chemical Equilibrium II. Applications

The applications of the principles of chemical equilibrium form a very important part of modern chemistry. In this chapter, the equilibria in a number of chemical systems will be treated in quantitative fashion. We will show how the form of the equilibrium constant and its magnitude are determined. The uses to which the equilibrium constant can be put, once its magnitude is known, will also be illustrated. It is assumed throughout that concentrations can be employed to represent activities; the limitations of this assumption are discussed in the preceding chapter.

Homogeneous Equilibria

15-1 Gaseous Equilibria

A homogeneous equilibrium is one in which all of the participating substances are present in the same phase, be it gaseous, liquid, or (rarely) solid. We will consider first the now familiar system for which the equilibrium expression is

$$H_2 + I_2 \rightleftharpoons 2HI$$

It is clear from the chemical equation that the equilibrium constant for this system is

$$K = \frac{[HI]^2}{[H_2][I_2]} \tag{15-1}$$

Equilibrium in this system can be reached by beginning with mixtures of hydrogen and iodine, with hydrogen iodide, or with a mixture of all three substances. The data in Table 15-1 were collected at 699°K; the starting materials were placed in glass tubes which were sealed and allowed to stand at the equilibrium temperature for up to thirty hours. At the end of this period, the tubes were rapidly cooled and the contents were analyzed. Since the reaction proceeds very slowly at lower temperatures, the system is effectively "frozen" in its equilibrium condition by the rapid cooling. The first five sets of data apply to tubes in which hydrogen and iodine were used as the starting materials. In the last three, hydrogen iodide alone was used.

Table 15-1 The hydrogen-iodine-hydrogen iodide equilibrium at 699°K.

H_2(moles/liter)	I_2(moles/liter)	HI(moles/liter)	K
1.831×10^{-3}	3.129×10^{-3}	17.67×10^{-3}	54.49
2.243×10^{-3}	2.336×10^{-3}	16.85×10^{-3}	53.97
2.907×10^{-3}	1.707×10^{-3}	16.48×10^{-3}	54.73
3.560×10^{-3}	1.250×10^{-3}	15.59×10^{-3}	54.62
4.565×10^{-3}	0.738×10^{-3}	13.54×10^{-3}	54.49
0.479×10^{-3}	0.479×10^{-3}	3.53×10^{-3}	54.4
0.495×10^{-3}	0.495×10^{-3}	3.66×10^{-3}	54.6
1.141×10^{-3}	1.141×10^{-3}	8.41×10^{-3}	54.4

From the excellent degree of constancy of the value calculated for K from use of Eq. (15-1), we may be sure that this expression does, in fact, represent the only equilibrium of any importance in the system. The constancy of the calculated values for K also provides assurance that equilibrium conditions were reached, particularly since the equilibrium point was approached from both directions. It should perhaps be noted, as a reminder, that the equilibrium constant for this particular system is a dimensionless quantity, so that the same value for K would have been obtained if the concentrations had been expressed in some other units, such as atmospheres of pressure.

With a knowledge of the magnitude of K, a variety of problems can be solved.

Example 15-1

Hydrogen gas at a pressure of 25 mm is allowed to mix with iodine gas at a pressure of 25 mm at 699°K. Calculate the concentration of hydrogen iodide at equilibrium, and the total pressure in the system.

Let the number of mm of hydrogen iodide pressure at equilibrium be x. Then, since two moles of hydrogen iodide are produced from the reaction of one mole of hydrogen and one mole of hydrogen iodide, we have for the pressures of all three at equilibrium

H_2	I_2	2HI
$25 - x/2$	$25 - x/2$	x mm

$$54.5 = \frac{x^2}{(25 - x/2)(25 - x/2)}$$

Solving the quadratic equation,

$$x = 39.34 \text{ mm} = \text{pressure of HI}$$

We could compute the total pressure by summing up the pressures of the three gases at equilibrium:

$$(25 - x/2) + (25 - x/2) + x$$

Of course, since the loss of one molecule of hydrogen and one molecule of iodine results in the production of two molecules of hydrogen iodide, the pressure is not dependent on the equilibrium, so that we know before solving for x that the total pressure must be just 50 mm.

Example 15-2

The equilibrium constant for the dissociation of hydrogen sulfide at 1065°C is 0.0118 atm. 0.112 mole of hydrogen sulfide is confined to a volume of 12.0 liters and heated to this temperature. What is the per cent of dissociation, and what is the total pressure?

The reaction is

$$2H_2S(g) \rightarrow 2H_2(g) + S_2(g)$$

Note that, at this high temperature, sulfur is a diatomic gas.

$$K = \frac{[S_2] \times [H_2]^2}{[H_2S]^2} = 0.0118 \text{ atm}$$

We first compute the initial pressure of the hydrogen sulfide under the conditions of the reaction, assuming no dissociation. From the ideal gas law,

$$P = \frac{0.112 \times 0.0821 \times 1338}{12.0} = 1.024 \text{ atm}$$

Let the pressure of the sulfur gas produced in the reaction be x atm. Then, from the balanced equation we see that the pressure of hydrogen gas must be $2x$, and the *loss* of pressure of the hydrogen sulfide must also be $2x$.

$$K = \frac{x(2x)^2}{(1.02 - 2x)^2} = 0.0118 \text{ atm}$$

This expression leads to a cubic equation in x which may be solved by a number of methods. The simplest procedure is to obtain an approximate value for x, and then determine a more exact solution by successive approximations. If we assume that the extent of the decomposition will be small, we can drop the $2x$ term out of the denominator. This leaves

$$0.0118 = 4x^3/1.02$$
$$x^3 = 0.00310 = 3.10 \times 10^{-3}$$
$$x = 0.146$$

If we now use this approximate value for x in the denominator and solve for a new x, we obtain

$$0.0118 = 4x^3/(1.02 - 0.29)^2$$
$$x = 0.116$$

Using this new value for x in the denominator, we have

$$0.118 = 4x^3/(1.02 - 0.23)^2$$
$$x = 0.122$$

This value is fairly close to the exact solution. It will be appreciated that we could obtain any desired degree of accuracy by this *method of successive approximations*, simply by applying it often enough.

Since $2x = 0.244$ atm is the amount of hydrogen sulfide which dissociated, the per cent dissociation is given by $(0.244/1.024) \times 100 = 23.8$ per cent. The total pressure is given by

$$\begin{array}{l} \quad 2H_2S \qquad 2H_2 \quad S \\ P^0 - 2x + 2x + x = P^0 + x = 1.024 + 0.122 = 1.146 \text{ atm} \end{array}$$

This problem differs from that of Example 15-1 in that the number of moles of gas in the products is not the same as that in the reactants. The final pressure is therefore dependent upon the extent of the reaction. In fact, the total pressure of the system might be measured to determine the extent of dissociation, and from this to calculate the equilibrium constant.

It is possible to evaluate the equilibrium constant from a knowledge of the standard free energies of the substances involved in a chemical reaction. As an example, consider the reaction

$$3H_2(g) + N_2(g) = 2NH_3(g)$$

From Eq. (14-13), we have

$$\log_{10} K_{eq} = \frac{-\Delta G^\circ}{2.3RT}$$

To obtain the value for the equilibrium constant of this reaction, we need to know the standard free energy of each substance in the balanced equation. These values are tabulated for a great many substances; the best single source is the National Bureau of Standards Circular No. 500. It is necessary to find the standard free energy for only one substance in this example, however, because *the standard free energy of pure elements in* *their standard states* at any temperature is defined to be zero.* The standard free energy of formation of ammonia at 25°C is listed as −3980 cal/mole. Since we require two moles from the balanced equation,

$$\Delta G^\circ = 2 \times (-3980) - 0 = -7960 \text{ cal}$$

$$\log_{10} K = \frac{7960}{2.303 \times 1.987 \times 298.2} = 5.835$$

$$K = 6.85 \times 10^5$$

Since the standard state for gases is defined in terms of atmospheres pressure, the units appropriate to K are atm^{-2}; 6.85×10^5 is the value of the equilibrium constant at 298°K, (25°C). To determine K at some other temperature in the same manner, we would need to know the free energy of formation of NH_3 at that temperature. The standard free energies of formation of some substances at 25°C are listed in Appendix E.

15-2 Solutions of Weak Acids

In the chapter on acids and bases it was pointed out that many acids, of which acetic acid is typical, react with water only to a slight extent. Strong acids such as hydrogen chloride, on the other hand, react with water almost completely. This means that the equilibrium constant for the reaction

$$HCl + H_2O \rightarrow Cl^- + H_3O^+ \tag{15-2}$$

is very large, and may in fact be assumed to be infinitely large. The equation representing the reaction of acetic acid with water is

$$HOAc + H_2O \rightarrow H_3O^+ + OAc^- \tag{15-3}$$

where, for simplicity, we use the symbol OAc^- for the acetate ion, CH_3COO^-. The equilibrium constant for this reaction is

$$K = \frac{[OAc^-][H_3O^+]}{[HOAc][H_2O]} \tag{15-4}$$

Interest in this type of equilibrium generally applies to solutions which are perhaps $0.1M$, rarely above $1M$. In a liter of pure water there is approximately 1000 g. The number of moles per liter of water is then $1000/18 = 55.5$. In a $0.1M$ solution this number will be very little different; even in a $1M$ solution of acetic acid it is still 52.2 mole/l. We can therefore, to a good degree of approximation, assume the concentration of water to be constant. Multiplying through by this constant, and making it part of the equilibrium constant, we have

$$K_{eq} \times [H_2O] = K_a = \frac{[OAc^-][H_3O^+]}{[HOAc]} \tag{15-5}$$

---

*The standard states for pure solids and liquids are the pure substances under 1 atm pressure. The standard state for a gas is defined to be 1 atm pressure.

K_a is called the *acid dissociation constant*. It is this value which is tabulated in handbooks, *etc*., for weak acids. The values of K_a for a number of weak acids are listed in Table 15-2. With these values, one can calculate the hydronium ion concentration under various conditions. Before the use of these equilibrium constants in calculation of hydronium ion concentration is illustrated, we must first consider the equilibrium occurring in the pure solvent.

Table 15-2 Dissociation constants of some weak acids in water at 25°C.

Acid	K_a	Ionization equilibrium
Acetic	1.76×10^{-5}	$CH_3COOH + H_2O \rightarrow CH_3COO^- + H_3O^+$
Benzoic	6.5×10^{-5}	$C_6H_5COOH + H_2O \rightarrow C_6H_5COO^- + H_3O^+$
Phenol	1.3×10^{-10}	$C_6H_5OH + H_2O \rightarrow C_6H_5O^- + H_3O^+$
Iodic	1.69×10^{-1}	$HIO_3 + H_2O \rightarrow IO_3^- + H_3O^+$
Nitrous	5.1×10^{-4}	$HNO_2 + H_2O \rightarrow NO_2^- + H_3O^+$
Hydrocyanic	4.9×10^{-10}	$HCN + H_2O \rightarrow CN^- + H_3O^+$
Formic	1.7×10^{-4}	$HCOOH + H_2O \rightarrow HCOO^- + H_3O^+$
Hydrofluoric	6.6×10^{-4}	$HF + H_2O \rightarrow F^- + H_3O^+$

15-3 The Autoionization of Water; pH

The equation which represents the autoionization of water is written

$$H_2O + H_2O \rightarrow H_3O^+ + OH^-$$

The corresponding equilibrium constant is

$$K = \frac{[H_3O^+][OH^-]}{[H_2O]^2} \tag{15-6}$$

From measurements of the electrical conductivity of very pure water, it is possible to determine the concentrations of hydronium and hydroxyl ions. At 24°C the concentration of each of these ions is 1×10^{-7} mole/l. The concentration of pure water in water is, as previously discussed, 55.5 mole/l.

$$K = \frac{(1 \times 10^{-7})^2}{(55.5)^2}$$

Since the denominator in this expression is effectively a constant under all conditions of interest, we may multiply through by it, and thus obtain

$$K_w = [H_3O^+][OH^-] = 1 \times 10^{-14} \tag{15-7}$$

K_w is called the *ion product for water*. It is a very important quantity, because it represents a condition which must exist in water and in *all* dilute aqueous solutions: The product of the hydronium and hydroxyl ion concentrations is a constant, 10^{-14}. This is true even in solutions of acids or bases in water. In pure water, the concentrations of hydronium and hydroxyl ions are equal; if an acid is added the hydronium ion concentration may increase tremendously, but the product of the hydroxyl and hydronium ion concentrations is still 10^{-14}.

Example 15-3

What are the concentrations of hydroxyl and hydronium ions in a $0.1M$ solution of hydrochloric acid?

Since hydrochloric acid is a strong acid, we know that the reaction represented by Eq. (15-2) goes essentially to completion. If the concentration of HCl is $0.1M$ therefore, the final concentration of Cl^- and H_3O^+ must be $0.1M$ each, since every HCl molecule is responsible for the formation of one chloride ion and one hydronium ion. Now, using Eq. (15-7),

$$K_w = 10^{-14} = 10^{-1} \times [OH^-]$$
$$[OH^-] = 10^{-13}$$

It is possible to obtain a simpler expression by taking the logarithm of both sides of Eq. (15-7) and multiplying through by -1.

$$-\log_{10}(10^{-14}) = -\log_{10}[H_3O^+] - \log_{10}[OH^-] \qquad (15\text{-}8)$$

We use base 10 logs in this equation, and all future discussion will be based on the use of this log scale. We define a new symbol pX, where p indicates the negative $\log_{10}$ of the quantity X. Applying this to the above equation, we have

$$14 = pH + pOH \qquad (15\text{-}9)$$

pH is the symbol universally used to represent $-\log_{10}[H_3O^+]$. As an example of the use of this equation, consider the two cases we have just discussed — pure water and a $0.1M$ solution of HCl. In the first of these, $[H_3O^+]$ and $[OH^-]$ are equal, both being 10^{-7}. We have therefore, $pH = pOH = -\log_{10}[10^{-7}] = 7$. In the $0.1M$ HCl solution $[H_3O^+]$ is 10^{-1}; pH is therefore 1, and

$$14 = 1 + pOH$$
$$pOH = 13$$

From these two examples we can see that the pH of a solution which contains more hydronium ion than hydroxyl ion is less than 7, while in

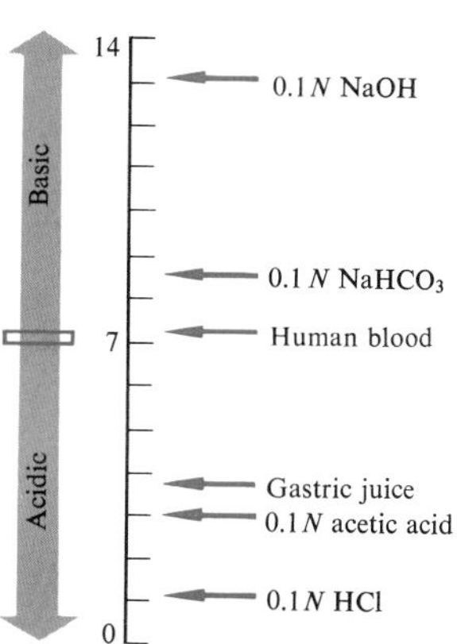

Fig. 15-1 A selection of representative pH values for aqueous solutions.

solutions containing an excess of hydroxyl ions the pH is greater than 7. The pH's of some solutions are indicated on the scale in Fig. 15-1. Since the pH is a quantitative measure of acidity of aqueous solutions, it is universally employed in medicine, biology, and chemistry.

Example 15-4

Returning to the consideration of weak acids, we now calculate the hydronium ion concentration and pH of a $0.1M$ solution of acetic acid. Using Eq. (15-5) and the value of K_a from Table 15-2, we have

$$1.76 \times 10^{-5} = \frac{[OAc^-][H_3O^+]}{[HOAc]} \qquad (15\text{-}10)$$

We will let x represent the concentration of acetate ion at equilibrium. Since each time an acetate ion is produced by the ionization of an acid molecule a hydronium ion is also produced, the concentration of hydronium ions which come from the ionization must also be x. There is a second source of hydronium ions which we must consider; *i.e.*, those produced by ionization of the solvent. However, we can ignore this source of hydronium ions, since it should be very small in comparison with that produced from the ionization of acetic acid. The justification for this must come with our final answer; we will have to be able to show that at least one hundred times as many hydronium ions are produced from the acid as are produced from ionization of the solvent.

The concentration of un-ionized acetic acid molecules at equilibrium is the initial concentration of the acid, less that which has been ionized. If we call the concentration of acetate ions produced from the ionization x, then the concentration of acetic acid molecules which have ionized is also x, since from the balanced equation there is a one-to-one correspondence between these two quantities. We have then

$$[OAc^-] = x$$
$$[H_3O^+] = x$$
$$[HOAc] = 0.1 - x$$

Inserting these in the equation, we have

$$\frac{x^2}{0.1 - x} = 1.76 \times 10^{-5}$$

This leads to a quadratic equation in which x is easily solved. There is a shortcut, however, which makes the problem even simpler. From the magnitude of the dissociation constant we can surmise that the percentage of acetic acid molecules which will dissociate is very small. This is equivalent to saying that x is small in comparison with 0.1. If we ignore the x which is to be subtracted from 0.1 in the denominator, we will therefore introduce a small error, and can write as a good approximation

$$\frac{x^2}{0.1} = 1.76 \times 10^{-5}$$
$$x^2 = 1.76 \times 10^{-6}$$
$$x = 1.33 \times 10^{-3}$$

Since $x = [H_3O^+]$,

$$pH = -\log [1.33 \times 10^{-3}]$$
$$= 3 - \log 1.33 = 2.87$$

$1.33 \times 10^{-3} M$ represents the concentration of acetate and hydronium ions produced by the ionization of the acid. It is indeed a large number in comparison with the hydronium ion concentration which is produced by ionization of water. Note, however, that it is a *small* number in comparison with 0.1, so our assumption that only a small fraction of acid molecules is ionized is satisfactory.

15-4 Solutions of Weak Bases

An entirely analogous approach is employed in solving problems involving bases in water. Strong bases such as sodium hydroxide — which are ionic substances — are assumed to be completely dissociated in water, so that the concentration of hydroxyl ion follows directly from the concentration of the salt.

Example 15-5

What is the pH of a $0.01M$ barium hydroxide solution?

The formula for barium hydroxide is $Ba(OH)_2$. In a $0.01M$ solution, there is, therefore, 0.02 mole/l of hydroxyl ion.

$$14 = pH + pOH = pH + (-\log 2 \times 10^{-2})$$
$$pH = 14 - 1.7 = 12.3$$

Weak bases, of which ammonia is typical, react with water to only a limited extent:

$$NH_3 + H_2O \rightarrow NH_4^+ + OH^-$$

$$K = \frac{[NH_4^+][OH^-]}{[NH_3][H_2O]} \tag{15-11}$$

Again, by removing the concentration of water from the denominator we obtain what is called the base dissociation constant K_b:

$$K_b = \frac{[NH_4^+][OH^-]}{[NH_3]}$$

The values of K_b for a few weak bases are listed in Table 15-3.

Table 15-3 Dissociation constants of some weak bases in water at 25°C.

Base	K_b	Ionization equilibrium
Ammonia	1.79×10^{-5}	$NH_3 + H_2O \rightarrow NH_4^+ + OH^-$
Diethylamine	9.6×10^{-4}	$(C_2H_5)_2NH + H_2O \rightarrow (C_2H_5)_2NH_2^+ + OH^-$
Pyridine	1.71×10^{-9}	$C_5H_5N + H_2O \rightarrow C_5H_5NH^+ + OH^-$
Aniline	4.0×10^{-10}	$C_6H_5NH_2 + H_2O \rightarrow C_6H_5NH_3^+ + OH^-$

Pyridine

Aniline

Example 15-6

What is the pH of a 0.05*M* solution of pyridine?

$$C_5H_5N + H_2O \rightarrow C_5H_5NH^+ + OH^-$$

$$K_b = \frac{[C_5H_5NH^+][OH^-]}{[C_5H_5N]} = 1.71 \times 10^{-9}$$

We make the following two assumptions, analogous to the two made in solving the acetic acid problem: (*a*) The amount of OH^- ion produced by the dissociation is large in comparison with that produced by ionization of water. (*b*) The fraction of pyridine which is ionized is very small. Then

$$[C_5H_5NH^+] = x$$

$$[OH^-] = x$$

$$[C_5H_5N] = 0.05 - x \approx 0.05$$

$$1.71 \times 10^{-9} = x^2/5 \times 10^{-2}$$

$$x^2 = 8.55 \times 10^{-11} = 85.5 \times 10^{-12}$$

$$x = 9.25 \times 10^{-6}$$

Since $x = [OH^-]$,

$$\text{pOH} = -\log(9.25 \times 10^{-6})$$

$$= 6 - \log 9.25 = 5.03$$

$$\text{pH} = 14 - 5.03 = 8.97$$

15-5 Common Ion Effect

We have so far considered solutions in which the weak acid or base is the only ionizing substance in the solution. If a salt which contains an anion or cation in common with the ionized acid or base is also present, it exerts a strong effect on the dissociation equilibrium. Consider as an example a 0.1*M* acetic acid solution which also contains 0.1*M* sodium acetate. The problem is to be solved as before: The concentration of hydronium ion produced by ionization of the acid is x. The concentration of acetate ions is the sum of the concentration produced by ionization of the acid — which is x — and that present from the sodium acetate. Since the salt is ionic, it is completely ionized in solution. The concentration of acetate ions from this source is therefore 0.1*M*. We have, then,

$$[H_3O^+] = x$$
$$[OAc^-] = x + 0.1$$
$$[HOAc] = 0.1 - x$$

We may examine Eq. (15-3) to learn, in a qualitative sense, what the effect of the added acetate ion on the equilibrium will be. A higher concentration of acetate ions increases the probability of a recombination of acetate and hydronium ions, and shifts the equilibrium to the left. We can say then that the fraction of acetic acid molecules which is ionized will be lower in this case than it is when only the acid is present. If it was permissible to ignore the x in comparison with 0.1 previously, it certainly is permissible now. Furthermore, we can say that the amount of acetate ion which is produced by the ionization of acetic acid is likely to be small in comparison with the quantity which is already present from the sodium acetate. We have, then,

$$\frac{x(0.1 + x)}{0.1 - x} = 1.76 \times 10^{-5} \approx \frac{0.1x}{0.1}$$
$$x = 1.76 \times 10^{-5}$$

Since x is the hydronium ion concentration, $\text{pH} = -\log 1.76 \times 10^{-5} = 4.76$. This is a much higher pH than that obtained for a $0.1M$ solution of acetic acid alone (Example 15-4). It is clear, then, that the addition of acetate ion has shifted the equilibrium to the left. A similar problem and method of solution applies in considering solutions of bases in which a salt with a common cation is present; *e.g.*, a solution containing ammonia and ammonium chloride.

15-6 Hydrolysis Equilibria

The common ion effect which acetate ion exerts on the dissociation of acetic acid takes on a somewhat altered aspect when considered from the viewpoint of the Brönsted acid-base theory. The acetate ion is a fairly good base, in the sense that it is a proton acceptor. Adding this basic ion to a solution of an acid — acetic acid — *should* and does result in a lowering of the hydronium ion concentration.

One might also expect that addition of sodium acetate to pure water would also result in a lowering of the hydronium ion concentration, or — what is the same thing — an increase in pH. This effect was discussed qualitatively in Chap. 12. We will now investigate it from a quantitative point of view.

The reaction of acetate ion with water is of the form

$$OAc^- + H_2O \rightarrow HOAc + OH^- \tag{15-12}$$

$$K = \frac{[HOAc][OH^-]}{[OAc^-][H_2O]} \qquad K_b = \frac{[HOAc][OH^-]}{[OAc^-]} \tag{15-13}$$

This reaction is no different from the reaction of any other base, such as ammonia or pyridine, with water; the fact that the acetate ion is charged is really incidental. We might then expect that one could evaluate the equilibrium constant by means of the same experimental techniques that are used in evaluating others. While this is correct, there is actually no need to perform these operations. By taking advantage of the fact that acetate ion is the conjugate base of acetic acid, K for the reaction of Eq. (15-12) can be calculated indirectly. We can make use of the following equilibrium constants:

$$K_w = [H_3O^+][OH^-]$$

$$[OH^-] = K_w/[H_3O^+]$$

$$K_a = \frac{[OAc^-][H_3O^+]}{[HOAc]}$$

$$\frac{1}{K_a} = \frac{[HOAc]}{[OAc^-][H_3O^+]}$$

We first substitute the expression for $[OH^-]$ into Eq. (15-13):

$$K_b = \frac{[HOAc]K_w}{[OAc^-][H_3O^+]} \tag{15-14}$$

Using the expression for $1/K_a$, we have

$$K_b = K_w/K_a$$

Stated verbally, the dissociation constant for a base in water is given by the ion product for water divided by the dissociation constant for the conjugate acid. Rearranging this equation a little, we have $K_a = K_w/K_b$. The dissociation constant for an acid in water is given by the ion product for water divided by the dissociation constant for the conjugate base. To confirm this last statement, the student should verify that the acid dissociation constant for the reaction

$$NH_4^+ + H_2O \rightarrow NH_3 + H_3O^+$$

is given by $K_w/1.79 \times 10^{-5}$.

Example 15-7

What is the pH of a solution which is $0.1M$ in sodium acetate?

The reaction of interest is given by Eq. (15-12). The dissociation constant is given by

$$K_b = \frac{10^{-14}}{1.76 \times 10^{-5}} = 5.68 \times 10^{-10} = \frac{[HOAc][OH^-]}{[OAc^-]}$$

Let x represent the concentration of un-ionized acetic acid formed. Then the concentration of OH^- formed by the reaction with water will also be x (we neglect the small amount produced from the ionization of water). The concentration of acetate ion is then $0.1 - x$.

$$5.68 \times 10^{-10} = \frac{x^2}{0.1 - x} \approx \frac{x^2}{0.1}$$

$$x^2 = 5.68 \times 10^{-11} = 56.8 \times 10^{-12}$$

$$x = 7.53 \times 10^{-6} = [OH^-] = [HOAc]$$

$$pOH = 5.12;$$

$$pH = 14 - 5.12 = 8.88$$

15-7 Buffer Solutions

In laboratory work, the necessity often arises of preparing solutions of known pH. These might bė prepared by adding a strong acid or base to water, but this is generally not a good method. If one wished, for example, to prepare a solution of pH 5 ($[H_3O^+] = 10^{-5}$), it would be necessary to make up a $10^{-5}N$ solution of a strong acid such as hydrochloric. Aside from the difficulties of doing this accurately, the problems involved in keeping the solution at this pH would be great. The slightest impurity might change the pH by an entire pH unit. These solutions are generally made up, therefore, by using more concentrated solutions of weak acids and bases, because these are not so susceptible to large pH changes.

Example 15-8

Prepare a solution of pH = 6.0 from pyridine and pyridinium chloride.

Pyridinium ion is the conjugate acid of the base, pyridine. Its dissociation constant, corresponding to the reaction

$$C_5H_5NH^+ + H_2O \rightarrow C_5H_5N + H_3O^+$$

is of the form and magnitude (Table 15-3)

$$K_a = \frac{[C_5H_5N][H_3O^+]}{[C_5H_5NH^+]} = \frac{K_w}{K_b} = \frac{10^{-14}}{1.7 \times 10^{-9}} = 5.85 \times 10^{-6}$$

By using this equilibrium expression it can quickly be seen that, if one had a $0.1M$ solution of pyridinium chloride, the pH would be below 6.0. The pH can be increased, however, by adding base to the solution. We begin with a $0.1M$ solution of pyridinium chloride, therefore, and add pyridine to bring the pH to 6.0

$$[H_3O^+] = 10^{-6}$$

$$[C_5H_5NH^+] = 10^{-1} \text{ (we neglect the small amount which ionizes)}$$

$$[C_5H_5N] = x$$

$$5.85 \times 10^{-6} = \frac{x \times 10^{-6}}{10^{-1}}$$

$$x = 0.585M$$

This is the concentration of pyridine which is required to raise the pH from whatever it is in a $0.1M$ solution of pyridinium chloride to 6.0. Because this solution contains relatively high concentrations of both the acid (pyridinium chloride) and its conjugate base, it is well able to absorb any small amounts of extraneous acid or base which may enter the solution. The pH would not change discernibly, for example, if a small amount of HCl vapor should happen to be absorbed into the solution.

15-8 Polyfunctional Acids and Bases

The acids and bases we have considered so far possess the capacity to release or attach one proton per molecule. One equilibrium and one equilibrium constant suffices, therefore, to describe their behavior. For some substances, more than one equilibrium constant is required to describe the acid-base behavior. Hydrogen sulfide, for example, ionizes according to the following equilibria:

$$H_2S + H_2O \rightleftharpoons HS^- + H_3O^+ \tag{15-15}$$

$$HS^- + H_2O \rightleftharpoons S^{-2} + H_3O^+ \tag{15-16}$$

The corresponding equilibrium constants are

$$K_{1_a} = \frac{[HS^-][H_3O^+]}{[H_2S]} = 1 \times 10^{-7} \tag{15-17}$$

$$K_{2_a} = \frac{[S^{-2}][H_3O^+]}{[HS^-]} = 1.2 \times 10^{-13} \tag{15-18}$$

From the magnitudes of these equilibrium constants one can see that hydrogen sulfide is a very weak acid. Furthermore, the second ionization is very weak in comparison with the first. The concentration of hydronium ions in a solution which contains just H_2S is therefore mostly dependent on the first equilibrium, with only a small contribution from the second. To demonstrate that this is true, let us calculate the pH of a $0.01M$ H_2S solution.

We let the concentration of HS^- ions produced by the ionization of H_2S equal x, neglecting for the moment the further ionization of these HS^- ions. Then, to a good approximation, the hydronium ion concentration is also x. We have then

$$K_{1_a} = \frac{x^2}{0.01 - x} = 10^{-7}$$

Neglecting the small amount of H_2S which ionizes in comparison with 0.01, we have

$$x^2 = 10^{-9} = 10 \times 10^{-10}$$
$$x = 3.1 \times 10^{-5}$$

We may now take this quantity as the starting point for the next ionization step, which follows Eq. (15-16). Let the concentration of S^{-2} produced by this ionization be y. Then y moles/l of hydronium ion is also produced by the same ionization, since for every sulfide ion produced a hydronium ion is also formed. But from the magnitude of K_{2a} we may expect that this quantity of H_3O^+ is trivial in comparison with that produced from the first ionization. We can therefore neglect it, and have

$$[H_3O^+] = 3.1 \times 10^{-5}$$
$$[S^{-2}] = y$$
$$[HS^-] = 3.1 \times 10^{-5} - y$$
$$\frac{3.1 \times 10^{-5} \times y}{3.1 \times 10^{-5} - y} = 1.2 \times 10^{-13}$$

If we assume that the fraction of HS^- which ionizes is small, y drops out of the denominator, the numbers in the numerator and denominator cancel, and we are left with

$$y = 1.2 \times 10^{-13}$$

This y is the concentration of sulfide ion produced from the ionization of HS^-. It is also the concentration of hydronium ion produced from this same ionization. We initially neglected this in comparison with that produced from the first ionization; this proves to have been a safe procedure, since 1.2×10^{-13} is very small in comparison with 3.1×10^{-5}. The pH is therefore $-\log_{10}(3.1 \times 10^{-5}) = 4.51$.

Table 15-4 Ionization constants for some polyprotic acids.

Acid	K_a	Equation
H_2CO_3	4×10^{-7}	$H_2CO_3 + H_2O \rightleftharpoons HCO_3^- + H_3O^+$
HCO_3^-	4.7×10^{-11}	$HCO_3^- + H_2O \rightleftharpoons CO_3^{-2} + H_3O^+$
H_2SeO_3	3×10^{-3}	$H_2SeO_3 + H_2O \rightleftharpoons HSeO_3^- + H_3O^+$
$HSeO_3^-$	5×10^{-8}	$HSeO_3^- + H_2O \rightleftharpoons SeO_3^{-2} + H_3O^+$
H_3PO_4	8×10^{-3}	$H_3PO_4 + H_2O \rightleftharpoons H_2PO_4^- + H_3O^+$
$H_2PO_4^-$	6×10^{-8}	$H_2PO_4^- + H_2O \rightleftharpoons HPO_4^{-2} + H_3O^+$
HPO_4^{-2}	1×10^{-12}	$HPO_4^{-2} + H_2O \rightleftharpoons PO_4^{-3} + H_3O^+$

A number of polyprotic acids are listed in Table 15-4 with the corresponding acid dissociation constants. The most important polyfunctional bases are the conjugate bases of the same acids — bases such as PO_4^{-3}, CO_3^{-2}, and S^{-2}. A solution of sodium carbonate is basic for the same reason that a solution of sodium acetate is basic: The anion is a base, and reacts with water, thus,

$$CO_3^{-2} + H_2O \rightleftharpoons HCO_3^- + OH^- \quad (15\text{-}19)$$

$$HCO_3^- + H_2O \rightleftharpoons H_2CO_3 + OH^- \quad (15\text{-}20)$$

By working through the equations, it can be seen that the same rule applies to these anions that applies to the acetate ion. The base dissociation constant is equal to the ion product for water divided by the acid dissociation constant for the conjugate acid. We have, then, for Eq. (15-19)

$$K_{b1} = 10^{-14}/4.7 \times 10^{-11} = 2.13 \times 10^{-4}$$

and for Eq. (15-20)

$$K_{b2} = 10^{-14}/4 \times 10^{-7} = 2.5 \times 10^{-8}$$

Example 15-9

As an illustration of the use of these equations, let us calculate the pH of a $0.02M$ solution of Na_2CO_3. First of all, it is clear from the relative values of the K's that the reaction represented by Eq. (15-19) will proceed to a much larger extent than that represented by Eq. (15-20). By proceeding as in the example with H_2S, it could be shown that the second equilibrium does in fact make a negligible contribution in this case to the pH of the solution.

Let the concentration of HCO_3^- produced by the first reaction equal x. This then is also the concentration of the OH^- ion produced by the same reaction. We have then

$$\frac{x^2}{0.02 - x} = 2.1 \times 10^{-4}$$

This is an instance in which it would not be safe to assume that the value of x is small in comparison with the initial concentration of reacting substance. Solving the quadratic equation yields

$$x = 1.95 \times 10^{-3} = [OH^-]$$
$$pH = 14 - pOH = 14 - 2.7 = 11.3$$

A somewhat more complex problem arises in considering the behavior of a solution of sodium bicarbonate, $NaHCO_3$. The bicarbonate ion is capable of acting as either acid or base. The two reactions and their corresponding dissociation constants are

$$HCO_3^- + H_2O \rightleftharpoons CO_3^{-2} + H_3O^+ \qquad K_{2a} = 4.7 \times 10^{-11}$$
$$HCO_3^- + H_2O \rightleftharpoons H_2CO_3 + HO^- \qquad K_b = 2.5 \times 10^{-8}$$

Since the equilibrium constant for the second reaction is about a thousand times larger than that for the first, it may be assumed to a good approximation that it is the important one, and the other equilibrium may be neglected entirely. If the two K's were of the same magnitude, however, the problem would be a good deal more complicated, though it could be solved by making use of all information known about the solution.

15-9 Acid-Base Titrations; Indicators

In Chap. 12 the neutralization reaction was considered from a qualitative point of view. We are now in a position to consider this important reaction quantitatively. The simplest case to consider is the addition of a strong acid solution to a strong base solution, or *vice versa*. Figure 15-2 shows the pH change which results from addition of 0.100 N NaOH solution to 50.00 ml of a 0.100 N HCl solution. It can be seen that the pH changes very little with addition of up to 49 ml of NaOH. The initial pH is, of course, 1. The total number of equivalents of acid is given by

$$\frac{0.1 \text{ equiv acid}}{\text{liter solution}} \times 0.0500 \text{ liter} = 0.00500 \text{ equiv acid}$$

After addition of 49 ml of NaOH, the total number of equivalents of base added is

$$\frac{0.1000 \text{ equiv base}}{\text{liter solution}} \times 0.0490 \text{ liter} = 0.00490 \text{ equiv base}$$

This quantity of base reacts with an equal quantity of acid, so that the quantity of acid remaining is then

$$0.00500 - 0.00490 = 0.00010 \text{ equiv}$$

The volume of solution at this point is 99.00 ml. The concentration of the remaining acid is then 0.00010 equiv/0.099 liters = 0.00101 equiv/liter. The pH is, therefore, approximately 3. Now, when exactly 50.00 ml of NaOH solution has been added, and the neutralization reaction is complete, the pH will be 7.0. During the addition of the last 1 ml of solution, therefore, the pH changes by 4 pH units — twice as large a change as that which occurred during addition of all the other NaOH. In a similar way (the student can verify this by a calculation like that given above), the pH continues to increase rapidly during addition of the first bit of NaOH solution in excess of that required for the neutralization.

The accurate and precise detection of the equivalence point (the point at which an equivalent amount of base has been added) is possible because the pH does change so rapidly in this range. It will be recalled from Chap. 12 that the endpoint of the neutralization reaction is detected by means of an indicator which is itself a weak acid or base (if the indicator is chosen correctly the endpoint and equivalence point are the same). Let us assume that the indicator to be used in the titration of HCl with NaOH solution is a weak base. It is added in very low concentration to the acid solution at the beginning of the titration. It then exists in the acid form HIN, where IN represents the indicator. HIN is itself a weak acid which ionizes according to the equation

$$\mathrm{HIN} + \mathrm{H_2O} \rightarrow \mathrm{H_3O^+} + \mathrm{IN}$$

where

$$K_{\mathrm{HIN}} = \frac{[\mathrm{IN}][\mathrm{H_3O^+}]}{[\mathrm{HIN}]} \tag{15-21}$$

We leave off any charge symbols on IN and HIN because the indicator might be an anion, a neutral molecule, or even a cation. Rearranging Eq. (15-21), we have

$$\frac{[\mathrm{IN}]}{[\mathrm{HIN}]} = \frac{K_{\mathrm{HIN}}}{[\mathrm{H_3O^+}]} \tag{15-22}$$

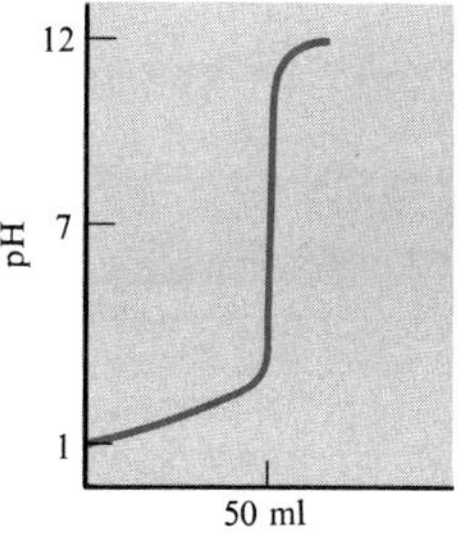

Fig. 15-2 pH *versus* volume of added 0.100*N* sodium hydroxide in the titration of 0.100*N* hydrochloric acid.

Now for all practical purposes, if the ratio of IN to HIN is less than 0.1, the color which the indicator imparts to the solution is that which is characteristic of HIN. On the other hand, if the ratio IN/HIN is greater than about 10, the color seen is due to IN. The endpoint occurs, ideally, when the ratio is 1. Taking the log of both sides of Eq. (15-22), we have

$$\begin{aligned}\log_{10} \frac{[\mathrm{IN}]}{[\mathrm{HIN}]} &= \log_{10} K_{\mathrm{HIN}} - \log_{10} [\mathrm{H_3O^+}] \\ &= \log_{10} K_{\mathrm{HIN}} + \mathrm{pH} \\ &= -\mathrm{p}K_{\mathrm{HIN}} + \mathrm{pH}\end{aligned} \tag{15-23}$$

The endpoint change occurs in the range of IN/HIN from 0.1 to 10. Since the logs of these ratios are -1 and 1, respectively, we see that the color change which is observed occurs over a pH range of about 2 units, possibly a little less (Fig. 15-3). From Eq. (15-23), when $[\mathrm{IN}]/[\mathrm{HIN}] = 1$, its log is zero, and $\mathrm{pH} = \mathrm{p}K_{\mathrm{HIN}}$ at the endpoint.

The titration of a weak acid such as acetic acid with a strong base such as sodium hydroxide provides an example of a somewhat more complicated system. Supposing that one begins with 50.00 ml of $0.1N$ acetic acid (Fig. 15-4); the pH of this solution as calculated on p. 358, is 2.87. The total number of equivalents of acid is

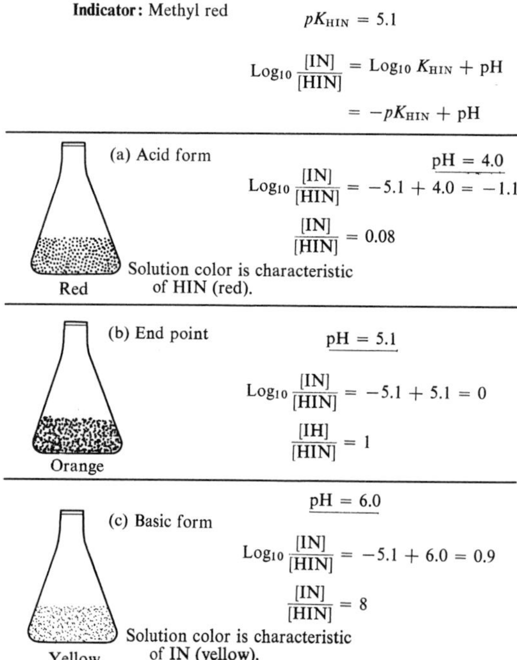

Indicator: Methyl red
pK_HIN = 5.1
Log10 [IN]/[HIN] = Log10 K_HIN + pH
= −pK_HIN + pH
(a) Acid form
pH = 4.0
Log10 [IN]/[HIN] = −5.1 + 4.0 = −1.1
[IN]/[HIN] = 0.08
Solution color is characteristic of HIN (red).
Red
(b) End point
pH = 5.1
Log10 [IN]/[HIN] = −5.1 + 5.1 = 0
[IH]/[HIN] = 1
Orange
pH = 6.0
(c) Basic form
Log10 [IN]/[HIN] = −5.1 + 6.0 = 0.9
[IN]/[HIN] = 8
Solution color is characteristic of IN (yellow).
Yellow

Fig. 15-3 Progress of an acid-base titration near the endpoint.

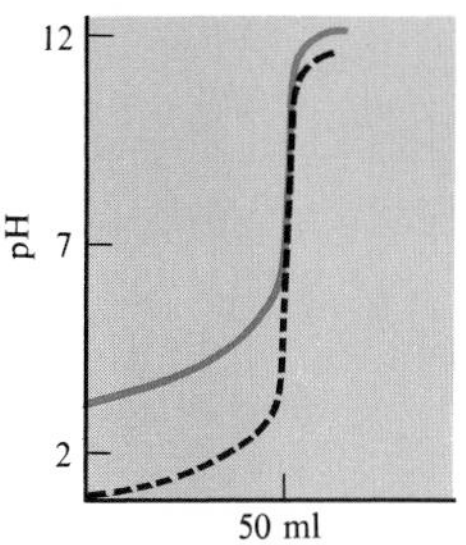

Fig. 15-4 Titration of 0.100*N* acetic acid with 0.100*N* sodium hydroxide. The dotted line represents the titration of a 0.100*N* strong acid with the same sodium hydroxide.

$$\frac{0.100 \text{ equiv acid}}{\text{liter solution}} \times 0.050 \text{ liter} = 0.0050 \text{ equiv acid}$$

49.0 ml of 0.100*N* NaOH represents

$$\frac{0.100 \text{ equiv base}}{\text{liter solution}} \times 0.049 \text{ liter} = 0.0049 \text{ equiv base}$$

The net amount of acid remaining after addition of 49.0 ml of sodium hydroxide solution is then 0.0001 equivalents. The concentration of the remaining acid is 0.0001 equiv/0.099 liter = 0.00101*N*. To determine the pH of the solution we must make use of the acid dissociation constant of acetic acid, and must take into account the fact that the neutralization produces sodium acetate as a product. The concentration of the latter is

$$0.0049 \text{ equiv}/0.099 \text{ liter} = 0.0495N$$

We have, then

$$HOAc + H_2O \rightarrow OAc^- + H_3O^+$$

$$K_a = \frac{[OAc^-][H_3O^+]}{[HOAc]}$$

Letting the concentration of H_3O^+ produced from the ionization be x,

$$[HOAc] = 0.00101 - x$$
$$[OAc^-] = 0.0495 + x$$
$$[H_3O^+] = x$$
$$1.76 \times 10^{-5} = \frac{(0.0495 + x)(x)}{0.001 - x}$$

Since x is likely to be small in comparison with both 0.0495 and 0.001, we drop it from the two terms involved, leaving

$$1.76 \times 10^{-5} = \frac{0.0495 \times x}{0.001}$$
$$x = 3.56 \times 10^{-7} = [H_3O^+]$$
$$pH = 6.45$$

When exactly 50 ml of NaOH has been added, no acid remains; the concentration of acetate ion is then 0.005 equiv/0.100 liter = 0.0500*N*. To calculate the pH of this solution we use the reaction

$$OAc^- + H_2O \rightarrow HOAc + OH^-$$

for which the dissociation constant is 5.68×10^{-10} (page 362). The pH is calculated to be 8.72 by the method employed on page 362.

The differences between the titrations of acetic and hydrochloric acids are: (*a*) The pH changes more gradually in titrating the weak acid. Thus, in adding the first 49.0 ml of base to acetic acid the pH changes by 6.45 − 2.87 = 3.58 pH units, whereas in titrating the hydrochloric acid the first 49 ml of NaOH produces a change of 2 pH units. The last 1 ml before the equivalence point causes a pH change of about 2.3 pH units with acetic acid and about 4 units with hydrochloric acid. (*b*) The equivalence point in the titration of the weak acid does not occur at a pH of 7, but at a higher pH. To choose an indicator whose color change occurs in the range of the equivalence point, therefore, we would need to use one such as phenolphthalein, with a pK_{HIN} of 9.1. If methyl red were used (see Table 15-5), the endpoint would come when the pH was about 5. But in titrating 0.1*N* acetic acid with 0.1*N* sodium hydroxide, a pH of 5 is reached before even 49 ml of base has been added; the endpoint and equivalence point would not be in good agreement. This example illustrates the importance of choosing a proper indicator for the titration at hand.

Table 15-5 Some acid-base indicators.

Indicator	pK_{HIN}	Color change	
		Acid	Basic
Thymol blue	2.0	red	yellow
Methyl orange	3.4	red	yellow
Bromocresol green	4.6	yellow	blue
Methyl red	5.1	red	yellow
Litmus	6.4	red	blue
Phenolphthalein	9.1	colorless	red
Alizarin yellow	11.0	yellow	red

If a weak base such as ammonia is to be titrated with a strong acid solution such as hydrochloric acid, the same considerations apply. In this case the equivalence point would occur in an acid solution, pH about 5.3. Methyl red would be a good indicator, since its p*K* is about the same as the pH at which the equivalence point occurs.

Titrations of weak acids with weak bases, or *vice versa*, are not commonly done. So long as the dissociation constants for the substances employed are known, however, the pH at which the equivalence point occurs could be calculated.

15-10 Complex Ion Equilibria

Metal complexes are formed when neutral base molecules or ions displace water from the immediate environment of a metal ion in solution. The groups, termed *ligands*, interact with the metal ion through a Lewis acid-base type of interaction, and form distinct chemical species. The number of ligands which are accommodated in the immediate spherical volume around the metal ion is termed the *coordination number*. The properties of metal complexes are discussed in detail in Chaps. 20 and 21. For now, we are concerned only with the equilibrium aspects of complex formation. As is usual in the case of aqueous solution equilibria, we omit water from the equations, since its concentration is, for all practical purposes, constant. Thus, the dissociations of copper tetrammine complex are written as

$$\begin{array}{lll} Cu(NH_3)_4^{+2} \rightleftharpoons Cu(NH_3)_3^{+2} + NH_3 & K_1 & \\ Cu(NH_3)_3^{+2} \rightleftharpoons Cu(NH_3)_2^{+2} + NH_3 & K_2 & \\ Cu(NH_3)_2^{+2} \rightleftharpoons Cu(NH_3)^{+2} + NH_3 & K_3 & (15\text{-}24) \\ Cu(NH_3)^{+2} \rightleftharpoons Cu^{+2} + NH_3 & K_4 & \\ \hline \text{over-all:} \quad Cu(NH_3)_4^{+2} \rightleftharpoons Cu^{+2} + 4NH_3 & K_d & \end{array}$$

Each of the successive dissociations involves displacement of an ammonia molecule by a water molecule, not shown. The inverse of the dissociation constant is often referred to as the stability constant, or formation constant. It corresponds to the reverse reaction of Eq. (15-24). The over-all dissociation constant for $Cu(NH_3)_4^{+2}$ is written as

$$K_d = \frac{[Cu^{+2}][NH_3]^4}{[Cu(NH_3)_4^{+2}]} \qquad (15\text{-}25)$$

K_d is the product of the successive constants

$$K_d = K_1K_2K_3K_4 = 2.6 \times 10^{-13}$$

Since so many different equilibria are involved, a large number of different species are present in the solution under any given set of conditions. The concentrations of all these species can be calculated by solving the cumbersome simultaneous equations which can be developed. This is a laborious job, and there are no simple shortcuts. Fortunately, however, we seldom need to know the concentrations of all of the intermediate species, but are interested only in the concentration of the most highly coordinated complex. We very often deal with solutions in which the ligand is present in excess, so that there is more than enough of it to form the highest complex in the series. We are then interested in the concentration of uncomplexed metal. For this purpose, we can employ the equilibrium expression and the equilibrium constant for the over-all process. As an example, let us suppose that we wish to know the concentration of ammonia-free copper ion in a solution which is $0.005M$ in Cu^{+2}, and $0.25M$ in ammonia. Since the equilibria shown above all have K values on the order of 10^{-3} or 10^{-4}, we can assume that each reaction will go far to the left, so that nearly all of the copper ion will be in the form of $Cu(NH_3)_4^{+2}$. Thus, if the initial Cu^{+2} concentration is $0.005M$, the concentration of $Cu(NH_3)_4^{+2}$ will be approximately $0.005M$ at equilibrium. The quantity of free ammonia is that initially present less that required to form the complex. The quantity of free Cu^{+2} we label x. Then

$$[Cu^{+2}] = x$$
$$[NH_3] \approx 0.25 - 4(0.005)$$
$$[Cu(NH_3)_4^{+2}] \approx 0.005$$
$$K_d = 2.6 \times 10^{-13} = \frac{x(0.23)^4}{5 \times 10^{-3}}$$
$$x = 4.65 \times 10^{-13} = [Cu^{+2}]$$

It is very important to note the limitations on the use of an over-all equilibrium constant expression such as that in Eq. (15-25). It is all right to use such an expression provided that one has in fact a reasonable basis on which to estimate the concentrations of the species which go into it. Thus, we have not assumed that the concentration of $Cu(NH_3)^{+2}$, for example, is zero, but we have assumed that nearly all of the copper (II) ion is converted into the 4:1 complex. In general, this will be true only when the formation constants are large (dissociation constants small), and when there is an excess of ligand.

Table 15-6 lists the dissociation constants for a number of metal complexes. These are often of value in calculating free metal ion concentrations to determine whether a precipitation or other reaction might occur.

Table 15-6 Dissociation constants and dissociation equilibria for some complex ions in water at 25°C.

Complex	K_d	Equilibrium expression
AlF_6^{-3}	1.45×10^{-20}	$AlF_6^{-3} \rightleftharpoons Al^{+3} + 6F^-$
$Co(NH_3)_6^{+2}$	1.8×10^{-5}	$Co(NH_3)_6^{+2} \rightleftharpoons Co^{+2} + 6NH_3$
$Co(NH_3)_6^{+3}$	6.2×10^{-36}	$Co(NH_3)_6^{+3} \rightleftharpoons Co^{+3} + 6NH_3$
$Fe(CN)_6^{-4}$	1×10^{-24}	$Fe(CN)_6^{-4} \rightleftharpoons Fe^{+2} + 6CN^-$
$Hg(NH_3)_4^{+2}$	4.0×10^{-20}	$Hg(NH_3)_4^{+2} \rightleftharpoons Hg^{+2} + 4NH_3$
$Hg(CN)_4^{-2}$	3×10^{-42}	$Hg(CN)_4^{-2} \rightleftharpoons Hg^{+2} + 4CN^-$
$Ag(NH_3)_2^+$	5.9×10^{-8}	$Ag(NH_3)_2^+ \rightleftharpoons Ag^+ + 2NH_3$
$Ag(CN)_2^-$	1×10^{-21}	$Ag(CN)_2^- \rightleftharpoons Ag^+ + 2CN^-$

Heterogeneous Equilibria

15-11 Solubility Product

A heterogeneous equilibrium is one in which the participating substances are present in more than one phase. Each of the following equations represents such an equilibrium.

$$Br_2(aq) \rightleftharpoons Br_2(CCl_4) \qquad (15\text{-}26)$$

$$2CH_3CO_2H(aq) \rightleftharpoons (CH_3CO_2H)_2(benzene) \qquad (15\text{-}27)$$

$$PbS(s) \rightleftharpoons Pb^{+2}(aq) + S^{-2}(aq) \qquad (15\text{-}28)$$

The first equation represents the distribution of bromine between two immiscible liquids, water and carbon tetrachloride. The second represents the distribution of acetic acid between water and benzene — this is a somewhat more complex system than the first because the acetic acid molecules are dimerized in the organic layer. The third equation represents the equilibrium between a solid and its ions in solution.

The equilibrium constants for these equilibria are set up in precisely the same manner as for homogeneous systems:

$$K_d = \frac{[Br_2]_{(CCl_4)}}{[Br_2]_{(aq)}}$$

$$K_d = \frac{[(CH_3CO_2H)_2]_{(benzene)}}{[CH_3CO_2H]^2_{(aq)}} \qquad (15\text{-}29)$$

$$K = \frac{[Pb^{+2}][S^{-2}]}{[PbS]}$$

The first two equilibria are no different in terms of how they are employed than the equilibrium constant expressions for homogeneous systems. In the last, however, a new element must be considered. The lead sulfide equilibrium involves two phases, solid lead sulfide and the aqueous phase. Now the concentration of lead sulfide in pure lead sulfide is a quantity which is not subject to any significant variation. As long as *any* quantity of this pure solid is present, the equilibrium is not affected by adding more to the system. Since the concentration term for the solid is effectively constant, it may be taken up into the equilibrium constant, with the result that the expression in Eq. (15-29) becomes

$$K[PbS]_s = K_{sp} = [Pb^{+2}][S^{-2}]$$

The modified constant is called the solubility product constant. Some examples of solubility equilibria, and the corresponding values for K_{sp}, are listed in Table 15-7.

Table 15-7 Values of some solubility products.

$Al(OH)_3 \rightleftharpoons Al^{+3} + 3OH^-$	1.4×10^{-34}
$BaCrO_4 \rightleftharpoons Ba^{+2} + CrO_4^{-2}$	1.2×10^{-10}
$CaCrO_4 \rightleftharpoons Ca^{+2} + CrO_4^{-2}$	7.1×10^{-4}
$CuS \rightleftharpoons Cu^{+2} + S^{-2}$	1×10^{-40}
$AgCl \rightleftharpoons Ag^+ + Cl^-$	1.8×10^{-10}
$MnS \rightleftharpoons Mn^{+2} + S^{-2}$	1×10^{-11}
$ZnS \rightleftharpoons Zn^{+2} + S^{-2}$	8×10^{-25}

The applications of the solubility product constant to chemical problems are numerous. In the following two examples we will consider determination of the solubility product from solubility data.

Example 15-10

The solubility of calcium fluoride in water at 18°C is given in handbooks as 0.00160 g per 100 ml of water. Calculate K_{sp}.

Since the solubility of the salt is quite low, we can assume that the volume of the saturated solution is essentially the same as that of the initial water, so the solubility is about 0.00160 g per 100 ml of solution.

$$\frac{0.00160 \text{ g } CaF_2}{0.100 \text{ liter solution}} \times \frac{1 \text{ mole } CaF_2}{78.1 \text{ g } CaF_2} = \frac{2.05 \times 10^{-4} \text{ moles } CaF_2}{\text{liter solution}}$$

The equilibrium in a saturated solution of CaF_2 is given by

$$CaF_2(s) \rightleftharpoons Ca^{+2}(aq) + 2F^-(aq) \qquad (15\text{-}30)$$

$$K_{sp} = [Ca^{+2}][F^-]^2$$

From Eq. (15-30) it can be seen that, for each mole of CaF_2 which dissolves, 1 mole of $Ca^{+2}(aq)$ and 2 moles of $F^-(aq)$ are produced. The concentrations of these two ions in the saturated solution at equilibrium are therefore

$$[Ca^{+2}] = 2.05 \times 10^{-4}$$

$$[F^-] = 4.10 \times 10^{-4}$$

$$K_{sp} = (2.05 \times 10^{-4})(4.10 \times 10^{-4})^2$$

$$= 3.45 \times 10^{-11}$$

It should be noted that in working this problem it was assumed that the only source of calcium and fluoride ion in the solution is the solid calcium fluoride. When this is the case, it necessarily follows that the concentration of fluoride is twice that of the calcium ion. That this need not always be the case is demonstrated by the next example.

Example 15-11

The solubility of barium fluoride in a 0.2*M* solution of sodium fluoride is 6×10^{-4} moles per liter. Calculate K_{sp} for BaF_2.

The solubility equilibrium expression is given by

$$BaF_2(s) \rightleftharpoons Ba^{+2}(aq) + 2F^-(aq)$$

Since the sole source of barium ion in the solution is dissolved barium fluoride, the concentration of barium ion at equilibrium is $6 \times 10^{-4}M$. The concentration of fluoride ion which is produced from the dissolving of the solid salt is, from the balanced equation, twice this, or 1.2×10^{-3} moles per liter. This, however, is not the only source of fluoride ion; the sodium fluoride which is also present contributes 0.2 mole/liter. The total fluoride ion concentration is then $0.2 + 0.0012 = 0.2012$. Since the quantity produced from the barium fluoride is small in comparison with that already present in the solution, it can be ignored; then

$$K_{sp} = [Ba^{+2}][F^-]^2 = (6 \times 10^{-4})(2 \times 10^{-1})^2$$

$$= 2.4 \times 10^{-5}$$

Using this value for K_{sp} we can now calculate the solubility of BaF_2 in water which contains no fluoride ion initially. Let the concentration of Ba^{+2} in the saturated solution be X. Then the concentration of F^- = is $2X$ since, from the balanced equation, two fluoride ions are produced in solution for each barium ion. Then

$$K_{sp} = (X)(2X)^2 = 2.4 \times 10^{-5}$$
$$4X^3 = 2.4 \times 10^{-5}$$
$$X = 1.8 \times 10^{-2}$$

Since X is the concentration of barium ion, and since 1 mole of barium ion corresponds to 1 mole of BaF_2, the solubility of BaF_2 is then $1.8 \times 10^{-2}M$. This is much greater than the solubility of the same salt in a solution which already contains some fluoride ion, as a comparison with the previous calculation shows. The effect of the fluoride ion in decreasing the solubility of BaF_2 is analogous to the effect of the acetate ion in reducing the extent of dissociation of acetic acid in solution. It is, indeed, another example of the common ion effect.

If a common ion effect is exercised by an anion which is the conjugate base of a weak acid, it is necessary to consider not only the solubility product equilibrium, but also the relevant weak acid dissociation equilibrium. As an example of this type of system we will consider the solubilities of sulfides in solutions which also contain hydrogen sulfide. Many procedures for separating and detecting the presence of metal ions in solution utilize sulfide precipitation.

Example 15-12

A saturated solution of hydrogen sulfide in water at room temperature contains about 0.1 mole per liter of H_2S. If a solution containing $0.1M$ Zn^{+2} is saturated with H_2S when the pH of the solution is 3.0, will ZnS precipitate from the solution? What will be the final concentration of Zn^{+2} in the solution if precipitation does occur?

The dissociation of H_2S proceeds stepwise according to the following equations:

$$H_2S + H_2O \rightleftharpoons H_3O^+ + HS^- \qquad (15\text{-}31)$$
$$HS^- + H_2O \rightleftharpoons H_3O^+ + S^{-2} \qquad (15\text{-}32)$$

In considering the extent to which these two equilibria contribute to the pH of a solution of H_2S, it was concluded that the second equilibrium is not important (p. 364). We are now interested, however, in quite another quantity — the sulfide ion concentration. It is true that the second dissociation step may occur to a very slight extent, but since it is the *only* source of sulfide ion, it must be considered. Furthermore, the pH in this problem is not an unknown quantity; by adding buffering substances to the solution the pH has deliberately been set at 3. We could calculate the sulfide ion concentration by setting up K_a for Eq. (15-31) and solving for $[HS^-]$. Using this value for $[HS^-]$ we could then set up K_a for Eq. (15-32) and solve for $[S^{-2}]$. The same value, 10^{-3}, would be used for $[H_3O^+]$ in both expressions.

There is, however, another and shorter means to the same end. If Eqs. (15-31) and (15-32) are added together, we obtain

$$H_2S + 2H_2O \rightleftharpoons 2H_3O^+ + S^{-2} \qquad (15\text{-}33)$$

The K for this expression is

$$K_{1,2} = \frac{[H_3O^+]^2[S^{-2}]}{[H_2S]} \qquad (15\text{-}34)$$

The student should verify that this expression is just the product of the K's for Eqs. (15-31) and (15-32) taken individually. The magnitude of $K_{1,2}$ is just $K_{a1} \times K_{a2}$. Using the values given on p. 364 we have $K_{1,2} = (1 \times 10^{-7}) \times (1.2 \times 10^{-13}) = 1.2 \times 10^{-20}$.*

In the present problem we have $[H_3O^+] = 10^{-3}$, $[H_2S] = 0.1$ (we neglect the small amount which ionizes). Then

$$1.2 \times 10^{-20} = \frac{[10^{-3}]^2[S^{-2}]}{10^{-1}}$$

$$[S^{-2}] = 1.2 \times 10^{-15}$$

The product of the zinc and sulfide ions after addition of the H_2S at pH = 3 is therefore

$$10^{-1} \times 10^{-15} = 10^{-16}$$

This is far in excess of the solubility product constant for ZnS (Table 15-7). Precipitation will therefore occur; if H_2S is continually bubbled in during precipitation to maintain the concentration of H_2S in the solution, the concentration of sulfide ion at the completion of precipitation will still be 1.2×10^{-15}. The solubility product for zinc sulfide is given by

$$K_{sp} = [Zn^{+2}][S^{-2}] = 8 \times 10^{-25}$$

since

$$[S^{-2}] = 1.2 \times 10^{-15}$$

$$[Zn^{+2}] = 7 \times 10^{-10}$$

From this result, it is clear that the addition of H_2S results in removal of essentially all of the zinc ion from the solution, provided that time is allowed for precipitation to occur, and provided that a sufficient quantity of H_2S is added to maintain the H_2S concentration.

All metal sulfides of low solubility do not precipitate under the conditions given in the above problem. K_{sp} for manganese sulfide, MnS, is 10^{-11}. If H_2S were added to a $0.1M$ solution of Mn^{+2} at a pH of 3, there-

*It is easy to fall into one very serious error in the use of Eq. (15-33) and the expression for $K_{1,2}$ given by Eq. (15-34). The concentration of hydronium ion is *not* twice that of the sulfide ion, even though the coefficients in the equation indicate that this is the case. It is true that, for every sulfide ion which is produced from H_2S, two hydronium ions are also produced. But it must be remembered that this is not the only source of hydronium ion. Even when no other substances are present in the solution, the hydronium ion concentration is much greater than twice the sulfide ion concentration, because of the first ionization step which produces HS^- and hydronium ion. Once the hydronium ion concentration in the solution is known, however, it can be substituted into Eq. (15-34), and the sulfide ion concentration solved for.

fore, the product $[Mn^{+2}][S^{-2}] = 10^{-16}$ would not exceed the solubility product, and MnS would not precipitate. It would therefore be possible, by adjustment of the pH to 3, to effect a separation of Zn^{+2} and Mn^{+2} ions by precipitating zinc as the sulfide, while Mn^{+2} would remain in solution.

Our final example involves a relationship between solubility product and complex formation.

Example 15-13

The solubility product for AgCl is 1.8×10^{-10}. Suppose that 0.01 mole of $AgNO_3$ is dissolved in a liter of 0.6*M* ammonia. Will AgCl precipitate if 0.01 mole of NaCl is added to the solution?

First compute the free Ag^+ concentration in the ammonia solution. From Table 15-6 we have

$$K_d = 5.9 \times 10^{-8} = \frac{[Ag^+][NH_3]^2}{[Ag(NH_3)_2^+]}$$

$$[Ag^+] = xM$$

$$[NH_3] = 0.60 - 2\,(0.01) = 0.58M$$

$$[Ag(NH_3)_2^+] \approx 0.01M$$

Then,

$$\frac{[x][0.58]^2}{[0.01]} = 5.9 \times 10^{-8}$$

$$x = 1.75 \times 10^{-9} = [Ag^+]$$

Precipitation of AgCl occurs when the product $[Ag^+][Cl^-]$ exceeds K_{sp}; *i.e.*, 1.8×10^{-10}. Since the Cl^- concentration is 0.01*M* after addition of NaCl, the maximum allowable Ag^+ concentration before precipitation is

$$Ag^+ = 1.8 \times 10^{-10}/10^{-2} = 1.8 \times 10^{-8}$$

The free Ag^+ concentration after addition of ammonia is less than this; AgCl does not precipitate from the solution.

Suggested Readings

Banks, J. E., "Equilibria of Complex Formation," *J. Chem. Educ.* 38, 391 (1961).

Bard, A. J., *Chemical Equilibrium*, New York: Harper and Row, 1966.

Butler, J. N., "An Approach to Complex Equilibrium Problems," *J. Chem. Educ.* 38, 141 (1961).

———, *Solubility and pH*, New York: Addison Wesley Publishing Co., 1965.

Petrucci, R. H., and P. C. Moews, "The Precipitation and Solubility of Metal Sulfides," *J. Chem. Educ.*, 39, 391 (1962).

Exercises

15-1. Write the equilibrium constant expression for the gas phase reaction

$$2NO + O_2 \rightleftharpoons 2NO_2$$

In what units may *K* be expressed? Does the numerical value for *K* depend on these units?

15-2. Oxygen gas is dissociated into atomic oxygen at very high temperatures. At 3000°C, and a total pressure of 1 atm, O_2 is 5.95 per cent dissociated. Calculate an equilibrium constant for the dissociation. Indicate the units employed, and write the balanced chemical equation for which the constant applies.

15-3. Is it possible to determine the dissociation equilibrium constant for a strong acid such as HCl in water? Explain your answer.

15-4. The first dissociation of H_2SO_4 is essentially complete in dilute aqueous solution. The proton in HSO_4^-, on the other hand, is only weakly ionized. How do you account for the difference?

15-5. Nitrogen dioxide exists at low temperatures as the dimer, N_2O_4. The dimer dissociates at higher temperatures according to the equation

$$N_2O_4 \rightleftharpoons 2NO_2$$

At a temperature of 60.2°C and a total pressure of 1 atm, the dimer is 53 per cent dissociated. Calculate a value for K in units of atmospheres. Can units of mole fraction be used?

The dimerization of NO_2 is a consequence of its electronic structure. Write the Lewis structure for the molecule and propose an explanation for the dimer formation.

15-6. Diethyl ether and boron trifluoride react to form an addition compound according to the equation

$$(CH_3CH_2)_2O + BF_3 \rightleftharpoons (CH_3CH_2)_2OBF_3$$

12.944 mm of BF_3 and 12.802 mm of diethyl ether vapor, both at 0°C, are mixed in a constant volume container. The pressure in the container at various temperatures is as follows: 69.9°C, 29.829 mm; 80.1°C, 31.540 mm; 94.7°C, 33.608 mm. Assuming ideal behavior, calculate K at each temperature for the above reaction in units of atm^{-1}. Judging from the variation of K with T, is the reaction exothermic or endothermic?

15-7. Using the data in Table 15-2, calculate the per cent of each of the following acids which is ionized in a $0.1M$ solution: nitrous, acetic, hydrocyanic. Repeat the calculations for a $0.03M$ solution. What general statement can be made about the variation in per cent dissociation with concentration for solutions of weak electrolytes such as these weak acids?

15-8. Calculate the pH of a $0.2M$ solution of benzoic acid. Repeat the calculation for $0.2M$ benzoic acid solutions containing the following concentrations of sodium benzoate: $0.05M$, $0.10M$, $0.20M$, $0.40M$. By rearranging the equilibrium constant expression, obtain an expression which readily gives the pH as a function of acid and added salt concentrations.

15-9. Write the equation for the autoionization of ammonia as a solvent.

15-10. The first ionization of sulfuric acid may be considered to go to completion. The dissociation constant for the hydrogen sulfate ion, however, is 1.0×10^{-2}. Using this value, calculate the pH of a $0.1M$ solution of sodium hydrogen sulfate; of a $0.1M$ solution of sodium sulfate.

15-11. It is apparent from the data in Table 15-4 that successive ionizations of protons from polyprotic acids are substantially less extensive. The following are the pK_a's for three diprotic acids of different structure. ($pK_a = -\log_{10}K_a$)

	H_2S	H_2SeO_3	$H_2C_2O_4$
pK_1	7.2	2.51	1.2
pK_2	14.9	7.3	4.2
ΔpK_a	7.7	4.8	3.0

Oxalic acid

$$\mathrm{HO(O{=})C{-}C({=}O)OH}$$

What relationship do you see between the variation in ΔpK and structure? How is the relationship explained?

15-12. Calculate $[H_3O^+]$ and pH in a $0.2M$ solution of HF (see Table 15-2), both by neglecting dissociation of the acid, and by taking account of it.

15-13. A $0.01M$ solution of potassium cyanide is known to be hydrolyzed to the extent of 3.7 per cent. Calculate the acid dissociation constant for HCN.

15-14. Calculate the pH of a $0.1M$ solution of sodium cyanide. Repeat for a $0.1M$ solution of sodium acetate. Account for the different pH's of the two solutions in terms of the Brönsted theory.

15-15. The dissociation constants for sulfurous acid, H_2SO_3, are 1.2×10^{-2} and 5.6×10^{-8}. What is the pH of a $0.1M$ solution of sodium sulfite, Na_2SO_3?

15-16. Calculate the pH of a $0.1M$ solution of aniline (Table 15-3).

15-17. A $0.15M$ solution of ammonia is to have its pH set at 7.8 by addition of ammonium chloride. Calculate the concentration of ammonium chloride required to do this. Does this calculated value exceed the solubility of ammonium chloride in water? (Consult a handbook.)

15-18. Amino acids are compounds in which there is a carboxylic acid group similar to that in acetic or benzoic acid, and an amino group, $-NH_2$, as in methylamine. If it is assumed that these two groups in the amino acid have approximately the same acid and base properties as in compounds where they are present separately, what is the state of the amino acid in a neutral water solution? In a basic solution? In an acid solution? As an example of an amino acid, consider glycine,

$$\mathrm{NH_2{-}CH_2{-}C({=}O){-}O{-}H}$$

15-19. Which of the following insoluble barium salts are also insoluble in strongly acidic medium, *e.g.*, in concentrated hydrochloric acid? Explain.

$Ba_3(PO_4)_2$ $BaSO_4$ $BaCO_3$

15-20. 50 ml of $0.1M$ solution of sodium cyanide is added to 50 ml of a $0.1M$ solution of acetic acid. When equilibrium is established, which solute species are present in greatest concentration?
[*Hint*: Set up the appropriate dissociation constant expressions, and determine the magnitude of the K for the reaction $CN^- + HOAc \rightleftharpoons HCN + OAc^-$.]

15-21. Calculate the pH of a $0.1M$ solution of sodium phenoxide, $NaOC_6H_5$; of sodium carbonate; of diethylamine hydrochloride, $(C_2H_5)_2NH_2Cl$.

15-22. What is a buffer solution? If you were given two solutions which have the same pH, one a buffer solution and the other not, how would you go about distinguishing them?

15-23. You are given $1M$ solutions of acetic acid and ammonia, and solid ammonium chloride and solid sodium acetate. What would you do to make up buffer solutions which have pH's of 6, 7, and 8, from these materials?

15-24. When solid calcium carbonate is placed in contact with water it does not dissolve. When the solution is made acid, however, it does so. Explain, using chemical equations to illustrate your answer.

15-25. K_{sp} for barium iodate, $Ba(IO_3)_2$, is 1.3×10^{-9}. What is the solubility of this salt in moles per liter? Barium iodate is placed in contact with a $0.025M$ solution of iodic acid, a strong acid. What is the solubility of the salt in this solution, in moles per liter?

15-26. An acid has the formula H_2B, where B is some divalent anion, B^{-2}. The first dissociation constant is 10^{-4}, the second is 10^{-8}.
(a) What is the pH of a $0.1M$ solution?
(b) 0.1 mole per liter of Na_2B is added to the $0.1M$ solution of the acid. What is the pH of the new solution?
(c) A $0.1M$ solution of Na_2B has what pH value?

15-27. Construct a titration curve for the titration of sodium cyanide. Begin with 25 ml of a $0.1M$ solution of sodium cyanide, and calculate the pH of the solution after adding the following amounts of $0.1M$ hydrochloric acid solution: 0, 20, 23, 24, 25, 26, 27 ml. Graph the calculated pH's *versus* the volume of acid added. What indicator (from those listed in Table 15-5) would you wish to employ in this titration?

15-28. Using the value for K_{sp} listed in Table 15-7, calculate the number of copper ions per liter in a solution which is saturated with CuS. Calculate the number of grams of CuS in a liter of the saturated solution.

15-29. K_{sp} for barium chromate is 1.2×10^{-10}. The acid H_2CrO_4 is a strong acid in the first ionization, but the dissociation constant for loss of the second proton is 3.0×10^{-7}. If a solution contains $0.01M$ sodium chromate and $0.01M$ barium chloride, at what pH will precipitation of barium chromate just begin? [*Hint:* Precipitation will just begin when the $[Ba^{+2}][CrO_4^{-2}]$ product is just 1.2×10^{-10}. $[Ba^{+2}]$ we know to be $0.01M$, but $[CrO_4^{-2}]$ is *not* $0.01M$ when an appreciable amount of acid is present. For precipitation to just begin, therefore, $[CrO_4^{-2}]$ must be $1.2 \times 10^{-10}/10^{-2} = 1.2 \times 10^{-8}$. What pH is required for this concentration of chromate ion?]

15-30. K_{sp} for calcium carbonate is 4.7×10^{-9}. Calculate the number of grams of this salt in a liter of saturated solution. Well water which has been in contact with deposits of limestone ($CaCO_3$) acquires calcium ion by dissolving $CaCO_3$. How does the amount of carbon dioxide (a weak acid) present in the solution affect the amount of calcium ion in the water at equilibrium?

15-31. The compound Ag_2CrO_4 has a solubility of 1.32×10^{-2} g/liter in water. What is K_{sp} for this salt?

15-32. In all of the problem solving we have done in this chapter we have employed concentrations in the equilibrium constant expressions. But the equilibrium constant is really constant only when activities are employed. The activity of an ion in solution is given by the product of the concentration times the activity coefficient: $a = \gamma c$. The activity coefficient in water solutions is a function of the ionic strength, as described for dilute solutions by the Debye-Hückel limiting law (see Problem 14-16). With these points in mind, explain why the solubility of copper (II) iodate in water at 25°C varies with the concentration of potassium chloride, as the following data show:

KCl conc (molal)	$Cu(IO_3)_2$ solubility (molal)
0.0000	0.003245
0.0050	0.00340
0.0100	0.00352
0.0200	0.00373

15-33. Calculate the concentration of free Hg^{+2} ion in a solution which is $0.01M$ in $Hg(NO_3)_2$ and $0.5M$ in cyanide ion, CN^-.

15-34. What explanation can you offer for the fact that the dissociation constant for the ammine complex of cobalt (III) is so much smaller than for cobalt (II), (Table 15-6)? (*Note:* This is discussed in detail in a later chapter, but you may be able to offer some suggestions at this point based on general considerations.)

15-35. The formation constant of the iron (III) oxalate complex, $Fe(C_2O_4)_3^{-3}$, is 1.5×10^{20}. Calculate the free iron (III) concentration in a solution containing a total iron (III) concentration of $0.02M$ and a total oxalate ion concentration of $0.15M$.

15-36. Solid $Ca(OH)_2$ and $CaSO_4$ are placed in contact with some water at 0°C, and allowed to come to equilibrium. From the data given below, determine the concentrations of Ca^{+2}, OH^-, and SO_4^{-2} in the saturated solution.

	Solubility in water (g/100g H_2O)
	0°C
$Ca(OH)_2$	0.185
$CaSO_4$	0.175

Chapter 16

Electric Cells

16-1 Introduction

Electrolyses of fused ionic substances and aqueous solutions of ionic substances illustrate the use of electrical energy for producing chemical change. By the same token it is also possible to convert chemical into electrical energy. This was first accomplished by Allesandro Volta in 1800.*

We will begin our discussion of electric, or voltaic, cells by considering one of the chemical reactions utilized by Volta. When a piece of zinc is immersed in a solution containing copper (II) ion, copper metal is deposited on the zinc bar. At the same time, zinc goes into solution as $Zn^{+2}(aq)$ ion.

$$Cu^{+2}(aq) + Zn \rightarrow Cu + Zn^{+2}(aq) \qquad (16\text{-}1)$$

This oxidation-reduction reaction involves a transfer of electrons from zinc metal to copper ions. We do not need to know the mechanism by which this occurs; the important thing is that there is a driving force which

*There is evidence that the Parthians, who lived in the region about Bagdad, employed electric cells as early as 250 B.C.

causes the reaction to occur spontaneously. Utilization of the chemical reaction in accomplishing useful electrical work requires simply that the spontaneous flow of electrons from reductant to oxidant be made to occur through an external path. When the reaction occurs in the beaker as described above, all of the energy difference between products and reactants appears as heat. By causing the electrons to flow through an external circuit, at least a part of the heat energy can be converted into useful work. The thermodynamic aspects of this system are discussed in Sec. 9-8.

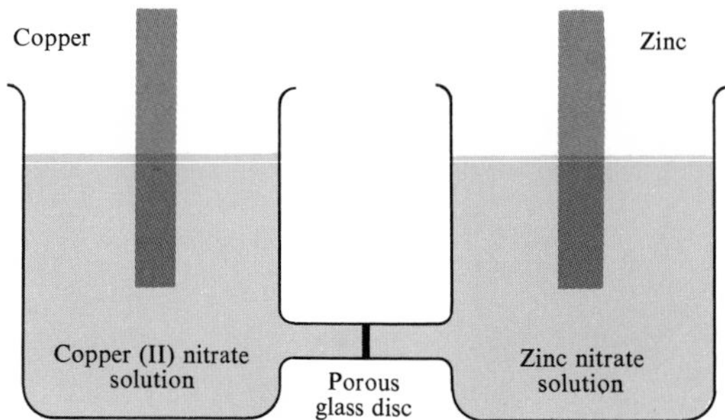

Fig. 16-1 A voltaic cell based on the reaction $Cu^{+2} + Zn = Cu + Zn^{+2}$.

A voltaic cell employing the copper-zinc reaction is shown in Fig. 16-1. In one compartment a copper electrode is immersed in a copper (II) nitrate solution at unit activity. In the other compartment a zinc bar is immersed in a zinc (II) nitrate solution, also at unit activity. The compartments are joined by a porous glass disc which, while it permits contact of the solutions, prevents extensive mixing.

If the metal electrodes were connected by a length of copper wire the chemical reaction described by Eq. (16-1) would proceed at a rapid rate. In the left-hand compartment, copper (II) ions would deposit on the electrode; in the compartment on the right, zinc would enter the solution as Zn^{+2} ion. The over-all chemical reaction can be written as the sum of two half-reactions:

$$\begin{aligned} &Cu^{+2}(aq) + 2e^- \rightarrow Cu \\ &Zn \rightarrow 2e^- + Zn^{+2}(aq) \end{aligned} \qquad (16\text{-}2)$$

Since electrons are released at the zinc electrode, and taken up at the copper electrode, electron flow in the external circuit is from right to left. The zinc bar is the anode — the electrode at which oxidation occurs — and the copper bar is the cathode. Note that the anode is negative in sign, as it is in all electric cells.

Progress of the chemical reaction in the cell is accompanied by migration of ions in the solution. It is clear that some movement of ions is necessary, since cations are being removed from the compartment on the left and added to the compartment on the right. If we could observe the flow of

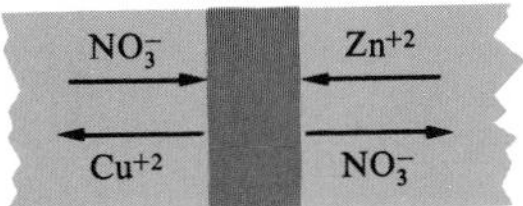

Fig. 16-2 The flow of ions through the porous glass disc during cell reaction.

ions through the porous glass disc (Fig. 16-2), we would find nitrate ions migrating toward the right, where new cations are being produced. Cations, on the other hand, would be observed to migrate toward the left, since there is a deficiency of cations here resulting from the reduction of copper ions. The relative numbers of ions which cross the disc in a period of time are governed by their relative mobilities (Chap. 11). The migration of ions must be such as to maintain electrical neutrality in both cell compartments.

16-2 Half-Cell Potentials

The voltage developed by the cell is measured by inserting a voltmeter in the external circuit. Just as the over-all cell reaction can be expressed as the sum of two half-cell reactions, so the over-all cell voltage can be expressed as the sum of two half-cell voltages. It is not possible to measure the voltage developed by an isolated half-cell, so we can never know more than the algebraic sum of two half-cell voltages. A scale of relative half-cell voltages can be developed, however, by arbitrarily assigning a value to one particular half-cell potential. The hydrogen-hydronium ion half-cell is assigned a potential of zero. When this half-cell is combined with the $Zn—Zn^{+2}(aq)$ half-cell, as in Fig. 16-3, the total cell potential is equal to the value of the half-cell potential for $Zn \rightarrow Zn^{+2}(aq) + 2e^-$. This is true because the over-all cell potential is the sum of the half-cell potentials.

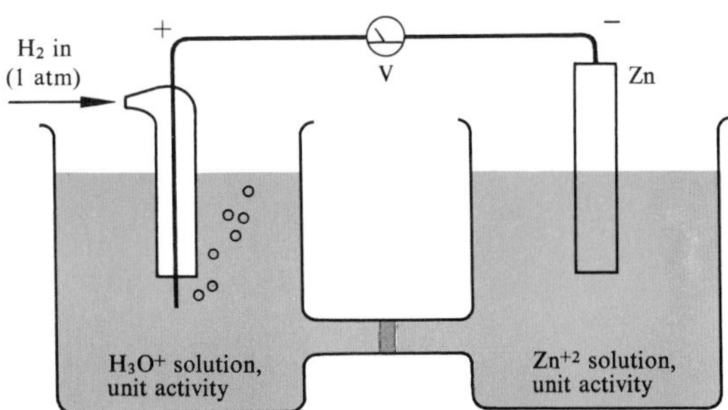

Fig. 16-3 Determination of the $Zn \rightarrow Zn^{+2} + 2e^-$ standard half-cell potential.

$$
\begin{array}{ll}
H_2 + 2H_2O \rightarrow 2H_3O^+ + 2e^- & \mathcal{E}_1^\circ = 0 \\
Zn \rightarrow Zn^{+2}(aq) + 2e^- & \mathcal{E}_2^\circ = X \\
\hline
2H_3O^+ + Zn \rightarrow Zn^{+2}(aq) + 2H_2O + H_2 &
\end{array}
\tag{16-3}
$$

$$
\begin{aligned}
\mathcal{E}_T^\circ &= -\mathcal{E}_1^\circ + \mathcal{E}_2^\circ \\
&= 0 + X
\end{aligned}
$$

Note that the half-cell reactions were both written as oxidation steps. One of them obviously must be reversed in the over-all reaction. To determine which, the polarity of the voltage measured must be known. The zinc electrode in this cell is negative (Fig. 16-3). This means that it is the anode, and that the reaction $Zn \rightarrow Zn^{+2}(aq) + 2e^-$ is the reaction which proceeds spontaneously. The complete cell reaction as written in Eq. (16-3) is therefore descriptive of the direction in which the cell reaction actually proceeds, and the measured voltage for the cell, 0.76 volt, is positive for the reaction as written. Since $\mathcal{E}_T^\circ$ is positive, the half-cell potential $\mathcal{E}_2^\circ$ must also be positive:

$$+0.76 = 0 + \mathcal{E}_2^\circ$$

It should be understood that the measured voltage, 0.76 volt, corresponds to standard conditions, as described in Fig. 16-3. Solutions of unit activity in the electrolytes would not, in general, be $1M$, since the activity coefficients would not generally be unity (see Table 14-1).

The cell described by Fig. 16-4 illustrates a slightly different situation. The two half-cell reactions in this instance are

$$
\begin{array}{ll}
H_2 + 2H_2O \rightarrow 2H_3O^+ + 2e^- & \mathcal{E}_1^\circ = 0 \\
Cu \rightarrow Cu^{+2}(aq) + 2e^- & \mathcal{E}_2^\circ = Y \\
\hline
2H_3O^+ + Cu \rightarrow Cu^{+2}(aq) + H_2 + 2H_2O &
\end{array}
\tag{16-4}
$$

$$
\begin{aligned}
\mathcal{E}_T^\circ &= -\mathcal{E}_1^\circ + \mathcal{E}_2^\circ \\
&= 0 + Y
\end{aligned}
$$

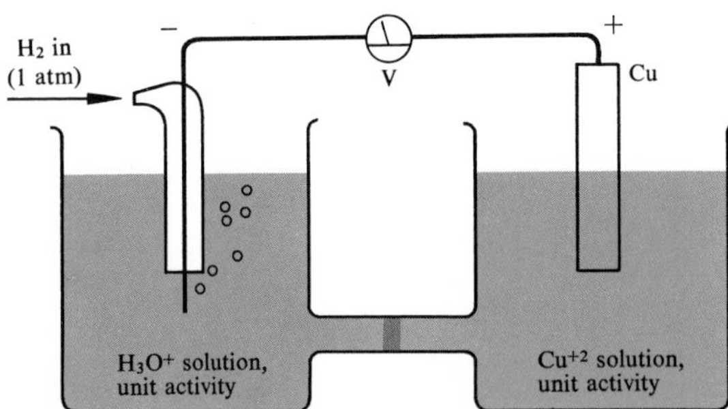

Fig. 16-4 Determination of the standard potential for the $Cu \rightarrow Cu^{+2} + 2e^-$ half-cell.

We now see that the measured potential, 0.34 volt, is of such a polarity as to make the metal electrode positive. From this we conclude that the reaction occurring spontaneously at the copper electrode is reduction:

$$Cu^{+2}(aq) + 2e^- \rightarrow Cu$$

Since the reaction as written in Eq. (16-4) involves oxidation of copper, this is clearly the reverse of the spontaneous reaction. The voltage for the reaction as written is therefore negative in sign. Since $\mathcal{E}_1^\circ$ is equal to zero, $\mathcal{E}_2^\circ$ (the half-cell potential corresponding to oxidation of copper) is -0.34 volt. By reversing the direction of the reaction we reverse the sign of the potential, so $\mathcal{E}^\circ$ for the reaction

$$Cu^{+2}(aq) + H_2 + 2H_2O \rightarrow Cu^{+2}(aq) + 2H_3O^+ \qquad (16\text{-}5)$$

is $+0.34$ volt.

A large number of half-cells can be compared under standard conditions with the hydrogen-hydronium ion half-cell in the manner illustrated by these two examples. From the results a list of standard electrode, or half-cell, potentials can be formed. Such a list has great utility, since any two half-cell reactions can be combined to produce a cell reaction represented by a complete oxidation-reduction equation. As an example, the cell illustrated in Fig. 16-1, and represented by the two half-cell reactions of Eq. (16-2), possesses under standard conditions a potential of $0.76 + 0.34 = 1.10$ volts.

Standard half-cell potentials are tabulated, as in Appendix B, for half-cell reactions written as oxidation processes.

The half-cell reactions may also be written as reduction steps; the half-cell potentials are then referred to as *reduction potentials.* The reaction described in the reduction half-reaction is the reverse of that described in the oxidation half-reaction. The reduction potential is thus the same numerically as the oxidation potential, but has an opposite sign.

To illustrate the use of the table, consider the reaction

$$2Ag + 2HClO + 2H_3O^+ \rightarrow 2Ag^+(aq) + Cl_2 + 4H_2O \qquad (16\text{-}6)$$

for which the two half-cell reactions are

$$2Ag \rightarrow 2Ag^+(aq) + 2e^- \qquad \mathcal{E}^\circ = -0.799$$
$$2e^- + 2HClO + 2H_3O^+ \rightarrow Cl_2 + 4H_2O \qquad \mathcal{E}^\circ = +1.63$$

(*Note:* The second reaction is the reverse of that listed in the Appendix. By reversing the direction of the reaction, the sign of the half-cell potential is also changed.) The total $\mathcal{E}^\circ$ for the cell is $-0.799 + 1.63 = +0.83$ volt.

It should be noted that the half-cell reaction representing the oxidation of silver is multiplied through by two in the balanced chemical equation. This has no effect, however, on the value for the potential. The voltage of an electric cell is what we term an *intensive*, rather than an *extensive*, property. An intensive property is related to the properties and behaviors of individual molecules; for example, the temperature or pressure of a

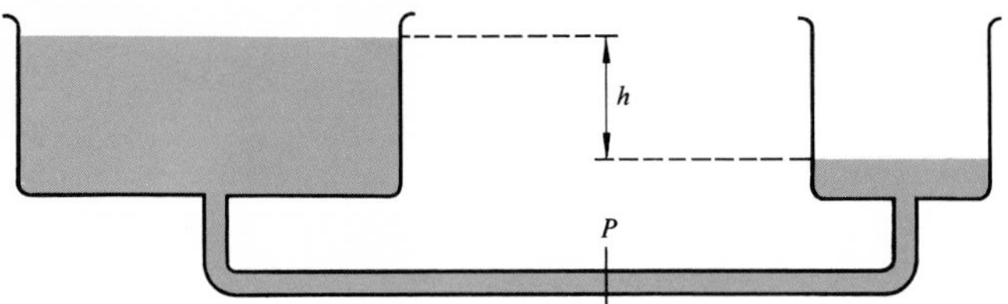

Fig. 16-5 Hydraulic analogue of the electric cell.

gas. An extensive property is related to the quantity of matter present, for example, total energy, volume, *etc.*

The point is perhaps most simply seen by making an analogy with a hydraulic system, as illustrated in Fig. 16-5. The pressure difference operating across the plane P in this system is dependent only on the difference in the heights of the water levels h and the gravitational constant, but not at all on the total amount of water present in the reservoirs. The amount of work which can be realized from the system, however, does depend both on the pressure and the total quantity of water. Similarly in the electric cell, if n equivalents of electrons are transferred under a potential of $\mathcal{E}$ volts, the maximum work which can be accomplished is the product of the total charge transferred and the potential:

$$W = n\mathcal{F}\mathcal{E} \tag{16-7}$$

When n is expressed as the number of equivalents, $\mathcal{F}$ — the value of the faraday (quantity of charge per equivalent) — in coulombs, and $\mathcal{E}$ in volts, the work is expressed in units of joules. To convert to calories, it is necessary to divide on the right by 4.184 (see p. 17); the formula for the work in calories is then

$$W = \frac{n \times 96{,}500 \times \mathcal{E}}{4.184}$$
$$= 23{,}060 n\mathcal{E}$$

16-3 The Nernst Equation

Suppose that the reaction

$$n\mathrm{A} + m\mathrm{B} \rightarrow r\mathrm{C} + s\mathrm{D}$$

can be made the basis of an electric cell. In Sec. 14-2, it is shown that the free energy change in the reaction can be described by the equation

$$\Delta G = \Delta G^\circ + RT \log_e \frac{a_\mathrm{C}^r \times a_\mathrm{D}^s}{a_\mathrm{A}^n \times a_\mathrm{B}^m} \tag{16-8}$$

If the cell reaction is allowed to proceed isothermally and reversibly, the work obtainable from the cell represents the free energy change in the cell reaction. From Eq. (16-7), therefore, in combination with Eq. (16-8), we can obtain

$$\Delta G = -W = -n\mathfrak{F}\mathcal{E} \tag{16-9}$$

The negative sign is in keeping with the convention that work done by the system is negative. Substitution of $\Delta G = -n\mathfrak{F}\mathcal{E}$, or $\Delta G^\circ = -n\mathfrak{F}\mathcal{E}^\circ$, into the above expression yields Eq. (16-10):

$$\mathcal{E} = \mathcal{E}^\circ - \frac{RT}{n\mathfrak{F}} \log_e \frac{a_C^r \times a_D^s}{a_A^n \times a_B^m} \tag{16-10}$$

The quantities in the log term are the activities of the reactant species. If these are all unity, the log term is zero, and $\mathcal{E} = \mathcal{E}^\circ$, as it should, since $\mathcal{E}^\circ$ is defined for the case where all activities are unity. When they are not all unity, $\mathcal{E}$ is in general not equal to $\mathcal{E}^\circ$. There are an infinity of states in which this may be true, but the state in which the system is in chemical equilibrium is of particular importance. *Under these conditions the cell voltage $\mathcal{E}$ must be zero;* if it were not, the reaction would proceed spontaneously in one direction or the other, contradictory to the definition of equilibrium. We have then

$$0 = \mathcal{E}^\circ - \frac{RT}{n\mathfrak{F}} \log_e \frac{a_C^r \times a_D^s}{a_A^n \times a_B^m} \tag{16-11}$$

But if the system is at equilibrium the ratio of terms in the log expression is equal to the equilibrium constant:

$$\frac{a_C^r \times a_D^s}{a_A^n \times a_B^m} = K_{eq}$$

$$\mathcal{E}^\circ = \frac{RT}{n\mathfrak{F}} \log_e K_{eq} \tag{16-12}$$

By means of Eq. (16-9), it is possible to calculate the equilibrium constant for a reaction from a knowledge of the required half-cell standard potentials. To illustrate, consider Eq. (16-6), for which $\mathcal{E}^\circ$ was determined to be 0.83 volt:

$$0.83 = \frac{0.0592}{2} \log_{10} K_{eq} \tag{16-13}$$

The value of $2.3RT/\mathfrak{F}$ is 0.0592 at 25°C. The factor 2.3 converts the log term from base e to base 10. Then

$$\log_{10} K_{eq} = 28.08$$

$$K_{eq} = \frac{a_{Ag^+}^2 \times a_{Cl_2}}{a_{HClO}^2} = 1.2 \times 10^{28}$$

Equation (16-10) is useful in calculating the potential for a cell in which all of the species are not at unit activity. Consider, for example, a cell represented by the equation

$$2Fe^{+2}(aq) + Cl_2(aq) \rightarrow 2Fe^{+3}(aq) + 2Cl^-(aq)$$

in which the following concentrations obtain:

$[Fe^{+2}] = 0.1M$; $[Fe^{+3}] = 0.01M$; $[Cl_2] = 0.002M$; $[Cl^-] = 0.001M$

The half-reactions and standard half-cell potentials are

$$Fe^{+2}(aq) \rightarrow Fe^{+3}(aq) + e^- \qquad \mathcal{E}^\circ = -0.771 \text{ volt}$$
$$Cl_2 + 2e^- \rightarrow 2Cl^-(aq) \qquad \mathcal{E}^\circ = +1.36 \text{ volts}$$
$$\mathcal{E}^\circ_T = +0.59 \text{ volt}$$

Approximating the activities by the concentrations, we have

$$\mathcal{E} = 0.59 - \frac{0.0592}{2}\log_{10}\frac{[10^{-3}]^2[10^{-2}]^2}{[2 \times 10^{-3}][10^{-1}]^2}$$
$$\mathcal{E} = 0.75 \text{ volt}$$

Since the activities of ions in solution are well approximated by the concentrations only in dilute solution, most measurements to determine values of $\mathcal{E}^\circ$ would actually be made in dilute solution.

The concentration cell is an interesting special case of a voltaic cell. Suppose that we have a cell such as that shown in Fig. 16-1, except that both electrodes are copper, and the electrolytic solutions in the two beakers are copper (II) nitrate; $0.01M$ in the left beaker, $0.1M$ in the right beaker. If we write the two half-reactions as though they were both oxidations, we have

$$Cu(s) \rightarrow Cu^{+2}(0.1M) + 2e^-$$
$$Cu(s) \rightarrow Cu^{+2}(0.01M) + 2e^-$$

Which of these half-reactions is more likely to occur? From entropy considerations we might conclude that the second, in which the copper goes into the more dilute solution, is favored. If one mole of copper is converted to copper (II) ion in a $0.01M$ solution, the ions are distributed throughout a larger volume than they are when a $0.1M$ solution is formed. The degree of randomness is thus larger, and ΔS should be more positive. We assume, therefore, that the over-all cell reaction proceeds in the direction:

$$Cu^{+2}(0.1M) \rightarrow Cu^{+2}(0.01M) \qquad (16\text{-}14)$$

Inserting the appropriate values into the Nernst equation, we have

$$\mathcal{E} = \mathcal{E}^\circ - \frac{0.0592}{2}\log_{10}\frac{[10^{-2}]}{[10^{-1}]}$$

$\mathcal{E}^\circ$ is, of course, zero, since the two half-reactions are identical when the activities of the ions are the same. Thus,

$$\mathcal{E} = -0.0295 \log_{10}(10^{-1})$$
$$= +.03 \text{ volt}$$

The cell voltage is positive for the reaction as written in the form of Eq. (16-14). This means that the copper electrode immersed in the $0.01M$ solution is the anode, and is negative. Concentration cells appear to be of importance in certain biological systems, where they may have something to do with transmission of nerve impulses. They are important also in metal corrosion.

16-4 Design of Electric Cells

Any oxidation-reduction reaction can, in principle, be made the basis of an electric cell. The physical arrangement required to produce a workable cell may, however, vary from one reaction to another. The most important variables involved are the choice of electrodes and the method of electrolytically joining the two half-cells. The use of cells as sources of electrical potential is discussed in Sec. 16-5. We are interested here in setting up cells for the purpose of measuring the potential. In making such measurements, it is necessary to draw as little current as possible from the cell. Special potentiometer circuits are used for this purpose. The variety of cell types encountered is exemplified by the following three oxidation-reduction equations:

$$Cu^{+2}(aq) + Zn \rightarrow Cu + Zn^{+2}(aq) \qquad (16\text{-}15)$$

$$2Ag(s) + Br_2(aq) \rightarrow 2AgBr(s) \qquad (16\text{-}16)$$

$$14H_3O^+ + 6Fe^{+2}(aq) + Cr_2O_7^{-2}(aq) \rightarrow 6Fe^{+3}(aq) + 2Cr^{+3}(aq) + 21H_2O \qquad (16\text{-}17)$$

The cell represented by the reaction of Eq. (16-15) is shown in Fig. 16-1. In this case, both electrodes enter into the chemical reaction. The half-cells are joined by a liquid junction at the porous disc. A suitable cell arrangement for the reaction represented by Eq. (16-16) is shown in Fig. 16-6. To see how this arrangement is arrived at, the over-all reaction should be looked at in terms of the half-reactions:

$$2Ag + 2Br^-(aq) \rightarrow 2AgBr(s) + 2e^-$$
$$Br_2(aq) + 2e^- \rightarrow 2Br^-(aq)$$

We see that bromide ion is involved in both half-cell reactions. The same electrolyte (*e.g.*, KBr) may, therefore, be used in both compartments. The half-cells need to be separated only to prevent bromine from diffusing to the silver electrode. The silver electrode is prepared by forming a coating of silver bromide on at least a part of its surface. The amount of AgBr

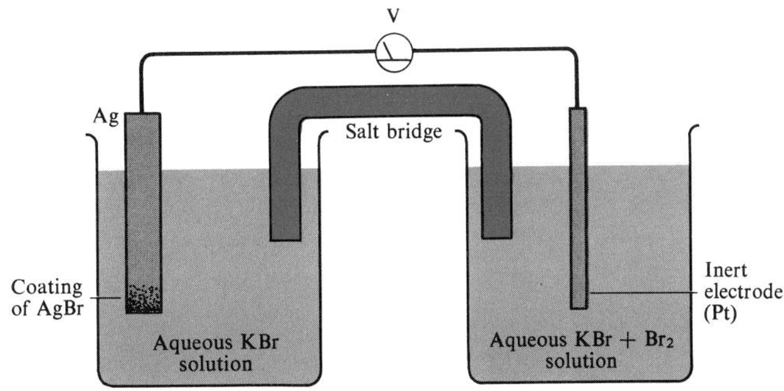

Fig. 16-6 A cell designed to operate according to the reaction $2Ag(s) + Br_2(aq) \rightarrow 2AgBr(s)$.

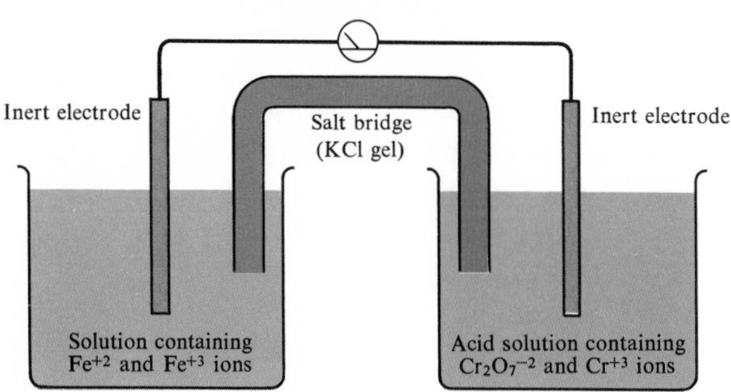

Fig. 16-7 A cell designed to operate according to the reaction expressed in Eq. (16-17).

formed is not important, since the activity of the pure solid is unity, no matter how much or how little is present. The two half-cells are joined by a salt bridge, which consists of an electrolyte (potassium bromide in this instance) in a gelatinous solution. It provides for electrolytic conduction, but mixing of the solutions is minimized. The electrode at which bromine is reduced is inert; it does not appear in the balanced chemical equation for the cell. Any suitably inert conductor such as a platinum wire or carbon rod will do.

Both electrodes in the cell for the reaction of Eq. (16-17) are inert (Fig. 16-7). A salt bridge rather than a liquid junction is required in this instance, since the components in solution must be kept separated. A liquid junction can be used only when the substances in solution do not react on contact.

16-5 Applications

Allesandro Volta first communicated the results of his experiments with electric cells to the Royal Society of London in 1800. He developed the "*voltaic pile*," which consists of pairs of metal plates (he preferred silver and zinc) separated by sheets of cardboard or other material soaked in water or salt solution. By stacking a number of such units together, a considerable voltage could be developed.

Volta's work came at a time when the scientific world was waiting — so to speak — for someone to develop a manageable source of electric energy. Within six weeks of the receipt of his letter by the Royal Society, Nicholson and Carlisle had constructed a voltaic pile and used it to electrolyze water for the first time. Humphrey Davy, using voltaic piles, had succeeded in producing the alkali metals by electrolysis by 1807. There were so many experiments to perform with the electricity that no one gave much thought to the voltaic cells themselves.

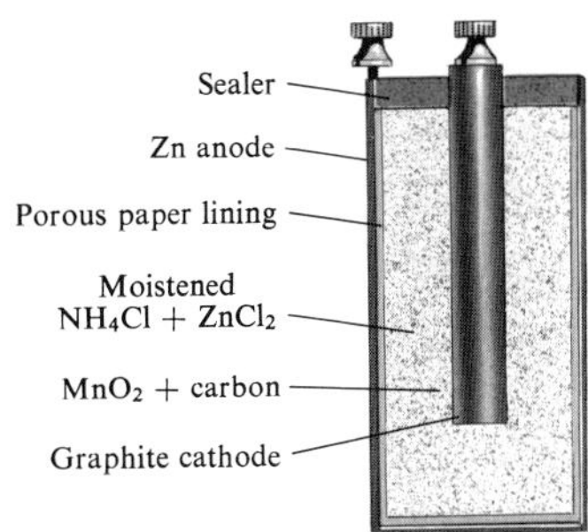

Fig. 16-8 The Leclanché dry cell.

The usefulness of electric cells as compact, portable sources of electrical energy has continued to the present. We may classify electrochemical power sources as primary cells or storage cells. In primary cells, chemical reactions are utilized without regard for their reversibility. Storage cells, on the other hand, are designed so that the chemical reaction can be reversed by application of an external potential.

Primary cells — The most familiar primary cell is the Leclanché dry cell. The electrodes are a zinc anode and a carbon rod cathode (Fig. 16-8). The substance reduced at the cathode is manganese dioxide. The half-cell reactions are (approximately)

$$2e^- + 2MnO_2 + 2H_3O^+ \rightarrow Mn_2O_3 \cdot H_2O + 2H_2O$$
$$4H_2O + Zn + 2OH^- \rightarrow Zn(H_2O)_4(OH)_2 + 2e^- \qquad (16\text{-}18)$$

The over-all reaction, after eliminating water from the equation, is

$$Zn + 2MnO_2 \rightarrow ZnO + Mn_2O_3 \qquad (16\text{-}19)$$

A number of substances are added to facilitate the reaction in one way or another. (They also serve to complicate the chemistry!) These include carbon black, ammonium chloride, and zinc chloride.*

The mercury dry cell uses a zinc anode and potassium hydroxide electrolyte. Mercuric oxide is reduced at the cathode. The over-all cell reaction is represented by the equation

$$Zn + H_2O + HgO \rightarrow ZnO + Hg + H_2O \qquad (16\text{-}20)$$

Depending upon the composition, the voltage of a mercury dry cell may vary from 1.10 to 1.25 volts. This cell has a much higher capacity per unit weight than the older Leclanché cell, and finds uses in hearing aids and similar devices where small size is particularly desirable.

---

*For a complete discussion of the Leclanché and other primary cells, see W. Vinal, *Primary Batteries*, New York: John Wiley & Sons, Inc., 1950.

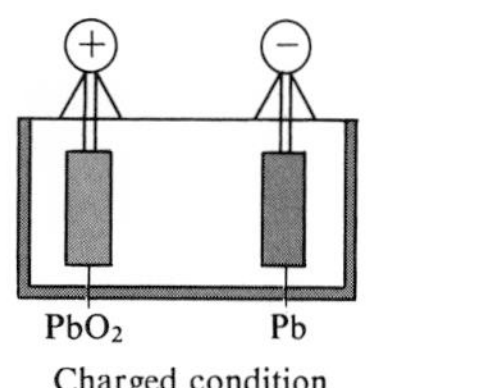

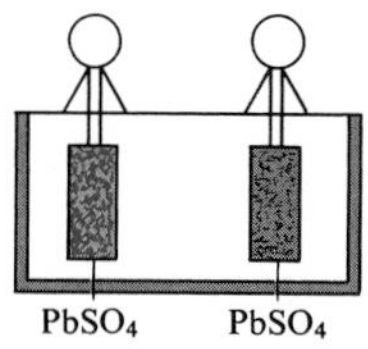

Fig. 16-9 A schematic illustration of the lead-acid storage cell.

Storage cells — The lead-acid cell, shown schematically in Fig. 16-9, is the most commonly employed storage cell. When in a fully charged condition, one electrode is a lead plate covered with a thick coating of lead dioxide, PbO_2. The other electrode is metallic lead; the electrolyte is sulfuric acid. The cell reaction is

$$PbO_2(s) + Pb(s) + 4H_3O^+ + 2SO_4^{-2} \rightarrow 2PbSO_4(s) + 6H_2O \quad (16\text{-}21)$$

Under the conditions normally encountered in automobile batteries,† the concentration of sulfuric acid in a fully charged cell is about $10.4N$. As the cell becomes discharged the concentration of electrolyte decreases. The reaction is reversible; in automobiles a generator unit provides the charging potential while the engine is running. A voltage regulator disconnects the generator when the battery voltage is up to that expected for full charge. The potential developed by the cell in Fig. 16-9 is about 2 volts. Six such cells are arranged in series in auto batteries to provide the 12 volts required.

The Edison cell is a storage cell which uses finely divided iron as the anode and hydrous nickel oxide, $NiO_2 \cdot xH_2O$, as the cathode. Both materials are supported on steel plates. The electrolyte is potassium hydroxide. The cell reaction is

$$Fe + NiO_2 + 2H_2O \rightarrow Fe(OH)_2 + Ni(OH)_2 \quad (16\text{-}22)$$

The cell voltage is about 1.4 volts. Despite its longer life expectancy under rough usage, the Edison cell has been too expensive to compete with the lead-acid cell for the large automobile battery market.

There has recently been an intensive interest in high energy density batteries which might be used to power an electric automobile. To achieve the desired goals, it is necessary to employ a cell reaction which is capable of generating a large voltage. This, in turn, means employing one of the more active metals; *e.g.*, lithium. Because lithium reacts rapidly with water, it is necessary to use a non-aqueous medium in the cell. This may be a

†The term battery refers, stricly speaking, to two or more cells connected together to give either greater current capacity or higher voltage. The distinction between a cell and a battery is usually not observed, however, in popular usage.

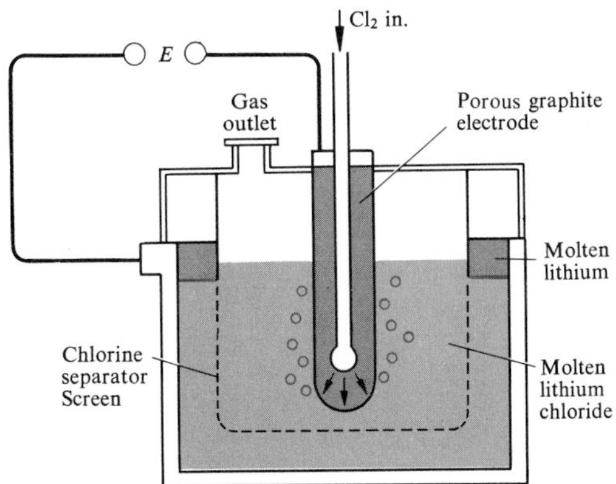

Fig. 16-10 A schematic illustration of a lithium-chlorine cell.

molten salt, or a polar organic liquid. One such cell, produced by the General Motors Corporation, utilizes the reaction

$$2Li(l) + Cl_2(g) \rightarrow 2LiCl(l)$$

The cell is illustrated schematically in Fig. 16-10. Chlorine is admitted through a porous graphite electrode. The electrolyte is molten lithium chloride (m.p. 613°C). Molten lithium metal, which floats on the molten chloride, is the anode. The cell is charged by electrolyzing the lithium chloride; *i.e.*, by reversing the above reaction. The chlorine gas which is produced is carried off to a storage compartment, to be returned under pressure through the graphite cathode when the cell reaction occurs. A screen is required in the cell to prevent the direct contact of chlorine with molten lithium metal. The cell has some obvious disadvantages for routine use. It must be maintained at a high temperature while in use, and the cell components are toxic and would thus be dangerous in the event of a leak or an accident. The cell is really a research prototype of a model which might some day be in common use.

Fuel cells — While it is rather early to make predictions, it seems quite probable that, during the next ten years or so, fuel cells will become an important method for converting the chemical energy of readily available fuels into electrical energy. Our present methods of converting the chemical energy of fuels such as petroleum or coal into electrical energy involve burning the fuel in air, and then using the heat which is liberated to drive a mechanical device such as a gas turbine. This in turn is coupled to a generator in which electrical energy is produced. The efficiency of the entire operation is rather low, and is limited in theory by the convertibility of heat into work (Chap. 9). In practice, efficiencies of perhaps 30 per cent are obtained.

In contrast, the free energy change accompanying fuel combustion can, in principle, be converted into electrical energy with 100 per cent efficiency,

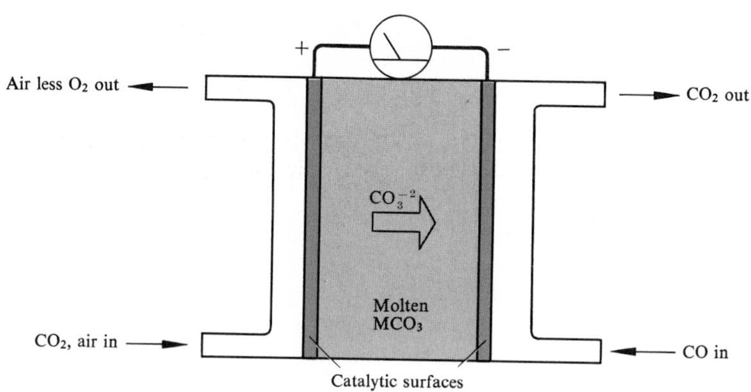

Fig. 16-11 A schematic illustration of a possible fuel cell.

if the reaction can be made the basis of an electric cell. Obviously this cannot be such a simple matter, or we would literally be surrounded with fuel cells. A number of difficulties need to be overcome before fuel cells become a practical reality, but a vigorous research effort is now under way in this area.

The essential components of a fuel cell are shown in Fig. 16-11. Suppose, for the sake of illustration, that the reaction to be utilized is the oxidation of carbon monoxide:

$$CO + \tfrac{1}{2}O_2 \rightarrow CO_2 \qquad (16\text{-}23)$$

The half-cell reactions are

$$\begin{aligned} CO + O^{-2} &\rightarrow CO_2 + 2e^- \\ \tfrac{1}{2}O_2 + 2e^- &\rightarrow O^{-2} \end{aligned} \qquad (16\text{-}24)$$

The half-cell reactions are made to occur at catalytically active surfaces which also permit the transfer of molecules or ions. The electrolyte is a molten carbonate; the oxide ions produced on the left react with CO_2 to form the carbonate ion, while on the right the carbonate ion combines with CO to yield carbon dioxide:

$$\begin{aligned} CO + CO_3^{-2} &\rightarrow 2CO_2 + 2e^- \\ 2e^- + CO_2 + \tfrac{1}{2}O_2 &\rightarrow CO_3^{-2} \end{aligned} \qquad (16\text{-}25)$$

Equations (16-25) differ from (16-24) only in the presence of CO_2, which acts as the oxide ion carrier. One obvious disadvantage of this fuel cell is that it must be operated at high temperatures, since molten metal carbonate is the electrolyte. Another is that the catalytically active electrode surfaces may be poisoned by impurities in the air or by carbon monoxide. It turns out in practice that, if high temperature systems are employed, a high catalytic efficiency is not required. There is then, however, the loss of efficiency in the heat required to maintain cell temperatures.

Suggested Readings

Bauman, R. P., *An Introduction to Equilibrium Thermodynamics*, Englewood Cliffs, N. J.: Prentice-Hall, Inc., 1966.
Egli, P. H., "Direct Energy Conversion," *American Scientist*, Vol. 48, 1960, p. 311.
Weissbart, J., "Fuel Cells," *J. Chem. Educ.*, Vol. 38, 1961, p. 267.

Exercises

16-1. The cell employed by the Parthians (see footnote on p. 383) consisted of a copper cylinder with a one-hole asphalt stopper through which projected an iron rod. What electrolyte they employed is unknown. Suggest some possible electrolytes for such a cell. Would it be better to use an ordinary aqueous solution or a gel of some type? Write the half-cell reactions and the over-all cell reaction, assuming that the electrolyte is dilute acetic acid.

16-2. Consider the cell illustrated in Fig. 16-7. Imagine a plane drawn through the midpoint of the salt bridge. Which ions are migrating in which direction across this plane when the cell is in operation? Explain.

16-3. (a) Draw a schematic diagram of a cell which would utilize the chemical reaction

$$5Mn + 2MnO_4^- + 16H_3O^+ \rightarrow 7Mn^{+2}(aq) + 24H_2O$$

(b) $\mathcal{E}°$ for this cell is 2.70 volts. $\mathcal{E}°$ for the half-reaction $Mn \rightarrow Mn^{+2}(aq) + 2e^-$ is 1.18 volts. Write the other half-cell reaction. Show how one would calculate $\mathcal{E}°$ for it.

(c) What would the voltage of this cell be if the concentrations were as follows: pH = 3, $[Mn^{+2}] = 10^{-1}M$, $[MnO_4^-] = 10^{-2}M$?

16-4. From the appropriate values of $\mathcal{E}°$, calculate the equilibrium constant for the reaction

$$Sn^{+2}(aq) + 2Fe^{+3}(aq) \rightarrow Sn^{+4}(aq) + 2Fe^{+2}(aq)$$

16-5. Calculate $\mathcal{E}°$ and K_{eq}, and write the equilibrium constant expression, for the reaction:

$$Br_2(l) + 2Fe^{+2}(aq) \rightarrow 2Br^-(aq) + 2Fe^{+3}(aq)$$

Calculate the relative concentrations of Fe^{+2} and Fe^{+3} ions in a solution made by adding 0.1ml of $1M$ Fe^{+2} solution to one liter of a solution which is saturated with respect to Br_2 (in contact with liquid Br_2), and which is $0.1M$ in Br^-.

16-6. Consider the following chemical reaction:

$$H_2O + Zn + HgO(s) \rightarrow Zn(OH)_2(s) + Hg$$

A cell can be designed for this reaction in which all of the materials are contained in one beaker. The electrolyte is sodium hydroxide solution. Draw a diagram for such a cell. Calculate the value of $\mathcal{E}°$ for the above reaction, and write the separate half-reactions.

16-7. Devise cells for each of the following chemical reactions:
(a) $Ce^{+4}(aq) + Ag(s) \rightarrow Ce^{+3}(aq) + Ag^{+}(aq)$
(b) $CuCl(s) + AgCl(s) \rightarrow Cu^{+2}(aq) + 2Cl^{-} + Ag(s)$
(c) $PbO(s) + H_2(g) \rightarrow Pb(s) + H_2O$

16-8. Two hydrogen-hydronium ion half-cells are connected together to make a single cell. The pH in one half-cell compartment is 1, in the other it is 6. Does this cell generate a potential? If so, calculate its magnitude.

16-9. Consider the cell diagrammed below:

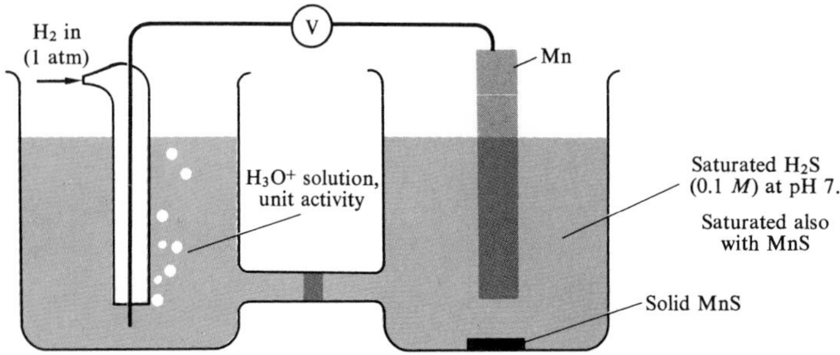

From the following data, calculate the potential of the cell.

K_{sp} for MnS is 10^{-11}
$Mn \rightarrow Mn^{+2} + 2e^{-} \qquad \mathcal{E}° = 1.18$ volts
$H_2 + 2H_2O \rightarrow 2H_3O^{+} + 2e^{-} \qquad \mathcal{E}° = 0$ volt
K_{1a} for $H_2S = 10^{-7}$, $\quad K_{2a} = 1.3 \times 10^{-13}$

16-10. Calculate the ratio $Fe^{+2}(aq)/Fe^{+3}(aq)$ at equilibrium in the system:

$$Ag(s) + Fe^{+3}(aq) \rightarrow Ag^{+}(aq) + Fe^{+2}(aq)$$

if $Ag^{+}(aq)$ at equilibrium is $10^{-2}M$.

16-11. Rearrange the Nernst equation so that the pH is given in terms of the variables in a cell for which the equation is:

$$Ag(s) + Cl^{-}(aq) + H_3O^{+} \rightarrow AgCl(s) + H_2O + \tfrac{1}{2}H_2(g)$$

Explain how a cell of this type might be used in carrying out an acid-base titration.

16-12. Measurement of the density of the solution in a lead-acid battery is often used to test for the state of charge of the battery. Explain.

16-13. When a lead-acid battery is charged too rapidly some of the lead dioxide which is formed may adhere to the plates poorly, and eventually fall to the bottom of the cell. What effect does this have on the properties of the battery (voltage, total capacity, *etc.*)?

16-14. 235 g of zinc is employed in the casing of one particular Leclanché dry cell. Assuming that this is all consumed in the cell reaction, how many grams of MnO_2 undergo reaction during the cell's lifetime? How many ampere-hours of electricity does the cell deliver during its lifetime?

16-15. From the values given in Appendix B for the half-cell reactions

$$Cu + Br^- \rightarrow CuBr + e^-$$
$$Cu \rightarrow Cu^+ + e^-$$

compute (using the Nernst equation) K_{sp} for CuBr.

16-16. Calculate from thermodynamic quantities the maximum voltage which the cell utilizing the reaction

$$MnF_2(s) + 2Li(s) \rightarrow Mn(s) + 2LiF(s)$$

might generate at 25°C, assuming that all substances involved are pure, and that the cell reaction is completely reversible.

Chapter 17

The Non-Metallic Elements

17-1 Introduction

Preceding chapters of this book have dealt with principles upon which one might base a study of chemistry. A number of important chemical systems have been dealt with in the course of discussing and illustrating these principles. In the next few chapters, we shall consider the chemical properties of elements and compounds in a more systematic fashion. Chemistry is, after all, the study of chemical change, the study of reactions. The principles which we have discussed are of value only insofar as they provide insight into what causes substances to behave as they do. Theories and "rules" may have another value, in that it is sometimes easier to remember chemical reactions, and to recognize similarities in different systems when such guidelines are available. It must not be thought that all observations made in the laboratory can readily be accounted for in terms of the concepts that we have so far dealt with; this is far from being the case. It is possible, however, to recognize and understand relationships among the physical and chemical properties of substances. This chapter deals with the properties of the non-metallic elements and some of their compounds.

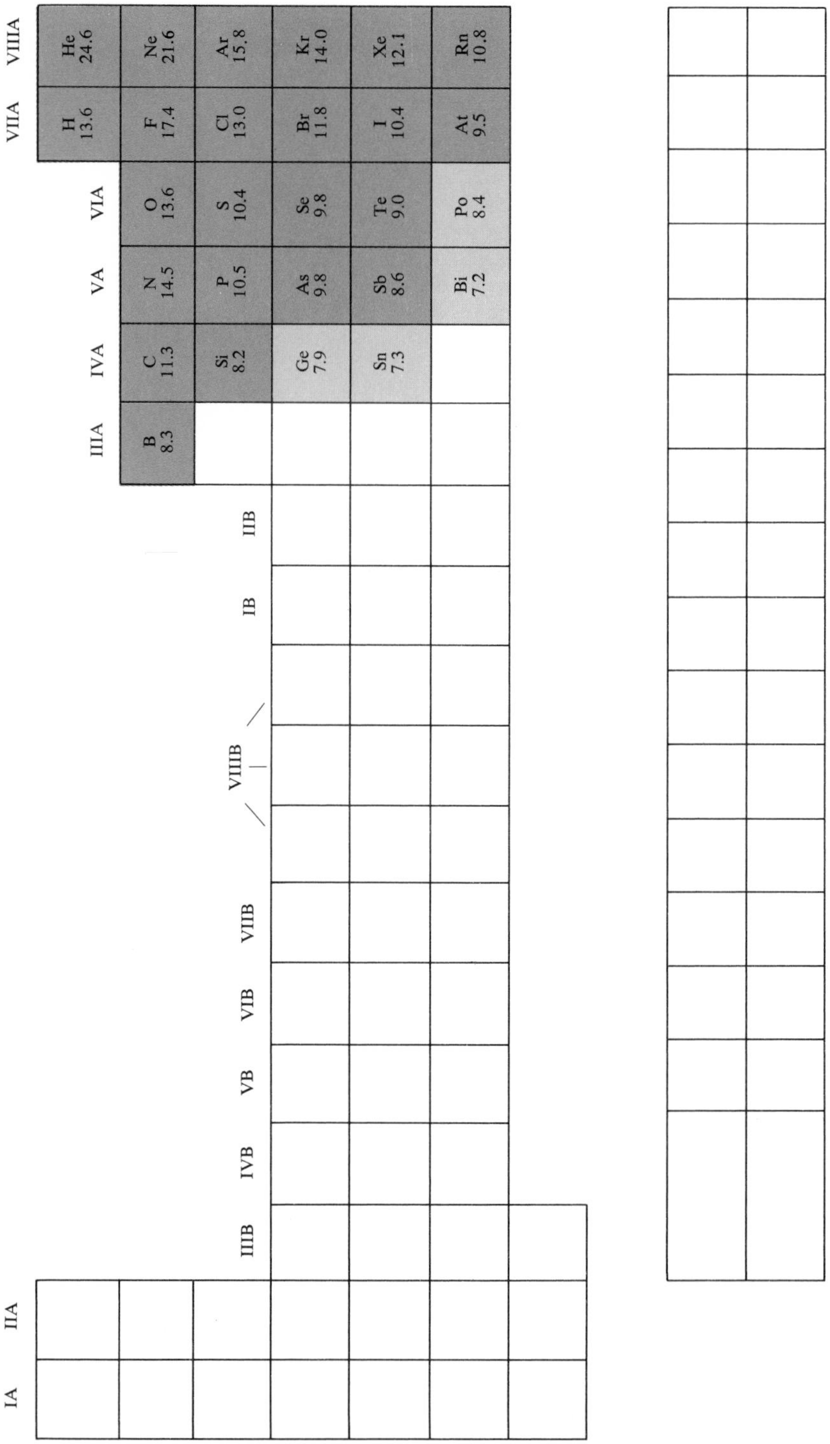

IA	IIA	IIIB	IVB	VB	VIB	VIIB	VIIIB	VIIIB	VIIIB	IB	IIB	IIIA	IVA	VA	VIA	VIIA	VIIIA
																H 13.6	He 24.6
												B 8.3	C 11.3	N 14.5	O 13.6	F 17.4	Ne 21.6
													Si 8.2	P 10.5	S 10.4	Cl 13.0	Ar 15.8
													Ge 7.9	As 9.8	Se 9.8	Br 11.8	Kr 14.0
													Sn 7.3	Sb 8.6	Te 9.0	I 10.4	Xe 12.1
														Bi 7.2	Po 8.4	At 9.5	Rn 10.8

Fig. 17-1 The location of the non-metallic elements in the periodic table. The first ionization potential (in electron volts) is listed for each non-metallic element.

The elements involved are shaded in the section of the periodic chart shown in Fig. 17-1.

The designation non-metallic is applicable to an element which exhibits all, or nearly all, of the following properties:

1. In the elemental state, the element lacks typical metallic properties such as lustre, ductility, metallic conductance, *etc.*
2. With metals, the element forms chemical compounds which are non-metallic in character; *i.e.*, which lack metallic properties such as those mentioned above.
3. With hydrogen, the element forms one or more compounds which are covalent and volatile.
4. The element forms covalent compounds with obviously non-metallic elements such as oxygen or fluorine.

Elements which border on the shaded area in Fig. 17-1 often exhibit one or more of the properties listed above. These elements are referred to as *metalloids*, to indicate that their properties are intermediate between those of metallic and non-metallic elements. The four elements which are more lightly shaded in Fig. 17-1, for example, are difficult to classify as either metallic or non-metallic since they exhibit many chemical properties characteristic of either.

The non-metals are the more electronegative elements. The characteristic properties listed above for non-metals can be accounted for in terms of the rather strong tendency for the non-metal to attract electrons in a chemical bond. The electronegativity (Sec. 8-4, Table 8-3) is a measure of non-metallic character. On this basis, fluorine is the most non-metallic element.

Certain trends in the properties of elements are very important for an understanding of how chemical behavior varies in a periodic fashion. There is an increase in electronegativity in proceeding from left to right in any horizontal row of the periodic table. This is the result of an increase in effective nuclear charge. The electrons in the outer shell of the atom are drawn in more tightly, creating a smaller atom. In keeping with this notion, the atomic radius decreases in going to the right in each row (Fig. 6-4). The second important trend is the increase in size, and decrease in electronegativity, in proceeding downward in any vertical column. The elements in each group (vertical column) tend to show the same valency, but because of variation in size and electronegativity, the heavier elements tend to exhibit higher coordination numbers. This means that iodine, for example, would tend to form a compound or ion with more atoms clustered about it than would chlorine. We shall see numerous examples of this size effect.

It is convenient to begin a discussion of the chemistry of the non-metals by considering the properties of the elements themselves. It is to be expected that, with the exception of the rare gases, the atoms of an element will be found in chemical combination of some sort. In the elemental substance the atoms are all alike, so that only non-polar bonds exist. The number of bonds per atom and the directions which the bonds assume will determine, therefore, the structures of the pure elements.

17-2 The Halogens

The halogens have enjoyed a prominent place in the history of chemistry. The first to be isolated as an element was chlorine, prepared by Scheele in 1774 — the same year in which Priestley discovered oxygen. Iodine was reported as a substance recovered from treatment of kelp by M. B. Courtois in 1813. Bromine was also first discovered as a constituent of kelp, by M. Balard in 1826. The preparation of fluorine, by electrolysis of anhydrous hydrogen fluoride, was reported by H. Moissan in 1886. Astatine, element 85, does not occur in nature. Its chemical properties were described by E. Segre and co-workers in 1940, on the basis of work performed with synthetically produced material.

Table 17-1 Properties of the halogens.

Element	Molecular form	Melting point (°C)	Boiling point (°C)	Solubility in water (moles/l at 20°C)	Bond dissociation energy (kcal)
Fluorine	F_2	−223	−187	—	37.8
Chlorine	Cl_2	−102	−34.6	0.090	58.0
Bromine	Br_2	−7.3	58.8	0.210	53.4
Iodine	I_2	114	183	0.00133	51.0

The electronic configuration of the halogens in the valence shell is described by the notation ns^2np^5, where n is 2, 3, 4, or 5.

ns	*np*
⇅	⇅ ⇅ ↑

Each halogen atom acquires a completed valence shell of electrons by forming a single bond to another atom. The halogens, in their stable states at room temperature, all exist in the form of diatomic molecules. Some important properties of the halogens are listed in Table 17-1. Note the much lower dissociation energy for F_2 than for the other halogens. This very low dissociation energy, coupled with the fact that fluorine forms unusually strong bonds to nearly all elements, is responsible for the great reactivity of fluorine and for the stability of fluorides. It is still not entirely clear why the F_2 bond energy is so much lower than that for the other halogens. One reason which is sometimes given is that the atomic orbitals on fluorine are quite highly contracted, so that the two atoms must approach quite closely for effective overlap. But at such close distances the nuclear-nuclear repulsion is large and cancels a larger part of the electronic binding energy than in most molecules. When fluorine forms compounds with other, less electronegative elements, it attracts charge to itself, and behaves in a fashion intermediate between fluorine atom and fluoride ion. The

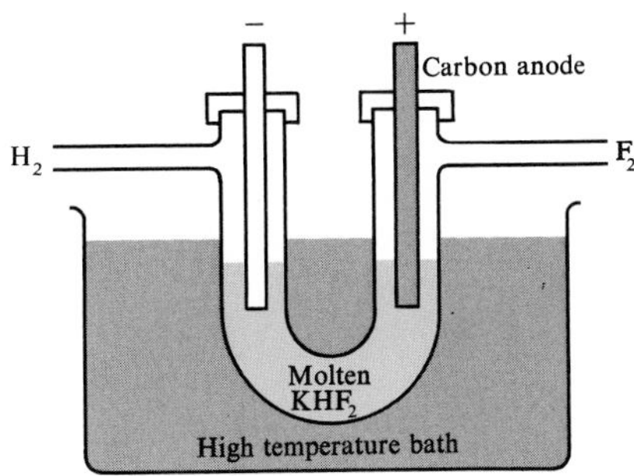

Fig. 17-2 Production of fluorine by electrolysis.

stability of the bonds is due in part to their polarity, a factor which is missing in F_2.

Fluorine — Fluorine gas may properly be described as the most reactive element. Because of its very high reactivity, it is prepared and stored only with great difficulty. The usual method of preparation involves electrolysis of a molten electrolyte containing fluoride ion. A mixture of potassium fluoride and hydrogen fluoride, which for simplicity we write as $KF \cdot HF$, or KHF_2, is commonly employed. Water must be rigorously excluded. The melt is electrolyzed at temperatures in the range of 75–250°C in a cell of the type illustrated in Fig. 17-2. The cell must be constructed of a material which does not continuously react with fluorine gas. The metals employed, usually copper, nickel, or steel, undergo an initial reaction which forms a coat of fluoride salt on the metal. This coat is relatively impervious to the passage of fluorine, so that further reaction is very slow. The anode, the electrode at which fluorine is generated, is compounded from graphite, which is moderately resistant to attack. The over-all equation for the electrolysis is

$$2KHF_2 \xrightarrow{\text{electrolysis}} H_2 + F_2 + 2KF \qquad (17\text{-}1)$$

Chlorine — Elemental chlorine is produced in present-day chemical industry in large quantities; in 1966, the production was about 7,000,000 tons. Chlorine gas, though not as reactive as fluorine, must nevertheless be handled with care. It liquefies upon compression at room temperature, and is stored and handled in steel cylinders as a liquid.

Nearly all chlorine is produced by electrolytic methods. The most important process involves electrolysis of aqueous sodium chloride solution. The over-all cell reaction* is

$$2NaCl(aq) + 2H_2O \rightarrow Cl_2(g) + 2NaOH(aq) + H_2(g) \qquad (17\text{-}2)$$

---

*The notation (aq) in a chemical equation is employed to indicate that the species which it follows is dissolved in aqueous solution.

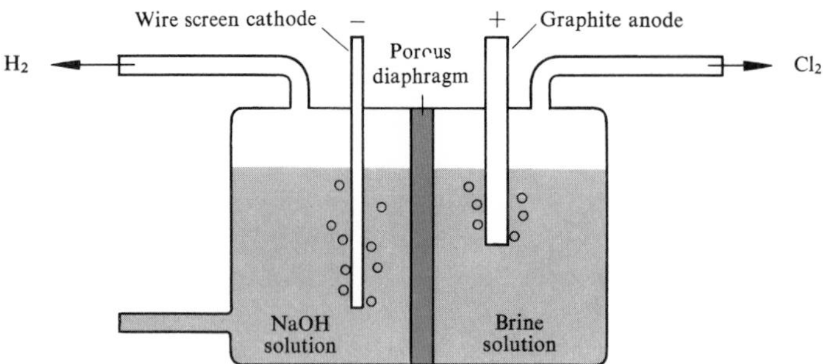

Fig. 17-3 Production of chlorine by electrolysis.

It can be seen from this equation that hydrogen gas and aqueous sodium hydroxide are also products of the electrolysis. If one is to obtain a high degree of purity in the products, the reaction must be conducted under carefully controlled conditions and in cells of special design. One of the important cell designs is shown in Fig. 17-3. Note that the anode and cathode compartments are separated by a diaphragm. The cell temperature must be carefully controlled to minimize reaction of chlorine with the solution, since this leads to products which contaminate the sodium hydroxide.

A second method of producing chlorine — electrolysis of molten metal chlorides — is also of considerable industrial importance. The principal products of these processes are the metals, but chlorine is a valuable by-product. The production of sodium metal and chlorine gas by electrolysis of molten sodium chloride is an example of the type of process. The overall cell reaction is

$$2NaCl(l) \rightarrow 2Na(l) + Cl_2(g) \tag{17-3}$$

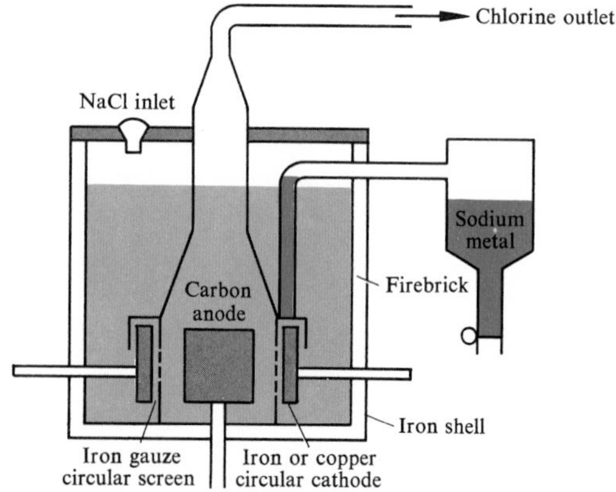

Fig. 17-4 Electrolysis of molten sodium chloride.

A cell in which this process is carried out industrially is illustrated in Fig. 17-4. The melt is not pure sodium chloride, but rather a mixture of sodium and calcium chlorides. The addition of the second salt lowers the melting point of the melt. Operation of the cell at a lower temperature results in a saving of fuel, and also reduces corrosion of the cell and electrodes.

Bromine — The most important sources of bromine are sea water and water from certain brine wells in Michigan and elsewhere. Bromine is present in these waters as bromide ion, and is removed by oxidation to bromine. This oxidation is generally effected by use of chlorine gas, which is relatively inexpensive:

$$2Br^-(aq) + Cl_2 \rightarrow Br_2 + 2Cl^-(aq) \quad (17\text{-}4)$$

(There is, of course, a positive ion which goes along with the negative ion in the above equation. It has been left out here because it is not a part of the chemical reaction.) This reaction illustrates the important point that the more electronegative element, chlorine, is more stable in the lower oxidation state. The bromine liberated by chlorine is removed from the system by a stream of air, and is subsequently purified.

Fig. 17-5 Sublimation of iodine.

Bromine is prepared in the laboratory by a method similar to that used in preparing chlorine:

$$2NaBr(aq) + 2H_2SO_4(aq) + MnO_2(s) \rightarrow$$
$$Br_2 + MnSO_4(aq) + 2H_2O + Na_2SO_4(aq) \quad (17\text{-}5)$$

Iodine — Iodine is also found in sea water, but its concentration is much lower than that of bromine. It is found concentrated in certain sea plants, from which it can be recovered by burning the dried plants and working up the residue. The methods are the same as those used in obtaining bromine. The iodine which is recovered is easily purified by sublimation; laboratory sublimation of iodine is illustrated in Fig. 17-5.

Iodine is found also in Chile in the form of iodate deposits. Iodates are generally quite soluble in water. The Chilean deposits have remained intact, however, because of an almost total absence of rainfall in the region. Iodine is liberated by treating the iodate in solution with bisulfite ion:

$$2H_2O + 2IO_3^- + 5HSO_3^- \rightarrow I_2 + 5SO_4^{-2} + 3H_3O^+ \quad (17\text{-}6)$$

In this reaction, the iodine is reduced from an oxidation state of +5 in the iodate ion to an oxidation state of zero in I_2. Sulfur is oxidized from +4 in the sulfite ion to +6 in the sulfate ion.

17-3 Group VI Elements

The electronic configurations of the oxygen group elements are described by the notation ns^2np^4, where n equals 2, 3, 4, or 5.

ns	*np*
⇅	⇅ | ↑ | ↑

The structures of the oxygen family elements vary considerably, but in all cases it is true that the bonding leads to a completed valence shell on each atom. The physical properties of the Group VI elements are listed in Table 17-2.

Table 17-2 Physical properties of Group VI elements.

Element	Molecular form	Melting point (°C)	Boiling point (°C)	Density of solid (g/cm³)
Oxygen	O_2	−219	−183	1.27
Sulfur	S_8	119	444	2.06
Selenium	Se_x	217	685	4.80
Tellurium	Te_x	450	1390	6.24

Subscript x indicates the chain-like, or "metallic" allotrope of the element.

Oxygen — This element is, of course, present in elemental form in the atmosphere, of which it constitutes about 20 per cent. The industrial preparation of oxygen involves liquifaction of air and subsequent fractional distillation.

Oxygen provides us with our first example of *allotropy*, which refers to the existence of an elemental substance in more than one form in the same physical state. The stable allotropic form of oxygen is O_2. This molecule is unusual in that it possesses two unpaired electrons (Sec. 8-2). The other allotropic form of the element is ozone, O_3. The structure of this species has been discussed earlier (p. 185). It is prepared by passing a silent electric discharge through oxygen. The energy absorbed from the electric discharge causes dissociation of an oxygen molecule; the oxygen atoms then combine with molecular oxygen:

$$O_2 + \text{energy} \rightarrow 2O$$
$$O_2 + O \rightarrow O_3$$

Ozone is a very reactive substance, and cannot be stored for long except at low temperatures. It is produced in the upper atmosphere through the action of high-energy ultraviolet solar radiation. If it were not for the removal of high-energy radiation from sunlight by this process, life as we know it on earth would be impossible (see Sec. 19-6).

Oxygen is the most abundant element in the earth's crust, and is second only to iron in abundance in the earth as a whole. In addition to its presence in the earth's atmosphere as molecular oxygen and — to a lesser extent — as O and O_3, oxygen is covalently bonded to hydrogen in water, which is very abundant on the earth's surface. It is also associated with a great many metals in the form of oxides (*e.g.*, corundum, Al_2O_3; cassiterite, SnO_2; spinel, $MgAl_2O_4$; hematite, Fe_2O_3). With many non-metals and metals it forms oxy-anions which are common in nature, such as sulfate, SO_4^{-2}; phosphate, PO_4^{-3}; iodate, IO_3^-; chromate, CrO_4^{-2}; molybdate, MoO_4^{-2}.

The quantity of atmospheric oxygen on earth is happily quite large; it sustains a prolific plant and animal life. It is not entirely clear to geologists why there should be so much free oxygen, since there is more than enough iron in the earth's crust in the form of iron (II) compounds to combine with all the oxygen of the atmosphere to form thermodynamically stable iron (III) substances. Atmospheric oxygen might have arisen from photodissociation of atmospheric water vapor, with subsequent escape of hydrogen from the earth's gravitational field. This would have occurred at a phase in the earth's development when the temperature at the surface was too low to permit rapid combination of oxygen with mineral substances.

Oxygen is involved in an organic cycle such as depicted in Fig. 17-6. Plants obtain their oxygen from carbon dioxide, through photosynthetic reduction of CO_2, and release oxygen gas. Animal life, and some higher forms of plant life, on the other hand, consume oxygen in respiration. Plant matter may be oxidized through combustion, as in the burning of coal, or through decay, liberating carbon dioxide and water.

Sulfur — Sulfur is a fairly abundant element in the earth's crust; it is estimated to average a little over one pound per ton. It exists mainly in the

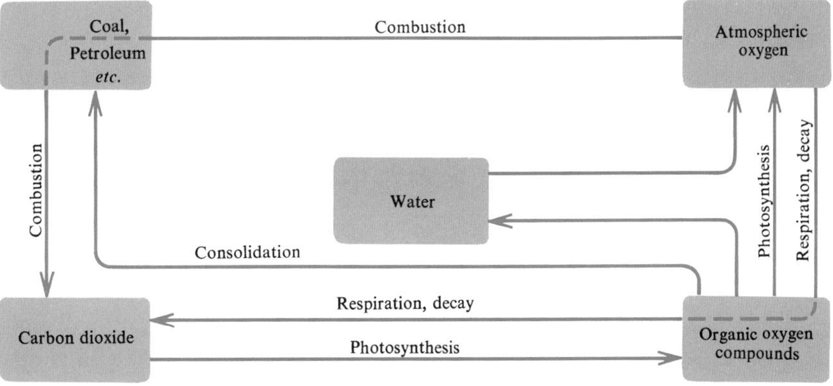

Fig. 17-6 Oxygen cycle in organic matter.

form of solid mineral sulfides. There are some metals which are found in nature mainly as oxides, and others largely as sulfides. Among the latter are copper, silver, zinc, cadmium, mercury, indium, thallium, arsenic, antimony, and molybdenum. Iron, which is commonly found in the form of oxide ores also occurs as pyrite, FeS_2.

Hydrogen sulfide may occur as a by-product of the formation of petroleum. There is one exceptional petroleum gas field in southern France which contains up to 18 per cent hydrogen sulfide by volume. The hydrogen sulfide is removed by absorption in a solvent under pressure, then later recovered from the solvent and catalytically decomposed to yield sulfur. The operation produces about 4000 tons of sulfur per day. Hydrogen sulfide is present in solution in waters from some deep mineral springs. Such waters have long been reputed to have therapeutic value.

Much of the sulfur present in the earth's crust is present in the form of sulfate. The sulfate ion, SO_4^{-2}, is the second most abundant anion in sea water. In addition, a number of solid sulfate minerals are known. Calcium sulfate, the best known of these, is found as the anhydrous salt or as gypsum, $CaSO_4 \cdot 2H_2O$.

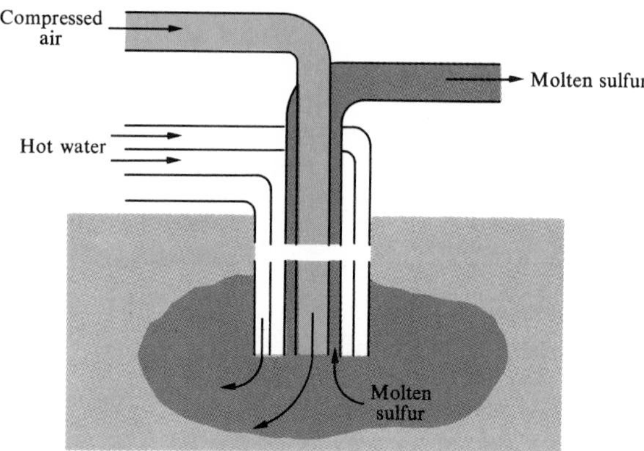

Fig. 17-7 Frasch process for mining of sulfur.

The extensive deposits of elemental sulfur found in various places throughout the world constitute the major commercial sources of sulfur. The vast deposits in Texas and Louisiana yielded about 6,500,000 tons of sulfur in 1966. The sulfur in these deposits is relatively free from impurities. Removal of the elements from the deposits is accomplished by the Frasch process, in which superheated water under pressure is forced down a concentric pipe, as shown in Fig. 17-7. The sulfur is melted by contact with the water, and a slurry of molten sulfur and water is pushed by the pressure up an inner pipe. The hot mixture is piped out into giant bins, where the water evaporates or runs off, leaving the solid sulfur behind. Many sulfur compounds of industrial importance are produced from this elemental sulfur, rather than from pyrite or sulfide ores, principally because of its higher degree of purity. In recent years, however, there has been a world shortage of sulfur, and more attention is being given to recovery of sulfur dioxide from roasting of sulfide ores. The stack gases resulting from combustion of coal and other sulfur-containing petroleum fuels contain sulfur dioxide, which is a major culprit in air pollution. A new process, developed in Japan, involves oxidation of the SO_2 to SO_3 and passage over aqueous ammonia, to form ammonium sulfate. This process has the double attractiveness of providing an additional source of sulfur compounds, and relieving a pollution problem.

Sulfur has a rather complicated allotropy. The most stable molecular form is the S_8 ring, depicted in Fig. 17-8. The rings may pack in the solid in different ways. At 1 atm pressure and at temperatures below 96°C, the stable crystalline form of sulfur is the rhombic. Above 96°C, the monoclinic form is stable. Sulfur may be prepared under various special conditions to yield still other crystalline forms of sulfur consisting of S_8 rings packed in still different ways. In one special case, the decomposition of thiosulfate ion under acid conditions (see Sec. 19-2), a form called *rho* sulfur is produced. This consists of S_6 rather than S_8 rings. It eventually converts to rhombic sulfur, but the process is slow in the solid state.

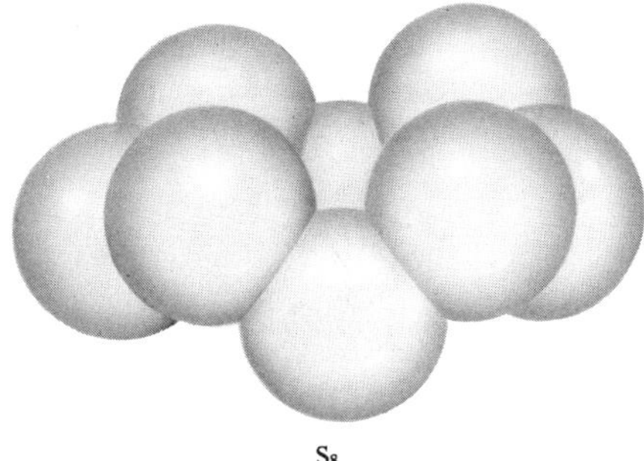

Fig. 17-8 The structure of elemental sulfur.

Rhombic sulfur melts at 112°C to form a pale, straw-colored liquid which pours easily. On further heating, above 160°C, the liquid becomes very viscous. It is believed that the increase in viscosity is due to opening of the S_8 rings to form long chains of sulfur atoms. When the hot viscous liquid is poured into cold water, plastic sulfur is obtained. Plastic sulfur is a metastable (unstable, but changing only slowly to a more stable state) form of sulfur, "frozen out" by the sudden cooling. On standing, it slowly reverts to rhombic sulfur. Other metastable forms of sulfur have been characterized by quenching (sudden cooling) of high temperature forms. As an extreme example, if sulfur gas at 1000°C is chilled on a surface at the temperature of liquid air, a purple solid is formed. The solid is thought to consist of S_2 molecules. On warming, it rapidly converts to a mixture of rhombic and plastic sulfur.

Selenium and Tellurium — If one restricts consideration to those forms which are easily obtained, selenium possesses a more complex allotropy than sulfur. At room temperature it exists in the solid state in the form of Se_8 rings; there are two common crystal modifications. Above about 60°C, however, the stable form is a more "metallic" variety of selenium, consisting of parallel, close-packed chains of atoms. This metallic form melts at about 220°C.

Tellurium does not form the eight-membered ring structure at all, but exists in the solid only in a form which is like the "metallic" form of selenium. Thus, we see a definite trend in the Group VI elements from a diatomic gas molecule for the lightest element in the series to the larger ring structure, and an increasing tendency to chain formation, with increase in atomic weight. The properties of the solid elements also reveal a trend toward increasing electrical conductivity, metallic luster, *etc.*, with increasing atomic weight.

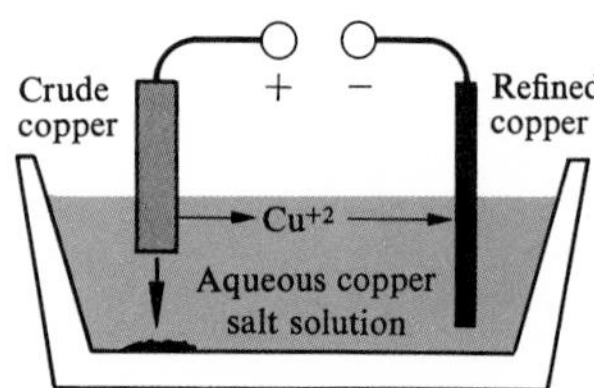

Fig. 17-9 Electrolytic refinement of copper.

The most important sources of selenium and tellurium are pyrite and sulfide ores which contain a small percentage of the two related elements. When copper ores are roasted, the copper is reduced to the metal. (Some oxide is formed, but this is reduced by further treatment.) The crude copper produced by these initial operations contains selenium and tellurium, as well as a number of other elements. The crude copper is refined by an electrolytic method, illustrated in Fig. 17-9. Impurities in the copper collect at the bottom of the bath. This so-called "anode mud" is a valuable source of precious metals such as gold, silver, and platinum; in addition, it contains selenium, tellurium, and possibly other elements. Although selenium and tellurium are not very important by-products of the treatment of this anode mud, a simplified description of the process for recovering selenium provides examples of interesting chemistry.

If the anode mud is treated with concentrated sulfuric acid at about 400°C, selenium is oxidized to selenium dioxide, which sublimes from the reaction mixture:

$$Se + 2H_2SO_4 \xrightarrow[\text{concentrated}]{\text{hot}} SeO_2 + 2SO_2 + 2H_2O \qquad (17\text{-}7)$$

The dioxide is recovered as a white powder from the upper furnace walls. It is then dissolved in dilute hydrochloric acid solution, in which it is quite soluble. Selenous acid, a weak acid, is formed:

$$SeO_2 + H_2O \rightarrow H_2SeO_3(aq) \qquad (17\text{-}8)$$

Sulfur dioxide is then bubbled into the solution, and an oxidation-reduction reaction occurs:

$$H_2O + H_2SeO_3(aq) + 2SO_2(g) \rightarrow Se(s) + 2H_2SO_4(aq) \qquad (17\text{-}9)$$

Tellurium is recovered from the anode mud by other procedures. Of course, the detailed procedures for recovery of these elements are more complex than described here, because it is necessary to make provision for separating closely related substances, and for recovering the more valuable metals.

17-4 Group V Elements

The Group V elements possess the electronic configuration ns^2np^3, where n is 2, 3, 4, 5, or 6.

ns	*np*
1l	1 | 1 | 1

The physical properties of the Group V elements in their common allotropic forms are given in Table 17-3. All of the Group V elements were discovered early in the development of chemistry. It seems rather curious from our present vantage point to notice that nitrogen was actually the last of these elements to be identified as such.

Table 17-3 Physical properties of Group V elements.

Element	Molecular form	Melting point (°C)	Boiling point (°C)	Density of solid (g/cm³)
Nitrogen	N_2	−210	−196	0.879
Phosphorus	P_4	44	280	1.82
Arsenic	As_x	814	—	5.7
Antimony	Sb_x	630	1380	6.7

Subscript x indicates the chainlike, or "metallic" allotrope of the element.

Among the Group V elements, nitrogen alone possesses a triple bond formed between two atoms in the stable state of the element. All of the other Group V elements exhibit allotropy.

Nitrogen — The free element is, of course, the major constituent of the earth's atmosphere. Nitrogen is found in inorganic materials in the form of nitrate (NO_3^-) compounds. It is widely and fairly heavily distributed in

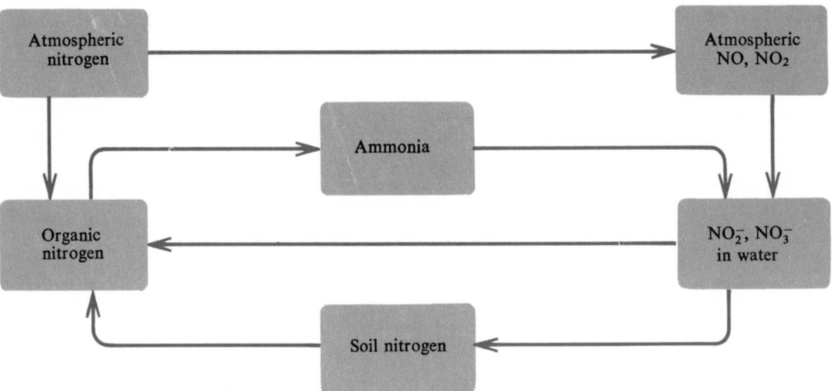

Fig. 17-10 The nitrogen cycle in nature.

plant and animal matter. The pure element is conveniently obtained by distillation of liquid air. Among the most important considerations in nitrogen chemistry are the nitrogen cycles in plants, animals, and in the earth's crust. These are rather complex, and will not be dealt with in detail here. The over-all character of the nitrogen cycle is shown in Fig. 17-10. The nitrogen of the atmosphere is converted in various ways into nitrogen compounds. This may occur as a result of solar radiation in the upper atmosphere, or of lightning storms. The nitrogen in such compounds is referred to as *fixed nitrogen.* Electrical discharges such as lightning cause the reaction

$$N_2 + O_2 \rightarrow 2NO \tag{17-10}$$

which is followed by the further oxidation:

$$2NO + O_2 \rightarrow 2NO_2 \tag{17-11}$$

The dioxide is washed down by rain, and so enters into the plant cycle. Rain may also contain ammonia, which enters the air as a result of decomposition of plant and animal material. It is estimated that the total precipitation of fixed nitrogen to the soil in the middle temperate zone of the U.S. is on the order of one ton per square mile per year.

Phosphorus — There are three major allotropic forms of phosphorus: white, black, and red. A great number of other, less easily obtained varieties have been reported. White phosphorus consists of tetrahedral P_4 molecules, as shown in Fig. 17-11. The P—P distance in the gaseous P_4 molecule has been found to be 2.21 Å. The P—P—P bond angle is 60°, as required by the geometry of the structure. The small bond angle is quite unusual, and it appears that there may be some strain in the molecule. If each phosphorus atom utilizes its three $3p$ orbitals in bonding to three other atoms, the un-

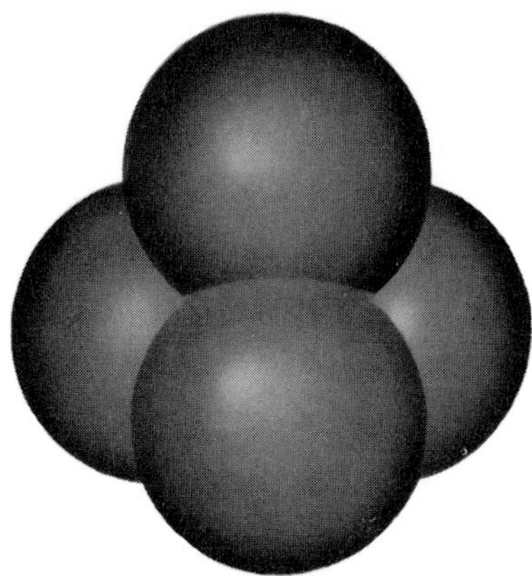

Fig. 17-11 Structure of white phosphorus.

strained bond angle at phosphorus would be 90°, as discussed in Sec. 8-5. Heating of P_4 vapor to temperatures above 900°C results in some dissociation to P_2 molecules. At 1700°C, there are about equal numbers of P_4 and P_2 molecules in equilibrium at a total pressure of about 1 atm. Study of the emission spectrum of the P_2 molecule has led to an estimate of the bond distance of 1.875 Å. This is considerably shorter than is the P—P distance in P_4, and it supports the idea of a triple bond in the P_2 molecule.

White phosphorus is a volatile, soft, low-melting solid. It is intensely reactive toward oxygen and inflames spontaneously in air. White phosphorus is a dangerous material which must be handled with great care. Phosphorus burns on the skin are painful and heal very slowly.

Black phosphorus is the densest and least reactive form of the element. It can be produced from white phosphorus by heating in the presence of a seed crystal of black phosphorus, with mercury as catalyst. It has a continuous layer structure, in which each phosphorus atom is bonded to three neighbor atoms. The layer structure gives black phosphorus a flaky character similar to that of graphite; it is a weak conductor of electricity. As one might expect of a substance which exists as a layer structure, black phosphorus is virtually insoluble in all solvents.

Red phosphorus is more common than the black variety, which it resembles in many ways. It is somewhat more reactive with oxygen than is the black form, but does not ignite on contact with air as does white phosphorus. It is considerably less poisonous than the white form. There is no one well-defined crystalline form for red phosphorus, but in general it appears to have a chain structure with each phosphorus bonded to three others in a manner similar to the bonding in black phosphorus. It is produced from white phosphorus by the action of heat or light, in the presence of a catalyst such as iodine.

Phosphorus is produced commercially by the reduction of a phosphate mineral with coke and silica in an electric furnace. The over-all process is described by the following equation:

$$2Ca_3(PO_4)_2 + 6SiO_2 + 10C \rightarrow P_4 + 6CaSiO_3 + 10CO \quad (17\text{-}12)$$

Phosphorus is distilled out as it is formed, and condensed under water as white phosphorus. This reaction probably occurs in two steps, the first being decomposition of the phosphate by silica, with formation of phosphorus pentoxide:

$$Ca_3(PO_4)_2 + 3SiO_2 \rightarrow 3CaSiO_3 + P_2O_5 \quad (17\text{-}13)$$

Calcium silicate forms a molten slag at the bottom of the furnace. Phosphorus pentoxide is a vapor at the furnace temperature, which is on the order of 1500° C; it is reduced by contact with the coke:

$$10C + 2P_2O_5 \rightarrow P_4 + 10CO \quad (17\text{-}14)$$

A great proportion of phosphorus chemistry begins with the white phosphorus produced in this manner. Total U.S. phosphorus production in 1966 exceeded 1,500,000 tons.

Arsenic — Arsenic is found in nature predominantly in sulfide ore bodies. For example, arsenopyrite, FeAsS, is the commonest arsenical mineral. It is found in pyrite (FeS_2) deposits. A great number of other arsenic-containing sulfide ores are known. When these ores are smelted as part of the process of recovering the metals, As_2O_3 is a by-product. It is, in fact, produced in quantities which far exceed the normal demand.

The free element is produced by reduction of the oxide with carbon:

$$As_2O_3 + 3C \rightarrow 2As + 3CO \quad (17\text{-}15)$$

Arsenic, like phosphorus, exists in allotropic forms. The most stable is the so-called "metallic" form (sometimes called grey arsenic), which has a structure similar to that of black phosphorus. Crystals of this form have a metallic luster; they are fairly good conductors of heat and electricity. On heating, arsenic volatilizes, and sublimes without melting at about 450°C to form a colorless vapor which consists of As_4 molecules. When heated under pressure in a sealed tube, it melts at 814°C (Table 17-3). Although the solid form of metallic arsenic does not readily oxidize, arsenic vapor burns in air at temperatures of about 400°C.

Yellow arsenic is an unstable allotrope of the element; it possesses the same structure as white phosphorus; *i.e.*, it consists of As_4 molecules. It is converted very readily into "metallic" arsenic. The conversion is catalyzed by light, even at temperatures as low as −180°C.

Antimony — Antimony is found in sulfide mineral deposits, in the same manner as arsenic, although it is much more scarce. It is also found as the mineral *stibnite*, Sb_2S_3, a black mineral, in a few rich deposits. The free element is recovered by one reduction process or another, depending upon the quality of the ore. For example, it might be reduced with metallic iron:

$$Sb_2S_3 + 3Fe \rightarrow 3FeS + 2Sb \qquad (17\text{-}16)$$

The stable allotropic form of antimony appears to be metallic, although it is brittle and easily powdered, only a fair conductor of heat and electricity, and exhibits magnetic properties characteristic of a non-metal. Other allotropic forms of the element are known, but they are very unstable.

The Group V elements furnish excellent examples of the trend toward metallic character with increase in atomic number in any one group. Where more than one allotropic form exists, one form can generally be associated with non-metallic properties, and the other form with metallic properties. We see in Group V an increasing relative stability of the metallic form with increasing atomic number. Furthermore, the metallic form appears more metallic in its properties as atomic number increases.

17-5 Group IV Elements

Group IV elements possess the valence shell electronic configuration ns^2np^2:

ns	*np*
↑↓	↑ | ↑ |

The diamond structure, shown in Fig. 17-12, is the allotropic form shared by all of the Group IV elements. The physical properties of the Group IV elements, including those elements that are classified as metallic, are given in Table 17-4.

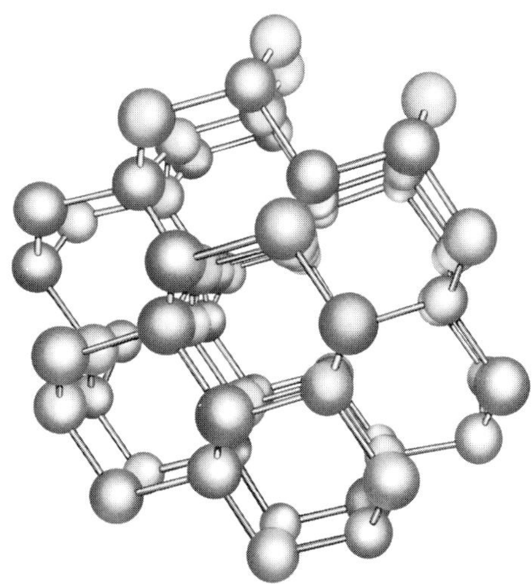

Fig. 17-12 Structure of diamond.

Table 17-4 Physical properties of the Group IV elements.

Element	Melting Point (°C)	Boiling Point (°C)	Density (g/cm^3)
Carbon[a]	3500	4220	3.51
Silicon	1420	350	2.49
Germanium	959	2700	5.36
Tin[b]	232	2260	7.29
Lead	327	1600	11.34

[a]Diamond
[b]White tin

Carbon — This element is found in the two crystalline allotropic forms, diamond and graphite. The graphite structure, shown in Fig. 17-13, consists of layers, within which the carbon atoms are strongly bonded to one another, with only weak bonding between layers. Each carbon atom is surrounded by three others at the corners of an equilateral triangle within the plane of the layer.

To understand the bonding in graphite, we must again have recourse to the concept of hybridization. We have seen that when one $2s$ and two $2p$ orbitals are hybridized, the three sp^2 orbitals which result are in a plane, and directed toward the corners of an equilateral triangle. We may begin our picture of the graphite structure, then, by assuming that the bonds between the carbon atoms are formed from the sp^2 hybrid orbitals. Each carbon then employs three of its valency electrons in forming three single electron pair bonds to the three neighboring carbon atoms. In addition, there is one electron in the $2p$ orbital which is perpendicular to the plane of the sp^2 hybrid orbital set. The $2p$ orbitals on adjacent carbon atoms

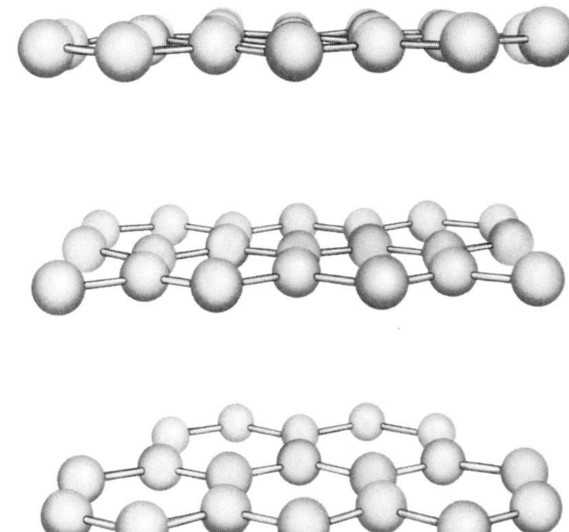

Fig. 17-13 Structure of graphite.

overlap to form π bonds, as explained in the earlier discussion of the benzene and ethylene structures, Sec. 8-6 and 8-7. Because each $2p$ orbital overlaps with more than one other $2p$ orbital, the electron pairs in the π bonds are not restricted to just one region of space between a specific pair of atoms. They are relatively free to move in the plane of the bonds, and are responsible for the good electrical conductivity and numerous other special properties of graphite. One might say that graphite is the "metallic" allotrope of carbon, and that diamond is the more characteristically non-metallic form.

The π electrons are uniformly distributed, on the average, among all of the possible π bonds in the plane. Since each carbon has one π bonding orbital to contribute, and since there are three neighboring atoms to which π bonding might occur, the average bond between any two atoms is a one-third bond. In addition, there is the full σ bond formed from the sp^2 hybrid orbitals, so the total bonding between any two carbon atoms is one and one-third bonds. In benzene, the total average bond order is one and one-half, as discussed in Sec. 8-7. It is of interest to compare the carbon-carbon distances in various carbon compounds with the formal bond order, which is the sum of the average number of bonding electron pairs, both σ and π. A few data are given in Table 17-5. It is clear that the carbon-carbon distance is shorter for higher carbon-carbon bond order.

Table 17-5 Comparison of carbon-carbon bond order and bond distance.

Compound	Average C—C bond order	Observed C—C distance (Å)
Diamond	1.00	1.542
Graphite	1.33	1.42
Benzene	1.50	1.39
Ethylene	2.00	1.353
Acetylene	3.00	1.207

When carbon-containing compounds are decomposed in the absence of air, as when sugar is strongly heated, a non-crystalline form of carbon results:

$$C_{12}H_{22}O_{11} \xrightarrow{800°C} 12C + 11H_2O \qquad (17\text{-}17)$$

Non-crystalline carbon, variously called coke, charcoal, soot, *etc.*, probably consists of microcrystals which possess the graphite structure.

At atmospheric pressure, graphite is thermodynamically more stable than diamond. For obvious reasons, considerable effort has been expended over the years in developing a method for converting graphite into diamond. Since the latter is the more dense substance (3.51 g/cc for diamond *vs* 2.25 g/cc for graphite), application of high pressure should shift the equilibrium in favor of the diamond form. High temperatures are also required

Fig. 17-14 Photograph of synthetic diamonds. (Courtesy of the General Electric Research Laboratory.)

in any practical process. Synthetic diamonds of industrial quality have been produced in recent years (Fig. 17-14).

Silicon — Compounds of this element are very abundant. Silicon dioxide, or silica, SiO_2, and its derivatives are so plentiful that silicon is, after oxygen, the most abundant element in the earth's crust.

The free element is produced by reduction of the dioxide. The reducing agent may be an active element such as aluminum, or carbon:

$$SiO_2 + 2C \rightarrow Si + 2CO \qquad (17\text{-}18)$$

The silicon which is produced reacts with the reducing agent to form a binary compound unless silica is present in excess. The element exhibits only one crystalline form, which has the diamond structure.

The production of ultra-pure silicon has been of great interest in recent years because of its electrical properties. It is termed a *semiconductor*, to indicate that its resistance to the flow of electric current is intermediate between that of an insulator and that of a good conductor of electricity, such as a metal. Controlled addition of "impurities" to the ultra-pure metal causes changes in the character of the electrical behavior which depend upon the chemical nature of the impurity. These properties form the basis for the use of silicon in transistors. The problems encountered in obtaining a really pure form of the element, and in then adding controlled quantities of another substance to it, have prompted much interesting research. The zone melting method of purification, which has been widely employed, is the most important method available for attaining high purity in elements such as germanium and silicon; it has also been applied with considerable success to purification of many of the metallic elements. Figure 17-15 shows a schematic diagram of a zone melting process. The substance to be purified is in the form of a solid rod, and is inside a container. The heater is annular, and fits closely around the tube. The temperature of the heater is adjusted so that the solid melts only in the area being heated. The heater is then moved slowly along the length of the

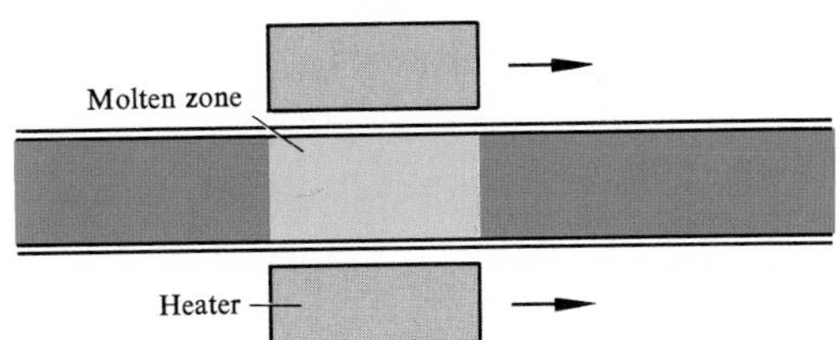

Fig. 17-15 Purification of a solid substance by zone melting.

rod. As it moves, solid melts on the leading edge, while molten material freezes out on the trailing edge. The melt may be thought of as a dilute solution of impurity substances in molten pure substance. The freezing which occurs on the trailing edge is analogous to that which takes place in the freezing point lowering experiment (p. 268). The solid which freezes out is nearly pure solvent, the solute remaining in solution. As the heater moves along the rod, therefore, impurities — which remain in the melt — move along with it and are carried to the end of the tube. The process may be repeated over and over again, to attain successively purer material. It is possible to obtain silicon and germanium which contain less than one part per billion of impurity.

Germanium — Germanium is not an extremely scarce element, in terms of over-all abundance, but it is not found in highly concentrated mineral deposits, and so is not easily accessible in large quantities. Until the recent advent of the transistor industry, there was very little demand for the element. Germanium is found in the ashes of some coals, and is extracted from these, where it is present as the oxide, GeO_2. One method consists in treating the residues with concentrated HCl, forming $GeCl_4$. This substance, which boils at 86°C, can be distilled off. The product is impure, since gallium, which also forms a volatile chloride, is also usually present.

Semiconductors

Pure silicon or germanium crystals are not conductors of electricity. All of the electrons present in the valence shell of each atom are involved in formation of covalent bonds with neighboring atoms in the diamond-type lattice. In the case of silicon, for example, the 3*s* and 3*p* orbitals of each atom are filled with shared electron pair bonds. There are other orbitals on the atoms which are completely vacant (*e.g.*, the 3*d*, 4*s*, *etc*). Now suppose that an arsenic atom is inserted in the lattice in place of a silicon or germanium atom. The arsenic atom must form bonds to its four neighbors, and so utilizes four of the five valence shell electrons. But after formation of the four bonds, there is no longer room in the valence shell of the arsenic atom for the extra electron. It must therefore be "promoted" out of the valence shell, into one of the higher, unused orbitals. But if it is to be located in one of the higher energy orbitals, there is no strong force which compels the electron to remain centered on the arsenic atom. The higher energy atomic orbitals of the arsenic overlap with similar, vacant orbitals of the surrounding silicon or germanium atoms. Indeed, there is an entire network of overlapping, vacant orbitals in the lattice. The extra electron is, therefore, more or less free to wander about in the lattice. Under the influence of an electric field in the solid, the electron migrates toward the positive pole. A Group IV element which has thus been "doped" with a Group V element, is termed an *n-type semiconductor*, to indicate that the charge species which moves in the lattice is a negative carrier of charge.

A p-type semiconductor is produced by adding a Group III element in small quantity to a Group IV element lattice. For example, if gallium is added to germanium, the gallium atom lacks one valence electron for bonding to the four neighboring silicon atoms. There is, therefore, a "hole" in the valence electron distribution, where there should be, but is not, a bonding electron. A bonding electron from a neighboring bond might jump into the hole to form a bonding electron pair, but in doing so, it leaves a hole behind. Under the influence of an electric field, the electrons move toward the positive plate by jumping from filled bonding orbitals into holes which are closer to the positive pole. This type of process is diagrammed in Fig. 17-16. If one looks at the hole, however, it appears that the hole is moving in the opposite direction from the electron migration. The hole thus seems to migrate toward the negative pole; in this sense, it behaves as a positive charge. A semiconductor which is produced by doping a Group IV element

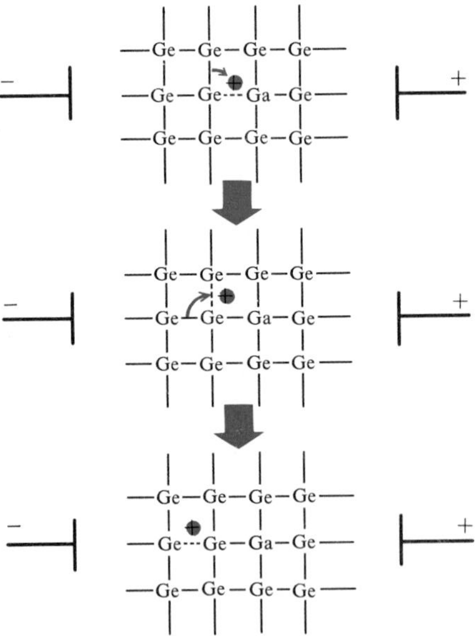

Fig. 17-16 Conductivity in a p-type semiconductor.

with a Group III element is called a *p-type semiconductor* to indicate that the species which appears to move under the influence of an electric field, the current carrier, is a positive hole. The two types of semiconductors can be combined in various ways to produce semiconductor diodes and other transistor configurations.

17-6 Group III Elements

Boron — The electronic configuration of boron, the only element of Group III which can be classified as a non-metal, is $2s^22p$:

2s	2p		
1l	1		

The element is difficult to prepare, particularly in a pure crystalline form. The best method, where only small quantities are required, involves the decomposition of the bromide on a hot tungsten wire:

$$BBr_3(g) \xrightarrow[\text{filament}]{\text{hot}} B(s) + \tfrac{3}{2}Br_2(g) \qquad (17\text{-}19)$$

The commercially important sources of boron are oxy-compounds such as Kernite, $Na_2B_4O_7 \cdot 4H_2O$, Borax, $Na_2B_4O_7 \cdot 10H_2O$ and Colemanite, $Ca_2B_6O_{11} \cdot 5H_2O$. The major deposits are found in the desert regions of California. Most of the material mined is used in the glass industry, as borate salts or boric oxide, B_2O_3.

Exercises

17-1. Consider the very simple reaction, $F_2(g) + Cl_2(g) \rightarrow 2ClF(g)$. The free energy of formation of ClF is given as $\Delta G_f(25°C) = -15.0$ kcal/mole. Calculate the equilibrium constant. What units does it have? What do you predict about the polarity of the product?

17-2. Calculate the quantity of sodium bisulfite required to free elemental iodine from 300 grams of potassium iodate.

17-3. Cite specific physical and chemical properties to show that selenium qualifies as a non-metal in terms of the criteria discussed in this chapter. Make use of material from at least one reference other than the text.

17-4. Fluorine cannot be prepared by electrolytic oxidation of fluoride ion in water. Why? Write a balanced chemical equation to illustrate your answer.

17-5. Describe experiments which you might perform to determine what allotropic form of sulfur is present in an unlabelled bottle of the element.

17-6. Draw Lewis structures for the following substances: S_8, P_4, S_6, P_2, graphite.

17-7. Discuss the relationship between ozone and sulfur dioxide, insofar as electronic structure is concerned. What implications can be drawn from the fact that the S_3 molecule is unknown?

17-8. Write balanced equations for each of the following:
(a) FeS_2 is roasted in air to give iron (III) oxide Fe_2O_3 and sulfur dioxide.
(b) Sodium chloride is reacted with manganese dioxide in sulfuric acid solution.
(c) Hydrogen sulfide is catalytically oxidized to free sulfur and water.
(d) Impure tellurium is treated with hot concentrated sulfuric acid.
(e) Germanium dioxide is reduced with coke.

17-9. The drawing shows a phase diagram for sulfur. The solid lines show the equilibrium values for phase changes; the dotted lines show the phase relationships which apply when the changes are made sufficiently rapidly so that phase changes in the solid do not have time to occur. S_α = rhombic sulfur, S_β = monoclinic sulfur. Identify the following points:
(a) A triple point involving solid, liquid, and vapor.
(b) The melting point of rhombic sulfur.
(c) The melting point of monoclinic sulfur.
(d) The highest pressure at which monoclinic sulfur can exist.
From the diagram, what can be said regarding the relative densities of rhombic and monoclinic sulfur; the relative densities of monoclinic and liquid sulfur? Explain.

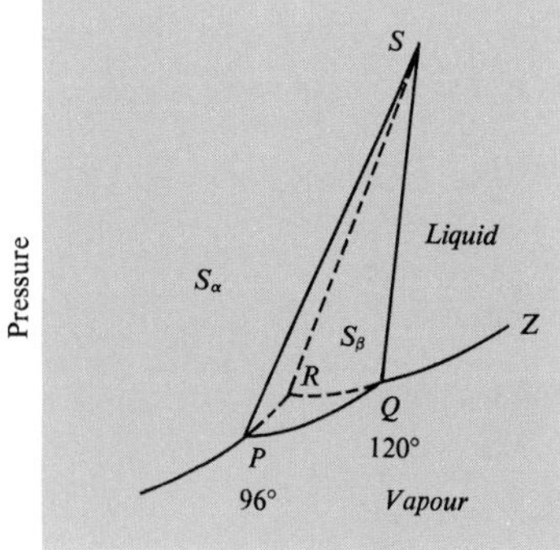

17-10. Suppose that the cost of electricity to a chlorine producer is 1.5 cents per kw-hr. What is the electrical cost involved in chlorine production, assuming 100 per cent efficiency in the cells, per pound of chlorine? The cells are operated at 10 v.

17-11. Suppose that you are given a vessel of about 1 l volume, painted black, which is said to contain the vapor of one of the following non-metallic elements: nitrogen, oxygen, chlorine, or argon. Cite chemical and physical tests which you might carry out to determine the nature of the gas.

17-12. The heats of formation of the gaseous atoms of some of the non-metallic elements are as follows (kcal/mole):

H	52.1	Si	105	As	60
C	171	P	75.3	Se	49
N	113	S	57	Br	26.7
O	59.5	Cl	28.9	Sn	72
F	18	Ge	89	Sb	61

Explain the trend in the series from C through F in terms of the structures of the elements. Do the same for the series Si through Cl. What is the significance of the variation among the Group IV elements?

17-13. Given that germanium possesses the diamond structure, from the density data in Table 17-4, from a knowledge of the carbon-carbon distance in diamond (Table 17-5), and from the relative atomic weights, calculate the covalent radius of germanium.

17-14. What do you expect would be the effect of adding small quantities of indium to an *n*-type semiconductor consisting of germanium doped with arsenic? Explain. To extend this idea further, what do you predict for the structure and properties of the stoichiometrically pure compound InAs? How might a semiconductor be produced from this substance?

17-15. Write balanced chemical equations for a method of preparing each of the following elements: phosphorus, bromine, fluorine, arsenic.

Chapter 18

The Chemistry of Hydrogen

18-1 Properties of the Element

Hydrogen, the lightest of the elements, possesses interesting properties in its own right, and forms important compounds with nearly all of the other elements in a variety of bonding situations. The element exists as the diatomic gas, H_2, which forms the simplest example of a molecular covalent bond. Hydrogen exists in nature in three isotopic forms. The most abundant, with nuclear mass one, is commonly referred to as hydrogen. The isotope of mass two is called deuterium (symbol D), and that of mass three is called tritium (symbol T). Deuterium (0.0156 per cent abundance) possesses one proton and one neutron in the nucleus; tritium (10^{-15} per cent abundance) possesses one proton and two neutrons. Because the percentage change in nuclear mass is so great among these isotopes, the properties of the hydrogen compounds composed from the different isotopes often differ significantly in both physical and chemical properties. This is particularly evident from an examination of the properties of H_2 and D_2, as shown in Table 18-1.

Table 18-1 Some comparative physical properties of H_2 and D_2.

Property	H_2	D_2
Melting point	−259.2°C	−254.4°C
Density of liquid	0.0710(−252.8°C)	0.163(−249.8°C)
Boiling point	−252.8°C	−249.5°C
Heat of dissociation (at 25°C, kcal/mole)	104.18	105.96
Heat of vaporization (cal/mole)	216	293

As a result of differences in the properties of H_2O and D_2O, it is possible to obtain quite highly enriched samples of D_2O, about 99 per cent, for rather low cost, despite the low abundance of deuterium in nature. Tritium is of such very low abundance that it is impractical to obtain it in any significant concentration from naturally occurring water. Tritium is an unstable nuclide, which decays to form a helium nuclide of mass three with a half-life of 12.5 years:

$$^{3}_{1}\mathrm{T} \rightarrow {}^{3}_{2}\mathrm{He} + {}^{0}_{-1}e \tag{18-1}$$

Because of this rather quick decay, there would be no tritium at all in nature if it were not for the continuous formation of new tritium by cosmic-ray-induced nuclear transformation in the upper atmosphere. Tritium is useful in research work for following the path of hydrogen in chemical systems by radiotracer counting techniques. For this purpose it is produced in nuclear reactors.

The element hydrogen ordinarily exists in the form of the H_2 molecule which possesses a substantial dissociation energy, 104 kcal/mole. If sufficient energy for rupture of the H—H bond is somehow supplied to the system, as in an electric discharge, atomic hydrogen is produced. If atomic hydrogen is produced in a dilute gas, *i.e.*, at a pressure of about 0.5 mm Hg, the rate of recombination of H atoms to form H_2 is remarkably slow, with a half-life of about 0.3 sec. From previous considerations of the rates of bimolecular reactions (Secs. 13-3, 13-4) it might be thought that the process $H + H \rightarrow H_2$ would have essentially no activation energy, and should therefore proceed at a rate not too much different from the calculated rate of bimolecular collisions of H atoms. If this were the rate of recombination, the half-life would be on the order of 10^{-8} sec. The slowness of the rate results from the need for a three-body, not a two-body, collision. If just two H atoms combine to form H_2, there is nowhere for the energy released in bond formation to go. It must appear, therefore, in the vibration of the H_2 molecule, which simply flies apart again. Only if there is a third body present at the moment of collision can the H_2 molecule transfer enough of the energy away, and survive. The third body could be another H atom, an H_2 molecule, or the wall of the container. Such three-body collisions are relatively rare, however, so the rate of H-atom recombination is much slower than one might otherwise expect.

Hydrogen gas is produced in large quantities as a by-product in industry. For example, it is formed in the catalytic cracking of hydrocarbons for gasoline use:

$$4C_2H_6 \rightarrow C_8H_{18} + 3H_2 \quad (18\text{-}2)$$

$$C_6H_{14} \rightarrow C_6H_6 + 4H_2 \quad (18\text{-}3)$$

It is evolved in the electrolysis of brine solutions to produce sodium hydroxide and chlorine (Sec. 11-4). It is formed in the reduction of water vapor by red-hot coke:

$$H_2O + C \rightarrow CO + H_2 \quad (18\text{-}4)$$

Hydrogen is used industrially in the Haber process for synthesis of ammonia, in hydrogenation of vegetable oils, in the production of methyl alcohol, and in other basic industrial processes. Industrial hydrogen production in the U.S. in 1966 is estimated to have been about 160 billion cubic feet (standard temperature and pressure), not counting that needed for synthesis of 11 million tons of ammonia.

Hydrogen forms chemical compounds with nearly all of the elements. These are classified into three groups: The covalent hydrides, the saline (or salt-like) hydrides, and the interstitial hydrides. The first classification is the most extensive; it includes such compounds as HCl, H_2O, *etc.*, as well as some exotic and interesting new compounds which have been prepared in recent years, in which there is a direct bond between a single hydrogen and a single transition metal atom. An example of the latter is manganese pentacarbonyl hydride, $Mn(CO)_5H$.

18-2 Salt-Like Hydrides

The salt-like hydrides include compounds such as sodium hydride, NaH. The compounds of this group display many characteristics of ionic compounds, in which there is a hydride ion, H^-, in an ionic lattice with a metal ion. The dividing line between ionic and covalent character is not very clear, however, and it is probably not safe to attempt a rigid classification of compounds such as MgH_2, BeH_2, *etc.*

A number of interesting hydride systems are formed combining two different metal hydrides, both of which are classifiable as ionic. For example, lithium hydride and aluminum hydride may be considered to react to form lithium aluminum hydride:

$$LiH + AlH_3 \rightarrow LiAlH_4 \quad (18\text{-}5)$$

(The compound is not actually prepared in this way.) The mixed hydride can be thought of as an Li^+ cation and an AlH_4^- anion.

All of the ionic hydrides are characterized as strong reducing agents. They react with water, or with any other compound containing a fairly positive hydrogen, to liberate hydrogen:

$$CaH_2 + H_2O \rightarrow Ca^{+2} + 2OH^- + 2H_2 \quad (18\text{-}6)$$

$$KH + NH_3 \rightarrow K^+ + NH_2^- + H_2 \quad (18\text{-}7)$$

When electrolyzed, the molten hydrides yield metal at the cathode and hydrogen at the anode:

$$2LiH(l) \rightarrow 2Li(s) + H_2(g) \qquad (18\text{-}8)$$

It is interesting to investigate the stability of a salt-like metal hydride lattice on the assumption that it is completely ionic, using the Born-Haber cycle. For example, for lithium hydride, we have the following data, all in kcal/mole:

Heat of formation	−22
Heat of sublimation of lithium	38
$\frac{1}{2}$ (Heat of dissociation of H_2)	52
Ionization potential of lithium	124
Electron affinity of hydrogen	16

Using the treatment of the Born-Haber cycle given in Sec. 7-3, we find that the lattice energy of lithium hydride is 220 kcal/mole. Clearly, its stability results from the very large energy terms involved in formation of the three-dimensional ionic lattice.

18-3 Interstitial Hydrides

The third class of hydrides is composed of systems which are often not really compounds at all, in the sense that a straightforward empirical formula can be written. For example, hydrogen reacts with palladium metal to form a species of variable composition, containing up to two hydrogen atoms for each palladium atom. In such systems, it appears that the hydrogen molecule is dissociated on the metal surface, and that the hydrogen atoms diffuse into the interior of the metal. They take up positions in the interstices of the metal lattice, and — through bond formation with the surrounding metal atoms — profoundly affect the properties of the metal. The interaction of hydrogen with metal surfaces may be utilized by using finely divided metals as hydrogenation catalysts. We will not discuss the properties of this type of hydride system any further at this point.

18-4 Hydrides of the Non-Metallic Elements

The non-metallic elements vary greatly in the ease and vigor with which they react with hydrogen, and in the stability of the resulting hydride. In general, one can say that the stability is greatest when the electronegativity of the non-metallic element differs most from that of hydrogen. In fact, the scheme originally devised by Pauling for determining electronegativities incorporated the X—H bond energy in determining the electronegativity of X. The standard enthalpies of formation of non-metal hydrides are given in Table 18-2. There is a fair degree of parallelism in the trends of hydride stability and the ease of formation by direct reaction with hydrogen.

Table 18-2 Standard enthalpies of formation of covalent hydrides (kcal/mole).

IV	V	VI	VII
CH_4	NH_3	H_2O	HF
−17.8	−11.0	−57.8	−64.2
SiH_4	PH_3	H_2S	HCl
−14.8	2.2	−4.8	−22.1
GeH_4	AsH_3	H_2Se	HBr
?	41.0	20.5	−8.7
	SbH_3	H_2Te	HI
	34	18	6.2

The hydrides of fluorine, chlorine and oxygen are all formed with great ease. The reaction with fluorine occurs with explosive violence upon contact of the two elements. The reaction involving oxygen or chlorine occurs with explosive violence if it is initiated by a photon of high energy, or by a spark.

$$\begin{aligned} H_2 + F_2 &\rightarrow 2HF \\ H_2 + Cl_2 &\rightarrow 2HCl \\ 2H_2 + O_2 &+ 2H_2O \end{aligned} \tag{18-9}$$

It will be noted that fluorine, oxygen, and chlorine are, in the order listed, the three most electronegative elements. The stability of the hydrides formed in the three reactions listed above is attested to by the fact that they are not appreciably dissociated at temperatures as high as 1000°C.

Moderately stable hydrides are formed by nitrogen, bromine, sulfur, iodine, and carbon. These elements form hydrides which are quite stable to dissociation at room temperature, but which are appreciably dissociated into the elements at temperatures of perhaps 600°C or so. Some of the formation reactions are as follows:

$$\begin{aligned} &N_2(g) + 3H_2(g) \rightarrow 2NH_3(g) && \text{(350°C, catalyst)} \\ &H_2(g) + Br_2(g) \rightarrow 2HBr(g) && \text{(rapid at 200-300°C with catalyst)} \\ &H_2(g) + S(g) \rightarrow H_2S(g) && \text{(600°C, catalyst)} \\ &H_2(g) + I_2(g) \rightarrow 2HI(g) && \text{(rapid at 600°C, incomplete)} \end{aligned} \tag{18-10}$$

It is not possible to make a detailed comparison of the case with which the above reactions proceed, since the form of the elements varies so widely. Thus, even though carbon does not react directly with hydrogen at all, methane, CH_4, is moderately stable when it is formed, so that carbon fits best in this class. The elements of this class are those which, though not the most electronegative, are clearly non-metallic in character. Hydrogen iodide is more easily dissociated than hydrogen bromide, a fact which further illustrates the relationship between electronegativity and stability.

On the basis of its electronegativity, one might expect that nitrogen would be in the first class of elements. The decomposition of ammonia, however, leads to the formation of N_2, which is unusually stable because of the triple bond between the nitrogen atoms. This factor makes ammonia relatively less stable with respect to decomposition than, for example, hydrogen chloride.

The remaining non-metallic elements — phosphorus, arsenic, selenium, tellurium, silicon, and boron—form hydrides of low stability. These elements do not react with hydrogen at room temperature, nor do they react at elevated temperatures, in some cases. The hydrides are appreciably dissociated at higher temperatures. It will be noted that the electronegativities of all of the elements in this class are approximately the same as that of hydrogen (Table 8-3).

Preparations — The direct reaction of the elements is very often not the best or most convenient method of producing the hydrides. The following reactions illustrate other methods which are employed.

1. $$H_2SO_4(conc) + NaCl(s) \rightarrow HCl(g) + NaHSO_4(s) \quad (18\text{-}11)$$

The treatment of an alkali metal salt with a strong, non-volatile acid at higher temperatures is employed in producing hydrogen fluoride, hydrogen chloride, hydrogen bromide, and hydrogen iodide. With the latter two elements, it is necessary to use another acid — usually phosphoric — because sulfuric acid reacts with the hydrogen bromide or iodide produced.

$$H_2SO_4 + 8HI \rightarrow 4I_2 + H_2S + 4H_2O \quad (18\text{-}12)$$

$$H_2SO_4 + 2HBr \rightarrow Br_2 + SO_2 + 2H_2O \quad (18\text{-}13)$$

These last two reactions illustrate a chemical property which the hydrides may exhibit; *i.e.*, an instability toward oxidizing agents. Since the non-metallic element is in a negative oxidation state, one might expect that the resistance to oxidation would be greatest for hydrides of the most electronegative elements. In a general sense, this is the case.

2. $$PBr_3(s) + 2H_2O(l) \rightarrow 3HBr(g) + H_3PO_3(aq) \quad (18\text{-}14)$$

This reaction illustrates the method most often employed in preparing laboratory quantities of hydrogen bromide.

3. $$H_2S(g) + I_2(s) \rightarrow 2HI(aq) + \tfrac{1}{8}S_8(s) \quad (18\text{-}15)$$

This reaction is carried out by bubbling hydrogen sulfide gas into an aqueous suspension of iodine. It is, in effect, a replacement reaction.

4. $$As_4O_6(s) + 12Zn(s) + 24H_3O^+ \rightarrow 4AsH_3(g) + 12Zn^{+2} + 30H_2O \quad (18\text{-}16)$$

This reaction is employed in reducing compounds of arsenic. In the Marsh test, the arsine, AsH_3, which is produced is passed over a hot surface. The compound decomposes, leaving a mirror of arsenic on the surface. Extremely small quantities of arsenic can be detected in this manner.

Physical and chemical properties — All of the non-metallic elements form binary hydrides which have formulas as expected from simple valence considerations. Many of the elements, however, most particularly boron, form more than one binary hydride. Formulas and a few physical properties of a number of the more complex hydrides are listed in Table 18-3.

Table 18-3 Formulas, and melting and boiling points of some higher hydrides of non-metallic elements.

Compound	Formula	Melting point (°C)	Boiling point (°C)
Hydrogen peroxide	H_2O_2	−0.4	150
Hydrogen disulfide	H_2S_2	—	74–75
Hydrogen trisulfide	H_2S_3	−52	45 at 4.5 mm pressure
Hydrazine	N_2H_4	2	113
Diphosphine	P_2H_4	—	64
Disilane	Si_2H_6	−132	−14
Trisilane	Si_3H_8	−117	53
Diarsine	As_2H_4	—	100
Diborane	B_2H_6	−165	−92.5
Tetraborane-10	B_4H_{10}	−120	16
Pentaborane-9	B_5H_9	−47	58
Pentaborane-11	B_5H_{11}	−123	65
Decaborane-14	$B_{10}H_{14}$	99.5	213

There is bonding between non-metal atoms in all of the more complex hydrides, so the ratio of non-metal to hydrogen is different from that in the simple hydride. In general, the more complex hydrides are considerably less stable than the simple hydride. Boron hydrides, discussed below, are an exception to this; in fact, the higher boron hydrides are produced by heating the simplest known compound, B_2H_6.

The physical properties of the simple hydrides are regular, except for those which exhibit anomalous properties due to hydrogen bonding, as discussed in Sec. 8-13. We will not discuss the physical properties further, except to note that the large relative mass difference between hydrogen and deuterium is reflected in the properties of the hydrides, as exemplified by the data in Table 18-4 for H_2O and D_2O. Mixtures of H_2O and D_2O

Table 18-4 Comparative physical properties of H_2O and D_2O.

Property	H_2O	D_2O
Density (25°C)	1.00000	1.10764
Temperature of maximum density	4.0°C	11.23°C
Boiling point	100°C	101.42°C
Freezing point	0°C	3.802°C
Heat of vaporization (kcal/mole)	9.700	9.960
Heat of fusion (kcal/mole)	1.436	1.510

possess physical properties which are intermediate between those for the pure components. The hydrogen and deuterium atoms of water molecules undergo extremely rapid exchange of hydrogen and deuterium from one molecule to another, so that in mixtures of H_2O and D_2O there is a predominance of HDO.

The hydrides may undergo double decomposition reactions, as in the following examples:

$$SiH_4(g) + Cl_2(g) \rightarrow SiH_3Cl(g) + HCl(g) \quad (18\text{-}17)$$

$$SiO_2(s) + 4HF(g) \rightarrow SiF_4(g) + 2H_2O \quad (18\text{-}18)$$

$$SiS_2(s) + 2H_2O(g) \rightarrow SiO_2(s) + 2H_2S(g) \quad (18\text{-}19)$$

$$TiCl_4(l) + 4HF(g) \rightarrow TiF_4(s) + 4HCl(g) \quad (18\text{-}20)$$

The direction of such reactions is determined by the thermodynamic stability of the products, and/or by the relative volatilities of the products as compared with those of the reactants. Some of these reactions (for example, the first listed) are quite violent.

The hydrides may be oxidized under various conditions, particularly in aqueous solutions. Even in the gas phase, reactions such as the following occur:

$$2HCl(g) + \tfrac{1}{2}O_2(g) = H_2O(g) + Cl_2(g) \quad (18\text{-}21)$$

$$H_2S(g) + \tfrac{1}{2}O_2(g) = H_2O(g) + S(s) \quad (18\text{-}22)$$

$$H_2S(g) + \tfrac{3}{2}O_2(g) = H_2O(g) + SO_2(g) \quad (18\text{-}23)$$

$$SiH_4(g) + 2O_2(g) = SiO_2(s) + 2H_2O(g) \quad (18\text{-}24)$$

Hydrogen chloride, hydrogen bromide, and hydrogen iodide react with water to produce strongly acid solutions (Chap. 12). Hydrogen fluoride reacts to produce a weakly acidic solution. Hydrogen sulfide, hydrogen selenide, and hydrogen telluride are very weakly acidic in water. Ammonia reacts with water to form a weakly basic solution; the other Group V hydrides are either *very* weakly basic or do not react at all.

A few miscellaneous properties of the hydrides are worth noting. The physiological action of these compounds varies widely. All of the hydrides, except water, are highly toxic. The Group V and VI compounds particularly, with the exception of water, are very toxic. Hydrogen sulfide, hydrogen selenide, and hydrogen telluride also have the dubious distinction of being among the worst-smelling compounds known.

The boron hydrides — The history of boron hydride chemistry is an interesting story in a number of ways. The earliest work of importance was performed by Albert Stock and his students during the years 1910 to 1930. Stock found that when the rather reactive magnesium boride, Mg_3B_2, was reacted with a 10 per cent hydrochloric acid solution, the products, in part, consisted of volatile compounds containing boron and hydrogen. In the course of some remarkable experimentation, Stock and his students succeeded in isolating and identifying a number of distinct substances. It is a tribute to the thoroughness of this early work that no new hydride of boron was discovered for many years afterward, despite a great deal of active research in this area.

The simplest hydride of boron might be expected to have the formula BH_3, since boron has three valence electrons in its valence shell. In point of fact, the simplest compound has the formula B_2H_6; it is called *diborane*. Table 18-2 lists a number of the other known compounds in the boron hydride series. The empirical formulas of these compounds cannot be explained by postulating boron-boron bonds, analogous to the carbon-carbon bonds in hydrocarbons. A compound containing two boron atoms, and with a single boron-boron bond, should have the formula $BH_2—BH_2$.

Since diborane is a rather difficult compound to work with (it inflames on contact with air, is a gas, and reacts rapidly with water), its structure remained unknown for some time. When the structure (Fig. 18-1) was finally determined, those chemists who were interested in valence theory had something very interesting to think about.

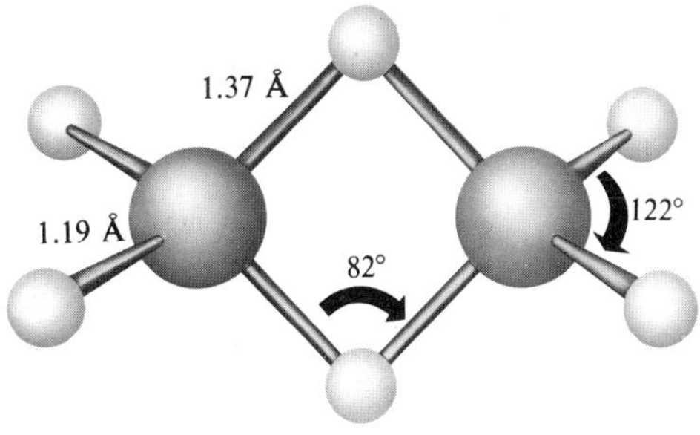

Fig. 18-1 The structure of diborane, B_2H_6.

There are three important things to note: (a) Each boron atom is apparently bonded to *four* hydrogens. (b) Two of the hydrogens are apparently bonded each to two boron atoms. These are called the *bridge hydrogens*, since they form a kind of bridge between the two boron atoms. The other four hydrogens are called *terminal hydrogens*. (c) The boron-bridge hydrogen distance is longer than the boron-terminal hydrogen distance. There is every indication that the bonds to the terminal hydrogens are normal, single electron pair bonds.

The total number of electrons in B_2H_6 is sixteen. Of these, each boron atom possesses a pair in the 1*s* orbital which is not used in bonding. Of the twelve remaining, eight are required for the normal boron-terminal hydrogen bonds, leaving four electrons for the bridge bonds. There are four boron-bridge hydrogen bonds, and only two electron pairs. We might attempt to describe the bonding in the bridges by writing a number of canonical resonance structures in which the two electron pairs are distributed variously among the possible bonds. The two most important structures are:

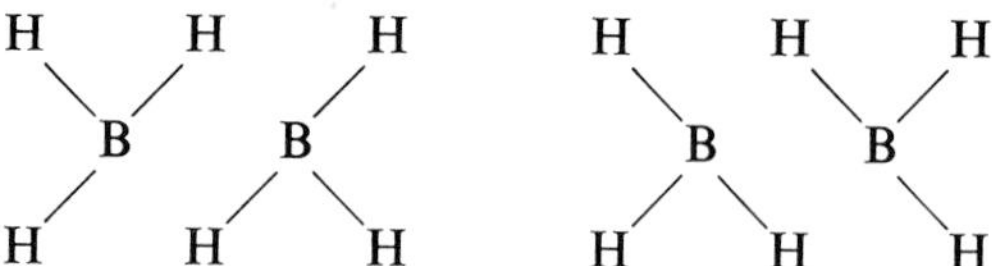

We see that each boron-bridge hydrogen bond possesses a pair of electrons in only half of the canonical structures, and might therefore be described as a half-bond. It is to be expected that the bond distance in a half-bond would be longer than in a normal, full bond, and this is found to be the case (Fig. 18-1).

The bridge bonds in diborane have also been described as three-center bonds. An ordinary covalent bond is a two-center bond, in the sense that the bonding electron pair is shared between two centers of positive charge. In the bridge bond, on the other hand, the single electron pair is shared between three nuclear centers: The two boron atoms and the bridging hydrogen atom. The bridge bonds are formed because the boron atom, after using all three of its valence shell electrons in bond formation, still has a vacant orbital in its valence shell. This orbital can be put to use only if bridge bonds are formed.

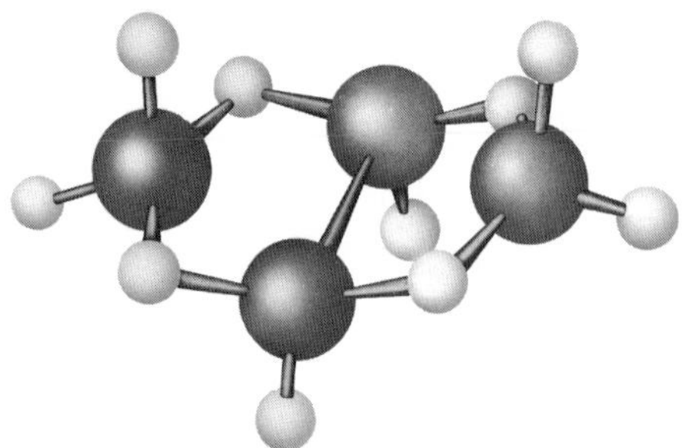

Fig. 18-2 The structure of tetraborane-10, B_4H_{10}.

Bridge bonds, as well as normal boron-hydrogen bonds, are also found in the higher boron hydrides. The structure of tetraborane-10 is shown in Fig. 18-2. It is obviously a complicated arrangement. It has only been in recent years — after a number of structures had been determined — that a consistent pattern has emerged. The most important work in this area has been done by Professor W. Lipscomb of Harvard University, and his students.

Some of the chemical properties of the boron hydrides have already been mentioned. They react readily with oxygen to form water and boric oxide, B_2O_3. They react with water at various rates, depending upon the compound, to give hydrogen gas and boric acid, H_3BO_3. With halogens, the products are partially halogenated boron hydrides and the corresponding hydrogen halide:

$$B_2H_6 + 4Cl_2 \rightarrow B_2H_2Cl_4 + 4HCl \qquad (18\text{-}25)$$

The products may vary in the extent to which the halogen replaces hydrogen, depending upon the reaction conditions.

Suggested Readings

Gould, E. S., *Inorganic Reactions and Structure* (2nd ed.), Henry Holt and Company, New York, 1962.

Jolly, W. L., *The Chemistry of the Non-metals*, Englewood Cliffs, N. J.: Prentice-Hall, Inc., 1966.

Siegel, B., "Hydride Formation by Atomic Hydrogen Reactions," *J. Chem. Educ.*, Vol. 38, 1961, p. 496.

Siegel, B., and J. L. Mack, "Boron Hydrides," *J. Chem. Educ.*, Vol. 34, 1957, p. 315.

Tyree, S. Y., and K. Knox, *Textbook of Inorganic Chemistry*, New York: The Macmillan Company, 1961.

Exercises

18-1. From the data given in Prob. 17-12 and Table 18-2, calculate the average thermochemical bond energies of as many non-metal-hydrogen bonds as the data permit. (See Sec. 9-4 for an example.) What significance can you attach to the trends observed in these data?

18-2. Each of the following chemical reactions leads to hydrogen gas as a product. Complete and balance each.

(a) $Fe + H_2O$ (steam, over red-hot iron) $\rightarrow Fe_3O_4 +$
(b) C (dull red heat) $+ H_2O(g) \rightarrow$
(c) $MgH_2 + H_2O \rightarrow$
(d) $Si + NaOH + H_2O \rightarrow Na_2SiO_3 +$
(e) $Zn + H_3O^+ \rightarrow$

18-3. Complete and balance each of the following. In cases where you are not sure of the products, assume reasonable products based upon your knowledge of the chemistry of the reactants.

(a) $TeO_2(s) + Zn + H_3O^+ \rightarrow$
(b) $NH_3 + O_2 \rightarrow NO +$
(c) $H_2S + H_2O_2 \rightarrow S +$
(d) $H_2S_3 + Cl_2 \rightarrow$
(e) $AsCl_3 + H_2O \rightarrow$
(f) $GeH_4 + HBr \rightarrow$
(g) $N_2H_4 + I_2 \rightarrow N_2 +$
(h) $Ca_2P_2 + H_2O \rightarrow P_2H_4 +$
(i) $H_3PO_3 \xrightarrow{\text{(heat)}} H_3PO_4 +$

18-4. Draw Lewis structures for the following hydrides: hydrogen peroxide, hydrazine, hydrogen trisulfide, disilane.

18-5. Draw a Lewis structure for tetraborane-10 (refer to Fig. 18-2). Indicate the bridge bonds by dotted rather than solid lines. Is it necessary to postulate a boron-boron bond in this molecule? Explain.

18-6. The equilibrium constants for the reaction $H_2S(g) \rightarrow H_2 + \frac{1}{2}S_2(g)$ as a function of temperature are as follows:

Temperature (°C)	750	830	945	1065	1132
K (atm$^{1/2}$)	1.06 $\times 10^{-6}$	4.2 $\times 10^{-6}$	24.5 $\times 10^{-6}$	107 $\times 10^{-6}$	226 $\times 10^{-6}$

(a) From these data, evaluate the heat of reaction. (b) Using the data of Problem 18-1, estimate a value for the S_2 bond energy.

18-7. The density of hydrogen fluoride gas at 25°C and 1 atm pressure is 2.34 g/l. Calculate the apparent molecular weight of the gas. Explain this value.

18-8. (a) How many grams of zinc are required for the preparation of 100 grams of arsine via Eq. (18-16)? (b) How many grams of chlorine would be required to completely convert 75 grams of silane to silicon tetrachloride?

18-9. The Marsh test is well known as a sensitive test for arsenic which can be performed with a very small amount of sample. Suppose that the film of arsenic which forms on the glass is two atomic layers thick, and must be at least 1 cm in diameter to be seen. Assume further that all of the arsenic in the sample is converted into arsine, but that only 10 per cent of the arsine is decomposed on the hot surface. What is the minimum amount of arsenic, in grams, that can be detected? Assume that each arsenic atom on the surface occupies a circular area of 2 Å radius, and that these circular areas are arranged in a rectangular array.

18-10. From the acid-base properties of the hydrides of the non-metallic elements, what rather crude generalization can be made regarding the electronegativity of an element and the acid-base property of its hydride in water? How is this to be explained in terms of structure? (Recall Sec. 8-5.)

18-11. Suppose that you are given three vessels, each containing a single substance, either hydrogen fluoride, nitrogen, or phosphine vapor. Cite physical and chemical tests which you could apply to determine which gas is contained in each vessel.

18-12. Describe one or more *chemical* tests which might be applied to distinguish one of the substances in each of the following pairs from the other.
(a) H_2S, HF.
(b) CH_4, PH_3.
(c) LiH, LiCl.
(d) Cl_2, HCl.
(e) SiH_4, H_2S.

18-13. The standard free energy of formation of the hydrogen atom at 25°C is given as 48.58 kcal/mole. The standard enthalpy of formation is

52.09 kcal/mole. What is the entropy change for the process $H_2(g) \rightarrow 2H(g)$ at 25°C? Is this result in accord with your expectations based on the discussion in Chap. 9? Assuming that the standard enthalpy of formation remains constant, at what temperature would you expect H_2 to be half-dissociated when total pressure is one atmosphere?

18-14. Diborane reacts with ammonia under carefully controlled conditions to yield a compound of the form BH_3NH_3. Indicate a pathway by which the reaction might proceed, and show a possible structure for the product. [*Hint:* This is a Lewis acid-base reaction.]

Chapter 19

Chemistry of the Non-Metals

19-1 Chemistry of the Halogens

For the elements to exist in a negative valence state, or in a negative oxidation state, they must acquire electrons. It follows, therefore, that there should be a good correlation between the electronegativity of an element and its stability in such a state. Of course, we must temper this optimism with the thought that "stability," however it is measured, is relative; the properties of the elements in other states, particularly as free elements, must also be considered.

In the halogen series — all members of which exist in the zero oxidation state as diatomic molecules — the stability of the -1 oxidation state varies in a simple way with the electronegativity. Thus, we find that the fluoride ion is very stable and not susceptible to oxidation by any other chemical agents, whereas the iodide ion is a rather good reducing agent; *i.e.*, the half-reaction

$$2I^-(aq) \rightarrow I_2 + 2e^-$$

can be made to occur easily. Chloride and bromide ions are intermediate

Table 19-1 Standard electrode potentials for oxidation of halide ions in water.

Reaction	Standard electrode potential
$F^-(aq) \rightarrow \frac{1}{2}F_2 + e^-$	−2.87 v
$Cl^-(aq) \rightarrow \frac{1}{2}Cl_2 + e^-$	−1.36 v
$Br^-(aq) \rightarrow \frac{1}{2}Br_2 + e^-$	−1.06 v
$I^-(aq) \rightarrow \frac{1}{2}I_2 + e^-$	−0.54 v

between these two extremes in the ease with which they are oxidized (Table 19-1).

It is of interest to examine the oxidation process in terms of a thermochemical cycle involving three separate steps:

$$\begin{array}{ccc} X^-(g) & \xrightarrow{\text{II}} & X(g) + e^- \\ \text{I}\uparrow & & \text{III}\searrow \quad \searrow \\ X^-(aq) & \xrightarrow{\text{A}} & \frac{1}{2}X_2(g) + e^- \end{array}$$

The free energy change for step A is thus equal to the sum of free energy changes for steps I, II, and III. There is in addition a term involving the electron, since the electron which is liberated in step A is in aqueous medium. We need not concern ourselves about the free energy value for the process of transferring the electron from the gas phase to the aqueous medium, since this step is identical for all of the halogens. Thus,

$$\Delta G(\text{A}) = \Delta G(\text{I}) + \Delta G(\text{II}) + \Delta G(\text{III}) + \text{constant}$$

Table 19-2 lists the values of ΔG for steps I, II, and III for the halogens. The large positive values for ΔG (total) are in keeping with the negative values for $\mathcal{E}°$, Table 19-1. If we compare the differences in ΔG(total) we

Table 19-2 Free energy values for processes involving the halogens (kcal/mole).

	F	Cl	Br	I
I. $X^-(aq) \rightarrow X^-(g)$ (− free energy of solution of the halide ion)	114	84	78	70
II. $X^-(g) \rightarrow X(g) + e^-$ (− ionization potential of halide ion*)	80	83	78	71
III. $X(g) \rightarrow \frac{1}{2}X_2(g)$ (− $\frac{1}{2}$ dissociation energy of the halide ion)	−19	−28	−23	−18
I + II + III = ΔG(total)	+175	+139	+133	+123

*This is also the negative of the electron affinity of halogen (see Sec. 6-12).

see that these are consistent with the observed differences in $\mathcal{E}°$. For example, the difference in ΔG(total) for fluorine as compared with bromine is 42 kcal/mole. Using the conversion factor, 23 kcal/volt, this implies a difference of 1.82 volts in $\mathcal{E}°$ values for the two elements, which is essentially identical with the observed difference.

By examining the values for the energy terms in Table 19-2, we can see that the single largest contributor to the large value of $\mathcal{E}°$ for fluorine is the free energy of hydration of the ion, which is larger for fluoride than it is for the other halogens. It also helps that the F—F bond dissociation energy is smaller than for the other halogens, but this is not the predominant factor. The higher free energy of solution of the fluoride ion in water as compared with that of the other halogens is the result of its smaller size (see Table 7-5, page 129). The fluoride ion is more strongly bound by the surrounding water molecules, and is thus further stabilized relative to F_2.

Oxyacids — The addition of a non-metal oxide to water results in an acid solution. Reaction of the oxide with water produces, in addition to the hydronium ion, an anion (or anions) characteristic of the particular compound. Many of the oxygen acids are not stable except in dilute solution. It is nevertheless possible to study them under these conditions with regard to acid strength and oxidation-reduction properties. Salts derived from the acids by neutralization with base are often more stable than the acids themselves, and are frequently obtainable from solution as crystalline solids.

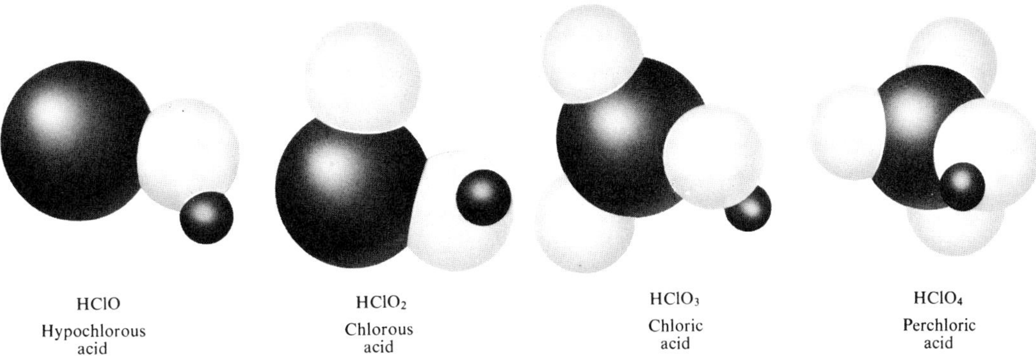

$HClO$ Hypochlorous acid | $HClO_2$ Chlorous acid | $HClO_3$ Chloric acid | $HClO_4$ Perchloric acid

Fig. 19-1 The oxyacids of chlorine.

Chlorine forms four oxyacids in aqueous solution (Fig. 19-1). It should be noted from the structures shown that in each case there is a hydrogen atom attached to oxygen, which is in turn bonded to chlorine. The anion related to each acid results from loss of the proton. The nomenclature employed in this series is general, and applies to all similar series of oxyacids. One particular oxidation state (in the case of chlorine, it is the +5 state) is selected as a sort of reference. The acid corresponding to this oxidation state is given a name which ends in *ic*. The next higher oxidation state is denoted by the prefix *per*, while the next lower is denoted by the changed suffix, *ous*, *etc.* Comparison of the chlorine oxidation states in these acids with those in the oxides discussed below shows that the anhydrides of chlorous and chloric acid are not known, while there are no acids which correspond to the oxides ClO_2 and ClO_3. When the latter two compounds are placed in contact with water they undergo the typical hydrolysis, but it is accompanied by disproportionation:

$$2ClO_2 + H_2O \rightarrow HClO_2 + HClO_3 \tag{19-1}$$

$$2ClO_3 + H_2O \rightarrow HClO_3 + HClO_4 \tag{19-2}$$

With increased numbers of oxygen atoms attached to the central atom, the chlorine exerts an increased attraction on the electrons in the bond to the hydrogen-bearing oxygen atom. The effect of this is to make the O—H bond itself more polar, so that the proton is more easily removed by a base. An increase in acid strength with increasing oxidation number of chlorine is therefore to be expected, and actually does occur — hypochlorous acid is very weak, while perchloric acid is one of the strongest acids known.

The ability of the acids to function as oxidizing agents is measured by the half-cell potential for the reduction processes:

$$\begin{array}{lr} & \mathcal{E}^\circ \\ 7e^- + ClO_4^- + 8H_3O^+ \rightarrow \frac{1}{2}Cl_2 + 12H_2O & 1.38 \text{ volts} \\ 5e^- + ClO_3^- + 6H_3O^+ \rightarrow \frac{1}{2}Cl_2 + 9H_2O & 1.47 \text{ volts} \\ 3e^- + HClO_2 + 3H_3O^+ \rightarrow \frac{1}{2}Cl_2 + 5H_2O & 1.63 \text{ volts} \\ e^- + HClO + H_3O^+ \rightarrow \frac{1}{2}Cl_2 + 2H_2O & 1.63 \text{ volts} \end{array} \tag{19-3}$$

It can be seen that, in a general way, the ability to function as an oxidizing agent decreases with increasing oxidation number of the chlorine. Although the cell potentials do not tell us anything about the rates with which the substances involved react, it is true in this case that the reactivity of the acids also decreases with increasing oxidation number. This generalization is subject to some uncertainty, however, because of the large number of variables involved (temperature, concentration, pH, *etc.*). In terms of half-reaction potentials, for example, perchloric acid in solution is capable of oxidizing water, with evolution of oxygen. Dilute perchloric acid solutions are quite stable, however, for indefinite periods of time.

In basic solution, all of the oxychlorine species exist as the oxyanions. The half-reaction potentials for reduction to chlorine are different from those in acid solution:

$$\begin{array}{lr} & \mathcal{E}^\circ \\ 7e^- + ClO_4^- + 4H_2O \rightarrow \frac{1}{2}Cl_2 + 8OH^- & +0.42 \text{ volt} \\ 5e^- + ClO_3^- + 3H_2O \rightarrow \frac{1}{2}Cl_2 + 6OH^- & +0.48 \text{ volt} \\ 3e^- + ClO_2^- + 2H_2O \rightarrow \frac{1}{2}Cl_2 + 4OH^- & +0.58 \text{ volt} \\ e^- + ClO^- + H_2O \rightarrow \frac{1}{2}Cl_2 + 2OH^- & +0.42 \text{ volt} \end{array} \tag{19-4}$$

It is evident from these data that oxychlorine species are more stable in basic than in acid solution.

Using the standard potentials given above, it is possible to calculate the standard potentials for other half-reactions. For example, suppose that we wish to know the standard oxidation potential for the half-reaction

$$\text{(A)} \qquad ClO_4^- + 4e^- + 2H_2O \rightarrow ClO_2^- + 4OH^- \qquad (19\text{-}5)$$

We have, from Eq. (19-4), the following two half-reactions:

$$\text{(B)} \quad ClO_4^- + 7e^- + 4H_2O \rightarrow \tfrac{1}{2}Cl_2 + 8OH^- \qquad \mathcal{E}^\circ = 0.42 \text{ volt} \qquad (19\text{-}6)$$

$$\text{(C)} \quad \tfrac{1}{2}Cl_2 + 4OH^- \rightarrow ClO_2^- + 2H_2O + 3e^- \qquad \mathcal{E}^\circ = -0.58 \text{ volt} \qquad (19\text{-}7)$$

Adding the two half-reactions together results in the desired half-reaction. It might be thought at first glance that we need only add the $\mathcal{E}^\circ$ values to obtain the correct $\mathcal{E}^\circ$ value for the new half-reaction. We can see that this is not the case, however, by recalling the thermodynamic significance of $\mathcal{E}^\circ$. The standard half-cell potential is related to the free energy change for the half-reaction by $\Delta G^\circ = n\mathcal{F}\mathcal{E}^\circ$, where n is the number of electrons transferred. In adding two half-reactions together we are adding the free energies, so we have

$$\begin{aligned} \Delta G^\circ(\text{A}) &= \Delta G^\circ(\text{B}) + \Delta G^\circ(\text{C}) \\ 4\mathcal{F}\mathcal{E}^\circ(\text{A}) &= 7\mathcal{F}\mathcal{E}^\circ(\text{B}) + 3\mathcal{F}\mathcal{E}^\circ(\text{C}) \\ \mathcal{E}^\circ(\text{A}) &= \frac{7(0.42) + 3(-0.58)}{4} = +0.30 \text{ volt} \end{aligned}$$

The student may wonder why this point was not raised in Chap. 16, in which half-reaction potentials were added together to give $\mathcal{E}^\circ$ values for cell reactions. When half-reactions are added or subtracted to give an over-all balanced equation, the number of electrons transferred is necessarily the same in each half-reaction, so that n is the same throughout. It is necessary to worry about the number of equivalents of electrons transferred in each half-reaction only when the net process is itself a half-reaction; *i.e.*, when there is a net number of electrons consumed or liberated in the final equation.

Chlorine provides examples of a reaction which is quite general for the non-metals — disproportionation in basic solution. The products of the reaction depend upon concentration and temperature as well as pH. The most readily observed reaction is the formation of hypochlorite:

$$Cl_2 + 2OH^- \rightarrow Cl^- + OCl^- + H_2O \qquad (19\text{-}8)$$

In *hot* aqueous solution, the reaction is:

$$3Cl_2 + 6OH^- \rightarrow 5Cl^- + ClO_3^- + 3H_2O \qquad (19\text{-}9)$$

The standard potential for either of these reactions can be calculated by proper combination of half-cell potentials. The over-all reaction in Eq. (19-9), for example, can be thought of as a sum of a reduction half-reaction:

$$\tfrac{1}{2}Cl_2 + e^- \rightarrow Cl^- \qquad \mathcal{E}^\circ = +1.36 \text{ volts}$$

and an oxidation half-reaction:

$$\tfrac{1}{2}Cl_2 + 6OH^- \rightarrow ClO_3^- + 3H_2O + 5e^- \qquad \mathcal{E}^\circ = -0.48 \text{ volt}$$

The over-all reaction has a positive $\mathcal{E}^\circ$ value, and the equilibrium lies far to the right in basic solution.

Hypochlorous acid can be obtained by bubbling chlorine into water and then adding a silver or mercury salt (usually the oxide) to precipitate the chloride ion:

$$\begin{aligned} Cl_2 + 2H_2O &\rightarrow Cl^- + H_3O^+ + HOCl \\ Ag_2O(s) + 2Cl^- + 2H_3O^+ &\rightarrow 2AgCl(s) + 3H_2O \end{aligned} \qquad (19\text{-}10)$$

Hypochlorous acid solutions are unstable; the acid decomposes, with the evolution of oxygen.

$$2HOCl \rightarrow 2HCl + O_2$$

Hypochlorites can be obtained by neutralizing a solution of hypochlorous acid with base, and then evaporating. They are made industrially by electrolyzing a sodium chloride solution, without a diaphragm separating the anode and cathode compartments. It will be recalled (p. 405) that electrolysis of aqueous sodium chloride solution, with separation of anode and cathode compartments, is used in production of chlorine and sodium hydroxide. When electrolysis is carried out with mixing of the cell contents, the chlorine produced at the anode reacts with hydroxide ion in solution:

$$Cl_2 + 2OH^- \rightarrow Cl^- + OCl^- + H_2O \qquad (19\text{-}11)$$

Sodium hypochlorite solution produced in this way is employed in bleaching.

"Bleaching powder" is a hypochlorite of calcium, formed by the action of chlorine on calcium hydroxide:

$$Ca(OH)_2(s) + Cl_2(g) \rightarrow Ca(OCl)Cl(s) + H_2O \qquad (19\text{-}12)$$

Chlorous acid solutions can be prepared by adding sulfuric acid to a solution of barium chlorite:

$$\begin{aligned} &2H_3O^+ + SO_4^{-2} + Ba^{+2} + 2ClO_2^- \rightarrow \\ &\qquad 2HClO_2 + 2H_2O + BaSO_4(s) \end{aligned} \qquad (19\text{-}13)$$

Note that this reaction proceeds by virtue of the insolubility of barium sulfate. Chlorous acid solutions are not stable.

Chlorite salts are usually obtained from chlorine dioxide; *e.g.*, sodium chlorite is prepared by passing chlorine dioxide into a solution of sodium hydroxide and hydrogen peroxide:

$$2ClO_2 + 2OH^- + H_2O_2 \rightarrow 2ClO_2^- + 2H_2O + O_2 \qquad (19\text{-}14)$$

The solid chlorites are dangerous substances. Many of them are likely to explode if heated, and they all react violently with organic substances, phosphorus, and sulfur.

Chloric acid solution is obtained by a reaction analogous to that used in producing chlorous acid solutions: Barium chlorate is substituted for barium chlorite. Dilute solutions of the acid are fairly stable. Chlorate solutions can be obtained by bubbling chlorine into a hot solution of sodium or potassium hydroxide:

$$3Cl_2 + 6OH^- \rightarrow ClO_3^- + 5Cl^- + 3H_2O \qquad (19\text{-}15)$$

The most important method used in industrial preparations is the electrolysis of chloride solutions at a temperature of about 70°C. Potassium chloride solutions are usually employed, since the potassium salt is less soluble in water and can therefore be more easily crystallized from solution after electrolysis. Chlorate salts are reasonably stable, but are vigorous oxidizing agents. They should be kept from contact with organic substances, sulfur, and powdered metals.

The anhydride of *perchloric acid* is chlorine heptoxide, but this compound is not employed as a route to the acid. Solutions of the latter are usually prepared by treating a perchlorate salt with strong acid and distilling. Solutions of up to about 72 per cent by weight of $HClO_4$ are obtained in this way.

Perchloric acid is a very strong acid; in acidic solvents such as acetic and sulfuric acids it is clearly a stronger acid than hydrochloric or nitric acids. Although in hot, concentrated solutions it is a powerful oxidizing agent, in dilute solutions at room temperatures its oxidizing reactivity is almost nil.

Perchlorates may be obtained by prolonged electrolysis of solutions of chlorates. Since the chlorates themselves are produced from electrolysis of chloride solutions, the starting material is the chloride. Potassium and ammonium perchlorates are more readily obtained than others because of their lower solubilities.

Perchlorates are employed in the explosives industry. They are vigorous oxidizing agents, though they do not generally react as readily as do chlorates. Both ammonium and lithium perchlorates have received attention in recent years as oxidizing agents in solid fuels for missiles. Anhydrous magnesium and barium perchlorates are powerful desiccants, and are often used in desiccators and drying towers.

Bromine forms only two oxyanions in aqueous solution, the hypobromite, OBr^-, and the bromate, BrO_3^-. Hypobromous acid is known; it is a strong oxidizing agent, and a weak acid, with a dissociation constant of 2×10^{-9}. Bromic acid, $HBrO_3$, is unstable. Acidification of an alkaline bromate solution results in formation of oxygen and bromine:

$$4H_3O^+ + 4BrO_3^- \rightarrow 6H_2O + 2Br_2 + 5O_2 \qquad (19\text{-}16)$$

Bromous acid and bromites, and perbromic acid and perbromate salts, are unknown, despite numerous attempts to prepare them.

Hypoiodous acid is very unstable; it decomposes in aqueous solution within a few minutes of formation. Iodic acid, HIO_3, is a white crystalline solid, very soluble in water. Periodic acid does not have the formula HIO_4, as we might expect, but H_5IO_6. This rather interesting compound, called para-periodic acid, calls our attention to the question of size relationships.

The size of the central atom — a factor which we have hitherto neglected — may be important in determining the stabilities of some oxy-compounds. For example, even if electronegativity considerations did not forbid it, the formation of FO_4^- ion would be difficult because of the small size of the fluorine atom. The oxygen atoms would be in contact with one another, and a repulsive interaction would develop to render the ion unstable.

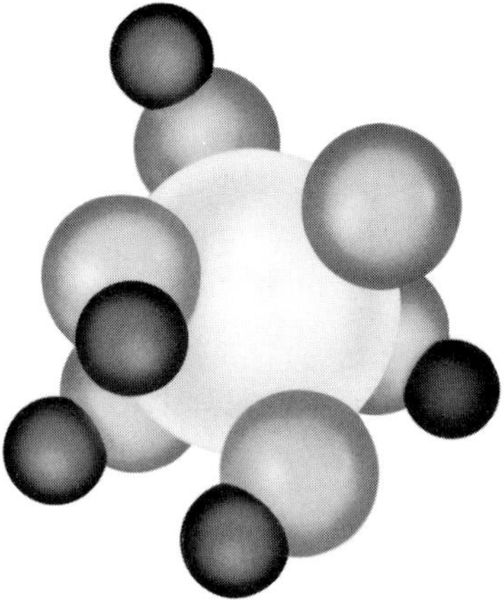

Fig. 19-2 Para-periodic acid, H_5IO_6.

The maximum number of oxygen atoms which can fit comfortably around a chlorine atom is four, as in ClO_4^-. Iodine, with a much larger atomic radius, can accommodate up to six oxygen atoms around it, and this number is present in para-periodic acid, Fig. 19-2. This compound can be thought of as formed from the compound HIO_4, meta-periodic acid, by addition of two moles of water. The six oxygen atoms are arranged octahedrally around the central iodine. The acid loses one proton quite easily ($K_{a1} = 10^{-2}$), and a second proton with less ease ($K_{a2} = 10^{-6}$). In considering the structure of periodic acid, H_5IO_6, it must be remembered that the iodine — to accommodate the six oxygen atoms — must have the orbitals available for them.

Halogen oxides — There are no oxyacids of fluorine, but fluorine does form compounds with oxygen. Oxygen difluoride (or fluorine monoxide), OF_2, is a pale yellow gas, with a boiling point of $-145°C$, which is capable of violent reaction when sparked in the presence of substances such as H_2, CH_4, or CO. It also reacts violently upon contact with any of the other halogens. OF_2 has the same structure as water; the O—F bond distance is 1.418 Å, and the F—O—F angle is 103°.

The great reactivity of OF_2 can be understood in terms of the general rule that the bonds between elements of similar electronegativity are not as stable as bonds between elements which differ significantly in electronegativity. We have already noted that the F—F bond is relatively weak. Since oxygen is close to fluorine in electronic configuration and atomic number, it is reasonable to suppose that the O—F single bond would not be very stable either.

OF_2 is readily hydrolyzed in basic solution:

$$OF_2 + 2OH^- \rightarrow O_2 + 2F^- + H_2O \qquad (19\text{-}17)$$

Other, less easily obtained oxygen compounds of fluorine are known. O_2F_2, analogous to hydrogen peroxide, is formed by passing oxygen and fluorine through an electric discharge in a quartz tube. It is a rather unstable red liquid, decomposing at −57°C.

Chlorine forms a number of oxides, none of which are very stable. In chlorine monoxide, Cl_2O, the chlorine may be assigned an oxidation number of +1. It can be prepared by passing chlorine gas over mercuric oxide:

$$2Cl_2(g) + HgO \rightarrow HgCl_2 + Cl_2O(g)$$

Cl_2O

The gas is explosively unstable, and must be handled with great care. It undergoes an interesting photochemical decomposition. It has been found that one photon causes the decomposition of two molecules of gas; this suggests that the reaction proceeds in a series of two steps such as the following

$$Cl_2O + h\nu \rightarrow Cl_2O^* \qquad (19\text{-}18)$$

$$Cl_2O^* + Cl_2O \rightarrow 2Cl_2 + O_2 \qquad (19\text{-}19)$$

(The asterisk denotes a photochemically excited molecule.) The number of molecules of reactant substance which are caused to react as a result of the absorption of one photon is termed the *quantum yield* of the reaction. In the above case, the quantum yield is two.

Chlorine dioxide, ClO_2, is also dangerously explosive; it has been used as a bleaching agent because of its powerful oxidizing properties. It is one of a relatively small number of substances known in which the molecules

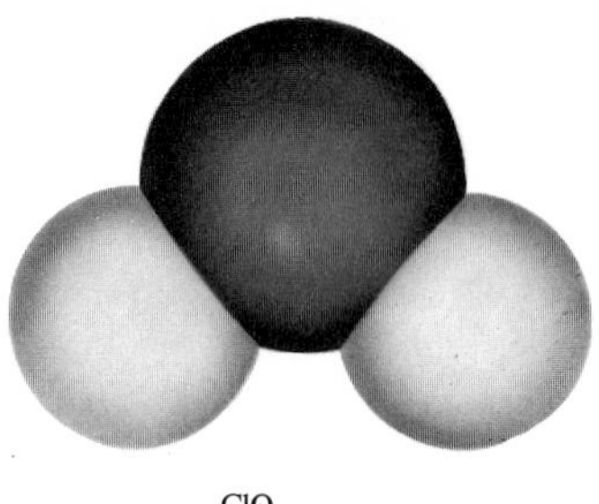

ClO_2

possess an odd number of electrons. The "odd," or unpaired, electron in ClO_2 is distributed throughout the molecule; this situation can be represented by writing resonant structures in which the electron is located on different atoms:

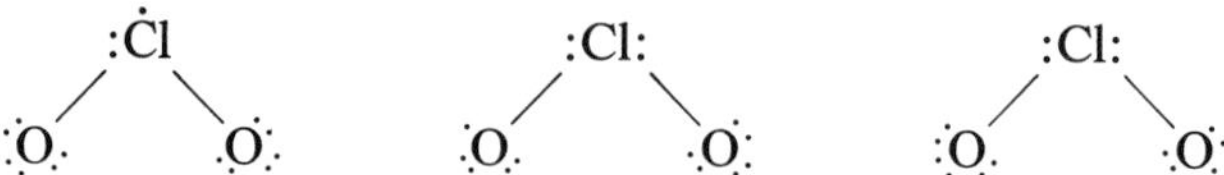

Molecules containing an unpaired electron are often colored; *e.g.*, chlorine dioxide is a yellow gas.

Other oxides of chlorine include ClO_3, which is also an "odd" molecule, and Cl_2O_7. Both are unstable and of little interest except for their relation to the oxyacids of chlorine to be discussed later.

Bromine and iodine also form a number of oxides, none of which are of special interest. They are in general more stable than the oxides of chlorine. For example, iodine pentoxide, I_2O_5, although a powerful oxidizing agent, is not explosively unstable; in fact, it can be made by heating iodine with oxygen gas under pressure at 250°C.

Interhalogen compounds — An interhalogen compound is a binary compound between two different halogens. Besides being of interest in themselves, these compounds provide examples of a diversity of valence types and geometries.

The simplest interhalogen compounds are diatomic molecules such as chlorine monofluoride, ClF, iodine monochloride, ICl, *etc.* Six such compounds are possible among the four halogen elements, F, Cl, Br, and I. Some of the six are not stable, and are known only as transient species in spectra. Those which are reasonably stable are not found to be unusual. The bond distance in each case is approximately the sum of the two covalent radii; the dipole moments increase with an increased electronegativity difference between the two atoms.

A second set of interhalogen compounds corresponds to the general formula XX'_3, where X represents chlorine, bromine or iodine, and X′ represents the second halogen, which is always lower in atomic number than is X. The only members of this group which have been studied to any extent are chlorine trifluoride, ClF_3, bromine trifluoride, BrF_3, and iodine trichloride, ICl_3. We will use chlorine trifluoride as an example in discussing the bonding in this group of compounds, since its structure is well known. The chlorine atom is bonded directly to three fluorine atoms by single covalent bonds. The total number of electrons in the valence shell of chlorine, therefore, is ten — seven from the chlorine itself and three from the fluorines. From previous acquaintance with expanded valence shells (p. 186), we expect that the hybrid orbitals employed about the chlorine atom are sp^3d. Into these five orbitals, we must place three electron pairs which form bonds to fluorine and two unshared electron pairs.

The geometry of the sp^3d hybrid orbitals is that of a trigonal bipyramid, as shown in Fig. 8-28. There is, however, more than one way in which the three bonding electron pairs and two unshared electron pairs might be placed. The structure which has been deduced from experimental studies is shown in Fig. 19-3. Because of the greater repulsion exerted by the unshared electron pairs (p. 175), the bond angles are somewhat distorted. In addition, the middle fluorine is at a different distance from the chlorine than the other two. BrF_3 and ICl_3 have similar structures.

The compounds bromine pentafluoride, BrF_5, and iodine pentafluoride, IF_5, are the known representatives of yet another group of interhalogen compounds. The central atom in these substances has twelve electrons in the valence shell; the expected set of hybrid orbitals is sp^3d^2, which has the symmetry of an octahedron (Fig. 8-27). In BrF_5, one of the six posi-

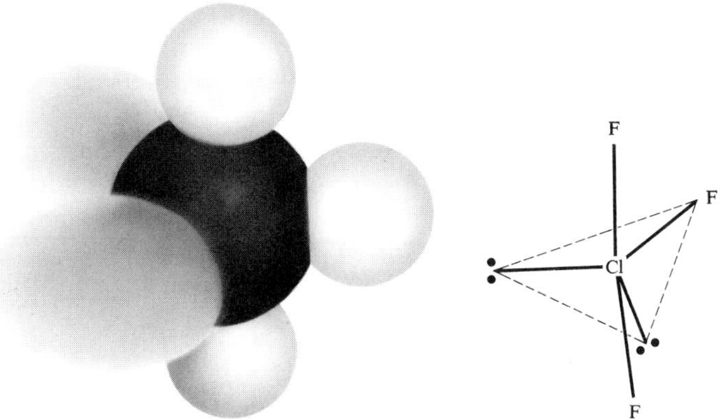

Fig. 19-3 The structure of chlorine trifluoride. The two large volume elements shown in the model represent the volume occupied by the unshared electron pairs.

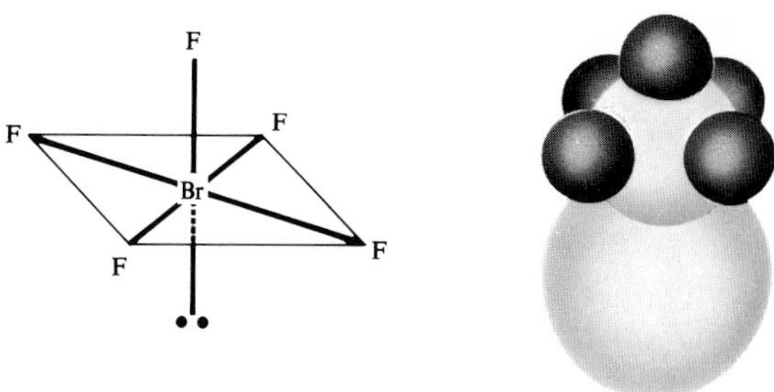

Fig. 19-4 The structure of bromine pentafluoride.

tions is occupied by an unshared electron pair, while the other five positions are the bonds to fluorine. Because of the greater repulsive effect of the unshared electron pair, the structure is distorted slightly with the central bromine atom just below the plane of the four fluorines, toward the unshared electron pair (Fig. 19-4). IF_5 is presumed to have a similar structure.

Iodine heptafluoride, IF_7, is the sole member of the final type of interhalogen. The structure of this compound is not known with certainty, but it is believed to have the form of a pentagonal bipyramid.

The fluorine-containing interhalogen compounds are powerful fluorinating agents. They react vigorously with organic matter, and must be handled with great care. The order of reactivity is $ClF_3 > BrF_5 > ClF > BrF_3 > IF_5 > BrF$. The compounds cannot be prepared or handled in glass apparatus, since they react with glass to form silicon tetrafluoride, SiF_4. The reaction of BrF_3 with silica is typical:

$$4BrF_3 + 3SiO_2 \rightarrow 3SiF_4 + 2Br_2 + 3O_2 \qquad (19\text{-}20)$$

The interhalogen compounds, in addition to their ability to act as fluorinating agents, are interesting in a number of other ways. Bromine trifluoride, for example, is of interest as a non-aqueous solvent. It is a colorless liquid (on the rare occasions when it is obtained in a pure state) with a boiling point of about 127°C at 1 atm of pressure. It is a rather polar substance with an unusually high conductivity. This is ascribed to the autoionization

$$2BrF_3 \rightleftharpoons BrF_2^+ + BrF_4^- \qquad (19\text{-}21)$$

Looked at from the Lewis acid-base theory, it is possible to view the BrF_2^+ ion as an acid species (since it can accept a pair of electrons) and BrF_4^- as the corresponding base species. Sodium fluoride would dissolve in this solvent to produce BrF_4^-

$$F^- + BrF_3 \rightarrow BrF_4^- \qquad (19\text{-}22)$$

and is, therefore, a characteristic base in this solvent. In fact, it is the analogue of NaOH in water. An acid (represented as A) in this solvent would have to be a substance which could react with the solvent to produce BrF_2^+:

$$A + BrF_3 \rightarrow AF^- + BrF_2^+ \tag{19-23}$$

Antimony pentafluoride is one of the few substances which does react in this manner:

$$SbF_5 + BrF_3 \rightarrow SbF_6^- + BrF_2^+ \tag{19-24}$$

19-2 Chemistry of the Group VI Elements

The Group VI elements are less electronegative than are the corresponding members of the halogen (Group VII) family. The most stable ionic form is that of the dinegative ions. This is particularly true of the oxide ion, as one might expect from the fact the oxygen is the most electronegative of the Group VI elements. The properties of metal oxides are treated in Chap. 20; the properties of the covalent non-metal oxides are discussed in connection with the chemistries of elements in this chapter. Metal sulfides are not so numerous nor so widespread as are oxides in nature, but they are, nevertheless, very common. Elements which tend to form sulfides are classified in geochemistry as *thiophiles.* All of the elements from copper through selenium on the periodic table, and all of those listed under these, are so classified. Selenium and tellurium occur with sulfur in ores which are predominantly sulfide.

The dinegative ions of Group VI as such do not exist to any degree in aqueous solution. Reaction of an ionic oxide such as CaO with water proceeds as

$$CaO(s) + H_2O \rightarrow Ca^{+2}(aq) + 2OH^-(aq) \tag{19-25}$$

A similar hydrolysis reaction occurs, though not so completely, with sulfide, selenide, and telluride salts.

In addition to the dinegative ions of Group VI elements, ions in which the elements have a -1 oxidation state are also found. The peroxide ion, O_2^{-2}, possesses an oxygen-oxygen bond, and has the structure.

$$[:\ddot{\underset{\cdot\cdot}{O}} - \ddot{\underset{\cdot\cdot}{O}}:]^{-2}$$

The corresponding hydride, hydrogen peroxide, is an interesting and important chemical. It can be prepared by the electrochemical oxidation of sulfate ion:

$$2SO_4^{-2} \rightarrow \left[\begin{array}{ccccccc} & O & & & O & \\ & | & & & | & \\ O- & S & -O-O- & & S & -O \\ & | & & & | & \\ & O & & & O & \end{array}\right]^{-2} + 2e^- \tag{19-26}$$

The peroxydisulfate ion, $S_2O_8^{-2}$, is subject to hydrolysis in acid solution:

$$H_2O + \left[\begin{array}{c} \quad O \qquad\quad O \\ \quad | \qquad\quad\; | \\ O\!-\!S\!-\!O\!-\!O\!-\!S\!-\!O \\ \quad | \qquad\quad\; | \\ \quad O \qquad\quad O \end{array}\right]^{-2} \rightarrow HSO_4^- + \left[\begin{array}{c} \qquad\quad O \\ \qquad\quad | \\ HO\!-\!O\!-\!S\!-\!O \\ \qquad\quad | \\ \qquad\quad O \end{array}\right]^{-} \qquad (19\text{-}27)$$

$$\left[\begin{array}{c} \qquad\quad O \\ \qquad\quad | \\ HO\!-\!O\!-\!S\!-\!O \\ \qquad\quad | \\ \qquad\quad O \end{array}\right]^{-} + H_2O \rightarrow H_2O_2 + HSO_4^-$$

Distillation from the aqueous solution leads to a 30 per cent by-weight solution of hydrogen peroxide. The pure material, which can be obtained by further distillation, resembles water in many ways. It is a strongly hydrogen-bonded liquid, as one might expect, and therefore has the rather high boiling point of 150°C. It melts at −0.4°C. Its dielectric constant at 20°C is 73, which puts it in the class of water insofar as polarity is concerned. Hydrogen peroxide differs from water in being a dangerously reactive substance. It is unstable toward decomposition, leading to water and oxygen. In aqueous solution it is a powerful oxidizing agent, as shown by the following oxidation-reduction half-reactions:

$$\begin{array}{lll} \text{Acid:} & 2H_3O^+ + H_2O_2 + 2e^- \rightarrow 4H_2O & \mathcal{E}^\circ = +1.77\text{ v} \\ \text{Base:} & H_2O_2 + 2e^- \rightarrow 2OH^- & \mathcal{E}^\circ = +0.87\text{ v} \end{array}$$

Despite the large driving force for reduction of H_2O_2 in acid solution with ordinary reducing agents, such reactions are not very rapid, and can usually be followed by ordinary kinetics procedures. The following reactions are typical:

$$3I^- + H_2O_2 + 2H_3O^+ \rightarrow I_3^- + 4H_2O \qquad (19\text{-}28)$$

$$Mn^{+2} + H_2O_2 + 2OH^- \rightarrow MnO_2 + 2H_2O \qquad (19\text{-}29)$$

$$SO_2 + H_2O_2 \rightarrow H_2SO_4 \qquad (19\text{-}30)$$

When hydrogen peroxide acts as a reducing agent, oxygen is a reaction product. The pertinent half-cell reactions and potentials are

$$\begin{array}{lll} \text{Acid:} & H_2O_2 + 2H_2O \rightarrow O_2 + 2H_3O^+ + 2e^- & \mathcal{E}^\circ = -0.67\text{ v} \\ \text{Base:} & H_2O_2 + 2OH^- \rightarrow O_2 + 2H_2O + 2e^- & \mathcal{E}^\circ = +0.08\text{ v} \end{array} \qquad (19\text{-}31)$$

The following reactions are typical of the behavior of hydrogen peroxide as a reducing agent:

$$\begin{array}{c} OCl^- + H_2O_2 \rightarrow H_2O + O_2 + Cl^- \\ Ag_2O + H_2O_2 \rightarrow Ag + H_2O + O_2 \\ PbO_2 + H_2O_2 + 2H_3O^+ \rightarrow Pb^{+2} + 4H_2O + O_2 \end{array} \qquad (19\text{-}32)$$

In certain instances, the peroxide ion can be produced easily in the solid state. For example, simply heating barium oxide in air causes formation of barium peroxide. Sodium peroxide is formed when sodium is burned in oxygen. A few of the other alkali metals also form peroxides with ease. When such salts are hydrolyzed, hydrogen peroxide may be recovered.

Oxyacids — Sulfurous acid, of which sulfur dioxide is the anhydride, is a rather weak acid. Sulfuric acid, on the other hand, is a strong acid, at least if one considers its first ionization constant. This comparison provides another demonstration of the rule which is applicable to the oxyacids of chlorine: *For acids containing the same central atom, the acid strength increases with increasing oxidation number of the central atom.* Although this statement is generally true, one must be careful in comparing apparently similar data in a quantitative sense. Thus, the apparent dissociation constant of the acid H_2SO_3 is $K_{a1} = 1.7 \times 10^{-2}$. But the extent of the reaction

$$SO_2 + H_2O \rightleftharpoons H_2SO_3 \tag{19-33}$$

for *dissolved* SO_2 is not really known, so that K_{a1} as given above is actually a minimum value; the actual value in terms of the correct concentration of H_2SO_3 may be much larger. It is certainly not as large as that for H_2SO_4, however, so our comparison is valid on a qualitative basis. In view of the foregoing remarks, it would be risky to pursue too far a comparison of the apparent acid dissociation constants of, for example, H_2SO_3 and H_2SeO_3. Suffice it to say that all of these acids appear to be of about the same acid strength. Sulfites are easily oxidized to sulfates; in fact, a solution of sodium sulfite which has been upon the laboratory shelf for a period of time is very likely to yield a positive test for sulfate — much to the dismay of the student learning qualitative analysis.

Sulfuric acid is, in terms of tonnage, the most important industrial chemical. Most of it is made by the so-called *contact process*, in which the oxidation of SO_2 to SO_3 is facilitated by a catalyst such as vanadium pentoxide. Sulfur trioxide is then absorbed into a concentrated sulfuric acid solution. Water is added as needed to form H_2SO_4:

$$2SO_2 + O_2 \rightarrow 2SO_3 \tag{19-34}$$

$$\begin{array}{c} SO_3 + H_2SO_4 \rightarrow H_2S_2O_7 \text{ (pyrosulfuric acid)} \\ H_2S_2O_7 + H_2O \rightarrow 2H_2SO_4 \end{array} \tag{19-35}$$

The reaction of SO_3 with sulfuric acid proceeds much more smoothly than does its reaction with water. Sulfuric acid, a strongly acidic medium, is self-ionized to an appreciable extent, according to the equation

$$\begin{array}{c} 2H_2SO_4 \rightleftharpoons H_3SO_4^+ + HSO_4^- \\ K_i = [H_3SO_4^+][HSO_4^-] = 2.4 \times 10^{-4}\ (25°C) \end{array} \tag{19-36}$$

Sulfur trioxide is a sufficiently strong Lewis acid (review Sec. 12-5) that, even in such a strongly acid medium as sulfuric acid, it reacts with the basic solvent species, HSO_4^-:

$$\begin{array}{ccc} \ddot{O}\ \ \ \ddot{O} & & \\ \diagdown\ /\!\!/ & & \\ S \leftarrow & \left[\begin{array}{ccc} & :\ddot{O}-H & \\ & | & \\ :\ddot{O}- & S & -\ddot{O}: \\ & | & \\ & :\ddot{O}: & \end{array}\right]^- & \rightarrow \left[\begin{array}{ccccc} :\ddot{O}: & & :\ddot{O}-H & \\ | & & | & \\ :\ddot{O}-S & -\ddot{O}- & S-\ddot{O}: \\ | & & | & \\ :\ddot{O}: & & :\ddot{O}: & \end{array}\right]^- \\ | & & \\ :\ddot{O}: & & \end{array} \tag{19-37}$$

The product is the anion (conjugate base) of pyrosulfuric acid, $H_2S_2O_7$. When one mole of SO_3 has been added for every mole of H_2SO_4, pyrosulfuric acid is the product. Dilution with water causes rupture of the central S—O—S grouping, to yield H_2SO_4.

The concentrated sulfuric acid of commerce is usually 98 per cent by weight, the remainder being water. This material boils at 340°C. It is a very strongly acid medium, and is also a powerful dehydrating agent. It is often used — particularly in organic reactions — to remove water as a reaction product from the system, and thus force the reaction further toward completion.

Sulfuric acid is a poor oxidizing agent at room temperature, but in hot, concentrated solution it does oxidize copper, as shown in Eq. (19-42).

The thiosulfate ion, $S_2O_3^{-2}$, is related to the sulfate ion; it is prepared by boiling an alkaline solution of sulfite with free sulfur:

$$SO_3^{-2} + S \rightarrow S_2O_3^{-2} \tag{19-38}$$

$S_2O_3^{-2}$

The thiosulfate ion may be thought of as a sulfate ion in which one of the four oxygen atoms has been replaced by sulfur. The thiosulfate ion is easily oxidized to the tetrathionate ion:

$$2S_2O_3^{-2} \rightarrow S_4O_6^{-2} + 2e^- \tag{19-39}$$

Thiosulfate solutions are frequently used in analyzing for iodine, since this element is rapidly and quantitatively reduced to iodide ion:

$$2S_2O_3^{-2} + I_2 \rightarrow S_4O_6^{-2} + 2I^- \tag{19-40}$$

When thiosulfate solutions are acidified, elemental sulfur and sulfur dioxide are produced. This particular reaction is a means of forming an allotropic modification of sulfur, called *rho* sulfur (Sec. 17-3), consisting of six-membered rings. The reaction is thought to proceed through a series of successive steps:

$$2S_2O_3^{-2} \rightarrow \underset{\text{intermediate}}{[S_4O_6]^{-4}} \xrightarrow{H^+} \left[\text{HS—S—}\overset{\overset{\text{O}}{|}}{\underset{\underset{\text{O}}{|}}{\text{S}}}\text{—O:} \right]^- + SO_3^{-2}$$

$$\begin{gathered} HS_3O_3^{-1} + S_2O_3^{-2} \rightarrow HS_4O_3^- + SO_3^{-2} \quad \textit{etc.} \\ H_2O + HS_7O_3^- \rightarrow S_6 + SO_3^{-2} + H_3O^+ \end{gathered} \tag{19-41}$$

The final reaction occurs when the chain of sulfur atoms has built up to such a length that a cyclic ring of sulfur atoms is possible. One way of looking at the final step is to consider that the end of the chain acts as a displacing group, and displaces the SO_3^{-2} group:

$$\begin{matrix} & SO_3^- & \\ & | & \\ & S & \\ S & & S{-}H \\ | & & | \\ S & & S \\ & S & \end{matrix} \rightarrow \begin{matrix} & S & \\ S & & S \\ | & & | \\ S & & S \\ & S & \end{matrix} + SO_3^{-2} + H^+$$

Oxides — The only oxides of importance among the Group VI elements are the di- and trioxides. In these compounds, the Group VI elements are in the +4 and +6 oxidation states, respectively.

Sulfur dioxide is the major product when sulfur is caused to react with an excess of oxygen. It may be prepared in the laboratory by reacting copper metal with hot concentrated sulfuric acid:

$$Cu + 2H_2SO_4 \rightarrow Cu^{+2} + SO_2 + SO_4^{-2} + 2H_2O \qquad (19\text{-}42)$$

or by adding acid to a sulfite salt:

$$SO_3^{-2}(aq) + 2H_3O^+(aq) \rightarrow 3H_2O + SO_2(g) \qquad (19\text{-}43)$$

Sulfur dioxide ordinarily behaves as a reducing agent, but in acidic solution it is itself reduced to free sulfur by ferrous and other ions.

Sulfur trioxide, SO_3, is of great importance because it is the anhydride of sulfuric acid. It is obtained by the oxidation of sulfur dioxide under the influence of a catalyst. The contact process for production of sulfuric acid involves oxidation of sulfur dioxide using a vanadium pentoxide catalyst (see p. 455).

In the gas phase, sulfur trioxide exists largely as single molecules which are planar and have the shape of an equilateral triangle. In the liquid state, it exists as an equilibrium mixture of monomer and trimer species. The trimer may be cyclic, and of the form shown in Fig. 19-5. Three

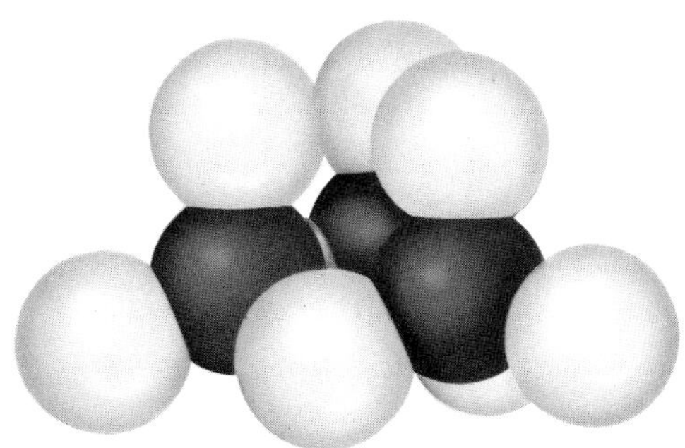

Fig. 19-5 Trimeric form of SO_3.

modifications of solid sulfur trioxide are known. These involve different groupings of the sulfur trioxide units to form polymers.

Only the dioxide of selenium is known to exist. Its properties are similar in many respects to those of sulfur dioxide, although it is more easily reduced to the free element. Tellurium forms both a dioxide and a trioxide.

Halides — The behavior of the other Group VI elements toward halogens is quite different from that of oxygen, described in Sec. 19-1. Although a compound such as S_2F_2, which would be analogous to one of the oxygen fluorides may possibly be prepared, it is certainly very unstable. The only reasonably stable fluorides of sulfur are SF_4, SF_6, and S_2F_{10}. Tellurium forms fluorides of the same stoichiometry, and selenium forms SeF_4 and

SF_4

SeF_6. In compounds of the XF_4 type, the Group VI element possesses ten electrons in the valence shell. This geometry can be visualized in terms of a sp^3d hybrid orbital set, as described in Sec. 8-9. One of the five hybrid orbitals, however, must contain an unshared electron pair. The structure is therefore intermediate between that for PF_5, a trigonal bipyramid, and that for ClF_3, in which two of the five hybrid orbitals contain unshared electron pairs. The unshared pair in the XF_4 compounds occupies one of the three positions in the plane of the bipyramid, as shown in Fig. 19-6.

The XF_6 species are octahedral, as already discussed in Sec. 8-9. It is rather interesting to compare the reactivity of SF_4 with that of SF_6. Ther-

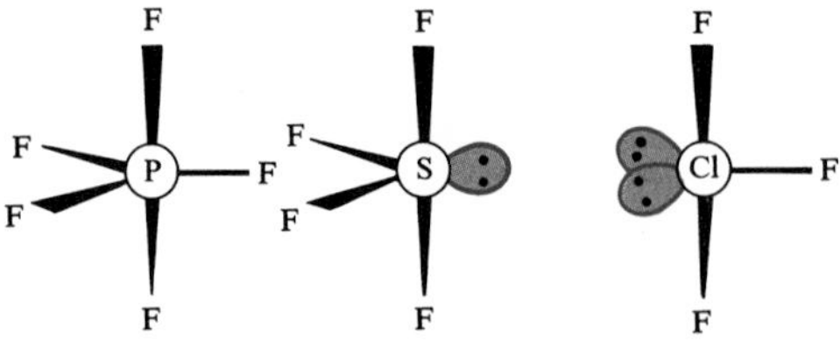

Fig. 19-6 Comparative structures of some molecules with expanded valence shell atoms.

mochemical data for heats of formation indicate that the average S—F bond energy in SF_4 is greater than that in SF_6. In spite of this however, SF_4 is a very reactive compound, whereas SF_6 is extraordinarily unreactive. SF_4 is instantly hydrolyzed by water to give HF and SO_2, whereas SF_6 is, for all practical purposes, totally unreactive. Since the difference in reactivities cannot be charged to the thermodynamic properties of reactants *vs* products, it must in some way be ascribed to the energetics involved in the reaction; *i.e.*, to the activation energies for reaction. The major factor seems to be the number of groups about the central atom. The six fluorines about the sulfur in SF_6 are closely packed, and there is not room for an attacking reagent to reach the central atom. This is not the case in SF_4, which has only five positions occupied (if we count the unshared pair as a position). In SeF_6 and TeF_6, this is not so important since the central atom is larger. Both these compounds are more reactive than is SF_6. TeF_6 is hydrolyzed by water over a period of about a day at room temperature.

Disulfur decafluoride is a by-product of the direct reaction of sulfur with fluorine, in which SF_6 is the major product. In this compound the coordination about sulfur is as in SF_6, but with one of the sulfur-fluorine bonds replaced by a sulfur-sulfur bond. It is nearly as unreactive as SF_6, probably because of the same steric factors.

Disulfur dichloride, S_2Cl_2, is an orange-colored liquid with a revolting smell. It has a structure analogous to that of hydrogen peroxide. It decomposes slowly in water to give a variety of sulfur-containing products. It can be chlorinated to yield SCl_2, which is unstable. At room temperature, SCl_2 is in equilibrium with chlorine and S_2Cl_2. Sulfur also forms a bromide S_2Br_2, but this is the extent of the well-characterized halogen chemistry of sulfur. By contrast, selenium and tellurium form XCl_4 species in addition to X_2Cl_2 and XCl_2 forms. Selenium forms a rather unstable $SeBr_4$, whereas the analogous compound of tellurium is quite stable, even above its melting point of 380°C. These comparisons illustrate the effect of the size of the central atom on stability. Sulfur is not large enough to accommodate four chlorines about it, whereas tellurium can accommodate four bromine atoms. The structures of the XCl_4 and XBr_4 species are presumably the same as that for SF_4, Fig. 19-6.

19-3 Chemistry of the Group V Elements

It might be expected that the accumulation of three negative charges on one atom, as is required to form the trinegative ions of Group V, would not occur readily. The nitride ion, N^{-3}, does occur in binary compounds with the more active elements, but does not exist as such in aqueous solution. Thus, lithium nitride is formed when nitrogen gas is passed over molten lithium:

$$6Li + N_2 \rightarrow 2Li_3N \qquad (19\text{-}44)$$

When this compound is placed in contact with water, the nitride ion (a strong Brönsted base) reacts to form ammonia:

$$Li_3N + 3H_2O \rightarrow NH_3 + 3Li^+ + 3OH^- \qquad (19\text{-}45)$$

Phosphorus unites with a number of the more active metals to form phosphides which are analogous to the nitrides, and are probably ionic in character. Hydrolysis of one such compound, aluminum phosphide, is a good method of preparing phosphine:

$$AlP + 6H_2O \rightarrow Al(H_2O)_3(OH)_3 + PH_3 \qquad (19\text{-}46)$$

The azide ion, N_3^- is an interesting species containing three nitrogen atoms in a linear structure:

$$[\ddot{\underset{..}{N}} = N = \ddot{\underset{..}{N}}]^-$$

Ionic azides are reasonably stable substances, but the compounds with heavy metals, and covalent azides, are explosively unstable. Azides can be prepared by the reaction of a nitrate salt with potassium amide in anhydrous ammonia:

$$KNO_3 + 3KNH_2 \xrightarrow[NH_3]{\text{liquid}} KN_3 + 3KOH + NH_3 \qquad (19\text{-}47)$$

Hydrazoic acid, HN_3, is a weak acid with an ionization constant in water of about 10^{-5}. It is extremely toxic and violently explosive, and must be handled with great care.

Oxyacids — There are three oxyacids of nitrogen which deserve mention: hyponitrous, nitrous, and nitric acids. Their structures and formulas are shown in Fig. 19-7.

Nitrous oxide is formally the anhydride of hyponitrous acid, since nitrogen exists in the +1 valence state in both compounds. It does not yield the acid, however, on contact with water. The usual method of preparing

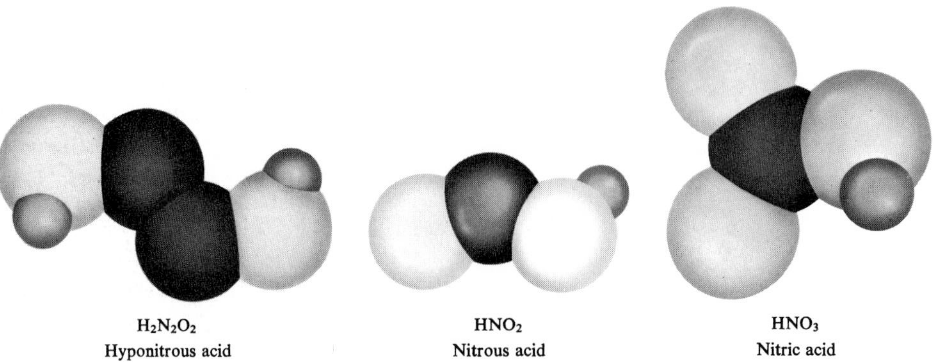

Fig. 19-7 The oxyacids of nitrogen.

hyponitrite ion, $N_2O_2^{-2}$, is the reduction of a solution of sodium nitrite by sodium amalgam. (An *amalgam* is a solution of another metal in mercury; this is used to moderate the activity of the active metal.)

The acid (in solution) is very easily oxidized to nitrate. The hyponitrite ion, which exists in alkaline solution, is also easily oxidized.

Nitrous acid is known only in solution. It is formed when equal quantities of nitric oxide and nitrogen dioxide are dissolved in water. It is a rather weak acid; K_a is about 10^{-4}.

Sodium nitrite, $NaNO_2$, is by far the most important compound in the nitrite class. It is used in large quantities in the dyestuffs industry, and is produced by passing an equimolar mixture of nitric oxide and nitrogen dioxide into sodium hydroxide solution.

Nitric acid is, of course, the most important oxyacid of nitrogen. It is produced in large quantities by a method described later, beginning with the catalytic oxidation of ammonia. Nitric acid is usually marketed as the concentrated acid, a 68 per cent-by-weight solution, of density 1.42. The pure, anhydrous acid boils at about 78°C at atmospheric pressure. It is not a stable compound at these temperatures, however.

Nitric acid is, in addition to being a strong acid, a powerful oxidizing agent. The amount and nature of the reduction products depends upon the reducing agent, the concentration of acid, and the temperature. The situation is complicated by the fact that the principal product (*e.g.*, nitrous acid) may decompose or disproportionate to yield still other products. To keep the balanced equations reasonably simple and still represent fairly well the major aspects of the reaction, the products are usually written as nitrogen dioxide for reactions involving concentrated acid, and as nitric oxide for those involving dilute acid:

$$\text{(concentrated)}\quad Cu + 2NO_3^- + 4H_3O^+ \rightarrow Cu^{+2} + 2NO_2 + 6H_2O \qquad (19\text{-}48)$$

$$\text{(dilute)}\quad 3Cu + 2NO_3^- + 8H_3O^+ \rightarrow 3Cu^{+2} + 2NO + 12H_2O \qquad (19\text{-}49)$$

With very strong reducing agents, the products may contain nitrogen in lower oxidation states; *e.g.*, N_2 or NH_3.

A number of nitrate salts are important industrial chemicals. Sodium nitrate, which occurs in large deposits in Chile, has for many years been the major source of naturally occurring nitrate. Ammonium nitrate is used as an explosive, and as a source of agricultural nitrogen. Since it contains nitrogen in the two extreme oxidation states of this element, it is not surprising that it should be easily decomposed into nitrous oxide and water. The reaction, which proceeds readily on heating, may become explosive.

The structural formulas and names of the three acids of phosphorus which we will discuss are shown in Fig. 19-8.

Hypophosphorous acid is obtained by reacting white phosphorus with alkali solution:

$$P_4 + 3NaOH + 3H_2O \rightarrow PH_3 + 3NaH_2PO_2$$

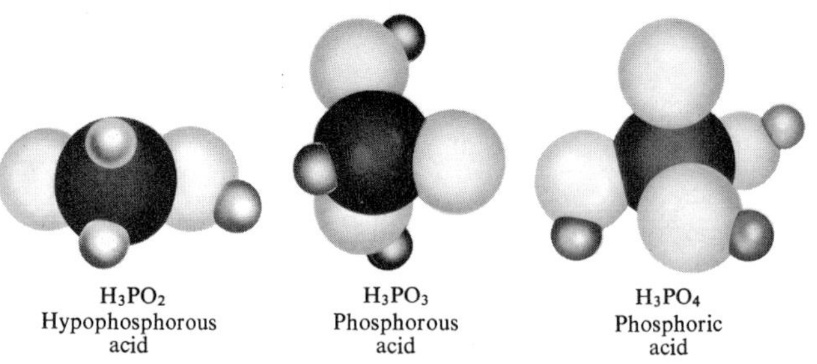

Fig. 19-8 The oxyacids of phosphorus.

If barium hydroxide is used, the excess alkali can be removed by bubbling in carbon dioxide:

$$\begin{aligned} CO_2 + 2OH^- &\rightarrow CO_3^{-2} + H_2O \\ Ba^{+2} + CO_3^{-2} &\rightarrow BaCO_3(s) \end{aligned} \qquad (19\text{-}50)$$

After filtering, the solution is evaporated until barium hypophosphite, $Ba(H_2PO_2)_2$, crystallizes out. This is redissolved in water, and an equivalent quantity of sulfuric acid is added to precipitate barium sulfate.

Only one of the three hydrogens in hypophosphorous acid is attached to oxygen, and is therefore replaceable. The acid is a powerful reducing agent; it will reduce sulfur dioxide to sulfur, and, with catalysis, reduces water:

$$H_3PO_2 + H_2O \rightarrow H_3PO_3 + H_2 \qquad (19\text{-}51)$$

Phosphorous acid is produced from the reaction of phosphorus trioxide or trichloride with water. It is a solid, melting at about 74°C, easily soluble in water. From the structural formula shown in Fig. 19-8, it can be seen that two of the hydrogens are attached to oxygens.

Orthophosphoric acid, representing phosphorus in the +5 oxidation state, is the product of complete hydrolysis of phosphorus pentoxide. It is distributed as an aqueous solution which is 85 per cent H_3PO_4 by weight. The three hydrogens are all ionizable, but phosphoric acid is a weak acid, even in the ionization of the first hydrogen (Table 15-4). It is not a strong oxidizing agent, even though the phosphorus is in its maximum oxidation state.

With the exception of the alkali metal and ammonium salts, orthophosphates are generally not soluble in water, but because the acid is a weak acid, they do dissolve in acidic solutions. Calcium phosphate occurs widely as a complex salt in minerals containing phosphate. It is the principal source of phosphorus and is the starting material for production of the element and other phosphorus compounds (Chap. 17). When treated with sulfuric acid, the calcium phosphate is converted to the dihydrogen salt, called "superphosphate":

$$Ca_3(PO_4)_2 + 2H_2SO_4 \rightarrow Ca(H_2PO_4)_2 + 2CaSO_4 \qquad (19\text{-}52)$$

In this form, it is easily soluble in water, and is therefore useful as a source of phosphorus in fertilizers.

Arsenic forms two oxyacids in water, arsenious acid, H_3AsO_3, and arsenic acid, H_3AsO_4. The last of these is analogous to phosphoric acid and behaves very much like it in solution. It can be crystallized from aqueous solution as a hydrate, $H_3AsO_4 \cdot H_2O$, which loses water and melts at about 100°C. Arsenious acid is not structurally identical with phosphorous acid, although it does have the same molecular formula. It appears that arsenious acid is a true trihydroxy compound, $As(OH)_3$, whereas phosphorous acid possesses a direct phosphorous-hydrogen link, as described above.

Oxides — Nitrogen forms a number of interesting and important oxides. Nitrous oxide, N_2O, in which nitrogen is in the $+1$ oxidation state, is a colorless gas with a slightly sweet taste. The name "laughing gas" for this substance comes from the fact that, when breathed in small amounts, it causes excitement. In larger dosages, however, it is a general anesthetic. It was the first substance to be so used, and is still employed to some extent for this purpose.

Nitrous oxide is prepared in the laboratory by the heating of ammonium nitrate:

$$NH_4NO_3 \rightarrow N_2O + 2H_2O \qquad (19\text{-}53)$$

Some care must be taken with this method of preparation, however, because if the temperature of the system rises too high an explosion may result.

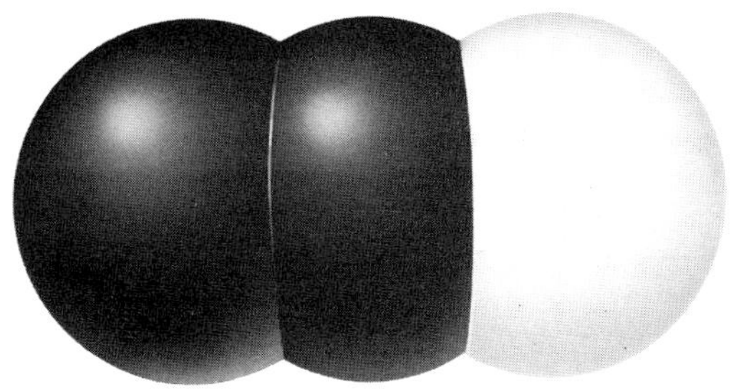

N_2O

Nitric oxide, NO, can be produced by the direct combination of nitrogen and oxygen:

$$N_2 + O_2 \rightarrow 2NO \qquad (19\text{-}54)$$

The process is endothermic, so the yield of product is greater at high temperatures. At 3000°C, the amount of nitric oxide in an equilibrium mixture of nitrogen and oxygen at one atm pressure is only 4 per cent. The compound is usually prepared in the laboratory by reduction of nitric acid with a reducing agent such as copper metal:

$$3Cu + 8H_3O^+ + 8NO_3^- \rightarrow 3Cu^{+2} + 6NO_3^- + 2NO + 12H_2O \quad (19\text{-}55)$$

The reaction does not yield pure nitric oxide; some nitrous oxide and oxygen is always present. Other reducing agents, such as ferrous ion, Fe^{+2}, may also be used in place of copper metal.

Nitric oxide is another "odd" molecule. The structure may be represented by the canonical forms

$$:\dot{N} = \ddot{\underset{\cdot\cdot}{O}} \qquad :\ddot{N} = \dot{\underset{\cdot\cdot}{O}}$$

Despite its being an odd molecule, nitric oxide is a colorless gas. As the major product produced in the catalyzed reaction of ammonia with oxygen, it is an intermediate in the production of nitric acid:

$$4NH_3 + 5O_2 \xrightarrow{\text{catalyst}} 4NO + 6H_2O \quad (19\text{-}56)$$

Nitric oxide reacts with oxygen in moist air to produce nitrogen dioxide, NO_2. The latter disproportionates on dissolving in water to produce nitrous and nitric acid:

$$2NO_2 + 2H_2O \rightarrow HNO_2 + NO_3^- + H_3O^+ \quad (19\text{-}57)$$

Nitrous acid, which is a weak acid in water, in turn disproportionates to nitrogen dioxide and nitric oxide:

$$2HNO_2 \rightarrow H_2O + NO + NO_2 \quad (19\text{-}58)$$

Since the reactions of Eqs. (19-57) and (19-58) occur concurrently, their sum describes the over-all reaction:

$$3NO_2 + 3H_2O \rightarrow 2H_3O^+ + 2NO_3^- + NO \quad (19\text{-}59)$$

Nitric acid solutions containing up to 55–60 per cent of HNO_3 by weight are produced in this process.

Nitrogen dioxide can be prepared by the action of concentrated nitric acid on copper, as in Eq. (19-48), or by heating lead nitrate:

$$2Pb(NO_3)_2 \xrightarrow{\text{heat}} 2PbO + 4NO_2 + O_2 \quad (19\text{-}60)$$

It is a red-brown gas (it is an odd molecule) which dimerizes at lower temperatures to form dinitrogen tetroxide, N_2O_4. This substance has a boiling point of 26°C, although at this temperature it is already appreciably dissociated into the monomeric form, as evidenced by the color of the liquid:

$$\underset{\text{colorless}}{2N_2O_4} \rightleftarrows \underset{\text{brown}}{2NO_2} \quad (19\text{-}61)$$

NO_2

Under a total pressure of one atm, the monomer NO_2 is present in the following pressures at various temperatures in the gas phase:

Temperature (°C)	27	70	100	135
P_{NO_2}	0.347	0.792	0.938	0.996

The structure of dinitrogen tetroxide is still not known with certainty, but it very probably contains a nitrogen-nitrogen bond.

Nitrogen pentoxide, N_2O_5, can be prepared by the dehydration of nitric acid. It is a crystalline solid which decomposes slowly at room temperature. This evidence of instability relative to the other oxides of nitrogen is to be expected, since the nitrogen in this compound is in a high (+5) oxidation state.

Phosphorus forms two easily obtained oxides, phosphorous trioxide, P_4O_6 and phosphorous pentoxide P_4O_{10}. The first of these is formed when phosphorus (usually white phosphorus) is oxidized in a restricted supply of air. In a plentiful supply of air, the pentoxide is formed. P_4O_{10} is called the pentoxide because the empirical formula is P_2O_5. The structures of these two oxides of phosphorus are interesting because of their relationship to the P_4 molecule (p. 415). P_4O_{10} is illustrated in Fig. 19-9.

Phosphorus trioxide, P_4O_6, is a reasonably stable compound, though it is easily oxidized to the pentoxide. The formula P_4O_{10} represents the most commonly obtained form of the compound with the empirical formula P_2O_5, but it is by no means the only one. On heating to about 425°C, it melts, then resolidifies at higher temperature as a glassy solid. This form is undoubtedly an extensively polymerized form of P_2O_5.

Arsenic and antimony also form oxides, which are analogous to the oxides of phosphorus. When arsenic or antimony is burned in air, however, the lower oxide, As_2O_3 or Sb_2O_3 is formed, whereas with phosphorus the higher oxide is formed.

Halides — Nitrogen forms trihalides with fluorine, chlorine, and iodine. In addition, a number of other nitrogen-fluorine compounds are known.

Nitrogen trifluoride is a colorless, inert gas, with a boiling point of −129°C, which can be prepared by electrolyzing an ammonium fluoride-hydrogen fluoride melt. The reaction probably proceeds in two steps:

$$\begin{aligned} F^- &\rightarrow \tfrac{1}{2}F_2 + e^- && \text{(oxidation of fluoride ion)} \\ 3F_2 + 2NH_4^+ &\rightarrow 2NF_3 + 3H_2 + 2H^+ && \text{(reduction of hydrogen)} \end{aligned} \qquad (19\text{-}62)$$

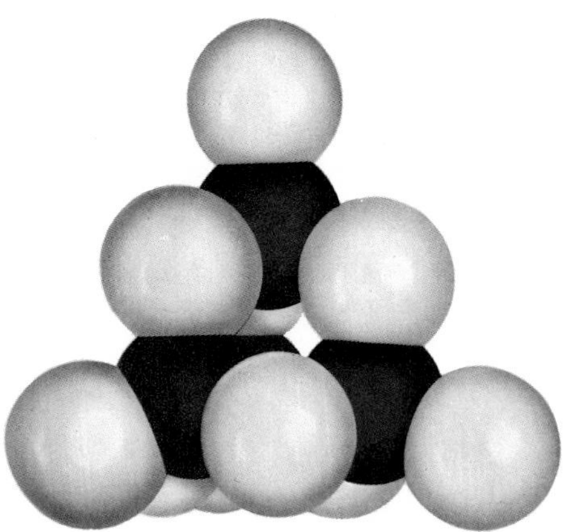

Fig. 19-9 Structure of P_4O_{10}. The structure consists of a tetrahedron of phosphorus atoms, with an oxygen atom bridging along each edge, and with a terminal P—O bond on each phosphorus. In phosphorus (III) oxide, P_4O_6, the terminal oxygens are absent.

Dinitrogen tetrafluoride, N_2F_4 is formed by action of fluorine gas on ammonia in a copper tube. This substance is a reactive gas, boiling point −73°C. It is the fluorinated analogue of hydrazine. On reaction with aluminum chloride, it is converted to the intensely reactive dinitrogen difluoride:

$$3N_2F_4 + 2AlCl_3 \rightarrow 3N_2F_2 + 3Cl_2 + 2AlF_3 \qquad (19\text{-}63)$$

This compound, also a gas, has the property of existing in two isomeric forms:

```
  N=N                 F
 /   \               /
F     F         N=N
               /
              F
   cis-          trans-
```

The interconversion of these two forms (the *trans-* is more stable) is restricted by the double bond between the nitrogen atoms. This is an example of *geometrical isomerism,* in which two compounds possess the same formula and the same bonds, but differ in the spatial arrangements of the atoms. Many examples of this form of isomerism are met with in organic chemistry.

Nitrogen trichloride is a yellow explosive oil, formed by the action of chlorine on anhydrous ammonia. It is an extremely dangerous substance, and its preparation should be avoided. It can be decomposed by hydrolysis. Related to the trichloride are monochloramine, NH_2Cl and dichloramine, $NHCl_2$.

Nitrogen tribromide, which has not been isolated as a pure substance, is also dangerously explosive. Nitrogen tri-iodide is obtained in the form of a black solid ammonia complex, NI_3NH_3, by the action of aqueous ammonia on iodine. This preparation, which is sometimes attempted by the unwary beginner in the chemistry laboratory, is extremely dangerous. The solid material is violently explosive.

This brief survey clearly indicates that among the trihalides only the trifluoride is reasonably stable. The driving force for decomposition in all cases is strong, since nitrogen gas, one of the products, has such a high stability. The standard enthalpy of formation of NF_3 is -26 kcal/mole, which means that it is stable relative to the free elements. (Ideally we should compare the standard free energy of formation, but this is not known with certainty. When ΔH_f is negative by this much, we can be sure that ΔG_f will be negative also.) By contrast, ΔH_f for NCl_3 is $+55$ kcal/mole, which means that there is a large thermodynamic driving force toward decomposition into the free elements.

Phosphorus forms stable halides with all of the halogens. With fluorine, both PF_3 and PF_5 are formed. PF_3 can be prepared by warming phosphorus tribromide with solid zinc fluoride

$$2PBr_3 + 3ZnF_2 \rightarrow 2PF_3 + 3ZnBr_2 \quad (19\text{-}64)$$

or by heating copper phosphide with lead fluoride. It is not an especially reactive gas; it does not fume in air, nor does it react with glass at room temperature as do many reactive fluorides, *e.g.*, Eq. (19-20). It is slowly hydrolyzed by water:

$$PF_3 + 3H_2O \rightarrow H_3PO_3 + 3HF \quad (19\text{-}65)$$

When phosphorus is burned in fluorine, PF_5 is formed. It is also formed by heating together a mixture of phosphorus pentoxide and calcium fluoride:

$$6P_2O_5 + 5CaF_2 \rightarrow 2PF_5 + 5Ca(PO_3)_2 \quad (19\text{-}66)$$

The pentafluoride is easily hydrolyzed, and fumes in air:

$$PF_5 + H_2O \rightarrow POF_3 + 2HF \quad (19\text{-}67)$$

Phosphorus forms both a trichloride and a pentachloride. The former is prepared by the action of dry chloride on either white or red phosphorus. When phosphorus is burned in an excess of chlorine, PCl_5 is produced. It is a white solid which sublimes easily, and which is dissociated in the vapor state into PCl_3 and Cl_2. Because of this relatively easy loss of chlorine, PCl_5 is a convenient chlorinating agent. It readily converts many metals into chlorides. Both PCl_3 and PCl_5 are hydrolyzed with great ease.

In the presence of a limited amount of water, PCl_5 may be hydrolyzed to the intermediate substance, phosphoryl chloride, $POCl_3$.

$$PCl_5 + H_2O \rightarrow POCl_3 + 2HCl \qquad (19\text{-}68)$$

Further hydrolysis results in the formation of orthophosphoric acid:

$$POCl_3 + 3H_2O \rightarrow H_3PO_4 + 3HCl \qquad (19\text{-}69)$$

Hydrolysis of PCl_3 results in formation of H_3PO_3.

PF_5 and PCl_5 in the vapor state have the trigonal bipyramidal structure, Fig. 19-6. In the solid state, and in certain polar organic solvents, PCl_5 exists as the ions PCl_4^+ PCl_6^-. Solid PCl_5 thus consists of an ionic lattice very much like the CsCl lattice. This is an interesting example of how the appearance of covalent or ionic character changes with the state of the substance. In general, ionic character is more highly developed in the solid state than in the gas phase or in solution in a non-polar solvent, because of the stabilizing influence of the lattice energy.

Phosphorus forms both a tri- and pentabromide which are much like the chlorides in properties. With iodine, it forms a tri-iodide and a substance of composition P_2I_4. This compound contains a phosphorous-phosphorous bond.

The trihalides of phosphorus have a pyramidal structure:

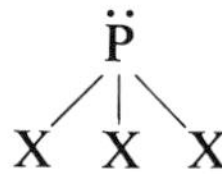

The oxyhalides, POX_3, have the same structure, except that an oxygen is bound to the phosphorus through a coordinate covalent bond.

The halides of arsenic and antimony are similar to the analogous phosphorus compounds.

19-4 Chemistry of the Group IV Elements

Carbon forms a number of binary compounds with metals, in which it appears that the carbon exists as a negatively-charged species. The carbides are usually formed when the oxide of the metal is reduced in an electric furnace in the presence of carbon. One group of compounds reacts with water to produce methane, acetylene, or other simple hydrocarbons. Examples include Li_2C_2, CaC_2, and Al_4C_3, which react with water as follows:

$$\begin{aligned} Li_2C_2 + H_2O &\rightarrow Li_2O + C_2H_2 \\ CaC_2 + H_2O &\rightarrow CaO + C_2H_2 \\ Al_4C_3 + 6H_2O &\rightarrow 2Al_2O_3 + 3CH_4 \end{aligned} \qquad (19\text{-}70)$$

The carbides of this type are referred to as salt-like, or ionic. The compounds formed with many other elements are, by contrast, notably unreactive, high-melting refractory solids. For example, TiC, TaC, Mo_2C, and many others, are in this refractory class. A third group, referred to as the interstitial carbides, is intermediate in character. Included in this group are, for example, Fe_3C and Cr_3C_2. The members of this group can be oxidized, and may react with acid solution, but are not as reactive as the salt-like carbides.

Oxyacids and oxides — Carbon forms three well-characterized gaseous oxides: carbon dioxide, CO_2, carbon monoxide, CO, and carbon suboxide, C_3O_2.

Carbon dioxide is a fairly abundant constituent of the atmosphere, about 0.03 per cent. It plays a vital role in photosynthesis in plants, and is important also in regulation of the temperature on the earth's surface (see Sec. 19-6).

Carbon dioxide is a stable substance. It does not support combustion except of the most active metals. Burning sodium, potassium, or magnesium continue to burn if plunged into CO_2; *e.g.*,

$$4K + 3CO_2 = 2K_2CO_3 + C \tag{19-71}$$

At high temperatures, CO_2 dissociates into CO and O_2. At a total pressure of one atm, the percentage dissociations as a function of temperature are:

Temperature (°K)	1000	2000	2500
Per cent dissociation	2.5×10^{-5}	2.05	17.6

Carbon dioxide dissolves in water to produce a weakly acidic solution. The acidity is presumed to arise from the dissociation of carbonic acid, H_2CO_3. The extent to which the dissolved carbon dioxide reacts with water to form carbonic acid is not known. The equilibrium $CO_2 + H_2O \rightleftharpoons H_2CO_3$ probably does not consume more than about 0.1 per cent of the dissolved CO_2. The *apparent* acid dissociation constant

$$H_2CO_3 + H_2O \rightarrow H_3O^+ + HCO_3^-$$

$$K_a = \frac{[H_3O^+][HCO_3^-]}{[H_2CO_3]} = 4 \times 10^{-7}$$

is based on the presumption that all the dissolved CO_2 is present as H_2CO_3. If the true acid concentration is only perhaps 1/1000 of this, the correct acid dissociation constant of carbonic acid is about 10^{-4}, which is more in keeping with the dissociation constants of carboxylic acids (Sec. 15-2). Although carbonic acid cannot be isolated as a pure compound, salts of the acid, carbonates (containing CO_3^{-2}) and bicarbonates (containing HCO_3^-) are numerous.

Carbon monoxide is formed from the combustion of coke or other fuels in a limited supply of air. It may also be formed by reduction of carbon dioxide by hot zinc or iron; *e.g.*,

$$CO_2 + Zn \rightarrow ZnO + CO \tag{19-72}$$

Carbon dioxide is also reduced to carbon monoxide fairly rapidly by carbon at temperatures above 1000°C. The physical properties of carbon monoxide are very similar to those of nitrogen. The resemblence results from the nearly identical molecular weights, and the low polarity of carbon monoxide, which has a dipole moment of only 0.12 Debye. In its chemical properties carbon monoxide differs dramatically from nitrogen. It is very poisonous for all forms of animal life which possess hemoglobin in the blood. The carbon monoxide molecule becomes attached to hemoglobin at the point which is normally employed in holding oxygen. The carbon monoxide compound formed is quite stable, with the result that the blood system loses its capacity for transfer of oxygen to the cells.

Carbon monoxide burns in air with a blue flame. Coke may be converted to a combustible gas by passing air through a red-hot bed of coke. At high temperatures, carbon monoxide is formed in high concentration. The action of steam on very hot coke also produces carbon monoxide and hydrogen. The resulting gas, called "water gas" is a cheap source of gaseous fuel. Carbon monoxide is oxidized by mercuric oxide at 200°C.

Carbon suboxide boils at 6°C and melts at −111°C. It is formed by dehydrating malonic acid, an organic acid, by heating under vacuum with phosphorus pentoxide at 300°C. The molecule, which is linear, may be described by the following Lewis structure:

$$\ddot{\underset{\cdot\cdot}{O}} = C = C = C = \ddot{\underset{\cdot\cdot}{O}}$$

A poisonous, pungent-smelling gas, it reacts explosively with oxygen to yield carbon dioxide.

Silica, SiO_2, is the only oxide of silicon of any importance. It exists in three polymorphic forms — quartz, tridymite, and cristobalite. (The term *polymorphic* refers to forms of a single substance which differ in the arrangement of the units in the solid structure.) Quartz, which is found in nature as rock-crystal and sand, is the common form. Silica is a very stable oxide, as we might expect. It melts at 1710°C, and becomes a transparent glass on cooling. Its transparency to radiation over a wide range of the spectrum is the basis for its use in construction of optical equipment such as prisms, sample containers, *etc.*

Halides — Carbon forms tetrahalides with all of the halogens. The stability toward decomposition decreases with the size of the halogen. Carbon tetraiodide is least stable, and cannot be distilled. The tetrafluoride is an extremely stable compound. Carbon tetrachloride is a commonly used solvent. It is photochemically unstable. In the presence of oxygen, and at high temperatures it may form phosgene, $COCl_2$, a poisonous gas. Carbon tetrachloride itself is a toxic substance, and should be handled with caution. It is rapidly absorbed into the body through the skin.

The other Group IV elements also form tetrahalides, although lead does so only with fluorine and chlorine. Silicon tetrafluoride is formed from the reaction of hydrogen fluoride with silica, or with silicates:

$$4HF + SiO_2 \rightarrow SiF_4 + 4HF \qquad (19\text{-}73)$$

When silicon is treated with chlorine, not only $SiCl_4$, but also higher chlorides, *e.g.*, Si_2Cl_6, *etc.*, are formed.

The very different behaviors of carbon tetrachloride and silicon tetrachloride toward water afford an interesting comparison. CCl_4 is practically inert toward water at room temperature, whereas $SiCl_4$ reacts violently, and fumes vigorously even in ordinary air. This great contrast is *not* due to thermodynamic factors, as the following data show:

$$\begin{array}{ll} CCl_4 + 6H_2O \rightarrow CO_2 + 4H_3O^+ + 4Cl^- & \Delta G^\circ = -90 \text{ kcal} \\ SiCl_4 + 6H_2O \rightarrow SiO_2 + 4H_3O^+ + 4Cl^- & \Delta G^\circ = -66 \text{ kcal} \end{array} \qquad (19\text{-}74)$$

There are two possible reasons for the difference in reactivity. On the one hand, the four chlorines about the rather small carbon atom probably use up the available space, so that an attacking water molecule would have to overcome sizable steric forces to penetrate to the central atom. This crowding should be less serious in the case of silicon. There is the further consideration that in the transition state, when the attacking water molecule has penetrated, the coordination about the central atom is, for the moment, five. Silicon can make use of a vacant 3*d* orbital in forming the extra bond, whereas the next available orbital for carbon is relatively much higher in energy. Whatever the actual cause, however, it is clear from the ΔG° values that the slower reaction of CCl_4 is due to a much higher activation energy than that for $SiCl_4$, and not to thermodynamic factors. As one might expect from these considerations, the tetrachlorides of the heavier Group IV elements are also susceptible to hydrolysis.

Selected Topics in Non-Metal Chemistry

19-5 Condensation Reactions

The formation of chlorine heptoxide from perchloric acid involves the removal of a molecule of water from between two acid molecules. The result is an oxygen bond between the two chlorine atoms:

$$\begin{array}{ccccccccc} & & :\ddot{O}: & & & & :\ddot{O}: & & \\ & & | & & & & | & & \\ :\underset{..}{\ddot{O}} & — & Cl & — & \underset{..}{\ddot{O}}—H + H— & \underset{..}{\ddot{O}} & — & Cl & —\underset{..}{\ddot{O}}: \rightarrow \\ & & | & & & & | & & \\ & & :\underset{..}{O}: & & & & :\underset{..}{O}: & & \end{array}$$

$$\begin{array}{ccccccccc} & & :\ddot{O}: & & :\ddot{O}: & & \\ & & | & & | & & \\ :\underset{..}{\ddot{O}} & — & Cl & —\underset{..}{\ddot{O}}— & Cl & — & \underset{..}{\ddot{O}}: + H_2O \\ & & | & & | & & \\ & & :\underset{..}{O}: & & :\underset{..}{O}: & & \end{array} \qquad (19\text{-}75)$$

This type of reaction is generally known as a *condensation reaction*, referring to the fact that two acid molecules are "condensed" into a single molecule. It is quite general in other series of acids as well. For example, the condensation of two molecules of sulfuric acid results in the formation of pyrosulfuric (disulfuric) acid.

$$
\begin{array}{c}
\begin{array}{ccccccccccc}
 & & :\ddot{O}: & & & & & & :\ddot{O}: & & \\
 & & | & & & & & & | & & \\
H- & \ddot{O}- & S & -\ddot{O}- & H & + & H- & \ddot{O}- & S & -\ddot{O}- & H \rightarrow \\
 & & | & & & & & & | & & \\
 & & :\ddot{O}: & & & & & & :\ddot{O}: & &
\end{array} \\
\begin{array}{cccccccc}
 & & :\ddot{O}: & & :\ddot{O}: & & \\
 & & | & & | & & \\
H_2O + H- & \ddot{O}- & S & -\ddot{O}- & S & -\ddot{O}- & H \\
 & & | & & | & & \\
 & & :\ddot{O}: & & :\ddot{O}: & &
\end{array}
\end{array}
\qquad (19\text{-}76)
$$

Condensation of a third molecule of sulfuric acid with pyrosulfuric acid would yield a compound of the form $H_2S_3O_{10}$; the process could be continued, yielding a whole series of so-called *isopoly* acids. The formulas of these would be $H_2S_4O_{13}$, $H_2S_5O_{16}$, *etc.* The general formula might be written as $H_2O\,(SO_3)_n$. As n becomes very large, the compound approaches the formula of the anhydride of sulfuric acid, SO_3.

The condensation products of orthophosphoric acid represent the most extensive known series of isopoly acids. Some of these are shown in Fig. 19-10. Note that, in addition to linear chains or polymers, cyclic structures are also formed. Although the compositions of many of the products of partial dehydration of orthophosphoric acid are quite complex, it has been possible in recent years to unravel these to a large extent. In some cases, polymers with molecular weights in the thousands are found. The ultimate composition of the linear polymer obtained by successive condensation of phosphoric acid is not the anhydride, but a chain with the formula $(HPO_3)_n$, termed *metaphosphoric acid.* The sodium salt of this polyacid, $(NaPO_3)_n$, is called *Graham's salt.* It is used for water softening because of its ability

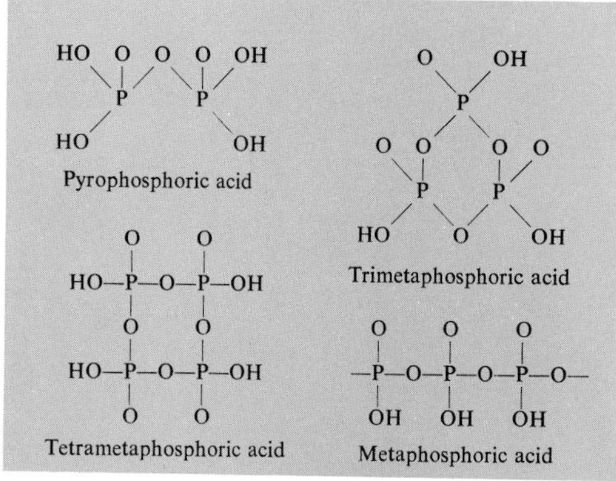

Fig. 19-10 Products of condensation reactions of orthophosphoric acid, H_3PO_4.

to form a complex with calcium ion and thus prevent its precipitation with soaps.

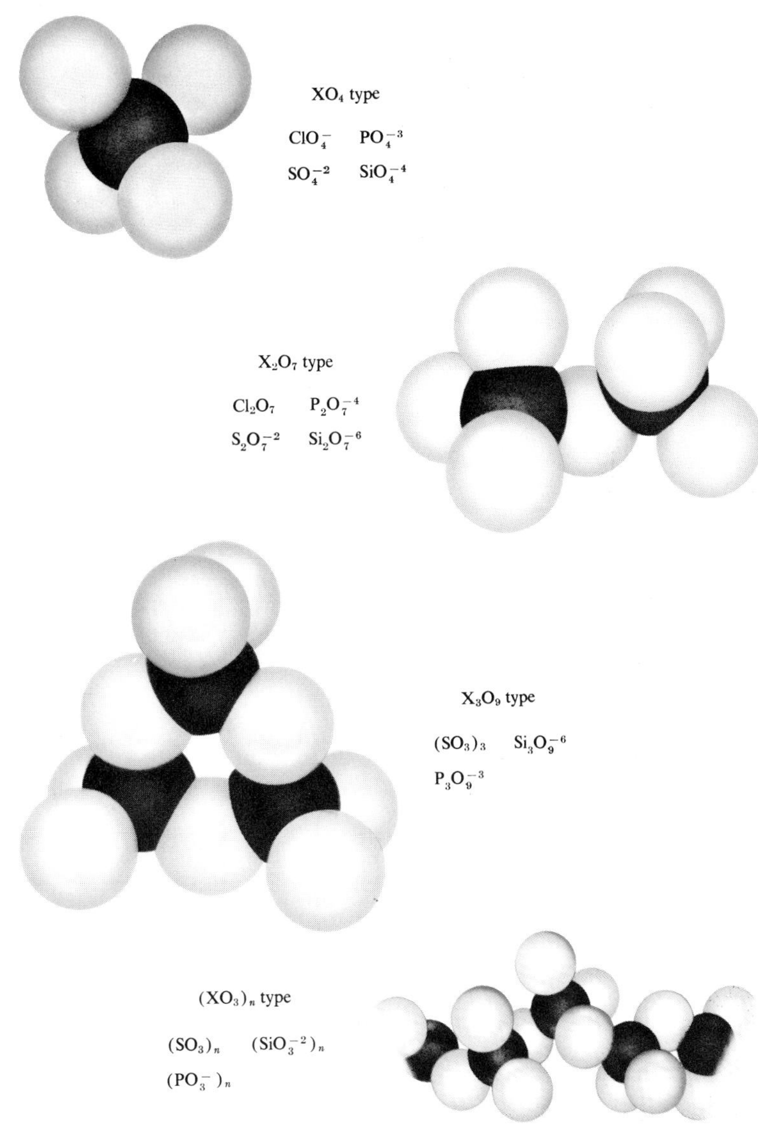

Fig. 19-11 Isoelectronic species among the oxy-anions and acid anhydrides. Although bond angles and relative bond distances may vary, the structures shown are generally representative of the species listed under each type.

The condensation products of orthosilicic acid, H_4SiO_4, are also important, since they are the basis for the whole range of silicate minerals found in nature.

The structural motif in all of the polyacids and polyanions we have discussed is the tetrahedron formed by the central atom and the four surrounding oxygen atoms. This is seen in the series ClO_4^-, SO_4^{-2}, PO_4^{-3}, and SiO_4^{-4}. These anions are isoelectronic, differing only in the nuclear charge on the central atom. Isoelectronic species are found also in the products of the condensation reactions (Fig. 19-11). Thus, for example, $(SO_3)_3$, $P_3O_9^{-3}$, and $Si_3O_9^{-6}$ are isoelectronic, and all possess a ring structure.

19-6 The Atmosphere

The earth's atmosphere is composed of a large number of substances (Table 19-3), although only a few are present to a large extent. The atmosphere is an extremely complex system because of variation in temperature and density with altitude and because of the effect of radiation from outer space, principally solar radiation. The sun's ultraviolet (high-energy) radiation breaks down molecules into atoms and ions. Oxygen is transformed into ozone, hydrogen is formed from the dissociation of water, and various other chemical changes occur. Species that are normally unstable are formed at very high altitudes, where the density of the atmosphere is so low that a long time may elapse before the unstable species undergo a recombination. For example, the mean free path of a gas molecule at a height of 500 kilometers is on the order of 20 kilometers! Solar radiation of wavelengths from about 3000 to 25,000 Å (near-ultraviolet, visible,

Table 19-3 Composition of dry air near sea level.

Component	Content (per cent by volume)	Molecular weight
Nitrogen	78.084	28.013
Oxygen	20.9476	31.998
Argon	0.934	39.948
Carbon dioxide	0.0314	44.0099
Neon	0.001818	20.183
Helium	0.000524	4.003
Krypton	0.000114	83.80
Xenon	0.0000087	131.30
Hydrogen	0.00005	2.0159
Methane	0.0002	16.0430
Nitrous oxide	0.00005	44.0128
Ozone	Summer: 0 to 0.000007	47.9982
	Winter: 0 to 0.000002	
Sulfur dioxide	0 to 0.0001	64.0628
Nitrogen dioxide	0 to 0.000002	46.0055
Ammonia	0 to trace	17.0306
Carbon monoxide	0 to trace	28.0105

near-infrared) is able to penetrate to the earth's surface. All other wavelengths, aside from a band in the radio wavelength region, are screened out. If it were not for this screening effect, life as we know it would be impossible on earth; the high-energy radiation of wavelength less than 3000 Å would destroy the complex molecular structures from which life systems are built.

Although the atmosphere may not appear to be dense, there is, in total, quite a bit of it — some 5500 trillion tons. The temperature of the atmosphere does not vary in a continuous way from the surface to outer space. The lowest temperatures, about −135°C, are reached at an altitude of 85 kilometers; the highest occur at heights of about 400 kilometers or more, and may reach 2000°K during periods of high solar activity. When it is said that an extremely high temperature prevails at high altitudes, hotness in the conventional sense is not meant. There are so few molecules or atoms per unit volume that a solid body placed in the environment would not attain the temperature indicated by a knowledge of the average kinetic energy of the gas and of extents of various dissociations. One can, however, speak of a *thermodynamic temperature*, as indicated by the extent to which the diatomic molecules are dissociated into atoms, and by the distribution of molecular and atomic speeds. It is in this sense that we say that very high temperatures prevail in the upper regions of the ionosphere.

The atmosphere is divided into vertical regions for the purpose of rough generalization and discussion. We will not discuss all of these but will consider those shown in Fig. 19-12. The *troposphere* extends to heights

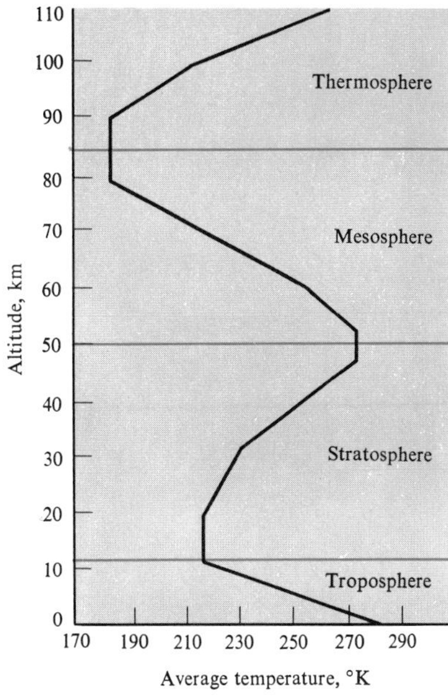

Fig. 19-12 Temperature variations in the atmosphere.

of about 11 kilometers. It contains all of the activity which we commonly think of as weather; rain, hail, lightning, hurricanes, *etc.* Most of the water vapor is in the troposphere, which also contains about 80 per cent of the total mass. The temperature of the atmosphere in the troposphere is determined largely by warming radiation from the earth's surface. For this reason, the temperature decreases more or less linearly with height, at the rate of about 6.5°C per kilometer.

Two minor constituents of the atmosphere, water and carbon dioxide, play a vital role in determining temperature in the troposphere, and thus in determining the conditions of life on earth. Both these substances are strong absorbers of infrared (*i.e.*, heat) radiation. They absorb infrared radiation from the earth's surface and reradiate it back toward earth, thus giving rise to what is frequently called the "greenhouse effect." This is a particularly important effect at present because man has, in recent times, brought about a significant change in conditions which have been operative for eons. The increased burning of coal, oil, and natural gas has brought about a large increase in the carbon dioxide content of the atmosphere. It is estimated that the average carbon dioxide content of the atmosphere has risen from an 1890 figure of about 0.0290 per cent to above 0.0315 per cent. This does not appear to be much of an increase, but it can, and almost certainly has, resulted in an increase in the average temperature on earth. If use of fossil fuels continues at the present rate, there will be 25 per cent more carbon dioxide in the atmosphere in the year 2000 than there is at present. Data obtained in a recent three-year period are shown in Fig. 19-13. An increase in the average temperature on earth sufficient to

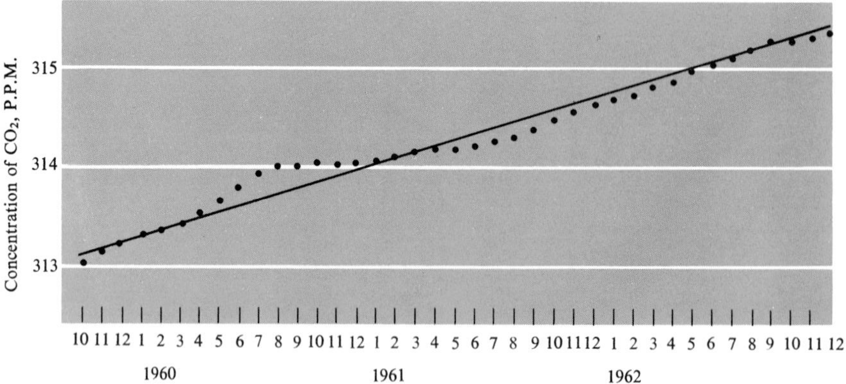

Fig. 19-13 Long-term increase of carbon dioxide in the atmosphere. Average monthly concentrations measured at the weather station on Mauna Loa in Hawaii (N. W. Rakestraw, *Chemistry*, Vol. 38, No. 4, p. 18).

cause drastic climactic changes may eventually occur if the present trend continues. For example, the change may be sufficient to completely melt the north polar ice cap, with disastrous consequences for many highly populated areas presently near sea level.

The *stratosphere* lies directly above the troposphere, and extends to about 50 kilometers. At the top of this region, the temperature may be as high as 270°K, not much different from that at the earth's surface. The composition of the stratosphere is essentially uniform, but there is a wide variation in the concentration of one minor constituent, ozone. In the center of the stratosphere region, the ozone concentration is about 10 parts per million (ppm), as compared with about 0.04 ppm in the troposphere. Ozone absorbs the sun's ultraviolet radiation in the 2100–2900 Å region. In addition to the vital role of screening out this high energy radiation, which would be harmful to life forms, the absorption also serves to heat the stratosphere, and thus maintain the heat balance in the troposphere. Ozone is formed in a two-step process:

$$\begin{aligned} O_2 + h\nu &\rightarrow O + O \\ O_2 + O + M &\rightarrow O_3 + M \end{aligned} \tag{19-77}$$

M represents any third body, such as a molecule of oxygen or nitrogen. The second reaction requires the third body to carry away the excess energy of combination. When it is not present, the O_3 molecule simply falls apart again after formation. Ozone is decomposed again by absorption of radiant energy

$$O_3 + h\nu \rightarrow O_2 + O$$

or by collision with an oxygen atom

$$O_3 + O \rightarrow 2O_2$$

The *mesosphere* and *thermosphere* are both part of the vast region known as the *ionosphere*, in which radiation from the sun causes formation of ions and free electrons from molecules. The two regions are distinguished by the way in which the temperature varies, as shown in Fig. 19-12.

A great number of chemical processes occur in the ionosphere, as a result of both solar radiation and galactic cosmic rays. The following processes are important in the production of free electrons in the ionosphere:

$$\begin{aligned} NO + h\nu &\rightarrow NO^+ + e^- \\ N_2 + h\nu &\rightarrow N_2^+ + e^- \\ O_2 + h\nu &\rightarrow O_2^+ + e^- \end{aligned} \tag{19-78}$$

The steps involving absorption of a photon are high energy processes and, therefore, require short wavelength radiation. The presence of free electrons in the ionosphere is responsible for reflection of radio waves; much worldwide radio communication is dependent upon ionospheric reflection.

Many of the reactions occurring in the thermosphere give rise to atoms, as in the dissociations of oxygen and nitrogen molecules:

$$\begin{aligned} O_2 + h\nu &\rightarrow O + O \\ N_2 + h\nu &\rightarrow N + N \end{aligned} \qquad (19\text{-}79)$$

The dissociation of oxygen molecules occurs at all altitudes, but recombination is fairly rapid at lower altitudes where the mean free paths are short. The probability of a three-body collision is extremely rare at an altitude of 120 kilometers, however, and there are on the average as many oxygen atoms as oxygen molecules at this height.

19-7 Compounds of the Noble Gases

The noble gases possess completed octets of electrons in the outermost orbitals of largest major quantum number. They are characteristically non-reactive, and, until recently, possessed a very limited known chemistry. We have already mentioned the helium hydride molecule-ion, HeH^+, the simplest heteronuclear diatomic molecule (Sec. 8-2). A number of other diatomic ions (*e.g.*, Ar_2^+, $HeNe^+$, *etc.*) are known from spectroscopic studies of electric discharges. Such species have only a transient existence, and do not afford much interesting chemistry. The first case of a *bona fide* compound of a noble gas element was reported in 1962 by Bartlett and Lochmann. They reasoned that transfer of an electron from xenon (ionization potential 12.1 e-v) to a species which would form a large anion, PtF_6, might lead to a lattice with sufficient lattice energy stabilization for the formation of a stable compound. The yellow compound $XePtF_6$ was prepared. Just afterwards it was discovered more or less simultaneously by more than one research group that xenon reacts with fluorine at elevated temperatures, about 400°C, to form a volatile, colorless solid, XeF_4, Fig. 19-14. Since then, a number of other compounds of xenon, involving fluorine and oxygen, have been prepared. In addition, compounds of radon

Table 19-4 Covalent compounds of the noble gases.

	Melting point (°C)	Remarks
XeF_2	140	Linear structure
XeF_4	114	$\Delta H_f^\circ(g) = -55$ kcal/mole; planar
XeF_6	46	$\Delta H_f^\circ(g) = -79$ kcal/mole
$XeOF_4$	−28	vapor pressure 29 mm at 23°C
XeO_3	—	$\Delta H_f^\circ(g) \approx +96$ kcal/mole
$Na_4XeO_4 \cdot 6H_2O$		Stable in basic solution, decomposes rapidly in acid
$K_4XeO_6 \cdot 2H_2O$		
KrF_2		
KrF_4	—	Sublimes at −30°C

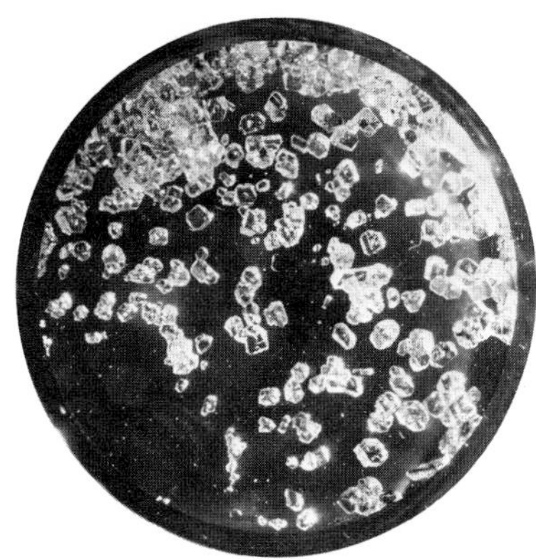

Fig. 19-14 Crystals of xenon tetrafluoride. The field of vision is 22 mm across (Figure by H. H. Claassen, Argonne National Laboratory, Argonne, Illinois).

and krypton have been discovered. Some of these are listed in Table 19-4, along with a few physical properties.

The structures of a number of these compounds have been determined by X-ray diffraction techniques. XeF_2 is linear, and XeF_4 consists of a xenon atom surrounded by four fluorine atoms in a plane. Xenon trioxide has a pyramidal structure, with the three oxygen atoms forming the base of a trigonal pyramid; in XeO_6^{-4}, the oxygen atoms are in an octahedral array about the xenon atom. The electronic structures can be most readily understood by considering the number of electrons which must be accommodated in the valence shell of xenon. XeF_2 is isoelectronic with species such as ICl_2^-, or I_3^-; the central atom must accommodate ten electrons, which leads to a prediction of sp^3d symmetry. The structure is thus similar to that for ClF_3, Fig. 19-6, except that the third fluorine, the one in the plane, is replaced by an unshared pair of electrons. Xenon tetrafluoride is isoelectronic in its outer electron arrangement with ICl_4^-, and BrF_4^-; there are twelve electrons in the valence shell of the central atom. This requires sp^3d^2 hybrid orbitals, which in turn implies an octahedral arrangement. The four positions in the plane are taken by fluorine atoms, the two axial positions by unshared electron pairs. On the basis of this type of argument, XeO_3 is isoelectronic with IO_3^-; it does in fact have the same type of structure, the pyramidal structure mentioned above. By proceeding along these lines, we can quickly conclude that XeF_6 should have 14 electrons in the valence shell of xenon. This implies a seven-coordinate arrangement. There is some controversy at present about whether this simple approach to the structures, analogous to that employed in considering the interhalogen compounds (Sec. 19-1), is valid. The structure of XeF_6 would provide a test of the model, since a seven coordinate arrangement would probably lead to some kind of distorted rather than a strictly symmetric, octahedral structure. The best evidence available at this time indicates that XeF_6 does indeed possess a distorted octahedral structure.

When the xenon fluorides are added to water, the following reactions may be observed:

$$XeF_2 + H_2O \rightarrow Xe + \tfrac{1}{2}O_2 + 2HF \qquad (19\text{-}80)$$

$$3XeF_4 + 6H_2O \rightarrow 2Xe + \tfrac{3}{2}O_2 + 12HF + XeO_3 \qquad (19\text{-}81)$$

$$XeF_6 + H_2O \rightarrow XeOF_4 + 2HF \qquad (19\text{-}82)$$

$$XeF_6 + 2H_2O \rightarrow XeO_2F_2 + 4HF \qquad (19\text{-}83)$$

$$XeF_6 + 3H_2O \rightarrow XeO_3 + 6HF \qquad (19\text{-}84)$$

Xenon trioxide may also be formed from the tetrafluoride through a disproportionation:

$$6XeF_4 + 12H_2O \rightarrow 2XeO_3 + 4Xe + 3O_2 + 24HF \qquad (19\text{-}85)$$

Although the trioxide is well behaved in aqueous solution, it is violently explosive as a dry solid. (Note that ΔH_f is positive for XeO_3, Table 19-4.) Exposure of either XeF_4 or XeF_6 to the air must be avoided, to prevent hydrolysis with formation of XeO_3.

The trioxide reacts with alkaline aqueous solutions to form perxenate ion, XeO_6^{-4}, with evolution of xenon. If the solution is made acid, the perxenic acid formed slowly decomposes to form XeO_3, with evolution of oxygen gas. The standard electrode potential for the Xe(VIII) → Xe(VI) half-reaction is estimated to be +3 volts.

Much work remains to be done on the chemistry of noble gas fluorides. XeF_4 appears to have promise as a fluorinating agent. A solution of XeF_4 in anhydrous HF attacks platinum, and apparently converts many organic compounds to their fluorinated derivatives. The water-soluble perxenate and xenate salts should provide interesting subjects for further study.

Suggested Readings

Cotton, F. A., and G. Wilkinson, *Advanced Inorganic Chemistry* (2nd ed.), New York, N. Y.: Interscience Publishers, Inc., 1966.

Hyman, H. (ed.). *The Noble Gas Compounds*. Chicago, Ill.: The University of Chicago Press, 1965.

Jolly, W. L., *The Chemistry of the Non-metals*. Englewood Cliffs, N. J.: Prentice-Hall, Inc., 1966.

Pauling, L., *The Chemical Bond*. Ithaca, N. Y.: The Cornell University Press, 1966.

Exercises

19-1. Using the method described in the text, write a balanced half-reaction for each of the following processes, and calculate the standard half-cell potential:

ClO_4^- to Cl^- , acidic solution
ClO_4^- to OCl^- , basic solution

19-2. Disproportionation reactions can be treated as the sum of two half-reactions. Calculate the standard free energy change for the reaction

$$Cl_2 + 2OH^- \rightarrow Cl^- + ClO^- + H_2O$$

19-3. Calculate $\mathcal{E}°$ values for the first and third reactions given in Eq. (19-32). Design cells which might be constructed using these reactions.

19-4. Can you suggest a reason why peroxides are generally more easily formed in the solid state with the heavier metals than with the light? [*Hint:* Consider the effect of radius ratio.]

19-5. Draw the Lewis structures for each of the following compounds: hydrogen peroxide, hydrazoic acid, chlorine trioxide, fluorine monoxide, metaperiodic acid, sulfur trioxide.

19-6. Which compound or compounds would produce in anhydrous ammonia a reaction which is analogous to the reaction of calcium oxide with water? Write a balanced chemical equation illustrating the reaction.

19-7. Complete and balance each of the following (it may be necessary in some cases to add hydroxide ion, hydronium ion or water):
(a) $As_2O_3 + H_2O_2 \rightarrow As_2O_5 +$
(b) $Fe^{+2} + H_2O_2 \rightarrow Fe^{+3} + H_2O$ (acidic)
(c) $Mn^{+2} + H_2O_2 \rightarrow MnO_2(s) + H_2O$ (basic)
(d) $As_2O_3 + C \rightarrow As + CO$
(e) $SF_4 + H_2O \rightarrow$
(f) $NO(g) + NO_2(g) + OH^-(aq) \rightarrow$
(g) $NO_2^- + Na$ (amalgam) $\rightarrow$ (basic)
(h) $Zn + NO_3^- + H_3O^+ \rightarrow N_2 +$

19-8. Write Lewis structures for each of the following:
(a) Nitrate ion.
(b) Disulfur decafluoride.
(c) Arsenious acid.
(d) Phosphoryl bromide.

19-9. Nitrogen trifluoride does not show any significant basic properties, in contrast with ammonia. Account for this difference.

19-10. Assuming complete conversion, what weight of lead nitrate is required for the preparation of 2.0 liters of dry NO_2 at 25°C, 1 atm pressure?

19-11. Write the Lewis structures for nitrous oxide, nitric oxide, nitrogen dioxide, and dinitrogen tetroxide. Indicate the need for resonance structures where applicable.

19-12. Calculate K for the equilibrium $N_2O_4 \rightarrow 2NO_2$ at 27, 70 and 100°C from the data listed on p. 465. (Optional) What is ΔH for this reaction? Is this a measure of the N—N bond strength?

19-13. Write the reactions for the electrolysis of a hot potassium chloride solution, assuming that the reaction proceeds in two steps:

(a) Evolution of chlorine gas.
(b) Reaction of chlorine with hydroxide.
Write an over-all balanced equation for the process.

19-14. By using your knowledge of the chemical properties of the substances involved, and by applying general principles, it should be possible for you to complete each of the equations listed below, or to indicate that no reaction occurs:

(a) $Cl_2O(g) + H_2O \rightarrow$
(b) $MnO_4^- + H_2O_2 \rightarrow$ (acid)
(c) $I_2O_5 + H_2O \rightarrow$
(d) $P_4O_6 + O_2 \rightarrow$
(e) $As_2O_3 + HOCl \rightarrow$
(f) $SO_2 + Fe^{+2} + H_3O^+ \rightarrow$
(g) $NaClO_3 + S \rightarrow$
(h) $HClO_2(aq) + Zn \rightarrow$
(i) $SO_3^{-2} + O_2 \rightarrow$

19-15. Suppose that the material which is to be oxidized in a solid fuel for a missile has the empirical formula SC_3H_6. Assuming that the products are SO_2, CO_2, H_2O, and NH_4Cl, calculate the weight of ammonium perchlorate required for complete reaction with 100 g of reductant. Assume that the oxidant is used with 100 per cent efficiency. Repeat the calculations for lithium perchlorate, assuming that LiCl is a product.

19-16. The species CO_2, N_3^-, and CNO^- are alike in being isoelectronic. Draw the structures for these ions or molecules.

19-17. It is often stated in textbooks that the structure of carbon dioxide is described in terms of the following canonical forms:

$$\ddot{\underset{..}{O}}{=}C{=}\ddot{\underset{..}{O}} \qquad :O{\equiv}C{-}\ddot{\underset{..}{O}}: \qquad :\ddot{\underset{..}{O}}{-}C{\equiv}O:$$

Which structure or structures are likely to make the major contribution to the actual structure? Explain.

19-18. Write a set of complete balanced equations for each of the following syntheses.

(a) A solution of hypophosphorous acid, beginning with white phosphorus.
(b) A solution of sodium nitrite, beginning with ammonia.
(c) Solid barium carbonate, beginning with graphite.
(d) Nitric acid, beginning with ammonia.
(e) Calcium sulfite, beginning with sulfur.
(f) Phosphorus (V) fluoride, beginning with phosphorus (III) fluoride and chlorine trifluoride.

19-19. Which set of hybrid orbitals would be employed by iodine in paraperiodic acid? By phosphorus in orthophosphoric acid? By iodine in ICl_3?

19-20. Draw a plausible structure for the tetrathionate ion, given the information that it contains a chain of four sulfur atoms.

19-21. The polyhalide ions are closely related structurally to the interhalogen compounds. When cesium iodide, for example, is treated with chlorine gas under pressure, the substance $CsICl_2$ is formed. It contains the ICl_2^- ion. Further treatment with chlorine yields $CsICl_4$. Discuss the polyhalide ions, ICl_2^- and ICl_4^-, from the viewpoint of their structural relationship to interhalogen compounds. Given that the ICl_2^- ion is linear, and that the ICl_4^- ion is planar, draw structures for these two species, showing the positions of the unshared electron pairs.

19-22. In connection with the reactivity of non-metal oxides, it was pointed out that the most reactive compounds are those in which oxygen is bonded to an element of high electronegativity. Compare the oxidizing properties of non-metal oxides with the fluorinating properties of non-metal fluorides in this respect.

19-23. Magnesium forms an unusual carbide of formula Mg_2C_3. Upon hydrolysis with water, the principal product is a hydrocarbon which analyses as 90.0 per cent carbon and 10.0 per cent hydrogen. The density of the hydrocarbon gas at 0.500 atm, 30°C, is 0.805 g/liter. What is the molecular formula of the gas? Write a balanced equation for the hydrolysis reaction. Can you suggest a possible structure for the compound?

19-24. By making an analogy with the behavior of chlorine, predict the formula and structure for a product of the reaction of argon with elemental fluorine.

19-25. The stepwise reaction of two moles of water with P_4O_{10} may give tetrametaphosphoric acid as one of the major products. Explain in detail how this product may be formed.

19-26. The anhydride of nitrous acid is dinitrogen trioxide, N_2O_3. Presuming that the substance does exist, suggest a possible structure.

19-27. It has been noted that PBr_5 exists in the solid state in the form of $PBr_4^+ + Br^-$ ions. Contrast this with the behavior of PCl_5, and account for the difference.

19-28. Antimony pentafluoride forms complex compounds of low solubility with KF and other ionic fluorides. How might these species be formulated as ionic species? Describe the structure of the anion which you visualize, and indicate the neutral non-metal fluoride with which it is isoelectronic.

19-29. Sulfur forms two oxychlorides, thionyl chloride, $SOCl_2$, and sulfuryl chloride, SO_2Cl_2. Draw Lewis structures for these two compounds, and discuss their probable geometry. (Note the relationship to SCl_2, similar to that which $POCl_3$ bears to PCl_3.)

19-30. Write the structural formula for the acid $H_2S_4O_{13}$. By generalizing on this, show the structure of polymeric SO_3.

19-31. The mineral thortveitite, $Sc_2Si_2O_7$, contains anions of the formula $Si_2O_7^{-6}$. The mineral wallastonite, $Ca_3Si_3O_9$, contains anions of the formula $Si_3O_9^{-6}$. Show the structures of these two anions. Is it possible to draw a linear structure for the second anion?

19-32. The ionization potential of the NO molecule is 15.5 e-v. What wavelength of photon is required to effect first the ionization process described by (19-78)?

19-33. Calculate the average thermochemical bond energy of the Xe—F bond in XeF_4 and XeF_6, from the data in Table 19-4 and other sources in the text.

19-34. Describe a single chemical test which would serve to distinguish the substances in each of the following pairs:
(a) H_2O_2 solution, Cl_2 solution.
(b) H_3PO_3 solution, H_2SO_3 solution.
(c) PF_3, PCl_3.
(d) XeF_4, HIO_3.
(e) Sodium thiosulfate solution, sodium perchlorate solution.
(f) Li_3N, CaC_2.
(g) CO, NO.

Chapter 20

Metallic Elements; Metal Ion Complexes

About three-fourths of the elements are classified as metallic. Their predominance is evident from Fig. 20-1, which shows the metallic elements as the shaded area. The metals are characterized by high thermal and electrical conductivity, luster, malleability, and ductility, reaction with aqueous acid resulting in the evolution of hydrogen, formation of basic oxides, and other properties. They differ, of course, in the degree to which they exhibit these evidences of metallic character. The elements which are on the borderline between the metals and non-metals may lack one or more of the above-mentioned properties.

20-1 Structures of Metals and Alloys

The metallic elements exhibit quite simple solid state structures. They are found as one of the two close-packed forms, hexagonal or cubic, or as a closely related structure, body-centered cubic. In the body-centered cube, each atom is surrounded by eight nearest neighbors at the corners of a surrounding cube, and by six others at slightly larger distance (Fig. 20-2). Recall that in the close-packed structures, each atom is surrounded by

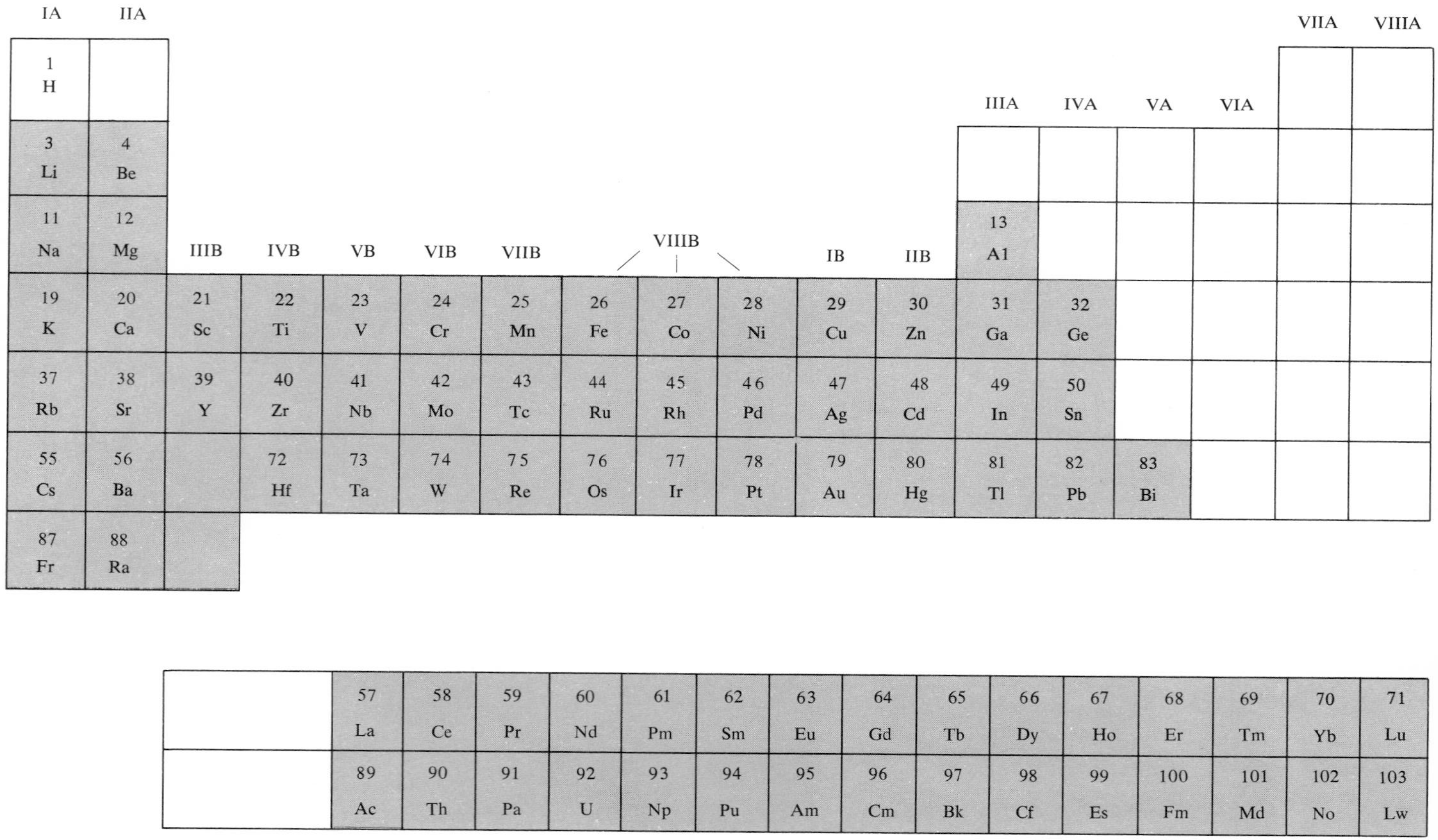

IA	IIA	IIIB	IVB	VB	VIB	VIIB	VIIIB			IB	IIB	IIIA	IVA	VA	VIA	VIIA	VIIIA
1 H																	
3 Li	4 Be																
11 Na	12 Mg											13 Al					
19 K	20 Ca	21 Sc	22 Ti	23 V	24 Cr	25 Mn	26 Fe	27 Co	28 Ni	29 Cu	30 Zn	31 Ga	32 Ge				
37 Rb	38 Sr	39 Y	40 Zr	41 Nb	42 Mo	43 Tc	44 Ru	45 Rh	46 Pd	47 Ag	48 Cd	49 In	50 Sn				
55 Cs	56 Ba		72 Hf	73 Ta	74 W	75 Re	76 Os	77 Ir	78 Pt	79 Au	80 Hg	81 Tl	82 Pb	83 Bi			
87 Fr	88 Ra																

	57 La	58 Ce	59 Pr	60 Nd	61 Pm	62 Sm	63 Eu	64 Gd	65 Tb	66 Dy	67 Ho	68 Er	69 Tm	70 Yb	71 Lu
	89 Ac	90 Th	91 Pa	92 U	93 Np	94 Pu	95 Am	96 Cm	97 Bk	98 Cf	99 Es	100 Fm	101 Md	102 No	103 Lw

Fig. 20-1 Location of the metals in the periodic table.

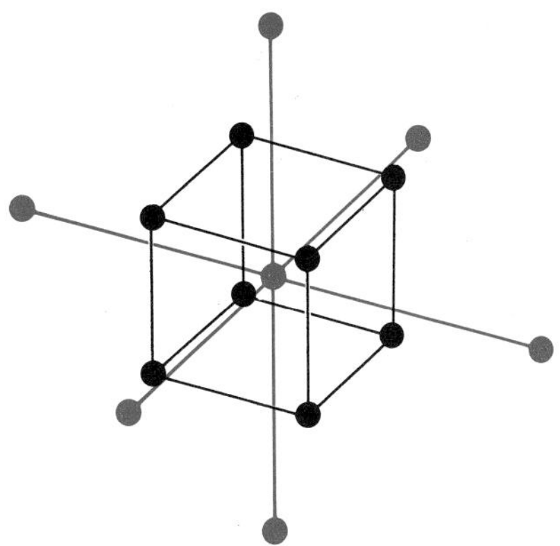

Fig. 20-2 The body-centered cubic structure, found for lithium and other metals.

twelve equivalent nearest neighbors (Sec. 7-6). The metals are thus characterized by large numbers of nearest neighbors for each atom.

This characteristic of metallic crystals is contrasted with the structures of non-metallic elements. In a sulfur crystal, for example, each sulfur atom has two nearest neighbors — the two adjacent atoms in the ring of which it is a part (p. 411). All other sulfur atoms in the structure are relatively far away, and there is no need to consider them insofar as possible chemical bonding is concerned. Ionic lattices, on the other hand, are similar to the metallic crystals in the arrangement of the units. The crystal structure of cesium chloride is the same as that of lithium metal, if one overlooks the difference between cation and anion. We have seen (Chap. 7) that the structures of ionic solids can be understood in terms of the potentials existing between the charged particles which make up the lattice. The same sort of model may be proposed for a metal. One can picture the lithium metal lattice, for example, as made up of a network of Li^+ ions, with electrons distributed throughout the lattice in such a way as to form a sort of "electrostatic glue" holding the lattice together. These electrons would be just the ones removed from the lithium atoms to form the Li^+ ions; there would be one, therefore, for each lithium ion. There is some virtue in this picture of a metal, but as stated it does not contain any real chemical insight. We will return later to a modified version which does have some connection with the more familiar ideas of chemical bonding.

The valence of the atoms in a metallic structure is reflected in the physical properties of the metal. Table 20-1 shows some typical properties for the series of metallic elements beginning with potassium. It is readily seen that, with an increase in the number of electrons beyond the argon configuration, the density and melting point increase.

The heats of atomization (*i.e.*, the heats of formation of the gaseous monatomic atoms) provide an indication of the total bonding stabilization in

Table 20-1 Physical properties and metallic valence in metallic elements.

Element	K	Ca	Sc	Ti	V	Cr	Mn	Fe	Co	Ni	Cu	Zn
Atomic number	19	20	21	22	23	24	25	26	27	28	29	30
Melting point (°C)	62	810	1200	1800	1720	1615	1260	1530	1480	1450	1083	420
Density	.84	1.6	2.5	4.5	5.9	6.9	7.2	7.9	8.9	8.9	8.9	7.1
Atomization energy (kcal/mole)	21	46	93	112	122	94	68	99	102	101	81	31

the solid. This increases regularly with the number of electrons available for bond formation, up to about five or six. Beyond this, some of the electrons in the valence shell are non-bonding. In fact, the number of bonding electrons appears to decrease for the heavier metals of the first transition series. The same general sort of behavior is observed also in the series beginning with rubidium and cesium.

Now in all cases, the number of electrons employed in bonding is far less than the number of nearest neighbors surrounding each atom. From this, we may infer that the bonding between a particular atom and one of its neighbors must be something less than a full bond. To return to lithium as an example: Each Li atom possesses a valence of one, but this single valence must be shared equally with eight nearest neighbors, and to a lesser extent with six next-nearest neighbors.

The nearest neighbor Li—Li distance in the metal is 3.03 Å, significantly larger than the 2.67 Å Li—Li distance in the Li_2 molecule. The individual Li—Li bonds in the metal are therefore probably much weaker than the Li—Li bond in the gaseous Li_2 molecule, for which the heat of dissociation is only 25 kcal/mole. This supports the idea that bonds between individual atoms in the metal are weak.

In a sense, then, the metallic crystal is a giant molecule with each atom bonded to its neighbors, which are in turn bonded to other atoms. The electrons — much fewer in number than the bonds — resonate among the many bonds so that a very large number of resonant structures are possible for the molecule. These electrons are for the most part paired in the bonds, but a small fraction of the electrons are unpaired, and are distributed among the various bonds as single electrons. The characteristic metallic properties of high thermal and electrical conductivity result from the extreme mobility of the valence electrons, which are free to move among the many available bonding sites.

The molecular orbital model has been successful in accounting for many characteristic properties of metals. The molecular orbital method was in fact first applied to metals, in a form which is sometimes referred to as *band theory*. To see how this model arises, let us consider some clusters of lithium atoms, as shown in Fig. 20-3. We will focus attention for the moment on the 2*s* orbitals of the lithium. We have already seen that two atomic orbitals can be combined under the appropriate conditions to form molecular orbitals (Sec. 8-2). When two atomic orbitals are combined, two molecular orbitals are formed. The energy separation between these orbitals is related to the degree of overlap of the atomic orbitals, and to the energy differences between them. (In the present case, all of the atomic orbitals are of the same energy.) When three atomic orbitals are combined, three molecular orbitals are formed. The energy difference between the upper and lower of these is not much different from that when two atoms are involved, but there is now a third molecular orbital of intermediate energy. With a further increase in the number of atoms which cluster together, the number of resulting molecular orbitals increases, always being

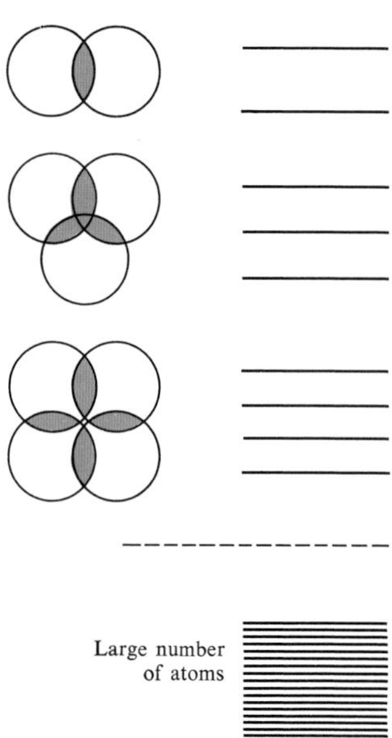

Fig. 20-3 Successive overlap of atomic orbitals to form molecular orbitals. In the limit of a very large number of contributing atoms, the individual molecular orbital energy levels are very close in energy and effectively form a continuum of allowed energies within the range from the highest to the lowest energy orbitals.

equal to the number of starting atomic orbitals. In lithium metal, the number of contributing atoms is extremely large, so the number of molecular orbitals formed is also extremely large. At the same time, the energy difference between the highest and lowest energy levels has not increased much. This means that the energy gap between successive levels is very small; so small, in fact, that the energy levels may be thought of as being continuous. When electrons are placed in the molecular orbitals, they occupy them in pairs in the order of increasing energy. When all of the electrons in lithium have been placed, however, half of the molecular orbitals still remain vacant. The lowest energy vacant levels are not separated from the highest energy filled levels by any significant gap in energy, so that the smallest push is sufficient to cause electrons in the highest energy filled levels to jump to the next higher levels, which are vacant. The electrons which have jumped are delocalized in the metal structure, and give rise to the high electrical and thermal conduction, high reflection of light, and other electrical, thermal, and magnetic properties of metals.

In beryllium metal, the overlap of the 2*s* orbitals of the Be atoms to form molecular orbitals should proceed as with lithium. In this case, however, each Be atom has two 2*s* electrons to contribute to the molecular orbitals which are formed. This number of electrons is just sufficient to completely fill the molecular orbitals. There are, therefore, no molecular orbitals formed from the 2*s* orbitals which are vacant, and which might function to impart metallic character to Be as described above for Li. How, then, does it happen that Be does possess metallic character? The answer lies in the fact that both lithium and beryllium have other atomic orbitals, in addition to the 2*s* orbitals, which overlap to form similar sets of molecular orbitals. The most important of these, in the case of lithium and beryllium, are the 2*p* orbitals. It happens that the spread of energy levels formed from the metal 2*p* orbitals in the metallic lattice (this is essentially a "band" of energy levels; hence, the term band theory) overlaps to some extent with the band formed from the 2*s* orbitals, as shown in Fig. 20-4. Because of this overlapping effect, the highest filled orbital in beryllium has just above it energy levels which are readily available. If the band of energies formed by the 2*s* orbitals did not overlap with that from the 2*p*, beryllium would not exhibit metallic properties, but would be an insulator, both electrically and thermally. It is possible to consider even a substance such as carbon in terms of this theory (Sec. 17-5). Let us suppose that in diamond all of

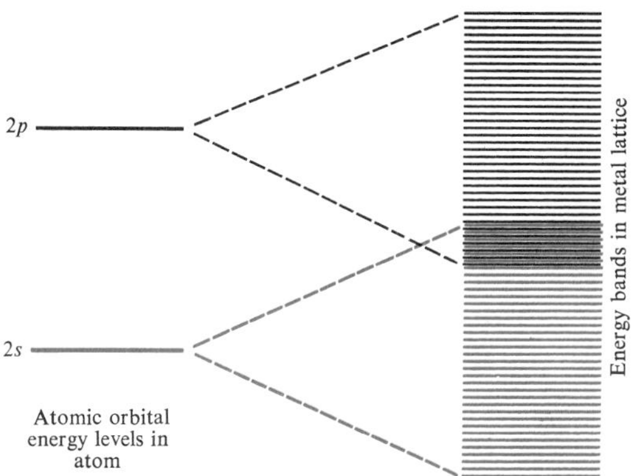

Fig. 20-4 Overlap of the allowed energy band formed from overlap of 2*s* atomic orbitals with that formed by overlap of 2*p* atomic orbitals. Such an overlap occurs when the energy separation between the 2*s* and 2*p* atomic orbitals is not too large. This separation is smallest for lithium, the most metallic element in the second row, and increases upon moving to the right in the periodic table.

the carbon 2*s* and 2*p* atomic orbitals interact to form molecular orbitals which are delocalized over the entire lattice. These energy levels would occur in the form of bands of allowed energy levels. The peculiar arrangements of the carbon atoms relative to one another in diamond is such that the electrons, when placed in the molecular orbitals, just fill an allowed band. Furthermore, there is a large energy gap to the next band of allowed energy levels, so diamond is an insulator. In graphite, on the other hand, the carbon 2*p* orbitals of the π variety overlap with one another to form a band of allowed energy levels which is only half-filled. The π electrons therefore have metallic properties such as high electrical and thermal conductivity, but only in the layer directions.

The free electron — or electron-in-a-well — model for metals is a simplified and easily understood version of the molecular orbital model. In this version, the electron (we will discuss the case of a single valence electron, as in lithium, for simplicity) is thought of as being located in a deep potential energy well. The actual shape of a potential well around a nucleus is something like that given by the shaded area in Fig. 20-5. We simplify this by assuming a square well, shown by the colored line in the figure. The depth of the well is actually equal to the ionization potential of the atom in question; for our purposes this factor is not important, however. Now the formation of a diatomic molecule in this model consists in the joining of the two potential wells of the constituent atoms. Whereas before, the two electrons were each in separate wells, they now share a single, larger well. The energy of the electrons in the potential wells is

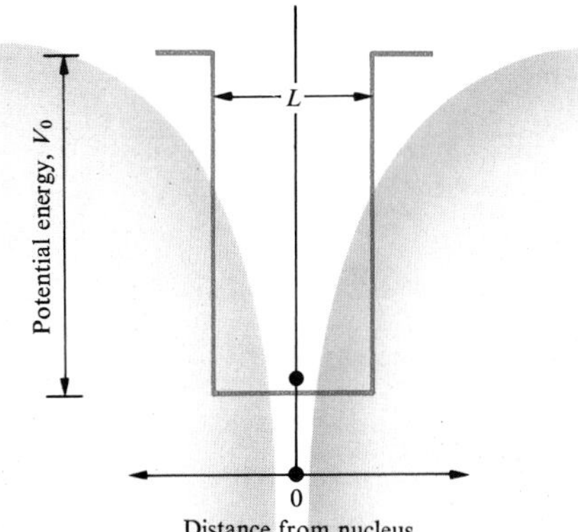

Fig. 20-5 Potential energy of an electron in the vicinity of the nucleus. The square well represents a simple approximation to the real potential, given by the shaded curve. The curve represents the manner in which the potential energy of the electron varies with distance from the nucleus, as measured on the horizontal axis.

quantized; the major factor in determining the magnitude of the quanta of energy is the width of the well, shown as L in Fig. 20-5. The formula for the energy of the electrons is

$$W = V_0 + \frac{h^2n^2}{8mL^2} \tag{20-1}$$

W is the total energy, V_0 is the potential energy (the depth of the well) and is always a large, negative number, h is Planck's constant, m is the mass of the electron, L is the width of the well, and n is the quantum number, which may take on the values $+1$, $+2$, $+3$. . . . Each value of n corresponds to an allowed energy state, just as with atomic orbitals. Each orbital here is capable of containing two electrons with paired spins.

The process of joining atoms together to form larger and larger potential wells is shown in Fig. 20-6. The energy levels of these potential wells are also shown in the figure, with the way in which they are occupied by the electrons. It is easy to see that, as the number of atoms joined together increases, the energy levels are brought closer and closer together, since L in Eq. (20-1) grows larger. When the number of atoms is very large, as

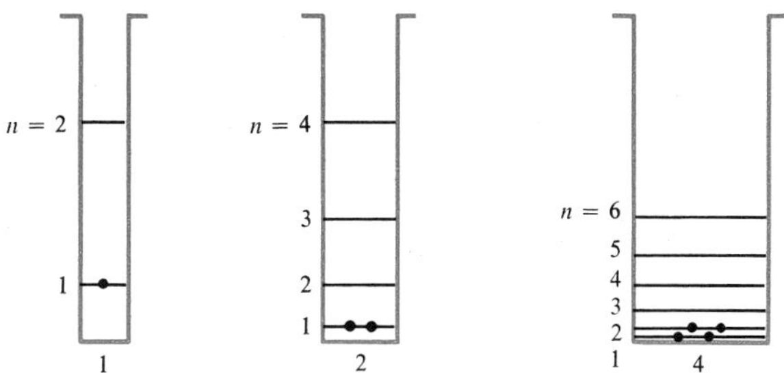

Fig. 20-6 An illustration of the simple free-electron model for a metal. Each atom is pictured as a potential energy well in which the electron is free to move. The joining together of atoms results in larger potential wells. The quantized energy levels are lowered in energy and become more closely spaced. Each allowed energy level is occupied by two electrons in accordance with the Pauli exclusion principle. As the number of atoms becomes very large, the energy levels become so very closely spaced that there is essentially a continuum of allowed energies.

in a metal crystal, the energies are so close together that they form a practically continuous sequence; very little energy is required to cause an electron to jump from the level it is occupying to the next higher level. At ordinary temperatures, there are a number of electrons in metals which exist in these higher levels, or which move into them under the influence of an applied field. These are the conduction electrons.

Metallic lustre is a consequence of the ability of a metal to absorb (and then re-radiate) any and all wavelengths of light. For any quantum of energy which excites an electron, there is an orbital at just the right energy for receiving that electron. When it returns to the lower level, the energy is re-emitted as reflected light. Metals which are very finely divided appear to be black, because the emitted light is scattered in all directions, very little going off in the direction from which it came.

Of course, it would be too much to expect that such a simple picture would explain all metallic properties. More elaborate versions of the molecular orbital theory have, however, been quite successful in this regard.

Since, as we have seen, bonding in metals is not specific in the sense of being directional, or limited to only one or two neighbors, it is not difficult to substitute one element for another in a metal, provided that certain conditions are met. The first condition is that the atomic sizes of the two kinds of atoms be nearly the same. This requirement is necessary to insure that an efficient packing arrangement is possible. The rule developed by Goldschmidt relating to isomorphous replacement in ionic crystals (p. 143) applies fairly well: The atomic sizes should not differ by more than about 15 per cent.

The second provision for simple substitution of one element for another is that the valence characteristics of the two elements be the same, or nearly the same. If they are not, special types of combinations leading to inter-metallic compounds may occur.

An *alloy* is a material which has metallic properties, and which is composed of more than one element. It need not be a distinct chemical compound, and, in fact, it seldom is. An alloy may consist of:

1. *Separate crystals of pure, or nearly pure, substances.* Tin and lead, for example, if melted together and allowed to cool form a solid in which the two elements are present as nearly pure, separate crystals, with only a little of one element dissolved in the other. The physical properties of this type of alloy depend upon the initial composition, and upon the manner in which the alloy is treated in respect to temperature, hammering, cold-rolling, *etc.*

2. *A solid solution of one metal in another.* This type occurs when the conditions mentioned above are met; *i.e.*, the atomic radii of the two elements are about the same, and the valence characteristics are not too different. Nickel and copper, or gold and silver, form solid solutions over the entire composition range. Depending upon the relative proportions of the constituents, the properties of these alloys are usually intermediate between those of the pure substances.

3. *A solid solution in which a second element — usually non-metallic — is located interstitially in the metallic lattice.* Formation of this type of alloy, an extremely important variety in metallurgy, requires that the second element introduced have a relatively small radius, so that it may fit into open spaces between spheres in the metallic lattice. Carbon, nitrogen, hydrogen, and boron all form alloys of this type with many metals. The interstitial atoms participate in bonding to neighboring metal atoms. As a result of these extra bonds, the metallic lattice becomes more rigid, more brittle, and melts at a higher temperature. Cast iron, which contains carbon alloyed with iron, is an example of this type of alloy. This type is actually a chemical compound but, since all of the available sites are not usually occupied, the alloys are non-stoichiometric, or of variable composition. However, pure carbides and nitrides of many metallic elements are known.

4. *Inter-metallic compounds between two or more metallic elements.* These occur when valences and size relationships are favorable. Although a great many inter-metallic compounds are known, their formulas are generally rather complex and not easy to understand in terms of simple valence theory. Some examples are Cu_5Zn_8, Cu_9Al_4, Fe_5Zn_{21}.

20-2 Metallurgy

Most metallic elements are found not in the free state in nature, but in the form of compounds such as silicates, oxides, or sulfides. The naturally occurring materials from which metals must be extracted are called *ores*. Metallurgy is the science (or art) of obtaining sufficiently pure metals from ores. The processes of metallurgy are, in effect, a series of concentration steps, in which the metal content of the worked material increases until, finally, a pure metal is obtained. The raw ore may be a far cry indeed from the final product; some copper ores which are now being worked successfully contain as little as 0.2 per cent copper.

The preliminary treatments of an ore are almost always physical processes, such as grinding and crushing, which put material dug from the earth into a workable state of subdivision.

The undesirable material (called gangue*) which accompanies the ore is removed, at least in part, by both physical and chemical methods. The physical methods may be based upon any significant difference between a physical property of the desired mineral and the gangue.

1. *Density differences.* Washing with water, or whirling finely ground ore in a "cyclone separator," serves to separate particles with large density differences.

2. *Wetting properties.* The froth flotation method, in which an ore is treated with an oil and water mixture, then thoroughly agitated with air,

*Pronounced "gang."

is a very important concentration method. The method has been applied most extensively to sulfide ores. These ores are not wetted by water, whereas the gangue — usually silicate or oxide minerals — *is* wetted. The agitation with air causes formation of a foam of oil-covered bubbles. Small particles of the desired constituent adhere to the bubbles and rise to the surface, whereas the gangue settles to the bottom. The foam containing the concentrated ore is skimmed off with paddles. Froth flotation methods have made it possible to work ore deposits of surprisingly low quality. As our richer mineral deposits become depleted and poorer sources must be resorted to, these methods will no doubt assume even greater importance.

3. *Miscellaneous physical properties.* These include electrostatic and magnetic properties, melting point differences and others.

Any chemical process which results in an increase in the weight percentage of metal in an ore may be considered a concentration step.

1. *Roasting.* This process, which applies generally to the treatment of sulfide ores, consists in their being heated in the presence of air. The product of the reaction may be the metal oxide, free metal, or metal sulfate. Some typical reactions are

$$\begin{aligned} 2ZnS + 3O_2 &\rightarrow 2ZnO + 2SO_2 \\ ZnS + 2O_2 &\rightarrow ZnSO_4 \\ 4FeS_2 + 11O_2 &\rightarrow 2Fe_2O_3 + 8SO_2 \\ HgS + O_2 &\rightarrow Hg + SO_2 \end{aligned} \tag{20-2}$$

Generally speaking, the free metal is formed only with the sulfide ores of metals which are lowest in the activity series.

2. *Calcination.* This term applies to the heating of a carbonate ore, or of a hydrate, with the evolution of carbon dioxide or water vapor. Some examples are

$$\begin{aligned} FeCO_3 &\rightarrow FeO + CO_2 \\ CaSO_4 \cdot 2H_2O &\rightarrow CaSO_4 \cdot \tfrac{1}{2}H_2O + \tfrac{3}{2}H_2O \\ 2Al(OH)_3 &\rightarrow Al_2O_3 + 3H_2O \end{aligned} \tag{20-3}$$

3. *Leaching.* It is sometimes advantageous to extract the valuable constituent from an ore by a chemical process which places the metal into solution. One of the most famous of leaching type processes is the cyanide process for recovery of gold and silver from low grade ores. The crushed ore is treated with an aqueous solution of sodium cyanide, with thorough aeration. In the presence of the strongly complexing cyanide ion, the free metal is oxidized:

$$4Ag + 8CN^- + O_2 + 2H_2O \rightarrow 4Ag(CN)_2^- + 4OH^- \tag{20-4}$$

Any silver present as silver chloride or silver sulfide is also dissolved, with formation of the dicyanatoargentate (I) ion, $Ag(CN)_2^-$. Other leaching operations of importance include treatment of ores containing zinc oxide with sulfuric acid. The free metal is generally recovered from solution by electrochemical means.

4. *Reduction to free metal.* For economic reasons, reduction of a metal compound (usually the oxide) to free metal must be accomplished with the cheapest possible reducing agents. For this reason many reductions are carried out with coke, which is an impure form of carbon obtained by heating coal in the absence of air. In most smelting operations, as these are called, the high temperatures necessary for reaction are obtained by combustion of coke. If the air supply is regulated properly carbon monoxide is produced, and is the substance which actually undergoes reaction with the oxide. Some typical reactions are

$$\begin{gathered} PbO + CO \rightarrow Pb + CO_2 \\ 2Fe_2O_3 + C \rightarrow CO_2 + 4FeO \\ FeO + CO \rightarrow Fe + CO_2 \end{gathered} \tag{20-5}$$

Although carbon is by far the most important reducing agent in the production of free metals, there are a number of metals for which such a reduction procedure either does not work at all, or for which the product is unsatisfactory. Vanadium pentoxide, for example, can be reduced by carbon, but the carbon in the product imparts undesirable properties to the metal. For such elements, other reducing agents must be found. Reduction with an active metal such as aluminum was developed by V. Goldschmidt as a means of preparing many metals. This method is still used extensively for a number of metals which are not needed in large quantities. Some examples of reactions of this type are

$$\begin{gathered} V_2O_5 + 5Ca \rightarrow 5CaO + 2V \\ Fe_2O_3 + 2Al \rightarrow Al_2O_3 + 2Fe \end{gathered} \tag{20-6}$$

The reaction with aluminum is called the *thermite* reaction.

In cases where the oxide of the metal does not reduce readily with carbon, it is sometimes possible to convert the oxide to another compound — usually a halide — which will react satisfactorily with a reducing agent. Titanium is a good example: Titanium dioxide reacts with chlorine in the presence of carbon to form titanium tetrachloride, $TiCl_4$:

$$TiO_2 + 2C + 2Cl_2 \rightarrow TiCl_4 + 2CO \tag{20-7}$$

Titanium tetrachloride is a volatile liquid which can be highly purified by fractional distillation. It is then reacted with sodium or magnesium in large reaction vessels under an argon atmosphere. Upon completion of the reaction, the vessel contains a hard, rock-like mixture of titanium metal and sodium chloride. This is broken up, and the sodium chloride is leached out with water. The dried residue is a relatively pure titanium sponge. The metal is not usable in this form; it must be melted under an inert atmosphere or in a vacuum, and cast into ingots.

When a very pure form of metal is desired, it is often necessary to resort to thermal decomposition of a volatile compound, usually the iodide. To produce pure vanadium metal, the crude metal is reacted with iodine vapor in a heated, evacuated chamber. In the center of this chamber is a loop of fine vanadium wire heated to incandescence by an electric current. Vanadium tri-iodide undergoes thermal decomposition on the hot filament, and vanadium metal is deposited:

$$VI_3 \rightarrow V + \tfrac{3}{2}I_2 \qquad (20\text{-}8)$$

Hydrogen is an excellent reducing agent, but for various reasons has not been much used in metallurgy. It is employed in the reduction of WO_3 to produce tungsten metal for use in light bulb filaments.

Depending upon the intended use for a metal, and upon the method used in preparing it, further purification, or refining, may be necessary. Purification by *electrolytic refining* involves using the relatively impure metal as an anode in an electrolytic bath containing a salt of the metal. The pure metal is then plated out on the cathode. The electrode reactions in the case of lead, which is often purified by this method, are

$$\text{Anode:} \quad Pb \rightarrow Pb^{+2} + 2e^-$$
$$\text{Cathode:} \quad Pb^{+2} + 2e^- \rightarrow Pb$$

The less active metals in the impure metal are not oxidized, and fall to the bottom of the cell to form a so-called anode sludge. The more active metals are oxidized at the anode but do not reduce at the cathode, and so remain in solution. Recovery of precious metals from the anode sludge is a profitable enterprise which generally makes up for the cost of the entire operation. Other refining methods include distillation, as with mercury and zinc, and zone refining, discussed in Sec. 17-5.

20-3 Solutions of Metal Ions

The formation of metal ions in aqueous solution can be viewed in terms of the following thermochemical cycle:

$$\begin{array}{ccccc} M(g) & \xrightarrow{B} & M^{+n}(g) & + & ne^-(g) \\ A\uparrow & & \downarrow C & & \downarrow D \\ M(s) & \xrightarrow{E} & M^{+n}(aq) & + & ne^-(aq) \end{array}$$

The free energy change in step E is expressed in terms of the standard electrode potential for the half-reaction. This is not, of course an absolute energy term, since the hydrogen-hydrogen ion half-cell potential is chosen as the arbitrary zero. The enthalpy and free energy values for step A are known. The enthalpies are listed for a number of metals in Table 20-1. The enthalpy change in step B corresponds closely to the ionization potential, or sum of successive ionization potentials if $n > 1$. The entropies of gaseous atoms or ions can be calculated from fundamental considerations, to yield estimates of the entropy change in step B. Step C involves hydration of the metal ion. There is no means of directly measuring the enthalpy and free energy changes for this step, but methods have been worked out for estimating the required values. The energetics of step D are not directly measurable either, but this step is presumably the same for all metal ions. We will confine our attention to comparing relative values in steps A, B, C, and E. Both steps A and B require energy, whereas step C is exothermic.

The values of $\mathcal{E}°$ for a number of metal-metal ion half-reactions are listed in Table 20-2. It is noteworthy that there is a gradual decrease in $\mathcal{E}°$ values for the formation of M^{+2} ions in the series of elements extending from Ca to Zn. The single most important factor in this trend is the gradual increase in ionization potential.

Table 20-2 Standard half-cell potentials in aqueous solution.

Half-cell reaction	$\mathcal{E}°$ (volts)
$Li(s) \rightarrow Li^{+}(aq) + e^{-}$	3.045
$Na(s) \rightarrow Na^{+}(aq) + e^{-}$	2.714
$K(s) \rightarrow K^{+}(aq) + e^{-}$	2.92
$Ca(s) \rightarrow Ca^{+2}(aq) + 2e^{-}$	2.87
$Ti(s) \rightarrow Ti^{+2}(aq) + 2e^{-}$	1.63
$V(s) \rightarrow V^{+2}(aq) + 2e^{-}$	1.18
$Cr(s) \rightarrow Cr^{+2}(aq) + 2e^{-}$	0.91
$Mn(s) \rightarrow Mn^{+2}(aq) + 2e^{-}$	1.18
$Fe(s) \rightarrow Fe^{+2}(aq) + 2e^{-}$	0.44
$Co(s) \rightarrow Co^{+2}(aq) + 2e^{-}$	0.277
$Ni(s) \rightarrow Ni^{+2}(aq) + 2e^{-}$	0.250
$Cu(s) \rightarrow Cu^{+2}(aq) + 2e^{-}$	−0.34
$Zn(s) \rightarrow Zn^{+2}(aq) + 2e^{-}$	0.76
$Al(s) \rightarrow Al^{+3}(aq) + 3e^{-}$	1.66
$Cr(s) \rightarrow Cr^{+3}(aq) + 3e^{-}$	0.74
$Fe(s) \rightarrow Fe^{+3}(aq) + 3e^{-}$	0.036
$Ce(s) \rightarrow Ce^{+3}(aq) + 3e^{-}$	2.48

Table 20-3 lists the ionization potential, or sum of ionization potentials, in units of kcal/mole, necessary to form the ion of indicated charge. At the same time the heats of hydration, corresponding to step C, are increasingly negative, which compensates for much of the variation in ionization potentials.

The heat of hydration of the metal ion is an important measure of the degree to which the metal ion interacts with surrounding water molecules. It is particularly important to note the very large differences in the values for metal ions of charges +1, +2, and +3, Table 20-3.

For ions of the same charge, the interaction with solvent increases with decreasing ionic size. The lithium ion, for example, interacts with water much more strongly than does the potassium ion. It is this factor which causes the half-cell potential for lithium to be greater than that for the other alkali metals, despite its higher ionization potential and sublimation

Table 20-3 Some thermodynamic and other properties of metal ions.

Ion	Total ionization potential (Step B)	Heat of hydration (Step C)	Ionic radius (Å)	Charge/Radius
Li^{+}	124	130	0.60	0.080
Na^{+}	118	104	0.95	0.0506
K^{+}	100	84	1.33	0.0361
Ca^{+2}	415	395	0.99	0.0970
Ti^{+2}	472	460	0.85	0.113
V^{+2}	480	—	0.80	0.120
Cr^{+2}	540	457	0.80	0.120
Mn^{+2}	530	455	0.78	0.123
Fe^{+2}	551	473	0.76	0.126
Co^{+2}	579	505	0.74	0.129
Ni^{+2}	595	519	0.73	0.131
Cu^{+2}	645	516	0.72	0.133
Zn^{+2}	631	503	0.74	0.129
Al^{+3}	1232	1135	0.52	0.277
Fe^{+3}	1257	1038	0.64	0.225
Cr^{+3}	1230	1044	0.62	0.232
La^{+3}	836	806	1.14	0.124

All energy values listed are in kcal/mole. The values in the last column on the right were obtained from the formula $n \times 4.8 \times 10^{-10}/r$, where n is the number of electronic charges, 4.8×10^{-10} is the magnitude of the electronic charge in electrostatic units, and r is the ionic radius in cm.

energy. A gradual increase in solvation energy with decreasing radius is also seen in the series extending from calcium to copper.

The major part of the solvation energy comes from the polarization of the surrounding solvent medium by the high field of the cation. Water molecules immediately adjacent to the ion are oriented so that the negative end of the molecular dipole is directed toward the cation. A sheath of tightly held water molecules, called the *primary hydration sphere*, is thus created about the ion. Other water molecules just outside the primary hydration layer may also be attracted to the cation, but the attractive forces are much weaker.

The interaction of water molecules with cations is a form of Lewis acid-base interaction, the water molecules acting as base toward the electron pair acceptor — the metal ion (p. 298). From a consideration of relative hydration energies (Table 20-3), we have seen that the binding of water molecules to the ion is favored by high charge and small size. The ratio of ionic charge to ionic radius should therefore be of value as a qualitative indication of the ability of a metal ion to coordinate water molecules. Some values of this ratio are shown in the last column on the right in Table 20-3.

One result of metal ion hydration is an increase in acidity of the hydrogens in the bound water molecules (Sec. 12-3). Accordingly, the acid dissociation constants of the hydrated metal ions might be taken also as a rough measure of the strength of hydration. Table 20-4 lists the first acid

Table 20-4 Acid dissociation constants for hydrated metal ions.

Metal ion	K_a	$\frac{\text{Charge}}{\text{Radius}}$
$Fe(H_2O)_6^{+3}$	1×10^{-3}	0.225
$Al(H_2O)_6^{+3}$	1.2×10^{-5}	0.277
$Cr(H_2O)_6^{+3}$	1.2×10^{-4}	0.232
$Sc(H_2O)_6^{+3}$	8×10^{-6}	0.178
$Pb(H_2O)_x^{+2}$	6×10^{-8}	0.114
$Cd(H_2O)_x^{+2}$	1×10^{-10}	0.099

dissociation constants for a few metal ions. Although the comparisons are rather crude, it is clear that the +3 ions have higher acid dissociation constants than do the +2 ions. There are, however, factors other than merely the charge/radius ratio involved, as evidenced by the results for the +3 ions. One factor which must be considered is that the hydroxide ion, which is formed in the primary hydration layer when ionization occurs, may interact more strongly with one metal ion than with another.

This view of solvation leads naturally to a more general notion of other bases interacting with the metal ion. In a solution of copper (II) ion in concentrated ammonia, water molecules in the primary hydration sphere are replaced by ammonia molecules:

$$Cu(H_2O)_x^{+2} + 4NH_3 \rightarrow Cu(NH_3)_4^{+2} + xH_2O$$

The change is easily observed, since the product is a much deeper shade of blue than is the hydrated copper ion. Similarly, addition of chloride ion to a copper (II) solution results in a marked color change — from blue to green — as a result of replacement of water molecules by chloride ions.

20-4 Coordination Compounds

It is possible to crystallize from solution solid compounds in which ions or molecules are retained about the central ion. Thus, in addition to the familiar hydrates of many salts, one may obtain $Cu(NH_3)_4SO_4$, K_2CuCl_4, and others. Compounds of this type were investigated in detail by Alfred Werner, beginning about 1890. Werner was able to show that many substances which at first glance possessed rather puzzling empirical formulas could be formulated as belonging to a type that he called *co-*

ordination compounds. He proposed that the metal ion in these compounds possesses both a primary and secondary valence. In the compound $Cr(NH_3)_6Cl_3$, the chromium possesses a primary valence of three, by which the chlorides are held, and a secondary valence of six, by which the ammonias are attached. The groups attached through the secondary valences are within the coordination sphere. When the compound is dissolved in water, it dissociates into three chloride ions and one complex ion:

$$Cr(NH_3)_6Cl_3 \rightarrow Cr(NH_3)_6^{+3} + 3Cl^-$$

The ammonia molecules are not immediately replaced by water, but remain attached to the metal ion.

Since Werner's work, the number of known coordination compounds has grown enormously. From many studies it appears that the bonding between the metal ion and the attached groups (called *ligands*) is in some cases simply electrostatic, in some covalent, and usually is a mixture of both electrostatic and covalent. We will attempt in this chapter to view the complexes in terms of electrostatic interactions, since this provides the simplest and most generally satisfactory approach.

If the interaction between metal ion and ligand is electrostatic in character, the attractive force is then of the ion-ion or ion-dipole variety. The potential energies in the two cases are of the forms

$$\text{ion-ion:} \qquad E = \frac{-Z^+Z^-e^2}{r}$$

$$\text{ion-dipole:} \qquad E = \frac{-Z^+e\mu}{r^2}$$

where

r = distance of separation of ion and ligand
Z^+, Z^- = number of electronic charges on cation and anion, respectively
e = magnitude of electronic charge, 4.8×10^{-10} esu
μ = magnitude of dipole moment in units of esu-cm (see p. 191)
E = potential energy (in ergs)

From these formulas, it is clear that a large value for Z^+ and a small value for r lead to increased potential energy of interaction.

The formulas are only approximately correct for the present problem, because other energy terms may also contribute. The field created by the cation is very large, and the ligands are distorted in it. This produces an added source of interaction energy.

Ligand size is also an important factor in coordination. The ligands exert a repulsive force on one another which acts to weaken ligand-ion binding. The number of ligands which can be accommodated about a metal ion is related, therefore, to the ratio of cation to anion radius (p. 139). For example, iron (II) ion accommodates six fluoride ions in FeF_6^{-4}, but only four chloride ions in $FeCl_4^{-2}$. There is not room for six chloride

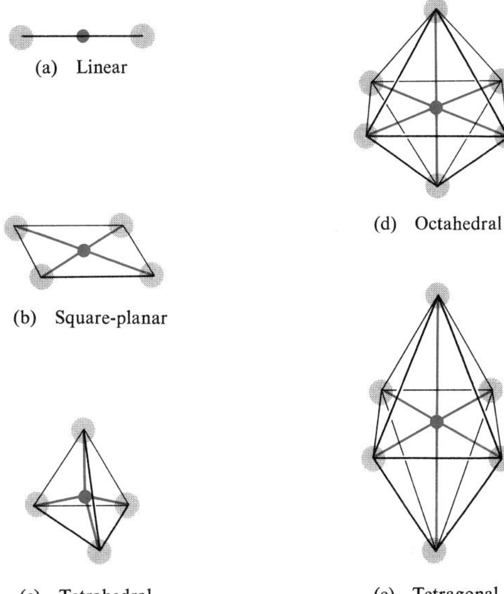

Fig. 20-7 Commonly observed coordination geometries in metal complexes. The tetragonal geometry is similar to the octahedral, except that the groups along the vertical axis are at a different distance from the central atom than are the other four.

ions about Fe^{+2} without development of unfavorable chloride-chloride repulsions.

The number of ligands surrounding the metal ion in a coordination compound is called the *coordination number*. The most commonly observed coordination number are two, four, and six. A coordination number of two results in a linear system, as shown in Fig. 20-7(a). With four ligands, either square-planar or tetrahedral arrangements are found [Fig. 20-7(b) and 7(c)]. Attachment of six ligands results in an octahedral arrangement, in which the ligands are equidistant from the metal ion [Fig. 20-7(d)]. In complexes of tetragonal symmetry two of the ligands are at a slightly greater distance along one of the axes, producing an axially distorted octahedral arrangement [Fig. 20-7(e)].

In addition to variations in the number of groups which may be coordinated, a considerable variety in the nature of the ligands is possible. In a given complex ion or coordination compound the ligands may differ chemically, and may be anions or neutral molecules. The complex as a whole may be negatively charged, neutral, or positively charged. Table 20-5 shows a series of complexes of platinum in which the complex ion —

Table 20-5 A series of chloro complexes of platinum (IV).

Compound	Number of dissociable chlorides	Molar conductivity*
$[Pt(NH_3)_6]Cl_4$	4	522.9
$[Pt(NH_3)_5Cl]Cl_3$	3	401
$[Pt(NH_3)_4Cl_2]Cl_2$	2	228
$[Pt(NH_3)_3Cl_3]Cl$	1	96.8
$[Pt(NH_3)_2Cl_4]$	0	0
$K[Pt(NH_3)Cl_5]$	0	108.5
$K_2[PtCl_6]$	0	256

*These are the conductivities of 0.1 mole per cent solutions in each case.

the portion within brackets — varies in charge from +4 to −2. The number of chloride ions in the coordination sphere can be inferred from the number of dissociated chloride ions in solution, determined by precipitation as silver chloride. Chlorides within the coordination sphere of the platinum are bound so tightly that they are not removed when silver ion is added to the solution. The number of ions formed in solution per mole of compound is reflected in the molar conductivities.

The ligands which have been mentioned up to this point are all alike in that they occupy only one coordination position. Some substances, however, are capable of coordinating to a metal ion at more than one point. One such is ethylenediamine:

$$NH_2—CH_2—CH_2—NH_2$$

It can be seen that, since there are two nitrogens in each molecule of this substance, it is possible (provided the molecule has the correct shape) that each would occupy a coordination position:

```
        H  H     H  H
         \ |     | /
   H      C——————C      H
    \    /        \    /
     N:              :N
    /  ·.          .·  \
   H     ·.      .·     H
            M
```

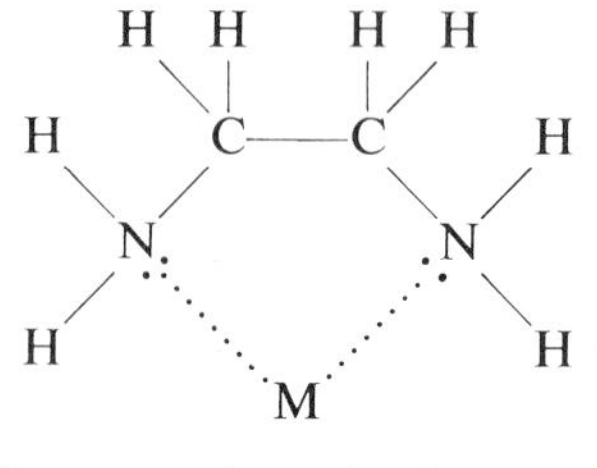

The compound does in fact coordinate in this way; it is termed a *bi*dentate ligand, as opposed to (*e.g.*) ammonia, which is a *mono*dentate ligand. The attachment of a single molecule at more than one site in the coordination sphere makes for unusually strong coordination. Thus, while one would not expect that ethylenediamine (en) would be much different from ammonia in its coordinating strength, complexes of (en) are much more stable in general than those of ammonia. Ligands are known which can coordinate to a single metal ion at three, four, five, and even six positions.

Because they coordinate so strongly, some of these have considerable practical importance.

Nomenclature — The naming of coordination compounds and complex ions is rather a complicated matter, since the name must impart so much information about the substance. The following rules are the most important, and suffice for most of the substances ordinarily encountered.

1. Name the positive ion in a coordination compound first. This may or may not be the complex ion.

2. In naming the complex ion or coordination compound, begin with the ligands; name the anions first, then the neutral molecules.

3. Indicate the number of ligands of a particular kind by the prefixes di-, tri-, tetra-, penta-, hexa- (or, in some cases, by the prefixes bis-, tris-, tetrakis-, when the use of the latin terms would cause confusion).

4. The names of anionic ligands end on -*o*: acetato, chloro, nitrato, *etc.*

5. Water as a ligand is called *aquo*. Ammonia is called ammine (note the double m in the spelling).

6. Following the names of the ligands, name the metal ion. If the complex is neutral or positively charged, the name of the element alone is sufficient. If the complex is negatively charged, the ending -*ate* is added.

7. The charge on the metal ion is denoted by Roman numerals in parentheses after the name of the metal.

Some examples:

$[Pt(NH_3)_4Cl_2]Cl_2$	Dichlorotetrammine platinum (IV) chloride
$K_2[PtCl_6]$	Potassium hexachloroplatinate (IV)
$[Cr(NH_3)_3Cl_3]$	Trichloro triammine chromium (III)
$K[Co(H_2O)(NH_3)Br_4]$	Potassium tetrabromo aquo ammine cobaltate (III)

Isomerism — In many of the complexes which contain more than one kind of ligand, the possibility of *geometrical isomerism* arises. Geometrical isomerism refers to the existence of two distinct compounds which possess the same molecular formula and the same kinds of chemical bonds, but which differ in the spatial arrangement of the atoms. For example, $[Pt(NH_3)_4Cl_2]Cl_2$ exists in two distinct forms, shown in Fig. 20-8. The

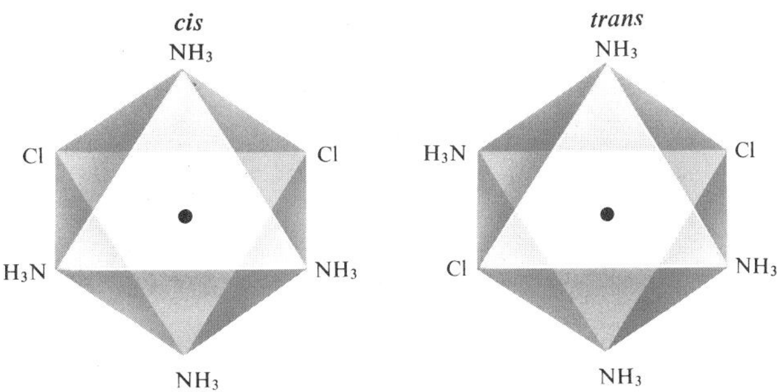

Fig. 20-8 *Cis-trans* geometrical isomerism.

chlorides in the *trans* isomer are opposite one another in the coordination sphere, whereas they occupy adjacent positions in the *cis* isomer. The two structures result in different solubility properties, and even different chemical behavior (one of the isomers may undergo a reaction at a much faster rate than the other). As the number of different ligands in a complex increases, the possibilities for geometrical isomerism increase, especially in octahedral complexes.

Coordination compounds may exhibit other types of isomerism besides geometrical. One of the more obvious is *structural, or coordination, isomerism.* Consider, for example, the complexes $[Pt(NH_3)_4Br_2]Cl_2$ and $[Pt(NH_3)_4Cl_2]Br_2$. These have the same empirical formula, but are clearly different compounds. In the first, dissolving in water would produce two equivalents of chloride ion in solution; in the second, two equivalents of bromide ion would be formed. There are also differences in other properties such as color, *etc.*

One of the most important forms of isomerism in metal ion complexes is *optical isomerism.* In order for this type of isomerism to exist, a molecule or ion must be missing certain symmetry properties: It must lack a center of symmetry or a plane of symmetry. Among the best examples of optical

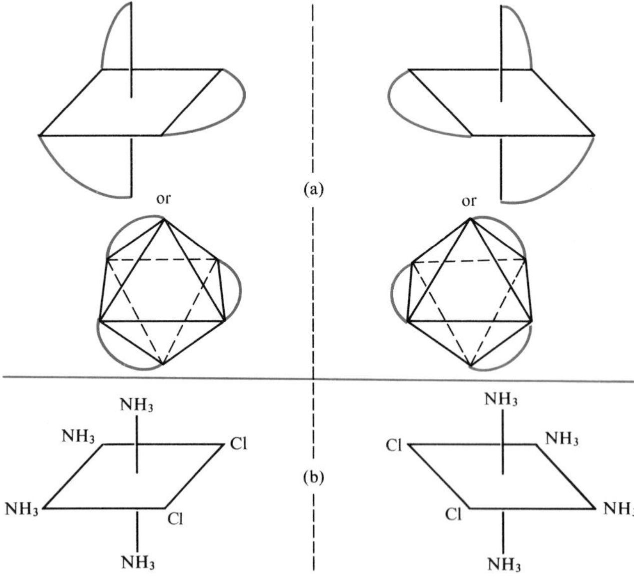

Fig. 20-9 Optical isomerism arising from the absence of a center or mirror plane of symmetry in a complex such as $Co(en)_3^{+3}$, (a). In (b), the complex does not exhibit optical isomerism, since it possesses a plane of symmetry.

isomerism are the *tris*-ethylenediamine complexes of transition metals such as cobalt (III), $Co(en)_3^{+3}$. These complexes, which are octahedral, can be represented as shown in Fig. 20-9, in which the bidentate nature of the ligand is indicated by the curved line which connects to the coordination sphere at two points. Consider the left-hand complex in Fig. 20-9(a), and imagine that the dotted line represents a mirror plane. Then the right-hand complex is the mirror image of the other. It is impossible to superimpose the two complexes on one another. By contrast, the mirror images of *cis*-$Pt(NH_3)_4Cl_2$, shown in Fig. 20-9(b), are superimposable. Note that *cis*-$Pt(NH_3)_4Cl_2$ has a plane of symmetry.

When $Co(en)_3^{+3}$ is prepared in the laboratory, equal quantities of the two optical isomers are formed. The two isomeric forms possess identical solubility properties, and are identical with respect to most physical and chemical properties. There are, however, methods for separating the optical isomers. Sometimes when a mixture of optical isomers (called a *racemic mixture*) is crystallized from solution, the two optical forms crystallize in separate crystals. It may then be possible to separate the two forms by examining the crystals under magnification, and looking for "right-handed" crystals as opposed to "left-handed" crystals. This laborious method was first employed by Pasteur. Other, more generally applicable methods are commonly used with coordination compounds.

The two optical forms, when separated, have identical properties in most respects. They differ, however, in their effect on polarized light. If a beam of plane-polarized light, such as might be generated by passing

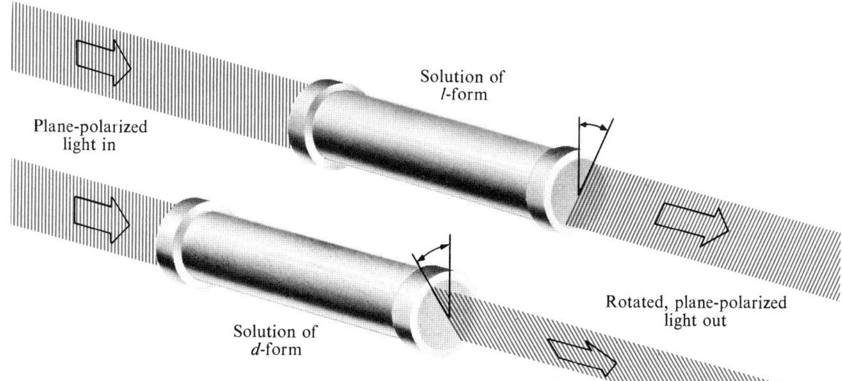

Fig. 20-10 Rotation of the plane of polarized light by solutions containing one isomer of an optically active complex. One of the optical isomers causes rotation of the plane of polarization in one direction. The other isomer causes a rotation of equal magnitude in the opposite direction.

light through a sheet of Polaroid film, is passed through a solution containing one optical isomer, the plane of the polarization is found to be twisted, or rotated (Fig. 20-10). One of the isomers rotates the plane of polarization in one direction, the other isomer rotates it an equal amount in the other direction. The two forms are designated *d*- or *l*-, indicating rotation of the light plane to the right (dextrorotatory) or left (levorotatory), respectively. It is this peculiar property with respect to light which gives rise to the name optical isomer. A racemic mixture of the two forms does not give rise to the effect, since the two optical isomers have equal and opposite effects on the light.

20-5 Ligand Field Theory

Although most of the common metal ions exhibit at least a limited degree of complex-forming ability, the major interest by far is in the complexes of the transition metals. From the values for the ratio of charge to ionic radius given in Table 20-3, it is evident that di- and trivalent transition metal ions are well suited for complex formation by this criterion. In general, compounds containing transition metal ions are colored. It has been known for a long time that the colors are related to the presence of *d* electrons in the ion. The behavior of the *d* electrons in the presence of the ligands which surround the metal ion in a complex can be understood in terms of an electrostatic model for the interaction between metal ion and ligands. We will confine the discussion for the present to the first transition series — the elements from titanium to zinc — and will limit ourselves to octahedral complexes.

The 3*d* atomic orbitals (five in number) are shown in one of their representations in Fig. 5-17. For this set of orbitals the principal quantum number is 3, the azimuthal quantum number $l = 2$ and the magnetic quantum number takes on the values -2, -1, 0, 1, and 2. In the representation shown in Fig. 5-17, two of the orbitals, d_{z^2} and $d_{x^2-y^2}$, possess regions of highest electron density along the coordinate axes x, y, and z. For the other three orbitals, d_{xy}, d_{xz}, and d_{yz}, the regions of highest electron density are found *between* the coordinate axes. For a free metal atom or ion in space, all five of the 3*d* orbitals are energetically equivalent; a single electron might be found in one as well as in another.

An octahedral complex of a metal ion with six ligands is viewed on the ionic model as an electrostatic interaction of a central, positively-charged ion with six negatively-charged particles surrounding it. The negatively-charged particles are used merely as an example; if the ligands are not charged, the interaction is of the ion-dipole type, with the negative ends of the dipole directed toward the metal ion, as in ammine or aquo complexes. The attractive force between the metal ion and the ligands tends to pull them as close in as possible. At the same time, however, the repulsive interactions between the similarly charged ligands tends to force them

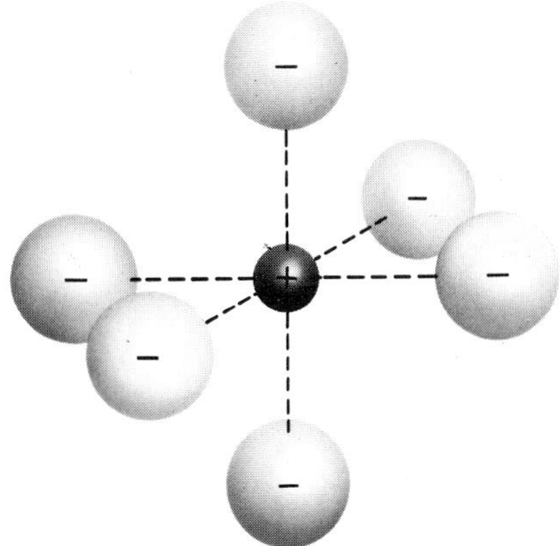

Fig. 20-11 Ligand field exerted by six ligands in an octahedral complex.

away from one another. The best compromise between the two opposing tendencies is reached when the ligands approach the metal ion from along the directions of the cartesian axes *x*, *y*, and *z*, as shown in Fig. 20-11.

The approach of the ligands has two distinct effects on the orbitals of the metal ion. The most important effect is one of raising the energies of the orbitals, especially of those orbitals which make up the outer periphery of the ion. The negatively-charged ligands exert a repulsive force on the electrons in the orbitals, thus increasing their energy. As a result, the energies of all five 3*d* orbitals are raised considerably in comparison with the free ion. The second effect, which is smaller than the first, arises from the fact that two of the 3*d* orbitals, the d_{z^2} and $d_{x^2-y^2}$, are pointed directly at the incoming ligands, whereas the other three are pointed between them. The former are therefore subject to stronger repulsion from the ligands; an electron located in either the d_{z^2} or $d_{x^2-y^2}$ orbital would be more strongly repelled by the ligand nearest it than if it were located in one of the other three 3*d* orbitals. Therefore, whereas the five 3*d* orbitals are energetically equivalent in the free metal ion, the presence of the ligands removes this equivalence; the five orbitals are now separated into two groups. The over-all effect of the ligands on the 3*d* orbitals is shown diagrammatically in Fig. 20-12.

The difference in energy between the two groups of 3*d* orbitals is known as the *ligand field splitting*. This quantity is of interest in the study of coordination compounds and complex ions, and is commonly designated by the symbol Δ.

If the energies of the five 3*d* orbitals are averaged, the resulting value is indicated by the short horizontal line labelled *s* in Fig. 20-12. The lowering of the *t* group orbitals with respect to this average value is called the *ligand field stabilization*. Since the average energy of all five *d* orbitals equals *s*, the *t* group orbitals must be 0.4Δ below *s*, and the *e* group orbitals 0.6Δ above *s* in energy.

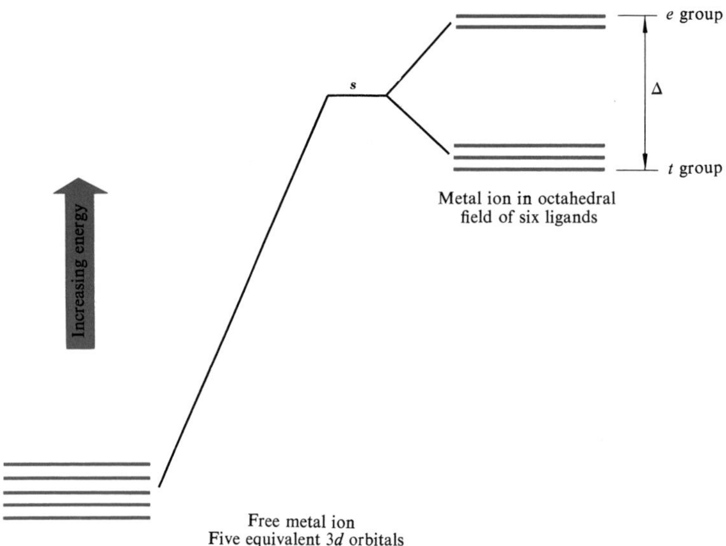

Fig. 20-12 The effect of the electrostatic field of six octahedrally placed ligands on the 3*d* energy levels in a metal ion. The energy levels of the 3*d* atomic orbitals are *all* raised by the repulsive effect of the ligands to the level indicated by *s*. There is a further splitting in the energies of the 3*d* orbitals which places them into two different groups, depending upon the orientation of the orbital with respect to the ligands.

The simplest example of a complex ion containing 3*d* electrons is the Ti^{+3} ion, with a single 3*d* electron. (It will be recalled that, in the ionization of a transition metal, the 4*s* electrons (or electron) are lost first. In forming ions of higher charge, electrons are removed from the 3*d* level.) In an octahedral complex such as $Ti(H_2O)_6^{+3}$, the electron is located in the lowest 3*d* level available; *i.e.*, it is in one of the three equivalent orbitals of the *t* group (Fig. 20-12).

A solution containing Ti^{+3} ion appears maroon in color when viewed in white light. The color is due to the solution's absorption of light energy from the visible portion of the spectrum. The wavelength absorbed corresponds to blue-green light. The removal of this light from white light results in a maroon color. The light absorption results from an electronic transition in which the single 3*d* electron is excited from the lower energy type *t* orbitals to one of the *e* group:

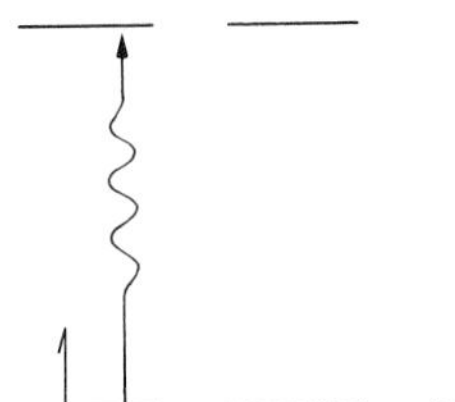

In aqueous solutions of Ti^{+3} there is an absorption at about 5000 Å wavelength, corresponding to the transition of the $3d$ electron. The energy of the photon required is then given by $\Delta E = h\nu = hc/\lambda = 6.63 \times 10^{-27} \times 3 \times 10^{10}/5 \times 10^{-5} = 3.98 \times 10^{-12}$ erg. On a molar basis, and converting to calories, we have:

$$(3.98 \times 10^{-12}\text{ erg/molecule}) \times (6.02 \times 10^{23}\text{ molecules/mole}) \times (1\text{ cal}/4.18 \times 10^{7}\text{ erg}) = 57{,}500\text{ cal/mole}$$

The ligand field stabilization energy (LFSE) for the single $3d$ electron of Ti^{+3} is then 0.4 of this, or about 23 kcal/mole. We may think of the LFSE as an extra gain in stability which comes from the ability of the t group electrons to avoid the ligands. An additional contribution to this stability comes from the fact that electrons in the t group orbitals do not shield the metal nucleus from the ligands as effectively as a spherically averaged $3d$ electron would, so there is more ligand-metal interaction, which leads to further stabilization. This extra stabilization should be manifested in thermodynamic quantities relating to formation of the hyhrated metal ion. After a few more points regarding ligand field theory have been discussed we will return to a consideration of the heats of hydration of the transition metal ions, and show that ligand field stabilization is evident in the data.

20-6 Coordinating Strength of Ligands

There are two $3d$ electrons in complexes of vanadium (III). These two electrons are in separate t group orbitals and have parallel spin (p. 106). Vanadium (II) and chromium (III) complexes possess three $3d$ electrons, each in a different t group orbital, with parallel spin. The complexes of chromium (III) have been rather extensively studied, and data on the wavelengths of light absorption for a number of the complexes are available. From these, it is possible to arrange the ligands in order of increasing energy difference between the e and t group orbitals, as shown in Fig. 20-13. The ligands that are most tightly bound to the metal ion create the strongest field, and cause the greatest splitting of the d orbital

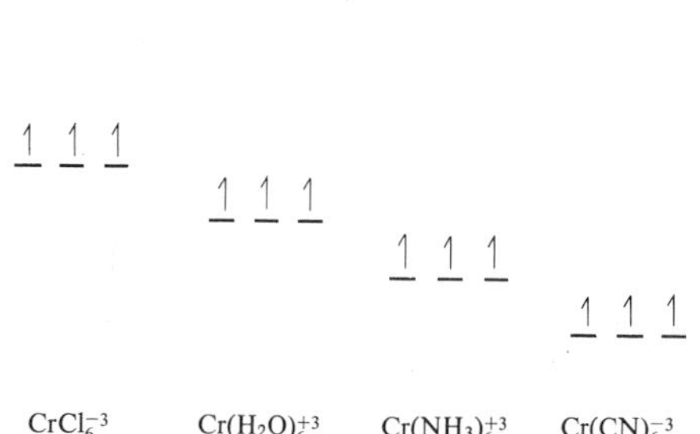

Fig. 20-13 Ligand field splitting in a number of octahedral chromium (III) complexes.

energy levels. The magnitude of the splitting, as revealed in the absorption spectra, is therefore a measure of the coordinating ability of the particular ligand. The relative order of ligand field splittings for different ligands in the chromium (III) complexes (Fig. 20-13) is observed in other transition metal complexes as well. We may conclude from this that the order of complexing strength in the ligands is $Cl^- < H_2O < NH_3 < CN^-$. By examining other complex ions, a more complete series of this type has been developed: $I^- < Br^- < Cl^- < F^- < OH^- < H_2O < NH_3 <$ en $< NO_2^- < CN^-$.

It is clear from this order that charge-charge interaction alone is not the principal factor determining the splitting, since neutral, polar molecules are among the stronger field ligands. There is good reason to believe that the simple ionic model is not complete, and that there is some covalent bonding between some of the ligands and the metal.

One might expect that the field created by a particular ligand will vary with the metal ion; this is indeed the case. In Fig. 20-14 the splittings in the hexa-aquo complexes of vanadium (II) and chromium (III) are shown. It is apparent that the water molecules create a stronger field in the chromium complex. Since the charge on the chromium ion is +3, and the charge on the vanadium ion is only +2, the water molecules are bound more tightly in the chromium complex. This comparison is valid, since the number of 3*d* electrons involved in both complexes is the same. One must be cautious, however, about making comparisons between complexes with different numbers of 3*d* electrons.

A new problem arises when we consider complexes of metal ions containing more than three 3*d* electrons. The fourth electron may go into one of the lower energy *t* group orbitals, and thus gain the ligand field

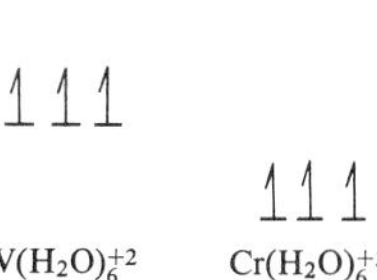

Fig. 20-14 The ligand field splitting in the hexa-aquo complex of vanadium (II) as compared with that in the iso-electronic chromium (III) complex.

stabilization energy. This requires, however, that two electrons occupy a single orbital, with opposed spins, and gives rise to a greater electron-electron repulsion than if the *e* group orbitals were utilized. We have, therefore, the following possible electron arrangements in a Mn^{+2} complex which is a $3d^5$ case:

$$\begin{array}{lcl} \underline{\uparrow}\ \underline{\uparrow}\ \ e & & \underline{\ \ }\ \underline{\ \ }\ \ e \\ & \text{or} & \\ \underline{\uparrow}\ \underline{\uparrow}\ \underline{\uparrow}\ \ t & & \underline{\uparrow\downarrow}\ \underline{\uparrow\downarrow}\ \underline{\uparrow}\ \ t \end{array}$$

Which of these two electron arrangements obtains in a particular complex of manganese (II) depends on the magnitude of the ligand field splitting. If the magnitude is great, the added electrons are found in the lower level orbitals despite the electron-electron repulsions. On the other hand, if the magnitude is not too great, the added electrons are in the higher level *e* group orbitals. The only complex of manganese (II) in which the electrons are paired in the lower level orbitals is $Mn(CN)_6^{-4}$. In all others, the ligand field is sufficiently small that the electrons are found in the upper *e* group orbitals.

Magnetic behavior — The disposition of the *d* electrons in coordination compounds is determined by measurement of the magnetic susceptibility of the substance. Ordinary substances, containing no unpaired electrons, are *diamagnetic*. This term is used to denote any substance which is less permeable than a vacuum to lines of magnetic flux. Substances which contain unpaired electrons, on the other hand, are *paramagnetic; i.e.*, they are more permeable than a vacuum to lines of magnetic flux. The magnitude of the paramagnetism is related to the number of unpaired electrons in a given weight of the sample. From a knowledge of the formula weight of a substance, it is then possible to calculate a molar magnetic susceptibility which is related in a straightforward way to the number of unpaired electrons. It is therefore possible to determine that there are five unpaired electrons per formula weight in a compound such as $Mn(H_2O)_6Cl_2$, whereas there is but one in $K_4Mn(CN)_6$.

One of the classical methods for measurement of magnetic susceptibility is the Gouy balance method. A tube containing the sample is suspended between the poles of an electromagnet. The pole faces are cut so that the magnetic field between the pole faces changes rapidly with vertical distance (Fig. 20-15). The weight of the sample is noted with the magnetic field off. The field is then applied and the weight again noted. A diamagnetic material is pushed upward and out of the field, so its apparent weight decreases, whereas a paramagnetic substance is pulled into the region of higher field, with a resultant increase in apparent weight. Calibration of the sample tube with substances of known magnetic susceptibility makes it possible to determine this quantity for an unknown material.

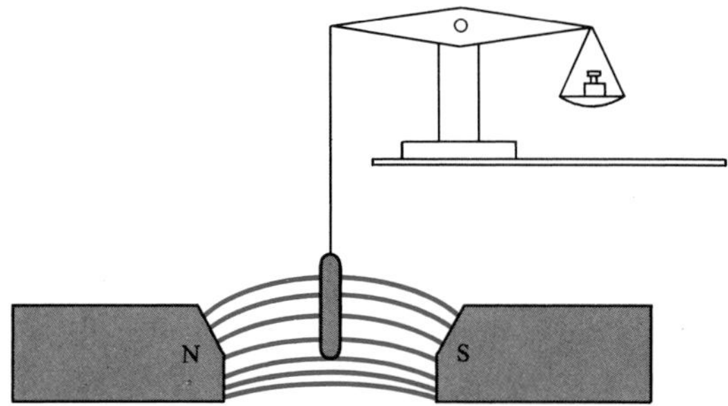

Fig. 20-15 An experimental arrangement for measuring paramagnetism.

The complexes of iron (II) and iron (III) ions afford some interesting comparisons in terms of the ligand field splitting. It is known that there are four unpaired electrons in the $Fe(H_2O)_6^{+2}$ ion and five in the $Fe(H_2O)_6^{+3}$ ion. The *d* electron configurations in these two ions must then be of the forms

$Fe(H_2O)_6^{+2}$	$Fe(H_2O)_6^{+3}$
↑ ↑	↑ ↑
↑↓ ↑ ↑	↑ ↑ ↑

The field created by the water molecules in these two complexes evidently does not cause a ligand field splitting great enough to force the electrons to pair in the lower level orbitals. Another fact worth noting about these two complexes is that the divalent ion is stable relative to the trivalent, since the electrode potential for the oxidation of iron (II) to iron (III) is negative:

$$Fe(H_2O)_6^{+2} \rightarrow Fe(H_2O)_6^{+3} + e^- \qquad \mathcal{E}^\circ = -0.77 \text{ volt}$$

With a much stronger coordinating group, such as cyanide ion, the situation is much different. In both the di- and trivalent complexes the field created by the ligands is sufficient to cause the electrons to pair in the lower energy orbitals:

$\overline{\quad}\ \overline{\quad}$ $\underline{1\!\downarrow}\ \underline{1\!\downarrow}\ \underline{1\!\downarrow}$ $Fe(CN)_6^{-4}$	$\overline{\quad}\ \overline{\quad}$ $\underline{1\!\downarrow}\ \underline{1\!\downarrow}\ \underline{1}$ $Fe(CN)_6^{-3}$

It is interesting to note that the oxidation of iron (II) to iron (III) is relatively easier in the cyanide than in the aquo complex:

$$Fe(CN)_6^{-4} \rightarrow Fe(CN)_6^{-3} + e^- \qquad \mathcal{E}^\circ = -0.36 \text{ volt}$$

The difference in the two values of half-cell potentials is related to the strengths of the two complexes. The iron (III) complexes are stronger than those of iron (II) in both cases, but the difference is much greater in the cyanide complex; perhaps the increased stability of the trivalent ion complex helps to compensate for the energy necessary to remove the electron.

The effect of complex ion formation on the relative stabilities of oxidation states is best exemplified by the case of cobalt. Cobalt (III) is not stable in solution as the aquo complex ion. The potential for the oxidation of cobalt (II) to cobalt (III) is large and negative:

$$Co(H_2O)_6^{+2} \rightarrow Co(H_2O)_6^{+3} + e^- \qquad \mathcal{E}^\circ = -1.84 \text{ volts}$$

Cobalt (III), in fact, oxidizes water with liberation of oxygen:

$$4Co(H_2O)_6^{+3} + 6H_2O \rightarrow 4Co(H_2O)_6^{+2} + 4H_3O^+ + O_2$$

It is apparent that, in the aquo complex at least, a third electron is lost from cobalt only with great difficulty.

In the presence of strong complexing agents, the relative stabilities of the (II) and (III) oxidation states are radically changed. Hexammine cobalt (III), for example, is nearly as stable as hexammine cobalt (II):

$$Co(NH_3)_6^{+2} \rightarrow Co(NH_3)_6^{+3} + e^- \qquad \mathcal{E}^\circ = -0.1 \text{ volt}$$

The explanation for this reversal in relative stability is found in the electron configurations of the complexes with strong coordinating groups. In the strong field of the ammonias, the electrons pair up as much as possible, leaving a single electron in the upper *e* level of the cobalt (II) complex. Loss of this electron leads to a much stronger field, not only because of the higher charge of the metal ion, but also because the orbitals which most strongly repel the ligands are then completely vacant. As a result, the cobalt (III) complex becomes relatively more stable.

With the above considerations in mind, we can now return to the hydrates of the divalent metals extending from Ca^{+2} to Zn^{+2}. The transition metal hydrates can be considered as hexacoordinated, octahedral complexes, of the weak ligand field type. The *d* electron configurations for this situation are shown in Table 20-6, with the net LFSE. The heats

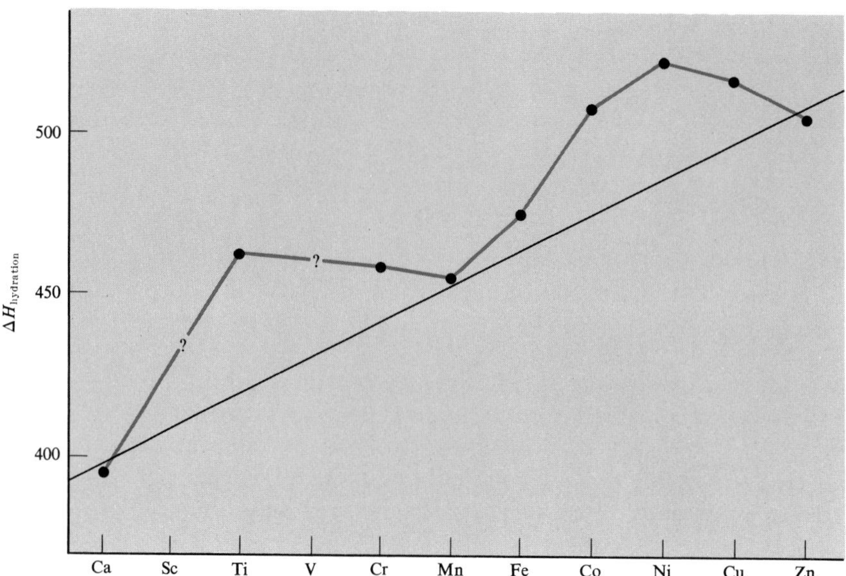

Fig. 20-16 Variation in the heat of hydration of the divalent metal ions of the first transition series, as a function of atomic number. The additional increments in ΔH above those expected from a gradual, smooth trend (black line) are ascribed to ligand field stabilization due to the field created by the coordinated water molecules.

of hydration of the divalent metal ions are graphed in Fig. 20-16. We expect that there should be a general, steady increase in the heat of hydration in passing from Ca^{+2} to Zn^{+2} because of a decrease in ionic radius with increasing atomic number. Superimposed on this general trend, however, is the LFSE. For the ions which are free of this effect, Ca^{+2}, Mn^{+2}, and Zn^{+2}, the heat of hydration *vs* atomic number data lie near the black line. *The heats of hydration of all of the other transition metal ions indicate an additional stability.* The vertical displacement of each of the other points from the dotted line is a measure of LFSE. The results are in accord with the expectations based on Table 20-6.

We have so far considered only the octahedral transition metal complexes. A great number of tetrahedral complexes (*e.g.*, $CoCl_4^{-2}$) are also known. It is interesting to compare the effect of tetrahedral coordination on the *d* orbital energy levels. The simplest way to visualize the tetrahedral complexes is to imagine that the ligands are at the alternate corners of a

Table 20-6 Ligand field stabilization energy, (LFSE), as a function of 3*d* electron configuration for divalent metal ions.

Number of 3*d* electrons	Ion	Electron configuration	LFSE
0	Ca^{+2}	O	0
1	Sc^{+2}	t^1	0.4Δ
2	Ti^{+2}	t^2	0.8Δ
3	V^{+2}	t^3	1.2Δ
4	Cr^{+2}	t^3e	0.6Δ
5	Mn^{+2}	t^3e^2	0 Δ
6	Fe^{+2}	t^4e^2	0.4Δ
7	Co^{+2}	t^5e^2	0.8Δ
8	Ni^{+2}	t^6e^2	1.2Δ
9	Cu^{+2}	t^6e^3	0.6Δ
10	Zn^{+2}	t^6e^4	0 Δ

cube. None of the *d* orbitals points directly at these corners, but a detailed examination of the geometry involved shows that the *t* group orbitals are pointed more directly at the ligands than are the *e* group orbitals. The resulting energy level diagram is thus very much like that for octahedral coordination, *except that the e and t group orbitals are reversed in relative energy.* For a given ligand, the energy separation in the two groups of orbitals is less for tetrahedral coordination than for octahedral; this means that the LFSE is smaller in tetrahedral coordination. For these reasons, tetrahedral coordination is not as common as octahedral. In those cases where strong tetrahedral complexes are formed, especially in the tetrachloro series, it may be that ionic size factors principally determine that only four chloride ions can be accommodated around the metal ion.

20-7 Reactions of Transition Metal Complexes

Transition metal complexes and complex ions may undergo a variety of reactions. Among the most common and important are substitution reactions of octahedral complexes, in which one ligand is replaced by another. In general, a complex of the form MY_6^n may undergo substitution by a new ligand Z in either of two ways:

$$MY_6 \rightarrow MY_5 + Y \qquad \text{(slow)} \qquad (20\text{-}9)$$

$$MY_5 + Z \rightarrow MY_5Z \qquad \text{(fast)} \qquad (20\text{-}10)$$

or

$$MY_6 + Z \rightarrow [Z\cdots MY_5\cdots Y] \rightarrow MY_5Z + Y \qquad (20\text{-}11)$$

We have omitted any indication of total charge in these equations, since Y and Z might be neutral or negatively charged. In the first mechanism, the rate is first order in MY_6, and is independent of the concentration of Z. In the second case, the reaction is first order in MY_6 and in Z. There are, of course, other possibilities in addition to these. For example, the solvent may be involved:

$$\begin{aligned} MY_6 + H_2O &\rightarrow MY_5(H_2O) + Y \qquad \text{(slow)} \\ MY_5(H_2O) + Z &\rightarrow MY_5Z + H_2O \end{aligned} \qquad (20\text{-}12)$$

Such possibilities can be tested for in some instances by study of MY_6 alone in the presence of solvent.

A complex that readily undergoes substitution is termed *labile;* one that does not is termed *inert.* In order to understand how chemical reactivity is related to $3d$ electron configuration, consider the bimolecular mechanism, Eq. (20-11). In the transition state the metal is seven-coordinate, since both Z and the leaving Y group must be bonded to M. Now Z will be able to intrude itself into the coordination sphere only if there is an orbital of fairly low energy available for accepting the electrons which Z has to donate. This will be the case if one of the t group orbitals is vacant. Another way of looking at it is that Z must enter the coordination sphere by forcing itself in between the Y ligands. But the t group d orbitals are directed in just these regions. If all of the t group orbitals are occupied by at least one electron, then electron-electron repulsions with the Z electrons operate to block entry of Z. We can state, therefore, that d^0, d^1, and d^2 complexes are labile, since they have a vacant t group orbital. Cr^{+3} complexes, on the other hand, which have a $3d^3$ configuration, are inert. Strong field Co^{+3} complexes, with a d^6 configuration, are even more inert. Complexes with one or more e group electrons tend not to react by bimolecular processes, since this would involve use of a high energy orbital in forming the transition state. The primary mechanism in substitution reactions is therefore the dissociative process, Eq. (20-9). In the dissociative process, Eqs. (20-9) and (20-10), the coordination number is five in the intermediate. In general, this means that the repulsion which raises the energy of the e group orbitals is lessened, although this may not always be the most important factor. As an example, $[Co(NH_3)_5Cl]^{+2}$ undergoes substitution by water with a rate constant of about 10^4 min^{-1} at room temperature in acid solution, pH $= 1$. The reaction is first order in the complex. It cannot be concluded from this, however, that the process

$$[Co(NH_3)_5Cl]^{+2} \rightarrow [Co(NH_3)_5]^{+3} + Cl^- \qquad (20\text{-}13)$$

is rate-determining. Since the reaction is carried out in water, the process might be bimolecular:

$$[Co(NH_3)_5Cl]^{+2} + H_2O \rightarrow [H_2O\cdots Co(NH_3)_5\cdots Cl]^{+2} \qquad (20\text{-}14)$$

On the basis of other evidence, however, it seems most likely that Eq. (20-13) represents the rate-determining step. The more strongly the ligand is bound to the metal, the less easily it will dissociate in the rate-determining step. Thus, as a rough rule, we might expect that weak field complexes will undergo dissociation more readily than will strong field complexes. In general, this seems to be the case. Another rule concerns the effect of metal charge. If the ligand which is expelled in the dissociation step is negatively charged, one might expect that it will more easily depart from a complex which has a low charge. Thus, complexes of chromium (III) tend to be more inert than those of chromium (II), because the +3 ion creates a stronger field, and because the potential energy of separating negatively charged ions such as Cl^- is greater for the +3 ion. Of course, this rule is not applicable if there are empty *t* group orbitals present, since this always leads to lability, regardless of charge.

Another type of reaction that has been studied is isomerization. For example, *cis*-$Co(NH_3)_4Cl_2^+$ isomerizes in aqueous solution to the *trans*-form. It is probable that the reaction proceeds through formation of a five-coordinate intermediate, which can recombine with chloride ion to form either the *cis*- or *trans*- form (Fig. 20-17). The *trans*- form is thermodynamically more stable, and is the predominant product.

$$\begin{array}{c} cis\text{-}[Co(en)_2Cl_2]^+ \rightleftharpoons [Co(en)_2Cl]^{+2} + Cl^- \\ \downarrow\uparrow \\ trans\text{-}[Co(en)_2Cl_2]^+ \end{array} \qquad (20\text{-}15)$$

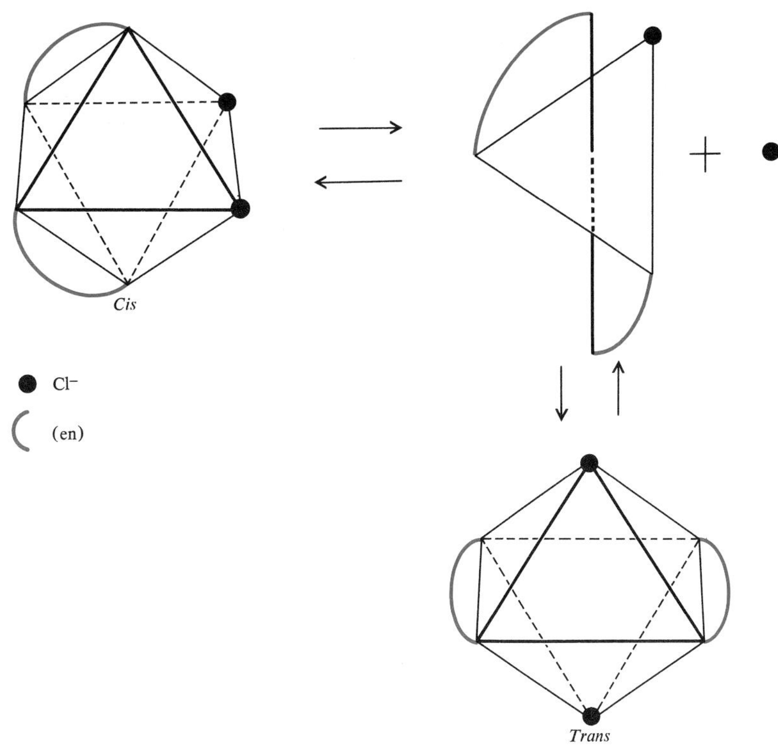

Fig. 20-17. A schematic illustration of the *cis-trans* isomerization of the $Co(NH_3)_4Cl_2^+$ ion. It is probable that the isomerization proceeds through a five-coordinate intermediate.

Oxidation-reduction reactions of complexes have been widely studied. The electron-transfer reaction is one of the simplest reaction types which one can have, since there is no net over-all change in concentrations of the different chemical species with time. For example, the reaction

$$\overset{*}{Fe}(CN)_6^{-3} + Fe(CN)_6^{-4} \rightarrow \overset{*}{Fe}(CN)_6^{-4} + Fe(CN)_6^{-3} \qquad (20\text{-}16)$$

involves simply the over-all transfer of an electron from the iron (II) to the iron (III) species. Since the coordination spheres around the two forms of iron are the same, there is no net change in the concentrations of the two chemical species present. It is believed that the reaction in Eq. (20-16) proceeds by a direct electron transfer. In other cases, there may be an atom transfer. A very well known case of this, first investigated at the University of Chicago by Professor Henry Taube, involves the electron transfer between chromium (II) and chromium (III) in aqueous solution. It was found that electron transfer between $Cr(H_2O)_6^{+2}$ and $Cr(H_2O)_6^{+3}$ is quite slow. On the other hand, transfer between $Cr(H_2O)_6^{+2}$ and $[Cr(H_2O)_5Cl]^{+2}$ is very fast. The latter complex is a chromium (III) complex, and is inert. Chromium (II) complexes are much more labile. If there were merely an electron transfer, then the reaction would proceed as follows:

$$\begin{array}{ll} Cr(H_2O)_6^{+2} + Cr(H_2O)_5Cl^{+2} \rightarrow Cr(H_2O)_6^{+3} + Cr(H_2O)_5Cl^{+} & \\ Cr(H_2O)_5Cl^{+} + H_2O \rightarrow Cr(H_2O)_6^{+2} + Cl^{-} \quad \text{(rapid)} & \end{array} \qquad (20\text{-}17)$$

That is, the chloride ion would be displaced from the chromium (II) chloro complex as rapidly as formed. But it was found that the chloride ion stays with the chromium (III). The transition state appears to involve a chlorine bridge:

$$(H_2O)_5\overset{*}{Cr}Cl^{+2} + (H_2O)Cr(H_2O)_5^{+2} \rightleftharpoons [(H_2O)_5\overset{*}{Cr} \overset{e^-}{\curvearrowright} :Cl: \; Cr(H_2O)_5]^{+4} + H_2O$$
$$\downarrow \uparrow$$
$$\rightleftharpoons \overset{*}{Cr}(H_2O)_6^{+2} + Cr(H_2O)_5Cl^{+2} \qquad (20\text{-}18)$$

The transition state, in brackets, can be thought of as a chloride ion bridge between chromium (II) and chromium (III) ions. When the bridge is symmetrical, the electron is free to jump back and forth rapidly, so that the chromiums appear to be equivalent. When the intermediate breaks up, the chloride ion goes one way, the electron goes the other. The over-all net effect is as though there were a transfer of a chlorine atom. Numerous other bridged intermediates of this type are known.

Since the chemical species do not appear to undergo change in concentration with time in Eqs. (20-16) and (20-18), some rather special techniques are necessary to follow the course of these and other electron transfer reactions. In some cases it is possible to use radioactive isotopes of the metals as tracers. The system described by Eq. (20-16), for example, was studied by using ^{55}Fe as tracer (see Sec. 23-2).

Suggested Readings

Bailar, J. C. Jr. (Ed.), *The Chemistry of Coordination Compounds*, New York, N. Y.: Reinhold Publishing Corporation, 1956, Chapters 1 and 3.

Busch, D. H., "The Coordinate Bond and the Nature of Complex Inorganic Compounds," *J. Chem. Educ.*, Vol. 33, 1956, pp. 376, 498.

Graddon, D., *Introduction to Coordination Chemistry*, New York, N. Y.: Pergamon Press, 1961.

Lewis, J. and R. S. Nyholm, "Modern Inorganic Chemistry," *Chem. Eng. News*, Dec. 4., 1961, p. 102.

Murmann, R. K., *Inorganic Complex Compounds*, New York, N. Y.: Reinhold Publishing Corporation, 1964.

Exercises

20-1. Write the electronic structures of each of the following:

(a) Al.
(b) Ca.
(c) Ti^{+2}.
(d) Co^{+3}.
(e) Sn^{+2}.
(f) Pt^{+4}.

20-2. What type of alloy would one expect for each of the following pairs of elements?

(a) Tungsten-carbon.
(b) Hafnium-zirconium.
(c) Molybdenum-aluminum.
(d) Palladium-hydrogen.

20-3. Give a definition of each of the following terms and an example of each:

(a) Molecular orbital.
(b) Coordination number.
(c) Calcination.
(d) Refining.
(e) Ore concentration.
(f) Intermetallic compound.

20-4. The entropy of hydration can be defined as the difference in entropy of a metal ion in the gas phase and in solution. Some entropies of hydration are as follows (units of cal/deg-mole):

Ion	ΔS_{hyd}	Ion	ΔS_{hyd}
Li^+	−32	Zn^{+2}	−70
Na^+	−24	Al^{+3}	−121
Mg^{+2}	−70	Fe^{+3}	−121
Ca^{+2}	−57	La^{+3}	−97

Explain why these values are negative. Explain the variations in ΔS among the metal ions in terms of properties of the ions.

20-5. Calculate the pH of a $0.01M$ solution of aluminum nitrate.

20-6. How many distinct compounds with formula $Pt(NH_3)_2Cl_4$ are possible? How many of the formula $Pt(NH_3)_4Cl_3Br$? Name all of the compounds you require in answering this problem.

20-7. What weight of each of the following reducing agents — assuming complete reaction — is required for reduction of 100 g of TiO_2?

(a) Aluminum. (b) Hydrogen. (c) Sodium.

20-8. For a large number of coordinating ligands, the tendencies of metal ions toward complex formation are in the following order:
(a) $Li^+ > Na^+ > K^+$.
(b) $La^{+3} > Ca^{+2} > Li^+$.
(c) $Ni^{+2} > Co^{+2} > Zn^{+2} > Fe^{+2}$.
Account for the observed trend in each case.

20-9. Which complex ion would you expect to exhibit the greater acidity in water, $[Co(NH_3)_5H_2O]^{+3}$ or $[Co(NH_3)_3Cl_2H_2O]^+$? Explain.

20-10. The oxalate ion has the structure $\left[\begin{array}{c} O \quad\quad O \\ \diagdown\!\!\diagdown \quad \diagup\!\!\diagup \\ C{-}C \\ \diagup \quad\quad \diagdown \\ O \quad\quad O \end{array}\right]^{-2}$. Draw a structural diagram of the *tris*-oxalatoferrate (III) ion. What form of isomerism do you expect this complex to exhibit?

20-11. Name each of the following complexes:
(a) $[Co(NH_3)_5H_2O]Br_3$.
(b) $[Ir(en)_2NH_3Br]Cl_2$.
(c) $K_2[ZnCl_4]$.
In each case, indicate the isomerism of which the complex is capable.

20-12. Draw an energy diagram showing the 3*d* electron disposition in the complex ion $Ti(H_2O)_6^{+2}$. What type of magnetic behavior should this compound exhibit? Would you expect the ligand field splitting in this ion to be greater or less than that in $V(H_2O)_6^{+3}$? Explain.

20-13. Compare the 3*d* electron dispositions in the octahedral complexes of chromium (II) and chromium (III).

20-14. Using the information given in the text about chromium and cobalt complexes, indicate what reasons we have for believing that cyanide ion is a stronger coordinating group than water.

20-15. On the basis of the ratio of charge to radius, which would be expected to form the stronger complexes with a given ligand, Al^{+3} or Cr^{+3}? Aluminum is in fact much more limited in its ability to form complexes than chromium. Suggest a plausible explanation for the discrepancy, if any, between the expectations and the facts.

20-16. From the half-cell potentials

$$V(H_2O)_6^{+2} \rightarrow V(H_2O)_6^{+3} + e^- \qquad \mathcal{E}° = 0.255 \text{ volt}$$
$$Cr(H_2O)_6^{+2} \rightarrow Cr(H_2O)_6^{+3} + e^- \qquad \mathcal{E}° = 0.41 \text{ volt}$$

calculate the equilibrium constant at 25°C for the reaction

$$V(H_2O)_6^{+2} + Cr(H_2O)_6^{+3} \rightarrow Cr(H_2O)_6^{+2} + V(H_2O)_6^{+3}$$

20-17. The equilibrium constants for the successive addition of ammonia to copper (II) ion in aqueous medium are: $K_1 = 1.4 \times 10^4$, $K_2 = 3.2 \times 10^3$, $K_3 = 7.8 \times 10^2$, $K_4 = 1.3 \times 10^2$. Example: $Cu(H_2O)_4^{+2}$ +

$NH_3 \rightarrow Cu(NH_3)(H_2O)_3^{+2} + H_2O$; $K_1 = 1.4 \times 10^4$. Calculate K for the reaction

$$Cu(H_2O)_4^{+2} + 4NH_3 \rightarrow Cu(NH_3)_4^{+2} + 4H_2O$$

Under what conditions could one use the K for this last reaction to calculate the concentration of $Cu(H_2O)_4^{+2}$? Explain.

20-18. From the spectral evidence it appears that (en) and ammonia create ligand field splittings of about the same magnitude in coordination compounds. Nevertheless, the equilibrium constants for formation of (en) complexes are much greater than for ammonia complexes. For example:

$$Cu(H_2O)_4^{+2} + (en) \rightarrow Cu(en)(H_2O)_2^{+2} + 2H_2O \qquad K = 3.5 \times 10^{10}$$

$$Cu(H_2O)_4^{+2} + 2NH_3 \rightarrow Cu(NH_3)_2(H_2O)_2^{+2} + 2H_2O \qquad K = 4.2 \times 10^7$$

Suggest an explanation for these results.

20-19. Summarize the evidence presented in this chapter and in other parts of the book that lithium ion is more strongly solvated in aqueous solution than the other alkali metal ions.

20-20. Werner and one of his students found that, when the compound $[Co(NH_3)_4Br_2]Br$ was placed in water, the molar conductivity changed with time as follows:

Time (minutes)*	Molar conductivity
0	191
5	288
10	326
15	341
20	348
40	364

*The first measurement was made as soon after dissolving the compound as possible.

Propose an explanation for the observed results. Write balanced chemical equations as part of the explanation. (It will be helpful to refer to the conductivity data in Table 20-5.)

20-21. As mentioned in the text, aqueous Ti^{+3} ion exhibits an absorption at 5000 Å due to a d electron transition from the t to the e group orbitals. What change would occur in the energy of this transition if the solution containing Ti^{+3} were subjected to an extremely high pressure? Explain.

20-22. What are the meanings of the terms, "strong-field complex," and "weak-field complex?" Can these terms be applied meaningfully to all transition metal complexes?

20-23. Which of the following complexes would you expect to be labile, and which to be inert?

(a) $V(H_2O)_4Cl_2^+$.
(b) $Mn(CN)_6^{-3}$.
(c) $CrCl_6^{-3}$.
(d) $Re(H_2O)_6^{+3}$.
(e) $Ti(NH_3)_6^{+2}$.

20-24. The complex $Co(NH_3)_4Cl_2^+$ undergoes substitution of both chlorides by water to yield finally, $Co(NH_3)_4(H_2O)_2^{+3}$. The rate at which the second chloride is replaced is, however, about 100 times slower than that of the first chloride. Explain.

20-25. $[Co(en)_2Cl_2]^+$ undergoes chloride ion exchange with radioactive chloride ion in water. The rate of exchange was found to be a first order process, independent of the chloride ion concentration. Write an appropriate equation or equations to describe the mechanism.

20-26. The complex *cis*-$Co(en)_2Cl_2^+$ is capable of existing in optically active forms. Show that this is the case by drawing pictures of the mirror image complexes. Suppose that the *d* form of the complex is isolated, and placed in aqueous solution in the presence of chloride ion. On the basis of the preceding problem, would you expect it to remain optically active? Aside from the exchange with chloride ion of the solution, what possibilities for isomerization are there? [*Hint:* There may be both optical and geometrical isomerization.] Draw some figures to illustrate your answer.

20-27. FeF_6^{-3} is labile toward substitution, whereas $Fe(CN)_6^{-3}$ is inert. How do you explain this?

Chapter 21

Chemistry of the Metallic Elements

21-1 Alkali Metals

The elements of Group IA are commonly referred to as the *alkali metals*. They are, as a group, the most metallic of the elements. They are found in nature as the +1 ion; their geochemistry is determined by the relative sizes of the ions. Table 21-1 lists a number of properties of the elements and the +1 ions. Because of its small size, lithium, Li^+, is not found in

Table 21-1 Comparative properties of the Group IA elements.

Element	Density	Melting point (°C)	Boiling point (°C)	I_1(e-v)	r^+	$\Delta H_{hydration}$	$\mathcal{E}^\circ$
Li Lithium	0.53	179	1317	5.39	0.60	130	3.04
Na Sodium	0.97	98	883	5.14	0.95	104	2.71
K Potassium	0.87	64	760	4.34	1.33	84	2.92
Rb Rubidium	1.53	38	668	4.18	1.48	78	2.92
Cs Cesium	1.87	28	705	3.89	1.69	70	3.08

association with the other alkali metals in minerals. It is found in a few aluminosilicate minerals (*e.g.*, Spodumene, $LiAlSi_2O_6$), which are commercial sources of the element. It is also present in the brines obtained from Searles Lake, California.

Sodium is found in wide distribution on the earth's crust. It is present in a large number of complex aluminosilicate minerals, in the oceans in the form of soluble sodium salts, and in a limited number of sodium halide deposits.

Potassium ion is sufficiently different in radius from Na^+ so that these two elements also do not tend to occur together in nature. Potassium is widely distributed in a variety of aluminosilicate minerals, particularly the *feldspars*, which have empirical formulas such as $KAlSi_3O_8$. These may be considered derivatives of silica, SiO_2, in which an Al^{+3} ion has substituted for a Si^{+4}. Charge balance is maintained by inclusion of K^+ in interstices of the close-packed oxygen lattice.

The alkali metal ions exhibit only the $+1$ oxidation state in aqueous solution. They are relatively poor complex formers, as one would predict from their low charge and rather large radius. Lithium ion is the most strongly hydrated of the group (Table 20-3); as a further consequence of this hydration, solid lithium salts have a stronger tendency to form hydrates than the corresponding salts of the other alkali metal ions.

It is apparent from the physical properties of the metals that the strength of metal bonding decreases regularly with increasing atomic weight. This is evident also in the ease of handling of the metals. Lithium is much harder than sodium, although it can be cut with a sharp knife. It can be handled with safety as long as water is absent. Lithium reacts vigorously with water, but the hydrogen produced is not ignited by the reaction. Sodium, on the other hand, is dangerously reactive with water, and there is almost always a fire. Potassium is much more reactive still, and may inflame simply upon contact with air.

In spite of these differences in physical properties and reactivity, the standard electrode potential for oxidation of lithium is the larger than for any of the other Group I elements except cesium. This is the result of the much greater degree of interaction of the Li^+ ion with the solvent, as a result of its smaller ionic radius (Sec. 20-3).

21-2 Alkaline Earth Elements

The Group IIA elements are commonly referred to as the *alkaline earths.* With the exception of beryllium, all of these elements are fairly abundant. Calcium is generally regarded as the fifth, and magnesium, the eighth, most abundant element in the earth's crust.

A number of properties of the Group IIA elements and their $+2$ ions are listed in Table 21-2. The trends in this group parallel those in the alkali metal series, Group IA. The stronger bonding of the metals as compared with that of the Group I elements is evident from the higher densities, and melting points.

Table 21-2 Some properties of the Group IIA elements and their ions.

Element	Density	Melting point (°C)	I_1	I_2	r^{+2}	$\Delta H_{\text{hydration}}$	$\mathcal{E}°$
Be Beryllium	1.85	1284	9.32	18.2	0.31	580	1.85
Mg Magnesium	1.74	651	7.64	15.0	0.65	474	2.37
Ca Calcium	1.54	851	6.11	11.9	0.99	395	2.87
Sr Strontium	2.6	770	5.69	11.0	1.13	354	2.89
Ba Barium	3.5	710	5.21	10.0	1.35	326	2.90
Ra Radium	5	960	5.28	10.1			2.92

The elements of the alkaline earth group are more active in forming complexes than are the alkali metal ions. Calcium and magnesium, particularly, form a number of complexes with strongly coordinating groups. Calcium ion is frequently found in fairly high concentration in water taken from underground water sources. Such water is called "hard water." When ordinary soaps that contain the sodium salts of long-chain fatty acids are used in hard water, insoluble calcium salts are formed. The precipitate formation can be avoided by adding to the water a *sequestering agent*, which complexes the calcium, thus lowering the concentration of free calcium ion. One such agent is ethylenediamine tetraacetic acid (EDTA), a hexadentate ligand (Fig. 21-1).

Calcium and magnesium are both found in plant and animal matter; the functions they perform in these systems are related in some instances to their ability to form complexes.

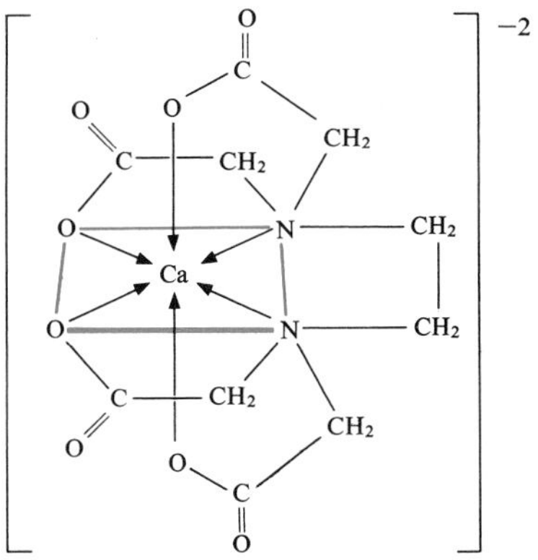

Fig. 21-1 The ethylenediamine tetraacetic acid (EDTA) complex of calcium ion.

Just as lithium is different from the other alkali metals in its properties, beryllium is quite different from the other alkaline earths. There is an interesting diagonal relationship rule which says that the lightest element in each of the first three groups behaves like the second element in the next group, *i.e.*, like the element that is located diagonally down and to the right in the periodic table. Therefore, in some respects, lithium resembles magnesium, beryllium resembles aluminum, and boron resembles silicon. To give just one example, the chlorides of all of the alkaline earth elements except beryllium behave as ionic substances. $BeCl_2$, however, is only slightly conducting when molten, and can be evaporated much more readily than the other alkaline earth chlorides. In these respects, it resembles $AlCl_3$. Because of its small size and +2 charge, Be^{+2} interacts very strongly with water molecules, and forms a rather strongly acidic species in solution. The chemistry of beryllium might have received much more attention than it has if the element were not so poisonous. Not only the free element, but most of its compounds, are highly toxic and must be handled with great care.

21-3 Group III Metals

We will consider in this section only the chemistry of the Group IIIA metals, Al, Ga, In, and Tl. Boron we have discussed briefly as a nonmetallic element. Aluminum is by far the most abundant of these elements. By contrast, gallium is not only rather rare, it is not found concentrated in any special minerals. A small amount of gallium is found in *bauxite*, a hydrated aluminum oxide, which is the chief commercial source of aluminum. The quantities of gallium recovered from this source, however, are small; not more than about 10 grams per ton of bauxite ore. Gallium is also found in small quantities in certain sulfide ores, notably zinc blende, ZnS. It is recovered from the flue dusts when the sulfide ore is roasted. Indium and thallium are also recovered from this source; indium is even scarcer than gallium, and about one-tenth as abundant as thallium.

Aluminum is very abundantly distributed in nature in the form of aluminosilicate minerals, but no full scale processes for winning the metal from these sources exist at present. The methods now in use all depend upon bauxite as the source. The major difficulty in obtaining pure aluminum metal is that it is quite high in the activity series; most other metals which are present as impurities in the ore to be reduced will therefore also be reduced with aluminum. For this reason, chemical purification of the bauxite ore before it is reduced is a necessity.

The purification step involves treating bauxite with sodium hydroxide solution to form the complex hydroxyaluminate (III) ion:

$$Al_2O_3 \cdot xH_2O + nOH^- \rightarrow [Al(OH)_4(H_2O)_2]^{-1} \qquad (21\text{-}1)$$

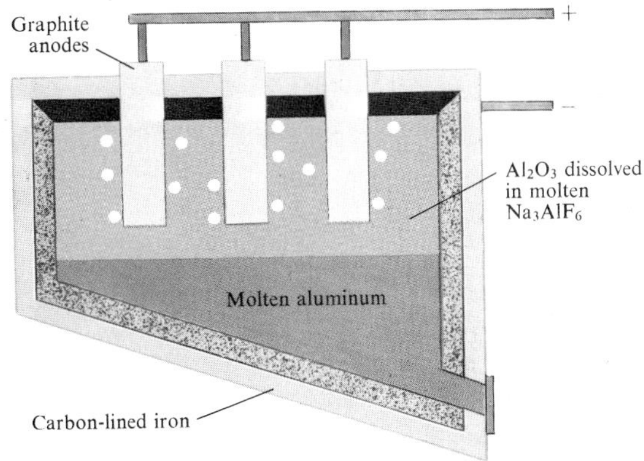

Fig. 21-2 Electrolytic preparation of aluminum.

The resulting solution is filtered to remove hydrous iron (III) oxide, $Fe_2O_3 \cdot xH_2O$, the major impurity. Dilution with water reverses the above reaction, precipitating pure hydrated aluminum hydroxide. The precipitate is calcined to remove all water, leaving aluminum oxide.

Aluminum cannot be obtained by electrolysis of an aqueous solution containing the aluminum ion because it is too high in the activity series; $\mathcal{E}°$ for reaction of aluminum ion with water is $+1.67$ volts. Since reduction by conventional chemical agents is also precluded by its high activity, electrolytic reduction in a non-aqueous medium is required. In the universally used Hall process, aluminum oxide is dissolved in fused *cryolite*, sodium hexafluoroaluminate (III), Na_3AlF_6. A diagram of the type of cell employed is shown in Fig. 21-2. The walls of the cell form the cathode at which the metal ion is reduced. The anodes are carbon. Ideally, the half-cell reactions would be

$$\begin{aligned} 2Al^{+3} + 6e^- &\rightarrow 2Al \\ 3O^{-2} &\rightarrow \tfrac{3}{2}O_2 + 6e^- \end{aligned} \qquad (21\text{-}2)$$

In practice, the carbon anodes are consumed, indicating that some of the oxidation reaction involves a process such as

$$O^{-2} + C \rightarrow CO + 2e^-$$

or

$$2O^{-2} + C \rightarrow CO_2 + 4e^- \qquad (21\text{-}3)$$

The temperature in the cell during electrolysis is quite high; the metal collects in the molten state, and can be periodically drawn off. It is 99.7 per cent pure, sufficiently pure for most purposes, so that a refining step is not necessary.

If the Group III metals formed simple, monomeric compounds with other elements such as the halogens there would be only six electrons in the valence shell of the metal. As a result, these compounds are strong

Lewis acids, and interact with a variety of electron pair donors. For example, aluminum chloride is found to form a stable adduct with trimethylamine, $AlCl_3 \cdot N(CH_3)_3$, and forms a chloro complex, $AlCl_4^-$. The trihalides of Al, Ga, and In with Cl, Br, and I exist in the vapor state and in non-coordinating solvents as dimers, which involve a halogen bridge; *e.g.*,

```
 ..      . .        ..
:Cl      .Cl.       Cl:
 ..\    /    \     / ..
    Al        Al
 ../   \    /   \ ..
:Cl     .Cl.      Cl:
 ..      . .       ..
```

In this structure, each metal atom makes use of all valence shell orbitals in bonding. The bridging chlorines may be thought of as forming a normal single bond to one aluminum atom, and acting as an electron pair donor to the other. The bridge structure is actually symmetric, so there is no real distinction between the two kinds of bonds.

Aluminum is the only one of the four elements in this group which does not possess a completed subshell of *d* electrons of major quantum number one less than that of the valence electrons. We have already seen that in the course of filling the *d* levels there is a gradual reduction in radius and increase in ionization potential (Sec. 6-9). A comparison of the properties of the elements, Table 21-3, shows that gallium, by these effects, is made more like aluminum in some respects and less like it in other respects. The ionic radii of Ga^{+3} and Al^{+3} are more alike than they would otherwise be. On the other hand, the $\mathcal{E}°$ values for the other elements are much lower than that for aluminum.

Table 21-3 Properties of the Group III metals and their ions.

Element	Density	Melting point (°C)	Boiling point (°C)	I_1	I_2	I_3	r^{+3}	$\mathcal{E}°$
Al Aluminum	2.70	660	2270	6.0	18.8	28.4	0.50	1.66
Ga Gallium	5.92	30	2070	6.0	20.5	30.7	0.62	0.53
In Indium	7.30	157	1450	5.8	18.9	28.0	0.81	0.34
Tl Thallium	11.8	304	1457	6.1	20.4	29.8	0.95	−0.74

This particular group of elements illustrates very well the increase in stability of a lower oxidation state with increase in atomic weight in a group. Aluminum has not been satisfactorily identified in anything other than a +3 oxidation state. At the other extreme, thallium is much more stable as Tl^+ than as Tl^{+3}. The $\mathcal{E}°$ value for the half-cell reaction $Tl^+ \rightarrow Tl^{+3} + 2e^-$ is -1.28 volts. Tl^+ closely resembles the alkali metal ions in its salt-forming properties. Gallium and indium also exhibit a tendency to form a +1 oxidation state under some circumstances. For example, a compound of empirical formula $GaCl_2$ has been known for some time. The composition of the compound has been shown to be $Ga^+GaCl_4^-$. That is, it is a mixed compound containing +1 and +3 gallium ions.

The +3 ions of aluminum, gallium, and indium all exhibit amphoteric behavior in aqueous solution. That is, the hydrous oxide which precipitates when an acid solution of the metal ion is neutralized redissolves upon addition of excess alkali (see Sec. 12-3).

21-4 Group IV Metals

Among the Group IVA elements, only tin and lead can be considered to be metallic. Tin is found highly concentrated in a few places in the form of complex sulfide ores, or as the mineral *cassiterite*, SnO_2. Lead is mined in the form of the mineral *galena*, PbS, *cerussite*, $PbCO_3$, or *anglesite*, $PbSO_4$. Their close relationship to the non-metals is seen in the fact that both SnO_2 and PbO_2 are distinctly acid in character. SnO_2 is insoluble in acid solution, but dissolves in base to form $Sn(OH)_6^{-2}$, The +4 oxidation state in aqueous solution is the most stable for tin, but lead is most stable as the +2 ion.

The behavior of tin toward the halogens is interesting. SnF_4 is a salt-like solid which sublimes only upon heating to high temperature, on the order of 700°C. $SnCl_4$, on the other hand, is a volatile liquid, boiling at 114°C. It reacts with water with great ease, and fumes vigorously upon contact with moist air. The $SnCl_4$ found on the reagent shelf in the undergraduate laboratory is the hexahydrate, $SnCl_4 \cdot 6H_2O$. Both SnF_4 and $SnCl_4$ form complex salts (*e.g.*, K_2SnF_6, $(NH_4)_2SnCl_6$), in which tin is octahedrally surrounded by halide ions.

Table 21-4 Properties of the Group IV metals.

Element	Density	Melting point (°C)	r^{+4}	$\mathcal{E}°$ ($M \rightarrow M^{+2}$)	$\mathcal{E}°$ ($M \rightarrow M^{+4}$)
Sn Tin	7.3	232	0.71	0.14	−0.007
Pb Lead	11.3	328	0.84	0.13	−0.66*
Ti Titanium	4.50	1670	0.68	0.83	0.89
Zr Zirconium	6.4	1855	0.74	—	1.53
Hf Hafnium	13.3	2200	0.75	—	1.70

*Product is $PbO_2(s)$

Titanium, zirconium, and hafnium are elements of Group IVB. A few characteristic properties of the elements are listed in Table 21-4. Note that hafnium has essentially the same radius as zirconium, even though it is in the next row of the table, and has an atomic number 32 higher than that of Zr. This effect is due to insertion of the 14 rare earth elements between hafnium and barium. In the elements 57 through 71, electrons are added to the 4*f* orbitals. These electrons do not completely screen the nucleus from the outer 6*s* and 6*p* electrons, so there is a gradual increase in effective nuclear charge in the rare earth series. When all fourteen 4*f* electrons have been added, the effect is just such that it cancels the effect of an increase in major quantum number. Hafnium and zirconium thus have about the same atomic radii. The equality in size of the second and third row transition elements in each group is commonly said to be due to the "lanthanide contraction," since the rare earth elements are often referred to as lanthanide elements.

In their highest oxidation state, +4, the Group IVB elements are devoid of *d* electrons in the shell just below the valence shell. By contrast, both Sn^{+4} and Pb^{+4} have these *d* levels filled. Titanium is found in nature in the form of TiO_2, and as titanate compounds ($CaTiO_3$, $FeTiO_3$, *etc.*). The metallurgy of this element is discussed briefly on page 497.

Although it is still not produced on a very large scale, zirconium has found new uses in recent years. Perhaps the most important one is based on its very low neutron capture cross-section, which makes it desirable as a structural material in nuclear reactors. The element is similar chemically to titanium; it occurs principally as the dioxide and the sulfate. At first, it might appear that the same metallurgical procedures employed in producing titanium would also be applicable to zirconium; *i.e.*, conversion to the tetrachloride, followed by reduction with an active metal (p. 497). The matter is greatly complicated, however, by the fact that zirconium is always accompanied in its ores by 0.5–3 per cent hafnium. In contrast to zirconium, hafnium has a high neutron capture cross-section; it must therefore be removed as completely as possible. Hafnium, which is in the same group of the periodic table as zirconium, also has the same atomic radius, so the two elements are very similar in their chemical and physical properties.

One method currently in use for producing zirconium involves separating zirconium dioxide from hafnium dioxide by a solvent extraction procedure which takes advantage of slightly differing solubility properties of the two compounds. The pure zirconium dioxide obtained is then converted to the tetrachloride by reaction with chlorine and carbon. The tetrachloride is reduced to the free metal with sodium or magnesium.

A new method, which holds considerable promise, involves converting the dioxides directly to the tetrachlorides. The mixture of zirconium and hafnium tetrachlorides is then reacted with zirconium dichloride, which selectively reduces zirconium tetrachloride to the trichloride.

$$ZrCl_4 \text{ (volatile)} + ZrCl_2 \text{ (non-volatile)} \xrightarrow{400°} ZrCl_3 \text{ (non-volatile)} \quad (21\text{-}4)$$

The volatile hafnium tetrachloride distills off, leaving $ZrCl_3$. This, when heated to 420–550°C, disproportionates to $ZrCl_2$ and $ZrCl_4$. The tetrachloride distills off, and is reduced as before to give a material containing only 0.01 per cent hafnium. The dichloride is used over again with a new batch of raw tetrachloride.

The most stable oxidation state of titanium is +4, but titanium (III) and titanium (II) compounds are known. Titanium dioxide is relatively acidic in character; it dissolves readily in aqueous base to form titanates

$$TiO_2 + 2OH^- \rightarrow TiO_3^{-2} + H_2O \qquad (21\text{-}5)$$

The TiO_3^{-2} ion is more faithfully represented by including the water of hydration

$$TiO_2 + 2OH^- + 2H_2O \rightarrow Ti(OH)_6^{-2} \qquad (21\text{-}6)$$

As one might expect, the Ti^{+4} ion is strongly hydrolyzed; $Ti(H_2O)_x^{+4}$ is found only in very acidic solution. Addition of complexing agents to a Ti^{+4} solution results in formation of octahedral complexes such as TiF_6^{-2} and $Ti(C_2O_4)_3^{-2}$, trisoxalatotitanate (IV). (Oxalate ion, $C_2O_4^{-2}$, is a bidentate coordinating ligand; see Problem 20-10.)

Titanium (III) ion in aqueous solution can be prepared by the electrolytic reduction of a titanium (IV) sulfate solution. The half-cell and overall cell reactions are

$$\begin{array}{l} 2Ti(H_2O)_x^{+4} + 2e^- \rightarrow 2Ti(H_2O)_x^{+3} \\ 3H_2O \rightarrow 2H_3O^+ + \tfrac{1}{2}O_2 + 2e^- \\ \hline 2Ti(H_2O)_x^{+4} + 3H_2O \rightarrow 2Ti(H_2O)_x^{+3} + 2H_3O^+ + \tfrac{1}{2}O_2 \end{array} \qquad (21\text{-}7)$$

The solution is stable for extended periods of time in the absence of oxygen or other oxidizing agents. It reacts quantitatively with reagents such as Fe^{+3}, MnO_4^-, or O_2. The divalent ion, Ti^{+2}, is not very well characterized in aqueous solution, since it reacts with water with evolution of H_2.

21-5 Group V Metals

The three elements of Group VB, vanadium, niobium, and tantalum, are characterized by a strong metallic bonding in the element, which leads to high melting and boiling points. Table 21-5 lists a few properties of these elements. An effect observed in Group IVB is seen here also; namely, the almost identical radii of the two heavier elements. Niobium and

Table 21-5 Properties of the Group IVB elements.

Element	Density	Melting point (°C)	Boiling point (°C)	Atomic volume*
V Vanadium	6.0	1917	3350	8.4
Nb Niobium	8.56	2470	4850	10.85
Ta Tantalum	16.6	3000	5300	10.9

*Atomic volume is atomic weight/density, a quantity proportional to volume per atom for elements of similar structure.

tantalum occur together in nature, and in general have rather similar chemical properties.

Vanadium forms compounds in which it is in the +2, +3, +4, and +5 oxidation states. In the +2 oxidation state, it is a powerful reducing agent. Compounds such as $VSO_4 \cdot 7H_2O$ are similar, insofar as the crystal structures are concerned, to the analogous compounds of other divalent metal ions; *e.g.*, $FeSO_4 \cdot 7H_2O$.

Vanadium (V) oxide, V_2O_5, is a yellow solid, soluble in both strongly basic and strongly acidic solutions. It exists as the orthovanadate ion, VO_4^{-3}, in very basic solution; as the solution is made more acidic a number of polyvanadates are formed, each species being most stable at a different value of pH. The compounds so formed are examples of the type of condensation reaction which was discussed on p. 471. Vanadium (V) forms complexes, all of which are notable for the presence of a V—O bond, as in $VOCl_4^-$.

Sulfur dioxide reduces a solution of V_2O_5 in sulfuric acid to form a blue solution of the vanadyl ion, VO^{+2}. On the other hand, V_2O_5 is reduced by magnesium metal in hydrochloric acid to the +3 state. Green vanadous hydroxide, $V(OH)_3$, is precipitated by addition of hydroxide.

Two lower oxides, VO_2 and V_2O_3, are formed from V_2O_5 by heating it in hydrogen. Vanadium is typical of a number of transition elements in that it forms more than one oxide. There is a general rule that in such series of oxides *the acid character increases with increasing oxidation number.* V_2O_5, which is the highest valence oxide of vanadium, is an acidic oxide. It is also noteworthy that the highest valence oxides of the transition metals are similar in many respects to the highest valence oxides of the non-metals in the corresponding Group A elements. For example, there are vanadate salts such as $Na_3VO_4 \cdot 12H_2O$, which are similar to the corresponding phosphate salts.

21-6 Group VI Metals

The Group VI elements and some of their properties are listed in Table 21-6. We see again the effect of the "lanthanide contraction" in the similar sizes of molybdenum and tungsten.

Table 21-6 Some properties of the Group VIB elements.

	Element	Density	Melting point (°C)	Boiling point (°C)	Atomic volume*
Cr	Chromium	7.14	1900	2500	7.28
Mo	Molybdenum	10.2	2600	4800	9.42
W	Tungsten	19.3	3400	5400	9.54

*Atomic volume is atomic weight/density, a quantity proportional to volume per atom for elements of similar structure.

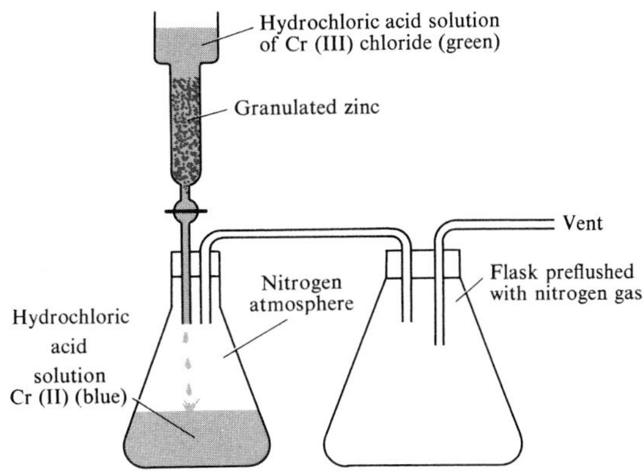

Fig. 21-3 Reduction of chromium (III) to chromium (II) with zinc in hydrochloric acid.

Chromium was first isolated by Vauquelin by reduction of green Cr_2O_3 with carbon at white heat. The commonest ore of the element is chromite, $FeCr_2O_4$. Reduction of this ore leaves an iron-chromium alloy, called *ferrochrome.* Alloy steels such as stainless steel (12-14 per cent Cr, about 0.5 per cent Ni) are made by adding ferrochrome, and whatever other metals are necessary, to iron. Chrome plating is carried out by electrolyzing a sulfuric acid solution of CrO_3. Chromium forms compounds in the +2, +3, and +6 oxidation states.

Chromium (II) ion is formed by reduction of an acid solution of chromium (III) with zinc, as shown in Fig. 21-3.

$$2Cr^{+3}(aq) + Zn \rightarrow Zn^{+2} + 2Cr^{+2}(aq) \qquad (21\text{-}8)$$

The solution is stable for a long time in the absence of oxygen and other oxidizing agents. The reaction with oxygen is sometimes used to remove traces of oxygen from gases when necessary:

$$4H_3O^+ + 4Cr^{+2} + O_2 \rightarrow 4Cr^{+3} + 6H_2O \qquad (21\text{-}9)$$

Chromium (III) is quite stable in aqueous solution and in the form of solid salts. Complexes of Cr^{+3} are generally inert to substitution, for the reasons given in Sec. 20-7. The color of a chromium (III) solution depends in some cases upon concentration, and upon whether various anions are present in excess. For example, in dilute solution $CrCl_3$ is a grey-violet color, denoting the presence of $Cr(H_2O)_6^{+3}$. In more concentrated solution, or in the presence of added chloride ion, the solutions are green, as a result of formation of $Cr(H_2O)_5Cl^{+2}$ and $Cr(H_2O)_4Cl_2^+$. It is actually something of a trick to prepare crystalline $CrCl_3 \cdot 6H_2O$, the species which crystallizes from aqueous solution, so that all three chloride ions are outside the coordination sphere. Whereas $[Cr(H_2O)_6]Cl_3$ is violet, $[Cr(H_2O)_5Cl]Cl_2 \cdot H_2O$

is green. These two compounds provide another example of coordination isomerism.

Chromium (III) sulfate is prepared as a violet crystalline solid by reacting dry chromium (III) hydroxide with concentrated sulfuric acid. It has the formula $Cr_2(SO_4)_3 \cdot 16H_2O$. The violet color suggests that the chromium is coordinated only by water. Water is lost by warming to 90°C. The crystals then turn green, and the composition changes to $Cr_2(SO_4)_3 \cdot 6H_2O$. In this form, the sulfate ions are coordinated to the chromium. Chromium (III) sulfate forms large crystals of *chrome alum* when a solution containing $Cr_2(SO_4)_3$ and an alkali metal sulfate is slowly evaporated. The *alums* have the general formula $AB(SO_4)_2 \cdot 12H_2O$, where A is an alkali metal, and B is a trivalent metal.

Chromium (II) oxide is distinctly basic in water. Cr_2O_3 is less basic; the corresponding hydroxide, $Cr(OH)_3$ is amphoteric. It dissolves in strongly basic solution, as a result of the formation of the complex hydroxide:

$$Cr(OH)_3(H_2O)_3 + OH^- \rightarrow Cr(OH)_4(H_2O)_2^- + H_2O \qquad (21\text{-}10)$$

CrO_3, a red solid, is very soluble in water, and forms strongly acidic solutions. This series again illustrates the rule that the acidity of the oxide increases with increase in metal oxidation number. CrO_3 is a powerful oxidizing agent. It reacts violently with many organic compounds; where the reaction can be controlled, it is a useful reagent for carrying out oxidations. In aqueous solution, Cr^{+6} exists as the chromate ion, CrO_4^{-2}, or a condensation product thereof. The chromate ion, CrO_4^{-2}, which is the stable form of chromium (VI) in basic solution, undergoes a series of condensation reactions with decreasing pH. The first of these leads to the dichromate ion:

$$2\left[\begin{array}{c} O \\ | \\ O\text{—}Cr\text{—}O \\ | \\ O \end{array}\right]^{-2} + 2H_3O^+ \rightarrow \left[\begin{array}{c} O \quad\quad O \\ | \quad\quad\; | \\ O\text{—}Cr\text{—}O\text{—}Cr\text{—}O \\ | \quad\quad\; | \\ O \quad\quad O \end{array}\right]^{-2} + 3H_2O$$

Trichromates and other, more highly condensed species exist in strongly acid solution. The dichromate ion is quite easily obtained; it is a strong oxidizing agent, as indicated by the half-cell potential:

$$Cr_2O_7^{-2} + 14H_3O^+ + 6e^- \rightarrow 21H_2O + 2Cr^{+3} \qquad \mathcal{E}^\circ = +1.33 \qquad (21\text{-}11)$$

By contrast, chromate ion, which is the stable form in basic solution, is not so easily reduced:

$$CrO_4^{-2} + 4H_2O + 3e^- \rightarrow 5OH^- + Cr(OH)_3(s) \qquad \mathcal{E}^\circ = -0.13 \qquad (21\text{-}12)$$

Insoluble chromate salts, which are a bright yellow, are widely used as paint pigments. Chrome yellow is the standard yellow of the artist.

Molybdenum occurs in nature mainly as the sulfide, MoS_2, which has a curious layer structure rather similar to that of graphite. It is, in fact, used in mixtures with oils and greases as a lubricant, just as graphite is. Roasting the sulfide in air results in formation of the trioxide, MoO_3. Reduction of this oxide with hydrogen at 1200°C leads to formation of the free metal. The trioxide may also be reduced with carbon, or with aluminum in a thermite reaction.

Molybdenum compounds are known in which the metal is in the +6, +5, +4, +3, and +2 oxidation states. Molybdic acid, H_2MoO_4, is known in the solid state. Other examples of +6 molybdenum are MoF_6, MoO_2Cl_2, and MoS_3. Molybdenum pentachloride, $MoCl_5$, is a volatile black solid which is formed by the action of chlorine on MoS_2. It is decomposed by water to form $MoOCl_3$. The element also forms a pentafluoride, MoF_5, by reaction of MoF_6 and Mo. Two lower valence chlorides, $MoCl_4$ and $MoCl_3$, are also known. In addition, other compounds of Mo in the +4 and +3 oxidation states are known. It is clear from these examples that molybdenum has a wider variety of reasonably stable oxidation states than chromium has.

Tungsten (called Wolfram in many parts of the world) is found as the mineral *Scheelite*, $CaWO_4$, named after Scheele who discovered the element. The metal is employed in lamp filaments, and in certain very hard alloys. Tungsten is rather similar to molybdenum in its chemistry.

21-7 Manganese, Iron, Cobalt, and Nickel

We shall discuss the chemistry of the next four elements of the first transition series collectively, rather than to continue with the group-by-group treatment as we have been doing. The heavier elements of Groups VIIB and VIIIB are quite different in chemical behavior from the lighter elements. Some properties of the four elements of interest are listed in Table 21-7.

Manganese is found as the ore *pyrolusite*, MnO_2, a black solid which is found in India, Russia, and parts of Africa. In addition, manganese nodules, large lumps of MnO_2 plus Fe_2O_3, are present in certain regions of the ocean bed. A great deal of manganese is used in steelmaking. For

Table 21-7 Some properties of the elements manganese, iron, cobalt, and nickel.

Element	Density	Melting point (°C)	Boiling point (°C)	Atomic volume
Mn Manganese	7.39	1240	2100	7.4
Fe Iron	7.86	1540	2450	7.1
Co Cobalt	8.8	1490	2900	6.7
Ni Nickel	8.8	1450	2900	6.7

these purposes, the oxide ore can be smelted with the iron ore in quantities to achieve the desired manganese content. The element has been identified in every oxidation state in the range 0 to +7. Among the oxides, MnO, Mn_3O_4, Mn_2O_3, MnO_2, MnO_3, and Mn_2O_7 are known. Mn_3O_4 can be thought of as a mixed oxide, $2MnO \cdot MnO_2$. MnO and Mn_3O_4 are basic oxides, Mn_2O_3 and MnO_2 are amphoteric, and the remaining higher oxides are increasingly acidic.

Manganese (II) salts and aqueous solutions are a pale pink. The Mn^{+2} ion has a d^5 configuration. In a weak field environment, which is what nearly all Mn^{+2} complexes appear to be, there is a precisely half-filled $3d$ subshell. There is, therefore, no ligand field stabilization, as noted in Sec. 20-4. Because of this, and because it has about the same ionic radius as Mg^{+2}, it resembles magnesium ion in many respects. Among the insoluble salts of Mn^{+2} are the hydroxide, $Mn(OH)_2$, carbonate, $MnCO_3$, oxalate, $MnC_2O_4 \cdot 2H_2O$, and sulfide, MnS.

Manganese exists in the +7 oxidation state as the purple permanganate ion, MnO_4^-, and in the +6 oxidation state as the green manganate ion, MnO_4^{-2}. The latter is stable only in basic solution. On acidification, it disproportionates into permanganate ion and manganese dioxide:

$$3MnO_4^{-2} + 4H_3O^+ \rightarrow 2MnO_4^- + MnO_2(s) + 6H_2O \qquad (21\text{-}13)$$

Permanganate ion is a strong oxidizing agent, particularly in acid solution. The product of its reduction in acid solution is manganese (II) ion; in basic solution, it is the insoluble black manganese dioxide:

$$\begin{array}{ll} \text{Acid:} & 8H_3O^+ + MnO_4^- + 5e^- \rightarrow Mn^{+2} + 12H_2O \\ \text{Basic:} & MnO_4^- + 2H_2O + 3e^- \rightarrow MnO_2(s) + 4OH^- \end{array} \qquad (21\text{-}14)$$

Iron occurs in the United States in the form of oxide ores, the most important being *hematite*, Fe_2O_3. It is also found as a carbonate, $FeCO_3$ (siderite), and as iron pyrite, FeS_2. Although pyrite deposits are very extensive, they are not at present employed as a source of iron for economic reasons. Other oxide ores of lesser importance than hematite are *magnetite*, Fe_3O_4, and *limonite*, $Fe_2O_3 \cdot xH_2O$. The great iron deposits in Minnesota, Michigan, and Wisconsin are largely hematite of varying quality. The accompanying gangue is largely clays, sand, and limestone. The quality of the ore in use at present is high enough so that the ore can be put through a reduction step immediately. This is done in a blast furnace, illustrated in Fig. 21-4.

The objectives of the blast furnace operation are twofold; to effect a reduction of Fe_2O_3 to metallic iron, and to provide for separation and removal of the gangue. The heat required for reduction is obtained from the combustion of coke. The coke is combusted to form carbon monoxide in the lower part of the furnace where temperatures are highest. As carbon monoxide rises in the furnace, it reduces hematite to metallic iron. The exhaust gases, which are quite hot, are used to preheat incoming air. Molten

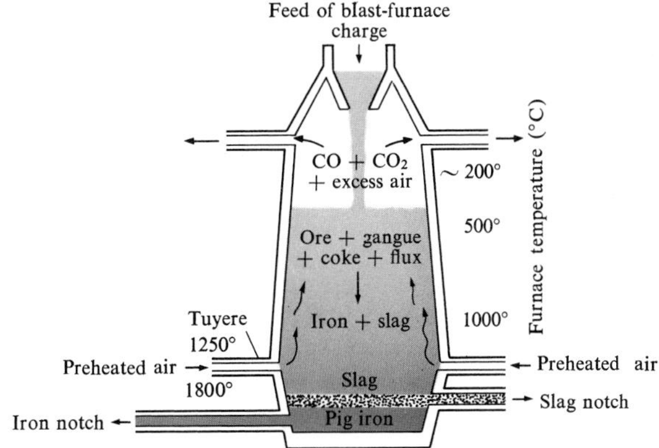

Fig. 21-4 Simplified illustration of a blast furnace used in reduction of iron ores.

iron collects at the bottom of the furnace and is periodically tapped off.

The gangue is removed by causing it to form a slag, a molten complex mixture of silicates. For this purpose limestone, $CaCO_3$, is introduced at the top of the furnace along with ore and coke. This undergoes calcination; the resulting CaO combines with SiO_2 to form the slag:

$$CaO + SiO_2 \rightarrow CaSiO_3$$

This reaction represents an idealized version of what happens; the slag is actually a complex mixture of silicates containing calcium, magnesium, aluminum, and small quantities of sulfur and manganese. The slag collects as a molten layer over the iron, and is withdrawn through a hole in the side of the furnace.

The product of the blast furnace, called *pig iron*, contains as impurities silicon, manganese, phosphorus, carbon, and sulfur. Of these, the carbon content is highest, being about 3.5-4.5 per cent.

Pig iron is not a usable form of the metal; the high carbon content is particularly undesirable, since it makes the metal extremely brittle. The conversion of pig iron to a more useful form is performed in either a Bessemer converter or an open hearth furnace. The latter process is the more important in the United States. An open hearth furnace is illustrated in Fig. 21-5. At the beginning of the process, the furnace is charged with a scrap iron and pig iron mixture (totaling perhaps 200 tons), and with some iron oxide ore and limestone. The iron oxide reacts with the carbon, silicon, manganese, and phosphorus in the pig iron to produce free metal and oxides, which in turn combine with the CaO to produce a slag which is run off. Heat for the process is provided by combustion of a gas in the open space over the charge. The hot exhaust gases are used to heat inlet gases to conserve fuel.

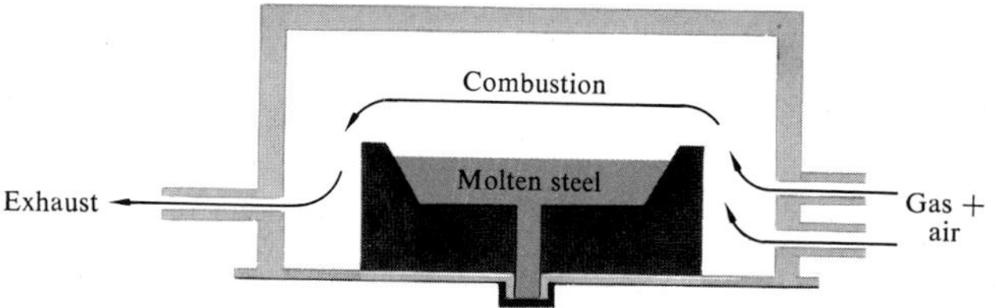

Fig. 21-5 Simplified diagram of an open-hearth furnace. The brickwork of the furnance is made up of either basic oxides (CaO, MgO) or an acidic oxide (SiO_2), depending upon the nature of the impurities in the raw iron.

The open hearth process is quite slow; the time for a 200-ton batch is on the order of 12 hours. The product of the process is steel, whose qualities are dependent principally upon the carbon content.

When special alloy steels requiring addition of other metals such as chromium or vanadium are desired, they are generally made up in electric furnaces. This procedure, although expensive, permits more exact control of alloy composition and properties.

Recently the so-called *basic oxygen process* has begun to replace the open hearth furnace as the major method of producing steel. In this process, high purity oxygen is blown directly into molten iron to remove carbon and other impurities by oxidation. The oxygen is obtained from giant liquid air plants, and piped directly to the furnaces. The method of introducing the oxygen depends upon the character of the ore and the intended use of the steel, but in general the process is very fast. A batch of raw iron can be processed in a 250-ton furnace in as little as 30 to 40 minutes.

Iron dissolves readily in dilute acid, producing ferrous salts. It dissolves in dilute nitric acid, however, to yield iron (III) nitrate in solution with evolution of nitric oxide:

$$Fe + 4H_3O^+ + NO_3^- \rightarrow Fe^{+3}(aq) + NO + 6H_2O \qquad (21\text{-}15)$$

Ferrous ion, Fe^{+2}, is quite easily oxidized to ferric ion, as evidenced by the half-cell potential:

$$Fe^{+2} \rightarrow Fe^{+3} + e^- \qquad \mathcal{E}^\circ = -0.77$$

The +3 ion forms very stable complexes with a number of ligands, particularly with EDTA (see Fig. 21-1). Strong coordinating ligands tend to stabilize the +3 state over the +2, although there is no case in which the $Fe^{+2} \rightarrow Fe^{+3}$ half-reaction has a positive $\mathcal{E}^\circ$. Iron (III) complexes are not as inert to substitution as Co^{+3} and Cr^{+3} complexes. With larger ligands, tetrahedral complexes are formed, as in $KFeCl_4$ and K_2FeCl_4.

Iron (II) chloride is a typically ionic salt; it can be prepared by heating iron in a stream of dry hydrogen chloride. Iron (III) chloride, formed by heating iron in the presence of chlorine, or by reducing ferric oxide with carbon in the presence of chlorine, is quite unlike the iron (II) compound. In many of its properties, it appears to be an essentially covalent compound. It is soluble in non-polar solvents, in which it is dimeric (Fe_2Cl_6), and it is fairly volatile, subliming at about 300°C. In these respects, it resembles aluminum chloride. Like aluminum chloride, also, it behaves as a Lewis acid and has been employed as a catalyst in organic reactions which are catalyzed by Lewis acids.

Cobalt is one of the rarer elements among the first row transition elements. It is sometimes found in usable quantities with copper, and is recovered as a by-product of copper mining. The element is found in the +2 or +3 oxidation state in its compounds. We have already seen that in the presence of strong complexing agents the +3 state is stabilized relative to the +2. We have also seen that complexes of the +3 ion are quite inert to substitution. Hexammine cobalt (III) ion is not attacked by concentrated hydrochloric acid, and is only slowly attacked in sodium hydroxide solution.

Cobalt (II) forms a number of tetrahedral complexes in addition to the many octahedral complexes already mentioned. Addition of concentrated hydrochloric acid to a pink solution of cobalt (II) ion produces the deep blue color of the tetrachlorocobaltate (II) ion:

$$Co(H_2O)_6^{+2} + 4Cl^- \rightarrow CoCl_4^{-2} + 6H_2O \qquad (21\text{-}16)$$

Nickel exhibits only the +2 oxidation state in aqueous solution. Although some of its complexes, like the deep blue $Ni(NH_3)_6^{+2}$ ion are six-coordinate, the majority are four-coordinate. Most of these are square-planar complexes.

Nickel is a much more abundant element than cobalt. It occurs in a number of well-defined minerals, mostly as sulfide ores (*e.g.*, NiS and Ni_3S_2), or as arsenosulfides. More than half of the world's production of nickel comes from the province of Ontario in Canada. Even though the raw ores from this area are relatively rich in nickel, they still contain only a few per cent Ni, and are concentrated by a froth-flotation process. The concentrated ore is roasted, reduced with coke and further treated in a converter to reduce carbon and other impurity levels. There is commonly a good deal of copper present in the ore, and at some stage a separation of nickel and copper is carried out.

Nickel exhibits only one common oxidation state, the +2. It should be noted that there is a progressive decrease in the stability of the higher oxidation states in moving across the first transition series. The effect is most pronounced with nickel and copper. It may be ascribed to the increase in effective nuclear charge with increasing atomic number. Aqueous nickel (II) ion is a bright green. Divalent nickel salts generally precipitate from solution as the green, hexahydrated nickel (*e.g.*, $NiSO_4 \cdot 7H_2O$). Nickel

(II) hydroxide, $Ni(OH)_2$, is precipitated from aqueous solution as a light green solid on addition of sodium hydroxide. It is not soluble in excess hydroxide, but does dissolve in excess aqueous ammonia. The difference in behaviors toward the two reagents is due to formation of the deep blue colored hexammine nickel (II) ion, $Ni(NH_3)_6^{+2}$:

$$Ni(OH)_2(s) + 6NH_3 \rightarrow Ni(NH_3)_6^{+2}(aq) + 2OH^- \qquad (21\text{-}17)$$

Nickel is commonly identified in qualitative analysis by formation of the characteristic bright red precipitate with dimethylgyloxime. The complex is planar, and involves two moles of this bidentate ligand:

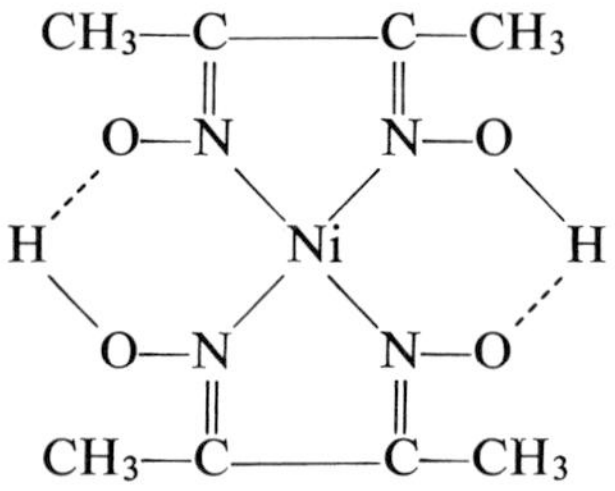

```
CH3—C————C—CH3
    ‖      ‖
  O—N      N—O
 .·   \   /   \
H       Ni      H
 \    /   \   ·'
  O—N      N—O
    ‖      ‖
CH3—C————C—CH3
```

The stability of the complex seems to be due in part to the formation of hydrogen bridge bonds between the two ligands, as shown by the dotted lines.

21-8 Group IB Elements

The elements of Group IB are the last group in the transition series. Copper itself actually has a completely filled 3*d* level; its electronic configuration is $3d^{10}4s^1$. Copper (II) ion, however, is colored; it has a $3d^9$ outer electron configuration.

The Group IB elements are quite different chemically from those of Group IA; they are much less reactive chemically, and exhibit a wider variety of oxidation states. The gradual increase in effective nuclear charge which occurs in the process of completing the *d* subshell in each period results in a gradual diminution in chemical activity.

Copper is a characteristic thiophilic (sulfur-loving) metal. It is found in extensive deposits as various kinds of sulfur compounds. Among these, *chalcocite* (Cu_2S), *covelline* (CuS), *chalcopyrite* ($CuFeS_2$), and *bornite* (Cu_5FeS_4) are of commercial importance. There are also extensive deposits of the native metal in certain places. Copper ores of very low copper content, on the order of 0.5 per cent, have been successfully worked using modern methods of enrichment such as froth flotation.

Compounds of both copper (I) and copper (II) are known. The +2 state is generally observed in aqueous solution, since copper (I) compounds are insoluble. Copper (I) chloride is precipitated as a colorless salt from a copper (II) chloride solution in the presence of HCl and copper metal:

$$Cu + CuCl_4^{-2} \rightarrow 2CuCl(s) + 2Cl^- \qquad (21\text{-}18)$$

It should be noted that copper (I) ion has a completed shell of 3*d* electrons; this accounts for the fact that the simple copper (I) compounds are colorless.

Copper (II) ion forms a number of complexes, most of which are four-coordinate, square-planar. The green tetrachloro cuprate (II) ion, $CuCl_4^{-2}$, is tetrahedral, but the blue tetrammine copper (II) ion is square-planar.

The hydroxide of copper(II) is amphoteric, dissolving easily in excess base:

$$Cu(H_2O)_2(OH)_2(s) + 2OH^- \rightarrow Cu(OH)_4^{-2} + 2H_2O \qquad (21\text{-}19)$$

The metallurgy of silver and gold are briefly mentioned in Sec. 20-2. Except for a few compounds with strongly electronegative elements (*e.g.* AgF_2), silver occurs in its compounds and in solution in the +1 oxidation state. In contrast with the alkali metal ions, a number of silver salts are insoluble in water. The list includes the chloride, bromide, iodide, and sulfide. Silver ion forms a number of complexes, in which its coordination number is two.

$Ag(CN)_2^-$	dicyanatoargentate (I) ion
$Ag(NH_3)_2^+$	diamminesilver (I) ion
$Ag(S_2O_3)_2^{-3}$	dithiosulfatoargentate (I) ion
$AgCl_2^-$	dichloroargentate (I) ion

The last of these is a rather weak complex. Silver chloride is precipitated by addition of dilute hydrochloric acid to a silver (I) solution, but partially redissolves if the solution is made concentrated in the acid.

Silver metal is relatively inert to non-oxidizing acids, but dissolves slowly in hot, concentrated sulfuric acid and readily in cold, dilute nitric acid.

Gold exists in both the +1 and +3 oxidation states in solution. Gold (III) is formed upon reaction of gold with *aqua regia*, a mixture of concentrated nitric and hydrochloric acids:

$$Au + 4Cl^- + 3NO_3^- + 6H_3O^+ \rightarrow AuCl_4^- + 3NO_2 + 9H_2O$$

Evaporation of the solution to dryness yields gold (III) chloride, which is dimeric in structure:

```
:C̈l       ˙C̈l˙       C̈l:
  ··\     /    \     /··
      Au          Au
  ··/     \    /     \··
:Cl       .Cl.       Cl:
  ··       · ·        ··
```

Gold (III) is readily reduced to a gold (I) compound, as exemplified by the reaction with iodide ion:

$$AuCl_4^- + 3I^- \rightarrow AuI(s) + I_2 + 4Cl^- \qquad (21\text{-}20)$$

Metallic gold is more inert to attack by acids than is silver. It does not dissolve in concentrated, hot nitric acid, but it does dissolve in aqua regia, as mentioned above.

21-9 Group IIB Elements

Zinc, cadmium, and mercury are post-transition elements, in that they possess completed 3*d* shells in the divalent state. They form a number of four-coordinate complexes with cyanide, iodide, ammonia, and other ligands. The hydroxides of both zinc and cadmium are amphoteric.

Both zinc and cadmium are found in solution only in the divalent state. Mercury is found in solution as both mercury (II) and mercury (I). In the latter oxidation state it exists as a dimeric ion, $(Hg—Hg)^{+2}$, with a single covalent bond between the mercury atoms. Mercury (I) is formed by adding mercury to a solution containing mercury (II).

$$Hg + Hg^{+2} \rightarrow Hg_2^{+2}$$

Other reducing agents such as tin (II) or sulfur dioxide may also be used. The chloride of mercury (I) is insoluble, whereas mercury (II) chloride is soluble. Although complexes of mercury (I) are not commonly observed, mercury (II) forms a number of four-coordinate complexes, including very stable species such as the tetraiodomercurate (II) ion, HgI_4^{-2}.

21-10 Platinum Metals

The six elements, ruthenium, rhodium, palladium, osmium, iridium, and platinum, are commonly referred to as the *platinum metals*. Although they are members of three different groups in the periodic classification, they share a number of common properties. The elements are all rare, and occur in small quantities with other elements. The nickel ores of Ontario, Canada are an important source of the platinum metals, which are recovered by a rather complicated metallurgical scheme.

With the exception of platinum, the metals are really not noble in the sense that gold is. They do form oxides and a number of other compounds quite easily. Osmium, for example, in a compact form, burns in oxygen at 400°C, forming OsO_4 vapor. Iridium is more inert; it is, in fact, insoluble in aqua regia, and is attacked by fluorine only at high temperatures. The standard meter in Paris is composed from a platinum-iridium alloy containing about 90 per cent platinum.

The platinum group elements possess a remarkably varied chemistry; we can mention here only a few of the more important facets. All of the elements form a variety of compounds with the halogens. All except palladium form a hexafluoride (palladium does not form any compounds in which it is in a +6 oxidation state). The highest known fluoride of this element is PdF_4. Ruthenium and osmium form tetroxides, RuO_4 and OsO_4. Both compounds are volatile. The ruthenium compound decomposes explosively to RuO_2 and O_2. The osmium compound, which melts at 41°C and boils at 131°C, is an important compound of osmium, but is evil-smelling, and very poisonous.

The platinum metals have a very interesting and complicated coordination chemistry. The platinum compounds are the most thoroughly studied of all coordination compounds. Much interesting work has been carried out in Russia, where there is an institute for the study of platinum metal chemistry. (Russia ranks third, behind South Africa and Canada, in platinum-metals production.) Platinum complexes exist for both the $+2$ and $+4$ oxidation states. The coordination number in complexes of the $+2$ state is four; in complexes of the $+4$ state it is six. The four-coordinate complexes are square-planar and the six-coordinate complexes are octahedral. The existence of *cis* and *trans* geometrical isomers of the complex $Pt(NH_3)_2Cl_2$ has been known for a long time. These two isomers have the structures shown in Fig. 21-6.

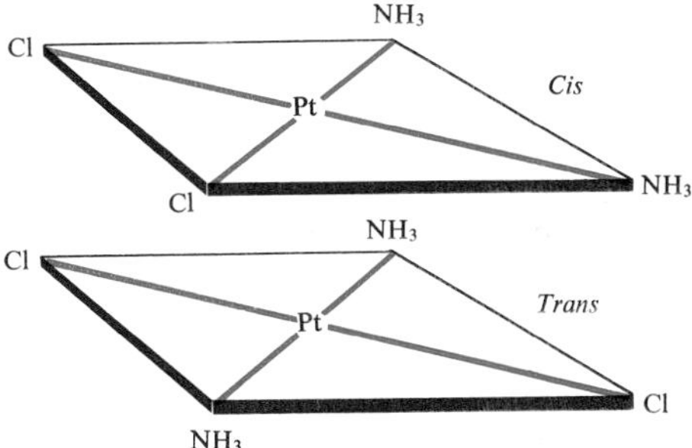

Fig. 21-6 Geometrical isomers of $Pt(NH_3)_2Cl_2$.

Complexes of the heavy metals such as palladium and platinum behave as strong field complexes. A square-planar complex does not have the same energy splittings as an octahedral complex, since the ligands on the z axis (Fig. 20-11) are missing. The d_{z^2} orbital is therefore not a high energy orbital, but is much lower in energy. This means that only one of the d orbitals, the $d_{x^2-y^2}$, is pointed directly at the ligands, and thus subject to a strong field. Both palladium (II) and platinum (II) complexes have a d^8 configuration. Since there are four d orbitals which are lower in energy than the $d_{x^2-y^2}$, the eight d electrons are paired up. In the $+4$ complexes of palladium and platinum, which are octahedral, the electron configuration is d^6. The six d electrons are paired in the lower energy t group orbitals, in the usual manner for a strong field complex. Thus, both the $+2$ and $+4$ complexes of palladium and platinum are diamagnetic, having no unpaired d electrons. Further, the complexes are relatively inert to substitution. This is particularly true of the $+4$ complexes.

21-11 Metal Carbonyls

One very interesting characteristic of the transition metals is their ability to form compounds in which the metal has a formal zero charge.

The most widely studied group which forms such compounds is carbon monoxide. It was discovered in about 1890 that finely divided nickel reacts with carbon monoxide at ordinary temperatures to form nickel tetracarbonyl, $Ni(CO)_4$, a volatile compound which melts at $-25°C$ and boils at 43°C. On heating, the compound decomposes to nickel metal and carbon monoxide. The ready formation of nickel tetracarbonyl, and its ready decomposition to free nickel, provides a means of purifying nickel by separating it from other metals such as copper which do not form a compound with carbon monoxide. The Mond process, which is still in use to some extent, makes use of this means of purifying nickel.

Subsequent investigation revealed that not only nickel, but many other elements form metal carbonyls. Table 21-8 lists some of the more common examples, with an indication of their properties.

Table 21-8 Some transition metal carbonyl compounds, and a few physical properties.

Compound	Melting point (°C)	Boiling point (°C)	Description
$Ni(CO)_4$	−25	43	Colorless, musty smell, very poisonous, burns readily, decomposes readily.
$Fe(CO)_5$	−20	103	Yellow liquid.
$Cr(CO)_6$	sublimes	...	Decomposes at 200°C, soluble in organic solvents.
$Mn_2(CO)_{10}$	155	...	Yellow solid, oxidizes in air, soluble in organic solvents.
$Co_2(CO)_8$	51	...	Orange solid, oxidizes rapidly in air.

In an earlier discussion of the chemical properties of carbon monoxide (Sec. 19-4), it was pointed out that CO is not a particularly reactive molecule. The CO molecule has the Lewis structure, :C≡O:, and is isoelectronic with N_2. It exhibits very little tendency to behave as an electron pair donor. There is no evidence, for example, that it reacts with the proton to form an HCO^+ species. Nevertheless, the formation of metal carbonyl compounds is best understood as a Lewis acid-base reaction between the carbon monoxide molecule which acts as the base, and the metal atom, which acts as the electron pair acceptor, *i.e.*, the acid. Why should a metal atom act as an acceptor toward CO when a strongly acid species such as H^+ does not? The answer involves the fact that carbon monoxide has orbitals which are not used in bonding (these are the anti-bonding orbitals of π symmetry in the C—O triple bond). When CO donates the unshared pair of electrons from carbon to the metal atom, there is a simultaneous back-donation of d electrons from the metal to the vacant, anti-bonding orbitals of the CO. There is a sort of push-pull process, as diagrammed in Fig.

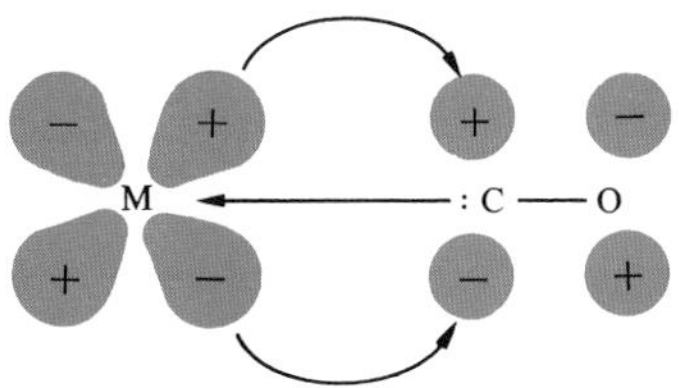

Fig. 21-7 A diagram of bonding between carbon monoxide and a metal atom. The CO molecule donates an electron pair on C to a vacant hybrid orbital on M. At the same time, the metal donates *d* orbital electrons to the CO group, by transferring electrons into the anti-bonding π orbitals of CO.

21-7. In this manner, carbon monoxide forms quite stable compounds with a number of the transition metals.

In order to understand the geometry of the complexes, it is convenient to apply hybrid orbital language to these compounds. Let us consider nickel tetracarbonyl as an example. Nickel in the zero charge state has 10 electrons in the orbitals outside the argon configuration. In the atom, the lowest energy configuration is $4s^2 3d^8$:

4s	4p			3d				
↑↓				↑↓	↑↓	↑↓	↑	↑

In forming the tetracarbonyl, each CO group donates one electron pair to the metal atom. Under the influence of the CO groups, the ten electrons of Ni^0 are paired up in the 3*d* orbitals:

4s	4p			3d				
				↑↓	↑↓	↑↓	↑↓	↑↓

This leaves the 4*s* and three 4*p* orbitals for bond formation. They form a $4sp^3$ hybrid orbital set, with tetrahedral geometry. Thus, we expect nickel tetracarbonyl to have a tetrahedral disposition of the CO groups about nickel, and we expect the compound to be diamagnetic.

Note that if we count two electrons from each CO group as entering the metal orbitals, the total number of electrons in the nickel valence orbitals is 18. Nickel has thus, in a manner of speaking, attained the krypton configuration in forming $Ni(CO)_4$. The "krypton rule" states that in a stable metal carbonyl compound, each metal atom has an 18-electron environment. We can apply this rule to iron, which has eight electrons beyond the argon configuration. We require 10 more electrons, which can be supplied by five CO groups. The formula for the carbonyl of iron should therefore be $Fe(CO)_5$, which indeed it is. The bonding scheme, in terms of hybrid orbitals, is as follows:

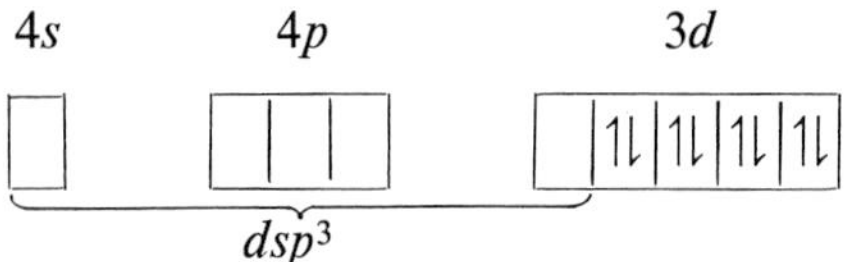

The use of dsp^3 hybrid orbital set suggests that $Fe(CO)_5$ should have a trigonal bipyramidal structure, similar to that for PF_5, *etc.* (Sec. 8-9), and it does.

Applying the same rule to the carbonyl of chromium, we conclude that it should have the formula $Cr(CO)_6$, and should have an octahedral arrangement of CO groups about chromium. These are, in fact, the formula and geometry of the compound. The carbonyl compounds of manganese and cobalt offer special problems; let us consider manganese. Mn possesses seven electrons beyond the inert gas configuration. Addition of five CO groups brings the total number of electrons in the valence orbitals to 17, one short of that required by the krypton rule. Addition of six CO groups would bring the total to 19, one over that called for by the rule. The latter possibility can be immediately ruled out since excess electrons would have to go into higher energy orbitals, and the compound would lack stability. On the other hand, $Mn(CO)_5$ would have an unpaired electron. Formation of an Mn—Mn bond, however, would provide an avenue for pairing of the odd electron on each $Mn(CO)_5$, and would effectively bring the total electrons per manganese atom to 18. Thus, we find that the molecular formula of the carbonyl compound of manganese is $Mn_2(CO)_{10}$. The structure is shown in Fig. 21-8.

Application of the same ideas to the carbonyl compound of cobalt suggests that we should find a $Co_2(CO)_8$ compound, which is in accord with the experimental facts. The structure of this compound is not so simple as we might predict from what has been said so far. For reasons which are still not entirely clear, two of the CO groups act as bridging groups, in which the CO bond is a double bond, and the carbon forms two electron pair bonds to the metals. The structure shown in Fig. 21-8 results.

A great number of other carbonyls are known. Most of the more complex ones, which contain more than one metal atom per molecule, can be understood in terms of the simple notion that each metal atom should have 18 electrons in its valence orbitals, although sometimes it is a bit difficult to keep track of where the electrons are.

Aside from their novelty as volatile compounds of the transition metals, the metal carbonyls undergo a number of interesting reactions. Other groups which have an unshared pair of electrons can replace carbon monoxide in a substitution reaction. Organic phosphorus compounds,

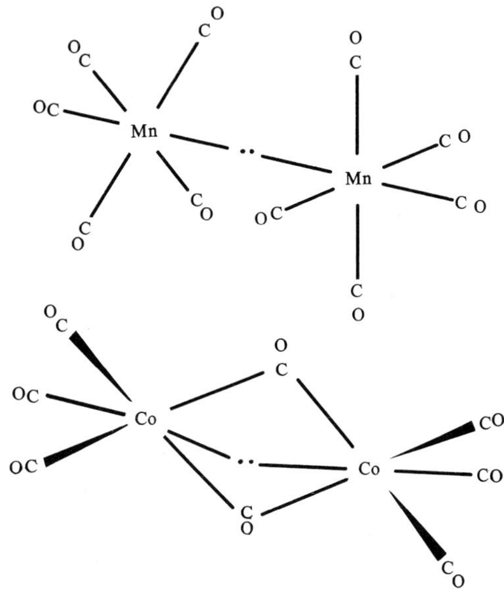

Fig. 21-8 The structures of $Mn_2(CO)_{10}$ and $Co_2(CO)_8$. In each case, there is an electron pair bond between the metals. In the cobalt compound, the structure is more complex; there are two CO groups acting as bridges.

such as triphenyl phosphine, $(C_6H_5)_3P\colon$, are capable of displacing CO to form a substituted compound:

$$Cr(CO)_6 + :P(C_6H_5)_3 \rightarrow Cr(CO)_5P(C_6H_5)_3 + CO \qquad (21\text{-}21)$$

A great variety of organic groups can be bonded to transition metals through reactions involving the metal carbonyls. The area of chemistry concerned with reactions and compounds of this type is called *organometallic chemistry*.

21-12 Chemistry of the Solid State

(The student should review Sec. 7-6 before making a serious attempt to learn the material in this section.)

The metal oxides and sulfides are an important class of substances. They are involved in practical considerations such as metal corrosion, and are employed in solid-state lasers, and as semiconductors. Metal oxide structures can be understood as close-packed oxide ion lattices, with the cations occupying regular interstices. Recall that there are one octahedral and two tetrahedral interstices for each metal atom. The particular type of hole the metal ion occupies is determined more by the radius ratio than any other single factor. In binary oxides such as TiO, MnO, and CdO, the metal ion occupies the octahedral holes. Since there is one metal ion for each oxide ion in the formula, all the octahedral holes are filled. In Li_2O, the cation is in tetrahedral holes, and — because there are two metal ions for each oxide ion — all the tetrahedral holes are filled. *Corundum*, a form of aluminum oxide, furnishes an example of incomplete filling of one type of hole; the Al^{+3} ions are in octahedral holes, and occupy two-thirds of them. Other oxides which have the same structure are α-Fe_2O_3 (the symbol α merely designates a particular crystal form of iron (III) oxide), V_2O_3, and Cr_2O_3.

Certain compounds which may be considered as mixed oxides furnish more complex examples of metal ion packing; for example, the spinels, which have the formula AB_2O_4. The mineral *spinel* has the formula $MgAl_2O_4$. A is always a divalent metal, and B is a trivalent metal. When the A ions occupy the tetrahedral holes and the B atoms occupy the octahedral holes, the compound is called a *normal spinel.* Examples of this are $MgAl_2O_4$, $FeAl_2O_4$, and $MgCr_2O_4$. If the divalent metal ions are in octahedral holes, and half of the trivalent ions are in tetrahedral holes, the compound is referred to as an *inverse spinel.* This situation occurs in $MgFe_2O_4$, and in Fe_3O_4.

Ionic lattices are never perfect, but contain flaws of various kinds. Dislocations are, as the name implies, irregularities of a mechanical sort in the layer structure. One type, the *edge dislocation*, can be visualized by imagining that the pages of this book are layers of ions in an ionic lattice. Now imagine that one of the pages is cut off so that it is an inch or so narrower than the others, and that the book is closed. The slight irregularity in the sheets along the edge of the short page is like an edge dislocation. A *screw dislocation* results when one lattice plane is twisted slightly with respect to another. The dislocations are an important aspect of the solid state, because absorbed substances, impurities, *etc.*, migrate very easily along dislocations. In addition, the ions which border on a dislocation are in a higher energy state than the others, and are often the first to react.

Whereas dislocations result from displacement of rather large sections of the solid structure with respect to other sections, *defects* result from the displacements or absences of individual ions or lattice units. We will be concerned here principally with defects in ionic lattices. If ionic lattices were free from defects, it would be impossible to pass an electric current through them. The conductivity of ionic lattices is usually not very large, but it is nevertheless possible to observe the passage of a current through them.

Defects in ionic lattices can be generally divided into three types. In the *Frenkel defect*, ions are removed from their normal sites, and displaced to higher-energy interstitial positions. This type of defect can be illustrated schematically in two dimensions as shown in Fig. 21-9. In silver bromide at 210°C, for example, 0.076 per cent of the silver ions are interstitially

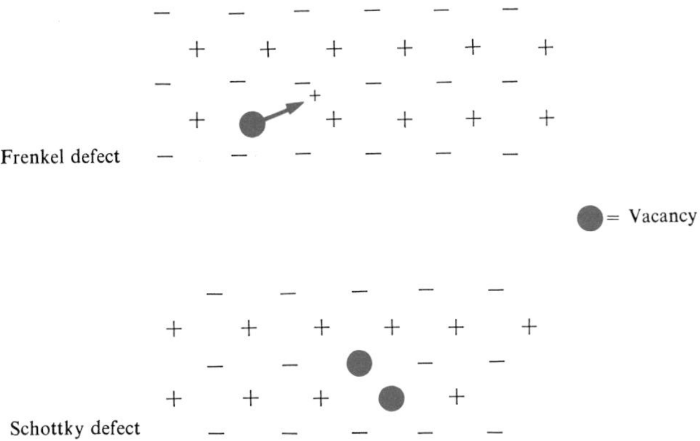

Fig. 21-9 Two types of defect found in ionic lattices.

located. Frenkel defects are generally found in lattices composed of small cations and relatively large polarizable anions. It is, of course, the small cations which are found in interstitial positions.

Schottky defects are characterized by the absence of cations and anions in equal numbers from their normal lattice positions. They are simply vacancies in the lattices created by these absences. Alkali halides exhibit such defects. The Schottky defect is illustrated in Fig. 21-9.

One of the more important types of defect, called an *impurity defect*, arises when an impurity cation of another charge type is found in the lattice; for example, $CaCl_2$ in a KCl lattice. This type of impurity defect is illustrated in Fig. 7-17.

The concentration of defects present in a given solid depends upon the temperature. As the temperature is raised, ions possess a greater average energy of vibration about their equilibrium lattice positions. There is, therefore, an increased probability that any one ion will acquire the energy necessary for a transition to the higher-energy interstitial position.

These ideas may be expressed more quantitatively by considering the equilibrium between ions in a normal site, and the higher-energy site. Let n be the number of ions in the normal site, and n^* the number in the high energy site. Then, $n \rightleftharpoons n^*$, and

$$K = \frac{n^*}{n}$$

Figure 21-10 shows the energy relationship between the ions in the two sites. Following the treatment in Sec. 14-4, we can obtain for the equilibrium constant

$$K = Be^{-\Delta E/RT}$$

From this relationship, we see that K increases with increasing temperature. For example, it was mentioned above that 0.076 per cent of the Ag^+ ions

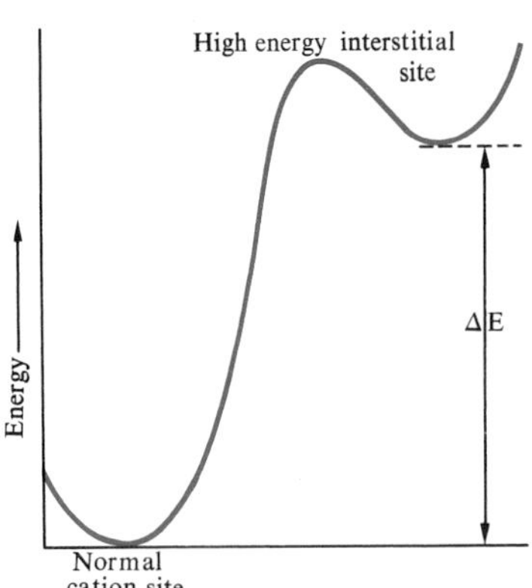

Fig. 21-10 Energy relationship between normal and interstitial cation sites in an anionic close-packed lattice. The normal cation sites are the regular holes in the close-packed lattice. The high energy sites are other holes which would force two cations to be closer than normal, or which would give an unfavorable radius ratio.

in AgBr are in high-energy interstitial positions at 210°C. At 300°C, 0.4 per cent are in such positions.

Non-stoichiometry — A stoichiometric compound is one which has a chemical composition of definite proportion by weight. But many ionic solids do not possess exactly the chemical formulas which we commonly write for them. For example, if we were asked to write the formula for zinc oxide we would write ZnO; for nickel (II) oxide, NiO. In fact, however, it is probably impossible to prepare either of these two oxides with precisely the formulas given. The correct formula for zinc oxide is generally of the form $Zn_{1+x}O$, and the correct formula for nickel oxide is of the form $Ni_{1-y}O$. The numbers written as subscripts in these two formulas are not exact, and the actual number varies from one preparation to another. Because there is not a precisely one-to-one relationship between the metal and non-metal in these compounds, they are called *non-stoichiometric*. Non-stoichiometry may arise from a number of causes:

1. *Excess cation due to anion vacancies.* Electrons are found in the anion vacancies. Some compounds exhibiting this type of non-stoichiometry are KCl, NaCl, and KBr. In the case of the alkali halides just mentioned, the compounds can be prepared with excess cation by allowing excess metal vapor to come in contact with the halogen vapor. This first type of non-stoichiometry is quite well known because the electrons which are in the vacancies (these electrons are necessary in order to preserve

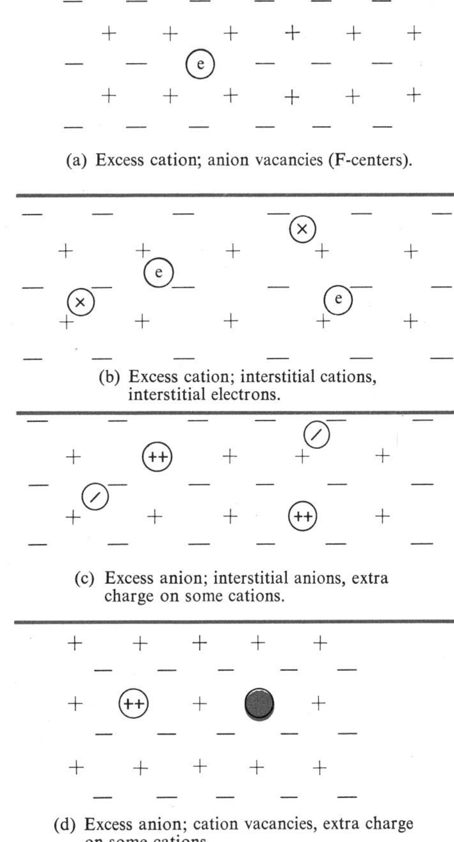

(a) Excess cation; anion vacancies (F-centers).
(b) Excess cation; interstitial cations, interstitial electrons.
(c) Excess anion; interstitial anions, extra charge on some cations.
(d) Excess anion; cation vacancies, extra charge on some cations.

Fig. 21-11 Schematic illustrations of various types of non-stoichiometry in ionic lattices.

electrical neutrality) give rise to intense color. They are, in fact, commonly referred to in solid state physics as *color centers*. Another name for them is *F-centers*. This first type of non-stoichiometry can be represented schematically as in Fig. 21-11(a).

2. *Excess cation due to interstitial cations.* The interstitial cations are accompanied by associated interstitial electrons. Examples of this type of non-stoichiometry are ZnO and CdO. Oxides of this variety are called *n-type oxides.* Because electrons are found in interstitial positions, and because these electrons migrate through the solid quite readily, these n-type oxides are moderately good conductors of electricity. Most of the electrical current is carried through the solid by the negatively charged particles. This type of non-stoichiometry is illustrated in Fig. 21-11(b).

3. *Excess anion due to interstitial anions.* There are cations in the lattice with extra positive charge. It is clear that this type of non-stoichiometry can arise only when the metallic element is capable of existing in more than one oxidation state. Furthermore, it is important only for large cations. An example of this type of non-stoichiometry is UO_2. See Fig. 21-11(c) for an illustration.

4. *Excess anion due to cation vacancies.* Other cations in the lattice carry an extra positive charge. Again, it is clear that this type of non-stoichiometry is found only when the metallic element is capable of existing in more than one oxidation state. Examples of this type are compounds such as Cu_2O, FeO, NiO, and FeS.

This type of non-stoichiometry is illustrated schematically in Fig. 21-11 (d). Iron sulfide prepared by heating iron and sulfur together is not stoichiometric, in general, but possesses a formula which ranges from $FeS_{1.00}$ to $FeS_{1.14}$. The excess of sulfur results from the fact that there are vacancies in the lattice where ordinarily there would be Fe^{+2} ions. Electrical neutrality is maintained by the presence of Fe^{+3} ions. The larger the excess of sulfur present in the formula of the compound, the lower the density of the FeS. There is, in fact, an experimentally observed decrease in density with increased sulfur content.

This fourth type of non-stoichiometry also results in a conducting substance. An electron may jump from a lower charged metal cation to a higher charged cation so that, in effect, there is a migration of the excess positive charge through the lattice. Oxides of the type under discussion are referred to as *p-type oxides.* Under the influence of an applied field, the electron, in jumping from the lower to the higher charged cations, migrates toward the positive electrode. It *appears,* however, in another way of looking at it, that the excess positive charge species is migrating. If we call the cation with excess positive charge a positive hole, it then appears as though the positive hole is migrating toward the negative electrode. Hence the name p-type semiconductor. The non-stoichiometric oxide and sulfide semiconductors are similar in many respects to the semiconductors prepared by doping silicon and germanium, as described in Sec. 17-5.

Corrosion — Chemical reactions which proceed as a result of the migration of ions through solid lattices are a very important part of chemistry. Perhaps the most important single instance of such reactions is that of metal corrosion. The rusting of iron is a deplorably common chemical process which costs the American economy billions of dollars every year. Despite this very high cost and despite a considerable amount of research, it is still not entirely clear how corrosion occurs. It is also not clear, therefore, how to prevent corrosion. We may imagine metal oxidation to occur in a stepwise manner. The first process is an initial chemisorption of O_2 on the surface of the metal. The oxygen may exist on the surface of the metal in one of a number of forms. For example it may exist as O_2^-, O^-, or O_2^{-2}. There are experiments which indicate that the O_2 molecule does dissociate upon adsorption. The energy of the adsorption process is quite high, on the order of ordinary chemical bond energies. For example, the heat of adsorption of the initial layer of oxygen on a fresh iron surface is on the order of 100 kcal/mole. Further layers of oxygen adsorb on top of this initial layer very rapidly up to perhaps eight or nine atomic layers. It is not entirely clear just how the metal oxide itself develops from this adsorbed oxygen layer. One theory is the so-called place exchange mechanism which is illustrated below:

```
                                               O2
                                              ↙
     (O)                                     M
     (M)                                     O
     (O)                                     M
----(M)----        ——————→             ----O----
      M                                      M
      M                                      M
```

This mechanism may predominate when the oxide layer is thin; *i.e.*, it consists of about eight to ten atomic layers. When the layer is thicker, the oxide film grows as a result of migration of cation through the oxide lattice. The particular mechanism which operates for a given metal depends upon the type of oxide which that metal forms. A general schematic illustration of the process of oxide growth can, however, be indicated and described in a stepwise manner as follows:

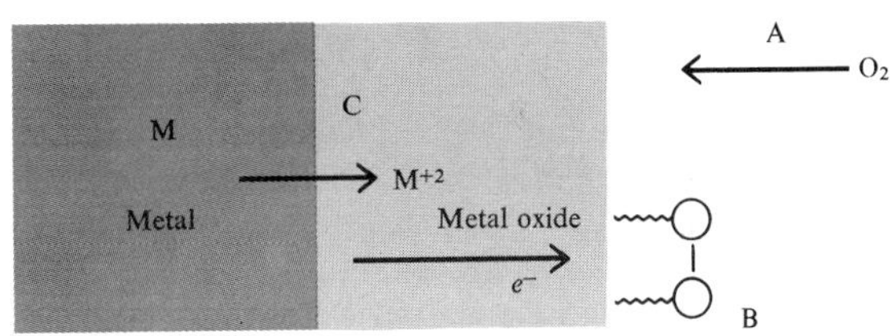

A. Oxygen is adsorbed on the surface of the oxide film.

B. Electrons which have migrated through the oxide lattice from the metal surface underneath are acquired by the adsorbed oxygen; the adsorbed oxygen thus becomes part of the close-packed oxide lattice.

C. Metal cations formed at the metal surface by the loss of electrons to the oxide lattice migrate through the oxide lattice by one of the mechanisms described above. For example, the cations may migrate as interstitial cations, jumping from one interstitial location to another until they arrive at the surface of the oxide film where the excess oxide ions are located. The rate-determining process in oxide film growth is thus generally the rate of diffusion of metal cations through the oxide lattice.

The view we have just presented is in fact drastically oversimplified. Numerous other problems in connection with oxide growth must be dealt with. For example, what are the effects of dislocations or macroscopic cracks in the oxide? What are the effects of impurity metals, or what are the effects of metals upon one another in a metal alloy? What are the effects of water or other chemical substances present in the environment to which a metal surface is commonly exposed? What chemical treatment may be applied to a metal surface to slow down or prevent corrosion? What particular combinations of metals in alloys best resist corrosion? These and many other such questions have not been completely answered.

Suggested Readings

Douglas, B. E., and D. H. McDaniel, *Concepts and Models of Inorganic Chemistry*, New York, N. Y.: Blaisdell Publishing Co., 1965.

Manch, W., and W. C. Fernelius, "Structure and Spectra of Nickel (II) and Copper (II) Complexes," *J. Chem. Educ.*, 38, 192 (1961).

McCarley, R. E., and T. M. Brown, "Tungsten Tribromide and Tungsten Tetrabromide," *J. Am. Chem. Soc.*, Vol. 84, 3216 (1962).

Sienko, M. J., R. A. Plane, and R. E. Hester, *Inorganic Chemistry*, New York, N. Y.: W. A. Benjamin, Inc., 1965.

Exercises

21-1. By considering the properties of the elements and their ions listed in Table 21-3, how do you account for the fact that aluminum forms the +3 state in water with a large $+\mathcal{E}°$, whereas thallium exhibits a negative value of $\mathcal{E}°$ for the analogous process?

21-2. What is the "lanthanide contraction"? Explain how the consequences of the contraction are seen in the properties of certain transition metals.

21-3. Make a table of metal oxides discussed in this chapter, classifying them according to whether they are acidic, basic, or amphoteric. What regularities can you observe in the data?

21-4. Write balanced chemical equations for each of the following reactions:
(a) Copper metal dissolves in an acid solution containing an excess of $NH_4Fe(SO_4)_2$, with no hydrogen evolution.
(b) An acid solution of V_2O_5 is reacted with SO_2 to yield a blue solution.
(c) Molybdenite, MoS_2, is treated with chlorine to form a volatile solid (you will have to make a reasonable guess about the fate of S in this reaction; see Chap. 19).
(d) A precipitate of nickel (II) hydroxide is dissolved in an excess of aqueous ethylenediamine solution.
(e) Solid hexammine cobalt (III) chloride is heated to over 200°C. One mole of ammonia is given off.

21-5. A solution containing both copper (II) and cadmium (II) ions is treated with concentrated hydrochloric acid. The pH is adjusted to about 1, and H_2S added. Only CuS precipitates. Upon addition of sodium acetate, followed by further addition of H_2S, CdS precipitates. Explain the observations, and write balanced net ionic equations for the important reactions which occur.

21-6. Can you provide an explanation of the fact that titanium tetrachloride hydrolyzes very rapidly, whereas carbon tetrachloride is quite stable to contact with water?

21-7. Write the equations representing the reaction of iron (III) oxide with carbon and with silicon in the open hearth furnace. Write an equation which shows the role of calcium oxide added in the open hearth furnace.

21-8. Calculate the weight of carbon anode consumed in the production of 1 mole of aluminum metal in the Hall process, assuming that carbon monoxide is the sole product at the anode.

21-9. Write a complete series of chemical equations representing the production of zirconium metal from zirconium dioxide by the process described in the text.

21-10. Write balanced chemical equations for the reaction of MnO, Mn_2O_3, and Mn_2O_7 with water.

21-11. In Chap. 19, the ability of the high oxidation state acids of non-metals to undergo condensation reactions was discussed. Write equations representing analogous reactions involving chromic acid, H_2CrO_4, and vanadic acid, H_3VO_4.

21-12. As a test of the assertion that the transition metals, in their higher oxidation states, compare well in their physical and chemical properties with the corresponding non-metals in the same oxidation states, look up the properties of the vanadium compound which corresponds to each of the following compounds of phosphorus, and make a comparison of the properties of the two. (It is just possible that the corresponding vanadium compound is not known; this is also relevant to the comparison we are making.) In addition to hand-books, the texts listed in Suggested Readings should be useful.

(a) P_2O_5.
(b) $POCl_3$.
(c) H_3PO_4.
(d) PCl_3.
(e) PCl_5.
(f) P_2O_3.
(g) PF_5.

21-13. Copper (I) forms a series of chloro complexes of the form $CuCl_2^-$, $CuCl_3^{-2}$ and $CuCl_4^{-3}$. Show how the equilibrium constant values for the equilibria

$$CuCl(s) + Cl^- \rightarrow CuCl_2^-$$
$$CuCl_2^- + Cl^- \rightarrow CuCl_3^{-2}$$
$$CuCl_3^{-2} + Cl^- \rightarrow CuCl_4^{-4}$$

could be obtained (at least in principle) from a study of the solubility of CuCl(s) in water containing varying concentrations of potassium chloride.

21-14. The Cd_2^{+2} ion has been formed by reaction of $CdCl_2$ with cadmium in molten $AlCl_3$. What is the electronic structure of the Cd_2^{+2} ion? Why is the preparation of the ion not carried out in aqueous solution? Predict what would occur on addition of $Cd_2[AlCl_4]_2$ to water.

21-15. Write balanced chemical equations representing each of the following observations:
(a) Addition of chlorine gas to a green manganate solution causes it to turn purple.
(b) Hydrogen sulfide is added to a basic permanganate solution, with resultant formation of a black precipitate.
(c) Iron (III) oxide is dissolved in acid solution.
(d) Addition of base to an acid solution containing zinc ion results in a precipitate. The precipitate dissolves on addition of excess base.
(e) Addition of mercury to a solution of mercury (II) chloride results in formation of a white precipitate.
(f) Addition of sodium cyanide solution to a precipitate of silver chloride causes it to dissolve.
(g) Copper (I) chloride dissolves in acidic aqueous hydrogen peroxide solution, with formation of a blue solution.

21-16. Write balanced chemical equations to complete each of the following:
(a) $Ti(H_2O)_6^{+2} + O_2 \rightarrow$
(b) $Cr(H_2O)_6^{+2} + Cr_2O_7^{-2} \rightarrow$
(c) $Co(H_2O)_6^{+3} + Br^- \rightarrow$ (acid)
(d) $Ag^+ + S_2O_3^{-2} \rightarrow$
(e) $VO_4^{-3} + H_3O^+ \rightarrow$ (first step)
(f) $Hg^{+2}(aq) + I^- \rightarrow$
(g) $Ag + H_3O^+ + NO_3^- \xrightarrow[\text{dilute}]{\text{cold}}$

21-17. Write an equation for the reaction of iron (III) chloride with a concentrated hydrochloric acid solution. [*Hint:* Compare with tin (IV) chloride.]

21-18. In one step of the metallurgy of vanadium, the compound NH_4VO_3 is roasted. What would you expect the product to be? Write a balanced chemical equation.

21-19. Propose a structure for the compound $VOCl_3$. What geometry should it have?

21-20. Write balanced net ionic equations for each of the following reactions:
(a) When manganese (III) fluoride is added to water, a black precipitate forms and the solution turns pink. No gas is evolved.
(b) When a green solution of chromium (III) chloride is poured over zinc in hydrochloric acid, the solution turns blue. Hydrogen is evolved.

(c) Titanium dioxide, TiO_2, does not dissolve in hydrochloric acid solution, but it does dissolve in hydrofluoric acid solution. Suggest a reason for this, and write an equation.

21-21. Vanadium forms a hexacarbonyl which does not combine with another molecule to form a dimer. Draw the expected structure for $V(CO)_6$. Indicate the electronic structure by drawing a box diagram representation of the vanadium atomic orbitals.

21-22. $Mn_2(CO)_{10}$ reacts with one mole of bromine in a polar organic solvent. No carbon monoxide is evolved. Indicate the geometry and electronic structure of the probable product.

21-23. Copper is known to exist in both the +1 and +2 oxidation states. On this basis, what type of non-stoichiometry would you expect copper (I) oxide to possess? Diagram schematically a metal oxide lattice of the type indicated by this non-stoichiometry.

21-24. A Nernst glower is a metal oxide which develops a low resistance to the passage of electric current when heated. It can thus be kept hot by passage of a large current, and serves as a source of infrared radiation. Why does the resistance decrease with increasing temperature, in contrast to metallic conduction, where the opposite is true?

21-25. Consider the following experiment: A block of silver and a block of sulfur are pressed together, with a series of fine platinum wires to serve as markers:

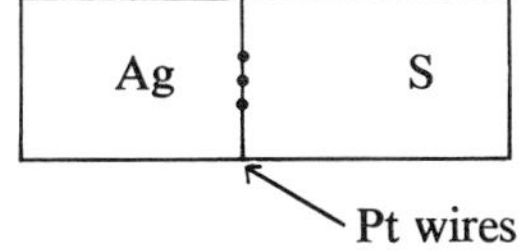

The assembly is placed in a hot environment for a period to speed chemical reaction. After a time, the assembly is found to be as follows:

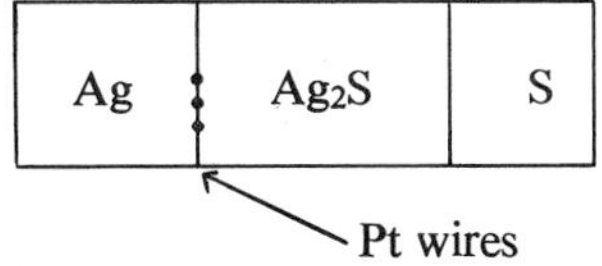

From the position of the marker wires relative to the compound formed, indicate a mechanism for the formation of Ag_2S.

21-26. It has been found that iron (II) oxide is non-stoichiometric, having a formula $FeO_{1.055}$ to $FeO_{1.19}$. Predict how the density of the compound might vary as a function of the formula. Explain your answer in terms of the type of defect structure you might expect.

21-27. Draw a qualitative picture of the d orbital energy splittings in a square-planar platinum complex. Indicate which d orbitals are assigned to each level.

Chapter 22

Organic and Biochemistry

22-1 Introduction

The distinction between organic and inorganic chemistry is, to some extent, an historical pheomenon. It was at one time held that the chemical substances which make up living matter are formed through the agency of some vital principle, and cannot be synthesized directly from substances which are obviously "inorganic." This so-called "vitalist" theory declined during the middle half of the nineteenth century, as powerful new methods for synthesizing compounds emerged. With these methods, it was possible to produce in the laboratory many compounds which it had been previously believed were formed only in living systems. Since that time, a great number of chemical substances found in living matter have been synthesized. As an example, one of the most recent achievements is the synthesis of chlorophyll, a plant pigment which plays an important role in photosynthesis.

Organic compounds are characterized chiefly by the presence of carbon-carbon and carbon-hydrogen bonds. The formation of a chain of chemically identical atoms through covalent bond formation is termed *catenation*. Carbon seems to be unique in its ability to form chains of literally unlimited length. Since carbon exhibits a covalency of four, other atoms may be bonded to the carbon atoms in the chain.*

22-2 Hydrocarbons

The simplest organic compounds are the hydrocarbons, which contain only carbon and hydrogen. The hydrocarbons are further subdivided according to the type of carbon-carbon bonding present.

Alkanes — The alkanes, or saturated hydrocarbons, contain only single carbon-carbon bonds. The carbon atoms may be joined in a straight chain, or if the number of carbon atoms is sufficiently large, may form a branched chain. The simplest saturated hydrocarbon is methane, CH_4. Compounds of increasing molecular weight are related to methane by the addition of CH_2 units to the molecular formula. The structures of some of the lower molecular weight compounds in the alkane series are shown in Fig. 22-1.

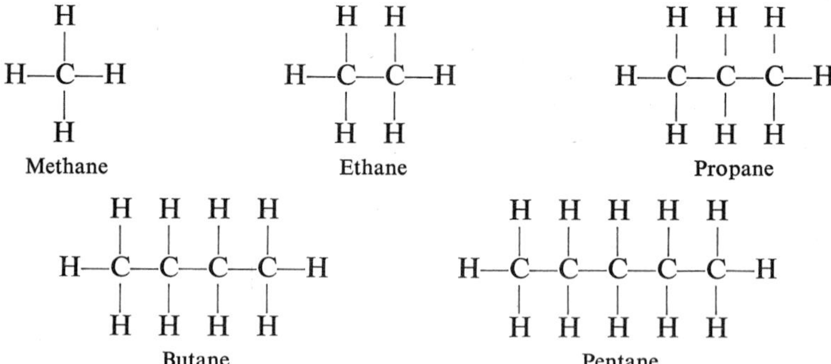

Fig. 22-1 Structural formulas for a number of straight-chain hydrocarbons.

The properties of a number of the compounds are listed in Table 22-1. The compounds in the table, related to one another by a succession of CH_2 groups, are members of a homologous series. The general formula for the homologous series of saturated hydrocarbons is C_nH_{2n+2}, where n = 1, 2, 3, ...

Isomerism. There are two compounds that possess the molecular formula C_4H_{10}:

---

*A distinction must be made between the linear carbon atom chains formed in organic compounds and other types of element-element bonding. In metals, a great many atoms of the same element are bonded together, but in a three-dimensional network. On the other hand, many elements are known to possess crystal structures in which the atoms are bonded in long chains; *e.g.*, tellurium. There are, however, no other kinds of atoms attached through covalent bonding to the atoms in the chain.

$$\begin{array}{ccccccccccc} & & H & & H & & H & & H & & \\ & & | & & | & & | & & | & & \\ H & — & C & — & C & — & C & — & C & — & H \\ & & | & & | & & | & & | & & \\ & & H & & H & & H & & H & & \end{array}$$

n-Butane

$$\begin{array}{ccccccccc} & & H & & H & & H & & \\ & & | & & | & & | & & \\ H & — & C & — & C & — & C & — & H \\ & & | & & | & & | & & \\ & & H & & H—C—H & & H & & \\ & & & & | & & & & \\ & & & & H & & & & \end{array}$$

iso-Butane

These two compounds represent examples of *structural isomerism.* The term structural isomerism applies whenever two or more compounds possess the same molecular formula, but differ in the atoms directly bonded to at least two of the atoms in the molecule. As the number of atoms in the hydrocarbon increases, the number of possible isomers also increases, and very rapidly.

The branched-chain hydrocarbons often have common names which are not very informative as to structure. However, they may be named systematically by considering them as derivatives of the straight-chain compound which corresponds to the longest straight chain in the molecule. For example, the compound with the common name *iso-octane* has the structure

$$\begin{array}{ccccccccccc} & & H & & CH_3 & & H & & CH_3 & & \\ & & | & & | & & | & & | & & \\ H & — & C & — & C & — & C & — & C & — & CH_3 \\ & & | & & | & & | & & | & & \\ & & H & & CH_3 & & H & & H & & \end{array}$$

Table 22-1 Physical properties of some straight-chain, saturated hydrocarbons.

Name	Boiling point (°C)	Melting point (°C)	Formula
Methane	−162	−183	CH_4
Ethane	−89	−172	C_2H_6
Propane	−42	−187	C_3H_8
Butane	−0.5	−135	C_4H_{10}
Pentane	36	−130	C_5H_{12}
Hexane	69	−94	C_6H_{14}
Heptane	98	−90	C_7H_{16}
Octane	126	−57	C_8H_{18}
Nonane	151	−54	C_9H_{20}
Decane	174	−29	$C_{10}H_{22}$
Undecane	196	−26	$C_{11}H_{24}$
Dodecane	216	−10	$C_{12}H_{26}$
Tridecane	232	−6	$C_{13}H_{28}$
Eicosane	—	37	$C_{20}H_{42}$
Heneicosane	—	40	$C_{21}H_{44}$
Docosane	—	44	$C_{22}H_{46}$
Triacontane	—	68	$C_{30}H_{62}$
Tetracontane	—	81	$C_{40}H_{82}$

The longest straight chain of carbon atoms in the compound is five; it is, therefore, a derivative of pentane. Numbering the carbon atoms along the chain, we signify the presence of the three CH_3 groups (called methyl groups) by the name 2,2,4-trimethylpentane.

Cycloalkanes — The cycloalkanes are characterized by the empirical formula C_nH_{2n}. The carbon atoms form a cyclic chain:

```
 H                   H                       H   H
  \                 /                        |   |
   C---------------C                 H      C---C       H
  / \             / \                 \    / |   | \    /
 H   \           /   H                 C    H   H    C
      \         /                     / \           / \
       \       /                     H   \         /   H
        \     /                           H-C-----C-H
           C                                |     |
          / \                               H     H
         H   H
```

Cyclopropane Cyclohexane

Rings containing less than six carbon atoms are strained, since the C—C—C angle must be substantially less than the $109\frac{1}{2}°$ tetrahedral angle. In cyclopropane, with an inside angle of only 60°, the strain is considerable; it is therefore a much more reactive substance than is cyclohexane.

Alkenes — Hydrocarbons containing at least one carbon-carbon double bond are termed *alkenes*, or *olefins*. The simplest member of this class is ethylene; the structure of this compound is discussed on p. 176. The carbon-carbon double bond may be described from either of two points of view. On the one hand, it may be envisaged as the overlapping of two tetrahedral orbitals from each carbon (Fig. 22-2). Some bending of the orbitals to give the best possible overlap is to be expected. Since energy is required to distort the orbitals, the double bond is rather reactive, and is easily opened through chemical reaction to give a carbon-carbon single bond plus two other single bonds to the carbon atoms.

On the other hand, one may assume that the carbon atoms are hybridized $sp^2 + p$, as in graphite or benzene (Sec. 8-6). The sp^2 orbitals are

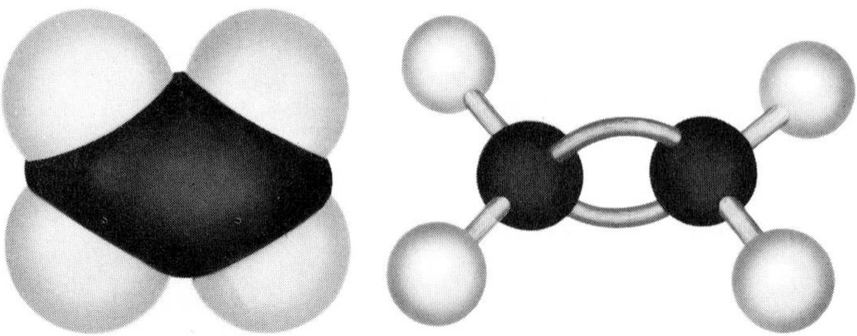

Fig. 22-2 Two representations of the carbon-carbon double bond in ethylene.

employed in forming single bonds to the hydrogen atoms and a single carbon-carbon bond. The *p* orbitals, by virtue of their sideways overlap, form a second carbon-carbon bond, which is of the π type. In this view, the carbon-carbon double bond consists of a σ type and a π type bond. The π bond is readily opened in the course of chemical reaction, and is the seat of the characteristic chemical properties of olefins.

Both models for the carbon-carbon double bond lead to the conclusion that the atoms in ethylene should lie in a plane (Fig. 22-2), in agreement with experimental observation.

Addition of CH_2 units to ethylene results in a homologous series of olefins of general formula C_nH_{2n}:

```
H         H          H     H    H
 \       /            \    |    |
  C═════C              C═══C────C─H
 /       \            /         |
H         H          H          H
   Ethylene              Propylene

H    H  H                 H  H          CH3         H
 \   |  |                 |  |             \       /
  C═C─C─CH3         H3C─C═C─CH3           C═════C
 /      |                                  /       \
H       H                               CH3         H
  1-Butene              2-Butene            Isobutylene
```

Note that, with chains of four or more carbon atoms, structural isomerism involving the location of the double bond is possible.

A further isomerism, classified as *geometrical isomerism*, is also found in many olefins. It arises because of restricted rotation about the double-bond axis. The rotation of groups about a carbon-carbon single bond, as in 1,2-dichloroethane, is not entirely free. Certain angular positions, such as those shown in Fig. 22-3, are more stable than others. In these positions,

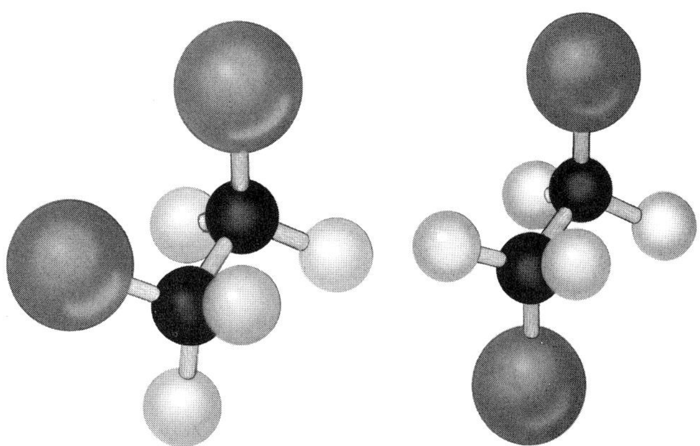

Fig. 22-3 Rotational forms of 1,2-dichloroethane.

the repulsions between groups attached to adjacent carbon atoms are minimized. The energy barrier which must be surmounted for the groups to twist into a different stable configuration is not large; at temperatures in the range of 25°C, the molecules pass from one configuration to another with great rapidity. The properties which we observe for 1,2-dichloroethane, therefore, are those of a mixture of the various possible forms, with the most stable ones predominating.

Rotation about carbon-carbon double bonds is, by comparison, very much restricted. 1,2-dichloroethylene, a planar molecule, exists in two isomeric forms which have different physical properties.

```
H         H              Cl         H
 \       /                 \       /
  C === C                   C === C
 /       \                 /       \
Cl        Cl              H         Cl

       cis                      trans
Boiling point = 60.1°C    Boiling point = 48.4°C
```

Conversion of one isomeric form to the other does not occur under ordinary conditions. Geometrical isomerism as found in coordination compounds is discussed on p. 505.

Alkynes — Hydrocarbons containing one or more carbon-carbon triple bonds are named *alkynes*, or *acetylenes*. The simplest member of the series is acetylene, or ethyne, $CH \equiv CH$. A homologous series of general formula C_nH_{2n-2} is formed by addition of CH_2 units.

Just as with the olefins, there are two alternative ways of viewing the bonding in alkynes (Fig. 22-4). In describing the bonding in terms of σ and π type bonds, it is necessary to invoke *sp* hybridization (Sec. 8-6). One of the *sp* orbitals is employed in bonding to the hydrogen atom, the other in forming the σ type carbon-carbon bond. The remaining $2p$ orbitals — two on each carbon — are perpendicular to the axis of the *sp* hybrid, and to each other. The two π bonds which are formed are thus at a 90° angle to one another.

Aromatic hydrocarbons — The aromatic hydrocarbons, exemplified by benzene, possess cyclic six-membered carbon rings with alternate carbon-carbon double bonds. The rings possess an unusual stability because of the particular character of the double-bonding (Sec. 8-7). The aromatic

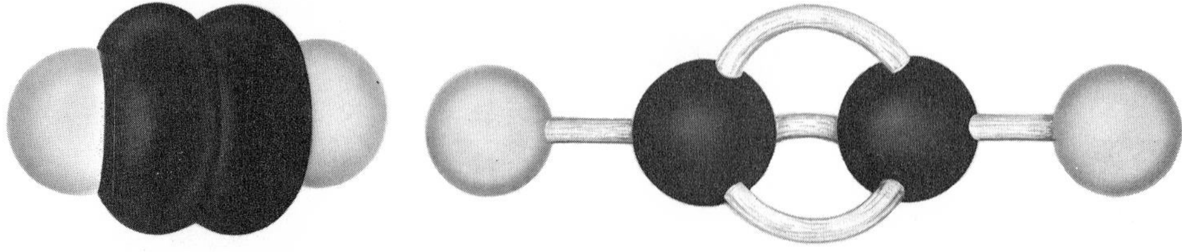

Fig. 22-4 Two representations of the bonding in acetylene.

compounds do not readily undergo a number of reactions which characterize the olefins, and so are rightly considered as a separate class. The common members of the aromatic family are all planar molecules. A few of these are shown in Fig. 22-5, with the symbolic notations often employed

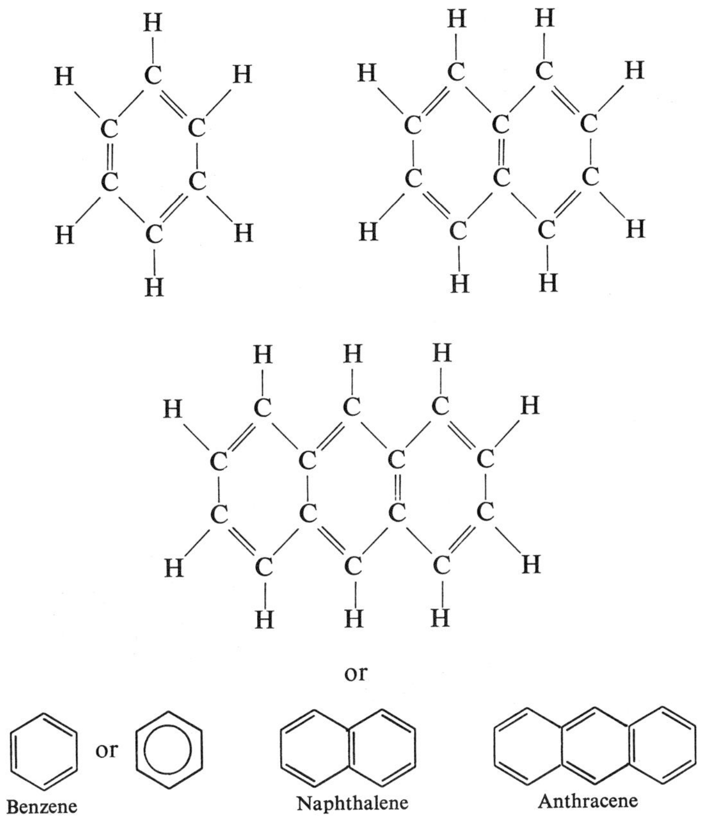

or

or

Benzene

Naphthalene

Anthracene

Fig. 22-5 Representations of the structures of simple aromatic compounds.

in representing them. The carbon-carbon bonds in the aromatic ring may be viewed in terms of canonical structures (p. 184), or in terms of π bonds constructed from carbon $2p$ orbitals. In any case, the extra energy of stabilization which the benzene ring seems to have is referred to as *resonance energy.* This extra stabilization energy can be estimated by utilizing thermochemical data. The heats of combustion of cyclohexane and cyclohexene have been accurately determined. By proceeding as described in Sec. 9-4, we can use the heat of combustion data to calculate the heats of formation of the two hydrocarbons, and thus to obtain the enthalpy change in the reaction

$$C_6H_{10} + H_2 \rightarrow C_6H_{12} \qquad \Delta H^\circ = -28.6 \text{ kcal/mole}$$

$$\text{cyclohexene} + H_2 \rightarrow \text{cyclohexane}$$

By proceeding in a like manner for cyclohexadiene, we ascertain that the enthalpy change for hydrogenating this compound is

$$C_6H_8 + 2H_2 \rightarrow C_6H_{12} \qquad \Delta H^\circ = -55.4 \text{ kcal/mole}$$

$$\text{cyclohexadiene} + 2H_2 \rightarrow \text{cyclohexane}$$

Since the enthalpy change for the second process is about twice that for the first, it would appear that there is a release of 28.6 kcal/mole for each double bond which is hydrogenated. If benzene could be thought of as cyclohexatriene, the heat of hydrogenation of benzene to cylcohexane would be estimated as $-3 \times 28.6 = -85.8$ kcal/mole of benzene. In fact, the heat of hydrogenation of benzene is only -49.8 kcal/mole. The heat released on hydrogenation of benzene is thus 36 kcal/mole *less* than expected. This amount of energy can be identified with the resonance energy, the extra stabilization which comes from the delocalization of the π electronic charge around the ring. Figure 22-6 shows a diagram of the energy relationships; note that benzene is lower in energy (more stable) than the hypothetical cyclohexatriene molecule, which would have no delocalization of its three double bonds.

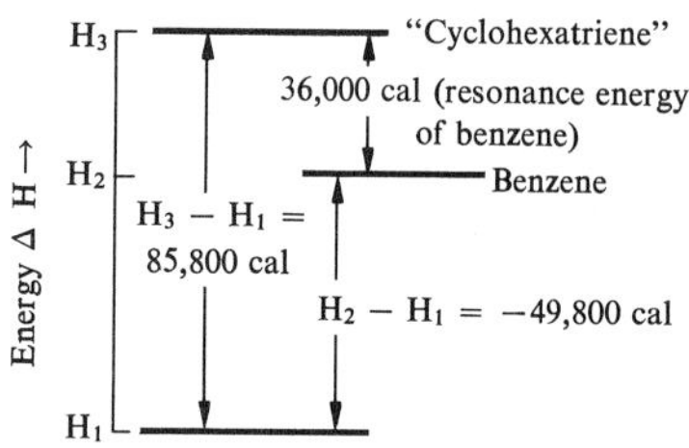

Fig. 22-6 The resonance energy of benzene.

22-3 Chemical Properties of the Hydrocarbons

The variability in carbon-chain length and the numerous possibilities for structural isomerism make for a large number of distinct hydrocarbon compounds. Each of these is capable of undergoing a number of chemical reactions leading to new organic compounds. The variety possible is indicated by the number of known organic compounds — at this time, about a million. In this brief introduction, we will consider only a few of the more common types of reactions which characterize the simple hydrocarbons and their derivatives.

Substitution reactions — Replacement of a hydrogen atom in a hydrocarbon by another atom or group is termed substitution. It occurs much more readily in the aromatic series than in the aliphatic series of compounds. When benzene is warmed in a strongly acid medium containing nitric acid, hydrogen is replaced by the nitro ($-NO_2$) group:

$$C_6H_6 + HNO_3 \xrightarrow[H_2SO_4]{conc} C_6H_5NO_2 + H_2O$$

The mechanism of the nitration reaction has been thoroughly studied. It has been established that the species which attacks the aromatic ring is the nitronium ion, NO_2^+, produced in sulfuric acid by the reaction

$$HNO_3 + 2H_2SO_4 \rightarrow NO_2^+ + H_3O^+ + 2HSO_4^-$$

The rate-determining step is the formation of a charged intermediate:

$$C_6H_6 + NO_2^+ \xrightarrow{slow} [C_6H_6NO_2]^+ \xrightarrow[HSO_4^-]{fast} C_6H_5NO_2 + H_2SO_4$$

More vigorous treatment results in the substitution of a second group:

$$C_6H_5NO_2 + HNO_3 \xrightarrow[H_2SO_4]{conc} C_6H_4(NO_2)_2 + H_2O$$

With more than one substituent group on the ring, the possibility of *positional isomerism* arises. The three possible dinitrobenzenes are

ortho-Dinitrobenzene *meta*-Dinitrobenzene *para*-Dinitrobenzene

Only the *meta* isomer is formed to any extent in the nitration of nitrobenzene.

Chlorination of aromatic hydrocarbons also proceeds relatively easily in the presence of iron as a catalyst:

$$C_6H_6 + Cl_2 \xrightarrow{Fe} C_6H_5\text{—}Cl + HCl$$

Chlorobenzene

$$C_{10}H_8 + Cl_2 \xrightarrow{Fe} C_{10}H_7Cl + HCl$$

α-chloronaphthalene

Introduction of alkyl groups into aromatic hydrocarbons is carried out via the Friedel-Crafts reaction (p. 329). The catalyst in this case is any strong Lewis acid.

$$CH_3CH_2\text{—}Cl + C_6H_6 \xrightarrow{AlCl_3} C_6H_5\text{—}CH_2\text{—}CH_3 + HCl$$

Ethylbenzene

The aliphatic hydrocarbons also undergo substitution reactions, but generally under much more rigorous conditions than those required for aromatic compounds. Chlorination occurs at high temperatures in the presence of ultraviolet light:

$$CH_3CH_2CH_3 + Cl_2 \xrightarrow[\text{ultra-violet light}]{\text{heat}} CH_3CHClCH_3 \quad \text{or} \quad CH_3CH_2CH_2Cl$$

Note that both possible isomers are produced. The reaction is difficult to control; the monochloro compounds may be further chlorinated to yield a variety of hard-to-separate isomers. Nitration of alkanes occurs only at elevated temperatures, and is accompanied by cleavage of the carbon chain.

Addition reactions — Addition reactions are characteristic of compounds containing carbon-carbon double or triple bonds. Addition to the multiple bonds in the aromatic series is also included in this classification, but it does not occur readily. The multiple bonding in the aromatic rings possesses unusual stability. The conditions required to get the reaction started are so rigorous that complete addition to all of the double bonds occurs after the process has once begun:

$$C_6H_6 + 3Cl_2 \xrightarrow{\text{ultra-violet light}} C_6H_6Cl_6$$

1,2,3,4,5,6-Hexachlorocyclohexane

By contrast, addition to olefins and acetylenes occurs very readily. Shaking an aqueous solution of bromine with an olefin results in addition to yield the dibromo compound

$$CH_2{=}CH{-}CH_3 + Br_2 \rightarrow CH_2Br{-}CHBr{-}CH_3$$

Acetylenes add up to 2 moles of bromine in similar fashion to yield tetrabromo derivatives.

Addition of hydrogen chloride to acetylene results in formation of vinyl chloride (chloroethene).

$$CH{\equiv}CH + HCl \rightarrow CH_2{=}CHCl$$

Addition of hydrogen to double bonds is an important reaction. Unsaturated fats from vegetable oils are hydrogenated in the presence of a catalyst to yield solid hydrogenated products.

Oxidation — The alkanes and aromatic hydrocarbons are relatively resistant to oxidation by reagents such as aqueous permanganate. They react readily with oxygen at elevated temperatures, however, to give (in excess oxygen) carbon dioxide and water:

$$C_3H_8 + 5O_2 \rightarrow 3CO_2 + 4H_2O$$
$$2C_6H_6 + 15O_2 \rightarrow 12CO_2 + 6H_2O$$

The reactions are highly exothermic, and the hydrocarbons are accordingly useful as fuels. Indeed, nearly all hydrocarbon material presently recovered from natural sources is eventually burned for its heat energy.

Olefins and acetylenes may also be combusted to give carbon dioxide and water. The heat released when acetylene is burned is particularly large; temperatures on the order of 3300°C are reached in the oxyacetylene torch, whereas maximum temperatures in an ordinary gas-air burner are about 1000°C.

Incomplete oxidation of hydrocarbons may lead to a variety of oxygen containing compounds. These will be discussed in the order of increasing degree of oxidation. It is important to recognize at the outset that, for the most part, the compounds to be discussed are not ordinarily produced by direct oxidation of simple hydrocarbons; much more efficient and otherwise superior methods for producing them are known.

22-4 Oxygen Derivatives of the Hydrocarbons

Alcohols — An alcohol is a hydrocarbon in which one of the hydrogens is replaced by an OH group. A few of the simple alcohols are shown in Fig. 22-7. In the most widely used naming system, the ending *-ol* is attached to the name for the hydrocarbon, after dropping the *-e*. Where structural isomerism is possible, the carbon atom carrying the alcohol group is designated by a number. The numbering is begun at the end of the chain nearest the OH group.

In a formal sense the alcohols may be considered as derivatives of water, an OH group being replaced by an O—R bond, where R represents an organic group. Intermolecular hydrogen bonding, though not as extensive as in water, is nevertheless important in determining the properties of the alcohols. The lower molecular weight compounds are relatively polar substances, completely miscible with water. With increasing size of the hydrocarbon group the properties of the alcohols become more hydrocarbon-like, as the solubility data in Table 22-2 indicate. The higher boiling points of the alcohols in general, as compared with those of the corresponding alkanes, are due to association through intermolecular hydrogen bonds.

Table 22-2 Some properties of the simpler aliphatic alcohols.

Alcohol	Melting point (°C)	Boiling point (°C)	Solubility in 100 g water
Methanol (CH_3OH)	−98	65	Completely miscible
Ethanol (C_2H_5OH)	−117	78	Completely miscible
1-propanol (C_3H_7OH)	−127	97	Completely miscible
1-butanol (C_4H_9OH)	−89	117	9 g at 15°C
1-hexanol ($C_6H_{13}OH$)	−52	156	Less than 1 g at 25°C

Methanol, CH_3OH, is sometimes referred to as wood alcohol because it was once produced mainly by the destructive distillation of hardwoods. The most important production method at present is the catalytic hydrogenation of carbon monoxide at elevated temperatures and pressures.

$$CO + 2H_2 \rightarrow CH_3OH$$

Methanol is used chiefly in the manufacture of formaldehyde, and as a radiator antifreeze. It is quite poisonous, and may cause death if taken internally.

Ethanol (ethyl alcohol), C_2H_5OH, is in many respects the most important of the alcohols. It has been produced for centuries by the fermentation of carbohydrates, and is the "active" ingredient in beer, wine, and many other beverages. The conversion of carbohydrates into alcohol is catalyzed by enzymes, high molecular weight biochemical catalysts. The starch molecules are first degraded into smaller sugar molecules in an enzyme-catalyzed hydrolysis reaction. The sugar (glucose) is then converted into ethanol and carbon dioxide by an enzyme produced by yeast cells:

$$\underset{\text{Starch}}{(C_6H_{10}O_5)_x} + xH_2O \rightarrow \underset{\text{Glucose}}{xC_6H_{12}O_6}$$

$$C_6H_{12}O_6 \xrightarrow{\text{Xymase}} 2C_2H_5OH + 2CO_2$$

Ethanol may be produced industrially by addition of water to an olefin, a method which is quite general in application. The addition of water is not accomplished directly, but through use of sulfuric acid:

$$CH_2{=}CH_2 + H_2SO_4 \rightarrow CH_3{-}CH_2OSO_2OH$$
$$CH_3{-}CH_2OSO_2OH + H_2O \rightarrow CH_3{-}CH_2OH + HOSO_2OH$$

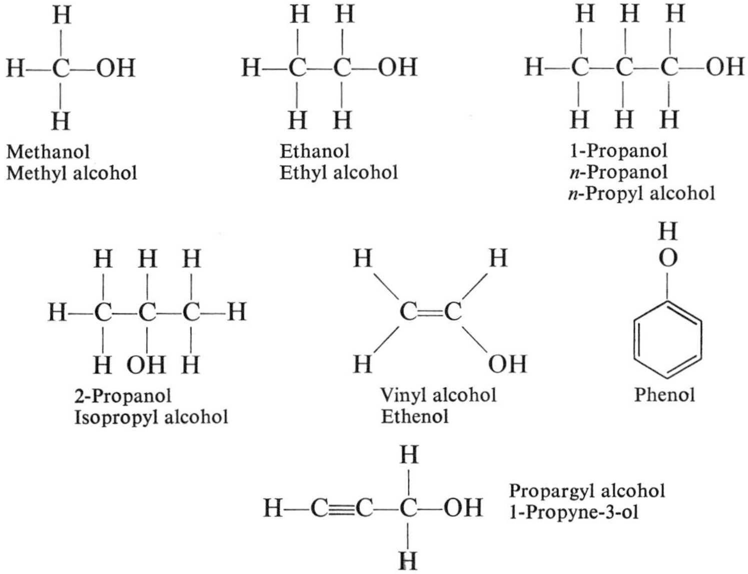

```
     H
     |
  H—C—OH
     |
     H
```

Methanol
Methyl alcohol

```
     H  H
     |  |
  H—C—C—OH
     |  |
     H  H
```

Ethanol
Ethyl alcohol

```
     H  H  H
     |  |  |
  H—C—C—C—OH
     |  |  |
     H  H  H
```

1-Propanol
n-Propanol
n-Propyl alcohol

```
     H  H  H
     |  |  |
  H—C—C—C—H
     |  |  |
     H  OH H
```

2-Propanol
Isopropyl alcohol

```
  H         H
    \     /
     C=C
    /     \
  H         OH
```

Vinyl alcohol
Ethenol

```
  H
  O
  |
 C6H5
```

Phenol

```
            H
            |
  H—C≡C—C—OH
            |
            H
```

Propargyl alcohol
1-Propyne-3-ol

Fig. 22-7 Structural formulas of some alcohols.

Polyhydroxy alcohols contain more than one hydroxy group per molecule. Ethylene glycol, 1,2-dihydroxyethane,

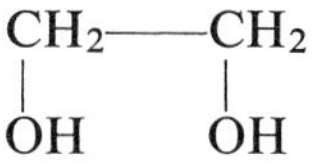

```
CH2——CH2
 |      |
 OH     OH
```

is used as an antifreeze in automobile radiators. Glycerine, or glycerol, is a trihydroxy alcohol:

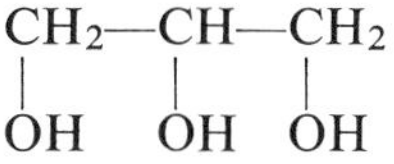

```
CH2—CH—CH2
 |    |    |
 OH   OH   OH
```

It occurs widely in nature as a component of fats.

The aromatic alcohols are typified by phenol, C_6H_5OH:

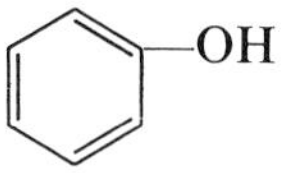

```
C6H5—OH
```

In addition to its use in dilute aqueous solution as an antiseptic wash, phenol is an important intermediate in the synthesis of dyes, plastics, and drugs. In contrast with the non-aromatic alcohols,* phenol is noticeably acidic in water. The origin of the much greater acidity of the aromatic compound is found in the anion which results when phenol ionizes. The

*The term *aliphatic* refers to all non-aromatic hydrocarbons except the cyclic ones; these are called *alicyclic*.

phenolate ion is stabilized by distribution of the negative charge throughout the entire molecule, instead of its being concentrated on one atom as it would be after ionization of an aliphatic alcohol.

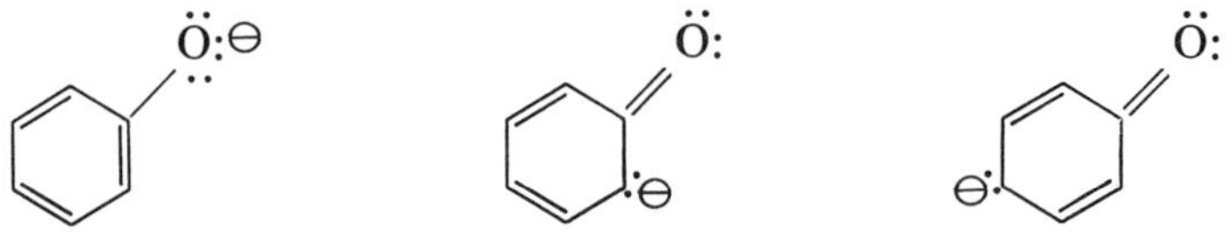

Alcohols undergo a number of chemical reactions. Alkoxide salts are produced with active metals, showing that, even though the alcohols do not show an easily measured acidity in water, the O—H bond is nevertheless quite polar.

$$2C_2H_5OH + 2Na \rightarrow 2C_2H_5ONa + H_2$$

Note the similarity of the alcohols to water in this reaction. Treatment with halogen acids under a variety of conditions leads to formation of alkyl halides.

$$CH_3CH_2CH_2OH + \text{Conc. } HCl \xrightarrow{ZnCl_2} CH_3CH_2CH_2Cl + H_2O$$

In the presence of a strong acid, such as H_2SO_4, the alcohols may lose water to form an alkene:

$$CH_3CH_2CH_2OH \xrightarrow[100°C]{H_2SO_4} CH_3CH{=}CH_2 + H_2O$$

A reaction of this type, in which a smaller molecule is expelled from a larger, is often referred to as an *elimination reaction.*

Aldehydes and ketones — Alcohols are easily oxidized. The product of the oxidation reaction depends on whether the alcohol is primary, secondary or tertiary; *i.e.*, whether the OH group is bonded to a carbon bearing one, two, or three carbon-carbon bonds. Primary alcohols are characterized by the grouping CH_2OH, secondary alcohols by CHOH, and tertiary alcohols by COH as follows:

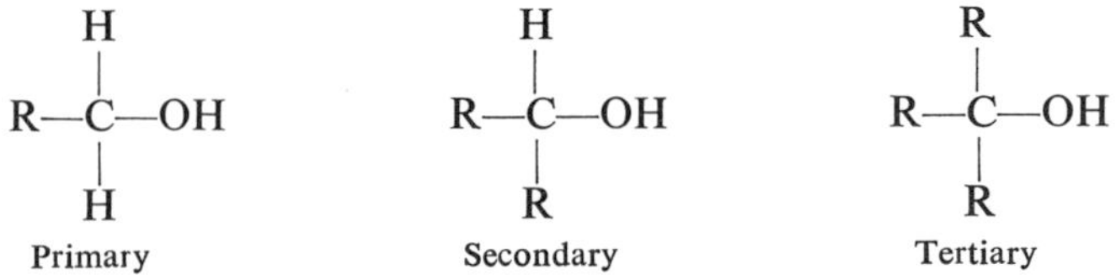

Primary	Secondary	Tertiary
R—C(H)(H)—OH	R—C(H)(R)—OH	R—C(R)(R)—OH

The first oxidation product of a primary alcohol is an *aldehyde:*

$$\underset{\text{1-Propanol}}{CH_3CH_2—CH_2—OH} + (O) \rightarrow \underset{\text{Propionaldehyde}}{CH_3CH_2—\overset{\overset{\displaystyle O}{\|}}{C}—H} + H_2O$$

The symbol (O) indicates oxygen from some oxidizing agent, in a form not specified. An aldehyde is a compound which possesses a CHO group.

If the alcohol is secondary, the product is a *ketone:*

$$\underset{\text{2-Propanol}}{CH_3\underset{\underset{OH}{|}}{C}HCH_3} + (O) \rightarrow \underset{\text{Acetone}}{CH_3\overset{\overset{O}{\|}}{C}CH_3} + H_2O$$

Ketones contain the R_1COR_2 group, where R_1 and R_2 are both hydrocarbon groups.

Tertiary and aromatic alcohols cannot undergo oxidation at the hydroxyl bearing carbon atom without cleavage of a carbon-carbon bond.

The lowest molecular weight compound in the aldehyde class is formaldehyde, HCHO, produced by oxidation of methyl alcohol in air over a catalyst at elevated temperatures:

$$2CH_3OH + O_2 \rightarrow 2H\overset{\overset{O}{\|}}{C}H + 2H_2O$$

It is employed (in the form of a 40 per cent aqueous solution called *formalin*) as an embalming fluid and as a preservative for biological specimens. The chief industrial use is in the formation of polymers.

Acetone, commercially the most important ketone, is used as a solvent. It should be noted that acetone and 1-propen-3-ol are structural isomers, since the molecular formula for both compounds is C_3H_6O. Structural isomerism is subdivided to distinguish cases in which the functional groups are different from those in which they are the same. Two compounds which differ only in the position of a common functional group are termed *positional isomers.* If they differ in the character of the functional group, they are termed *functional-group isomers:*

$$CH_3CH_2CH_2OH \qquad CH_3-\underset{\underset{OH}{|}}{\overset{\overset{H}{|}}{C}}-CH_3$$

Positional isomerism

$$\underset{\text{1-Propen-3-ol}}{CH_2{=}\overset{\overset{H}{|}}{C}-CH_2-OH} \qquad \underset{\text{Acetone}}{CH_3-\overset{\overset{O}{\|}}{C}-CH_3}$$

Functional-group isomerism

Carboxylic acids — Reaction of two atoms of oxygen with a primary alcohol results in formation of a carboxylic acid:

$$\underset{\text{1-Propanol}}{CH_3CH_2CH_2OH} + 2(O) \rightarrow \underset{\text{Propionic acid}}{CH_3CH_2\overset{\overset{O}{\|}}{C}OH} + H_2O$$

In laboratory work the oxidizing agent may be dichromate or permanganate. Catalyzed air oxidation of alcohols is employed industrially, as in the preparation of acetic acid from ethanol.

Many of the acids have been known for a long time as naturally occurring substances, and non-systematic names have become attached to them. Formic acid, HCOOH, first was obtained from the destructive distillation of ants.

Table 22-3 Some long-chain fatty acids.

Formula	Name
$CH_3(CH_2)_{10}COOH$	Lauric acid
$CH_3(CH_2)_{14}COOH$	Palmitic acid
$CH_3(CH_2)_{16}COOH$	Stearic acid
$CH_3(CH_2)_7CH{=}CH(CH_2)_7COOH$	Oleic acid
$CH_3(CH_2)_7CH{=}CH(CH_2)_{11}COOH$	Erucic acid
$CH_3(CH_2)_4CH{=}CHCH_2CH{=}CH(CH_2)_7COOH$	Linoleic acid

Acetic acid, CH_3COOH, results from the enzyme-catalyzed oxidation of alcohols; it is present in soured fruit juices and vinegar. Butyric acid, $CH_3CH_2CH_2COOH$, imparts a rather indelicate aroma to rancid butter and limburger cheese. Long chain carboxylic acids (Table 22-3) occur in animal and vegetable fats as the esters of glycerol. Note that some of the acids contain double bonds. Fats derived from these are usually liquid, and are termed unsaturated fats.

The simplest carboxylic acid representative from the aromatic series is benzoic acid:

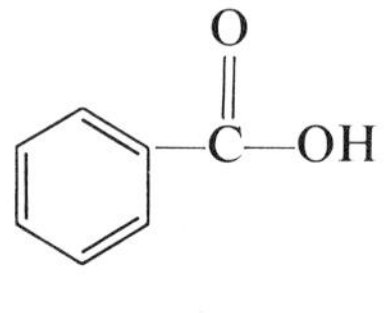

O
‖
C_6H_5—C—OH

A number of acids found in plants and animals contain more than one carboxyl (COOH) group. Some examples are shown in Fig. 22-8.

Although the carboxylic acids vary widely in acid strength, very few of them are strong acids. For the majority, K_a is in the range 10^{-2} to

Table 22-4 Effect of substituents on carboxylic acid strength.

Acid	Formula	K_a(25°C)
Acetic acid	CH_3COOH	1.8×10^{-5}
Monochloroacetic acid	$CH_2ClCOOH$	1.5×10^{-3}
Dichloroacetic acid	$CHCl_2COOH$	5.0×10^{-2}
Trichloroacetic acid	CCl_3COOH	2.0×10^{-1}
Trifluoroacetic acid	CF_3COOH	5.8×10^{-1}

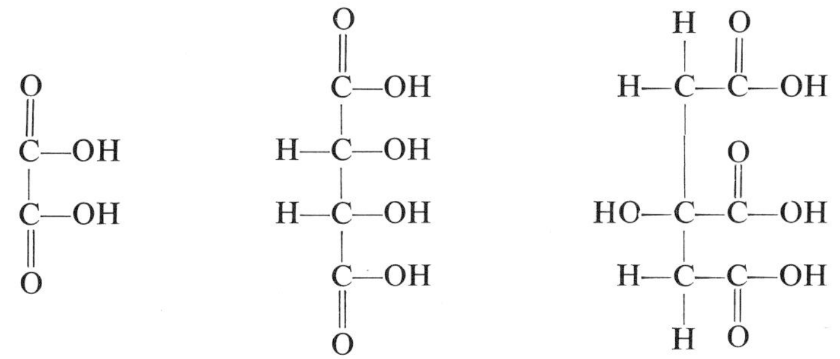

$$\begin{array}{c} O \\ \| \\ C{-}OH \\ | \\ C{-}OH \\ \| \\ O \end{array}$$

Oxalic acid

$$\begin{array}{c} O \\ \| \\ C{-}OH \\ | \\ H{-}C{-}OH \\ | \\ H{-}C{-}OH \\ | \\ C{-}OH \\ \| \\ O \end{array}$$

Tartaric acid

$$\begin{array}{c} H\quad O \\ |\quad \| \\ H{-}C{-}C{-}OH \\ | \\ \quad O \\ \quad \| \\ HO{-}C{-}C{-}OH \\ | \\ H{-}C{-}C{-}OH \\ |\quad \| \\ H\quad O \end{array}$$

Citric acid

Fig. 22-8 Structural formulas of some common organic acids containing more than one carboxyl group.

10^{-5}. If, however, strongly electron-attracting groups are attached to the carboxyl carbon atom, loss of the proton is made easier, and the acid strength increases. Attachment of electron-attracting halogen atoms to the CH_3 group in acetic acid brings about an increase in acid strength, as shown by the data in Table 22-4.

The condensation of an acid with an alcohol, with the splitting out of water, results in formation of an ester:

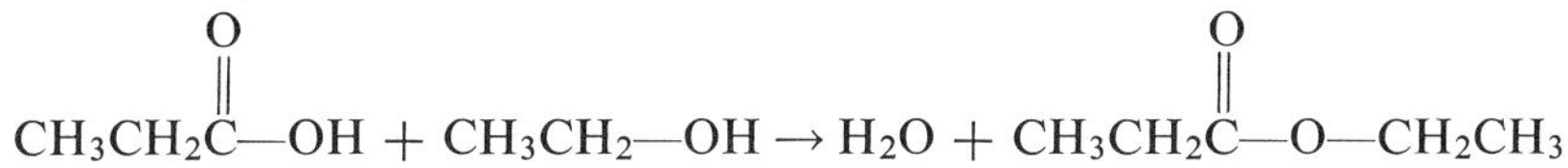

$$\underset{\text{Propionic acid}}{CH_3CH_2\overset{\overset{O}{\|}}{C}{-}OH} + \underset{\text{Ethanol}}{CH_3CH_2{-}OH} \rightarrow H_2O + \underset{\text{Ethyl propionate}}{CH_3CH_2\overset{\overset{O}{\|}}{C}{-}O{-}CH_2CH_3}$$

Esters are named by first naming the group derived from the alcohol, and then the group derived from the acid (*e.g.*, ethyl butyrate, $C_3H_7COOC_2H_5$; phenyl acetate, $CH_3COOC_6H_5$). Some of the lower molecular weight esters contribute to the pleasant odors of fruits and flowers. The high molecular weight esters are waxy substances found in beeswax, carnauba wax, *etc.*

Methyl acetate and ethyl acetate are low-boiling liquids which are widely used as solvents for fast-drying paints and cements.

Condensation reactions are quite general, and may be applied to other than carboxylic acids. The phosphate esters, compounds of the form

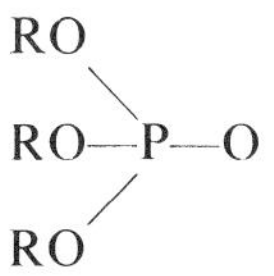

$$\begin{array}{l} RO \\ \quad\diagdown \\ RO{-}P{-}O \\ \quad\diagup \\ RO \end{array}$$

are derived (formally at least) from phosphoric acid and the alcohol ROH.

Fats are esters of glycerol with long-chain carboxylic acids:

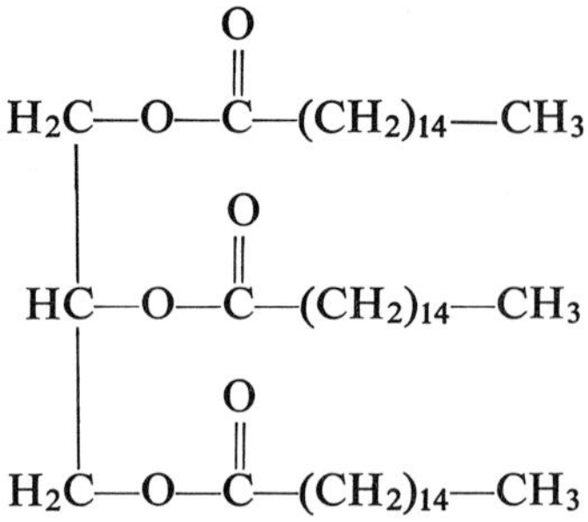

$$\begin{array}{l} H_2C-O-\overset{\overset{\large O}{\|}}{C}-(CH_2)_{14}-CH_3 \\ \;| \\ HC-O-\overset{\overset{\large O}{\|}}{C}-(CH_2)_{14}-CH_3 \\ \;| \\ H_2C-O-\overset{\overset{\large O}{\|}}{C}-(CH_2)_{14}-CH_3 \end{array}$$

Glyceryl tripalmitate

The hydrolysis of esters, with formation of the acid and alcohol, is catalyzed by basic solutions. Fats such as the one shown above are rapidly hydrolyzed to yield glycerol and the acid anion, when heated with sodium hydroxide or sodium carbonate solution. Soap is obtained by saturating the solution with sodium chloride to cause precipitation of the sodium salt of the carboxylic acid. The composition of the soap depends on the particular fats used in its preparation.

Ethers — The condensation of two alcohol molecules with a splitting out of water leads to the formation of ethers. Treatment of ethyl alcohol with concentrated sulfuric acid at 150°C leads to dimethyl ether. The reaction is essentially a dehydration process, catalyzed by the great avidity of concentrated sulfuric acid for water.

$$CH_3CH_2-O\boxed{H + HO}-CH_2CH_3 \xrightarrow{H_2SO_4} CH_3CH_2-O-CH_2CH_3 + H_2O$$

Ethers may be considered as disubstituted water molecules. With loss of both OH groups all possibility of hydrogen bonding disappears. The ethers are accordingly not very polar in character, and are immiscible with water. They do act as Lewis bases, however, as the reaction of dimethyl ether with boron trifluoride illustrates.

$$CH_3-\ddot{\underset{\cdot\cdot}{O}}-CH_3 + BF_3 \rightarrow (CH_3)_2\ddot{O}-BF_3$$

22-5 Petroleum, Polymers, and Plastics

Petroleum is a complex liquid solution of gaseous, liquid, and solid hydrocarbons. It is organic in origin, but the nature of the organic material and the manner of its conversion into petroleum are uncertain. The original material apparently lay in beds under salt water at one stage in the process. The petroleum, formed in a series of reactions, migrated through porous strata, finally to accumulate in pockets which happened to be favorable for such accumulations. The entire process required a very long time — perhaps thousands of centuries. It gives one pause to consider that mankind, in all likelihood, will have completely undone nature's handiwork in perhaps two centuries.

For purposes of handling and transportation, petroleum deposits may be classified as liquid or gaseous. The natural gas deposits consist largely of methane, with some ethane, propane, and other hydrocarbons of low molecular weight also present. (There is also some helium, which is removed under the auspices of the Federal government.) The liquid fraction may consist of both alkanes and cycloalkanes. The first step in refining the liquid fraction is accomplished in huge fractionating towers by separating it into batches with different boiling ranges. The important fractions obtained are listed in Table 22-5.

Table 22-5 Petroleum fractions.

Fraction	Size range	Boiling point range (°C)
Gas	C_1–C_5	(−160)–(−30)
Petroleum ether	C_5–C_7	30–90
Gasoline	C_5–C_{12}	30–200
Kerosene	C_{12}–C_{16}	175–275
Fuel oil, Diesel oil	C_{15}–C_{18}	250–400
Lube oils	C_{16}–up	350–
Paraffin waxes	C_{20}–up	Melting point 50–up
Residues (pitch, tar)		

The need for motor fuels creates the largest demand for petroleum products. It is desirable, therefore, that as much crude petroleum as possible be turned to this use. The gasoline fraction in the initial fractionation, called *straight-run gasoline*, consists largely of unbranched alkanes. As it happens, the branched-chain compounds possess superior burning characteristics in modern, high-compression automobile engines. The straight-run gasoline is therefore often put through an isomerization operation, in which — under the influence of a catalyst — the straight-chain compounds are isomerized to branched-chain compounds.

The higher molecular weight fractions (kerosene, diesel, and fuel oil) are converted into lower boiling fractions by a thermal cracking operation. Hydrocarbon vapors are fed into a cracking chamber at high temperatures (400-600°C) under high pressure. The catalysts (certain clays and aluminosilicate minerals) are fed in as a dust which remains suspended in the gas stream. The products of the cracking operation are lower molecular weight, highly branched hydrocarbons — ethylene, propylene, and hydrogen. The olefins are used in the production of plastics. The hydrogen which is formed may be combined with nitrogen in the Haber process for forming ammonia. The petrochemical industry provides some interesting examples of how a new process — developed to make use of a material available in large supply — leads to new byproducts, for which, in turn, new processes are developed. The result of this kind of development has been an ever-widening scope for synthetic chemistry in meeting the needs of modern society.

Polymers and plastics — A polymer is a high-molecular-weight molecule possessing a characteristic repeating unit. The repeating units may extend in a long chain, in two or three dimensions. Polymers range in molecular weight from a few thousand to perhaps ten million units. They are of great importance, both as synthetic products and as naturally occurring substances. Linear polymers are the most important type, although in some cases connections between chains (cross-links) exist at random points.

Natural rubber is a hydrocarbon polymer extracted from the rubber tree, which grows in tropical climates.

$$\begin{array}{cccccccccccccc} & H & & & H & H & & & H & H & & & H & \\ & | & & & | & | & & & | & | & & & | & \\ - & C - & C = & C - & C - & C - & C = & C - & C - & C - & C = & C - & C & - \\ & | & | & | & | & | & | & | & | & | & | & | & | & \\ & H & CH_3 & H & H & H & CH_3 & H & H & H & CH_3 & H & H & \end{array}$$

Natural rubber

Natural rubber is not useful as such; it is soft and sticky, and possesses little resiliency. It is converted into useful form by heating with sulfur, a process called *vulcanization*. The sulfur adds to some of the double bonds along the polymer chain. By joining carbon atoms in two adjacent chains, the sulfur atoms act as bridging groups which reduce the flexibility of the polymers and add strength to the material (Fig. 22-9).

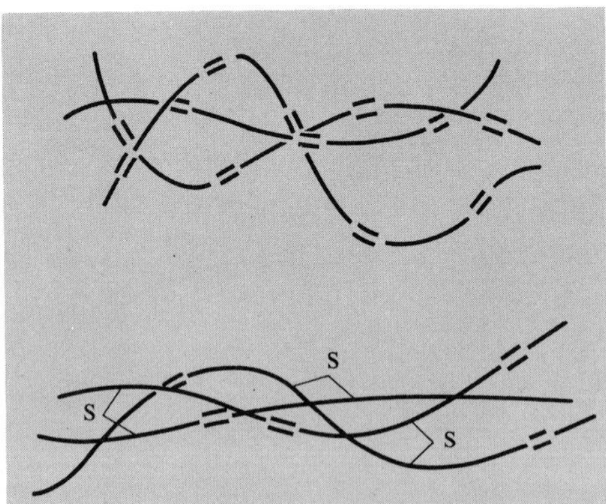

Fig. 22-9 Schematic representation of vulcanization of naturally occurring polyisoprene to produce commercially useful rubber. Sulfur atoms (or short chains of sulfur atoms) connect the polyisoprene chains, imparting hardness and greater elastic strength.

Synthetic polymers which are somewhat analogous to natural rubber have been produced in increasing quantities since World War II. Neoprene is a polymer of chloroprene, which is formed by addition of two molecules of acetylene with one of hydrogen chloride:

$$2CH{\equiv}CH + HCl \rightarrow CH_2{=}\underset{\displaystyle Cl}{\underset{|}{C}}{-}\overset{\displaystyle H}{\overset{|}{C}}{=}CH_2$$

Chloroprene (2-chloro-1,3 butadiene)

$$-\underset{\underset{\displaystyle H}{|}}{\overset{\overset{\displaystyle H}{|}}{C}}-\underset{\underset{\displaystyle Cl}{|}}{C}=\underset{\underset{\displaystyle H}{|}}{C}-\underset{\underset{\displaystyle H}{|}}{\overset{\overset{\displaystyle H}{|}}{C}}-\underset{\underset{\displaystyle H}{|}}{\overset{\overset{\displaystyle H}{|}}{C}}-\underset{\underset{\displaystyle Cl}{|}}{C}=\underset{\underset{\displaystyle H}{|}}{C}-\underset{\underset{\displaystyle H}{|}}{\overset{\overset{\displaystyle H}{|}}{C}}-\underset{\underset{\displaystyle H}{|}}{\overset{\overset{\displaystyle H}{|}}{C}}-\underset{\underset{\displaystyle Cl}{|}}{C}=\underset{\underset{\displaystyle H}{|}}{C}-\underset{\underset{\displaystyle H}{|}}{\overset{\overset{\displaystyle H}{|}}{C}}-$$

Neoprene

Polymers such as Neoprene are termed *addition polymers*, since the monomers merely add together to form the chain; aside from the catalyst no other substance is involved. The addition of the monomer units to one another is of a special kind, referred to as 1,4 *addition*. This indicates that the added groups are attached to the end carbons of the monomer unit. If the addition occurred across just one of the double bonds it would be labelled a 1,2 addition. In a sense, one addition reaction occurs across two double bonds, with the net result that one double bond remains after the addition.

Polyethylene is another important addition polymer; the monomer unit is ethylene, a byproduct of the thermal cracking process in the petrochemical industry. In recent years the methods employed for polymerization of ethylene have been vastly improved, and it has become the most commonplace of the synthetic plastics. Polyvinyl chloride (from vinyl chloride, $CH_2{=}CHCl$) and Dacron and Orlon (from acrylonitrile, $CH_2{=}CHCN$) are examples of addition polymers formed from ethylene derivatives as monomers.

Condensation polymerization involves the joining of monomer units with splitting out of a smaller molecule, usually water. A number of important polymers, including naturally occurring substances, are of this type (Table 22-6).

The properties of materials fabricated from polymer substances are, of course, dependent upon the structure of the polymer molecule, but they also depend upon the treatment of the raw polymer material in the fabrication process. Since the formula for a polymer chain is written as a long line of symbols, it is easy to envisage the molecules themselves as long straight chains. In actuality, however, polymers may twist and coil in a great variety of ways. Sometimes special effects operate to stabilize one configuration relative to others. In proteins, for example, intramolecular hydrogen bonds may stabilize a helical coil configuration. Cellulose is

Table 22-6 Some examples of condensation polymers.

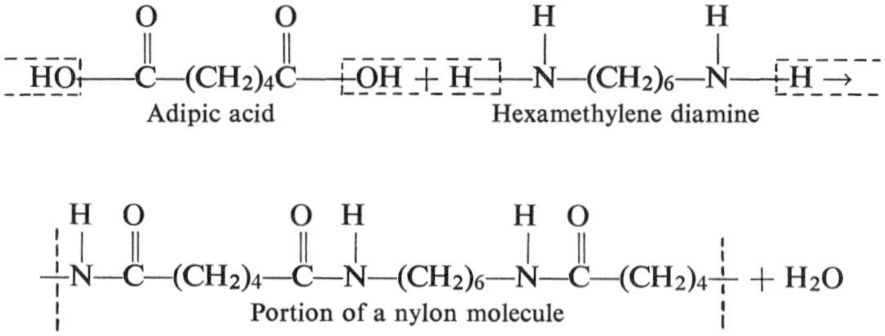

$$HO-\overset{O}{\overset{\|}{C}}-(CH_2)_4-\overset{O}{\overset{\|}{C}}-OH + H-\overset{H}{\overset{|}{N}}-(CH_2)_6-\overset{H}{\overset{|}{N}}-H \rightarrow$$

Adipic acid   Hexamethylene diamine

$$-\overset{H}{\overset{|}{N}}-\overset{O}{\overset{\|}{C}}-(CH_2)_4-\overset{O}{\overset{\|}{C}}-\overset{H}{\overset{|}{N}}-(CH_2)_6-\overset{H}{\overset{|}{N}}-\overset{O}{\overset{\|}{C}}-(CH_2)_4- + H_2O$$

Portion of a nylon molecule

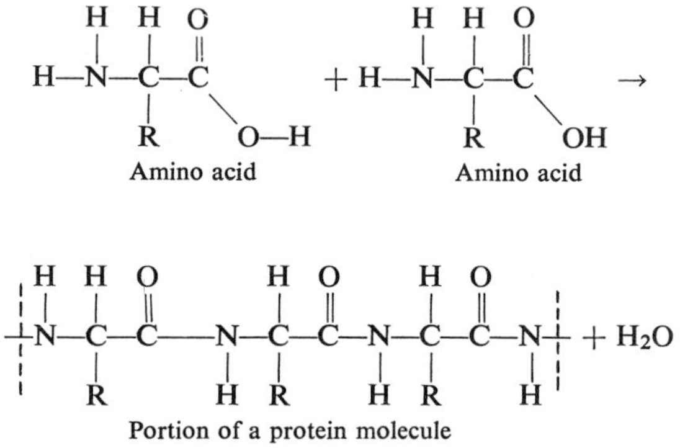

$$H-\overset{H}{\overset{|}{N}}-\underset{R}{\overset{H}{C}}-\overset{O}{\overset{\|}{C}}-O-H + H-\overset{H}{\overset{|}{N}}-\underset{R}{\overset{H}{C}}-\overset{O}{\overset{\|}{C}}-OH \rightarrow$$

Amino acid   Amino acid

$$-\overset{H}{\overset{|}{N}}-\underset{R}{\overset{H}{C}}-\overset{O}{\overset{\|}{C}}-\underset{H}{N}-\underset{R}{\overset{H}{C}}-\overset{O}{\overset{\|}{C}}-\underset{H}{N}-\underset{R}{\overset{H}{C}}-\overset{O}{\overset{\|}{C}}-\underset{H}{N}- + H_2O$$

Portion of a protein molecule

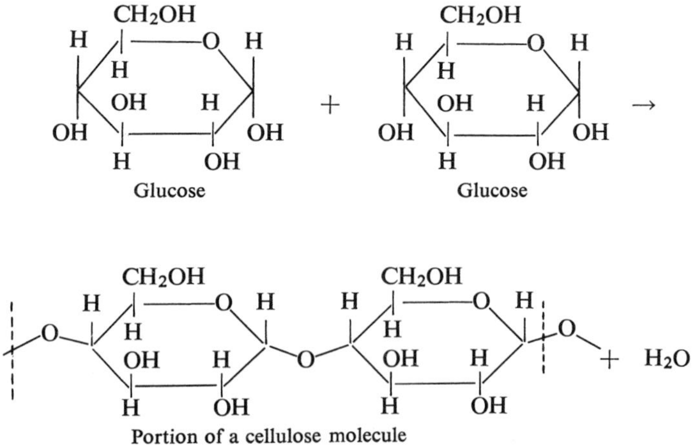

CH_2OH, H, O, H, H, OH, H, OH, OH, H, OH + CH_2OH, H, O, H, H, OH, H, OH, OH, H, OH →

Glucose   Glucose

CH_2OH, H, O, H, H, OH, H, H, OH, —O—, CH_2OH, H, O, H, H, OH, H, H, OH, —O— + H_2O

Portion of a cellulose molecule

Note: Each unlabelled corner of the hexagon represents a saturated carbon atom. The structure of cellulose is similar to that for starch.

found in plants as long filament-like molecules which lie in parallel array in the plant structure.

It is possible to cause at least a partial alignment of linear polymer molecules by imposing an external force. If a solution containing polymer molecules is evaporated on a glass slide, and at the same time the drying material is stroked in a given direction, the molecules align with long axes parallel to the direction of the stroke. Nylon thread is formed by extruding the molten polymer through very small orifices and cooling it. The filaments which form are then stretched, to cause the molecules to align along the thread axis.

22-6 Biochemistry

Biochemistry, which is the study of the chemistry of plant and animal systems, is tremendously varied. The molecules involved may be as small as water molecules or as large as protein molecules with a molecular weight on the order of a million. Complexity arises in biochemistry partly because of this great variety in size. There is also great variety in the functional groups involved in reactions. In this very brief introduction to the subject, we shall mention only three major classes of substances — proteins, carbohydrates, and fats — which are important in human nutrition, and then consider the molecular structures and properties of a few familiar compounds.

Amino acids and proteins — Proteins are large molecules, rather high in nitrogen content, found in the cells of plants and animals. The name protein was suggested by Berzelius in about 1835. All proteins contain carbon, hydrogen, nitrogen, oxygen, and (usually) sulfur. Nitrogen is present to the extent of about 16 per cent by weight. Under conditions similar to those which cause hydrolysis of esters, the proteins yield as a major product a mixture of simpler substances, the *amino acids.* The hydrolysis is effected by boiling with about $6N$ hydrochloric acid. Base hydrolysis is also effective in breaking up the protein, but it destroys some of the amino acid constituents.

The amino acids are the basic structural units of protein. With only two exceptions, the amino acids derived from protein are primary alpha amino acids; *i.e.*, they contain a carboxyl and an amino group bound to the same carbon:

$$\begin{array}{ccccc} & & H & & O & & \\ & & | & & \| & & \\ R & — & C & — & C & — & OH \\ & & | & & & & \\ & & NH_2 & & & & \end{array}$$

There are some twenty different amino acids known to occur in protein. With the two exceptions mentioned above, they differ only in the nature

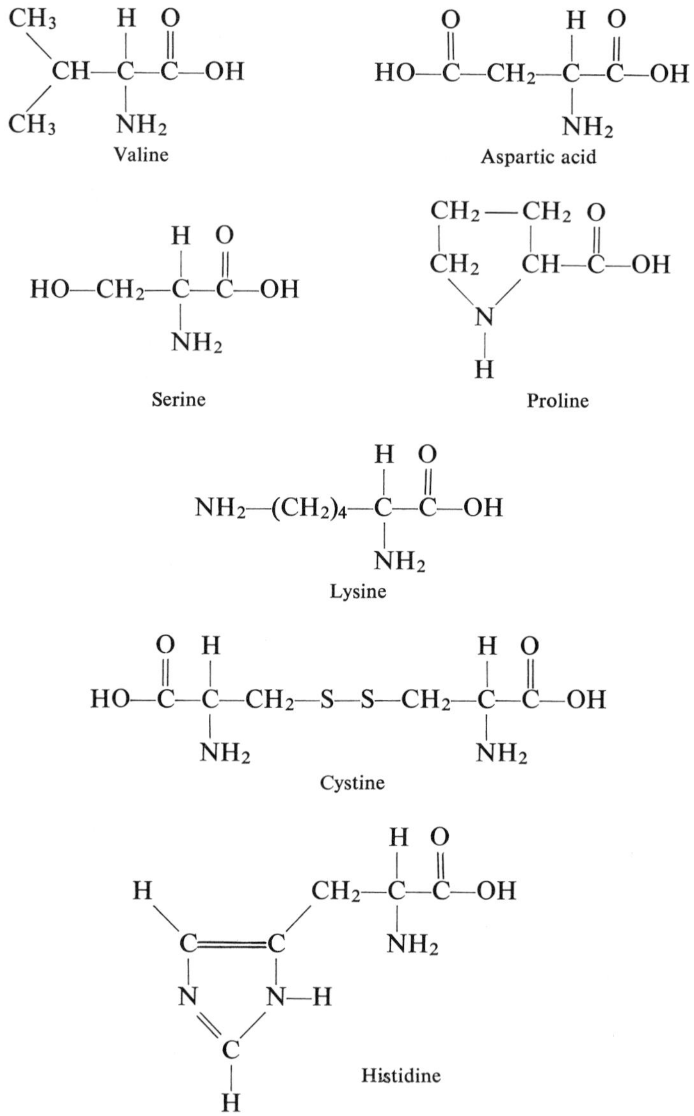

$(CH_3)_2CH{-}CH(NH_2){-}C(=O){-}OH$

Valine

$HO{-}C(=O){-}CH_2{-}CH(NH_2){-}C(=O){-}OH$

Aspartic acid

$HO{-}CH_2{-}CH(NH_2){-}C(=O){-}OH$

Serine

$CH_2{-}CH_2{-}CH_2{-}N(H){-}CH{-}C(=O){-}OH$ (ring: $CH_2{-}CH_2$, CH_2, CH, $N{-}H$)

Proline

$NH_2{-}(CH_2)_4{-}CH(NH_2){-}C(=O){-}OH$

Lysine

$HO{-}C(=O){-}CH(NH_2){-}CH_2{-}S{-}S{-}CH_2{-}CH(NH_2){-}C(=O){-}OH$

Cystine

$H{-}C{=}C(CH_2{-}CH(NH_2){-}C(=O){-}OH){-}N{-}H{-}C(H){=}N$ (imidazole ring)

Histidine

Figure 22-10 Some naturally occurring amino acids.

of the group R. A few of the more common acids are shown in Fig. 22-10. Note that some of the acids contain a second acid group, while others contain a second amino group. The presence of these groups has an important effect on the acid-base behavior of the compound.

The amino acids behave both as weak acids and as weak bases, since they contain both an acidic and basic group. Each group has its characteristic acid or base dissociation constant. A monoamino-monocarboxylic acid compound such as glycine is internally ionized in solution:

$$\begin{array}{ccccc} & H & & O & \\ & | & & \| & \\ H- & C & - & C & -O^- \\ & | & & & \\ & NH_3^+ & & & \end{array}$$

A dipolar ion of this type is termed a *zwitterion.* For a given amino acid in solution, there is a particular value of pH — called the *isoelectric point* — at which the number of protonated amino groups is just equal to the number of ionized carboxyl groups. The value of pH at the isoelectric point can be determined from a knowledge of the acid and base dissociation constants, as illustrated below:

Acid dissociation: $HA + H_2O \rightarrow H_3O^+ + A^-$

$$K_a = \frac{(H_3O^+)(A^-)}{(HA)}$$

Base dissociation: $B + H_2O \rightarrow BH^+ + OH^-$

$$K_b = \frac{(BH^+)(OH^-)}{(B)}$$

Since B and HA are the same molecule, (B) = (HA)

$$\frac{(H_3O^+)(A^-)}{K_a} = \frac{(OH^-)(BH^+)}{K_b}$$

Since, at the isoelectric point, there are equal numbers of ionized amino and carboxyl groups, $[A^-] = (BH^+)$; using $K_w = (OH^-)(H_3O^+)$

$$(H_3O^+)^2 = K_w \times \frac{K_a}{K_b} \qquad pH = 7 + \frac{pK_a - pK_b}{2}$$

With the exception of glycine, all of the naturally occurring amino acids are capable of existing as optical isomers. This is possible because there is at least one carbon atom (the one bearing the amino group) which has four different groups attached to it. The amino acid molecule thus does not possess a center or plane of symmetry, and can exist in either of two non-equivalent mirror image forms. With the exception of glycine, *all of the amino acids found in living systems are optically active.* They consist of just one of the two possible isomers; furthermore, all of the different acids possess the same spatial arrangement of groups about the carbon atom. The particular configuration found in nature is labelled the L form. Living systems are very particular about this matter of isomerism, even though it involves merely relative arrangement of the same groups in space. The D form of a necessary amino acid does not satisfy the nutritional requirements of an animal system.

Peptides are formed from the condensation of two or more molecules of amino acid, with the splitting out of water. Glycylglycine is an example of a dipeptide:

$$\mathrm{H-\underset{\underset{NH_2}{|}}{\overset{\overset{H}{|}}{C}}-\overset{\overset{O}{\|}}{C}-\underset{\underset{H}{|}}{N}-CH_2-\overset{\overset{O}{\|}}{C}-OH}$$

Hofmeister and Fisher both proposed in 1902 that the peptide bond (enclosed within the dotted lines above) is the main type of linkage between amino acids in proteins. This hypothesis has been firmly established by a large body of evidence. The major chain, or skeletal backbone, of a protein chain is thus as shown in Table 22-6. The side groups, which differ for each amino acid, are very important in determining the structure.

The proteins range in molecular weight from perhaps 6000 to many millions. Many of them have been obtained as homogeneous, crystalline substances, but the majority are known only by their properties in solution. Molecules of very high molecular weight behave in many respects more like colloidal suspensions than homogeneous solutions, since a single molecule is large in terms of ordinary molecular dimensions. They are amphoteric substances, just as their constituent amino acids are, and are often drastically altered by changes in pH.

The proteins are classified in a number of ways, depending upon chemical composition, physical properties, or occurrence in biological systems. They range from water-soluble substances such as egg albumin to tough, insoluble materials such as hair and fingernails. The proteins may contain any of the 20 amino acids known in protein work, and these may be connected in any sequence. Even if the over-all amino acid composition of a protein is known, therefore, the structure of the protein is still not determined, since the sequence in which the acids occur in the protein is not known. It is no simple task to unravel the protein, bit by bit, to discover the acid sequence. At the present time, only a few proteins have been completely worked out.

Determination of the amino acid composition and sequence is not the entire story in determining protein structure. A complete understanding of the protein involves knowledge of:

1. *The primary structure.* The amino acid composition and sequence.

2. *The secondary structure.* The long protein chain may possess a number of possible configurations. Internal hydrogen bonding is important in determining the spatial arrangement of the bond angles in the chain. On the other hand, a polar solvent such as water also interacts with the atoms along the chain, and is important in determining the chain configuration. Intramolecular hydrogen bonding from an N—H bond to the carbonyl oxygen of another link along the chain gives rise to a helical configuration (Fig. 22-11). If the intramolecular hydrogen bonds are broken by

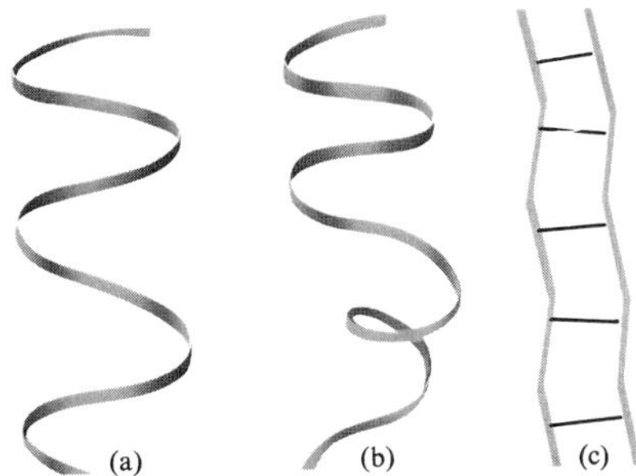

Fig. 22-11 Diagrammatic representation of possible configurations of protein (polyamino acid) chains: (a) helical, (b) random coil, (c) extended.

adding to the solution some reagent which is a very strong hydrogen bonding substance, the protein may assume a random coil configuration. If the side groups along the chain possess charged sites, the helical configuration may be unstable because of the repulsions between like charges. The protein then may assume an extended configuration (Fig. 22-11) in which form it may be hydrogen bonded to adjacent, parallel chains.

3. *The tertiary structure.* Proteins, in addition to interacting *intra*molecularly through hydrogen bonding, also interact *inter*molecularly through hydrogen bonding, electrostatic forces, van der Waals' forces, *etc.* The intermolecular interaction leads to the so-called tertiary structure, in which the proteins are assembled end to end or laterally. It is even possible to speak of a quaternary structure, which is a grouping of the tertiary structures to form a crystalline or quasi-crystalline structure, such as occurs in the virus particles and in crystalline proteins. From these brief remarks on the nature of proteins — one of the basic types of molecular substance in living systems — it should be apparent that life is very highly organized. Each living system contains a great number of proteins — each different from the others, each performing some highly specific role in the dynamic chemical system.

The *enzymes*, substances which act as catalysts for biochemical reactions, are proteinic in nature. Many enzymes are extremely specific in character, operating only with a single chemical substance and no other. On the other hand, some enzymes catalyze reactions of a particular type, and operate on any substance which conforms to a rather general requirement. Certain enzyme-catalyzed reactions require — in addition to the enzyme itself — the presence of one or more smaller molecules or ions, called cofactors, or coenzymes. Elucidation of the mode of action of enzymes is a challenging and important research goal. Although considerable progress has been made in recent years, it cannot be said at this time that we possess a satisfactory understanding of all aspects of any enzyme-catalyzed reaction.

Carbohydrates — The carbohydrates have the general formula $C_x(H_2O)_x$. They are thus formally "hydrates of carbon," although there is no structural significance to this notion. All of the carbohydrates are derived from *saccharides*, polyhydroxy compounds with either an aldehyde or ketone group. One such unit is called a *monosaccharide*. The saccharides may be joined together to form disaccharides, trisaccharides, and other polysaccharides. The structural formulas of a number of monosaccharides are shown in Fig. 22-12. The convention which is used in this type of representation is that the carbon atoms which form the backbone of the saccharide molecule are not shown explicitly, but are at the intersections of the vertical line with the horizontal lines. Thus, ribose has a five-carbon chain, glucose has a six-carbon chain, *etc.* Since there are carbon atoms which have four different groups attached in the molecules, there is the possibility of optical isomerism. Only one of the two optical isomers is found in living systems. For example, in the case of glucose, only D-glucose is found.

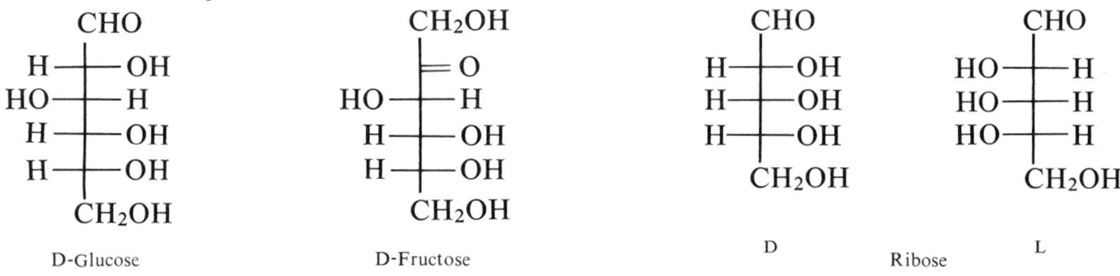

Fig. 22-12 Structural formulas of three monosaccharides.

The saccharides may exist in another form from that shown in Fig. 22-12, and in fact the alternative form is the one in which they are found in existence in polysaccharides. By rearranging the bonding slightly, it is possible to form a six-membered ring involving five carbon atoms of the chain and the oxygen atom of the aldehyde or ketone group:

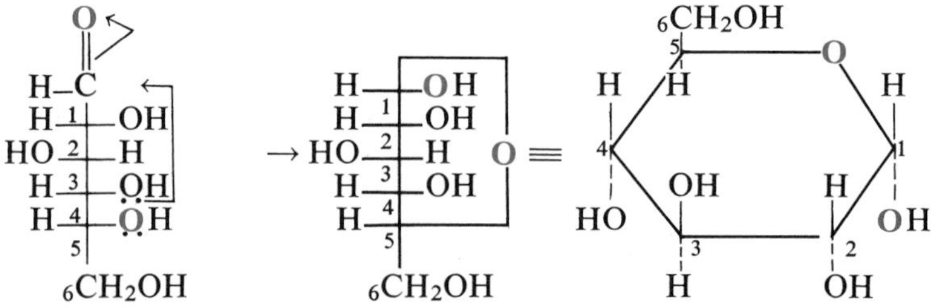

When the polysaccharides are formed from monosaccharides in this form, linear chains of the six- (sometimes five-) membered rings are formed, as exemplified by the structures for cellulose shown in Table 22-6.

The carbohydrates are a relatively direct product of photosynthesis in plants, for which the over-all reaction is

$$CO_2 + H_2O \rightarrow C_x(H_2O)_x + O_2$$

Starch, which is found mainly in the seeds or tubers of plants, and *cellulose*, which forms the walls of cell plants, are both polysaccharides of D-glucose. Cellulose molecules are generally of higher molecular weight than starch. There is also a slight difference between cellulose and starch in the geometrical relationships of the one of the OH groups to the others. The difference, though slight, is important for human nutrition, since starch is metabolized in the body (*i.e.*, used as a source of chemical energy), whereas cellulose is not.

The ordinary sugar of commercial importance is *sucrose*, a disaccharide of glucose and fructose. A number of other sugars, some monosaccharides and other disaccharides, are found in fruits and other foods. Not all sugars have sweet taste to the same degree as sucrose; in general, however, they are all soluble in water, and are readily converted into chemical energy in the body.

Fats — Animal fats such as lard and butter, and vegetable fats such as corn and soybean oils, are composed principally of esters of glycerol, as discussed on page 578. A great variety of long-chain carboxylic acids are found in the form of these esters, called *glycerides.* A few of the more common acids are listed in Table 22-3. In general, the acids tend to have an even number of carbon atoms in the chain. The glycerides formed from the acids containing double bonds are often referred to as *unsaturated fats.* Lower melting vegetable fats such as corn or cottonseed oil are "hardened" by hydrogenation to form higher-melting saturated fats which are used in such commercial products as Crisco and Spry.

Fats are an important source of energy in human nutrition. The complete metabolism (*i.e.*, chemical breakdown of a complex molecule, with release of energy) of fats results in the release of twice as many calories as are released by the metabolism of an equal weight of carbohydrate.

Selected compounds of biochemical interest — A great many substances and elements are essential to long term health in animal systems. Whereas fats and carbohydrates are important principally as a source of heat energy, human need for protein is more specific. Nine of the amino acids are essential; they are not manufactured by the body. Many inorganic substances, particularly quantities of various metal ions, are essential. Although iron, for example, is required in rather large quantities, other elements such as cobalt or molybdenum are required only in trace quantities. A number of organic substances, called *vitamins*, are also required for human health. The vitamins are all substances of moderately low molecular weight (200-300 or less). They serve a number of different purposes in the body, and a deficiency of any one gives rise to characteristic symptoms. The structural formulas of three of the vitamins are shown in Fig. 22-13, principally for the purpose of showing that they are not unusually complex molecules. The structure for vitamin B_1 also exhibits another characteristic of biologically interesting compounds, the prevalence of *heterocyclic rings.* A heterocyclic ring is one in which one or more of the carbon atoms is replaced by nitrogen, sulfur, or some other element.

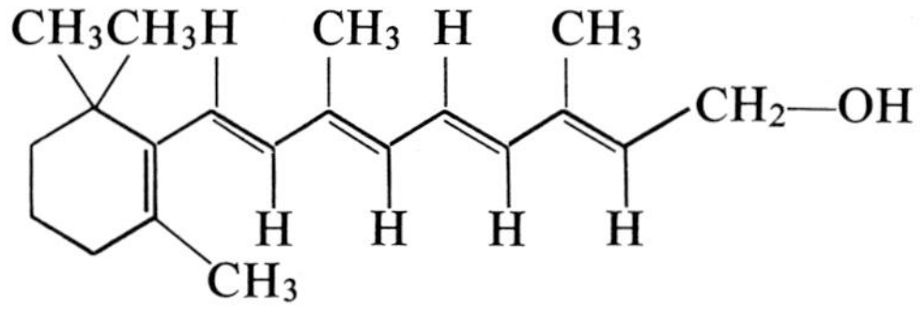

Vitamin A

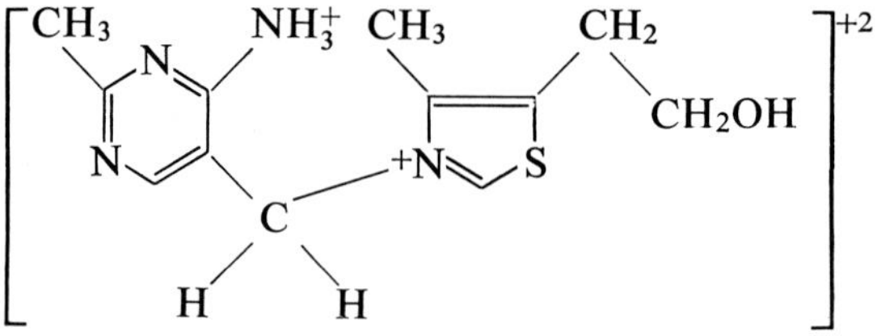

Vitamin B_1

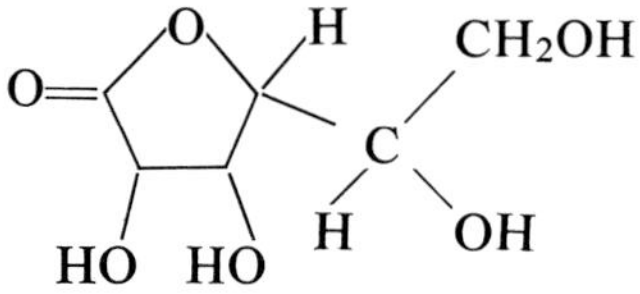

Vitamin C

Fig. 22-13 Structural formulas of vitamins A, B_1 and C.

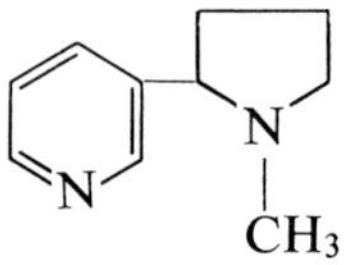

Nicotine

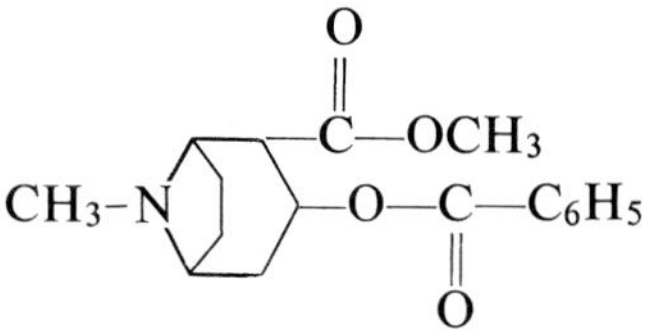

Cocaine

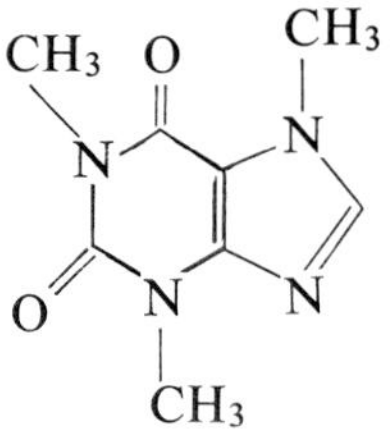

Caffeine

Fig. 22-14 Structural formulas of three physiologically active compounds.

Many of the heterocyclic systems of importance involve nitrogen and are aromatic in character. The nucleic acids, which constitute the genes, are composed of heterocyclic rings connected through sugar (ribose) and phosphate groups.

Many compounds which occur in plants produce marked physiological activity when ingested into the human system. Such compounds have been used for relief of pain, for lowering of blood pressure, and for many other purposes. The structures of a few such molecules are shown in Fig. 22-14.

Suggested Readings

Pauling, L., "The Stochastic Method and the Structure of Proteins," *American Scientist*, Vol. 43, 1955, p. 285.

Shaw, W. H. R., "Kinetics of Enzyme Catalyzed Reactions," *J. Chem. Educ.*, Vol. 34, 1957. p. 22.

Smith, W. B., *A Modern Introduction to Organic Chemistry*, Columbus, Ohio: Charles E. Merrill Books, Inc., 1961.

White, E. H., *Chemical Background for the Biological Sciences*, Englewood Cliffs, N. J.: Prentice-Hall, Inc., 1964.

Exercises

22-1. Write structural formulas for all of the structural isomers of C_7H_{16}.

22-2. How many different products might result from the following reaction?

$$C_4H_{10} + Cl_2 \rightarrow C_4H_9Cl + HCl$$

22-3. Write the structural formulas and names for all compounds of molecular formula C_4H_8.

22-4. Graph the melting and boiling points of the straight-chain alkanes *vs* the number of carbon atoms in the chain. Are there any peculiarities in this graph? How do you explain the general trend? How do you explain any peculiarities you find?

22-5. Consider the following values of K_a for a series of carboxylic acids:

		K_a
α-chlorobutyric acid	$CH_3CH_2CHClCOOH$	1.45×10^{-3}
β-chlorobutyric acid	$CH_3CHClCH_2COOH$	8.80×10^{-5}
Butyric acid	$CH_3CH_2CH_2COOH$	1.50×10^{-5}

Why is the K_a for α-chlorobutyric acid greater than that for butyric acid? Explain the variation in K_a with the location of the chlorine atom.

22-6. A compound of formula C_4H_6 reacts with 2 moles of Br_2 to give $C_4H_6Br_4$. Is this information above sufficient to determine the structure of the compound? Explain.

22-7. The standard heats of combustion of ethane and ethylene are -373 and -337 kcal/mole, respectively. The standard heats of formation of $H_2O(l)$ and $CO_2(g)$ are -68 and -94 kcal/mole, respectively. From these data, calculate $\Delta H°$ for the reaction

$$C_2H_4 + H_2 \rightarrow C_2H_6$$

How does this compare with the heat of hydrogenation of cyclohexene?

22-8. Draw structural formulas for, and name, all of the possible isomers of the aromatic compound, trichlorobenzene.

22-9. Explain what is meant by the concept of resonance energy. What experimental evidence exists in support of the concept?

22-10. The following statement has appeared in the chemical literature: "The pK_b of 2-aminofluorene

—NH_2

in 70 per cent ethanol at 25°C is 9.30. At $HClO_4$ concentrations between 0.02 and 0.05*M*, with 0.005*M* amine, the amount of free amine should vary from 0.01 per cent to 0.38 per cent of the total 2-aminofluorene present." Perform the calculations necessary to substantiate this result. Are the statements correct? [*Note:* The fact that the solvent is 70 per cent ethanol, 30 per cent water is not important to the calculations, except that K_w should be corrected, Sec. 15-3.; assume that perchloric acid is a strong acid in the mixed solvent.]

22-11. Glacial acetic acid is sometimes employed as a non-aqueous solvent medium (p. 300). Employing Brönsted acid-base theory, write the reactions of perchloric acid, and of a base, acetamide, $CH_3C(=O)—NH_2$, with acetic acid as solvent.

22-12. Note the melting points listed below for some di-substituted benzene derivatives. Suggest an explanation for the *para*-substituted compounds' melting at a higher temperature.

	Melting points (°C)		
Substituents	*ortho-*	*meta-*	*para-*
Methyl-methyl	−25	−47	13.2
Methyl-chloro	−34	−47	7.5
Methyl-bromo	−27	−40	28
Methyl-nitro	−10	16	51
Chloro-chloro	−18	−25	53
Bromo-bromo	2	−7	87
Nitro-nitro	118	90	174

22-13. Consider a compound of molecular formula $C_3H_8O_3$. Suggest a reasonable structure; indicate which functional groups can be ruled out immediately on the basis of the formula alone.

22-14. A gaseous hydrocarbon is found to have the empirical formula C_2H_3. From a measurement of gas density at known pressure and temperature, it is found that the molecular formula is C_4H_6. Draw the structural formulas of the compounds it might be, and name each one.

22-15. Describe a simple test which might be applied semiquantitatively to commercial cooking oils to determine the presence and amount of unsaturation.

22-16. The acid dissociation constant for alanine, CH_3CHNH_2COOH, is 1.35×10^{-10}, while the base dissociation constant is 2.24×10^{-12}. What is the pH of the isoelectric point for this compound in water? Alanine is dissolved in water and the pH is adjusted to 7.0. A pair of electrodes is inserted into the solution and a field applied. Toward which electrode will the amino acid migrate? Explain.

22-17. Draw a portion of the polymer chain formed from polymerization of acrylonitrile, $CH_2{=}CHCN$. (The cyano group, $C{\equiv}N$, is not involved in the polymerization.)

22-18. It is proposed that one determine the molecular weight of a protein by freezing point lowering experiment. To determine the feasibility of this approach, calculate the freezing point lowering for a 1 per cent-by-weight solution of a protein of MW = 100,000 in water. The cryoscopic constant for water 1.86°C/mole/1000 g solvent.

22-19. The ease with which carbon rings are formed seems to be maximized at six carbons. What reasons can you advance for the decreased ease of formation of both smaller and larger rings? Discuss both aromatic and saturated rings.

22-20. Write the most important resonance structures for naphthalene. Would you expect naphthalene to have a greater or lesser "resonance energy" than benzene? Explain.

22-21. Arrange the following amino acids in the order of increasing pH for a 0.1*M* solution of the acid: valine, lysine, aspartic acid.

22-22. Two compounds with the same empirical formula, C_3H_5Cl, are obtained by a chemist. On reaction with bromine, both compounds yield the same product, of formula $C_3H_5Br_2Cl$. What are the probable structures of the two compounds, if cyclopropane derivatives are excluded?

22-23. Write structural formulas for all of the possible oxidation products of ethylene glycol.

22-24. Write a balanced chemical equation for each of the following reactions:

(a) Oxidation of *n*-butanol to butyric acid by permanganate in acid solution.
(b) Oxidation of naphthalene to carbon dioxide and water.
(c) Preparation of isopropyl benzene from benzene.
(d) Mild oxidation of 2-pentanol.
(e) Formation of a dipeptide from two molecules of serine.
(f) Addition of a mole of bromine to propyne.
(g) Mild oxidation of 2-butanol.
(h) Friedel-Crafts reaction between benzene and acetyl chloride,

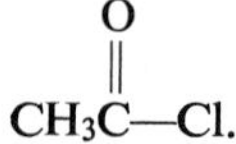

Chapter 23

Nuclear and Radiochemistry

On the first page of text, we made the point that the lines dividing the physical sciences are not generally well defined. The subject matter of this final chapter provides an excellent illustration of a borderline area in which major advances have been achieved by both chemists and physicists.

The theory dealing with the nature of the nucleus, its structure, and the types of particles found therein, has become a primary concern of physicists. On the other hand, chemists have played important roles in solving problems connected with nuclear fission and the synthesis of new elements. In fact, the nuclei which are the products of any nuclear reaction — whether a fission reaction or one produced in a high energy particle machine — are often identifiable only by chemical characterization of the corresponding atoms.

23-1 Nuclear Fission

The term fission is normally applied to a process in which a nucleus splits into two main fragments of comparable mass. *Spallation* refers to reactions in which the nuclei lose a number of nucleons (protons or neutrons) but otherwise remain intact. *Fragmentation* refers to processes in which the products of the nuclear decomposition are both large and small. Both spallation and fragmentation are brought about by bombardment of lighter-weight elements in high energy particle machines. We will discuss only the fission reaction, since it is historically the most interesting and important.

Before describing the fission reaction and some of the historical background relating to it, let us first consider some of the more obvious facts about nuclei. We notice first of all that, with an increase in atomic weight, there is a tendency for the ratio of neutrons to protons to increase in naturally occurring nuclei. This can be seen in Fig. 23-1, which shows a graph of the atomic number for known stable nuclides as a function of mass number. Secondly, we may note that any combination of neutrons and protons which results in a stable nucleus involves a great deal of stored energy. The mass of a nuclide should (if there were no stabilization energy) be just the sum of the masses of the particles which make it up. If, on the other hand, there is a large stabilization energy involved, the mass of the nuclide will be less than the expected sum; the mass difference is related to the stabilization energy through the Einstein equation

$$E = mc^2$$

This is usually expressed as the binding energy per nucleon:

$$\text{binding energy} = \frac{\text{mass of nuclide} - Zp - (A - Z)n}{A} \times c^2$$

Z = atomic number $\quad p$ = mass of proton
A = mass number $\quad n$ = mass of neutron

The binding energy per nucleon, given in units of million electron volts (mev), is shown graphed as a function of mass number in Fig. 23-2. One

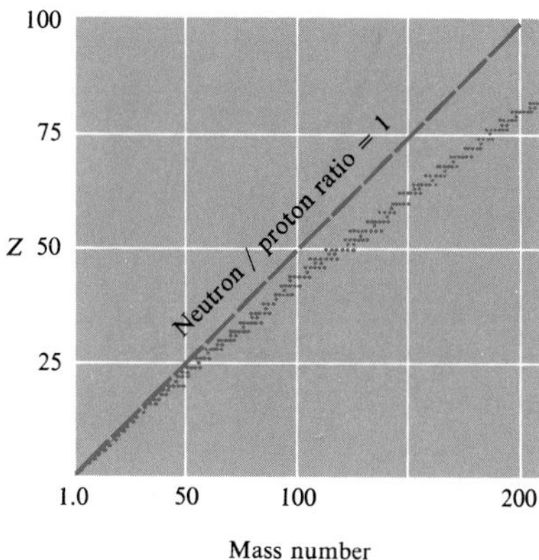

Fig. 23-1 A graph of mass number versus atomic number for the stable isotopes of the elements.

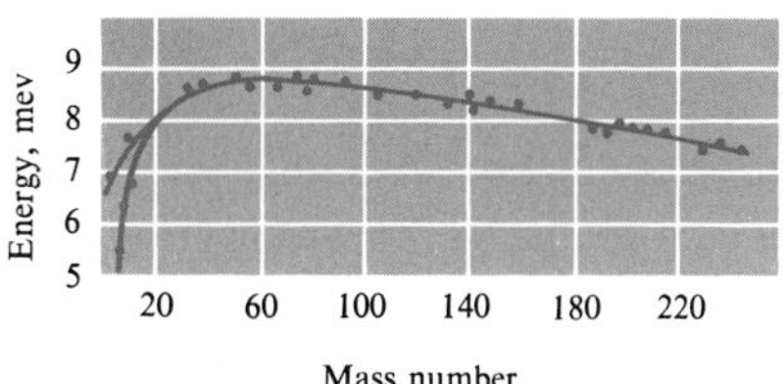

Fig. 23-2 Binding energy per nucleon as a function of mass number.

can see from this that, if the heavy nuclei were to be broken into two medium-weight nuclei, there would be an increase in the total binding energy. Furthermore, one can predict that if such a change were to occur, there would be quite a few excess neutrons left over, since the nuclei of medium mass number have lower neutron/proton ratios than do the heavier ones.

Hahn and Strassmann postulated in 1939 that the uranium nucleus could divide into two parts upon bombardment with neutrons. This conclusion was the culmination of five years of intensive research effort in their own and other laboratories, looking for something quite different! Enrico Fermi had suggested in 1934 that a uranium nucleus which was bombarded with neutrons might, after absorbing a neutron, lose one or more electrons to become a nucleus of an element with atomic number greater than 92.

$$^{1}n + {}^{n}_{92}\mathrm{U} \rightarrow {}^{n+1}_{92}\mathrm{U} \rightarrow {}^{\ 0}_{-1}e + {}^{n+1}_{93}\mathrm{X}$$
$${}^{n+1}_{93}\mathrm{X} \rightarrow {}^{\ 0}_{-1}e + {}^{n+1}_{94}\mathrm{Y}$$

(The notation for nuclear reactions is discussed in Chap. 2). It is at this point that chemistry enters the picture. The new elements were predicted to have the chemical properties of the corresponding transition elements. For example, element number 93 would be similar to the Group VIIB elements, manganese, technetium, and rhenium.

The nuclide formed by neutron capture might also decay by other means. It might emit a helium nucleus, as do many of the naturally occurring radioactive nuclides of heavy elements. This would lead to an element of lower atomic number:

$${}^{n+1}_{92}\mathrm{U} \rightarrow {}^{4}_{2}\mathrm{He} + {}^{n-3}_{90}\mathrm{Th}$$

The next few years following 1934 saw a number of papers in which Fermi's predictions were confirmed. As an example of the experimental approach taken, a solution of a uranium salt was irradiated with neutrons; added manganese was then precipitated as the dioxide. The product of the nuclear reactions — if a member of Group VIIB — was expected to coprecipitate with the manganese. (It would have been futile to attempt to precipitate the new element by itself because of the extremely small quantities involved.) Radioactivity was noted in the precipitated material, and this was taken as evidence that the new element of atomic number 93 was present.

As the work progressed, a number of inconsistencies were noted which cast some doubt on the assumption that the new elements should be analogues of the transition elements; it was suggested that perhaps the new elements would be analogous to the rare earths. Finally, it began to look as though some of the new substances giving rise to radioactivity in the products of nuclear reaction were not heavy elements at all, but were elements of much lighter atomic weight. Hahn and Strassmann found that three of the products were isotopes of barium, and another was an isotope of technetium, while still others were isotopes of krypton and xenon. They were thus led to their celebrated conclusion that at least some of the uranium nuclei had undergone fission. Soon after Hahn and Strassmann's paper appeared, it was suggested that the fission reaction was feasible as a source of energy; in 1941 the massive secret wartime effort to develop the atomic bomb began in the United States.

Before going on, a word on the nature of the fission process itself is in order. The nucleus can be thought of as a charged liquid drop. The drop is assumed to have a high surface tension which holds it together against the mutual repulsive forces of the protons. By making some reasonable assumptions about the dependence of the repulsive forces upon the number of protons and upon the volume of the drop, it can be shown that nuclei of mass numbers in the range of uranium should become unstable if given an initial distortion. The energy necessary to give the initial distortion is called the *fission threshold.* For some nuclei this distorting input may not even be necessary, and the nucleus undergoes spontaneous fission. Californium 254, for example, undergoes spontaneous fission with a half-life of 56 days.

Naturally occurring uranium consists of about 0.7 per cent ^{235}U, the remainder being ^{238}U. The fission threshold for the lighter nuclide is quite low, whereas for the mass-238 nuclide it is high. Capture of a slow (thermal) neutron by ^{235}U, therefore, leads to fission. Two nuclei of lighter weight and a number of neutrons are produced. Capture of a neutron by ^{238}U leads to ^{239}U, which decays in two successive electron emissions to give $^{239}_{94}Pu$:

$$^{1}_{0}n + ^{238}_{92}U \rightarrow ^{239}_{92}U \rightarrow ^{0}_{-1}e + ^{239}_{93}Np \rightarrow ^{0}_{-1}e + ^{239}_{94}Pu$$

Plutonium 239 is of particular interest because it has a low fission threshold. In a similar way, absorption of neutrons by ^{232}Th eventually results in ^{233}U, which also has a low fission threshold. The first reactor, called an *atomic pile* at the time, was constructed by Fermi and his co-workers on the University of Chicago campus, and became operative on December 2, 1942. Fermi reasoned that, if the neutrons produced by a ^{235}U nuclide fission could be captured, they could be used to convert ^{238}U to fissionable material. He began, therefore, with uranium containing the isotopes in their natural abundances. This was mixed with graphite, which acted to slow down the "hot" neutrons produced in the fission reaction and so to increase the probability of their capture by another uranium nucleus. Once the fission process began — initiated by a stray neutron from outside — the neutrons would be produced in sufficient quantity to keep ^{235}U nuclides fissioning, producing still more neutrons, *etc.* The nuclear reaction would therefore be self-sustaining. To keep it from getting out of hand, the pile was constructed so that cadmium rods could be inserted; cadmium nuclei have a large cross-section for neutron capture,* so the reaction could be controlled by their insertion into the pile.

A nuclear reactor which operates in the manner of Fermi's first atomic pile is known as a *breeder reactor.* Not only is a great deal of heat evolved from the fission processes, but the excess neutrons are used to convert otherwise non-fissionable material into fissionable material.

The chemist plays an important role in nuclear technology in a number of ways. The heat exchangers which are employed to carry off the enormous energies released in fission must be able to operate under conditions of high temperature and high radiation density. Structural material used in the reactor must have, in some cases, other special properties (see the discussion of the metallurgy of zirconium for an example, p. 532). One of the more challenging problems is the separation of the fission products from the reactor after it has been running for a time. A good deal of effort was devoted during the war to working out the chemistry of plutonium so that it could be separated efficiently from uranium. To add to the difficulties of working with small quantities, the substances are radioactive and intensely poisonous.

A fissioning nucleus tends to break into two masses of comparable mass. Since the neutron/proton ratio in the region of mass from about 90 to 150 is lower than at mass 240, the nuclides which are produced do not require all of the neutrons present in the original nuclide. They are therefore released in the fission process, as we have already seen. The masses of the fission products actually cover a wide range. A fission yield curve, such as the one shown in Fig. 23-3, describes the abundances of the various nuclides as a function of mass. It is interesting that the fissioning nucleus does not usually break up to give two products of equal mass; if this were the case, the curve would have a single maximum at half of the original mass. Instead there is a minimum at this point. The details of the fission yield curve depend upon the nuclide undergoing fission, and upon the means by which the fission threshold is surmounted, but for elements of atomic number 92 or higher they are always similar to the one shown in Fig. 23-3. The determination of the fission yield curve is essentially a chemical problem, and one which taxes the ingenuity and patience of the investigator.

---

*The capture cross-section is a measure of the probability of neutron capture by a given nuclide.

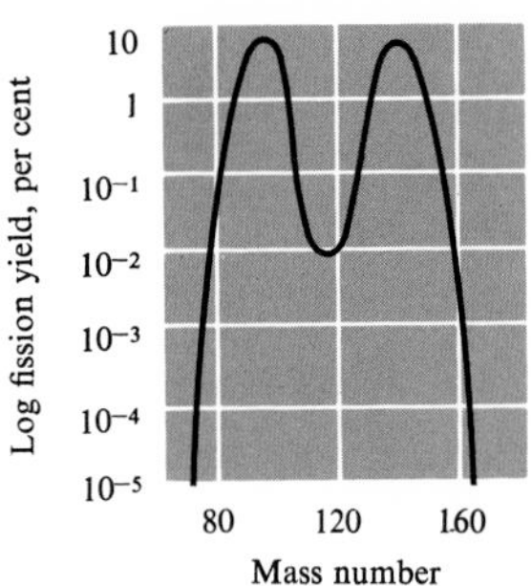

Fig. 23-3 A fission yield curve similar to that obtained for ^{235}U.

23-2 Radioactivity

Nuclides which possess an unfavorable neutron/proton ratio may decompose spontaneously, emitting a single particle and producing a new nuclide. The decay process follows a first-order rate law (p. 312); the rate of decomposition is proportional to the number of decomposing nuclei present

$$\frac{-dN}{dt} = kN$$

From which we obtain

$$N = N_0 e^{-kt}$$

N represents the number of decomposed nuclei which remain at time t, and N_0 represents the number present at the start of the time period. The decay rates of unstable nuclei are expressed in terms of half-life, the period of time required for one half of the nuclei present at a given time to decay.

Radioactivity was discovered by Becquerel in 1896. The Curies searched for radioactivity in naturally occurring substances and isolated the elements polonium and radium. Rutherford determined that the radiation emanating from a radioactive mineral was of three types: *alpha*, *beta*, and *gamma* rays. Alpha rays are helium nuclei, beta rays are electrons, and gamma rays are high-energy photons. Alpha rays have very low penetrating power: 5 to 7 cm of air serves to completely absorb most of them. On the other hand, a typical beta ray is about half absorbed after passing through 25-50 cm of air, while gamma rays may be only half absorbed after passage through about 100 meters of air.

There are nine radioactive nuclides of the elements which have half-lives comparable in time to the known age of the earth (about 3.5×10^9 years). The elements which make the greatest contribution to the natural radioactivity of the earth's crust are — in the order of their importance — uranium (90 per cent), thorium (5.0 per cent) and potassium (4.9 per cent).

Natural radioactivity is confined principally to the heavy elements. Uranium, thorium, and actinium are all the "parent" elements of three different radioactive decay series which are found in nature. In order to follow a natural radioactive decay scheme, it is necessary to note the effect of each of the three types of radioactive emissions. *Emission of an alpha particle* from the nucleus causes a decrease of two in atomic number, and a decrease of four in atomic weight. A *beta emission* causes an increase of one in atomic number, and no change in nominal atomic weight. *Gamma-ray emission* has no effect on either the atomic number or atomic weight. Figure 23-4 shows a grid chart with atomic number plotted against atomic weight. The figure shows a diagram of one of the natural radioactive-decay schemes, that of ^{238}U. This nuclide of uranium has a half-life of 4.49×10^9 years, *i.e.*, in that period of time, half of the uranium present in the initial sample has decayed. The decay path may branch along the way, but in any case the end product is ^{206}Pb. The half-life for each nuclide produced in the decay path is indicated in the box, in terms of years, days, *etc.*, as appropriate. Note that a process which moves one place to the right, with no change in mass, represents a beta decay, whereas

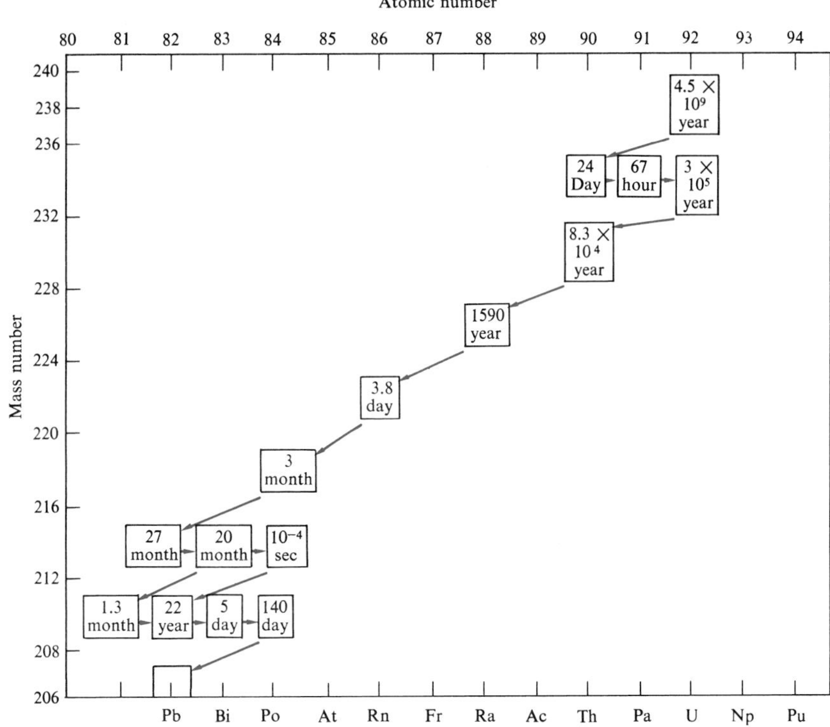

Fig. 23-4 Natural radioactivity decay series for ^{238}U.

a process moving four mass units down and two places to the left is an alpha emission.

The production of radioactive isotopes has increased enormously since World War II because of the ready availability of neutron sources. Their usefulness in chemical and biological research has also increased enormously, partly because they are more readily available, but also because of the great advances in the instruments used in measuring radioactivity and in handling radioactive materials.

Neutron absorption by a stable nuclide results in a higher neutron/proton ratio. If the resulting nucleus is unstable, it is likely to decay by emission of a beta particle, since this results in a lower neutron/proton ratio. Thus, if a solution of iodine is irradiated with slow neutrons, ^{127}I is converted to ^{128}I, which decays by emission of a beta particle, with a half-life of about 25 minutes.*

It would be impossible to discuss here all of the interesting aspects of the use of radiochemicals in research, or the techniques involved in working with such materials. The list of references at the end of the chapter will serve as a guide to the student who wishes to do further reading in this area. We will conclude this chapter by discussing a few chemical problems which have been attacked by use of radiochemical techniques.

The structure of the thiosulfate ion which is generally accepted shows that the two sulfur atoms in the ion are not equivalent. The correctness of this structure can be demonstrated by dissolving ^{35}S in a basic sulfite solution. After a period of time the solution is made acid, causing the precipitation of free sulfur. If the sulfur atoms in the thiosulfate were equivalent, one would expect that half of the original sulfur activity would remain in the sulfite ion in solution.

$$S^* + SO_3^{-2} \rightarrow S^*SO_3^{-2}$$

$$S^*SO_3^{-2} + 2H_3O^+ \rightarrow H_2SO_3 + 2H_2O + S^*$$

or

$$S^*SO_3^{-2} + 2H_3O^+ \rightarrow H_2S^*O_3 + 2H_2O + S$$

The sulfite in the acid solution can be precipitated as barium sulfite, and thus separated completely from the sulfur. When the experiment is performed, it turns out that there is essentially zero activity in the barium sulfite, indicating that the two sulfur atoms in thiosulfate ion are not equivalent. The sulfur which is removed on acidification is the same sulfur originally added to form the thiosulfate (see p. 456).

When a solution of iron (III) perchlorate containing radioactive ^{55}Fe is mixed with a solution of iron (II) perchlorate, the radioactivity begins to appear in the iron (II) form. The experiment is performed by withdrawing aliquots from the solution at known times after the initial mixing, and then precipitating the iron (III) ion as the hydroxide. The precipitate

*It is fortunate that beta particle emission is the predominant mode of decay, because beta particles are more conveniently counted than either alpha particles or gamma rays.

is dried and its activity determined. The activity is observed to decrease steadily during the first hour or so after mixing. From the results, it is apparent that a reaction of the form

$$Fe^*(H_2O)_6^{+3} + Fe(H_2O)_6^{+2} \rightarrow Fe(H_2O)_6^{+3} + Fe^*(H_2O)_6^{+2}$$

is occurring. This is an example of an *electron exchange*, or *electron transfer, reaction.* It represents a simple system from the chemical point of view, since the composition of the system does not change with time. The only event occurring is the transfer of an electron from the iron (II) to the iron (III) species, so the over-all concentrations of all species remain unchanged with time. Quantitative studies of the rates and mechanisms of electron transfer reactions have contributed a great deal to our understanding of oxidation-reduction processes (p. 520).

Carbon 14 has been a very useful radioactive isotope in organic chemical research. As an illustration of its use, consider the following reaction:

$$R{-}\overset{\overset{\displaystyle O}{\|}}{C}{-}\overset{\overset{\displaystyle O}{\|}}{C}{-}OC_2H_5 \xrightarrow{\text{heat}} R{-}\overset{\overset{\displaystyle O}{\|}}{C}{-}OC_2H_5 + CO$$

To determine which CO group is lost as carbon monoxide, the compound was synthesized with ^{14}C in the position adjacent to R. It was found that none of the radioactivity was present in the CO evolved. It is thus clear that the carbonyl group lost on heating is the one attached to the OC_2H_5 group:

$$R{-}\overset{\overset{\displaystyle O}{\|}}{C}{}^*{-}\overset{\overset{\displaystyle O}{\|}}{C}{-}OC_2H_5 \xrightarrow{\text{heat}} R{-}\overset{\overset{\displaystyle O}{\|}}{C}{}^*{-}OC_2H_5 + CO$$

The method of dating archeological objects and other subjects of historical interest by measurement of ^{14}C activity was developed by Professor W. F. Libby and his co-workers at the University of Chicago. The interested student is referred to the works on "Radiocarbon Dating" cited below.

Suggested Readings

Asimov, I., "Relative Contributions of Various Elements to the Earth's Radioactivity," *J. Chem. Educ.*, Vol. 31, 1954, p. 70.

Drouin, J. S. St., and L. Yaffe, "The Half-life of ^{41}Ca," *Canadian Jour. of Chem.*, Vol. 40, 1962, p. 833.

Hamill, W. H., R. R. Williams, and R. H. Schuler, "Laboratory Exercises in Nuclear Chemistry. I. Principles," *J. Chem. Educ.*, Vol. 26, 1949, p. 210.

Hindman, J. C., "Neptunium and Plutonium," *J. Chem. Educ.*, Vol. 36, 1959, p. 22.

Libby, W. F., "Developments in Peacetime Uses of Atomic Energy," *J. Chem. Educ.*, Vol. 36, 1959, p. 627.

Libby, W. F., *Radiocarbon Dating*, Chicago, Ill.: University of Chicago Press, 1954.

Libby, W. F., "Radiocarbon Dating," *Proc. Chem. Soc.* (London), 1960, p. 164.

New Chemistry (A *Scientific American* book), Simon & Schuster, New York: W. F. Libby, "Hot Atom Chemistry"; I. Perlman and G. T. Seaborg, "Synthetic Elements, I"; A. Ghioroso and G. T. Seaborg, "Synthetic Elements, II."

Overman, R. T., *Basic Concepts of Nuclear Chemistry*, Reinhold Publishing Corporation, New York, 1963.

Quagliano, J. V., *Chemistry*, Prentice-Hall, Inc., Englewood Cliffs, N. J., 1958, Chapt. 32.

Exercises

23-1. The rate of any particular nuclear decay process is found to be independent of external variables such as pressure or temperature. What conclusions can be drawn from this fact?

23-2. How is the atomic number of an element changed when an alpha particle is emitted from the nucleus? When a beta particle is emitted? When a gamma ray is emitted?

23-3. ^{232}Th is the parent of a natural decay series which terminates with ^{208}Pb. How many beta particle emissions, and how many alpha particle emissions, are there in this series?

23-4. When ^{242}Cm undergoes fission, the most probable masses of the products are 134 and 106. If these are the product masses, how many neutrons are released per curium nucleus? If one of the fission products from a particular nucleus is antimony ^{134}Sb, what might the other mass product be? Is either of these two fission products likely to be stable? [*Hint:* Compare with the graph in Fig. 23-1.]

23-5. Write a set of nuclear equations showing the formation of ^{233}U from neutron bombardment of ^{232}Th.

23-6. Occasionally a nucleus is found which undergoes radioactive decay by capturing an electron from the space outside the nucleus (usually an electron from the $n = 1$ level). What is the effect of this change on the atomic number of the element?

23-7. The radioactivity from a sample of ^{128}I is being counted. The radioactivity is measured by counts per second on a radiation detection device. The following data are obtained.

Time (minutes)	Activity (counts/sec)
3.00	4780
5.00	4530
8.00	4165
15.00	3440
25.00	2610
34.50	2015
43.00	1590

Compute the half-life of ^{128}I.

Appendix A

Oxidation-Reduction Equations

A chemical equation is a shorthand notation for representing a set of observations about a chemical system. It possesses a qualitative significance in that the chemical species involved are shown. It also possesses a quantitative significance in that mass balance is maintained and the composition of each substance is indicated. The most important part of the equation for most purposes is its expression of the chemical entities which are the reactants and products. It is from this information that we make deductions and draw conclusions about valence and other aspects of chemical behavior. Of course, the requirement of mass balance is important as well, because once the products of a reaction are known, the mass balance determines the relative quantities required to carry out the reaction. In what follows we will discuss the methods employed in balancing oxidation-reduction reactions, assuming that the products are known. It should be kept in mind that the balancing operation is purely mechanical and has nothing to do with the mechanism of reaction, with the rate at which it proceeds or even with valence.

An oxidation-reduction reaction is not easy to define, because our notion of what constitutes oxidation or reduction is itself not easily defined. We think of oxidation as a "loss" of electrons, and reduction as a "gain" of electrons. In dealing with some ionic species the gain or loss is well defined, but this is not the case with covalent substances. For this reason, we define an oxidation-reduction equation as one in which at least some of the reactants undergo a change in oxidation number. We then define an oxidation number in accordance with a rigid set of rules, so that there is no difficulty in determining this quantity for any atom in a given situation. We must emphasize that the oxidation number is not the same thing as the valence. It is, in fact, merely a bookkeeping tool. To ease the burden of remembering a set of rules, the oxidation number is made to follow as much as possible from accepted valence ideas, but this cannot always be done.

Every atom which appears in a chemical equation can be assigned an oxidation number, in accordance with the following set of rules:

1. The algebraic sum of the oxidation numbers of all the atoms in a neutral compound must equal zero; for an ionic substance, it must equal the charge.

2. The oxidation number of all elements in the free state is zero. When an atom is bonded to another chemically identical element this bond makes no contribution to the oxidation number of either atom.

3. The alkali metals assume a $+1$ oxidation number in their compounds; the alkaline earth elements assume an oxidation number of $+2$.

4. The oxidation state of oxygen is -2, except in peroxides, in which case it is -1. (It must be -1, for example, in Na_2O_2, by application of rules 1 and 3.)

5. Hydrogen has an oxidation number of $+1$ when bonded to nonmetallic elements, but -1 in the ionic metallic hydrides such as NaH.

6. A single bond between two atoms for which the above rules are not applicable contributes a -1 to the oxidation number of the more electronegative atom. A double bond contributes a -2 to the oxidation number of the more electronegative atom, *etc.*

The rules are best appreciated by looking at some examples:

P_2O_5 — By rule 4, the oxidation number of phosphorus is $+5$.
H_2CO — By rules 4 and 5, the oxidation number of carbon is zero.
$SOCl_2$ — Since chlorine is more electronegative than sulfur, and by rule 4, the oxidation number of sulfur is $+4$.
$HClO_2$ — By rules 4 and 5, the oxidation number of chlorine is $+3$.
MnO_4^{-2} — By rules 4 and 1, the oxidation number of manganese is $+6$.

It should be clear that the oxidation number is not identical with valence, except coincidentally.

When the reactants and products of an oxidation-reduction reaction are known and written down, the next step is to determine which atoms undergo a change in oxidation number. For the most part, it is best to work with the net ionic equation, although — if it seems easier — one may work with a complete equation and later reduce this to a net ionic equation.

Let us take a simple example: It is observed, that, when manganese dioxide is treated with hydrochloric acid, chlorine gas is evolved. Further testing discloses that manganese (II) ion is present in solution. We can begin, therefore, by writing

$$MnO_2(s) + Cl^- \rightarrow Mn^{+2} + Cl_2(g) \qquad (1)$$

We identify the oxidation numbers of manganese and chlorine, the two elements which undergo change in oxidation number:

$$\overset{+4}{Mn}O_2 + \overset{-1}{Cl^-} \rightarrow \overset{+2}{Mn^{+2}} + \overset{0}{Cl_2} \qquad (2)$$

The increase in oxidation number of the material being oxidized* must equal the total decrease in oxidation number of the material being reduced. After multiplying by the proper coefficients to ensure that this is the case, we have

$$MnO_2(s) + 2Cl^- \rightarrow Mn^{+2} + Cl_2(g) \qquad (3)$$

We now balance the total charge on both sides of the equation. Since the reaction is carried out in acid medium we know that the charged solvent species present in the hydronium ion. We add four hydronium ions on the left to balance the charge. Mass balance therefore requires that we add six water molecules on the right:

$$4H_3O^+ + MnO_2(s) + 2Cl^- \rightarrow Mn^{+2} + Cl_2(g) + 6H_2O \qquad (4)$$

Written in the molecular form, the equation looks like this:

$$MnO_2(s) + 4HCl \rightarrow MnCl_2 + Cl_2(g) + 2H_2O \qquad (5)$$

Let us work through another example: When lead dioxide is treated with a solution of manganese (II) nitrate in the presence of sulfuric acid, lead (II) sulfate precipitates, and permanganate ion is formed in solution:

$$PbO_2(s) + Mn^{+2} + SO_4^{-2} \rightarrow PbSO_4(s) + MnO_4^- \qquad (6)$$

Since the sulfate ion is well known to have a charge of -2, we can say without further ado that the oxidation number of lead in $PbSO_4$ is $+2$. The oxidation number of manganese in permanganate ion is $+7$. Balancing the changes in oxidation number (two for the lead, five for the manganese), we have

$$5PbO_2(s) + 2Mn^{+2} + 5SO_4^{-2} \rightarrow 5PbSO_4(s) + 2MnO_4^- \qquad (7)$$

We now require four hydronium ions on the left for a charge balance:

$$5PbO_2(s) + 2Mn^{+2} + 5SO_4^{-2} + 4H_3O^+ \rightarrow 5PbSO_4(s) + 2MnO_4^- \qquad (8)$$

*Oxidation is defined as an increase in oxidation number, reduction as a decrease in oxidation number.

To balance either oxygen or hydrogen, six water molecules are required on the right:

$$5PbO_2(s) + 2Mn^{+2} + 5SO_4^{-2} + 4H_3O^+ \rightarrow 5PbSO_4(s) + 2MnO_4^- + 6H_2O \quad (9)$$

The question of what to do with the ionic components of water (H_3O^+ or OH^-) continually recurs in dealing with equations for reactions in aqueous medium. We have already seen what needs to be done in acid solution. In a basic solution we can have either water or hydroxyl ion appearing in the balanced equation. We would not have hydronium ion present to any extent since it would react with hydroxyl ion to form water.

Example

Chromium (III) ion is oxidized to chromate ion in basic solution by hydrogen peroxide. From a knowledge of the chemistry of chromium (Chap. 21) we know that chromium (III) in basic solution should be represented as the complex ion, $Cr(OH)_4(H_2O)_2^-$.

$$Cr(OH)_4(H_2O)_2^- + H_2O_2 \rightarrow CrO_4^{-2} + H_2O$$

The two entities which undergo a change in oxidation number are the chromium (+3 to +6) and oxygen (−1 to −2). A balance of the changes in oxidation number leads to

$$2Cr(OH)_4(H_2O)_2^- + 3H_2O_2 \rightarrow 2CrO_4^{-2} + H_2O \quad (10)$$

We balance the charge by adding two hydroxyl ions on the left:

$$2OH^- + 2Cr(OH)_4(H_2O)_2^- + 3H_2O_2 \rightarrow 2CrO_4^{-2} + H_2O \quad (11)$$

To balance hydrogens and oxygens on both sides we require 12 molecules of water on the right:

$$2OH^- + 2Cr(OH)_4(H_2O)_2^- + 3H_2O_2 \rightarrow 2CrO_4^{-2} + 12H_2O \quad (12)$$

Balancing equations by the method of half-reactions — It is possible to balance oxidation-reduction equations by treating each over-all equation as the sum of two half-equations which add together to give the over-all expression. Some advantage is claimed for this procedure in that it relates the over-all reaction to cell reactions occurring in electrochemical cells (Chap. 16). As we have already pointed out, however, the balancing operation is more or less mechanical, and any method which works may be properly used. The method chosen should be the one which can be used most rapidly and with least difficulty. We illustrate the method of half-reactions with the first example discussed above. The two half-reactions are, in balanced form,

$$4H_3O^+ + MnO_2 + 2e^- \rightarrow Mn^{+2} + 6H_2O \quad (13)$$
$$Cl^- \rightarrow \tfrac{1}{2}Cl_2 + e^-$$

It should be noted that each half-reaction must be balanced for mass and for the same total charge on each side of the equation. In order to add these two half-reactions together to obtain a complete reaction, we must multiply through the second reaction by two, so that the electrons transferred in each case balance. After doing this and adding, the over-all reaction is the same as that given in Eq. (4).

Let us do another example by this method: When copper (II) sulfate is reacted with iodide ion in solution, copper (I) iodide forms as a solid, and iodine is liberated. We have, then

$$I^- + Cu^{+2} + e^- \rightarrow CuI(s) \qquad (14)$$

$$2I^- \rightarrow I_2 + 2e^-$$

Note that we include iodide ion in the first half-reaction not as an oxidant or reductant, but because it forms CuI with copper (I). To balance the half-reactions for electron transfer, the first reaction must be multiplied through by two. Adding, we have then

$$4I^- + 2Cu^{+2} \rightarrow 2CuI(s) + I_2 \qquad (15)$$

Finally, let us do an example involving basic solution: Consider the oxidation of hexacyanomanganate (II) to hexacyanomanganate (III) by molecular oxygen in basic solution. We have

$$Mn(CN)_6^{-4} \rightarrow Mn(CN)_6^{-3} + e^- \qquad (16)$$

$$4e^- + O_2 + 2H_2O \rightarrow 4OH^-$$

In the second half-reaction, we require that the oxygen appear in the -2 oxidation state. Since we are working with a basic solution, the only choice is hydroxyl ion; we could hardly have water as the product, since this would require that we have hydronium ion present initially, and this could not be the case in basic solution. Multiplying through the first reaction by four, and adding

$$4Mn(CN)_6^{-4} + O_2 + 2H_2O \rightarrow 4Mn(CN)_6^{-3} + 4OH^- \qquad (17)$$

Oxidation-reduction (redox) titrations, normality — Oxidation-reduction reactions which proceed rapidly in solution form the basis of a number of quantitative analytical procedures. As an example, an iron ore sample is dissolved in acid and all of the iron in solution converted to iron (II) ion by passing the solution over granulated zinc. The resulting solution is then titrated with a potassium permanganate solution of known concentration. The reaction which occurs is

$$8H_3O^+ + 5Fe^{+2} + MnO_4^- \rightarrow 5Fe^{+3} + Mn^{+2} + 12H_2O \qquad (18)$$

An indication of the equivalence point in the titration would be provided by the reagent itself. Permanganate solutions are a deep purple, whereas solutions of Fe^{+3} or Mn^{+2} are not strongly colored.

From the volumes of a permanganate solution of known concentration, the total quantity of iron (II) in the solution could be determined. To make such a calculation, however, account must be taken of the fact that five moles of iron (II) ion is required for each mole of permanganate. We could say that there are five equivalents of permanganate per mole of permanganate. This is perhaps made more evident by writing the half-reactions for Eq. (18):

$$Fe^{+2} \rightarrow Fe^{+3} + e^-$$
$$8H_3O^+ + MnO_4^- + 5e^- \rightarrow Mn^{+2} + 12H_2O$$

We define the normality of a solution to be employed in an oxidation-reduction reaction as the molarity times the number of moles of electrons released or taken up per mole of reagent. In the iron (III) — iron (II) pair, molarity and normality are the same, whereas for the permanganate, normality is five times molarity.

Example

A sample of sodium oxalate, $Na_2C_2O_4$, weighing 0.2633 g, was dissolved in water and titrated with a permanganate solution: 34.87 ml was required. From these data, calculate the concentration of the permanganate solution.

The balanced equation for the reaction is

$$16H_3O^+ + 2MnO_4^- + 5C_2O_4^{-2} \rightarrow 10CO_2 + 24H_2O + 2Mn^{+2}$$

The half-reactions are

$$C_2O_4^{-2} \rightarrow 2CO_2 + 2e^-$$
$$MnO_4^- + 8H_3O^+ + 5e^- \rightarrow Mn^{+2} + 12H_2O$$

The calculation of the permanganate normality is as follows:

$$\frac{0.2633 \text{ g } Na_2C_2O_4}{0.3487 \text{ liters solution}} \times \frac{1 \text{ mole } Na_2C_2O_4}{134.0 \text{ g } Na_2C_2O_4} \times \frac{2 \text{ equiv } C_2O_4^{-2}}{1 \text{ mole } Na_2C_2O_4}$$
$$\times \frac{1 \text{ equiv } MnO_4^-}{1 \text{ equiv } C_2O_4^{-2}} = \frac{0.1125 \text{ equiv } MnO_4^-}{1 \text{ liter } MnO_4^- \text{ solution}}$$
$$= 0.1125N$$

To make up a liter of this solution, one would have had to weigh out $\frac{1}{5} \times 0.1125 = 0.0225$ mole $KMnO_4$.

Example

A sample of iron ore weighing 0.3043 g is dissolved in acid, passed through a column of granulated zinc, and then titrated against the permanganate solution (Example), 40.32 ml being consumed. Calculate the percentage of iron in the sample.

$$\frac{0.04032 \text{ liter } MnO_4^- \text{ solution}}{0.3043 \text{ g sample}} \times \frac{0.1125 \text{ equiv } MnO_4^-}{1 \text{ liter solution}} \times \frac{1 \text{ equiv } Fe^{+2}}{1 \text{ equiv } MnO_4^-}$$
$$\times \frac{1 \text{ mole Fe}}{1 \text{ equiv } Fe^{+2}} \times \frac{55.85 \text{ g Fe}}{1 \text{ mole Fe}} = \frac{0.833 \text{ g Fe}}{1 \text{ g sample}} = 83.3 \text{ per cent Fe}$$

Note that in the calculations the conversion from permanganate to iron (II) ion is made by recognizing that one equivalent of the one reagent equals one equivalent of the other. This is indeed the very meaning of normality — that solutions of equal normality are equal to one another in oxidizing or reducing capacity.

Exercises

1. Compute the oxidation number of the element in boldface in each of the following:

   $NaBrO_3$ Na_2FeO_4 $POCl_3$ ICl Bi_2O_3
   $CuCl_4^{-2}$ $Na_3H_2IO_6$ $NaNO_3$ N_2O_4 Fe_3O_4

2. Complete and balance each of the following (it may be necessary to add water, hydroxyl or hydronium ion to some of the equations, in a manner consistent with the conditions under which the reaction occurs):
   (a) $H_2S + I_2 \rightarrow S(s) + I^-$ (acid)
   (b) $H_2S + SO_2 \rightarrow H_2O + S(s)$
   (c) $IO_3^- + SO_3^{-2} \rightarrow I_2 + SO_4^{-2}$ (acid)
   (d) $CuCl(s) + O_2 \rightarrow Cu^{+2} + Cl^-$ (acid)
   (e) $Cl^- + NO_3^- \rightarrow NOCl + Cl_2$ (acid)
   (f) $KClO_3(s) \rightarrow KCl(s) + KClO_4(s)$
   (g) $KMnO_4(s) \rightarrow K_2MnO_4(s) + MnO_2(s) + O_2$
   (h) $Cl_2 + OH^- \rightarrow Cl^- + ClO_3^-$ (basic)
   (i) $NiS(s) + NO_2^- \rightarrow Ni^{+2} + NO_2 + S(s)$ (acid)
   (j) $NO_3^- + Al \rightarrow Al(OH)_6^{-3} + NH_3$ (basic)
   (k) $Bi + NO_3^- \rightarrow Bi^{+3} + NO + H_2O$ (acid)
   (l) $P_4(s) + OH^- \rightarrow PH_3 + H_2PO_2^-$ (basic)
   (m) $I^- + Cr_2O_7^{-2} \rightarrow I_2 + Cr^{+3}$ (acid)
   (n) $SeCl_2 + H_2O \rightarrow Cl^- + H_2SeO_3 + Se(s)$
   (o) $C_2H_2 + O_2 \rightarrow C_2H_4O_2 + H_2O$

3. A sample of an ore containing chromium is fused with sodium peroxide to convert all of the chromium present to chromate ion, CrO_4^{-2}. The sample is dissolved in water and the solution boiled to remove any excess peroxide. It is then made acidic, (chromate, CrO_4^{-2}), is converted to dichromate, $Cr_2O_7^{-2}$, and titrated with an iron (II) sulfate solution. The iron (II) sulfate solution was previously standardized by titrating a known quantity of potassium dichromate, $K_2Cr_2O_7$. From the data given below, calculate the percentage of chromium in the ore.
   Weight of potassium dichromate used in standardization: 0.2200 g.
   Volume of iron (II) sulfate solution needed for titration: 45.65 ml.
   Weight of sample of unknown ore: 0.1325 g.
   Volume of iron (II) sulfate solution needed in titrating unknown sample: 33.65 ml.
   The half-reactions required for the calculations are:

$$14H_3O^+ + Cr_2O_7^{-2} + 6e^- \rightarrow 2Cr^{+3} + 21H_2O$$
$$Fe^{+2} \rightarrow Fe^{+3} + e^-$$

Appendix B

Oxidation Potentials

The table which follows contains the values for a number of standard oxidation potentials at 25°C (298°K). The values are taken from W. M. Latimer, *The Oxidation States of the Elements and Their Potentials in Aqueous Solutions* (2nd ed.), Prentice-Hall, Inc., New York, 1952. A number of the oxidation potentials for metals are given in Table 20-2, p. 499.

The following table is by no means complete. Other standard potentials may be found in Latimer's book, or in one of the following references:
Handbook of Chemistry and Physics, Chemical Rubber Publishing Co., Cleveland, Ohio.
Inorganic Chemistry, Kleinberg, J., W. J. Argersinger, and E. Griswold, D. C. Heath & Co., Boston, 1960, pp. 106-116.

It should be noted that the acidic species is written simply as H^+ rather than as hydronium ion, H_3O^+, and that the metal ions are shown without the hydrating water molecules. The half-cell reactions are frequently written as reduction steps rather than as oxidations. Thus, for example, the first half-reaction in the table would be listed as

$$AlF_6^{-3} + 3e^- \rightarrow Al + 6F^- \qquad \mathcal{E}^\circ = -2.07$$

The potentials for the half-reactions written in this manner are termed reduction potentials. The sign for the reduction potential is the opposite of the sign for the oxidation potential (Chap. 16).

Acid solutions

Couple	$\mathcal{E}°$ (volts)
$Al + 6F^- \rightarrow AlF_6^{-3} + 3e^-$	2.07
$Hf \rightarrow Hf^{+4} + 4e^-$	1.70
$B + 3H_2O \rightarrow H_3BO_3(s) + 3H^+ + 3e^-$	0.87
$Tl + I^- \rightarrow TlI(s) + e^-$	0.75
$AsH_3 \rightarrow As + 3H^+ + 3e^-$	0.60
$Cr^{+2} \rightarrow Cr^{+3} + e^-$	0.41
$Cd \rightarrow Cd^{+2} + 2e^-$	0.403
$Tl \rightarrow Tl^{+1} + e^-$	0.336
$V^{+2} \rightarrow V^{+3} + e^-$	0.255
$Sn + 6F^- \rightarrow SnF_6^{-2} + 4e^-$	0.25
$Cu + I^- \rightarrow CuI(s) + e^-$	0.185
$Sn \rightarrow Sn^{+2} + 2e^-$	0.136
$Cu + Br^- \rightarrow CuBr(s) + e^-$	−0.03
$H_2S \rightarrow S + 2H^+ + 2e^-$	−0.14
$Sn^{+2} \rightarrow Sn^{+4} + 2e^-$	−0.15
$H_2SO_3 + H_2O \rightarrow SO_4^{-2} + 4H^+ + 2e^-$	−0.17
$Ag + Cl^- \rightarrow AgCl + e^-$	−0.222
$Bi + H_2O \rightarrow BiO^+ + 2H^+ + 3e^-$	−0.32
$Cu \rightarrow Cu^{+2} + 2e^-$	−0.34
$S + 3H_2O \rightarrow H_2SO_3 + 4H^+ + 4e^-$	−0.45
$Cu \rightarrow Cu^+ + e^-$	−0.52
$MnO_4^{-2} \rightarrow MnO_4^- + e^-$	−0.564
$Au + 4CNS^- \rightarrow Au(CNS)_4^- + 3e^-$	−0.66
$H_2O_2 \rightarrow O_2 + 2H^+ + 2e^-$	−0.682
$Se + 3H_2O \rightarrow H_2SeO_3 + 4H^+ + 4e^-$	−0.74
$2CNS^- \rightarrow (CNS)_2 + 2e^-$	−0.77
$Fe^{+2} \rightarrow Fe^{+3} + e^-$	−0.771
$2Hg \rightarrow Hg_2^{+2} + 2e^-$	−0.789
$Ag \rightarrow Ag^+ + e^-$	−0.799
$Hg_2^{+2} \rightarrow 2Hg^{+2} + 2e^-$	−0.920
$HNO_2 + H_2O \rightarrow NO_3^- + 3H^+ + 2e^-$	−0.94
$2Cl^- + \frac{1}{2}I_2 \rightarrow ICl_2^- + e^-$	−1.06
$2Br^- \rightarrow Br_2(l) + 2e^-$	−1.065
$H_2SeO_3 + H_2O \rightarrow SeO_4^{-2} + 4H^+ + 2e^-$	−1.15
$ClO_3^- + H_2O \rightarrow ClO_4^- + 2H^+ + 2e^-$	−1.19
$\frac{1}{2}I_2 + 3H_2O \rightarrow IO_3^- + 6H^+ + 5e^-$	−1.195
$HClO_2 + H_2O \rightarrow ClO_3^- + 3H^+ + 2e^-$	−1.21
$2H_2O \rightarrow O_2 + 4H^+ + 4e^-$	−1.23
$Mn^{+2} + 2H_2O \rightarrow MnO_2 + 4H^+ + 2e^-$	−1.23
$HClO_2 \rightarrow ClO_2 + H^+ + e^-$	−1.28
$2Cr^{+3} + 7H_2O \rightarrow Cr_2O_7^{-2} + 14H^+ + 6e^-$	−1.33
$2Cl^- \rightarrow Cl_2 + 2e^-$	−1.36
$Pb^{+2} + 6H_2O \rightarrow PbO_2 + 4H_3O^+ + 2e^-$	−1.455

Couple	$\mathcal{E}°$ (volts)
$Mn^{+2} \rightarrow Mn^{+3} + e^-$	−1.51
$\frac{1}{2}Cl_2 + H_2O \rightarrow HClO + H^+ + e^-$	−1.63
$Ni^{+2} + 2H_2O \rightarrow NiO_2(s) + 4H^+ + 2e^-$	−1.68
$2F^- \rightarrow F_2 + 2e^-$	−2.87
$2HF(aq) \rightarrow F_2 + 2H^+ + 2e^-$	−3.06

Basic solutions

Couple	$\mathcal{E}°$ (volts)
$Mg + 2OH^- \rightarrow Mg(OH)_2(s) + 2e^-$	2.69
$Al + 4OH^- \rightarrow H_2AlO_3^- + H_2O + 3e^-$	2.35
$Cr + 3OH^- \rightarrow Cr(OH)_3(s) + 3e^-$	1.3
$Zn + 2OH^- \rightarrow Zn(OH)_2(s) + 2e^-$	1.245
$Fe + 2OH^- \rightarrow Fe(OH)_2 + 2e^-$	0.88
$Cd + 4NH_3 \rightarrow Cd(NH_3)_4^{+2} + 2e^-$	0.597
$S^{-2} \rightarrow S + 2e^-$	0.48
$Cu + 2CN^- \rightarrow Cu(CN)_2^- + e^-$	0.43
$Ag + 2CN^- \rightarrow Ag(CN)_2^- + e^-$	0.31
$SeO_3^{-2} + 2OH^- \rightarrow SeO_4^{-2} + H_2O + 2e^-$	−0.05
$Hg + 2OH^- \rightarrow HgO(s) + H_2O + 2e^-$	−0.098
$ClO_2^- + 2OH^- \rightarrow ClO_3^- + H_2O + 2e^-$	−0.33
$ClO_3^- + 2OH^- \rightarrow ClO_4^- + H_2O + 2e^-$	−0.36
$4OH^- \rightarrow O_2 + 2H_2O + 4e^-$	−0.40
$Ni(OH)_2(s) + 2OH^- \rightarrow NiO_2(s) + 2H_2O + 2e^-$	−0.49
$MnO_2(s) + 4OH^- \rightarrow MnO_4^{-2} + 2H_2O + 2e^-$	−0.60
$Br^- + 6OH^- \rightarrow BrO_3^- + 3H_2O + 6e^-$	−0.61
$ClO^- + 2OH^- \rightarrow ClO_2^- + H_2O + 2e^-$	−0.66
$Cl^- + 2OH^- \rightarrow ClO^- + H_2O + 2e^-$	−0.89

Appendix C

Mathematical Operations

In this section we will discuss some of the mathematical techniques and operations employed in the text.

Logarithms — The logarithm (log) of a positive number is simply the power to which another number must be raised to equal it. If

$$Y^p = X$$

then the log of X, to the base Y, is p:

$$\log_Y X = p$$

The bases in common use are ten and e. Tables of logarithms are readily available for these two bases. We will discuss the base ten, or common, logarithms. Logs to the base e, called *natural logarithms*, are obtained by multiplying the base-ten log values by 2.303.

The log of a number contains two parts, the *characteristic* and the *mantissa*. The characteristic is determined by the position of the decimal point. The mantissa is determined by the sequence of digits in the number. The log of 1 is zero, since 10^0 is one. The log of ten is 1, since 10^1 equals 10. The log of a number which lies between one and ten will, therefore be some number less than one. Any number can be written as a number between one and ten times ten to some whole-number power.

Examples

$$137 = 1.37 \times 10^2 \qquad 1{,}300{,}000 = 1.3 \times 10^6$$
$$6735 = 6.735 \times 10^3 \qquad 0.075 = 7.5 \times 10^{-2}$$

The log of two numbers which are multiplied together is the sum of the logs of each number. The logs of the above numbers are, therefore,

$$\log_{10} 137 = 2 + \log_{10} 1.37 \qquad \log_{10} 1{,}300{,}000 = 6 + \log_{10} 1.3$$
$$\log_{10} 6.735 = 3 + \log_{10} 6.735 \qquad \log_{10} 0.075 = -2 + \log_{10} 7.5$$

The logs of the numbers involved can be found in a log table containing a sufficient number of places, or may be read from a slide rule if the accuracy required is not great. Since the number in each case lies between one and ten, the characteristic is zero. We have, then, for the four numbers above,

$$\log_{10} 137 = 2 + 0.137 = 2.137$$
$$\log_{10} 6735 = 3 + 0.828 = 3.838$$
$$\log_{10} 1{,}300{,}000 = 6 + 0.114 = 6.114$$
$$\log_{10} 0.075 = -2 + 0.875 = -1.125$$

Logs are particularly useful when one wishes to take the root of a number. In general, for X^n, $\log X^n = n \log X$. For example, the fifth root of 6735 is found as follows:

$$\log_{10} 6735 = 3.828$$
$$\log_{10} (6735)^{1/5} = (\tfrac{1}{5}) \log_{10} 6735 = 0.766$$

The number whose $\log_{10}$ is 0.766 is found by finding 0.766 in the body of the table, and reading out the number which corresponds to it. We see that it is a number between one and ten from the fact that the decimal point immediately precedes the log: $(6735)^{1/5} = 5.85$.

Further Examples

(a) Find $(5.23)^4$.

$$\log_{10} 5.23 = 0.718$$
$$4 \times 0.718 = 2.870$$
$$\text{antilog} = 741$$

(b) $X^4 = 0.035 \times 10^{-17}$. Find X.

$$0.035 \times 10^{-17} = 3.5 \times 10^{-19}$$
$$\log_{10}(3.5 \times 10^{-19}) = -19 + 0.544 = -18.456$$
$$\log_{10} X = \frac{-18.456}{4} = -4.614 = -5 + 0.386$$
$$X = 2.44 \times 10^{-5}$$

The straight line and linear relationship — There are distinct advantages to be gained from expressing two related quantities in terms of a linear relationship, if one can be found. The straight line requires only two points for its definition, so a minimal amount of data is required to establish the relationship. The most general equation for a straight line is

$$U = mY + a \tag{1}$$

where U and Y represent the variables, and m and a are constants. By setting Y equal to zero, we see that a is the point at which the line intersects the U-axis (Fig. C-1). The second constant m corresponds to the slope of

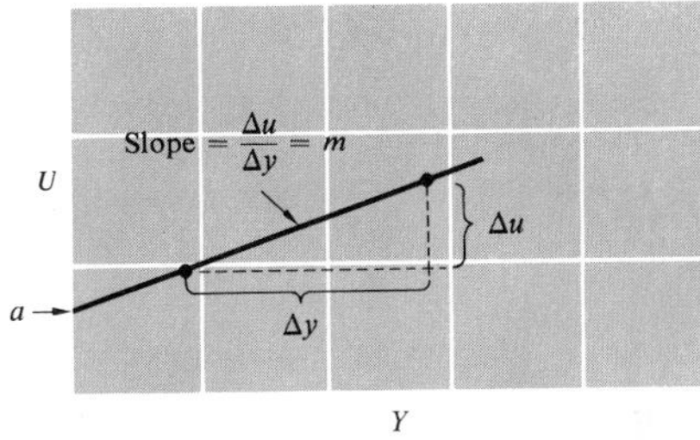

Fig. C-1

the line. Consider two points on the line, u_1, y_1 and u_2, y_2. We have for these

$$\begin{aligned} u_1 &= my_1 + a \\ u_2 &= my_2 + a \end{aligned} \tag{2}$$

Taking the difference between these we have

$$\begin{aligned} u_2 - u_1 &= m(y_2 - y_1) \\ \Delta u &= m\Delta y \end{aligned} \tag{3}$$

A change in the value of u, given by Δu, corresponds to a change of Δy multiplied by m. One can say that the value for u changes m times as rapidly as the value for y. m may be a number zero to infinity, and may be either positive or negative in sign. The intercept a may be of either sign, or may be zero, in which case the line passes through the origin.

Example

The following values were obtained for the pressure of a gas held at constant volume as the temperature was changed. Assuming that the relationship between pressure and temperature is linear, determine the equation of a straight line relating the pressure to the temperature. (The temperature scale in this problem is not one of those in common use.)

Pressure	Temperature
1.273 atm	17.00
1.462 atm	53.50

If the problem were to be solved graphically, we could plot the data and draw a straight line. We can also determine the equation for the line by analytic means. Using Eq. (3), we have

$$\Delta P = m\,\Delta T$$
$$0.189 = m(36.50)$$
$$m = 0.00518$$

We have then, using Eq. (1), $P = 0.00518T + a$. Substituting this value for P at the first temperature, we have

$$1.273 = 0.00518(17.0) + a$$
$$a = 1.185$$

Knowing both m and a, we can determine that, when $P = 0$,

$$T = -1.185/0.00518 = -230$$

This is the value of absolute temperature on the scale we are working with. The graph of the straight line is shown in Fig. C-2.

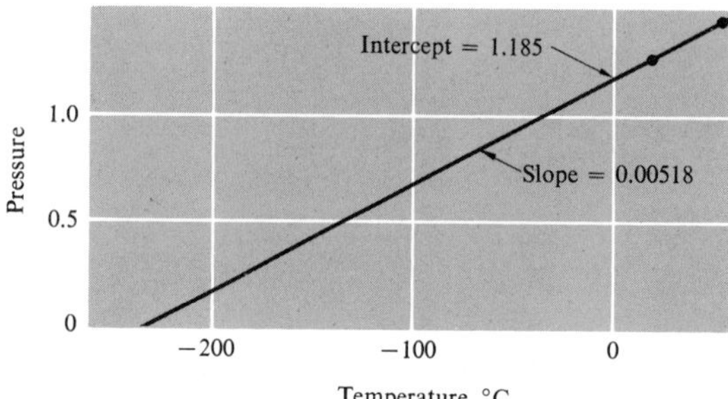

Fig. C-2

It is often possible to obtain a linear relationship between two variables by plotting some special function of them. For example, suppose that a quantity is defined with respect to time t by the equation

$$Z = Ae^{-kt}$$

where A and k are constants. By taking the log of both sides, we have

$$\log_e Z = \log_e A - kt$$

or

$$\log_{10} Z = \log_{10} A - kt/2.303$$

This is an equation of the same form as Eq. (1), as we can see by the following substitutions:

$$\log_{10} Z = U$$
$$\log_{10} A = a$$
$$-k/2.303 = m$$
$$t = Y$$

If data are obtained for Z as a function of time, a graph of $\log_{10} Z$ *vs* time will yield a line of slope $-k/2.303$. A straight line relationship of this form is used in handling data obtained for first-order rate processes such as radioactive decay.

Similarly, an equation of the form

$$S = Ae^{-k/t}$$

reduces to a straight line function in which one variable is log S and the other is $1/t$, since

$$\log_{10} S = \log_{10} A - \frac{k}{2.3}(1/t)$$

Exercises

1. Determine the base ten logs of the following numbers:

8235	4.3×10^{-6}
7.34	3.82×10^{4}
0.0362	0.000273

2. Calculate each of the following quantities by use of logarithms:

$(3.271)^{1/3}$	$(5.62 \times 10^{3})^{-3}$
$(2.176)^{1/4}$	$(0.0267)^{-1}$
$(5260)^{7}$	

3. Determine the slope and the intercepts for the straight line functions relating the two variables in each of the following equations. Indicate which functions of the variables are employed in the linear relationship.
(a) $3216X + 0.0367Y = -27.52$.
(b) $X = 37.2y^2 + 6.2$.
(c) $U = 0.32e^{-0.032t}$.
(d) $P = -3200/1.67t + 36.2$.
(e) $36.2XY = 18.72$.

Appendix D

The Electronic Configurations and First Ionization Potentials of the Elements

Z	Element	1	2		3			4				5				6			7	I_1(e.v.)
		s	*s*	*p*	*s*	*p*	*d*	*s*	*p*	*d*	*f*	*s*	*p*	*d*	*f*	*s*	*p*	*d*	*s*	
1	H	1																		13.595
2	He	2																		24.58
3	Li	2	1																	5.39
4	Be	2	2																	9.32
5	B	2	2	1																8.30
6	C	2	2	2																11.26
7	N	2	2	3																14.53
8	O	2	2	4																13.61
9	F	2	2	5																17.42
10	Ne	2	2	6																21.56
11	Na	2	2	6	1															5.14
12	Mg	2	2	6	2															7.64
13	Al	2	2	6	2	1														5.98
14	Si	2	2	6	2	2														8.15
15	P	2	2	6	2	3														10.48
16	S	2	2	6	2	4														10.36
17	Cl	2	2	6	2	5														13.01
18	Ar	2	2	6	2	6														15.76

Z	Element	1	2		3			4				5				6			7	I_1(e. v.)
		s	*s*	*p*	*s*	*p*	*d*	*s*	*p*	*d*	*f*	*s*	*p*	*d*	*f*	*s*	*p*	*d*	*s*	
19	K	2	2	6	2	6		1												4.34
20	Ca	2	2	6	2	6		2												6.11
21	Sc	2	2	6	2	6	1	2												6.54
22	Ti	2	2	6	2	6	2	2												6.82
23	V	2	2	6	2	6	3	2												6.74
24	Cr	2	2	6	2	6	5	1												6.76
25	Mn	2	2	6	2	6	5	2												7.43
26	Fe	2	2	6	2	6	6	2												7.87
27	Co	2	2	6	2	6	7	2												7.86
28	Ni	2	2	6	2	6	8	2												7.63
29	Cu	2	2	6	2	6	10	1												7.72
30	Zn	2	2	6	2	6	10	2												9.39
31	Ga	2	2	6	2	6	10	2	1											6.00
32	Ge	2	2	6	2	6	10	2	2											7.88
33	As	2	2	6	2	6	10	2	3											9.81
34	Se	2	2	6	2	6	10	2	4											9.75
35	Br	2	2	6	2	6	10	2	5											11.84
36	Kr	2	2	6	2	6	10	2	6											14.00
37	Rb	2	2	6	2	6	10	2	6			1								4.17
38	Sr	2	2	6	2	6	10	2	6			2								5.69
39	Y	2	2	6	2	6	10	2	6	1		2								6.38
40	Zr	2	2	6	2	6	10	2	6	2		2								6.84
41	Nb	2	2	6	2	6	10	2	6	4		1								6.88
42	Mo	2	2	6	2	6	10	2	6	5		1								7.10
43	Tc	2	2	6	2	6	10	2	6	5		2								7.28
44	Ru	2	2	6	2	6	10	2	6	7		1								7.36
45	Rh	2	2	6	2	6	10	2	6	8		1								7.46
46	Pd	2	2	6	2	6	10	2	6	10										8.33
47	Ag	2	2	6	2	6	10	2	6	10		1								7.57
48	Cd	2	2	6	2	6	10	2	6	10		2								8.99
49	In	2	2	6	2	6	10	2	6	10		2	1							5.78
50	Sn	2	2	6	2	6	10	2	6	10		2	2							7.34
51	Sb	2	2	6	2	6	10	2	6	10		2	3							8.64
52	Te	2	2	6	2	6	10	2	6	10		2	4							9.01
53	I	2	2	6	2	6	10	2	6	10		2	5							10.45
54	Xe	2	2	6	2	6	10	2	6	10		2	6							12.13
55	Cs	2	2	6	2	6	10	2	6	10		2	6			1				3.89
56	Ba	2	2	6	2	6	10	2	6	10		2	6			2				5.21
57	La	2	2	6	2	6	10	2	6	10		2	6	1		2				5.61
58	Ce	2	2	6	2	6	10	2	6	10	2	2	6			2				6.5
59	Pr	2	2	6	2	6	10	2	6	10	3	2	6			2				5.7

Z	Element	1	2		3			4				5				6			7	I_1(e. v.)
		s	*s*	*p*	*s*	*p*	*d*	*s*	*p*	*d*	*f*	*s*	*p*	*d*	*f*	*s*	*p*	*d*	*s*	
60	Nd	2	2	6	2	6	10	2	6	10	4	2	6			2				5.7
61	Pm	2	2	6	2	6	10	2	6	10	5	2	6			2				—
62	Sm	2	2	6	2	6	10	2	6	10	6	2	6			2				5.64
63	Eu	2	2	6	2	6	10	2	6	10	7	2	6			2				5.67
64	Gd	2	2	6	2	6	10	2	6	10	7	2	6	1		2				6.16
65	Tb	2	2	6	2	6	10	2	6	10	9	2	6			2				6.7
66	Dy	2	2	6	2	6	10	2	6	10	10	2	6			2				6.8
67	Ho	2	2	6	2	6	10	2	6	10	11	2	6			2				—
68	Er	2	2	6	2	6	10	2	6	10	12	2	6			2				—
69	Tm	2	2	6	2	6	10	2	6	10	13	2	6			2				—
70	Yb	2	2	6	2	6	10	2	6	10	14	2	6			2				6.2
71	Lu	2	2	6	2	6	10	2	6	10	14	2	6	1		2				5.0
72	Hf	2	2	6	2	6	10	2	6	10	14	2	6	2		2				—
73	Ta	2	2	6	2	6	10	2	6	10	14	2	6	3		2				7.88
74	W	2	2	6	2	6	10	2	6	10	14	2	6	4		2				7.98
75	Re	2	2	6	2	6	10	2	6	10	14	2	6	5		2				7.87
76	Os	2	2	6	2	6	10	2	6	10	14	2	6	6		2				8.7
77	Ir	2	2	6	2	6	10	2	6	10	14	2	6	7		2				9.0
78	Pt	2	2	6	2	6	10	2	6	10	14	2	6	9		1				9.0
79	Au	2	2	6	2	6	10	2	6	10	14	2	6	10		1				9.22
80	Hg	2	2	6	2	6	10	2	6	10	14	2	6	10		2				10.43
81	Tl	2	2	6	2	6	10	2	6	10	14	2	6	10		2	1			6.11
82	Pb	2	2	6	2	6	10	2	6	10	14	2	6	10		2	2			7.42
83	Bi	2	2	6	2	6	10	2	6	10	14	2	6	10		2	3			7.29
84	Po	2	2	6	2	6	10	2	6	10	14	2	6	10		2	4			8.43
85	At	2	2	6	2	6	10	2	6	10	14	2	6	10		2	5			9.5
86	Rn	2	2	6	2	6	10	2	6	10	14	2	6	10		2	6			10.75
87	Fr	2	2	6	2	6	10	2	6	10	14	2	6	10		2	6		1	3.83
88	Ra	2	2	6	2	6	10	2	6	10	14	2	6	10		2	6		2	5.28
89	Ac	2	2	6	2	6	10	2	6	10	14	2	6	10		2	6	1	2	6.9
90	Th	2	2	6	2	6	10	2	6	10	14	2	6	10		2	6	2	2	—
91	Pa	2	2	6	2	6	10	2	6	10	14	2	6	10	2	2	6	1	2	—
92	U	2	2	6	2	6	10	2	6	10	14	2	6	10	3	2	6	1	2	—
93	Np	2	2	6	2	6	10	2	6	10	14	2	6	10	4	2	6	1	2?	—
94	Pu	2	2	6	2	6	10	2	6	10	14	2	6	10	5	2	6	1	2?	5.1
95	Am	2	2	6	2	6	10	2	6	10	14	2	6	10	7	2	6		2?	6.0
96	Cm	2	2	6	2	6	10	2	6	10	14	2	6	10	7	2	6	1	2?	—
97	Bk	2	2	6	2	6	10	2	6	10	14	2	6	10	8	2	6	1	2?	—
98	Cf	2	2	6	2	6	10	2	6	10	14	2	6	10	9	2	6	1	2?	—
99	Es	2	2	6	2	6	10	2	6	10	14	2	6	10	10	2	6	1	2?	—
100	Fm	2	2	6	2	6	10	2	6	10	14	2	6	10	11	2	6	1	2?	—
101	Md	2	2	6	2	6	10	2	6	10	14	2	6	10	12	2	6	1	2?	—
102	No	2	2	6	2	6	10	2	6	10	14	2	6	10	14	2	6		2?	—
103	Lw	2	2	6	2	6	10	2	6	10	14	2	6	10	14	2	6	1	2?	—

Appendix E

Values of Thermodynamic Functions for Selected Substances at 298°K*

Inorganic substances

Substance	ΔH_f°	ΔG_f°	S°
Al(s)	0.00	0.00	6.77
Al_2O_3(s)	−399.09	−376.77	12.19
Ag(s)	0.00	0.00	10.21
Ag^+(aq)	25.31	18.43	17.67
Ag_2O(s)	−7.31	−2.59	29.09
AgCl(s)	−30.36	−26.22	22.97
As(g)	60.64	50.74	41.62
$AsCl_3$(g)	−71.5	−68.5	78.2
As_4O_6(s)	−313.9	−275.4	51.2
Ba(s)	0.00	0.00	15.1
Ba^{+2}(aq)	−128.67		3.0
Br(g)	26.71	19.69	41.81
Br^-(aq)	−28.90	−24.57	19.29

*ΔH°_f and ΔG°_f are in units of kcal/mole, at 298°K. S° values are in units of cal/deg-mole, at 298°K.

Inorganic substances

Substance	ΔH_f°	ΔG_f°	S°
Br_2(g)	7.34	0.75	58.64
Br_2(l)	0.00	0.00	36.4
Ca(s)	0.00	0.00	9.95
Ca^{+2}(aq)	−129.77	−132.18	−13.2
$CaCO_3$(calcite)	−288.45	−269.78	22.2
$CaCO_3$(aragonite)	−288.49	−269.53	21.2
CaO(s)	−151.9	−144.4	9.5
$Ca(OH)_2$(s)	−235.80	−214.33	18.2
Cl(g)	29.01	25.19	39.46
Cl^-(aq)	−40.02	−31.35	13.17
Cl_2(g)	0.00	0.00	53.29
Cr(s)	0.00	0.00	5.68
Cr^{+3}(aq)	−61.2	−51.5	−73.5
Cu(s)	0.00	0.00	7.96
Cu^{+2}(aq)	15.39	15.33	−23.6
CuCl(s)	−32.2	−28.4	21.9
Fe(s)	0.00	0.00	6.49
Fe^{+2}(aq)	−21.0	−20.30	−27.1
Fe^{+3}(aq)	−11.4	−2.52	−70.1
Fe_2O_3(s)	−196.5	−177.1	21.5
Fe_3O_4(s)	−267.0	−242.4	35.0
H(g)	52.09	48.58	27.39
H^+(aq)	0.00	0.00	0.00
H_3O^+(aq)	−68.32	−56.69	16.72
HBr(g)	−8.66	−12.72	47.44
HCl(g)	−22.06	−22.77	44.62
HF(g)	−64.2	−64.7	41.47
HI(g)	6.20	0.31	49.31
H_2(g)	0.00	0.00	31.21
H_2O(g)	−57.80	−54.64	45.11
H_2O(l)	−68.32	−56.69	16.72
H_2O_2(l)	−44.84	−27.24	
H_2O_2(aq)	−45.68	−31.47	
H_2S(g)	−4.82	−7.89	49.15
HS^-(aq)	−4.22	3.01	14.6
S^{-2}(aq)	10.0	20.0	5.3
Hg(g)	14.54	7.59	41.80
Hg(l)	0.00	0.00	18.5
Hg_2Cl_2(s)	−63.32	−50.35	46.8
I(g)	25.48	16.77	43.18
I^-(aq)	−13.37	−12.35	26.14
I_2(g)	14.88	4.63	62.28
I_2(s)	0.00	0.00	27.9
K(g)	21.51	14.62	38.3
K(s)	0.00	0.00	15.2
K^+(aq)	−60.04	−67.47	24.5
KCl(s)	−104.18	−97.59	19.76

Inorganic substances

Substance	ΔH_f°	ΔG_f°	S°
KNO_3(s)	−117.76	−93.96	68.87
Li(g)	37.07	29.19	33.14
LiF(s)	−146.3	−139.5	8.57
Mg(s)	0.00	0.00	7.77
Mg^{+2}(aq)	−110.41	−108.99	−28.2
$MgCl_2$(s)	−153.40	−141.57	21.4
Mn(s)	0.00	0.00	7.59
Mn^{+2}(aq)	−52.3	−53.4	−20
MnF_2(s)	−189	−179	
MnO_2(s)	−124.2	−111.1	12.7
MnO_4^-(aq)	−129.7	−107.4	45.4
N(g)	112.98	108.88	36.62
NH_3(g)	−11.04	−3.98	46.01
NH_4^+(aq)	−31.74	−19.00	26.97
NO(g)	21.60	20.72	50.34
NO_2(g)	8.09	12.39	57.47
N_2(g)	0.00	0.00	45.77
N_2O(g)	19.49	24.76	52.58
N_2O_4(g)	2.31	23.49	72.73
Na(g)	25.98	18.67	36.7
Na(s)	0.00	0.00	12.2
Na^+(aq)	−57.28	−62.59	14.4
NaCl(s)	−98.23	−91.78	17.30
$NaHCO_3$(s)	−226.5	−203.6	24.4
Na_2CO_3(s)	−270.3	−250.4	32.5
Ni(g)	101.61	90.77	43.59
Ni(s)	0.00	0.00	7.20
NiF_2(s)	−159.5		
ONCl(g)	12.57	15.86	63.0
O(g)	59.16	55.00	38.47
OH^-(aq)	−54.96	−37.60	−2.52
O_2(g)	0.00	0.00	49.00
O_3(g)	34.0	39.06	56.8
P(s) (white)	0.00	0.00	10.6
P(g)	75.18	66.71	48.90
P_4O_{10}(s)	−720	—	—
PCl_3(g)	−73.2	−68.42	74.49
PCl_5(g)	−95.35	−77.59	84.3
Pb(s)	0.00	0.00	15.51
$PbCl_2$(s)	−85.85	−75.04	32.6
S(s) (rhombic)	0.00	0.00	7.62
S(s) (monoclinic)	0.071	0.023	7.78
SO_2(g)	−70.96	−71.79	59.40
SO_3(g)	−94.45	−88.52	61.24
SO_4^{-2}(aq)	−216.90	−177.34	4.1
$SbCl_3$(s)	−91.34	−77.62	44.5
Zn(s)	0.00	0.00	9.95

Inorganic substances

Substance	ΔH_f°	ΔG_f°	S°
Zn^{+2}(aq)	−36.43	−35.18	−25.4
$ZnCl_2$(s)	−99.40	−88.26	25.9
ZnO(s)	−83.17	−76.05	10.5

Carbon Compounds

Substance	ΔH_f°	ΔG_f°	S°
C(g)	171.70	160.84	37.76
C(diamond)	0.45	0.68	0.58
C(graphite)	0.00	0.00	1.36
CCl_4(g)	−25.5	−15.3	73.95
CCl_4(l)	−33.3	−16.4	51.25
$CHCl_3$(l)	−31.5	−17.1	48.5
CH_4(g)	−17.89	−12.14	44.50
CO(g)	−26.42	−32.81	47.30
CO_2(g)	−94.05	−94.26	51.06
CO_3^{-2}(aq)	−161.63	−126.22	−12.7
H_2CO_3(aq)	−167.0	−149.0	45.7
HCO_3^-(aq)	−165.18	−140.31	22.7
C_2H_2(g)	54.19	50.00	48.00
C_2H_4(g)	12.50	16.28	52.45
C_2H_6(g)	−20.24	−7.86	54.85
C_3H_8(g)	−24.82	−5.61	64.5
CH_3OH(l)	−57.02	−39.73	30.3
C_2H_5OH(l)	−66.36	−41.77	38.4
CH_3COOH(l)	−116.4	−93.8	38.2
C_6H_6(l)	11.72	29.76	41.3
C_6H_6(g)	19.82	30.99	64.34

Appendix F

Values of Selected Fundamental Constants

Standard acceleration due to gravity	980.665 cm/sec^2
Standard atmosphere of pressure	1,013,250 dynes/cm^2
Standard thermochemical calorie	4.1840 joules
Velocity of light, c	2.9979×10^{10} cm/sec
Planck constant, h	6.6252×10^{-27} erg-sec
Avogadro constant, N	6.0229×10^{23} mole^{-1}
Faraday constant, $\mathfrak{F}$	96,491 coulombs/equiv.
Electronic charge, e	1.6021×10^{-19} coulomb
	4.8029×10^{-10} esu
Gas constant, R	1.9872 cal/°K-mole
Boltzmann constant, k	1.38044×10^{-16} erg/°K

Appendix G

Answers to Selected Exercises

Chapter 1

1-2. Heat is a content term; it refers to a quantity of energy. Temperature refers to the average value of a quality possessed by all the parts of something. We refer to heat as an *extensive* property, one which is proportional to the total quantity of matter present. Temperature is an *intensive* property that is not related to the total quantity of matter present, but to the average energy of all of the components which make up the system.

1-3. (a) 25.5°C.

(c) 307.6°K.

(e) 84°K.

1-6. Initial kinetic energy ($\frac{1}{2}\ mv^2$) = 4.84×10^{10} erg. When this is all converted to potential energy, $gmh = 4.84 \times 10^{10}$ erg, $h = 27.4$ kilometers.

1-8. 1.16×10^{13} ergs converts to 2.77×10^5 cal.

1-11. A small hole has the properties of a black body surface, because each photon which "hits" the hole goes into the enclosure and is absorbed somewhere inside. The surface of the hole thus acts as a perfect absorber of incident light. Photons are coming out of the hole, too, of course, but these are the ones produced by vibrations of the atoms and molecules of the walls inside the box, and are not the ones which strike the hole from outside. The surface of the hole thus acts as a source of photons determined only by the inside temperature.

1-12. 3×10^9 cycles/sec, or 3,000 megacycles.

1-16. 2.84×10^{-12} erg and 4.97×10^{-12} erg, respectively.

1-17. (a) cm/sec.
(b) g/cm^3.
(c) $g\text{-}cm^2/sec$.
(d) cm/sec^2.

Chapter 2

2-2. Yes. The electrons must have long free paths in the tube, so that deflections incurred by the fields will not be disturbed as the electrons travel to the point where they are detected.

2-4. The repulsive forces arising from the electrostatic interaction of the positively charged helium and gold nuclei are very large at small distances. These forces vary inversely with distance. When the distance between a helium nucleus and a gold nucleus is such that the repulsive force is of the order of magnitude of the kinetic energy of the helium nucleus, deflection will occur.

2-5. (a) The energy of the electron that has fallen through the 100-volt potential field is $100 \times 1.6 \times 10^{-12}$ erg. Kinetic energy is expressed as $\frac{1}{2} mv^2$, where v is the velocity in cm/sec. The energy the electron has gained by falling through the potential field is in the form of kinetic energy.

$$\begin{aligned} \tfrac{1}{2} mv^2 &= 1.6 \times 10^{-10} \text{ erg} \\ v^2 &= 3.2 \times 10^{-10} \text{ erg}/9.107 \times 10^{-28} \text{ g} \\ v^2 &= 3.52 \times 10^{17} \text{ cm}^2/\text{sec}^2 \\ v &= 5.93 \times 10^8 \text{ cm/sec} \end{aligned}$$

(b) 1.38×10^7 cm/sec.

2-7. Suppose that the applied electric field is in a direction such that the upper plate is positive. Drops with a positive charge will then be accelerated in their downward motion, while those with a negative charge will be decelerated, or will move upward.

Chapter 3

3-1. 174.8 cm.

3-3. 1.33 g.

3-5. 113 cm^3 at STP. 0.00505 moles regardless of gas composition, assuming ideal behavior.

3-6. 188°C.

3-9. 1.183 g/liter.

3-12. 20.192.

3-14. Helium has a much higher average molecular speed than does a typical atmospheric gas such as O_2. It diffuses fairly readily even through glass and causes gas failure of vacuum tubes. It also has a very high thermal conductivity since its mean free path is longer than average and since it undergoes frequent gas collisions because of its high average molecular speed.

3-16. (a) 114 g/mole.
(b) 1. Calculated MW is high.
2. Calculated MW is high.
3. Essentially no effect.

3-19. Let the volume of the bottle be V cm^3. The weight of the solution is then $V/2$ g, and the weight of H_2O_2 is 0.025V g. The number of moles of H_2O_2 is 0.025V/34. The number of moles of O_2 is half of this, or 0.0125 V/34. The volume of the gas space, in liters, is V/2000, since V was defined as the total volume of the bottle in cm^3. Using the ideal gas law, we then have

$$P = \frac{(0.0125\ V/34) \times 0.08206 \times 300}{V/2000}$$
$$= 18.1 \text{ atm}$$

3-22. l at 1 atm $= 1.38 \times 10^{-5}$ cm
l at 1 mm $= 0.0105$ cm
l at 10^{-6} mm $= 1.05 \times 10^{4}$ cm

3-24. Multiplying both numerator and denominator in the exponent of Eq. (3-26) by N,

$$f_j = e^{-\frac{1}{2}Mu_j^2/RT}$$

from Eq. (3-27), if u_j is set equal to the r.m.s. speed, u, then

$$u_j^2 = \frac{3\ RT}{M}$$

On insertion in the expression for f_j, we obtain

$$\begin{aligned} f_j &= e^{-3/2} \\ \log_e f_j &= -1.5 \\ 2.3 \log_{10} f_j &= -1.5 \\ \log_{10} f_j &= -0.652 = -1 + 0.348 \\ f_j &= 0.223 \end{aligned}$$

The fraction f_j is clearly a constant for all gases, independent of molecular weight or temperature.

3-27. 0.69 liters.

3-28. The corrected volume is 2.449 liters, as compared with 2.451 liters from the ideal gas equation. The error in estimating the volume is thus on the order of 1 part in 1,000. For ordinary analytical purposes, this is not significant.

3-30. 3.24×10^9 molecules/cm^3.

Chapter 4

4-2. 26.0 g ClCN.

4-4. 0.047 moles SO_2.

4-7. $FeCr_2O_4$.

4-9. $0.684 \text{ g } SnO_2 \times \dfrac{118.7 \text{ g Sn}}{150.7 \text{ g } SnO_2} = 0.539 \text{ g Sn}$

$$\frac{0.539 \text{ g Sn}}{1.356 \text{ g sample}} \times 100 = 39.7 \text{ per cent Sn.}$$

4-12. 23.2 liters CO_2.

4-14. CH_2Cl_2.

4-16. 36.9 liters H_2.

4-19. $SnCl_2$.

4-22. 107.88 g/mole.

Chapter 5

5-1. Mendeleev's periodic table was based on the assumption that the atomic numbers of the elements could be assigned to them on the basis of a continuous increase in the atomic weight with atomic number. There is no doubt about the proper places of potassium and argon in the table. The reversal in order of atomic weights with the potassium-argon pair is an important indication that atomic weight is not the property that relates fundamentally to atomic number.

5-10. (b) The hydrogen atom, and hydrogen-like ions are unique in being one-electron systems. For these, the energy of the electron depends only upon n, the major quantum number.

(c) For any multi-electron atom, the electron energy depends not only on n, but also on l, the azimuthal quantum number.

5-14. (a) 6.

(b) 14.

(c) 18.

(d) 2.

5-15. 2s depends on r only.

$2p_z$ depends on r and θ only.

$d_{x^2-y^2}$ depends on r, θ, and ϕ.

p_x depends on r, θ, and ϕ.

Chapter 6

6-4. The effective nuclear charge acting on the 5s electron is much greater for silver. The electrons added to the 4d orbitals as the nuclear charge is increased from 37 to 47 do not completely screen the 5s electron from the additional increase of ten in nuclear charge.

6-5. (a) Increase.

(b) Decrease.

(c) Remain the same, insofar as the valence shell electrons are concerned.

(d) Increase.

6-7. Li, Cu, N, Ne.

6-10. The second electron removed from sodium is a 2p electron. These electrons "see" a large effective nuclear charge, because the six 2p electrons do not shield one another from the nucleus very well. The second electron removed from magnesium is another 3s electron which *is* very effectively shielded from the nucleus by all the other electrons.

6-13. Pb $1s^2\ 2s^2\ 2p^6\ 3s^2\ 3p^6\ 3d^{10}\ 4s^2\ 4p^6\ 4d^{10}\ 4f^{14}\ 5s^2\ 5p^6\ 5d^{10}\ 6s^2\ 6p^2$.

6-17. (a) Cl, since the electron added to form Cl^- experiences a substantial nuclear attraction due to incomplete shielding by the other 3p electrons.

(b) Ca^{+2}. Same nuclear charge, but fewer electrons.

6-18. About +2.2 e-v, or +50 kcal/mole (23 kcal = 1 e-v). The positive sign indicates that energy is *required* to form Na^- from Na.

Chapter 7

7-3. (a) CaO smaller cation, closer approach of oppositely charged centers.

(b) CaO. Iso-structural, but charges are +2 in this ion as opposed to +1 in the KCl lattice.

(c) Na_2S. Both cation and anion are smaller, allowing for closer approach of oppositely charged ions.

7-5. For LiCl:

$$V = \tfrac{1}{2} D + S + I + E + U$$
$$= 28 + 38 + 124 - 83 - 192$$
$$= -85 \text{ kcal/mole}$$

For LiI:

$$V = 18 + 38 + 124 - 71 - 170$$
$$= -61 \text{ kcal/mole}$$

The lower heat of formation of LiI is due primarily to the lower lattice energy, since the variations in $\frac{1}{2}$ D and E for the halogens approximately cancel.

7-9. There are four LiF per unit cell. The volume per LiF unit is 16.1927 $Å^3$. The volume of a mole of LiF is 25.937/2.662 = 9.7434 cm^3/mole LiF. From these, $N = 6.017 \times 10^{23}$ LiF molecules/mole.

7-10. At 22° 58′ and 51° 17′.

7-12. Co^{+3}: [Ar] $3d^6$.

Ga^{+3}: [Ar] $3d^{10}$.

Fe^{+2}: [Ar] $3d^6$.

7-13. (a) 6.

(c) 4.

(e) 6.

7-16. 1.53 hr.

7-19. NiO NaCl structure

CoO NaCl structure

FeO NaCl structure

NiS Zinc blende structure

CoS Zinc blende or NaCl structure

FeS Zinc blende or NaCl structure

The last two examples are very near the dividing line, but in all likelihood would be found to have the zinc blende arrangement.

7-21. The ions of the Group IB and IIB elements have smaller ionic radii than their Group A counterparts. This is consistent with the higher ionization potentials of the Group B elements.

Chapter 8

8-4. Be_2^{+2} is isoelectronic with Li_2. It should be less stable than Li_2, since the greater nuclear charges in Be_2^{+2} cause contraction of the electron densities around the nuclei, preventing effective overlap. Another way of looking at it is that there are no more bonding electrons than in Li_2, but the nuclear-nuclear repulsion is greater in Be_2^{+2}.

8-7. The energy separation would be smaller. The interaction between orbitals on the two centers is decreased as the internuclear distance increases. In the limit, as r approaches large values, the interaction decreases to zero. The separation between the bonding and anti-bonding orbitals is determined by the extent of interaction between orbitals on the two centers.

8-10. N—Cl < N—Br < N—F

Se—H < S—H < O—H

The bond strength generally decreases as the electronegativity difference between the bonded atoms decreases, provided that the bond order (*i.e.*, the number of electron pairs in the bond) remains the same.

8-12. Electron-volts can be converted to kcal by multiplying by 23.05. The result of applying Mulliken's formula to the data is listed below, with the values taken from Table 8-3 for comparison.

	Mulliken	Table 8-3
F	3.75	3.9
Cl	3.0	3.0
Br	2.75	2.8
I	2.44	2.5
O	2.71	3.5

Mulliken's method is really not very good for an element such as oxygen.

8-16. (a) p^3.

(b) sp.

(c) sp.

(d) sp^3.

(e) sp^2.

8-22. $SnCl_4 + 2Cl^- (aq) \rightarrow SnCl_6^{-2} (aq)$

sp^3 $sp^3 d^2$

We would not expect similar behavior for CCl_4, since there are no low-lying *d* orbital levels available for carbon to use to expand its valence shell. Alternatively, we might argue that carbon would not exhibit this expansion of its valence shell because of steric limitations. There simply isn't room for six chlorine atoms about the carbon atom.

8-25.

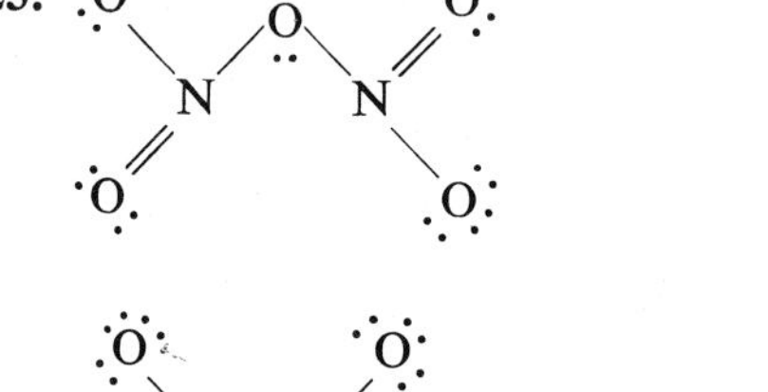

Only one resonance structure is shown in each case.

Chapter 9

9-2. One simple way of doing this experiment is to measure the heat produced by a set of paddles turning in a water-filled calorimeter and powered by a falling weight:

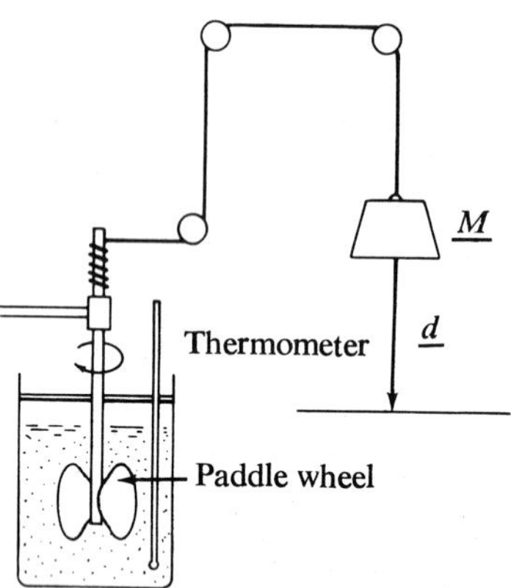

Measurement of the mechanical equivalent of heat. The decrease in the potential energy of the weight M occasioned by its being lowered through a distance d is seen as an increase in the temperature of the calorimeter containing the paddle.

Potential energy decrease $= -Mgd$ where g is the gravitational constant.

Heat appearing $= \Delta t \times C_c$ where Δt is the temperature rise, and C_c is the heat capacity of the calorimeter.

This is the method employed by Joule. Other substances may be used in place of water in the calorimeter.

9-4. (a) No work done, since the external pressure is zero.

(b) Since the external pressure is constant, we can use $w = P(V_2 - V_1) = 1(5) = 5$ liter-atm. Note that the quantity of gas involved does not enter into this work expression.

9-6. (a) $w = P(V_2 - V_1) = 17.5$ liter-atm $= 424$ cal
$\Delta E = 0,\ q = w$

(b) $w = \int P dv = nRT \ln\frac{V_2}{V_1} = 1040$ cal
$\Delta E = 0,\ q = w$

9-9. The higher heat capacity of iodine indicates that the vibrational energy levels of the iodine are more closely spaced. This means that there are more lower-lying levels which can be populated by I_2 molecules at room temperature.

9-11. $\Delta H^\circ_f = -57.07$ kcal.

9-13. Hydrogen bonding is less extensive in water at higher temperatures, so the heat required to vaporize the water is reduced. The variation in ΔH_v between water, methyl alcohol and carbon tetrachloride can be ascribed to hydrogen bonding effects.

9-15.

P-Cl	78.1
As-Cl	73.1
Sb-Cl	74.3
Bi-Cl	67.1

Estimated thermochemical value for N-Cl, based on available data for NCl_3, is ~46 kcal/mole.

9-18.

	Molar heat of combustion	Heat evolved per kg
$C_2H_2(g)$	−300.1	11,550 kcal
$C_2H_4(g)$	−316.2	11,300 kcal
$C_2H_6(g)$	−341.3	11,380 kcal

9-22. $V_{Na} = 23/0.95 = 24.2\ cm^3/mole.$
$V_K = 39.1/0.85 = 46.0\ cm^3/mole.$

Total volume on mixing one mole of each is 70.2

$$\Delta S = \Delta S_{Na^+}\, \Delta S_K = 0.23 \times 1.987 \left(\log_{10} \frac{70.2}{24.2} + \log_{10} \frac{70.2}{46.0}\right)$$
$$= 0.296 \text{ cal/°K-mole}$$

9-23. (a) $\Delta S < 0$.

(b) $\Delta S > 0$.

(c) $\Delta S < 0$.

(d) $\Delta S > 0$.

9-25. (a) False. Heat cannot be entirely converted into work.

(d) True. An isolated system is defined as one which exchanges neither heat nor work with the surroundings. By the first law, this requires $\Delta E = 0$ for any process.

(i) False. There is no necessary connection between the rate of a process and the value for the free energy change accompanying the process.

9-27. $\Delta G^\circ = -32.81 + 48.96 = +16.1$ kcal.

$\Delta H^\circ = -26.42 + 53.5 = +27.1$ kcal.

$\Delta S^\circ = (27.1 - 16.1)/298 = +36.9$ cal/°K-mole.

9-33. The vaporization of a mole of gallium may be thought of as a two-step process:

Ga(l) $\rightarrow$ Ga(g) $p = 760$ mm Hg

Ga(g) ($p = 760$ mm) $\rightarrow$ Ga(g) ($p = 6 \times 10^{-4}$ mm Hg)

An expansion of the gallium vapor from 760 mm to 6×10^{-4} mm Hg at constant temperature, 1210°K may be carried out with an entropy change, if the vapor were ideal, of

$$\Delta S = 2.3 \times 1.987 \log_{10} \frac{760}{6 \times 10^{-4}}$$
$$= 27.9 \text{ (cal/°K-mole)}$$

The total entropy change for the equilibrium vaporization is

$$\Delta S = \frac{63.8 \times 10^3}{1210} = 52.7 \text{ cal/°K-mole}$$

As a rough estimate for the entropy change on vaporization to 1 atm of gas, we have

$$\Delta S = 52.7 - 27.9 = 24.8 \text{ cal/°K mole}$$

Chapter 10

10-1. (a) $m = 1.08$, $M = 1.05$.

(b) $m = 1.85$, $M = 1.68$.

10-4.

	Aqueous solution	Benzene solution
M	0.0250	0.0250
m	0.0251	0.0285
f	4.52×10^{-4}	27.2×10^{-4}

10-6. In the vicinity of room temperature, a $1M$ solution. Since the density of water is about one, there is less water in a liter of $1M$ solution than in a $1m$ aqueous solution. If the solvent density is greater than one, molal solutions are more concentrated than molar solutions. If solvent density is less than one, the reverse is true.

10-8. The apparent vapor pressure is too high at all temperatures. The air has the effect of lowering the apparent heat of vaporization. It contributes relatively less to the apparent vapor pressure at higher temperatures, so the effect is to lower the apparent degree of variation in vapor pressure with temperature.

10-10. $M = 6.4 \times 10^{-3}$.

10-12. Molecular weight = 144 g/mole.

10-16. $$K = \frac{1 \text{ atm HCl}}{0.055 \text{ mole fraction}} \times \frac{760 \text{ mm HCl}}{1 \text{ atm HCl}}$$
$$= 1.38 \times 10^4 \frac{\text{mm HCl}}{\text{mole fraction}}$$

The large variation in HCl solubility shown in Table 10-3 is probably due mostly to hydrogen bonding of HCl to solvent molecules. This is particularly evident for the last three solvents listed, all of which have lone pair electrons on oxygen atoms.

10-20. Weight of a single particle $\approx 4.2 \times 10^{-19}$ g.
Molecular weight of particles $\approx$ 250,000.

Chapter 11

11-1. $HF > C_2H_5OH > CH_3OCH_3 > CHCl_3$.
This order is based on the idea that higher dielectric constant means more effective separation of charges.

11-4. It is sufficient to determine whether the dissolving of sodium perchlorate is exothermic or endothermic at the temperature of interest. If the solution process is endothermic, the solubility increases with increasing temperature. If solution is accompanied by release of heat (exothermic), solubility decreases with increasing temperature.

11-6. 29 g Pb.

11-10. $HCl \approx LaCl_3 > CaCl_2 \approx KOH > NaBr \approx LiCl$.

11-11. The heats for step 3 calculated by setting up the cycle

$$\begin{array}{ccc} M(g) & \xrightarrow{②} & M^+(g) + e^- \\ {}^1\uparrow & & \downarrow{}^3 \\ M(s) & \xrightarrow{④} & M^+(aq) + e^- \end{array}$$

are: Li^+, 229 g kcal/mole; Na^+, 201 kcal/mole; K^+, 182 kcal/mole. The decrease in these heats with increasing ionic size is due to the poorer solvation of the larger ions.

11-14. Λ versus $N^{1/2}$ yields a slightly curved line. One can extrapolate the data to $N \rightarrow 0$, to yield an estimated $\Lambda°$ of 126.3. A graph of Λ versus N, on the other hand is quite markedly curved, and does not extrapolate easily.

Chapter 12

12-2. The H_3O^+ ion is the stronger acid. As an illustration of this, if NH_3 is added to an aqueous solution of H_3O^+, there is extensive transfer of H^+ to the NH_3, forming NH_4^+. This occurs despite the much higher concentration of H_2O, which might be thought of as competing with NH_3 for the proton. We can say that NH_3 is a stronger proton acceptor than H_2O. The conjugate acid, NH_4^+, is correspondingly weaker than H_3O^+, the conjugate acid of H_2O.

12-4. $\underset{\text{acid}}{HCN} + \underset{\text{base}}{H_2O} \rightarrow \underset{\text{conjugate acid}}{H_3O^+} + \underset{\text{conjugate base}}{CN^-}$

HCN is a weak acid.

12-6. HI is an acid.

$HSeO_4^-$ might function as either an acid or base.

PO_4^{-3} can function only as a base.

Fe^{+3} acts as an acid in the sense that we consider its hydration sphere, *i.e.*, $Fe(H_2O)_n^{+3}$.

12-9. About 4.75 ml concentrated H_2SO_4.
0.3346*N*.

12-10. KCl, $BaCl_2$, $NiCl_2$, $FeCl_3$. This is the order of increasing cation hydrolysis which is in turn related to small ionic size and high charge.

12-12. 114 ml.

12-14. 14.45 per cent N.

Chapter 13

13-3. $\dfrac{-dP_{SO_2}}{dt} = k\, P_{SO_2} \times P_{H_2}$

k would have units of $atm^{-1}\ sec^{-1}$.

13-5. It might be followed by observing a volume change, keeping the system at constant pressure.

13-6. Make a graph of $[I^-]$ *versus* time. Draw in the tangents to the curve at $t = 4$ and $t = 8$ sec. Determine the slopes of these lines; the slopes are equal to the rates at the two times.

At $t = 4$ sec, rate $= -0.00122$ M/sec.

At $t = 8$ sec, rate $= -0.000750$ M/sec.

If the rate expression is of the form $-$ rate $= k[I^-][OCl^-]$, then

$$k = \frac{-\text{rate}}{[I^-][OCl^-]}.$$

Since $[I^-] = [OCl^-]$, we can test this at $t = 4$ and $t = 8$ sec, using the graphically estimated rates. At $t = 4$ sec, from the graph, $[I^-] =$ 0.00144 M. Then

$k(4 \text{ sec}) = 1.22 \times 10^{-3}/(1.44 \times 10^{-3})^2 = 5.9 \times 10^2 \text{ M}^{-1} \text{ sec}^{-1}$

$k(8 \text{ sec}) = 7.50 \times 10^{-4}/(1.12 \times 10^{-3})^2 = 6.0 \times 10^2 \text{ M}^{-1} \text{ sec}^{-1}$

The near-identity of these values for k confirms that the reaction is probably first order in each reagent; at any rate, it *is* second order.

13-10. Under the conditions described, only the forward process in the equilibrium

$$X(s) \rightleftarrows X(\text{solution})$$

need be considered. The rate of this forward step will be proportional to surface area. The surface area of a sphere is given by $4\pi r^2$, the volume by $\frac{4}{3}\pi r^3$. Let us say that we have a total y volume of the solid in each case. Then in the first case, radius $= r_1 = 0.1$, we have a total number of spheres given by $y/(\frac{4}{3})\pi r^3$. The total area is this number times the area per sphere:

$$4\pi r_1^2 \times \frac{y}{(\frac{4}{3})\pi r_1^3} = \frac{3y}{r_1}$$

Similarly the surface area in the second case is $3y/r_2$, and the ratio of surface area is

$$\frac{S(0.1 \text{ mm})}{S(2.5 \text{ mm})} = \frac{2.5}{0.1} = 25$$

The rate is thus inversely proportional to the radius of the spheres, assuming constant total volume (or weight) of solid.

13-13. $k = 4.68 \times 10^{-3}$ sec^{-1}.

E_a is about 63 kcal/mole.

13-15. Average $k = 3.4 \times 10^{-4}$ liter/mole-sec.

13-19. Using $A/A_o = e^{-kt}$.

$$\begin{aligned} \log_{10}(A/A_o) &= -kt/2.3 \\ &= -0.535 \times 0.50/2.3 \\ &= -0.1161 \\ A/A_o &= -1 + 0.884 \\ &= 0.764 \end{aligned}$$

We see, therefore, that 76.4 per cent remains after 30 minutes. The half-life is given by $t_{1/2} = 0.693/k = 1.29$ hr.

Chapter 14

14-2. (a) $K = \dfrac{P_{SO_2} \times P_{H_2O}}{P_{SO_3} \times P_{H_2}}$ units: none.

(b) $K = \dfrac{P_{NO}^4 \times P_{H_2O}^6}{P_{NH_3}^4 \times P_{O_2}^5}$ units: atm.

14-6. One possible starting composition is pure $POCl_3$. Another is $POCl_3$ with either of the reactants. Another is a mixture in any proportion of the two reactants. Finally, one could begin with an initial composition including all three ingredients in any proportion.

14-9. We must first write an equilibrium constant:

$$SO_2(g) \rightleftarrows SO_2(solu)$$

$$K = \frac{[SO_2(solu)]}{P_{SO_2}}$$

In order to obtain the heat of solution, we need to graph $\log_{10} K$ *vs* $1/T$. It doesn't matter what units we employ for K, since the units would merely be a constant factor which drops out in taking the log. We can, therefore, merely graph the log of the solubility, as given in the table, *vs* $1/T$. (The denominator is constant, as stated in the problem). The slope of the $\log_{10} K$ *vs* $1/T$ plot is $-\Delta H/2.3R$. From the graph, we obtain a slope of $+1{,}320$. ΔH is therefore -6.1 kcal/mole.

The dissolving of SO_2 in water is thus an exothermic process. As the temperature increases, the magnitude of the equilibrium constant decreases.

14-11. (a) Shift to right.

(b) No change.

(c) Shift to right.

(d) Shift to left.

(e) Shift to right.

14-15. $K_1 = 1.855.$

$K_2 = K_1^2 = 3.44.$

Chapter 15

15-1.

$$K = \frac{P_{NO_2}^2}{P_{O_2} \times P_{NO}^2}$$

K has units of atm^{-1} if pressures are expressed in atm. The units employed (atm^{-1}, mm^{-1}, liter/mole, *etc.*) determine the numerical value obtained for K.

15-3. It is not possible to measure an equilibrium constant for HCl in water. The dissociation of HCl is so nearly complete that there are no measurable quantities of undissociated HCl present. Attempts to work at high HCl concentrations meet with the difficulty of correcting for non-ideal solution behavior.

15-4. The loss of the second proton, from HSO_4^-, is more difficult because the proton feels the attractive force of the negative site left by the first proton.

15-6. One first computes what the pressures of BF_3 and ether would be at the higher temperatures if there were no reaction. The observed total pressure in the system, P_T, is equal to $P^\circ_{BF_3} + P^\circ_{ether} - P_x$, where P_x is the pressure of product. Compute P_x, then $P^\circ_{BF_3} - P_x$ and $P^\circ_{ether} - P_x$, which give the actual pressures of BF_3 and ether in the system at equilibrium. Compute K in units of mm^{-1}, multiply by 760 to convert to atm^{-1}.

Temperature (°C)	K (atm^{-1})
69.9	10.17
80.1	5.97
94.7	2.71

The decrease in K with increasing temperaure shows that the reaction is exothermic in the forward direction.

15-10.

Solution	pH
0.1M H_2SO_4	0.96
0.1M $NaHSO_4$	1.57
0.1M Na_2SO_4	7.49

15-11. The effect of the first ionization in reducing the ionization of the second proton depends upon how close the proton is to the atom which previously held the first proton. As this distance increases in going from H_2S to H_2SeO_3 to $H_2C_2O_4$, the *difference* in ionization constants decreases. Note that we compare pK_a values pK_a = $-\log K_a$, and $\log K$ relates directly to the standard energy difference

$\Delta G°$. We are saying, therefore, that there is less free energy difference between first and second ionizations when the distance between the ionizing sites increases. The second proton is not so strongly attracted to the negative center left by the first proton.

15-14. pH of $0.1M$ NaCN solution is 11.15
pH of $0.1M$ CH_3COONa solution is 8.88

Cyanide ion is a stronger base than acetate ion; it is the conjugate base of a weaker acid.

15-15. pH = 10.1. Only the first hydrolysis reaction need be considered.

15-19. $BaSO_4$ is insoluble in acid. The other two are soluble, since the acids from which the anions derive are weak acids. Carbon dioxide is, of course, evolved on adding acid to $BaCO_3$.

15-20. K for the reaction shown is just K_a for acetic acid divided by K_a for hydrocyanic acid: 3.6×10^4. The reaction thus proceeds far to the right.

15-24. The dissolution is the result of reaction of the solid with hydronium ion to form the weak acid, H_2CO_3, which decomposes to liberate CO_2:

$$CO_3^{-2} + 2H_3O^+ \rightarrow [H_2CO_3] + 2H_2O$$

$$H_2O + CO_2 \uparrow \quad \hookleftarrow$$

15-27. The end point occurs at a pH of about 5.3. Methyl red would be a good indicator.

15-29. pH = 0.6. Above this value, the barium chromate is insoluble.

15-31. $K_{sp} = 2.56 \times 10^{-13}$.

15-32. The thermodynamic solubility product for copper (II) iodate is

$$K = a_{Cu^{+2}} \times a^2_{IO_3^-}$$

where each activity is given by $a = \gamma c$. Since the activities must remain constant in order that K remain a constant, any change in the activity coefficients γ are accompanied by an equivalent change in the concentrations. From the Debye-Hückel limiting law (Problem 14-16) we see that an increase in ionic strength results in a decrease in the value for γ. But the ionic strength depends on the *total* ionic composition of the solution, so when KCl is added the ionic strength increases, and γ correspondingly decreases. This decrease in turn is compensated for by an increase in the equilibrium concentration of the copper(II) iodate. It is possible to use the Debye-Hückel limiting law to calculate what the change in solubility ought to be for a given concentration of added KCl. Rather than go through these calculations, which are rather lengthy, you may wish to see whether the solubility of $Cu(IO_3)_2$ varies in a linear manner with the square root of the total ionic strength, as it should if the limiting law is obeyed.

15-35. The net oxalate ion concentration, assuming complete complex formation, is $0.15 - 3(0.02) = 0.09M$. The concentration of complex is approximately $0.02M$. Then,

$$1.5 \times 10^{-20} = \frac{X(0.09)^3}{0.02}$$

$$X = 4 \times 10^{-19}M$$

Chapter 16

16-2. Potassium ion migrates to the right; *i.e.*, toward the half-cell in which reduction occurs. Chloride ion moves in the opposite direction.

16-4. $K = 10^{21}$.

16-8. The cell described is a concentration cell.

$$H_3O^+ (a = 10^{-1}) \rightleftarrows H_3O^+ (a = 10^{-6})$$

$$\mathcal{E} = -0.0592 \log_{10} 10^{-6}/10^{-1}$$

$$\mathcal{E} = +0.296$$

16-10. We have

$$K_{eq} = \frac{(Ag^+)(Fe^{+2})}{(Fe^{+3})}$$

$$\mathcal{E}^\circ = -0.028 = \frac{RT}{F} \log K_{eq} = 0.0592 \log_{10} K_{eq}$$

$$\log_{10} \frac{(Fe^{+2})}{(Fe^{+3})} = \mathcal{E}^\circ/0.0592 - \log_{10} (Ag^+)$$

$$= 1.526$$

$$(Fe^{+2})/(Fe^{+3}) = 33.6$$

16-13. Lowers total charge capacity. Voltage is only slightly affected. Eventually, the battery may short-circuit internally.

16-16. $\mathcal{E}^\circ = 2.46$ volts.

Chapter 17

17-2. 365 g $NaHSO_3$.

17-5. Among the more important, a careful molecular weight measurement of the sulfur in a non-polar solvent such as benzene. Solubility, density, and color are all relevant properties.

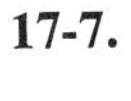

17-7.

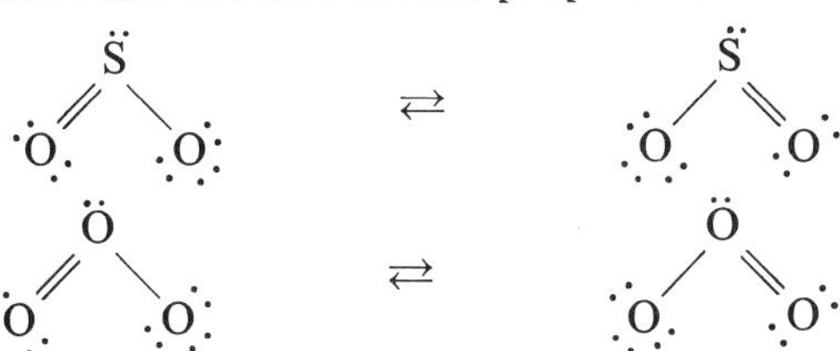

If one considers only the valence shell electrons, these two substances are "iso-electronic." The non-existence of S_3 is related to the poor ability of non-metals from the second full row and below to form multiple bonds with one another.

17-8. (a) $4FeS_2 + 11O_2 \rightarrow 2Fe_2O_3 + 8SO_2$

(b) $4H_3O^+ + 2Cl^- + MnO_2(s) \rightarrow Mn^{+2} + Cl_2(g) + 6H_2O$

(e) $GeO_2 + 2C \rightarrow Ge + 2CO$

17-10. We first need the quantity of charge required:

$$454 \text{ g } Cl_2 \times \frac{1 \text{ mole } Cl_2}{70.9 \text{ g } Cl_2} \times \frac{2 \text{ equiv}}{1 \text{ mole } Cl_2} \times \frac{96{,}500}{1 \text{ equiv}} \text{ amp-sec} = 1.23 \times 10^6 \text{ amp-sec}$$

A watt is a volt-amp. The power consumption in producing the chlorine is $V \times I \times t$:

$$\begin{aligned} W &= 1.23 \times 10^6 \times 10 = 1.23 \times 10^7 \text{ watt-sec} \\ &= 1.23 \times 10^7/3600 = 3.42 \times 10^3 \text{ watt-hour} \\ &= 3.42 \text{ kilowatt hour} \end{aligned}$$

The electricity for a pound of chlorine thus costs $3.42 \times 1.5 =$ 5.1 cents.

17-12. In the series C through F, the number of valence electrons decreases. The same applies to the series Si through Cl. Since the heats of formations of the elements in their standard states are arbitrarily set at zero, the ΔH_f° of the gaseous atoms is simply a measure of the total energies of the bonds holding each atom in the free element.

The decrease in ΔH_f° with increasing atomic weight is an indication of weaker bonding between atoms, since the structures all have the same (diamond) structure.

17-14. The indium has the effect of neutralizing the arsenic, because it provides holes for the extra electrons. InAs, as a stoichiometrically pure substance, would thus be an insulator, just as pure germanium. An excess of In or As would lead to a *p*- or *n*-type semiconductor.

Chapter 18

18-1. Average Thermochemical Bond Energies

CH	NH	OH	FH
99	93	111	134
SiH	PH	SH	ClH
82	76	83	103
	AsH	SeH	BrH
	58	67	88
	SbH		
	61		

18-3. (a) $TeO_2 + 2Zn + 4H_3O^+ \rightarrow Te + 2Zn^{+2} + 6H_2O$.

(b) $4NH_3 + 5O_2 \rightarrow 4NO + 6H_2O$.

(e) $AsCl_3 + 3H_2O \rightarrow H_3AsO_3 + 3HCl$.

(f) $GeH_4 + HBr \rightarrow GeH_3Br + H_2$.

(This reaction would probably require a catalyst, *e.g.*, $AlBr_3$, to proceed.)

18-6. The data should be treated graphically, by plotting $\log_{10} K$ *vs* $1/T$. The slope of the line which results is equal to $\Delta H/2.3R$. We obtain $\Delta H = 40.2$ kcal/mole.

We can write the following:

1. $H_2S\ (g) \rightarrow 2H(g) + S(g)$.
2. $2H\ (g) \rightarrow H_2(g)$.
3. $S(g) \rightarrow \frac{1}{2}\ S_2(g)$.
4. $H_2S(g) \rightarrow H_2(g) + \frac{1}{2}\ S_2(g)$.

We have then that 4. = 1. + 2. + 3., or 3. = 4. − 1. − 2. = 40 − 166 + 104 = −22. The S_2 bond dissociation energy is just twice the negative of this, 44 kcal/mole. This is a rough value only, because we are mixing values of ΔH which are applicable at 25°C with a ΔH value for a temperature range much higher.

18-8. 252 g Zn to produce 100 g AsH_3

662 g Cl_2 to convert 75 g silane to $SiCl_4$

via

$$SiH_4 + 4Cl_2 \rightarrow SiCl_4 + 4HCl$$

18-12. (c) Reaction with water; LiH yields H_2 gas, basic solution.

(d) Add to water; HCl yields acid solution. Cl_2 reacts with bromide or iodide solution to yield Br_2, I_2 respectively, easily identified.

Chapter 19

19-1. $\mathcal{E}° = [7(1.38) + 1.36]/8 = 1.38$

$\mathcal{E}° = [7(0.42) - 0.42]/6 = 0.42$

19-5. H—Ö—Ö—H   H—N̈=N=N̈

:Ö: | :Ö—Cl—Ö: ; :Ö: | :Ö—Cl—Ö·   :F̈—Ö—F̈:   :Ö: | :Ö—I—Ö—H | :Ö: :Ö: | S / \\ :O: :O:

19-7. (c) $Mn^{+2} + H_2O_2 + 2OH^- \rightarrow MnO_2(s) + 2H_2O$.

(g) $2H_2O + 2NO_2^- + 4Na(Hg) \rightarrow N_2O_2^{-2} + 4Na^+(aq) + 4OH^-$

19-10. $2Pb(NO_3)_2 \xrightarrow{\Delta} 2PbO + 4NO_2 + O_2$

13.5 g $Pb(NO_3)_2$ required.

19-12. For example, $K = P^2_{NO_2}/P_{N_2O_4} = (0.347)^2/0.653$

$= 0.184$ atm at 27°C.

From $\log_{10} K$ *vs* $1/T$ graph, slope equals $\Delta H/2.3R$. $\Delta H = 13.4$ kcal/mole. This is indeed a measure of the N—N bond strength.

19-15. $4SC_3H_6 + 11NH_4ClO_4 \rightarrow 4SO_2 + 12CO_2 + 12H_2O + 11NH_4Cl$

$4SC_3H_6 + 11LiClO_4 \rightarrow 4SO_2 + 12CO_2 + 12H_2O + 11LiCl$

435 g NH_4ClO_4, or 394 g $LiClO_4$ required.

19-18. (b) $4NH_3 + 5O_2 \rightarrow 4NO + 6H_2O$

Half the NO is oxidized; then,

$$NO + \tfrac{1}{2}O_2 \rightarrow NO_2$$

$$NO + NO_2 + 2OH^-(aq) \rightarrow 2NO_2^-(aq) + H_2O$$

(f) $PF_3 + ClF_3 = PF_5 + ClF$

19-20.

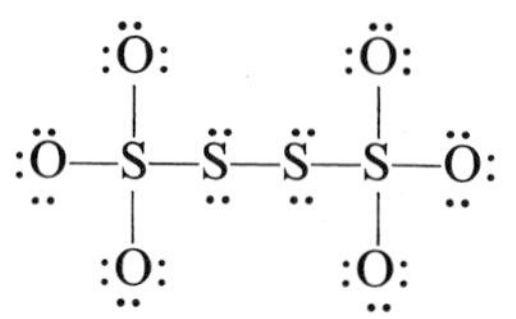

```
          ..        ..
         :O:       :O:
          |         |
    ..    |   ..  ..|   ..
   :O—————S———S———S—S———O:
    ..    |   ..  ..|   ..
          |         |
         :O:       :O:
          ..        ..
```

19-24. By analogy with ClF_3, one expects ArF_2.

19-27. PCl_5 forms $PCl_6^- + PCl_4^+$. The different behavior of the bromide may be ascribed to the instability of PBr_6^-, probably because six bromide atoms cannot fit around the phosphorus atom.

19-31.

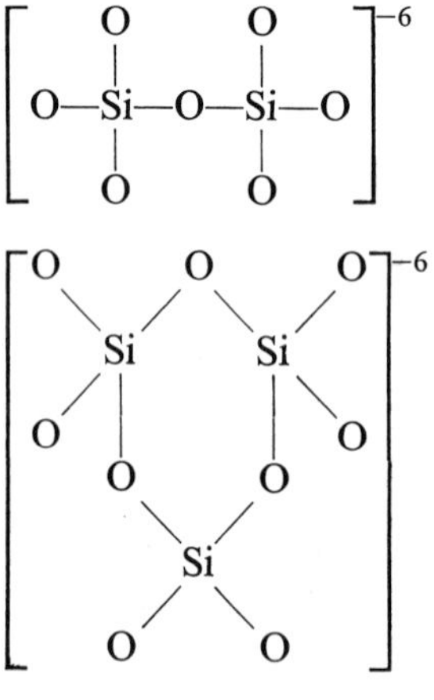

```
[     O       O     ]-6
      |       |
  O—Si—O—Si—O
      |       |
[     O       O     ]
```

```
[O       O       O]-6
  \     / \     /
   Si        Si
  /  |       |  \
 O   |       |   O
     O       O
      \     /
        Si
      /    \
[    O      O     ]
```

Chapter 20

20-2. (a) Interstitial.

(b) Solid solution.

(c) Either intermetallic or separate crystals.

(d) Interstitial.

20-4. The process involved is of the form

$$Li^+(g) + nH_2O(l) \rightarrow [Li(H_2O)_n]^+ (aq)$$

The transfer of metal ion from the gas phase to the liquid phase is accompanied by the addition of some number of water molecules to the coordination sphere around the metal. Since these water molecules are strongly held in position, they lose motional freedom. There is a consequent loss in entropy in the system. The loss in entropy is proportional to the number of water molecules bound, and the firmness with which they are held. This in turn should be proportional to metal charge, and inversely proportional to radius. Compare ΔS with charge/radius, Table 20-3.

20-6. Two isomers of $Pt(NH_3)_2Cl_4$ are possible, *cis* and *trans*.

$Pt(NH_3)_4Cl_3Br$ may possess coordination isomers:

1. $[Pt(NH_3)_4Cl_2]$ ClBr, dichlorotetrammine platinum (IV) chlorobromide. This species may possess either *cis* or *trans* isomerism.
2. $[Pt(NH_3)_4ClBr]$ Cl_2, chlorobromotetrammine platinum (IV) chloride. This compound may exist in either the *cis* or *trans* form. Neither form can exhibit optical activity.

20-11. $[Co(NH_3)_5H_2O]Br_3$ aquopentammine cobalt (III) bromide. (No isomers possible.)

$[Ir(en)_2NH_3Br]Cl_2$ bromoammine bis ethylenediammine iridium (III) chloride. (Either *cis* or *trans* isomers. *Cis* isomer also capable of optical isomerism. Coordination isomerism, involving chlorine-bromine exchange, also possible.)

$K_2[ZnCl_4]$ Potassium tetrachlorozincate (II). (No isomers possible.)

20-15. Since Al^{+3} has a very favorable charge/radius ratio, its much more limited ability to form coordination compounds must arise from the absence of *d* level electrons. It is unlikely that crystal field stabilization is the major factor in the difference between Al^{+3} and the transition metal series. It appears that covalent bond formation, involving the *d* orbitals, make a major contribution to ligand-metal bond stability in the transition metal series.

20-17. K (overall) is the product of the four K_i given: $K = 4.5 \times 10^{12}$. The overall K is useful when NH_3 is present in excess, so that nearly all Cu^{+2} present is in the form of $Cu(NH_3)_4^{+2}$.

20-21. It should increase in energy, since the molecules creating the electrostatic interaction giving rise to the splitting are forced closer to the metal ion.

20-25.

$$[Co(en)_2Cl_2]^+ \rightarrow [Co(en)_2Cl]^{+2} + Cl^- \qquad \text{slow}$$

$$[Co(en)_2Cl]^{+2} + Cl^- \rightarrow [Co(en)ClCl]^+ \qquad \text{fast}$$

There may also be fast intermediate steps involving coordination of water and then displacement by Cl^-. The important point is, dissociation is rate-determining.

Chapter 21

21-1. Tl^{+3} is a larger ion, therefore not as well solvated in solution. Furthermore, the total energy required to remove three electrons from Tl is 3.1 electron-volts. These two factors more than make up for the fact that it is easier to get the metal atom from the solid to the gas phase (Section 20-3).

21-5. K_{sp} for CuS is much smaller than for CdS. At pH = 1, $[S^{-2}]$ is extremely low, since H_2S is a weak acid. Only CuS precipitates because the solubility product for CdS is not exceeded. Addition of acetate ion results in an increase in pH, thus increasing the concentration of free S^{-2} ion. Now K_{sp} for CdS is exceeded, and precipitation occurs.

21-7.

$$Fe_2O_3 + C \rightarrow 2Fe + 3CO$$

$$2Fe_2O_3 + 3Si \rightarrow 4Fe + 3SiO_2$$

$$CaO + SiO_2 \rightarrow CaSiO_3$$

21-10.

$$MnO + H_2O \rightarrow Mn^{+2} + 2OH^-$$

$$Mn_2O_3 + H_2O \rightarrow H_3MnO_3 \text{ [or } Mn(OH)_3]$$

$$Mn_2O_7 + 3H_2O \rightarrow 2H_3O^+ + 2MnO_4^-$$

21-14. The Cd_2^{+2} ion is the analogue of the Hg_2^{+2} ion, and contains a single metal-metal bond between Cd atoms: $[Cd—Cd]^{+2}$. It should be a powerful reducing agent, and would undergo immediate reaction with water:

$$Cd_2^{+2} + 2H_2O \rightarrow 2Cd^{+2} + H_2 + 2OH^-$$

21-15. (c) $Fe_2O_3(s) + 6H_3O^+ \rightarrow 2Fe^{+3} + 9H_2O$.

(e) $Hg^{+2} + Hg(l) + 2Cl^- \rightarrow Hg_2Cl_2(s)$.

(g) $2H_3O^+ + 2CuCl(s) + H_2O_2 \rightarrow 2Cu^{+2} + 4H_2O + 2Cl^-$.

21-16. (b) $14H_3O^+ + 6Cr(H_2O)_6^{+2} + Cr_2O_7^{-2} \rightarrow 8Cr(H_2O)_6^{+3} + 9H_2O$.
(f) $Hg^{+2} + 4I^- \rightarrow HgI_4^{-2}$ (aq).

21-18. Assuming that the nitrogen is reduced to N_2, the most probable reaction is

$$2NH_4VO_3 \rightarrow N_2 + 2VO + 4H_2O$$

VO is a black substance which dissolves in acid to yield V^{+2} solutions.

21-22. $Mn(CO)_5Br$. Octahedral, with 5 CO groups and 1 Br about Mn. Br_2 simply cleaves the Mn—Mn bond.

21-23. Excess O (oxygen) due to cation vacancies. Charge neutralized by Cu^{+2} sites in lattice. See Fig. 21-11 for an example of a schematic representation.

Chapter 22

22-2. Two, the 1-chloro and 2-chloro compounds.

22-3. Cyclobutane.
1-butene.
2-butene.
2-methylpropene.

22-7. The standard enthalpy change in the reaction $C_2H_4(g) + H_2(g) \rightarrow C_2H_6(g)$ is -32 kcal/mole, somewhat larger than the -29 kcal/mole observed for cyclohexene.

22-10. We have $RNH_2 + H_2O \rightarrow RNH_3^+ + OH^-$
from which we derive

$$\frac{(RNH_3^+)}{(RNH_2)} = \frac{K_b(H^+)}{K_w}$$

K_w is not simply $(H^+)(OH^-) = K_w = 10^{-14}$, since we do not have pure water any more, but only 30 per cent water. K_w should therefore be multiplied by 0.3 as a rough approximation. Then $(OH^-) = 3 \times 10^{-15}/(H^+)$. Since $pK_b = 9.30$, $K_b = 5 \times 10^{-10}$.

Then,

$$\frac{(RNH_3^+)}{(RNH_2)} = 1.7 \times 10^5 (H^+)$$

For the (H^+) values given, the ratios are about 3.4×10^3 and 8.5×10^3. The reciprocals times 100 and 0.03 per cent are 0.011 per cent, in reasonable agreement with the reported values.

22-13. Since the ratio of H to C is the same as that for a saturated alkane, the oxygens must not take up carbon valences at the expense of C—H bonds. This rules out aldehydes, ketones, and acids. The oxygens might be ether or alcohol groups, *e.g.*,

```
        H                            H       H
        |                            |       |
  H2C—C—CH2             H3C—O—C—O—C—OH
   |    |    |                       |       |
  OH  OH  OH                        H       H
```

22-24. (a) $12H_3O^+ + 5C_3H_7CH_2OH + 4MnO_4^- \rightarrow 23H_2O + 5C_3H_7COOH + 4Mn^{+2}$.

(e) The product is: $CH_2OHCHNH_2CONHCH(CH_2OH)COOH$.

(h) $C_6H_6 + CH_3COCl \xrightarrow{AlCl_3} C_6H_5COCH_3$.

Chapter 23

23-2. Alpha emission lowers the atomic number by two. Beta emission increases it by one. There is no change with gamma emission.

23-4. Probably two neutrons per Cm nucleus. A second possible product would be ^{106}Rh. This species decays by beta emission to give ^{106}Pd, a stable nucleus. ^{134}Sb decays to give ^{134}Te, which in turn decays to ^{134}I. This again emits a beta particle to give the stable ^{134}Xe. We have, therefore, three successive beta decays.

23-6. The atomic number is lowered by one. The process is referred to as *K* capture.

INDEX

Index